DATE			

TO RENEW THIS BOOK
CALL 275-5367

© THE BAKER & TAYLOR CO

UNDERSTANDING GENETICS

UNDERSTANDING GENETICS

Second Edition

NORMAN V. ROTHWELL

Long Island University, The Brooklyn Center

NEW YORK   OXFORD UNIVERSITY PRESS

Copyright © 1979 by Oxford University Press, Inc.

Library of Congress Cataloging in Publication Data

Rothwell, Norman V
  Understanding genetics.

  Includes bibliographies and index.
  I. Genetics. I. Title.
QH430.R67 1979   575.1   78-6203
ISBN 0-19-502440-0

Third printing, 1980

Printed in the United States of America

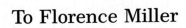

To Florence Miller

PREFACE

This second edition of *Understanding Genetics,* as in the case of the first edition, has been designed to serve the student whose background includes little more than the genetic information which is ordinarily encountered in a general biology course. It presupposes only a passing familiarity with elementary genetic principles and little contact with molecular biology. During 21 years of teaching genetics, I have tried several approaches in my undergraduate course. I am still convinced that the best is the one which assumes no more than a nodding acquaintance with the language of genetics. Frequently advanced biology students, even those on the graduate level, admit a complete lack of understanding of such elementary concepts as linkage and crossing over. I have found that such extremely critical topics as these have often been slighted as the student is plunged at the outset of his or her first genetics course into the details of chemical genetics and the intricacies of sophisticated experimental designs. Such a popular approach is based on the supposition that the student has been provided with a strong foundation which permits a ready insight into refined research. No doubt there are students who can do this; however, I have not encountered them, and I have no reason to believe that the students who come to me are atypical. I have met many graduate students from a variety of schools who are familiar with specific details of molecular biology but who have no inkling of the general biological significance of such knowledge. It seems to me that only by a clear, concise approach which encompasses the details of classical genetics can the excitement of today's research be appreciated. Moreover, the great strides in human genetics which present hope and challenge to today's society demand a firm grasp of basic genetic principles along with the elements of probability and statistics.

Therefore, the sequence of topics in this book, as in the first edition, starts from the "beginning" and leads to some of the latest advances reported in the literature at the time of writing. Wherever possible throughout the development of basic topics, elementary or historical points are associated with modern ideas and related to current problems. The reader will find a great deal of emphasis placed on the concept of allelism, to me the single most misunderstood point in genetics and the one so essential to an insight into the significance of viral and bacterial research. Certain topics in the first edition have been eliminated in order to direct attention to the new material which has been incorporated and which seems at this time to reflect significant advances in our understanding of genetic phenomena. Certain chapters have been rewritten, extended, and presented in a new sequence to provide a more cohesive treatment of the gene at the molecular level. I believe the present sequence of chapters yields a good bridge between classical genetics and more modern concepts derived from molecular biology.

A solid foundation in the basics of molecular genetics is essential for any biology student today, and I believe sufficient depth is given here to the topic to serve as a point of departure for perusal of more detailed and technical presentations in the literature. Certain items have been omitted, as in the previous edition, as they seem superfluous to the presentation of topics. For example, only those basic mathematical approaches are discussed which the student will encounter most frequently. On the other hand, parts of the text may appear somewhat repetitive or to overemphasize certain basic matters. This has been done intentionally to stress those principles which I find confusing to many students who are often embarrassed to ask for clarification of points they feel to be too rudimentary. The final chapter of this edition is essentially new and attempts to bring together some of the provocative aspects of present-day research against the background of human behavior.

The book can be used for a one-semester course or as a guide for two semesters, depending on the specific goals and the preparation of the students. The following chapters would appear essential to a one-semester presentation: 1–8, selected parts of Chapter 10, and Chapters 11–15. If Chapters 1–7 are to be assigned only for review, the course could in essence begin with Chapter 8. I would recommend inclusion of Chapter 8 regardless of the background of the students, since the fundamentals of linkage and crossing over are so critical to genetic analysis.

Essential to the understanding of scientific concepts in a textbook are the figures and other illustrative material. I have been most fortunate in this respect to work along with Mrs. Diane Abeloff, who has been in charge of the art work. In addition to her attention to detail in the depiction of more familiar subjects, she has contributed original ideas to clarify the presentation of some of the abstract material. This has required extra diligence in addition to talent as an artist. Her efforts form an integral part of the completed work.

Special recognition and appreciation are due to Dr. Robert F. Lewis, my friend and colleague of many years. His encouragement plus his valuable aid and suggestions throughout the project are in large part responsible for the completion of the manuscript.

This book has not been written for an advanced level course, but I feel that the treatment of most topics can be of use to a graduate student in biology, as well as an undergraduate. I am confident that the text can serve most biology students as I have come to know them throughout the years. While I do not consider this work to be a technical reference, I am confident that it can aid both the student and teacher who want to clarify certain genetic points.

NORMAN V. ROTHWELL
Brooklyn, New York
1978

CONTENTS

UNDERSTANDING
GENETICS

FOUNDATIONS OF GENETICS

The accomplishments of many brilliant biologists throughout the years have established the foundations upon which classical genetics, modern genetics, and the science of molecular biology rest. These contributions tell the story of the logical development of genetic principles. Familiarity with some of the major efforts is needed to provide a background in the concepts of heredity. Such knowledge can provide insight into the challenge which genetic phenomena present to us today.

ACHIEVEMENTS BEFORE 1900. Before 1900, the history of the science of heredity was closely interrelated with that of cytology, the discipline concerned with cellular structure and function. The origin of both sciences can be traced to the discovery of the cell. This landmark is credited to Robert Hooke, who in 1665 described cellular entities in sections of cork. In the following decade, Leeuwenhoek described many interesting cell types, recognizing many free-living forms in addition to those cells which are associated to form tissues. However, more than a hundred years were to pass before any other significant observations would be made. We can appreciate the main reason for this standstill when we consider the fact that before the nineteenth century, there were no notable advances in the development of optical tools. Very little could actually be seen inside the cells. As a result, cell walls and cell boundaries were stressed rather than contents of the cell. It was essential to distinguish the subcellular components before the location of the genetic material could be established.

In 1831, Brown called attention to a body within the cell which he recognized as a regular, constant cellular element. This structure was the nucleus. A few years later (1838-1840), another important contribution was made, the proposal of the "cell theory" by Schleiden and Schwann. This concept is considered by many to be the most significant, all-encompassing generalization in biology: all living things are composed of one cell or more and their products. Schleiden and Schwann were not the first to recognize the cellular nature of organisms; however, they were able to present the idea very convincingly by compiling their own observations with those of others. Once the cell was accepted as the unit of life, several biological disciplines emerged. In addition to cytology, such fields as physiology, embryology, and pathology were to undergo rapid advances.

Even though cytologists were now concentrating on cellular components, not too much was seen before 1850. This resulted from the still limited optical instruments and the lack of appropriate stains and fixatives. Most of the important improvements in this area occurred in Germany after a government subsidy of the dye industry. In the 1870's, development of the aniline dyes led to better staining methods. About this time, Abbé developed the condenser and the oil immersion lens. Aided by these advances, microscopists were able to see and describe in some detail the nucleus and changes associated with it.

At this time a large number of highly significant descriptions of the chromosome and its behavior suddenly appeared. Several investigators noticed the presence of threads in the nucleus of the cell and reported that these threads became split. Flemming, a zoologist, reported that the halves of the split threads separate, and in 1882, he gave the name "mitosis" to the process. The botanist Strasburger was the first to give a complete description of mitotic events. During this period, some biologists were paying special attention to the germ cells. By 1879, fertilization had been recognized in plants by Strasburger and in animals by Hertwig and Fol.

The observations made on mitosis eventually led to the establishment of the "chromosome theory of heredity." In 1884-1885, four different investigators (Hertwig, Kolliker, Weismann (all zoologists), and the botanist Strasburger) concluded independently that the physical basis of inheritance is in the nucleus and that the hereditary material is in the substance which makes up the chromosomes. To these workers as well as others, it was apparent that the separation of nuclear threads at mitosis is a very accurate procedure. In contrast to this, other portions of the cell are not so precisely distributed at cell division. Moreover, the cells uniting at fertilization, sperm and egg, differ immensely in size, primarily due to non-nuclear material. So, although cytoplasm can vary greatly in amount, the nuclei of gametes are quite similar. The larger sized female gamete does not contribute more nuclear material than the male. Inasmuch as heredity appeared to be equal from both parents, it seemed logical to assume that the hereditary material was in the nucleus, carried in the chromosomes. Although the chromosomes seemed to disappear in the "resting phase," they were still somehow carried along from fertilization on and distributed accurately from one cell to the next when the nucleus divided.

Therefore, in the 1880's, there were strong suggestions that the chromosomes are the carriers of hereditary material, because only they are quantitatively divided. The word "chromosome" was coined in 1888 by Waldeyer. However, there was still no real proof of the chromosome theory. Chromosomes can be seen only when the cell is dividing; therefore, it was essential to obtain evidence that they persist throughout all stages of the cell cycle, even when they are not visible. Also needed was some evidence to correlate chromosome behavior with the inheritance pattern of specific traits. Such confirmation was to be presented in the next few decades through the work of several outstanding scientists.

Of the four original proponents of the chromosome theory, Weismann was the one who recognized its additional implications. He reasoned that if the chromosome theory were correct, then the germ cells of one generation must supply the chromosomes for both the body cells and the germ cells of the next generation. These germ cells in turn give rise to still other body cells and germ cells (Fig. 1-1A). From such reasoning, Weismann formulated the idea of the "continuity of the germ plasm." According to this concept, it is the germ cells which connect one generation to the next. Although they provide the body cells as well as other germ cells with chromosomes, the body cells (or somatoplasm) do not form a continuum from one generation to the next; they are a "dead end," so to speak. Any hereditary effect, therefore, must come through the germ plasm, *not* the body cells. Such an idea argued against acquired characteristics, a theory which had been proposed in 1809 by Lamarck and which has been a popular one even in the twentieth century. According to this concept, a change in the body, resulting from an environmental influence during an individual's lifetime, can be passed on to the next generation. Most of us have heard the familiar story of the lengthening of the neck of the giraffe as the result of the animal's constant stretching to reach leaves in the trees. If the neck should somehow become physically stretched in a parental animal, the theory of acquired characteristics would maintain that this increase in length may be inherited directly by the offspring. But according to the theory of the continuity of the germ plasm, this would not be so, because body cells of one generation do not bridge the gap from one generation to the next. Any inherited change affecting the body must come through the germ cells of the previous generation. To support his idea, Weismann cut off the tails of mice for over 20 generations and demonstrated that the practice had no effect in altering the tail length at the end of the study, even though gametes came from tailless mice (Fig. 1-1B).

Furthermore, Weismann realized that if there is continuity of the germ plasm, there must be some mechanism to prevent the doubling of the chromosome number in each new generation over that in the previous one. The body cells cannot have the same chromosome number as the germ cells because each new generation would have twice the chromosome number of the parents. Weismann proposed that a reduction division must occur, a device which would reduce the number of chromosomes in the germ cells to one-half that of the body cells (Fig. 1-1C). He began searching for evidence, and he did notice that during parthenogenesis in aphids, only one polar body is formed. He assumed that reduction division was eliminated here to make up for the lack of fertilization. Although Weismann never found cytological evidence from chromosome behavior to support the actuality of a reduction division, his suggestions directed the attention of other workers to this problem. Cytologists then began to observe the process we know today as meiosis. The details of this divi-

sion which is so critical to all sexually reproducing life forms were not worked out until after 1900.

The turn of the century marks a very important date for the science of heredity, for it was in 1900 that the work of Gregor Mendel was discovered. Mendel had actually read his paper at Brunn, Czechoslovakia, 35 years earlier, before any detailed observations had been made on the cell nucleus and the chromosomes. The lack of cytological information accounts in part for the fact that Mendel's report lay unappreciated and poorly disseminated among scientists until its significance was pointed out independently by three biologists: DeVries, Correns, and

FIG. 1-1. Weismann's concepts. *A.* The germ plasm (gametes) forms a continuous line connecting one generation to the next; the somatoplasm (body cells) does not link one generation to the other; thus, changes in it cannot influence the next generation. *B.* Amputation of the tails of mice *before* mating, generation after generation, does not influence the length of the tails of the offspring. The tails will not disappear if it is the germ cells and not the body cells which supply the genetic information for the next generation *C.* If the germ line is continuous, the gametes must have one-half the number of chromosomes present in the body cells. A reduction in number must occur in each generation before gametes are formed; otherwise, the chromosome number would increase with each generation.

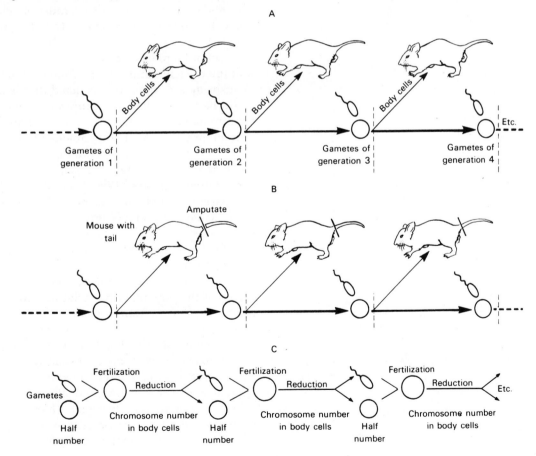

FOUNDATIONS OF GENETICS

5

von Tschermak. All three worked with plant material, and their results confirmed those of Mendel. By 1900, a receptive climate prevailed for the recognition of Mendel's principles. For, at that point, some hereditary concept was needed to tie in with the observations which had been made on cell structure and cell division by pioneer cytologists. Moreover, the recognition of Darwin's theory of natural selection called for a solid explanation of the basis of genetic variation.

Before 1900, some breeding work had been performed, but it contributed little to an understanding of the mechanism of inheritance. Several nineteenth-century biologists had actually obtained the same results as Mendel. However, none were able to arrive at a correct interpretation of their findings. One of them actually described the phenomenon of segregation, the basis of Mendel's first law. But, unlike many of these earlier investigators, Mendel most likely had formulated a working hypothesis which enabled him to present a valid interpretation of this results. The nineteenth-century biologists knew that the embryo developed from the fusion of sperm and egg. They realized that all hereditary potential must be transmitted this way and that the nuclear contributions from each parent to offspring are equal. It was noted that hybrids might be intermediate or at times resemble one parent and not the other. The predominant concept of inheritance in the nineteenth-century was a blending theory. According to this view, the hereditary material was pictured as some sort of fluid, perhaps even blood. The fluids of two parents would come together and blend in some fashion. It would be impossible to separate these hereditary fluids, just as one would fail in an attempt to separate a mixture of red and white paints.

A contrasting viewpoint was one which conceived of the hereditary material as particulate in nature. Francis Galton pointed out that particles must be handed down. He based this on the perceptive observation that one parental trait in a pedigree may seem to disappear, only to show up again in a later generation. Galton approached heredity in a statistical way by measuring individuals for a variable characteristic, such as height. He compared the measurements which he made on populations from one generation to the next. Galton must be credited with the establishment of biometry, the valuable science of statistics as applied to biological problems. However, he reached erroneous conclusions at times, and his investigations, like those of many others of his day, did not reveal what was really happening when contrasting traits were followed in a family history or lineage.

MENDEL AND THE CHOICE OF A GENETIC TOOL. The great significance of Mendel's work stems from the fact that it established the particulate concept of heredity and replaced the blending theory. Whereas many before him had failed, Mendel was able to obtain clear-cut genetic results through his recognition of several requirements which are critical to genetic analysis. The garden pea plant with which he worked enabled him to meet these needs. First of all, it provided several pairs of contrasting traits for study. The plants varied in such characteristics as height, seed texture, seed color, and flower color. Some sort of variation is essential if anything at all is to be learned about the inheritance of any character. Suppose, for example, that all pea plants were the same height and had the same flower color generation after generation. Obviously, no information could be gained from following plant height and flower color in genetic studies, as every individual would continue to look alike in these respects. A characteristic must have alternative traits or variant forms which can be followed if an insight is to be gained into its inheritance.

In all, Mendel was able to study seven distinct characteristics, each with two well defined traits (Fig. 1-2). Moreover, in order to follow a characteristic clearly from one generation to the next, its traits must be easy to score. The characteristic height in the pea plant illustrates this point: the varieties which Mendel used exhibited contrasting traits which were easily recognized as either tall or dwarf. There were no continual grades of height between the two extremes to obscure the picture. This should be contrasted with height in humans or some other animal where the range varies gradually from very short to very tall with no distinct categories. The inheritance of height in these cases involves more complexities than in Mendel's peas. The use of a highly variable characteristic with ill-

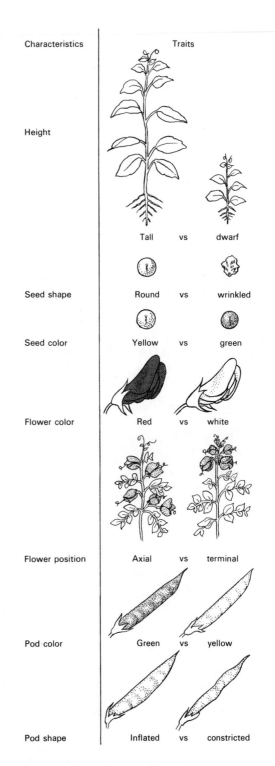

FIG. 1-2. Seven characteristics in the pea plant followed by Mendel. Each characteristic has two well-defined traits which are easily recognized. This feature permits ready identification of the traits as they are followed from one generation to the next.

defined traits in a pioneer genetic investigation is almost certain to lead to confusion.

Mendel also recognized the need to work with a species which could provide a large number of offspring and several generations in a short period of time. Observations made on a few individuals in just one or two generations yield insufficient data for formulating valid conclusions. The elephant would be a very poor animal to select for basic genetic studies. Not only is the gestation period of the elephant 20 months, but the reward is just a single offspring. At the end of a lifetime of crossing elephants, a worker would accumulate too few data to be meaningful. As we will see more clearly in Chapters 6 and 7, large numbers are usually essential for scientific significance. Basing conclusions on small samples is unreliable and increases the probability that the observations may be distorted by chance factors. The garden pea completely satisfied Mendel's need to follow more than one characteristic through many generations composed of large numbers of individuals.

An additional advantage is that the investigator can quite easily manipulate the pea plant to obtain the exact type of cross or mating he desires. Peas are normally self-pollinated; therefore two genetically identical individuals can be crossed. This is so because self-pollination is equivalent to mating one individual with another one exactly like itself. This feature can be put to even further advantage. One may allow plants with a certain set of traits to self-pollinate through one generation or more. If the alternative traits fail to appear among hundreds of offspring, one can feel certain that the strain is pure breeding. In addition, the pea plant can be handled to permit cross-pollination. Consequently, a plant from a particular variety can be used either as a female parent (by removing its stamens and placing pollen from another variety on the stigmas of its flowers) or a male parent (by placing its pollen on the stigmas of a different plant whose stamens have been removed). Therefore, with the common pea plant, the geneticist can perform crosses in the way he elects, and he can easily establish a great deal about the lineage or pedigree of each parent in a particular cross. This is a far cry from the situation in human populations where an investigator cannot dicate the mating between two different types and where he may know little or nothing about the family histories of the individuals in question. The results obtained by many geneticists before 1900 were ambiguous because the species they employed lacked one or more of the features found in the garden pea. Other organisms (the fruit fly, corn, bacteria, and others) which have also contributed a wealth of information to the science of genetics possess this same combination of desirable attributes: characteristics which show variations but whose traits are well defined and easily scored, the production of appreciable numbers of offspring in a relatively short generation time, and a situation that allows crosses to be controlled with ease.

MENDEL'S MONOHYBRID CROSSES. Part of Mendel's genius lay in his ability to formulate a scientific problem. The aim of his experiments was to study the numbers and kinds of offspring produced by hybrid individuals and to determine from the observations whether any statistical relationship existed among these offspring. Mendel undertook the study of the seven characteristics and their variant forms one at a time: tall vs. dwarf, red flower color vs. white and yellow seeds vs. green, etc. He performed *monohybrid crosses*, crosses in which only one pair of contrasting or alternative traits is being followed. In his study of the inheritance of height (Fig. 1-3), Mendel knew that his tall plants and his dwarf ones were all pure breeding, because he had allowed tall and dwarf types to self-pollinate for two generations. When he crossed the pure breeding tall forms with the pure breeding dwarf (the P_1 or first parental generation), the offspring (the F_1 or first filial generation) were all tall. The same result was obtained whether he used the tall plants as male parents or female parents. Mendel realized, however, that these F_1 tall individuals were not the same as the P_1 tall parents because they had a dwarf parent. He followed these F_1 hybrids to the next generation (the F_2 or second filial generation) by allowing them to self-pollinate. In one cross, he actually obtained 787 tall offspring and 277 short, a ratio that is approximately 3:1. He followed these second-generation plants still further by allowing them to self-pollinate. He found

out that the short plants continued to produce only short ones. On the other hand, the tall plants were of two types: approximately 1/3 of them produced only tall offspring upon self-fertilization; the remaining 2/3 gave rise to both tall and short, again in a ratio of approximately 3:1 (Fig. 1-3).

Comparable results were obtained when monohybrid crosses were performed for each of the other six characteristics. For example, in the case of red vs. white flower color, a cross between the two types of pure breeding varieties yielded all red-flowered plants in the first generation. When these were allowed to self, the second generation gave 705 red offspring and 224 white, again close to a ratio of 3:1. The white-flowered plants produced only white upon self-pollination. The red-flowered ones, as in the case of tallness, were shown to consist of two types: 1/3 pure breeding for red flowers; the other 2/3 producing red and white in a ratio of 3:1. The red trait thus behaved similarly to tall in the first example and white similarly to dwarf.

MENDEL'S LAW OF SEGREGATION. It is remarkable to realize that Mendel knew nothing of mitosis or meiosis, that he never heard of chromosomes or the units of heredity, the genes, which they carry. Yet his explanation of his results reads almost as if he sensed that such phenomena did indeed exist. His knowledge of probability and his understanding of the goal of his investigations undoubtedly account in large measure for the perception which is evident in his conclusions. From the data of his monohybrid crosses, he was able to recognize a pattern and formulate what is called "Mendel's first law" or the "law of segregation." In essence, it states that the hereditary characteristics are determined by particulate units or factors. These unit factors occur in pairs in an individual, but in the formation of germ cells, these entities are segregated so that only one member of the pair is transmitted through any one gamete. When the male and female gametes unite, the double number of factors is restored in the offspring. Substituting the word "gene" in place of "factor," the statements read almost as if the physical basis of heredity had been known to Mendel at the time the experiments were performed.

Mendel coined the term *dominant* for traits

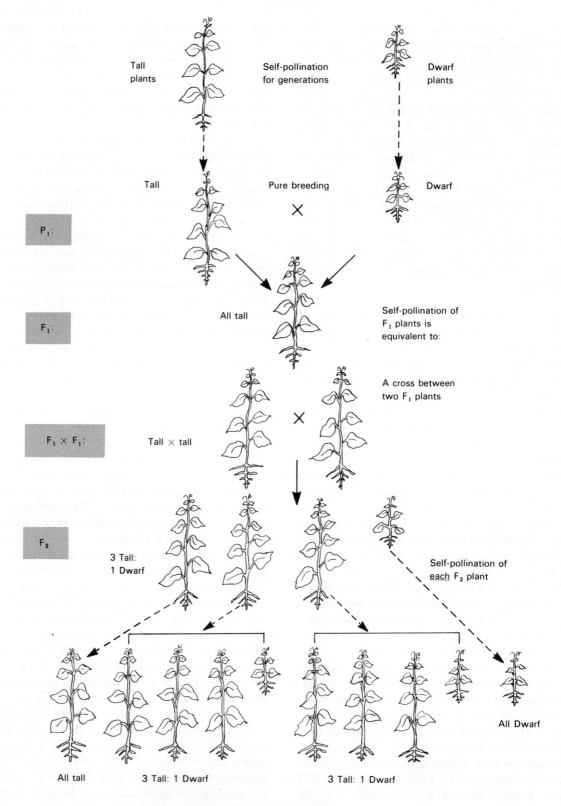

FIG. 1-3. Monohybrid cross in the pea plant followed through three generations. The results of the reciprocal cross, dwarf × tall, are identical.

such as tallness or red flower color because they expressed themselves when present with the factors for the contrasting traits and thus seemed to dominate them. He used the word *recessive* for traits like dwarfness and white flower color; these traits are not expressed when present in the hybrid with factors for the dominant traits. Although Mendel coined the terms *dominant* and *recessive* to designate traits, throughout the years these expressions have been widely used to refer to the genetic factors associated with the traits as well as to the traits themselves. Throughout our discussions, *dominant* and *recessive* will be used both ways.

We can now reexamine the cross of the tall and dwarf plants in light of Mendel's concepts (Fig. 1-4). It should be noted here that when letters are assigned to a pair of contrasting factors (such as those for tallness and dwarfness), the same letter of the alphabet is used. The capital is reserved for the dominant factor (T for tallness) and the small letter for the recessive (t for dwarfness). This serves to avoid confusion when two or more pairs of factors are being followed at the same time in a cross. Still another accepted way of designating genetic factors will be presented at length at the end of Chapter 4.

In Figure 1-4, the Punnett square method, also known as the checkerboard method, is used to bring together the gametes from both parents in all possible combinations. We will see later in this chapter that other devices may be used to achieve the same thing. As the figure shows, each parent in the F_1 generation is a hybrid and can contribute to the F_2 offspring *either* a factor for tallness (T) *or* a contrasting factor for dwarfness (t). A sperm with the factor for tallness (T) can unite with an egg bearing the same factor or it can just as well unite with an egg carrying the contrasting factor for dwarfness (t). Similarly, a sperm with the recessive factor for dwarfness (t) has an equal chance of combining with the "T" egg or the "t" egg. Inasmuch as large numbers of gametes would actually be involved, the resulting combinations in the F_2 occur in a ratio of 1 TT: 2 Tt: 1 tt. Due to the fact that tallness is dominant, the ratio of different physical types is 3 tall: 1 dwarf. However, it can be seen that of these F_2 tall plants, only one-third of them (the TT types) will breed true when self-pollinated to give all tall offspring. The other two-thirds are hybrids (Tt). When these are selfed, it is the same as crossing Tt × Tt, and tall and dwarf offspring will be produced in a ratio of 3:1. The plants which are pure breeding for tallness (TT) have identical genetic factors for that trait. We say they are *homozygous*. The dwarf plants are also homozygotes, individuals with identical genetic factors for a trait. Indeed, the dwarf plants *must* be homozygous (tt) because the presence of the dominant factor for tallness would cause a plant to be tall. We see, therefore, that tall plants may be of two types: TT or Tt. The latter are not pure breeding and are said to be *heterozygous*, meaning that they have contrasting genetic factors (T vs. t) for a particular characteristic. As we will see more clearly later (Chap. 14), contrasting factors such as these are really different forms of the *same* gene which result in contrasting detectable effects. Contrasting forms of a gene (T vs. t; R vs. r) are called *alleles*. So the homozygous individual, that is TT or tt, has only one form of a gene. The same allele is represented twice in each homozygote. On the other hand, the heterozygote possesses both forms—contrasting forms of the same gene. Therefore, the heterozygote (Tt) contains a pair of contrasting alleles for the characteristic, height.

The term *allele*, therefore, is used to indicate a specific form of a given gene. In this case, we are discussing the *gene* for the characteristic of "height" in the pea plant, and we are following the transmission of two specific forms of that gene, the allele "T" for tallness and the allele "t" for dwarfness.

We cannot tell simply by looking at a tall plant whether it is homozygous or heterozygous. All that we can say is that it is tall in contrast to a plant which is dwarf. We refer to the physical appearance or the detectable attributes of an individual as its *phenotype*. A knowledge of the phenotype does not necessarily tell us anything conclusive about the genetic endownment of the individual. The term *genotype* is used to describe the genetic composition or the kinds of genetic factors carried by an individual. If we know that an allele is recessive, such as the ones for dwarfness or white flowers, then simply by observing a dwarf or white-flowered phenotype, we know the genotype in regard to the particular trait.

However, looking at a tall or a red phenotype does not tell us whether the genotype is homozygous or heterozygous. We may gain this information only from further knowledge about the ancestry or about the offspring of the individuals expressing a dominant allele.

BLENDING INHERITANCE DISPROVED. An important point to appreciate is that regardless of the dominant or recessive nature of an allele, blending inheritance does not occur. In the case of flower color in Mendel's peas, the alleles for redness and whiteness were in no way contam-

FIG. 1-4. Monohybrid cross of tall and dwarf pea plants. By convention, the female parent is written on the left. The results of the reciprocal cross are the same. Compare with Fig. 1-3.

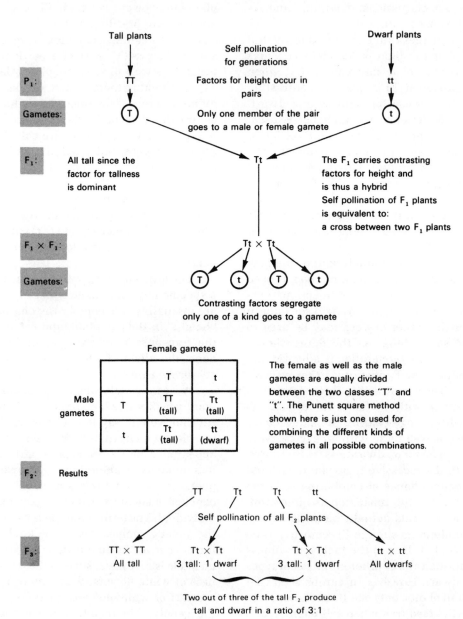

UNDERSTANDING GENETICS

inated or altered by being present together in the F_1 hybrids, the heterozygous Rr individuals. Although white seemed to disappear, it *did* show again in the F_2, and these white-flowered plants were identical phenotypically to the original whites in the P_1, the first parental generation. Likewise, the red factor was in no way diluted by the presence of the contrasting allele, the factor for white. This is certainly not to be expected if the fluids from two organisms are mixing, because we would then expect the original qualities to be lost or at least permanently altered in some way. The concept of blending is still entertained by those who are naive about genetic principles.

Let us turn our attention for a moment to another plant species which was not studied by Mendel, but in which the expression of flower color is somewhat different than in the garden pea. The common horticultural plant known as the four-o'clock may also have red or white flowers. However, if pure breeding white plants are crossed with pure breeding red ones, the F_1 offspring are intermediate in color, pink (Fig. 1-5). When two pink-flowered plants are crossed, the F_2 does not fall into a phenotypic ratio of 3:1 but rather into a ratio of 1 red: 2 pink: 1 white. Has blending inheritance occurred in this case because the intermediate pink shade has appeared? Far from contradicting Mendelian principles, such an example lends strong support to the concept of particulate inheritance. Neither the factor for redness (R) nor the one for whiteness (r) has in any way been altered by being present together in the cells of an individual. The F_2 reds and whites are the same as the original red and white P_1's. The cross illustrates Mendel's first law beautifully and differs from that of the pea plant only in relationship to the question of dominance. In this case, neither form of the gene is dominant to the other. Each expresses itself in the presence of its allele to produce an intermediate effect. We say that alleles such as these show *incomplete dominance*. As Figure 1-6 shows, we may diagram the cross in the four-o'clock in exactly the same way as for the pea plant. It does not matter if we choose to represent the allele for red color as "R" or "r," as long as we remember that the heterozygote is an intermediate. An important point which should be evident from this example is

that the 3:1 phenotypic ratio, which typically results from a cross of two monohybrids, has been modified to a phenotypic ratio of 1:2:1. This is not surprising. In the case of the pea plant, where red (R) is dominant to white (r), the heterozygotes (Rr) are classed as red because they cannot be distinguished from the homozygotes (RR). But in the case of the four-o'clocks, the heterozygotes can be easily recognized phenotypically. Inasmuch as we *can* distinguish them from the homozygous red plants, the phenotypic ratio becomes 1:2:1. It should be noted that this is the same as the *genotypic ratio* resulting when any two monohybrids are crossed. So when dominance is incomplete, a cross of two monohybrids (such as Rr × Rr pink plants) gives a phenotypic ratio (1 red: 2 pink: 1 white) which is identical to the genotypic ratio (1 homozygous RR: 2 heterozygous Rr: 1 homozygous rr). Where complete dominance is operating in a monohybrid cross, the phenotypic ratio (3:1) and the genotypic ratio (1:2:1) are different.

DOMINANCE IS NOT ALWAYS ABSOLUTE. Although Mendel noted dominance and recessiveness among the seven pairs of alleles which he followed, he did not say that dominance must always pertain. A survey of a variety of organisms, including the human, shows that incomplete dominance is common in living things. Indeed, cases of complete dominance may prove to be the exception rather than the rule. For today, we are often able to examine individuals in more detail than was formerly possible. When heterozygotes are examined closely, they frequently reveal some detectable difference from the homozygous dominant types. Therefore, while the heterozygotes may superficially appear to be the same as the homozygotes, the so-called recessive allele may actually be producing some effect. The expression "incomplete dominance" is also used to refer to these situations where both alleles are expressing themselves in the hybrid but where the effect of one allele appears greater than that of the other. In the absence of a refined examination, one allele may even appear to be completely dominant to the other. Another frequently encountered term is "codominance," and this is also applied in cases where both alleles produce an effect in the heterozygote. However, in its strict sense, codominance implies that a definite product or substance controlled by each allele can be identified. As we will see in Chapter 14, the genetic factors which govern the A and B blood types in man are alleles. Each controls the formation of a different red blood cell protein or antigen, antigen A in one case, antigen B in the other. Neither allele is dominant to the other. The individual that is blood type AB is a heterozygote and has both the allele for antigen A and the allele for antigen B. Both of these proteins, A and B, are easily detected in equal amounts in the red cells. This

FIG. 1-5. Monohybrid cross: red × white-flowered four-o'clocks. The F_1 is intermediate in color. A cross of two of these pink-flowered plants yields red, pink, and white in a ratio of 1:2:1. This same pattern of inheritance is seen in other plant species; for example, flower color in snapdragons.

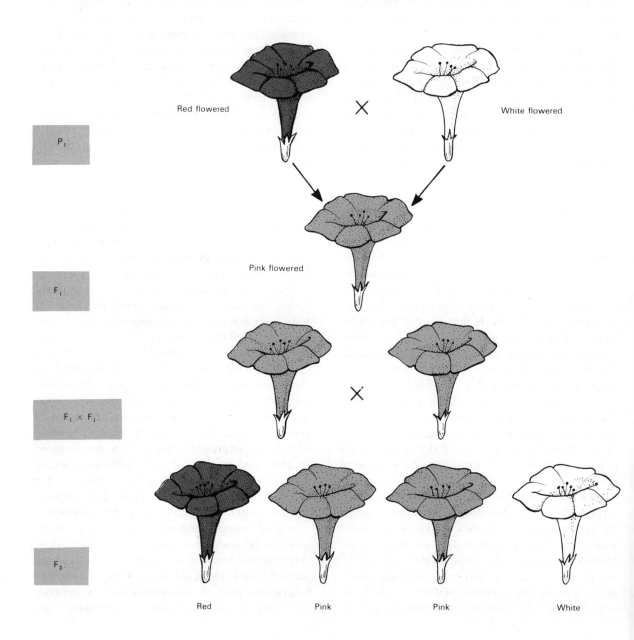

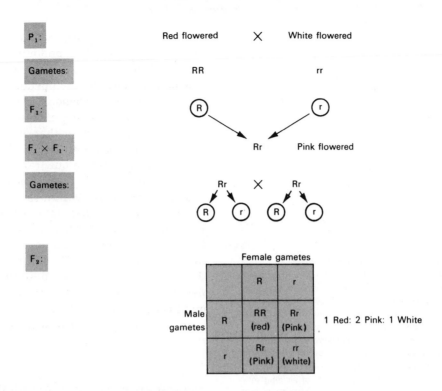

P₁: Red flowered   ×   White flowered

Gametes: RR   rr

F₁:

Rr   Pink flowered

F₁ × F₁:

Gametes: Rr   ×   Rr

F₂:

Female gametes

Male gametes		R	r
	R	RR (red)	Rr (Pink)
	r	Rr (Pink)	rr (white)

1 Red: 2 Pink: 1 White

FIG. 1-6. Incomplete dominance in the four-o'clock. Neither of the two alleles, R or r, is dominant. Consequently the F₂ phenotypic ratio is modified from 3:1 to 1:2:1, which is identical to the ratio of genotypes.

is a clear-cut case of codominance between a pair of alleles.

In other situations it may not be so easy to distinguish codominance from incomplete dominance because use of the proper term depends on our ability to detect in the heterozygote distinct substances controlled by each member of the allelic pair. A case from man illustrates the difficulty often involved in a precise distinction between codominance and incomplete dominance. Many of us are familiar with the unfortunate blood abnormality known as sickle-cell anemia, a fatal disorder in which an abnormal hemoglobin is present (p. 382). It is so named because the red blood cells tend to assume a sickle shape in the blood vessels (Fig. 1-7), resulting in a clogging of the capillaries, an assortment of dire symptoms, and eventual death. The lethal condition stems from the presence in the homozygote of a genetic factor which governs the formation of the abnormal hemoglobin. Completely normal persons possess only the typical hemoglobin in their red blood cells. Heterozygous persons appear completely normal under usual conditions. We can represent the alleles responsible for the normal hemoglobin and the sickle cell hemoglobin as "S" and "s", respectively. Superficially, it would appear that SS and Ss individuals are identical because neither appears to differ from the other, whereas the homozygote, ss, presents a very abnormal picture. However, the heterozygote, Ss, *is* actually different from the SS normal homozygote. This can become apparent at high altitudes where the heterozygotes may suffer ill effects due to the lower oxygen tension. Moreover, they may be recognized when their blood is examined microscopically because a certain percentage of their red cells will sickle under reduced oxygen tensions. The factor for the sickle-cell hemoglobin is obviously not completely recessive. Is the normal allele S incompletely dominant to the abnormal one, or are both alleles codominant? In the sense that the normal S results in a heterozygote which is more normal than abnormal, the allele would appear to be an incomplete dominant. However, refined techniques demonstrate that two different proteins, normal hemoglobin controlled by gene form S and sickle-cell hemoglobin controlled by its allele s, are both present in the blood. So based on this criterion, the alleles are codominants. It is evident that the choice of words may depend on the level at which we describe the phenotype of the heterozygote; it also depends on the techniques available to us to detect chemical differences. However, it is important that we avoid getting lost in a maze of terminology. The expression we use may depend on our frame of reference. The alleles for the A and B blood groups are clearly codominants. We may say that the alleles, R and r, in the four-o'clock are incompletely dominant because the heterozygote is intermediate and no distinct substances specific to each have been recognized. The normal and the sickle-cell allelic pair may be described as incompletely dominant at one level of reference (a small percentage of sickling) or as codominants at another (presence of two specific protein products).

MISCONCEPTIONS TO BE AVOIDED. To avoid erroneous interpretations of Mendelian principles which have been discussed so far, it is essential to pause and reexamine the implications of certain basic ideas. Several important points concern the concept of dominance, which unfortunately is often misunderstood. Although Mendel showed that the factor for red flower color in the pea plant is dominant to its allele for white, the genetic factors for red and white flower in the four-o'clock are incompletely dominant. This example was given to illustrate the fact that a genetic factor which produces a certain phenotypic effect (red, in this example) in one species is not the same as one in a different species, even though the two genetic factors appear to produce identical phenotypes (red flower color in both the pea and the four-o'clock). Simply because a particular factor in one group of animals or plants expresses itself as a dominant is no reason to assume that a different factor with a similar or even identical expression must also be dominant in another group. In animals, a factor for black fur may be dominant in one species (as in guinea pigs) but recessive in another (as in sheep). This presents no contradiction to Mendelian principles which nowhere imply that similar phenotypes must be due to the presence of similar genes.

FIG. 1-7. Blood smear from a person with sickle-cell anemia showing typical sickle-shaped cells. A person homozygous for the allele for normal hemoglobin (SS) will have disk shaped cells under all levels of oxygen concentration. The homozygote for sickle-cell anemia (ss) carries red blood cells which assume an abnormal sickling shape under the lower oxygen concentrations in the blood vessels. The heterozygotes (Ss) may show some degree of sickling in the blood vessels, although these persons usually do not develop anemia. However, in environments where the oxygen concentration is very low, a sufficient amount of sickling may occur and cause some anemia. (Courtesy of Carolina Biological Supply Company.)

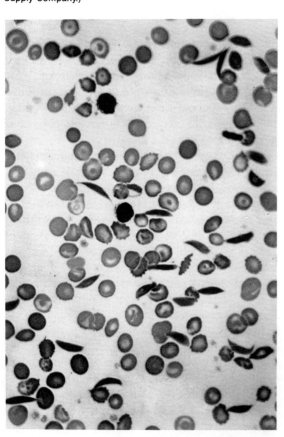

Another very common misunderstanding of dominance concerns the frequency of dominant alleles in a group or population. Too often, the student assumes that simply because a particular factor is dominant it must be more abundant in the population than its recessive allele. Many wonder why there are not more curly-haired people in the population, because the gene form for that phenotype is inherited as a dominant or incomplete dominant. Implicit in this erroneous idea is the belief that a dominant factor *must* increase in frequency over its recessive allele. A moment's reflection will tell us in the simple case of eye color in humans that this is not so. Although the factor for brown eye color (B) behaves as a dominant to its allele for blue (b), there may be more blue-eyed people in a population than brown-eyed ones. In Scandinavia this would be so, whereas in Mediterranean countries, we would expect more people to be brown eyed. As we will learn in Chapter 16, the frequency or abundance of an allele is not the same as its degree of dominance. The allele for brown eye color is dominant to the one for blue, and this means just what the definition of dominance says: the factor for brown eyes expresses itself in the presence of its allele, whereas the factor for blue does not. Which allele, the one for brown or the one for blue eye color, will be the most frequent is a very different matter. The frequency of any allele, dominant or recessive, is related to the survival value that it imparts to a population and not just to its dominance or recessiveness.

Nor does the fact that an allele is dominant mean that it must necessarily be better or that it will somehow produce a stronger or more desirable effect. A long list of undesirable dominant alleles in various species could be presented. One need only reflect on the fact that in humans the following seem to depend on dominants: extra fingers or toes, a form of muscular dystrophy, a type of dwarfism, a form of progressive nervous deterioration resulting in death, and several others. Such conditions are far from desirable and certainly do not make the afflicted person stronger. It is also apparent that the conditions, and hence their associated dominant alleles, are fortunately not as common in the population as the normal conditions and the normal, but recessive gene forms. Moreover, these undesirable dominants are *not* increasing

to the disadvantage of the whole population. If their dominance made them increase, everyone would eventually become afflicted.

INTERPRETATIONS OF CROSSES AND RATIOS. It is also essential at the outset of a study of genetics to appreciate the correct use of the Punnett square (Figs. 1-4 and 1-6) and the interpretation of genetic ratios. We have so far become acquainted with two important ratios: 3:1 and 1:2:1. It will be recalled that one reason for Mendel's achievement was that he conducted his studies with large numbers of plants. His F_1 and F_2 generations did not consist of just a few individuals. Few of us would expect a cross between two heterozygous tall plants (Tt × Tt) to yield exactly three tall and one dwarf if only four offspring were obtained. What does the 3:1 ratio (or any other) tell us? It tells us the *probability* of getting a certain phenotype or genotype when a particular cross is made. In his experiments, Mendel obtained numbers which were very close to 3:1 because he worked with large numbers of plants. But even then, the numbers were not exactly 3:1. We instinctively know that if we toss a coin a large number of times we will obtain heads in approximately 50% of the tosses. We also realize that there is a 50% chance for a boy to be born at any birth and a 50% chance for a girl. We do not consider it amazing to find a family in which all five children are girls. Nor are we very surprised if five tosses of a coin yield all heads. Only after hundreds of coin tosses or after counting the children in hundreds of families do we come close to realizing 50% heads and 50% tails or equal numbers of boys and girls. Knowing this, we see how illogical it is to assume that any four offspring from a pair of monohybrid parents must be three of one kind and one of another. This would be the same as expecting a girl baby to be followed by a boy who in turn would be followed by a girl, and so on, just because the chances are equal for the arrival of a boy or a girl. With this in mind, we can avoid absurd replies in situations illustrated by the following. In humans, normal skin pigmentation depends on the presence of a dominant factor, "A". Its recessive allele, "a," results in albinism, a deficiency of melanin pigment. Suppose that two people are known to be monohybrids for this character, Aa. They already have three normal children and inquire about the appearance of their next child. Many times, the novice will reply that the child will be an albino. Much confusion and anxiety can result from misuse of genetic concepts. The answer in this case can be given only in terms of probability. *No matter* how many offspring the couple has, the chance is 3:1 *at each birth* in favor of a normal child. In a similar family, it would not be unusual to find three albinos and one normal, even though there is only one chance out of four for an albino at each conception. It must be kept in mind when ratios are discussed that they do not tell us what the exact results will be from a cross. Instead, they give the chance *at any one time* that a certain event will occur.

THE TESTCROSS AND ITS VALUE. With these points clarified, we can proceed to further experiments of Mendel. Once he formulated the law of segregation, Mendel performed other kinds of crosses to see if the rules in his hypothesis were followed. One of the first types of crosses he made is what we know today as the *testcross*. To perform this, Mendel took his F_1 hybrids and mated them with recessives, such as a cross between hybrid tall plants and dwarf ones (Fig. 1-8). Mendel realized if the law of segregation were correct that the monohybrid tall plants should produce two classes of gametes in equal amounts, one-half with the factor for tall (T) and one-half with the factor for dwarf (t). A dwarf plant, carrying only the recessive allele for height, can produce only one class of gametes for this characteristic, those carrying the allele, t. Consequently, genetic factors for height coming from the dwarf parent will not mask any gene forms for height contributed by the hybrid parent. As a result, it becomes possible to determine the kinds of gametes formed by the hybrid as well as the proportion in which they are formed. To do this it is only necessary to observe the kinds of offspring in the testcross and the frequency in which they occur. As Figure 1-8 shows, one-half of the offspring from the testcross should be tall and the other half dwarf, a ratio of 1:1. This is so because the two alleles (T and t) are segregating in the hybrid. They are produced in equal amounts, and it is a matter of chance whether a gamete carrying T or one

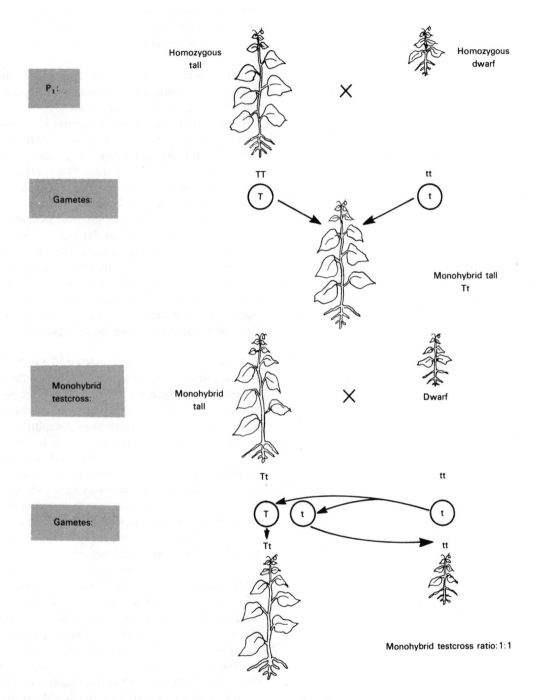

P₁:
Homozygous tall   ×   Homozygous dwarf

TT   tt

Gametes:
T   t

Monohybrid tall
Tt

Monohybrid testcross:
Monohybrid tall   ×   Dwarf

Tt   tt

Gametes:
T   t   t

Tt   tt

Monohybrid testcross ratio: 1:1

FIG. 1-8. Monohybrid testcross. The testcross ratio, 1 tall:1 dwarf, is exactly the same as the ratio of the two types of gametes produced by the monohybrid, one kind with allele T, the other with allele t. This is so, because the testcross parent (tt) carries only the recessive allele for the character, height, and cannot mask any factors segregating in the hybrid parent. The testcross shown here is also a backcross, because "dwarf" is one of the P₁ types.

carrying t combines with the gamete from the recessive parent. Mendel performed many testcrosses of monohybrids and obtained results very close to a 1:1 ratio in each case, thus supporting the ideas expressed in his first law. The testcross is such a valuable procedure in genetic analysis that we will frequently refer to it throughout the text.

The term "testcross" is often used interchangeably with the expression "backcross." However, the two are not necessarily the same. We have just seen that a testcross is a cross of any individual to one which is recessive for a trait which is being followed in a cross. A backcross, on the other hand, implies a mating of an individual to a type like one of the parents. In our example (Fig. 1-8) we have *both* a testcross and a backcross. The monohybrid is crossed to the recessive, but the recessive phenotype is also one of the parental types. If the hybrid had been crossed instead to a homozygous tall (TT), we would have an example of a backcross but not a testcross. Since a backcross may have special value in some genetic analyses, its distinction from the testcross should be recognized.

DIHYBRID TESTCROSSES AND INDEPENDENT ASSORTMENT. Mendel's work with monohybrids indicated clearly that a 3:1 ratio is to be expected from a cross of two monohybrids and that a 1:1 ratio will follow from a testcross of a monohybrid. Mendel's further studies entail the behavior of two or more pairs of alleles and led to the formulation of Mendel's second law. A question which occurred to Mendel was how two or more pairs of factors might behave in relationship to one another when followed at the same time in a cross. From his monohybrid crosses, Mendel knew that the color of the pea seed can be either yellow or green and that the factor for yellow (G) is dominant to green (g). When plants homozygous for yellow seeds are crossed with those having green seeds, the F_2 generation contains the yellow and green phenotypes in a ratio of 3:1. Likewise, the shape of the seed, round (W) vs. wrinkled (w), depends on a pair of alleles with the factor for the former being dominant. The F_2 ratio is thus three round to one wrinkled. But if both characteristics, seed color and seed shape, are followed at the same time, how will the factors for the two of them behave?

Suppose a plant which is pure breeding for yellow, round seeds is crossed to one from a green, wrinkled variety (Fig. 1-9A). From Mendel's first law, we would expect two factors for each character to be present in each parent. Knowing which genes are dominant, we may represent the genotypes of the parents as GGWW (yellow, round) and ggww (green, wrinkled). The F₁ dihybrid between them will receive one factor for each character from each one of the parents and will have the genotype GgWw. The F₁'s are all found to have yellow, round seeds. This is not surprising because yellow and round have been shown to be dominant. The question which now arises is, "What will happen when the F₁ dihybrids are followed further?" Will the factors for yellow and round stay together through later generations because they were together in one of the original parents? The same question arises for green and wrinkled. It is also possible that

FIG. 1-9. Origin of a dihybrid and its testcross. A. Cross producing a dihybrid. The yellow, round phenotype is to be expected because the factor for yellow (G) is dominant to its allele (g), and the one for round (W) is dominant to wrinkled (w). B. The dihybrid is testcrossed. The phenotypic ratio 1:1:1:1 is exactly the same as that in which the four classes of gametes are produced in the dihybrid: 1GW:1Gw:1gW:1gw.

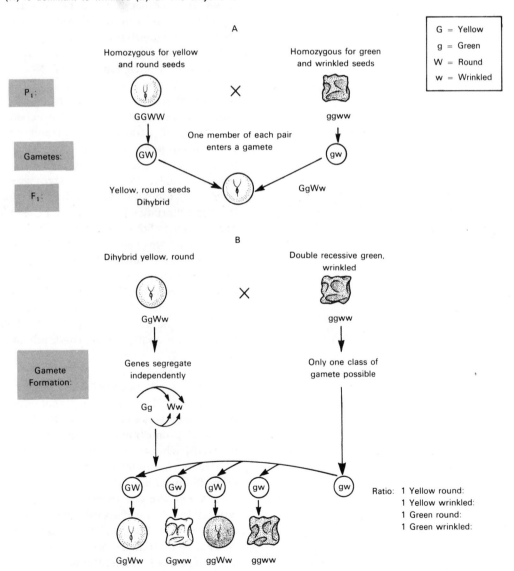

the original gene combinations are not required to remain together but that they are free to enter into new combinations. The most direct way to settle the matter is to take the F_1 dihybrid and perform a testcross because a testcross can tell us the kinds of gametes produced by an individual and the frequency of the different kinds. As Figure 1-9B shows, the offspring resulting from this testcross are of four types, and they occur with equal frequencies, a ratio of $1:1:1:1$. This tells us directly that the dihybrid is forming four different classes of gametes. This in turn means that the factors for seed color and those for seed shape are behaving independently. Just because the factors for yellow and round (G and W) and those for green and wrinkled (g and w) were together in the original parents did not mean they had to stay together. In addition to the old combinations, GW and gw, the F_1 dihybrid also produces new ones, Gw and gW. Nothing seems to tie the factors together. They are free to form new combinations. We say that they "assort independently." As a result, the expected dihybrid testcross ratio is $1:1:1:1$, a reflection of the types of gametes and their proportions. We see here, as in the case of the monohybrid, that the testcross parent (double recessive "ggww") produces only one kind of gamete in relationship to the characteristics we are following. Since all the gametes carry both recessives, "g" and "w," they cover up nothing contributed by the other parent. Thus, we can estimate the kinds of gametes and their proportions produced by this other parent simply by counting the kinds of offspring. Figure 1-10 shows one other way to envision the independent behavior of two pairs of alleles in a dihybrid and to determine the kinds of gametes formed.

INDEPENDENT ASSORTMENT AND THE F_2 DIHYBRID RATIO. Knowing that any dihybrid individual forms four different kinds of gametes in equal proportions, we can depict diagrammatically what to expect in a dihybrid cross (one in which two pairs of contrasting factors are being followed) carried through the F_2 generation (Fig. 1-11). We see from the figure that the F_2 will fall into four different phenotypic categories in a ratio of $9:3:3:1$. This is exactly what is to be expected from the independent behavior of the two pairs of alleles. When Mendel performed

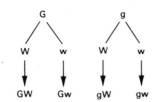

The alleles for color segregate from each other.

The alleles for seed shape also segregate, but they do so independently of those for seed color. Therefore—

the allele for yellow has an equal chance of entering a gamete with either the allele for round or the one for wrinkled. The same is true for the allele for green seed.

FIG. 1-10. Branching method. One way to determine the different types of gametes which can be formed by a dihybrid.

the cross, the actual frequencies which he obtained were: 315 round yellow, 108 round green, 101 wrinkled yellow, and 32 wrinkled green. Statistical analysis (Chap. 6) shows that these figures are compatible with a 9:3:3:1 ratio. This classic dihybrid ratio is so fundamental to genetic studies and so frequently encountered that we must be able both to recognize it and also to predict the phenotypes which are to be expected from a cross of any two dihybrids. Note (Fig. 1-11) that while only four different phenotypes result when dominance is operating in both pairs of factors, there are actually nine different genotypes which arise from the independent assortment of the genes. We can think of a cross between any two dihybrids in the following general way: AaBb × AaBb. When dominance is involved, the expected results are: 9A__B__ (those showing both dominant traits; the dashes indicate that either the dominant or the recessive allele may be present); 3A__bb (those expressing the dominant A and the recessive b); 3aaB__ (individuals expressing the recessive a and the dominant B); and 1 aabb (double recessives).

Other crosses of pea plants in which the remaining five pairs of alleles were followed produced comparable results and established the 9:3:3:1 ratio as the one to be expected when two dihybrids are crossed and 1:1:1:1 as the dihybrid testcross ratio. From such data, Mendel formulated his second law, *the law of independent assortment,* which states what we have just seen: the members of one pair of factors (alleles) segregate independently of members of other pairs at the time of gamete formation (pair Gg segregates independently of pair Ww). This rule holds for two, three, or more pairs, but we will see that the law pertains only to those genes

located on different pairs of chromosomes (Chap. 3). Mendel was fortunate in encountering seven genes which were located on different chromosomes, each gene having two distinct allelic forms. The complication of linkage (genes found together on the same chromosome) was not involved in Mendel's crosses to cloud the picture of independent assortment.

SEPARATE CHROMOSOMES AND INDEPENDENT ASSORTMENT. Knowledge of independent assortment enables us to predict the kinds of offspring in any crosses involving two or more pairs of factors on separate chromosomes. The Punnett square method can aid us in visualizing this; it also provides us with a picture of the different genotypes. However, the method may be cumbersome at times, particularly when more than two pairs of alleles are involved. We can avoid the Punnett square by following an easier way to diagram crosses. This requires nothing more than a grasp of a simple mathematical principle which can be stated as follows: if we know the chances of two independent events happening separately, the chance of the two occurring together is the product of the separate probabilities. A dihybrid cross can be treated as if it were two separate crosses. Returning to the dihybrid pea plants whose seeds were round and yellow (GgWw × GgWw), we may think of the cross as two monohybrid crosses. We can consider it to be a cross between the one kind of monohybrid (Gg × Gg) and also a cross between the other kind (Ww × Ww). Our knowledge of monohybrid crosses tells us that in the first case we would expect 3/4 yellow and 1/4 green and in the second cross 3/4 round and 1/4 wrinkled. We know that the 3:1 ratio is an expression of

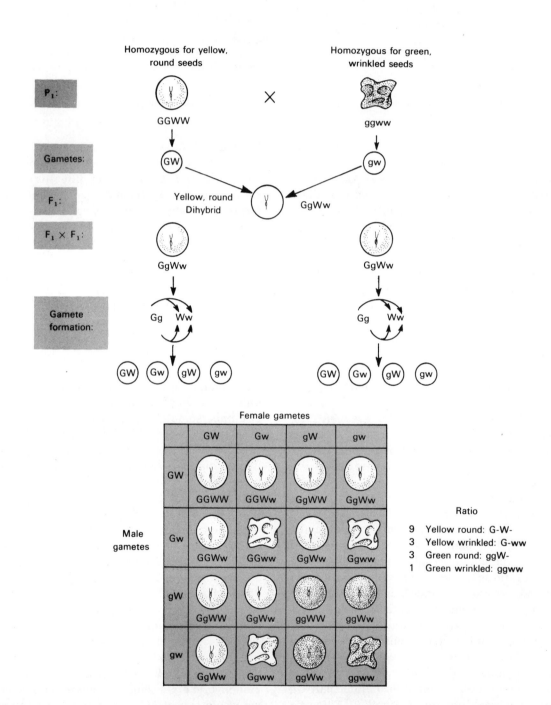

FIG. 1-11. Dihybrid cross followed through second generation. Inspection of the Punnett square will show that nine different genotypes are found among the F₂ offspring. Because of domi- nance in one member of each pair of alleles, only four phenotypes are expressed.

probability, the chance of a certain event happening at any one time. So in this dihybrid cross, we know the separate chances for round vs. wrinkled at any one time and similarly for the independent occurrences of yellow vs. green. To determine the different combinations which are possible in a dihybrid cross which involves the traits for seed color and for seed texture, we simply multiply together the separate probabilities:

3/4 round; 1/4 wrinkled
3/4 yellow; 1/4 green

9/16 round yellow: 3/16 wrinkled yellow: 3/16 round green: 1/16 wrinkled green

This is a much more direct and time-saving method than the Punnett square and accomplishes the same thing in telling us the different phenotypes to be expected and their frequencies. Suppose that we were also following a third characteristic at the same time, height, and wanted to know the results of a cross between two trihybrids: GgWwTt × GgWwTt. We obtain the answer quickly by proceeding as if we were dealing with three separate monohybrid crosses. We simply multiply the results of the dihybrid cross above by the chances for tall vs. dwarf:

9/16 round yellow: 3/16 wrinkled yellow: 3/16 round green: 1/16 wrinkled green
3/4 tall: 1/4 dwarf

27/64 tall round yellow: 9/64 tall wrinkled yellow: 9/64 tall round green: 3/64 tall wrinkled green: 9/64 dwarf round yellow: 3/64 dwarf wrinkled yellow: 3/64 dwarf round green: 1/64 dwarf wrinkled green

Although the Punnett square will confirm this, it is evident that an appreciable number of squares is required. Sixty-four are needed because each trihybrid forms eight different classes of gametes as a result of independent assortment of three pairs of factors. The testcrosses of the trihybrid (GgWwTt × ggwwtt) shows us this. The trihybrid testcross itself may be treated as if it were three separate monohybrid testcrosses: Gg × gg, Ww × ww, and Tt × tt. Inasmuch as the expected ratio for each is 1:1, the trihybrid testcross becomes:

1/2 yellow: 1/2 green
1/2 smooth; 1/2 wrinkled

1/4 yellow smooth: 1/4 green smooth: 1/4 yellow wrinkled: 1/4 green wrinkled
1/2 tall: 1/2 dwarf

1/8 tall yellow smooth: 1/8 tall green smooth: 1/8 tall yellow wrinkled: 1/8 tall green wrinkled: 1/8 dwarf yellow smooth: 1/8 dwarf green smooth: 1/8 dwarf yellow wrinkled: 1/8 dwarf green wrinkled.

The above trihybrid testcross ratio of 1:1:1:1:1:1:1:1 is obtained because eight different types of gametes are possible from a trihybrid. Figure 1-12 depicts how the branching method may also enable us to derive the different combinations and to reach the same results. The independent behavior of the pairs of alleles permits us to estimate the number of different kinds of gametes which will be produced by any hybrid. From the information presented on testcrosses, we have seen that a monohybrid (Gg) forms two different kinds of gametes, the dihybrid and trihybrid (GgWw and GgWwTt), four and eight different types, respectively. A testcross of a tetrahybrid can result in the formation of 16 gene combinations. The number of kinds of possible combinations which can be formed by individuals of different degrees of hybridity forms a geometric progression. This fact can be summarized by the very helpful expression: 2^n. This tells us the number of different kinds of gametes which can be formed by any hybrid, allowing "n" to stand for the number of hybrid gene pairs. Table 1-1 summarizes a great deal of valuable information which can be derived from the basic genetic principles discussed so far.

We can easily illustrate some of the points contained in Table 1-1. A monohybrid tall plant (Tt) forms *two* classes of gametes: T and t. If two such monohybrids are crossed, the number of phenotypic classes is again *two*: tall and dwarf. The number of phenotypic classes in the monohybrid testcross also equals *two*: tall (Tt) and dwarf (tt). However, the number of different genotypes which can arise when two monohybrids are crossed is three: TT, Tt, and tt. If there is no dominance, as in the case of the four-o'clocks, the number of phenotypic classes will be equal to the number of genotypic classes because the heterozygotes can be distinguished from the homozygotes (RR, red; Rr, pink: rr, white). If dihybrid and trihybrid crosses, etc. are considered in the above context, it will be apparent that the simple expression 2^n is indeed a valuable one to keep in mind. It also tells us something about our own species. Humans possess 23 pairs of chromosomes. The human is also highly heterozygous. For any one pair of chromosomes, man is hybrid for many, many allelic pairs. But to keep the example as simple as possible, let us assume that man is hybrid for only one pair of alleles on each pair of chromosomes. We can now consider the possibilities which can arise from independent assortment alone. Table 1-1 tells us that any one parent would form 2^n or 2^{23} different kinds of gametes. From two such parents, at least that many phenotypic classes can appear. However, many more kinds of genotypic classes are possible, 3^{23}. In actuality, this illustration is simplified to the point of absurdity, because countless other allelic pairs on the same chromosome would be involved. We will learn that reassortment among genes on the same chromosome may take place, and this increases the possibilities many times more (Chap. 8). We are also ignoring lack of dominance which will add to the number of phenotypic classes. At any rate, although 2^{23} is the minimal figure for the number of kinds of gametes and phenotypic classes, it is so immense that it defies the imagination. We should appreciate why no two people on earth, except for

FIG. 1-12. Trihybrid testcross. Branching method is used here to derive the different kinds of gametes formed by the trihybrid.

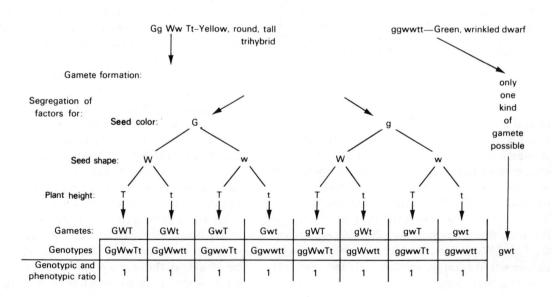

TABLE 1-1 Summary of important consequences of segregation and independent assortment

NO. OF ALLELIC PAIRS IN HYBRID	NO. OF KINDS OF GAMETES WHICH HYBRIDS CAN FORM	NO. OF PHENOTYPIC CLASSES WHEN TWO SUCH HYBRIDS ARE CROSSED (ASSUMING DOMINANCE)	NO. OF PHENOTYPES IN TESTCROSS OF THE HYBRID	NO. OF GENOTYPES WHEN TWO SUCH HYBRIDS ARE CROSSED (ALSO THE NO. OF PHENOTYPES WHEN NO DOMINANCE PERTAINS)
1	2	2	2	3
2	4	4	4	9
3	8	8	8	27
4	16	16	16	81
5	32	32	32	243
6	64	64	64	729
n	2^n	2^n	2^n	3^n

identical twins, are genetically identical. Each of us is unique. The chance for another to be exactly like us, even from our own parents, is so improbable that the idea can be dismissed. Each of us owes his distinct nature in a large degree to independent assortment. In the two chapters which follow, we will see that this is a consequence of chromosome behavior and the nature of the sexual process.

RECOGNITION OF THE SIGNIFICANCE OF MENDELIAN PRINCIPLES. Before the discovery of Mendel's paper in 1900, several investigators had almost deduced Mendelian principles from their own experiments. De Vries, one of the three biologists who found Mendel's paper, had made many of the same observations in the course of his many studies. In 1900, he reported that Mendelian principles apply to approximately a dozen very different groups of seed plants. De Vries also made an important contribution to genetics as a result of his investigations with the evening primrose, a species which seemed to violate Mendelian rules. Actually, the odd results derive from an unusual chromosome situation in this plant and present no real exception to the chromosome theory (Chap. 10). From the unusual results obtained from work with the evening primrose, De Vries formulated the mutation theory, now a fundamental concept in genetics. He thought that the unexpected appearance of certain phenotypes among his plants resulted from mutations, or sudden inheritable changes. Although the unusual plants were later shown to arise for other reasons, the mutation theory of

De Vries focused the attention of geneticists on the fact that sudden inheritable changes may occur and that these may provide the source of the genetic variation required for evolutionary progress.

Correns, another of Mendel's discoverers, realized the full significance of the data on the garden pea. In 1902, he connected the behavior of Mendel's factors with the chromosomes and was thus the first to attribute the operation of Mendelian laws to chromosome behavior. After 1900, Mendel's principles were confirmed and extended to include many other species, animals as well as plants: Cuenot in France with mice, Castle in the United States studying rodents, and Davenport (U.S.) and Bateson (England) working with poultry. The latter was another geneticist who was very close to deducing Mendelian principles just from his own breeding work. Among his many contributions to genetics is the very name of the science itself, which he coined along with "heterozygous," "homozygous," and "allelomorph" (which we shorten to allele). We owe the expressions "gene," "genotype," and "phenotype" to the Danish botanist, W. Johannsen. In one of his classic studies, Johannsen devised a method to distinguish the variation in beans which was due to environment from the variation contributed by genetic factors. Many of his ideas greatly influenced genetic thought and will be discussed in later chapters. Many of the investigators, including Mendel, appreciated the need to consider the environment when undertaking genetic analysis.

At the turn of the century, the features of meiosis were being unraveled. Montgomery postulated that the chromosomes occur in corresponding pairs, one member of each pair contributed by each parent. Other workers besides Correns then began to point out that the chromosomes might provide the physical basis for Mendel's factors. In 1902, Sutton gave perhaps the most convincing arguments, building on the ideas of earlier biologists and his own observations. He reasoned that since the gametes are the only connection between the generations, all the hereditary material must be carried in them. The egg contributes a great deal of cytoplasm at fertilization, whereas the sperm sheds most of its cytoplasm as it matures. Since genetic experiments indicated that the hereditary contribution is equal from both parents, the factors must reside in the nucleus. The only visible parts of the nucleus which are accurately divided when the cell divides are the chromosomes; therefore, the genes must be carried on them. The chromosomes of an individual occur in pairs; so do Mendelian factors. In the formation of gametes, the chromosomes segregate; Mendelian factors also segregate at some time prior to gamete formation. Mendel's factors segregate independently. The chromosomes seemed to do this also, but there was still no definite proof of the point, which was to come through the studies of Carothers in 1913. Nevertheless, the observations of cytologists before this date gave no reason to support the idea that the chromosomes are tied together in some way which can prevent their independent assortment.

Therefore, before 1910 there was no conclusive proof that the hereditary factors, the genes, were on the chromosomes, but all the evidence favored this interpretation. Perhaps the strongest argument came from the parallel between chromosome behavior and the behavior of the genetic factors as expressed in Mendel's laws. We will examine these points in some detail in the two chapters which follow.

REFERENCES

Dunn, L. C. *A Short History of Genetics*, McGraw-Hill, New York, 1965.

Mendel, G. *Experiments in Plant Hybridization.* Translated in Classic Papers in Genetics, J. A. Peters (ed.), pp. 1–20, Prentice-Hall, Englewood

Cliffs, N.J., 1959. [Also appears in C. Stern, and E. R. Sherwood, 1966 (below).]

Stern, C. and E. R. Sherwood, (eds.), *The Origin of Genetics, A Mendel Source Book*. W. H. Freeman, San Francisco, 1966.

Strickberger, M. W. *Genetics*, Macmillan, New York, 1976.

Sturtevant, A. H. *A History of Genetics*, Harper and Row, New York, 1965.

REVIEW QUESTIONS

1. In tomatoes, round fruit (O) is dominant to oblong fruit (o). Write as much as possible of the genotypes of:

    A. A plant from a homozygous round-fruited stock.

    B. A plant from an oblong strain

    C. A round-fruited plant which resulted from a testcross

    D. A oblong-fruited plant which resulted from the cross of two round-fruited ones

2. Using information from the above question, give the genotypic and phenotypic ratios to be expected in regard to fruit shape from each of the following crosses:

    A. homozygous round × heterozygous round

    B. heterozygous round × heterozygous round

    C. oblong × oblong

    D. heterozygous round × oblong

3. Two short-haired female cats are mated to the same long-haired male. Several litters are produced. Female No. 1 produced eight short-haired and six long-haired kittens. Female No. 2 produced 24 short-haired ones and no long-haired. From these observations, what deductions can be made concerning hair length inheritance in these animals? Assuming the allelic pair S and s, give the likely genotypes of the two female cats and the male.

4. In cattle, the hornless condition (H) is dominant to the horned condition (h).

    A. A horned bull is mated to a hornless cow that is heterozygous for the condition. What kinds of offspring are to be expected and in what ratio?

    B. If the cow is next mated to a hornless bull that is a heterozygote, what is the chance that the first calf will be horned?

    C. Assuming that the first calf born has horns, what is the chance that the second calf will be hornless?

5. In cattle, the alleles for red coat (R) and white coat (r) behave as codominants. Both red and white hair are produced so that the heterozygote is intermediate or roan colored.

    A. Give the phenotypic and genotypic ratios to be expected among the offspring from a cross of two roan animals.

    B. What are the expected genotypic and phenotypic ratios from a cross of a roan animal and a white one?

6. Write as much as possible of the following genotypes:

    A. A horned white bull.

    B. A hornless roan cow.

    C. A hornless red cow whose male parent was horned and roan.

7. In guinea pigs, short hair (L) is dominant to long (l). Black fur (A) is dominant to albino (a). A female from a strain which is pure breeding for black fur and short hair is mated to a male from a strain pure breeding for the albino condition and long hair.

    A. What will be the phenotypes of the F_1?

    B. If members of the F_1 are mated among themselves, what percentage of offspring can be expected to be homozygous for both traits? Give the genotypes and phenotypes of these homozygotes.

8. In certain breeds of chickens, the factor (B) is responsible for the production of black feathers and its allele (b) produces feathers that are basically white (except for flecks of black). The heterozygote is blue-feathered. Another factor (F) produces straight feathers, whereas its allele (f) results in the frizzled condition, giving brittle

feathers. The heterozygote is mildly frizzled. What is to be expected from the following crosses? Give the phenotypic ratios.

A. A black, frizzled hen × a white, straight rooster.
B. A white, frizzled hen × a blue, straight rooster.
C. A blue, mildly frizzled hen × a white frizzled rooster.
D. A blue, mildly frizzled hen × a blue, mildly frizzled rooster.

9. In the human, the allele for normal skin pigmentation (A) is dominant to (a) for the absence of pigment, the albino trait. A certain type of migraine headache (M) behaves as a dominant to the normal condition (m), absence of headache. Write as much as possible of the genotypes of the following three persons:

A. A phenotypically normal individual whose mother was an albino with migraine.
B. A person suffering from migraine who has normal skin pigmentation. Both parents have normal skin pigmentation, and one of them suffers from migraine.
C. An albino person who does not have headaches, whose normally pigmented parents both suffer from migraine.

10. A person who is dihybrid with respect to both the skin pigmentation and headache characteristics marries a person who is heterozygous at the skin pigmentation locus and who does not have headaches.

A. Give the genotype of each and the kinds of gametes each will produce.
B. What kinds of offspring are to be expected and in what phenotypic ratio?

11. The factors for normal hemoglobin (S) and for sickle-cell hemoglobin (s) may be considered codominants. A blood examination can detect the heterozygote, since a small percentage of the cells will show sickling. A man whose cells show no sickling upon examination marries a healthy woman who is found to have a certain percentage of sickle cells.

A. What is the chance of their having a baby with the severe anemia?

B. The woman later marries a healthy man whose blood examination reveals sickle cells. They have three children free of sickle-cell anemia. What is the chance that the next child will have the severe disorder?

C. What is the chance that any one of the healthy children is a carrier of the sickle-cell allele?

12. Assume that medical science finds a treatment for sickle-cell anemia. When instituted in infancy, it permits an otherwise doomed person to live a normal life span without any serious effects of the anemia. If two such persons, saved from the fatal effects of the disease, become parents, what will be the genotypes of their offspring? Would they require treatment? Explain. Also give an explanation which would be expected from a person who supports the Lamarckian concept.

13. In the human, free ear lobes (F) are dominant to attached lobes (f). Using this information and that on sickle-cell hemoglobin in Question 11, answer the following:

A. Give the genotypes of the following persons:
   (1) A healthy woman with attached ear lobes whose red blood cells show some sickling.
   (2) A man with free ear lobes whose mother also had free lobes but whose father had attached ones. This man's red blood cells show some sickling.

B. What kinds of offspring are to be expected if the man and woman marry, and in what ratio?

14. The fatal disorder, Huntington's disease, results in progressive nervous deterioration and death. It usually strikes well after the age of puberty so that afflicted persons may have produced offspring. The genetic factor responsible for the disorder is a dominant one (H), which is very rare in the population and does not seem to be on the increase.

A. Offer an explanation for the fact that more and more persons do not seem to be suffering from the disorder resulting from this dominant.

B. If a young person had one parent who died of the disorder and another who is apparently free of it, what does he or she know about the chances of developing the disorder?

15. Give the number of phenotypic classes expected from each of the following:

A. Testcross of AaBbCC.
B. Testcross of AaBbCcDd.
C. Testcross of AaBbCcDdEE.

16. Give the number of phenotypic classes to be expected when the following crosses are made and when there is dominance:

A. Cross of two individuals who are AaBbCcDd.
B. Cross of two individuals who are AaBbCcDDEE.
C. Cross of two individuals who are AaBbCcDdEe.

17. Give the number of genotypic classes to be expected in A, B, and C of the above question.

In the pea plant, tallness (T) is dominant to shortness (t). Yellow seed color is dominant (G) to green (g), and round seed shape (W) is dominant to wrinkled (w).

18. Show the different kinds of gametes which can be formed by individuals of the following genotypes: (1) TtGG, (2) TtGGWw, (3) TtGgWw.

19. What would be the expected phenotypic ratio if each of the above three plants were testcrossed?

20. Suppose plants (2) and (3) in Question 18 were crossed. What would be the expected phenotypic ratio among the offspring?

CHROMOSOMES AND THE DISTRIBUTION OF THE GENETIC MATERIAL

Many descriptions of the cell which were given by the pioneer cytologists in the early part of this century are known for their accuracy and attention to detail. Much of our knowledge today about chromosome structure and behavior at the time of cell division is based on the accurate accounts of these keen scientists. Throughout the years, improvements in cytological methods have made possible an even greater resolution of nuclear features and have enabled us to ascertain the fine structure of many cytoplasmic elements. The electron microscope has been one of the most important tools in this area of cell study. Electron microscopy and the advances in molecular biology in the past 20 years have greatly enhanced our appreciation of the dynamic interplay which takes place among the many parts of the cell, both nuclear and cytoplasmic. Before concentrating on some of the details of the chromosome and its behavior, based in large part on the work of the early descriptive biologists, let us review some general concepts which are recognized by the student of the cell.

ALL CELLS MUST SOLVE CERTAIN PROBLEMS. Early in the evolution of life, long before the origin of the first cell, primitive systems composed of aggregates of macromolecules and simpler substances probably settled out of the primordial seas. The only systems able to survive were those which by chance came to incorporate several features that guaranteed their persistence and their continuity. Such systems were able to maintain themselves and eventually gave rise to the first primitive cells. From these, the familiar cell types of today have evolved. We find in all of them, from the bacterial cell to the most specialized ones in higher organisms, those features which the ancestral cell types acquired as they solved the problems of survival and reproduction. All cells have some way of separating themselves from the external environment. This is accomplished almost universally by the presence of a cell membrane, which in turn may be surrounded by a wall (typical of plants) or a pellicle. However, this separation from the external environment must not be one of complete isolation because the continuation of internal cellular activities depends on communications with the outside. Substances must enter the cell when new building blocks are required; useless or toxic substances from cellular metabolism must pass out of the cell. Therefore, the barrier between internal and external environment must not be complete, nor may it be haphazard. Remarkable features of the cell membrane solve this dilemma and enable it to *selectively* control the entry and exit of substances to and from the internal cellular environment. This property of selective permeability possessed by the cell membrane is not just a passive process but depends on living cellular activities, most of which require an energy source. And so we find that all cells have some means to release and transfer energy. This energy is needed if the cell is to grow, but it is also required just to maintain stable metabolic processes. In most plant and animal cells, the release of energy and its transfer are mediated largely through the activities of certain intracellular bodies (organelles), the mitochondria, which can be seen with the light microscope but whose morphology is resolved only by the electron microscope (Fig. 2-1). These organelles, which may be very abundant in the cytoplasm, contain all the enzymes required for aerobic respiration and the manufacture of adenosine triphosphate (ATP), the molecule in which high

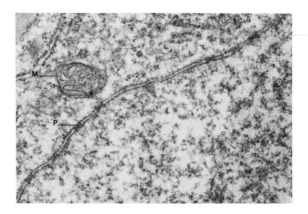

FIG. 2-1. Electron micrograph of portion of a cell of the grass, *Panicum*, showing mitochondrion (*M*) in the cytoplasm. Note pores (*P*) in nuclear membrane. (Courtesy of R. F. Lewis.)

energy is stored in most cells. Energy release through anaerobic processes also takes place in the cytoplasm, but the mitochondria are by far the most important and efficient cell components in the manufacture of high-energy bonds. In addition to mitochondria, certain plant cells may also contain chloroplasts, chlorophyll-containing organelles in which energy of the sun may be trapped and converted to ATP. This trapped energy may then be utilized to manufacture food in the chloroplast. As we will learn in Chapter 20, both the mitochondria and the chloroplasts contain genetic information which may play a role in extrachromosomal inheritance.

However, not all cells possess mitochondria for the formation and transfer of energy. Cells may be classified into two general categories. Those which compose the most familiar plants and animals are termed *eukaryotic*—cells which contain an assortment of intracellular organelles in which certain specific processes are packaged and carried out. Eukaryotic cells also possess a distinct nucleus with a membrane which surrounds the genetic material. On the other hand, bacterial cells and blue-green algae lack true membrane-bound nuclei and never contain certain organelles, such as the mitochondria and chloroplasts. Cells of this type are called *prokaryotic*. Although prokaryotes are faced with the same problems as eukaryotes, they solve them in a somewhat different fashion, which will be considered more fully in later chapters. The discussion here will be confined mainly to eukaryotes which include the bulk of the organisms used in classical genetic studies.

If a cell is to persist from one generation to the next, it must have some mechanism to store and transfer information. Otherwise, newly formed cells would lose the ability to build essential cellular products and to carry on metabolic activities in a coherent manner. Moreover, cellular identity would be lost. Such chaotic systems would probably collapse and cease to exist. All cells have solved this problem through the coding of information in certain large molecules. These are the nucleic acids, deoxyribonucleic acid (DNA) and ribonucleic acid (RNA). The former is the chemical substance of the gene and is confined primarily to the nucleus. RNA, however, occurs as several molecular types which are widely distributed throughout the nucleus and cytoplasm.

COMMUNICATION BETWEEN NUCLEUS AND CYTOPLASM. The nuclear region of eukaryotes is bounded by an envelope. The electron microscope clearly shows that this is not a membrane which seals the nucleus off from the cytoplasm (Fig. 2-1). Instead, it is perforated by pores through which certain products may pass. Again, this is a selective process which depends on living activities. Communication between nucleus and cytoplasm can be visualized by electron micrographs which demonstrate that the nuclear membrane is continuous with the membranes of the endoplasmic reticulum (ER). The latter (Fig. 2-2) is an intracellular system which performs several functions. Not only does it supply channels through which nuclear and cytoplasmic products may pass, it also allows communication with the outside environment. Technically the channels of the ER are outside the cell. This membrane-bound network furnishes a great deal of added surface area for interactions between the cytoplasm and external environment. Of utmost importance is the fact that these membranes themselves contain enzymes and provide surfaces upon which many of the cell products may be constructed. In many active cells, the membranes of the ER are typically covered with particles which can be resolved easily with the electron microscope. These are the important ribosomes, found in all kinds of cells (Fig. 2-2). In eukaryotes, they may be seen lying free in the cytoplasm or attached to the ER, depending on the particular cell. Ribosomes are critical to the life and economy of the cell because it is through their activities that the messages coded by DNA and RNA are translated into the diversity of proteins associated with a given cell type. These little ribosomal bodies, composed of protein and RNA, are actually quite complex in their organization, the details of which will be examined later (Chap. 13). However, one point must be stressed here. The ribosomes depend on the DNA of the nucleus which contains the coded information needed both for ribosomal construction and the formation of cellular proteins; nevertheless, the DNA can accomplish nothing without the ribosomes and other components of the cytoplasm. Although it is the chromosomes with their DNA which contain the information required to perpetuate the cell, the genetic information by itself becomes meaningless if there is no way for coded messages to be interpreted and put to constructive use. In addition, the DNA depends on several devices for its orderly distribution to new cells at the time of cell division.

THE DISTRIBUTION OF THE GENETIC MATERIAL.   Mitosis is the nuclear division which ensures that two newly formed cells will contain genetic information identical to that in the parent cell. Several mechanisms were perfected during cellular evolution to make mitosis accurate and precise. While we examine some of these, we will also review Mendelian concepts in relationship to their physical basis in the cell. A good understanding of genetic phenomena depends on a comprehension of the details of chromosome structure and behavior.

When nondividing cells are stained with certain dyes, the nuclei become quite distinct. This is due to a reaction of the coloring agent with the nuclear material which then assumes a granular or netlike appearance. The name *chromatin* was given to this material in the nucleus which has an affinity for dyes, especially those basic in nature (Fig. 2-3A). As a nucleus starts to divide, one of the first detectable changes is seen in the chromatin, which gradually becomes more and more distinct. This is a consequence of the condensation of the chromatin into the bodies we call the *chromosomes*.

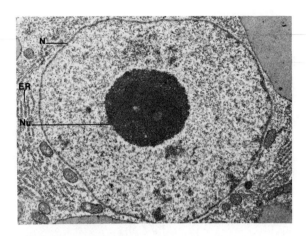

FIG. 2-2. Electron micrograph of portion of a cell of *Panicum* showing endoplasmic reticulum (*ER*) covered with ribosomes and the nucleus (*N*). Note conspicious nucleolus (*Nu*). (Courtesy of R. F. Lewis.)

It is important to realize that the term "chromatin" was not coined to define a specific kind of chemical material. Analysis of the chromatin has shown it to consist of several diverse substances: DNA, histone and nonhistone proteins, and RNA. Therefore, "chromatin" refers to an assortment of macromolecules, whereas "chromosomes" are the bodies evident at nuclear divisions and which are composed of chromatin.

Very characteristic changes in the chromosome and other parts of the cell accompany the process of mitotic division. These are so typical of cells in general that it has been possible to assign names to representative stages: prophase, metaphase, anaphase, and telophase (Fig. 2-3 B-E). However, remember that nuclear division is an uninterrupted process, one in which the various activities proceed continuously without abrupt breaks between the successive stages.

CHROMOSOME MORPHOLOGY.   When describing chromosome morphology, cytologists refer to the "mature" or metaphase chromosome. This is customary because the chromosome at metaphase is in a compact state which can be easily studied with the light microscope. We can see clearly that the chromosome is composed of two halves or two sister chromatids. The expression "sister" implies that both structures are identical and have resulted from the replication of an original chromosome strand (Fig. 2-4). This double nature of the chromosome is apparent even

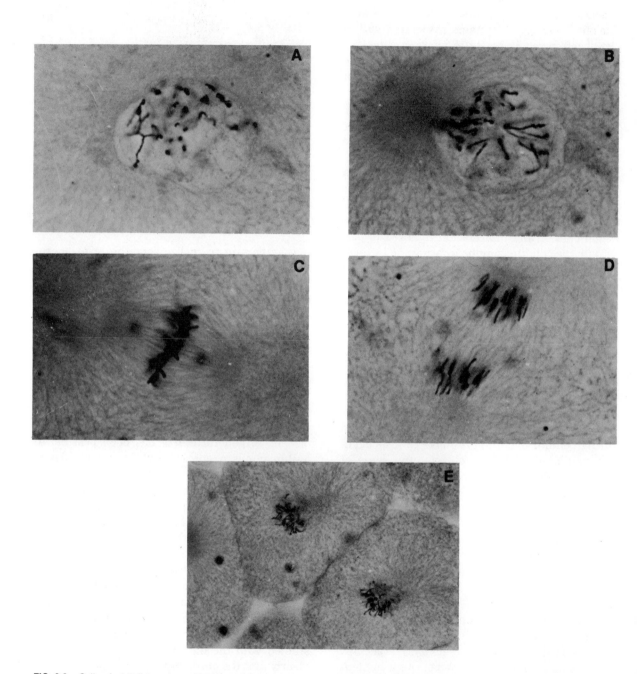

FIG. 2-3. Cells of whitefish embryo showing nuclear changes at mitotic division. *A.* Nucleus of cell in which the chromatin is just starting to condense as cell enters prophase (*B*). During this stage, chromosomes become more and more evident. The dense region at the left of the nucleus is the centrosome which will divide and eventually mark the poles of the spindle. At metaphase (*C*) the chromosomes are arranged at the equator of the spindle. Anaphase (*D*) is recognized by the movement of chromosomes to opposite poles of the spindle. At telophase (*E*) chromosomes become less and less evident as the nuclear membranes reform. Constriction of the cytoplasm in this species achieves the division into two cells.

at early prophase, the first stage of mitosis. It is a direct consequence of the fact that the chromosome material duplicates during *interphase,* the portion of the cell cycle between divisions and before the onset of prophase. This double nature extends the length of the chromosome, except at one region, the *centromere.* It is also called the "primary constriction," due to the characteristically constricted appearance of the two sister chromatids which are held together at this point (Figs. 2-4 and 2-5*A*). The centromere is the dynamic center of the chromosome and is responsible for many of the chromosome movements at mitosis. It is the division of the centromere which helps trigger the movement at anaphase when the sister chromatids of each chromosome separate and move to opposite poles. Once the centromere divides, each chromosome is composed of only one chromatid (Fig. 2-3*D* and 2-5*B*). A chromosome, therefore, may be composed of one *or* two chromatids, depending on the stage of the nuclear cycle. We must adopt the habit of thinking in terms of the chromatid as the physical basis of genetic phenomena because the fundamental genetic principles can be explained by its behavior when the nucleus divides.

When we observe the mature chromosome complements of many species, we can often see that the chromosomes have different appearances. (Note the size differences of human chromosomes shown in Fig. 2-4). They may range in size from small to large. Another reason for their variation is related to the position of the centromere. It may be located more or less in the center of the chromosome, in which case the centromere is in a "median position." Consequently, the arms of such a metacentric chromosome appear approximately equal, and during its movements, the chromosome may assume a V shape (Fig. 2-6). If the centromere is just off to one side of center, it is then considered to be "submedian" in position. This kind of chromosome is submetacentric and will have one arm distinctly shorter than the other and may resemble an L figure at anaphase. When the centromere is way off to one side, giving the chromosome a large arm and one which is very small or inconspicuous, the chromosome is termed "acrocentric" and may have the shape of a J or I (Fig. 2-6). In normal chromosomes, centromeres

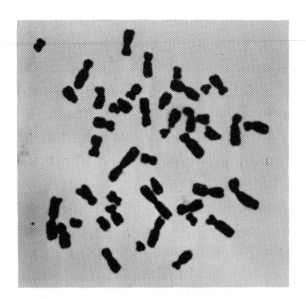

FIG. 2-4. Human chromosomes in white blood cell from a human male. The cell was actually in metaphase but was swollen by a treatment which separates the chromosomes so that they may be counted and their morphology observed. Note that each chromosome is composed of two chromatids and is double throughout its length. The chromatids are held together by the centromere which is evident here as a constricted region. Human chromosomes range in size from large to small and differ in the position of the centromeric region.

are not believed to occur at the extreme end. However, when the centromere appears to be at an extremity and the chromosome has only one evident arm, the expression "telocentric" is commonly used.

In addition to the narrowing attributed to the centromere, some chromosomes may possess still other constrictions, known as *secondary constrictions.* These may be so pronounced that nothing more than a filament appears to connect the two portions of the chromosome. In such a case the smaller portion is called the satellite (Fig. 2-7). Satellites frequently are associated with the formation of the nucleolus. The nucleolus does not simply form anywhere in the nucleus nor does it float about in the nuclear sap. Instead, it arises at a special site on a specific chromosome. In a diploid organism, at least two such chromosomes are present. The special sites are the *nucleolar organizing regions,* and these are generally found in the vicinity of the secondary constrictions or satellites. When nucleoli reform at telophase, it is at these sites. They will remain attached to the nucleolar organizing chromosomes until the end of the next prophase. Even in the nondividing cell, the nucleolus remains associated with the chromosomal region on which it was formed. The number of nucleoli present will vary from one species to another, depending on the number of nucleolar organizing chromosomes. Since two or more nucleoli may fuse, the number per cell can vary, even within an organism. Although the nucleolus does not influence chromosome movement and is not involved in the segregation of genetic factors, its role in the economy of the cell is very significant. The nucleolus is essential for the assembly of the ribosomes; therefore, protein synthesis depends on the presence of normal nucleoli. On the level of molecular events, the nucleolus is a critical part of the cellular machinery essential for the translation of the information stored in the genes.

HOMOLOGOUS CHROMOSOMES AND GENETIC LOCI. When the chromosomes are studied at metaphase, it becomes evident that they exist in pairs. For any one chromosome of a particular length and shape, another can usually be found to match it. We say that the chromosomes exist in *homologous pairs*. This is so, because any sexually reproducing organism receives one set of chromosomes from each parent. Therefore, a

FIG. 2-5. Chromosome and chromatid. *A.* The chromosome exists as one chromatid before it duplicates. The chromosome is already replicated before the onset of prophase, so that the prophase chromosome is composed of two identical sister chromatids. At metaphase, these can be clearly distinguished. Both chromatids are held together by the centromere. *B.* Division of the centromere triggers anaphase movement. The sister chromatids move to opposite poles. Each chromosome, double at metaphase, is now composed of only one chromatid.

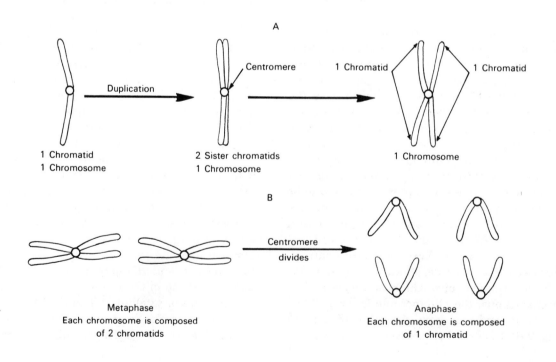

A

Duplication

Centromere

1 Chromatid   1 Chromatid

1 Chromatid
1 Chromosome

2 Sister chromatids
1 Chromosome

1 Chromosome

B

Centromere
divides

Metaphase
Each chromosome is composed
of 2 chromatids

Anaphase
Each chromosome is composed
of 1 chromatid

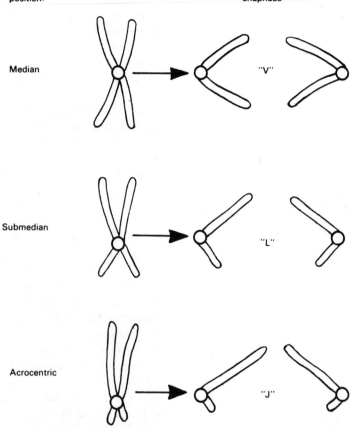

Centromere position:

Median

Submedian

Acrocentric

Shape at anaphase

"V"

"L"

"J"

FIG. 2-6. Chromosome morphology. The shape of the chromosome at metaphase depends largely on the position of the centromere. At anaphase, chromosome shape is the result of the arm length and the fact that the centromere is leading in the movement to the poles.

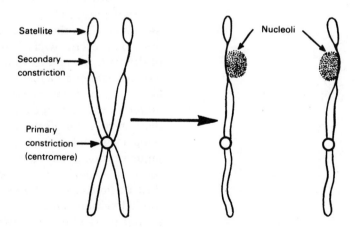

Satellite

Secondary constriction

Primary constriction (centromere)

Nucleoli

FIG. 2-7 Secondary constrictions. These occur on some chromosomes and may be so pronounced that a portion of the chromosome appears to be set off as a satellite. These secondary constrictions frequently contain nucleolar organizing regions at which nucleoli reform at telophase. Every typical normal diploid possesses at least two chromosomes with nucleolar organizing regions.

CHROMOSOMES AND DISTRIBUTION OF GENETIC MATERIAL

particular kind of chromosome from the maternal parent should also be found in the chromosome contribution from the paternal parent. An exception to this occurs in the case of those organisms which have sex chromosomes. In such species, the sex chromosomes received from the two parents may be quite different in appearance and may contain extensive regions which are not homologous. In some species, the sex chromosome may be completely unpaired, so that one sex has one more chromosome than the other. There is no real contradiction to the statement that chromosomes exist in pairs if we consider the sex chromosomes to be a special kind of chromosome pair. The sex chromosomes are discussed more fully in Chapter 5.

All of our genetic evidence tells us that genes occur in a linear order on the chromosomes. The position of a specific gene on a chromosome is called the *locus* of that gene. A locus is a constant site which does not move around but has a characteristic location on a chromosome. Since chromosomes exist in homologous pairs, each chromosome in a diploid cell has a mate which corresponds to it locus for locus. For example, in the fruit fly there is on chromosome III a locus where a gene occurs that affects eye color and which is called the "sepia locus" (the naming of loci is discussed at the end of Chap. 4). Therefore, any fly would have this locus for eye color represented twice in its body cells, one on the chromosome III received from the maternal parent, the other on the chromosome III from the paternal parent (Fig. 2-8). However, this does not mean that the identical form of the gene must be present twice. Red eye color is normal in the fruit fly, and any one fly could have two identical genetic factors for red eye color at this locus: a gene form for red pigmentation on the maternal chromosome and the same one also on the paternal. Similarly, another fly could have sepia eyes and have the gene form for sepia pigmentation at that locus on both members of the homologous pair. We would say that these flies are *homozygous* for eye color, meaning that the individuals have identical forms of the gene at that particular locus, two forms for red eyes in the one case, and two for sepia in the other.

Since flies with red eyes can be easily distinguished from those with sepia coloration, two distinct phenotypes may be recognized on the

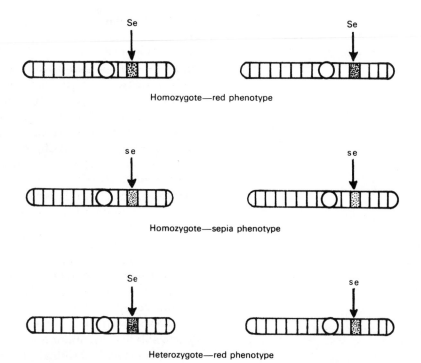

Homozygote—red phenotype

Homozygote—sepia phenotype

Heterozygote—red phenotype

FIG. 2-8. Loci on chromosomes. The figure depicts in a highly schematic way one pair of chromosomes, Chromosome III of *Drosophila*. The diploid receives one such chromosome from each parent. The two correspond locus for locus. The "sepia" locus is, therefore, present twice in every diploid cell, once on each chromosome of the homologous pair. The individual represented in the top part of the figure has the same gene form on each of the two chromosomes, the allele for red eyes, whose symbol is "Se." The fly is homozygous for the allele (Se Se) and would have a red-eyed phenotype. A fly with two recessive alleles for sepia (middle) at the sepia locus is also homozygous (se se) but will show sepia eye color. A fly with contrasting genetic factors at the sepia locus (Se se) is a heterozygote with red eyes. The factor for red eye color (Se) is dominant to its allele or alternative form (se) for sepia eye color.

basis of eye color. We learned in the last chapter that a knowledge of the phenotype may give us an indication of the genotype or kinds of genetic factors the individual carries for a particular character. Due to the fact that the sepia form of the gene is recessive to the form for red pigmentation, we would know that the sepia flies are homozygous at the sepia locus. However, we could not infer the genotype of a red fly simply by observing the phenotype. A red fly could receive a factor for red pigmentation from one parent and the alternative gene form for sepia pigmentation from the other parent. This fly would not be homozygous for eye color but would be *heterozygous,* having a pair of contrasting forms of the eye color gene at that particular locus which affects eye pigmentation (Fig. 2-8). Therefore, any locus has contrasting possibilities because different forms of a gene can be found at that locus, and these may produce very dif-

ferent phenotypic effects. In the last chapter, we also learned that contrasting forms of the same gene are called "alleles." We now see that any pair of alleles is associated with a specific locus on the chromosome. The factor for red eye and the one for sepia are alleles; they produce contrasting effects and are associated with variation at the same chromosome location or locus.

CHROMOSOME, CHROMATID, AND THE DISTRIBUTION OF GENETIC INFORMATION. DNA is a major constituent of the chromatin and thus of the chromosome. The chromosome, therefore, may be thought of as the vehicle for distribution of the genetic material at nuclear divisions. Since it transports the substance of heredity, its behavior directly determines the genetic composition of newly formed cells. Any abnormalities in chromosome distribution or any damage to the chromosomes through environmental influences can produce serious effects. A normal cell contains at least one complete balanced set of instructions coded in the DNA of its chromosomes. When the cell divides, it is critical for each of the two new cells to receive information identical to that present in the parent cell which gave rise to them. The elimination of information or the addition of extra copies can cause an unbalanced genetic condition resulting in abnormality or cellular death. It is the mitotic process with its characteristic stages which guarantees the transmission of a balanced set of genetic information from one cell generation to another.

Portions of the complete set of instructions are packaged in the separate chromatids of the cell. Any one chromatid contains a fixed unit of genetic information which is part of the total amount coded in the DNA of the cell. All the chromatids with their units of information are responsible for distributing a complete and balanced set to newly forming cells. For simplicity, we may think of the chromatid as a body containing a thread along whose length the different genes are arranged in a linear order. Various genetic phenomena can be explained by a model of the chromatid as a single fiber of DNA with associated proteins and RNA. However, this picture is undoubtedly an oversimplification. In spite of the great strides which have been made in cytology and molecular biology, there still exists a large gap in our knowledge of chromosome structure. Evidence gathered from electron microscopy, autoradiography, and procedures utilizing highly refined techniques of physical chemistry indicate that only one DNA molecule is actually present per chromatid. However, it is certain that this molecule does not exist in the form of a long, straight fiber. The evidence from electron microscopy indicates that a very long fiber is present which runs the length of the chromatid and folds back upon itself repeatedly. Regardless of the exact details of the arrangement of the fiber, a most provocative question remains to be answered: "How can such a long, folded thread replicate and segregate accurately at mitosis to ensure equal DNA distribution?" Nevertheless, as far as genetic observations are concerned, it is still the behavior of the chromatid which accounts for the equal distribution of genes at mitosis.

CHROMOSOMES IN THE NONDIVIDING NUCLEUS. Although discrete chromosomes are not ordinarily detectable in the nondividing nucleus, it is most important to realize that they are indeed present and have not lost their individuality. They are invisible to us largely because the chromatin fibers of each chromosome are in a highly uncoiled state. The narrow, extended threads are greatly elongated, making visualization difficult and the identification of entire, separate chromosomes impossible. It is not generally appreciated that the chromosomes in this extended interphase condition are very active metabolically. Indeed, it is in the nondividing cell when the chromosomes are *not* condensed that the major molecular activities are occurring involving DNA, RNA, and protein interactions. On the other hand, a chromosome in the condensed state, typical of division stages, is quite inactive in the dynamics of cellular metabolism. The activities which we see at these times are concerned mainly with packing the chromosome material and distributing it; little else is going on. Therefore, it is incorrect to think of the nondividing cell as "resting" and doing nothing. It is wrong to believe that chromosomes are active only when we see them at mitosis or meiosis. A better term than "resting" for a nondividing nucleus is "metabolic" because it reflects the true activities of that nuclear stage. "Interphase" may also be used to describe a metabolic nucleus of a cell between divisions.

CHANGES AT MITOSIS. The interphase chromosomes begin to contract with the onset of prophase and become progressively more condensed and conspicuous as this stage progresses (Fig. 2-3*B*). By the end of prophase, we can distinguish the two chromatids and the centromere position of each chromosome. With the disappearance of the nucleoli from the nucleolar organizing regions and dissolution of the nuclear membrane, the chromosomes arrange themselves at the middle of the spindle, marking the beginning of metaphase (Fig. 2-3*C*). The spindle has been forming throughout prophase. Electron microscopy has shown that each fiber in the spindle is a microtubule, a hollow cylinder composed of linear arrays of units. Each of these units is composed of a single kind of protein or protein monomer. In many animal and lower plant cells,

the spindle forms in association with the *centrioles*, bodies which come to mark the poles of the spindle at metaphase. The centriole area with associated microtubules is often referred to as the *aster*. Since asters and centrioles are absent from higher plant cells which still manage to form spindles, their exact role in spindle formation is unclear. At any rate, in all eukaryotic cells, the spindle fibers which are composed of protein microtubules are organized during prophase and are essential to the orderly distribution of the chromosomes.

At metaphase of mitosis, each chromosome is lined up at the midplane or equator of the spindle. Actually, it is the centromere of each chromosome which is at the equatorial plane of the spindle, and this accounts for the fact that the chromosomes when viewed with the microscope do not appear as they do in diagrammatic representations. The arms of the chromosomes project from the centromeric regions in various ways; they are being carried along passively by the dynamic center, the centromere, and do not all lie parallel to the equator of the spindle. The midplane of the spindle is often called the *equatorial plate*. It is not an actual structural component of the cell or the mitotic apparatus. As they line up at the middle of the spindle at metaphase, the centromeres become attached to certain of the fibers composing the spindle. These are the *chromosomal fibers* and are distinct from those known as the *continuous fibers* which run from pole to pole without any evident connection to the chromosomes (Fig. 2-9).

One of the most critical events occurring during the mitotic process is the division of the centromere. Recall that the chromosome has been double except for this region. With division of the centromere, metaphase comes to an end, and anaphase movement is initiated (Fig. 2-3*D*). Normally, centromeric duplication occurs about the same time in all the chromosomes. At anaphase, it is most apparent that the centromere is the part of the chromosome active in movement. The two sister chromatids which composed a chromosome move to opposite poles of the spindle. As they travel, the chromatids appear in various configurations of V, J, and I (Fig. 2-6). This results from the position of the centromere and the fact that the centromere is the actively leading point of every chromosome.

Each of the original sister chromatids may now be called a chromosome. As mentioned earlier, a chromosome may be composed of one or two chromatids, depending on the stage of the nuclear cycle. It is most important to keep in mind the following fact: Since each of the two sisters composing a chromosome moves to opposite poles at anaphase, identical genetic material will be found at opposite ends of the cell when this stage is completed. By the end of telophase, the spindle disappears, nuclear membranes and nucleoli are reconstituted, and each new (daughter) nucleus assumes the appearance characteristic of interphase (Fig. 2-3E).

Although mitosis is commonly thought of as a cell division, it should be kept in mind that it is basically a division of the nucleus. *Cytokinesis,* or division of the cytoplasm, may or may not take place. What is important from a genetic point of view is that the two daughter nuclei which have resulted from the mitotic division of an original nucleus are identical insofar as their chromosome complements are concerned. Thus, mitosis is a process which ensures continuity of the identical genetic material from one nuclear division to the next. Barring abnormalities, the same genes will be present on the chromosomes of all those cells which can be traced back to the same original cell, even though many mitotic divisions may have intervened. No new combinations of genetic factors result from this process. The genetically significant result is that one kind of gene arrangement is preserved from one cell generation to the next. The importance of such a process in any multicellular organism is obvious. Every normal cell composing the body will contain a full complement of genetic material characteristic of that species. Any departures may bring about abnormalities because the cell may not have a normal or balanced set of genes necessary for the production of a normal phenotype.

Variations from the normal genetic constitution may result from spontaneous aberrations arising during a nuclear division. For no apparent reason, a chromosome may lag at anaphase, fail to reach the poles, and disintegrate in the cytoplasm (Fig. 2-10). The result is a cell with one less chromosome than normal. Several other types of irregularities may take place and lead to the origin of atypical cells. Although such

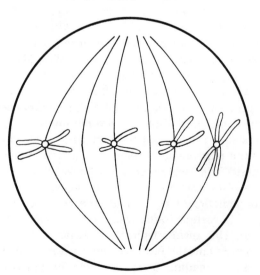

FIG. 2-9. Metaphase orientation. Two pairs of homologous chromosomes are represented, one pair with median centromeres; the other is acrocentric. It is the centromeres of the chromosomes which are oriented on the equatorial plate. Certain spindle fibers (chromosomal fibers) attach to the centromeres. Others are continuous and run from pole to pole.

A

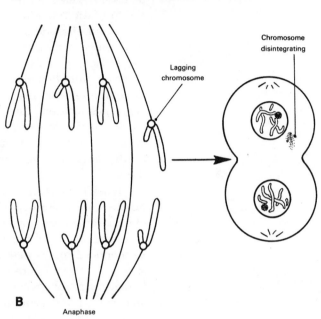

B

Anaphase

FIG. 2-10. Mitotic irregularity. *A.* Lagging chromosome in a root cell of *Vicia* (× 1050). *B.* A chromosome which lags behind the others may fail to reach the pole and may be left out of the nucleus at telophase. It will disintegrate in the cytoplasm, leaving one nucleus lacking in genetic information.

accidents apparently occur from time to time in all living things and may even take place in the course of aging, mitosis is still a remarkably accurate process. A multicellular organism may contain trillions of cells produced from countless mitotic divisions which continue throughout the lifetime of the individual in certain tissues. In this light, mitosis is seen to be an exceptionally precise event, usually free of errors. Irregularities in chromosome distribution at mitosis may also be induced in various ways by radiations and chemicals. The problems which this poses for today's society will be presented in later chapters.

THE HUMAN CHROMOSOME COMPLEMENT. The correct number of chromosomes in the human was established in 1956. The great strides which permitted the accurate description of the human chromosome complement were made possible largely through the perfection of a simple procedure in which a small sample of blood is taken and cultured. The cultured cells are exposed to a mitotic stimulator which is derived from the castor bean plant. This exposure triggers certain white blood cells to undergo mitosis. These cells are then trapped in mitotic metaphase through the application of the drug colchicine, a chemical which disrupts the organization of the mitotic spindle and thus prevents anaphase movement. The cells are then subjected to a chemical fixative (killing agent) and exposed to a hypotonic solution which swells them and disperses the chromosomes so that they become separated from one another. Finally, the cells are attached to slides and stained for observation with the microscope. The complement of the human cell can then be photographed (Fig. 2-4). The term "karyotype" may be used to refer to the chromosomes of a single cell or of an individual. Each chromosome may be cut out of a photograph and then placed in a systematic arrangement (Fig. 2-11). When a karyotype is prepared, the chromosomes are assembled into homologous pairs and then arranged in order of decreasing size. In any karyotype preparation, the centromeres are placed on the same line or level so that centromere position and length of chromosome arms can be easily compared from one chromosome pair to the next.

When the chromosome complements of most humans are examined, it is seen that 23 pairs of chromosomes are present and that 22 of these pairs are identical when the chromosome complements of males and females are set side by side (Fig. 2-11). There is, however, a difference in one pair, the pair designated the *sex chromosomes*. It is seen that a female karyotype contains two sex chromosomes which are of appreciable size and identical appearance. These are the X chromosomes. On the other hand, the male karyotype shows one X chromosome and, in addition, a very small chromosome, the Y. All the chromosomes in the human complement other than the sex chromosomes are called the *autosomes*. The human therefore has 22 pairs of autosomes plus one pair of sex chromosomes (2 X chromosomes in the female and 1 X and 1 Y in the male).

Table 2-1 describes the major groups of human chromosomes. By agreement among cytologists, the autosomes are serially numbered 1 to 22 in descending order of length. The sex chromosomes continue to be called X and Y. Following routine staining procedures with common dyes such as Giemsa, the chromosomes can be easily classified into seven groups. Within a group, further identification of individual chromosomes can sometimes be made. However,

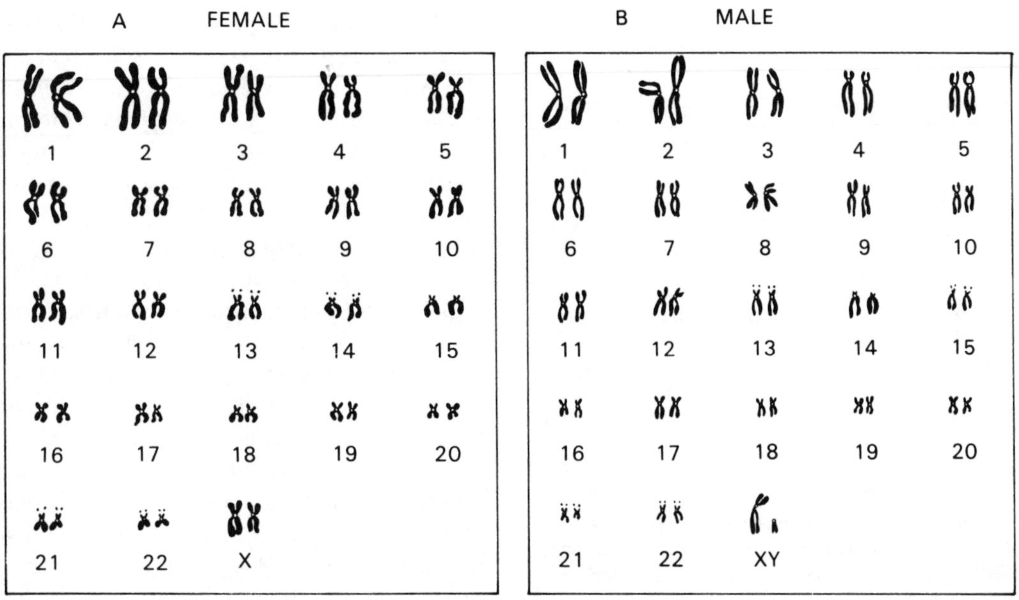

FIG. 2-11. Metaphase mitotic chromosomes of a human female (*A*) and a male (*B*) arranged in homologous pairs. (Reprinted with permission from V. A. McKusick, *Human Genetics,* Prentice-Hall, Englewood Cliffs, N. J., 1964.)

TABLE 2-1   Classification of human mitotic chromosomes*

GROUP	DESCRIPTION
1-3 (A)	Large chromosomes with approximately median centromeres. The three chromosomes are readily distinguished from each other by size and centromere position.
4 and 5 (B)	Large chromosomes with submedian centromeres. The two chromosomes are difficult to distinguish, but Chromosome 4 is slightly longer.
6-12 (C)	Medium-sized chromosomes with submedian centromeres. The X chromosome resembles the longer chromosomes in this group, especially Chromosome 6, from which it is difficult to distinguish. This large group is the one which presents major difficulty in identification of individual chromosomes.
13-15 (D)	Medium-sized chromosomes with nearly terminal centromeres (acrocentric chromosomes). Chromosome 13 has a prominent satellite on the short arm. Chromosome 14 has a small satellite on the short arm. Chromosome 15 was later found to possess a satellite as well. Satellites may be difficult to detect in preparations.
16-18 (E)	Rather short chromosomes with approximately median (in Chromosome 16) or submedian centromeres.
19 and 20 (F)	Short chromosomes with approximately median centromeres.
21 and 22 (G)	Very short, acrocentric chromosomes with satellites on the short arms. The Y chromosome resembles these chromosomes and is placed in this group. It cannot be distinguished from the other members in many cases.

*Reprinted with permission from the *J.A.M.A., 174:* 159–162, 1960.

distinctions are frequently difficult or uncertain within a size group, as in Group C which contains the X chromosome and in Group G which contains the smallest autosomes as well as the Y. Fortunately, since 1970, techniques have been perfected which bring out distinctions among all the chromosomes, even those which may appear identical when stained with common dyes. In these more refined procedures, chromosomes are exposed to fluorescent dyes, such as quinacrine mustard. Techniques of this type bring out characteristic bands on the chromosome (Fig. 2-12). For a given kind of treatment, the banding pattern is distinctive and constant for a given chromosome. The bands vary in width and amount of fluorescence along the length of a chromosome as well as from one chromosome to another. The regions of fluorescence can be visualized by a light microscope fitted with appropriate lenses. Not only have the fluorescent

FIG. 2-12. Human chromosome complement showing banding patterns after a combined chemical and immunochemical procedure. (Reprinted with permission from Schreck, Warburton, and Miller *et al., Proc. Natl. Acad. Sci., 70:* 804-807, 1973.) (Courtesy of Dr. O. J. Miller.)

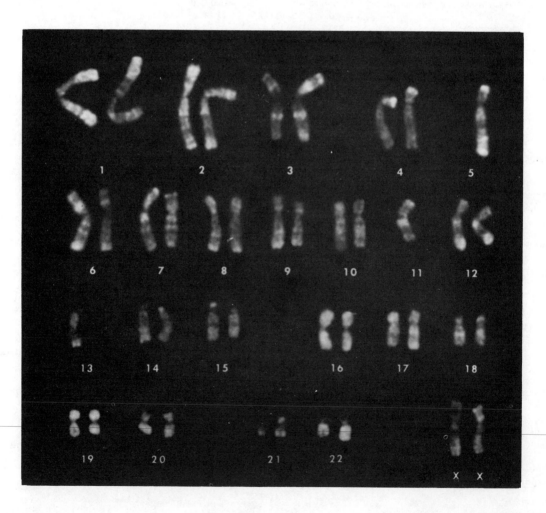

staining procedures permitted precise identification of each human chromosome, they have also revealed that nondividing cells from males contain a fluorescent body which is absent from cells taken from females. This body apparently represents the Y chromosome.

GIANT CHROMOSOMES. No discussion of chromosomes would be complete without reference to the unusual chromosomes found in several groups of flies. One of the many features of the fruit fly *Drosophila* that has contributed to its value as a genetic tool is its giant chromosomes found in the immense nuclei of the larval salivary gland cells. The larval stage of the fly is primarily concerned with food getting; efficient salivary glands are important to an insect which literally eats its way through its food. These active glands, as well as certain other organs found in some of the dipteran insects (flies), achieve their growth largely by increase in cell mass and volume rather than by increase in the number of cells. After a certain number of cells is established through mitosis, further cell divisions cease. Growth, however, continues by cell enlargement. As the nuclear size increases, the chromosomes duplicate over and over again without an accompanying mitotic division. These extra chromosome sets are probably necessary for the increased activities of cells of greater size and mass. Another unusual feature of these cells is the phenomenon of somatic chromosome pairing. Typically, homologous chromosomes pair only at meiosis (next chapter). But in the salivary glands, not only do homologous chromosomes pair, but the chromatids of each chromosome duplicate over and over again throughout the larval stage. Approximately 11 rounds of duplication take place until thousands of individual chromatids are formed. The very strong forces of pairing which seem to operate in these cells prevent the threads from separating. The end result is that each nucleus comes to contain a haploid number (four in *Drosophilia melanogaster*) of giant chromosomes. Each, of course, is in actuality composed of the many separate chromatids of the two homologues. All the threads held together side by side as a large compound unit form a giant structure, a *polytene chromosome,* a chromosome composed of many threads. This growth of

the salivary glands by increase in cell size rather than cell number may be a very efficient mechanism for the best operation of the enzymatically active cells during the period of active food getting. The glands will finally be resorbed when the larva is transformed into the pupa.

The large salivary cells are not destined to divide again; therefore, the large polytene chromosomes are not in any stage of mitosis. Actually, they are in an extended interphase. They are stretched out almost to their maximal length, a condition which makes them superb subjects for cytogenetic studies. They provide an unusual advantage for assigning genes to definite chromosome regions. When the polytene chromosomes are stained, they display deeper staining regions, bands of varying widths which alternate with lighter stained regions, the interbands (Fig. 2-13). The width and arrangement of the bands is not random but is so characteristic of every chromosome region that a small chromosome segment by itself may be identified by its banding pattern. The deeply staining bands are also called *chromomeres.* Comparable deep-staining regions are very evident along the length of the chromosomes at early meiosis (next chapter). There is reason to believe that in the region of the bands the chromosome threads are more tightly folded or packed than they are in the interbands. As a result, a greater concentration of hereditary material would occur in the band region, and this seems to be the case, as indicated by stains which are specific for DNA. Since thousands of separate chromatids are associated side by side to form a polytene chromosome, the bands or chromomeres would necessarily stain more deeply than the interbands. Specific genes have been associated with specific bands, although it is still unsettled whether one gene or more is located in a distinct band.

A chromosome may become altered in its structure, and a gene may actually be lost or deleted. Such a loss often can be detected by the genetic effects which ensue. Observation of the giant chromosomes may then show that a loss of a certain band accompanies the loss of the gene. Other kinds of changes besides gene deletion may also alter chromosome structure, such as inversions in which the order of the genes becomes reversed, or translocations in which chromosome arms may be shifted to new locations. We will see (Chap. 10) that the genetic effects which accompany these changes can be correlated with corresponding changes in the bands. This enables the cytogeneticist to assign genes in *Drosophila* to definite chromosome locations. The polytene chromosomes are also of immense value in studies of cell differentiation and gene action (Chap. 19). We will continue to refer to these giant structures throughout our discussions.

REFERENCES

Cold Spring Harbor Symposium on Quantitative Biology. *Chromosome Structure and Function,* vol. 38. Cold Spring Harbor Laboratory, New York, 1973.

Du Praw, F. J., *DNA and Chromosomes.* Molecular and Cellular Biology Series, Holt, Rinehart and Winston, New York, 1970.

Felsenfeld, G. Chromatin. *Nature, 271*: 115, 1978.

Mazia, D. Mitosis and the physiology of cell division. In *The Cell,* vol. III, J. Brachet and A. E. Mirsky (eds.). p. 77–412, Academic Press, New York, 1961.

Mazia, D. The cell cycle. *Sci. Am. (Jan):* 54, 1974.

Ruddle, F. H. and R. S. Kucherlapati, Hybrid cells and human genes. *Sci. Am. (July):* 36, 1974.

Thomas, C. A. Jr. The genetic organization of chromosomes. *Ann. Rev. Genet., 5:* 237, 1971.

Yunis, J. J. (ed.). *Human Chromosome Methodology.* Academic Press, New York, 1974.

Yunis, J. J. and O. Sanchez. G-banding and chromosome structure. *Chromosoma: 44,* 15, 1973.

REVIEW QUESTIONS

1. Name the stage of mitosis or the cell cycle at which:

    A. The nucleolus disappears.

    B. Nuclear membranes reform.

    C. Centromeres are aligned on the equatorial plate.

    D. Microtubules are associating to form the spindle fibers.

    E. The DNA of each chromatid undergoes duplication.

    F. Chromatids move to opposite poles.

2. The normal chromosome number of a body cell in the human is 46.

    A. In a human cell, how many chromatids are present at: (1) prophase? (2) telo-

FIG. 2-13. The salivary gland chromosomes of *Drosophila melanogaster.* (Reprinted with permission *Hered, J. 25:* 464-476, 1934.)

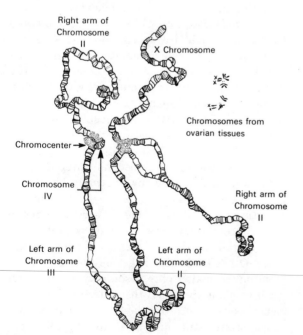

Right arm of Chromosome II

X Chromosome

Chromosomes from ovarian tissues

Chromocenter →

Chromosome IV

Right arm of Chromosome II

Left arm of Chromosome III

Left arm of Chromosome II

phase? (3) interphase, following DNA replication?

B. Give the chromosome number at: (1) prophase (2) telophase (3) the interphase following DNA replication.

3. In the human:

A. How many autosomes are present in body cells of a male?

B. How many autosomes are present in body cells of a female?

C. How many sex chromosomes are present in a male?

D. How many sex chromosomes are present in a female?

E. How many major groups of chromosomes can be recognized on the basis of size and shape?

4. Allow "A" to designate one entire haploid set of human autosomes, and allow "X" and "Y" to represent the sex chromosomes.

A. How can one represent the chromosome constitution of the body cells of a female in regard to autosomes and sex chromosomes?

B. Answer the above for a male.

5. From the column on the right, select the letter of the term which applies best to each of the following statements. A term may be used more than once or not at all.

1. Does not occur in a cell having two nuclei.	A. chromomere
2. Its position determines the length of the chromosome arms.	B. cytokinesis C. chromatin D. chromosome
3. Marks the poles of the spindle in some types of cells.	E. centromere
4. Deeply staining band or region of a polytene chromosome.	F. centriole G. chromatid H. satellite
5. Holds the chromo-	

some halves together at prophase and metaphase.

6. Nuclear material composed of several kinds of large molecules.

7. Its division ends metaphase and initiates anaphase.

8. Associated with the formation of the nucleolus.

9. Separates from its sister at anaphase.

6. Give at least three features of eukaryotic cells which are absent in prokaryotes.

7. What term applies to:

1. The specific position on a chromosome which can be occupied by a particular gene.

2. Chromosomes which correspond in size, shape, and genetic regions?

3. An individual possessing two identical forms of a gene at a specific genetic region which is under consideration?

4. Alternate forms of a gene?

5. The actual chemical substance composing the gene?

6. A V-shaped chromosome having its centromere in the middle?

7. A chromosome composed of many separate chromosome threads which are intimately associated?

8. The chromosome constitution of a single cell or individual?

9. That part of some chromosomes appearing as an appendage due to a secondary constriction?

10. An I-shaped chromosome having one very short arm?

11. The drug which prevents anaphase movement and which is used in chromosome analysis?

3

CHROMOSOMES AND GAMETE FORMATION

How remarkable that Mendel, totally unaware of any descriptions of cell division, was able to make the deductions which formed the basis of his two laws of inheritance! Perhaps even more incredible is the fact that the intricacies of mitosis and meiosis were being unraveled while Mendel's paper lay ignored. For *there* in the report of Mendel's experiments was the genetic evidence for the existence of factors whose behavior parallels that of the chromosomes. The beginning of the twentieth century was to bring not only a recognition of Mendel's paper but also independent genetic and cytological evidence to support it.

Although Weismann predicted some sort of reduction division in the life cycle of sexual creatures, the details were to be slow in coming. Before 1900, the half number of chromosomes had been reported in the gametes of animals (Van Beneden with *Ascaris*) and in plants (Strasburger). But the many changes taking place in the appearance and behavior of the meiotic chromosomes must have defied interpretation. One problem was the difficulty in recognizing homologous chromosomes. In 1901, Montgomery postulated homologues, one set of chromosomes provided by the female parent, the other by the male. It is these which pair at meiosis; pairing is not a random process. Sutton, a student of Wilson, confirmed Montgomery's findings and said further that the paired homologues must somehow separate. Such behavior, he noted, would provide the physical basis for both of Mendel's laws. Conclusive proof of these points was to take several more years. Further genetic and cytological studies were needed to place Mendel's factors in the chromosomes and to verify the chromosome theory of inheritance. At this point, let us examine the main features of the meiotic process to appreciate the problems which confronted cytologists and to enable us to grasp other genetic phenomena which were soon to be discovered.

MEIOSIS ENTAILS TWO NUCLEAR DIVISIONS. Like mitosis, meiosis is fundamentally a nuclear event, and many features of the two processes are identical. However, there are several distinctions between them which produce very different genetic results. In the last chapter, we saw that mitosis preserves the genetic identity between the parent cell and the two resulting daughter cells. The effect of meiosis is quite the opposite, for it is a process in which the chromosome number of the parent cell is reduced by one-half. Each meiosis involves not just one but two nuclear divisions. As in mitosis, it is again the chromatid which acts as the agent to distribute the genetic information. But at meiosis the chromatid behavior is such that *new* genetic combinations are formed. Since two nuclear divisions take place, the expected number of nuclei upon completion of meiosis is four. The four nuclei which are typically formed are referred to as the *tetrad,* the immediate products of a meiotic division. Cytoplasmic division usually, but not always, accompanies the meiotic process so that a tetrad of haploid cells is the characteristic end product. Exactly what these four cells will become after meiosis (sperms, eggs, spores) depends both on the species and the sex of the organism. Meiosis and all of its complications are an integral part of the sexual process and appeared early in the evolution of living things.

THE MORPHOLOGY OF THE EARLY MEIOTIC CHROMOSOME. Most of the significant differences between mitosis and meiosis occur during *first prophase.* The duration of prophase I of meiosis is quite prolonged in comparison to that of mitosis and entails pronounced changes in chromosome appearance and behavior. At least five substages are generally recognized. At the earliest of these, *leptonema,* the chromosomes are very stretched out, more so than at any other of the meiotic stages. Individual chromosomes cannot be identified, and the nucleus appears to contain a network of thin, entangled threads. The thin chromosome thread is designated the *chromonema* (Fig. 3-1A and B). Although chromonemata can be demonstrated at mitosis by chemical treatment of prophase and metaphase chromosomes, it is during early prophase of meiosis that they are most apparent. Indeed, we can think of all the changes in chromosome appearance during nuclear divisions as a result of changes in the coiling of the chromosome thread. In many species, the chromonemata exhibit deeper staining regions along their lengths, giving the threads the appearance of strings of beads. These intensely staining areas are the *chromomeres* and are comparable to deep-staining bands along the lengths of the polytene chromosomes (Chap. 2). As mentioned in relationship to the giant chromosomes of the *Drosophila* salivary glands, the chromomeres have been thought by some to represent the sites of individual genes, but this idea is controversial. Studies with the electron microscope support their interpretation as coils in the chromosome thread. Whether or not they represent the sites of single genes, their positions are not random but appear at characteristic locations along a particular chromosome. Certain genes in some organisms have actually been associated with specific chromomeres.

Although the filament at leptonema is very thin and appears to be single, arguments have raged for years concerning the number of strands which actually compose each chromosome at this early stage of meiosis. It would be irrelevant here to enter into this problem. Genetic interpretations are compatible with the concept of the chromonema of early meiosis as a single thread. Figure 3-1C is a diagram of an imaginary cell from an organism with a diploid chromosome number of four, where one pair of chromosomes is larger than the other. At leptonema, these two homologous pairs, although long and extended,

FIG. 3-1. Leptonema. *A.* Microspore mother cell of the lily. *B.* The chromosomes are in a greatly extended condition and appear as slender threads, the chromonemata. *C.* Chromosomes number is four. Chromatid number is four. Two pairs of homologous chromosomes, one pair with the alleles A and a; the other with the alleles B and b. The chromosomes are unassociated and distributed at random in the nucleus.

A

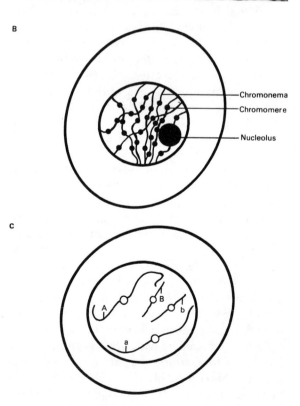

B

Chromonema
Chromomere

Nucleolus

C

are separated and are not associated in any prescribed way. The cell in the figure depicts a heterozygous individual, a dihybrid carrying two pairs of contrasting genetic factors. These are the alleles, "A" and "a", which occupy a locus on the larger of the chromosomes and the alleles "B" and "b," located at a site on the smaller ones.

THE HOMOLOGOUS CHROMOSOMES PAIR. One of the most critical differences between meiotic and mitotic chromosome behavior involves the pairing or *synapsis* of homologous chromosomes. The act of pairing marks the onset of *zygonema*, which is defined as the stage of active pairing (Fig. 3-2*A*). This synapsis is very exact and can be seen to take place chromomere for chromomere. If we recall that genetic loci are distributed along the length of the chromosome, we can envision the homologues pairing locus-for-locus throughout zygonema until all the corresponding loci on all the chromosomes are intimately associated. Therefore, at zygonema the alleles in a heterozygote would be closely paired and no longer apart in the nucleus (Fig. 3-2*B*). Although the nature of the synaptic force is not understood, it serves to bring the homologous chromosomes together in a close union even though they may be widely separated in the nucleus before the onset of meiosis. Whatever its exact basis may prove to be, the force is an exact one which operates over large distances on the level of the nucleus.

EACH CHROMOSOME IS DOUBLED. After zygonema, the chromosome threads become more conspicuous, and the next stage of meiotic prophase I can be recognized, *pachynema* (Fig. 3-3*A*). The thickening which becomes evident is due partly to a contraction of the chromonemata and partly to the fact that each thread has become detectably double. Exactly when this doubling first occurs is still in question. Studies of the incorporation of radioactive deoxyribonucleic acid (DNA) precursors indicate that replication is completed well before the onset of leptonema. It is at pachynema, however, that the double nature of each chromosome thread can be clearly demonstrated. At this stage, the homologous chromosomes are intimately paired throughout their lengths, and each individual chromosome is composed of two chromatids (Fig. 3-3*B*). Therefore, at this stage separate chromosomes are not distributed throughout the nucleus, as is true in mitotic prophase. What appears to be an individual chromosome is actually an association of two, each of which is double. This association of two homologues is termed a *bivalent*. The expression "tetrad," referring to the four threads which are present, is also used, but many prefer to restrict usage of the term to the immediate products of meiosis after telophase II. Thus, the nucleus at pachynema contains a number of bivalents which corresponds

to the haploid chromosome number for the species. Figure 3-3*B* shows that although four chromosomes are present, they are arranged as bivalents. Since each chromosome thread with all its loci is definitely double by this time, two "A" and two contrasting "a" genetic factors are present. The same is true for the other pair of alleles, "B" and "b." It is very important to note that as far as the number of *chromatids* is concerned, the total is the same in both the mitotic and the meiotic prophases. For at prophase of *mitosis,* each chromosome is also composed of two chromatids. The dihybrid represented in Figure 3-3*B*

FIG. 3-2. Zygonema. *A.* Microspore mother cell of the lily. *B.* Chromosomes number is four. Chromatid number is four. This is the stage of active pairing of the homologous chromosomes. The corresponding loci become intimately associated during this stage.

FIG. 3-3. Pachynema. *A.* Microspore mother cell of the lily. *B.* Chromosome number is four. Chromatid number is eight. Each chromosome thread is now detectably double. The number of bivalents shown here is two, and each is composed of four threads. Each locus is represented four times: A, A, a, a, and B, B, b, b.

A

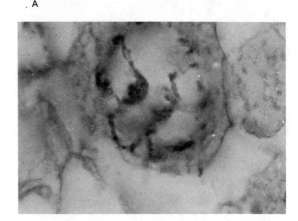

A

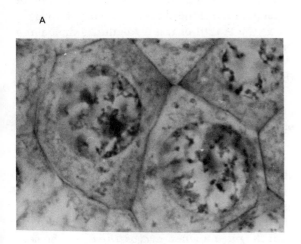

B

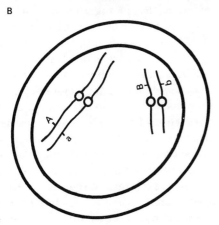

B

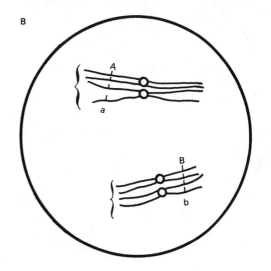

has eight, the same number that would be present if the cell were undergoing mitosis. Therefore, at both the mitotic and the meiotic prophases, the same situation pertains regarding the number of chromosome threads because each chromosome in each of the processes is composed of two halves. The main difference is the *distribution* of the chromosomes in the nucleus. At mitotic prophase, the chromosomes are separate and unassociated. At prophase I of meiosis, the homologues are paired as bivalents, and it is these which are found in the haploid amount.

THE CHROMATIDS SEPARATE. Once intimate pairing is achieved and the chromosomes appear doubled, another force begins to operate which seems to oppose that of pairing. We recognize *diplonema* of prophase I when the threads composing each bivalent appear to repel (Fig. 3-4A). Diplonema is often referred to as the stage in which "opening out occurs," a recognition of the separation of the chromatids which is evident along the length of the bivalent. As diplonema proceeds, this becomes more pronounced, particularly at the region of the centromere where the forces of repulsion seem to be strongest (Fig. 3-4B). Indeed, the bivalent association would probably fall apart were it not for the presence of certain regions where the homologous chromosomes still remain in intimate contact. These regions are called the *chiasmata* because at each chiasma the threads of the separate chromatids appear to cross. Although their formation is still another matter for debate among cytogeneticists, the chiasmata do seem to be associated with the phenomenon of *crossing over*, the separation of genes found together on the same chromosome. A full discussion of this topic is presented in Chapters 8 and 9, but a few points about its physical basis are in order here. Genetic analyses have firmly established that homologous chromosomes exchange segments during meiotic prophase. In certain organisms, the frequency of these exchanges is closely correlated with the frequency of chiasma formation. A one-to-one correspondence is not evident in all species, but the consensus is that the chiasmata do reflect a physical exchange of segments between homologues. Exactly when the exchange occurs and the precise mechanism involved are still

FIG. 3-4. Diplonema. *A.* Microspore mother cell of the lily. *B.* Chromosome number is four. Chromatid number is eight. The number of bivalents is two. The chromosomes composing a bivalent begin to repel each other at the beginning of this stage, and the forces of repulsion progress throughout. The chiasmata (*arrows*) hold the four chromatids together in a bivalent, which would otherwise fall apart. Only the nucleus is represented here.

A

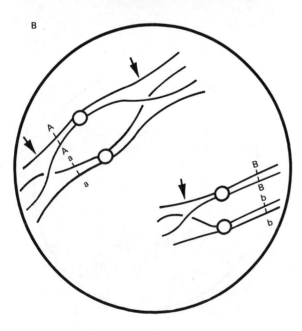

B

Point of reciprocal
exchange

FIG. 3-5. A single crossover event following genes on the same chromosomes: Cc and Dd. In the bivalent, two of the four threads participate in any *one* crossover (*left*). However, more than one crossover may take place in a bivalent, and any two homologous threads may be involved. In a *single* crossover, a reciprocal exchange of chromosome segments occurs between two homologous chromatids. After the event, a new combination of alleles results. On the right, we see that in each chromosome one chromatid shows a new arrangement (C-d in the upper; c-D in the lower). This is the direct result of the reciprocal exchange of homologous chromosome segments past the point of the crossing over. (*Arrow* indicates the point where the reciprocal exchange occurred between the two chromatids.)

matters for lengthy discussion. Evidence from studies with microorganisms in the past 10 years strongly supports the concept of an actual breakage of chromosome threads. In this process, broken chromatid ends would join up reciprocally with chromatid segments of the homologous chromosome (Fig. 3-5). Since this rejoining is reciprocal, the final outcome is an exchange of corresponding blocks of genes between homologous chromosomes. Many cytologists believe that this exchange takes place sometime between late pachynema and the onset of diplonema. The process of crossing over is of utmost importance in the formation of new combinations between genes linked together on the same chromosome. It is an integral part of the normal sexual process, and its role in adding to the number of new combinations of the genetic complement of a species cannot be overemphasized. We will reexamine this aspect of meiosis in later chapters in relationship to a variety of topics.

The next clearly defined stage of prophase I is *diakinesis*, marked by rather pronounced changes in the appearance and distribution of the chromosomes within the nuclear membrane (Fig. 3-6). The chromosomes contract markedly, and the chiasmata become increasingly evident as the stage progresses. The bivalents themselves tend to become widely spaced, as if some other force of repulsion were operating to cause them to separate from one another. Conse-

quently, this stage is an excellent one for chromosome counting because the haploid number is clearly indicated by the number of contracted and widely separated bivalents. However, nothing of genetic significance is actually taking place to alter the chromosomes or rearrange the hereditary material. We may consider this stage to be an exaggeration of diplonema.

CHROMOSOME BEHAVIOR AT MEIOSIS AND MENDEL'S LAWS. As in mitosis, prophase ends with dissolution of the nuclear membrane and the disappearance of the nucleoli. Common to both processes, metaphase movements begin, and the chromosomes arrange themselves at the midplane of the spindle. However, a major distinction is found between the metaphase of mitosis and that of the first meiotic division. In the former, the centromeres of *individual* chromosomes are arranged at the equatorial plate (Chap. 2, Fig. 2-9), whereas at metaphase I, it is the centromeres of *bivalents* which are found at the equator of the spindle (Fig. 3-7 *A* and *B*). Thus, the main difference is that at mitotic metaphase, *single* chromosomes are oriented on the spindle, whereas *pairs* of chromosomes occupy the equator in the meiotic cycle. The centromeres of the individual chromosomes of every bivalent have been repelling each other and are now clearly pointed toward opposite poles of the spindle. It is particularly important at this point to note the arrangement of the two pairs of chromosomes in Figure 3-7B. The diagram shows the "A" allele of one chromosome and the "B" allele of the other directed to the lower pole and the "a" and "b" toward the upper. However, there is nothing which dictates such an orientation requiring factors "A" and "B" to go to the same pole. It is equally possible for the chromosome with allele "A" and the chromosome with allele "b" to face the same pole and those with "a" and "B" to face the opposite one. Assuming that the chromosomes with the "A" and the "B" genetic factors came from the maternal parent and those with the "a" and "b" factors from the paternal one, we can see that there is no reason to suppose that "A" and "B" *must* travel together as well as "a" and "b." The two loci are on completely different chromosomes. How the four chromosomes become arranged in relationship to the poles of the spindle is a matter of chance and forms the foundation of Mendel's law of independent assortment. It is here at metaphase I and the ensuing anaphase I that we find the basis of both of Mendel's primary laws, *segregation* as well as *independent assortment*.

Anaphase I, like anaphase of mitosis, is recognized by the movement of chromosomes to opposite poles of the spindle (Fig. 3-8A). It will be recalled that in mitosis the centromere divides, initiating anaphase, and sister chromatids separate. This is *not so* at first anaphase of meiosis where the centromere remains undivided. The individual centromeres of homologous chromosomes in each bivalent repel each other and travel to opposite poles. Consequently, each chromosome is still composed of two chromatids (Fig. 3-8A and B).

Telophase I involves the reformation of nuclear membranes and nucleoli, as in telophase of mitosis (Fig. 3-9A). However, at meiotic telophase, the haploid number of chromosomes is present in the newly formed nuclei. In our example (Fig. 3-9B), each nucleus contains two chromosomes instead of the diploid number of four. However, each of these chromosomes, unlike those of mitotic telophase, is composed of two chromatids. If we count the chromatid number, in this case four, we see that true reduction *has not yet* taken place. To bring about reduction in chromatid as well as chromosome number requires the second division of meiosis.

FIG. 3-6. Diakinesis in the lily. At this stage, the chromosomes are extremely contracted, chiasmata are very evident, and the bivalents tend to be separated from each other in the nucleus. The relationship of the chromatids, however, is basically the same as in the previous stage, diplonema.

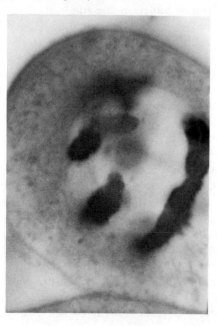

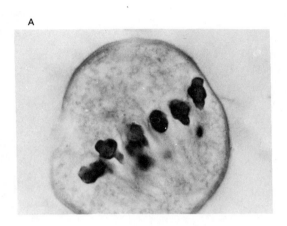

A

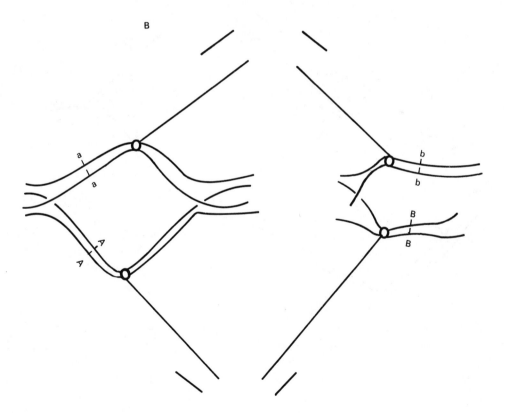

B

FIG. 3-7. Metaphase I. *A.* Meiotic cell in the lily. *B.* Chromosome number is four. Chromatid number is eight. The two bivalents are arranged at the equator of the spindle. The chromosomes with alleles A and B are directed toward the lower pole, but it is just as possible to find other arrangements, because the separate nonhomologous chromosomes are not tied together in any way.

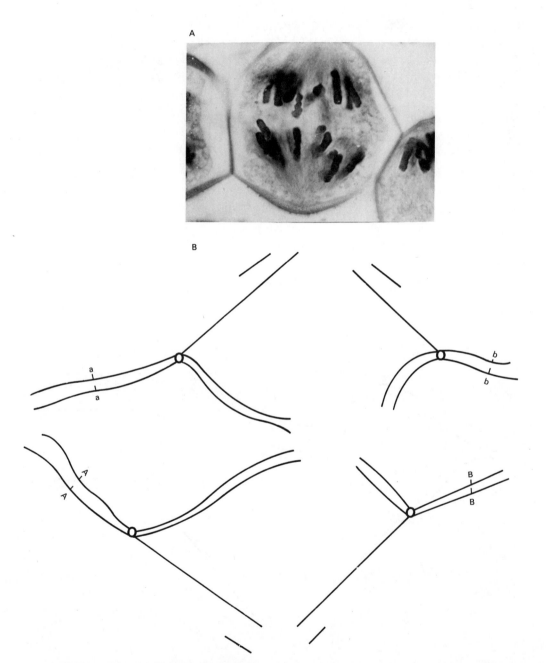

FIG. 3-8. Anaphase I. *A.* Meiotic cell in the lily *B.* The centromere of each chromosome remains undivided so that each chromosome moving to a pole is composed of two chromatids.

SECOND MEIOTIC DIVISION EFFECTS TRUE REDUCTION. The onset of the second meiotic division varies with the species involved. Prophase II may begin after a distinct interphase period. In some organisms, the interphase may not occur, so that an abrupt transition is found from telophase I to prophase II and in certain cases even to metaphase II. At any rate, the second division stages of meiosis resemble those of mitosis when viewed with the microscope (Figs. 3-10A and 3-11A). Unfortunately, this has led to a hasty description of the second meiotic division as a mitotic one. Such a designation is quite incorrect for several reasons. Most obvious is the fact that the double nature of the chromosome at mitotic prophase results from the duplication of the genetic material in the preceding interphase. Although the prophase II chromosome is also double, no duplication of the chromatin immediately preceded it. Less apparent, but equally important, is the fact that the chromatids composing the *mitotic* chromosome are sisters, meaning that they are identical throughout their lengths. Those at prophase II, however, are not truly sisters because crossing over probably occurred somewhere in each chromosome. Only portions of the chromatids of a single chromosome would be identical. We have noted (Fig. 3-5) that after the point of a crossover, a segment of a chromatid is now associated with a block of genes from the homologous chromosome. Second division of meiosis is necessary to separate the dissimilar chromatids of each chromosome. This critical division will achieve complete reduction in the amount of genetic material and bring about new combinations of genes from the male and female parent of the preceding generation. The second meiotic division is essential to the sexual process and should not be considered equivalent to a mitosis.

At metaphase II, single chromosomes, each composed of two chromatids, are at the equator, and their appearance is similar to those of mitosis (Fig. 3-10A and B). As in mitosis, the centromere then divides, and anaphase II is initiated (Fig. 3-11 A and B). It is *only now* with the division of the centromere that true reduction is achieved. The nuclei of telophase II (Fig. 3-12 A and B) contain chromosomes each composed of one chromatid. Figure 3-12B shows the reduction in both chromosome number (2) and the number

A

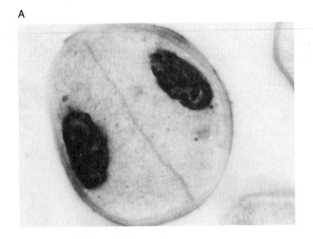

B

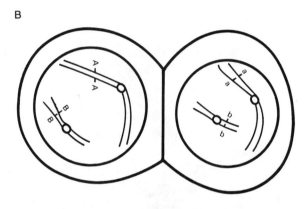

FIG. 3-9. Telophase I. A. Meiotic cells in the lily. At this stage, the nuclear membranes reform. B. Chromosome number per nucleus is two. Chromatid number per nucleus is four. Each sister nucleus is genetically different; the top one is ab, the lower AB. Reduction in chromosome number has occurred but not in chromatid number.

of chromatids (2). The figure also shows that the nuclei of the tetrad are genetically of two types; two of the resulting cells are "AB," having received the chromosome with allele "A" and the other with allele "B," and the other two are "ab." It is essential here to reexamine metaphase I (Fig. 3-7B). It was mentioned that the orientation of the bivalents at this stage is not mandatory. The factors on separate chromosomes segregate independently. If the "A" and the "b" chromosomes face the one pole and the "a" and "B" chromosomes the other, the final products at the completion of meiosis would be nuclei with different genetic arrangements from those in Fig-

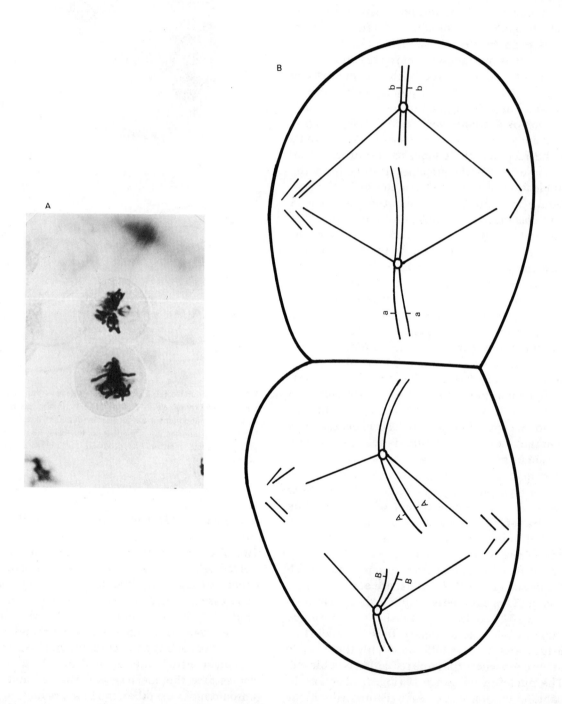

FIG. 3-10. Metaphase II. *A.* Meiotic cells in the lily. *B.* The single chromosomes in each sister cell are at the equatorial plate. Contrast this with metaphase I (Fig. 3-7) where bivalents are at the equator.

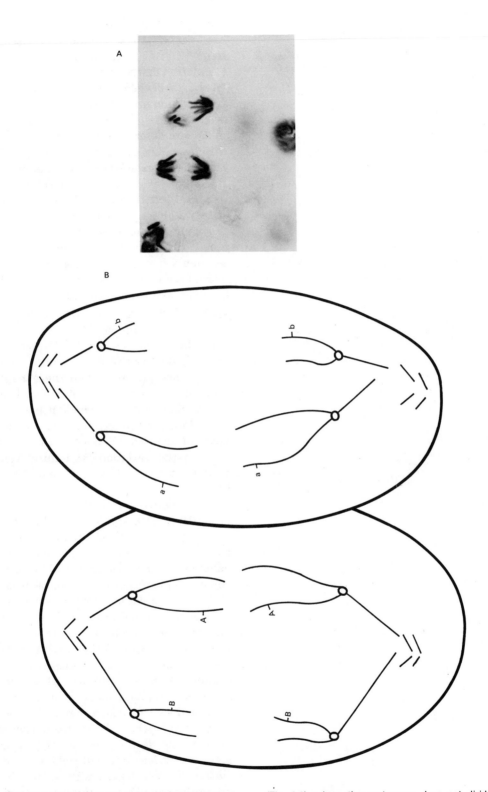

FIG. 3-11. Anaphase II. *A.* Cells in the lily. *B.* The centromere of each chromosome now divides so that the two chromatids of each chromosome are separated. Contrast this stage with anaphase I (Fig. 3-8), where the centromere does not divide, and each chromosome remains double.

ure 3-12B: "Ab" and "aB," instead of "AB" and "ab." Since both possibilities can arise with equal frequency as the result of chance chromosome orientation on the spindle, meiosis in a dihybrid, "AaBb," will yield four different kinds of gametes in equal proportions: AB, Ab, aB, and ab. So the origin and frequencies of different kinds of gametes are determined by the physical events of meiosis which shuffle the maternal and paternal genetic factors into new combinations. Although the details of linkage and crossing over will be left until later, it should be appreciated at this point that even those genes which are linked on the same chromosome become shuffled as a result of the physical exchange of chromosome segments between homologues (Fig. 3-5). The overall significance of crossing over, like independent assortment, is the formation of new combinations of genetic material in the gametes.

We see that unlike mitosis, the nuclei formed from the meiotic divisions of a parent nucleus are very different. Almost limitless new combinations are possible when these haploid cells unite with others at fertilization. The meiotic event, along with independent assortment and recombination through crossing over, is the focal point of the sexual process and has supplied most of the variation for natural selection to work on in the evolution of the diverse forms of life.

MEIOSIS IN THE MALE.   All sexually reproducing organisms contain meiotic cells at some portion of their life cycle. Once chromosome behavior during meiosis is fully grasped, it becomes easy to recognize the characteristic stages of the two nuclear divisions of meiosis in specific cells of the animal or plant. In the testis of the mature male animal, a variety of cell types is found in the wall of the seminiferous tubule. Among these is the *spermatogonium,* which is capable of mitotic division. However, it may enter the prophase stage of first meiosis, and once it does so, it is recognized as a primary spermatocyte (Fig. 3-13). Each primary spermatocyte is destined to complete both of the meiotic divisions and will yield a tetrad of *spermatids*. Each of these in turn will finally undergo dramatic cytoplasmic transformations into a sperm (*spermiogenesis*) without further significant changes in the genetic complement. The

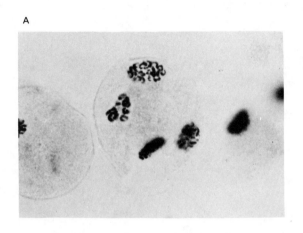

A

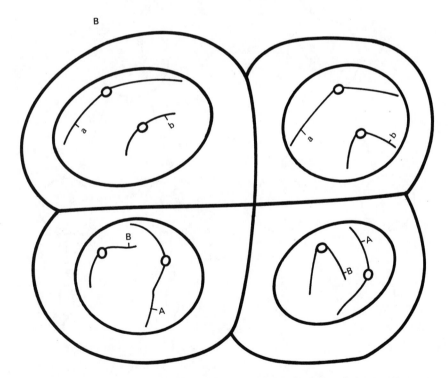

B

FIG. 3-12. Telophase II. *A.* Tetrad of microspores in lily. *B.* Chromosome number per nucleus is two. Chromatid number per nucleus is two. Cells of the tetrad have the true reduction in both chromosome and chromatid numbers. They are also unalike genetically as a result of segregation and independent assortment. In this example, one-half of the nuclei are AB and the rest ab.

CHROMOSOMES AND GAMETE FORMATION

65

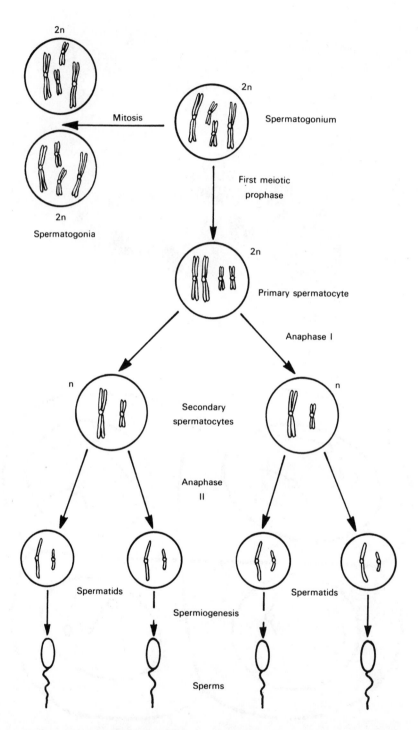

FIG. 3-13. Diagram of spermatogenesis. Among the various types of cells in the seminiferous tubules is the spermatogonium, which may divide mitotically or which may undergo the meiotic divisions. The tubules are always filled with cells in various stages of gamete formation. The spermatids are the most numerous. Note that the eventual products of one primary spermatocyte are four functional sperms.

entire series of events beginning with meiosis and resulting in the formation of male gametes is known as *spermatogenesis*. In the human, it has been estimated that approximately 64 days are required for spermatogenesis, starting with a spermatogonium and ending with the development of mature sperms. Any slide of a testis squash or section shows a diversity of cell types because different meiotic stages occur simultaneously along the length of the tubules. The most abundant cell type is the spermatid which assumes an assortment of shapes as it transforms into a mature sperm.

The grasshopper will be used as the species of reference in the following discussion of *spermatogensis* since meiotic behavior in the male grasshopper is typical of male animals, and its chromosome number is low, making observations quite easy. Moreover, it has an unpaired X chromosome; no Y is present at all. In this so-called "X-O condition," the identification of the X chromosome at meiosis is simpler than in the more common X-Y situation found in man and other mammals.

The characteristic stages of first meiotic division during spermatogenesis all take place in the primary spermatocyte which, as noted above, is distinguished from a spermatogonium by the onset of meiosis. By the time of pachynema, the nuclei of the primary spermatocytes of the grasshopper show an identifiable X chromosome (Fig. 3-14). At early prophase stages, it is usually more deeply stained and contracted than any other member of the chromosome complement; by diplonema, it is very well defined (Fig. 3-15 *A* and *B*). In contrast, the other chromosomes appear fuzzy in outline. The cell at diakinesis typically contains a sex chromosome off to one side of the bivalents. When first metaphase is reached, it is often off the equatorial plate (Fig. 3-16). At metaphase I, the unpaired X, which has been the most distinct chromosome of the complement, begins to appear fuzzy and less intensely stained, resembling the early prophase stages of the rest of the chromosomes. This differential reaction to staining, often displayed by the sex chromosomes, is called *heteropyknosis*. The phenomenon may also be characteristic of certain regions of other chromosomes. *Positive heteropyknosis* designates the more deeply stained condition of early

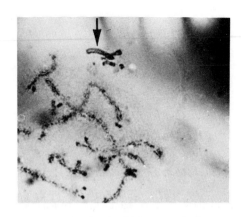

FIG. 3-14. Pachynema in primary spermatocyte of grasshopper. The X chromosome is much more condensed and stains more deeply that the other chromosomes in the complement (*arrow*).

FIG. 3-15. Primary spermatocytes of grasshopper. These cells show 11 bivalents and an unpaired X (*arrow*). The X stains more deeply than the other chromosomes at diplonema (*A*). By diakinesis (*B*) all of the chromosomes are highly contracted and react similarly to stains.

A

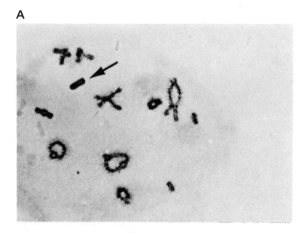

B

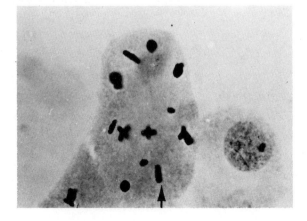

FIG. 3-16. First metaphase in primary spermatocyte of grasshopper. Note that the X chromosome is off by itself and is now staining more faintly than the rest of the chromosomes.

FIG. 3-17. First meiotic anaphase in the male grasshopper. The chromosomes at each pole will become incorporated in nuclei of cells which will develop into secondary spermatocytes. Note that 11 chromosomes are at one pole and 12 at the other due to the fact that the latter contains the X chromosome.

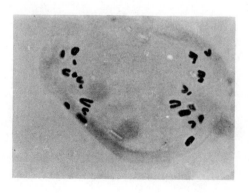

meiosis and *negative heteropyknosis* the less intensely stained appearance often observed at first metaphase. Difference in staining behavior of the sex chromosomes as opposed to the autosomes (all the other chromosomes in the complement) is typical of most animals having the X-O and X-Y mechanisms of sex determination, although details vary among the different groups.

At anaphase I, the negatively heteropyknotic X chromosome passes undivided to one of the poles (Fig. 3-17). After reconstitution of the nuclear membranes at telophase I, two cells are formed, the *secondary* spermatocytes. These will be quite different genetically (refer back to Fig. 3-9*B*). A major difference is found in regard to the X chromosome. Since it passed undivided to one pole at anaphase I, one of the two secondary spermatocytes in the grasshopper will contain one less chromosome than the other. (In animals where both an X and a Y are present, one of the cells will contain the X and the other the Y because the X separates from the Y at first division.) Secondary spermatocytes will be mixed with the many other cells of the testis. Their identification is often difficult because they are easily confused with young spermatids; both cell types are undistinguished looking cells with large nuclei.

Each one of the two secondary spermatocytes formed from first meiotic division will in its turn undergo the second stages of meiosis. The sex chromosome, displaying negative heteropyknosis, may be recognized at both second metaphase and anaphase. The secondary spermatocyte with the X will produce two spermatids, each with an X. The spermatocyte lacking the X will give rise to two spermatids with no X chromosome (but with a Y if the species contains both X and Y). The meiotic end product of each primary spermatocyte in an animal is a tetrad of cells, two which contain an X and two which lack it (the latter two will contain a Y in those males that are XY). The significance of this in sex determination is a major subject for discussion in Chapter 5.

MEIOSIS IN THE FEMALE.   The nuclear events in both sexes are essentially the same. There are, however, some different aspects between them in regard to the meiotic process. In females

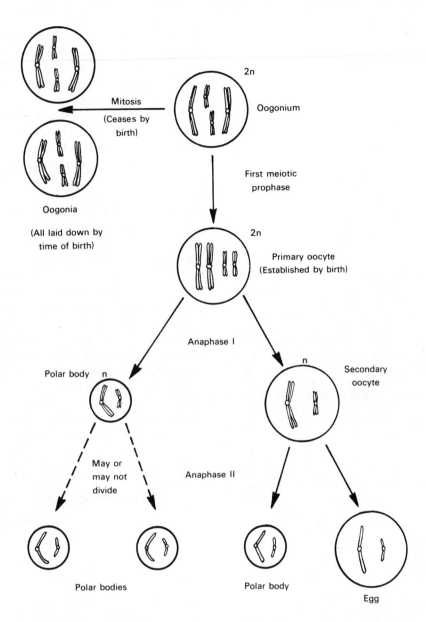

FIG. 3-18. Diagram of oogenesis. The meiotic stages of oogenesis are comparable to those of spermatogenesis (Fig. 3-13). The same chromosome behavior is involved, although cytoplasmic events are very different in the two processes. Note that the eventual product of one primary oocyte is one functional egg.

of higher animals, meiosis does not take place continually throughout the age of reproductive maturity as it does in the male. In female mammals, for example, the mitotic divisions of *oogonia*, the counterparts of the spermatogonia, cease during embryological development (Fig. 3-18). Therefore, the entire complement of *primary oocytes* is established by the time of birth. A primary oocyte thus remains in the first meiotic prophase stages for many years. The first meiotic division is not resumed until the egg is about to mature in the follicle at the age of sexual maturity. When the primary oocyte does finally divide, it gives rise to two cells, but these are unequal in size. This results from the fact that the larger one, the *secondary oocyte,* receives most of the cytoplasm. The smaller cell, the *polar body,* may or may not divide again. The secondary oocyte completes the second meiotic division only if a sperm enters the cytoplasm.

The result of the division is again two cells of unequal size, a large one which will become the egg and another small polar body. All polar bodies eventually disintegrate. Throughout oogenesis, the process in the female which produces a gamete from the maturation of an immature germ cell, both X chromosomes stain to the same degree, in contrast to the single X of the male which shows heteropyknosis in the spermatocytes.

MEIOSIS IN THE HIGHER PLANT. The sexual process undoubtedly arose early in the evolution of eukaryotic cells, long before the divergence of the plant and animal kingdoms. We therefore find comparable meiotic stages in both plants and animals. Plant life cycles, however, are very diverse and often are quite complicated. Meiosis generally takes place in specific plant organs, but the direct result of the process usually is *not* gamete formation as it is in animals. Instead, haploid cells called *spores* are formed. These then undergo a series of mitotic divisions which eventually leads to the origin of sex cells. In the flowering plant (Fig. 3-19), meiosis occurs in two parts of the flower. On the male side, the anther

FIG. 3-19. Meiosis in the flowering plant. Meiotic divisions leading to the formation of male gametes occur in the microspore mother cells of the anther. Four microspores, each of which will become a pollen grain, arise from each mother cell. The haploid nucleus of the pollen divides mitotically and yields two haploid nuclei. One of these divides again to produce two male gametes, one of which will fertilize an egg. The ovary contains ovules, and in each of these, one cell, the megaspore mother cell, undergoes meiosis. Of the four haploid megaspores formed, only one survives. It enlarges to give rise to the embryo sac. Mitotic divisions follow in the embryo sac, and one of the nuclei becomes the egg. Growth of the pollen tube down the style delivers one male nucleus in each pollen tube to an egg in the embryo sac of an ovule.

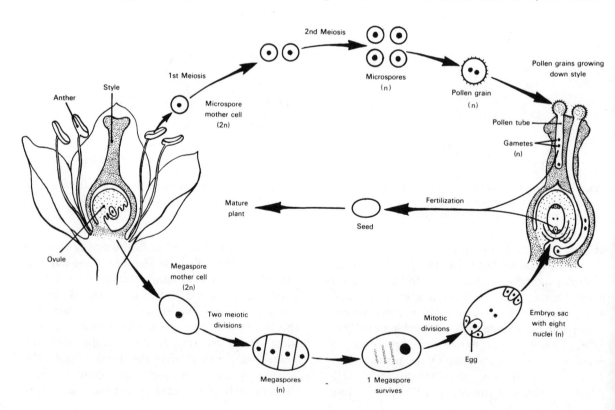

is the organ involved. Special cells, *microspore mother cells,* arise from the anther wall, and these will undergo the meiotic divisions called *microsporogenesis.* (Figures 3-1 to 3-12 show the meiotic divisions and products of microspore mother cells.) The end product of meiosis of each microspore mother cell is a tetrad of four haploid cells, the *microspores.* Each one of the microspores will mature into a pollen grain. During this maturation, each haploid nucleus of a microspore divides *mitotically* to produce two haploid nuclei per microspore or pollen grain. Of the two haploid nuclei, one divides again to form two male gametes, one of which will fertilize the egg. Therefore, all the stages of meiosis which bring about segregation and independent assortment are completed by the time of microspore formation, well before the production of the male gametes.

Inside the ovary of the flower are found the *ovules,* or immature seeds. In each ovule, *megasporogenesis* will occur, the meiotic divisions which precede the formation of the female gamete. In each ovule, one cell enlarges, the *megaspore mother cell.* It undergoes the first meiotic division and produces two cells. Each of these in turn undergoes the second meiotic division, with the formation of a tetrad of four haploid *megaspores.* Of these four cells, three disintegrate, leaving a single cell which goes on to enlarge. The nucleus of this remaining megaspore divides mitotically. Typically, two more mitotic divisions follow so that a group of eight haploid cells is formed. Only one of these is destined to act as the egg and be fertilized if pollination is successful. It is easy to recognize a parallel between megasporogenesis in plants and oogenesis in animals. In both processes, a mother cell (megaspore mother cell or primary oocyte) undergoes meiosis and only one functional cell results. On the male side in both plants and animals, each tetrad of cells forms four products (microspores or spermatids) from each of which a functional male gamete is derived.

MITOSIS AND MEIOSIS ARE COMMON TO ALL EUKARYOTES. Regardless of differences in detail among species, the meiotic process is remarkably similar in all sexual organisms. This undoubtedly reflects its value and early selection by the forces of evolution, which have refined it since its origin in ancestral sexual cells. By increasing the speed with which a wider variety of new genetic combinations can be formed, meiosis gives a decided advantage to sexual forms over those which are exclusively asexual. Sexual reproduction with its meiotic mechanism makes possible an increase in the rate of evolutionary progress. It makes available more diverse kinds of living things which can explore a wider range of environments. Mitosis and meiosis are both fundamental features of eukaryotic cells, and the many genetic phenomena which occur in higher organisms are related directly to both processes. If the essentials of mitosis and meiosis presented in Chapters 2 and 3 have been grasped, the genetic principles encountered in the ensuing discussions should be easily followed.

REFERENCES

De Robertis, E. D. P., F. A. Saez, and E. M. F. De Robertis, *Cell Biology,* 6th ed. W. B. Saunders, Philadelphia, 1975.

Henderson, S. A. The time and place of meiotic crossing over. *Ann. Rev. Genet., 4:* 295, 1970.
Stern, H., and Y. Hotta, Biochemical control of meiosis. *Ann. Rev. Genet.,* 7: 37, 1973.

REVIEW QUESTIONS

1. Name the precise stage of meiosis in which the following take place.

   A. The chromosome threads in the bivalent appear to repel, and chiasmata become evident.

   B. Pairing of homologous chromosomes takes place.

   C. The chromosomes are greatly contracted, and the bivalents are widely separated in the nucleus.

D. The chromonema with obvious chromomeres is greatly elongated and appears as a single filament.
E. Chromatid separation occurs following division of the centromere.
F. Chromosomes composed of two chromatids are aligned separately on the equatorial plate.

2. The chromosome number for the human is a diploid one of 46. Give the chromatid number, the chromosome number, and the bivalent number for a cell or nucleus at each of the following stages: (1) pachynema; (2) diplonema; (3) diakinesis; (4) telophase I; (5) prophase II; (6) telophase II.

3. How can one distinguish cytologically:

A. A cell in mitotic metaphase from one at first meiotic metaphase?
B. A cell at mitotic anaphase from one at first meiotic anaphase?
C. A mitotic prophase from a first meiotic prophase?

4. A cell is dihybrid for the allelic pairs Aa and Bb. The two loci are found on nonhomologous chromosomes. What are the possible genetic combinations which can arise:

A. When this cell undergoes mitosis?
B. When a cell of the same genotype undergoes meiosis?

5. A human body cell normally contains 46 chromosomes. Give the number of chromosomes in each of the following:

A. A cell resulting from the mitotic division of a spermatogonium.
B. A primary oocyte.
C. A secondary spermatocyte.
D. The polar body formed along with a secondary oocyte.
E. A spermatid.
F. A cell formed after spermiogenesis.

6. Allow "A" to stand for one set of autosomes in the human, and let "X" and "Y" represent the sex chromosomes.

A. What will be the chromosome constitution of eggs with regard to autosomes and sex chromosomes?

B. Answer the above question in regard to sperms.
C. What will be the chromosome constitution in a spermatogonium?
D. What will be the chromosome constitution in an oogonium?

7. In which of the cell types in Question 5 are bivalents present?

8. A. How many sperms or male nuclei will arise from each of the following?
    (1) 1000 primary spermatocytes.
    (2) 1000 secondary spermatocytes.
    (3) 1000 spermatids.
    (4) 1000 microspore mother cells.

    B. How many egg cells will arise from each of the following?
    (1) 1000 primary oocytes.
    (2) 1000 secondary oocytes.
    (3) 1000 megaspore mother cells.
    (4) 1000 megaspores which are the immediate products of meiosis?

9. In certain insects such as the grasshopper which has the X-O condition, the male has only one sex chromosome, an X, in contrast to the female which has 2 X chromosomes. Assuming the females of such a species have a diploid chromosome number of 12, answer the following questions.

    A. How many autosomes will there be in the wing cells of a female?
    B. Answer the above for a male.
    C. What will be the total chromosome number in the wing cells of a male?
    D. How many chromosomes will there be in the sperm cells which bear an X chromosome?
    E. How many chromosomes will there be in the non-X bearing sperms?

10. In corn, there are ten pairs of chromosomes in the somatic cells. What would be the expected chromosome number in:

    A. A pollen tube nucleus?
    B. A cell in a petal?
    C. A cell in the embryo of the seed?
    D. A pollen mother cell?
    E. A megaspore?
    F. A cell of the embryo sac?

4

GENIC INTERACTIONS

GENOTYPE, PHENOTYPE, AND CHARACTERIS-TIC. Throughout the early decades of this century, the particulate nature of inheritance was found to apply to a large number of different plant and animal species. It therefore became essential to define the nature of these factors which are transmitted from one generation to the next. Many of the early twentieth-century geneticists were vague or even confused on the distinction between the unit factors of Mendel and their relationship to the actual visible characteristics shown by an individual. No one person merits more credit on the clarification of this critical relationship than the Danish botanist, W. Johannsen. In his first use of the words "phenotype" and "genotype," he clearly distinguished between the assortment of observable characteristics which an individual possesses (the phenotype) and the genetic constitution (the genotype). The genotype is composed of elements (genes) which the individual receives through the gametes. It establishes the foundation from which development of the individual proceeds. The final phenotype in turn depends on the interaction of these genetic elements with factors in the environment. Johannsen clearly pointed out that *no one* kind of gene corresponds directly to a particular characteristic in the phenotype of an individual. For the phenotype is *not* directly inherited. It results from a complex interplay of genetic determinants with the various aspects of the environment. Failure to envision this critical distinction between genotype and phenotype, between gene and character, has persisted to the present time and is responsible for many absurd conclusions concerning inheritance and the environment. How ridiculous it would be to diagram a pair of chromosomes showing a blue eye at one locus and a brown eye at the corresponding locus on the homologous chromosome! Yet in effect, this summarizes the naive concepts which imply that the final character seen in an individual corresponds exactly to what is inherited and that one gene is responsible entirely for a finished characteristic. Some of the aspects of the inheritance of feather color in poultry may help us to avoid such erroneous ideas.

INTERACTION OF TWO OR MORE PAIRS OF ALLELES. The early work of Bateson with poultry demonstrated that *not* just one pair of Mendelian factors but that at least two could interact to affect the color of the feathers. It was found when certain different races of white birds were crossed that the F_1 offspring were not white but colored! The explanation put forth by Bateson (the correct one) was that two separate pairs of Mendelian units are involved in such cases and that these can interact to produce an effect which is distinct from that of either one by itself. We can visualize this by assuming two different dominant genetic factors for pigment formation (C and O) and their recessive alleles for no pigment (c and o). If *any* color at all is to be produced, *both* dominants must be present. In the absence of either dominant allele, no pigment can be formed and the feathers are therefore completely white (Fig. 4-1A). Consequently, two different races may show the same white phenotype but actually possess different genotypic constitutions. Race A may be white because it is ccOO, whereas race B is white because it is CCoo. A cross between members of the two races brings together the two dominants, one from each race, and the offspring can have colored feathers. *Exactly what* the color will be, however, depends on still other genes.

Another locus is present which can affect the *type* of pigmentation. We may let B represent the determinant which governs the formation of black pigment. Its allele b results in white feathers flaked with pigment (white splashed). There is lack of dominance between these two alleles so that the heterozygote, Bb, is blue feathered. Any bird may be black (BB), blue (Bb), or white splashed (bb), *but only if both* "C" and "O" are present at the other loci (Fig. 4-1B). For example, the genotype BbCcOo would produce blue feathers, but Bbccoo can only result in pure white. This example clearly shows that there is not *a gene* for the color of the plumage. At least three separate loci are involved (really more), the interactions of which may produce different effects. The example should also eliminate the notion that a unit character, such as color, is somehow carried in the cell and transmitted as such to the next generation. We cannot, without being silly, say that the genotype is "black" or "white feathered" or "one-half black feathered" and "one-half white." Rather, the genetic component in this case includes certain genes which can determine the presence or absence of specific pigments in the feathers. The final color which is realized phenotypically will depend on the interaction of these several genetic elements, and, as we will see, upon environmental factors as well.

APPRECIATION OF GENIC INTERACTION AND THE NATURE OF WILD TYPE. When we are following a simple monohybrid or dihybrid cross, how do we take into consideration the fact that many genes are actually involved in the expression of a particular character? To answer this, let us turn our attention next to the many pairs of loci which are involved in the production of coat color in the mink. In this animal, at least a dozen genes are known to influence the color of the fur. The standard coat color (wild type) results from a certain combination of genes. Minks which are pure breeding for wild fur color must be homozygotes of the following constitution: PP IpIp AlAl BB BgBg BiBi CC OO ss ff ebeb cmcm.

Substitution at just one of these loci may result in a phenotype quite distinct from the standard. For example, the presence of two recessives, pp, in place of the dominant P will cause the fur to be platinum. Replacing the dominant allele B with the recessive condition bb will change the fur to brown, even though two doses of the dominant allele for wild coat color are present at many of the other loci. A substitution of one dominant Eb for just one recessive eb in the genotype can alter the coat color from wild to ebony. How then are we to diagram through the F_2 generation the following monohybrid cross: wild mink × platinum? It would be ridiculous to write more than is shown by the simple Punnett square method in Figure 4-2. The reason for this is that *only one* locus is under consideration; only one pair of alternatives at a specific locus is being followed. Although we must appreciate that any character such as coat color involves many loci, a particular cross does not necessarily involve more than one or two of them. In any diagram, we symbolize only those loci which vary in the particular cross. In our example, only the variation at the platinum locus concerns us. Although we should appreciate that the genes at the other loci are also exerting an effect, we need not represent these

FIG. 4-1. Gene interaction in poultry. *A.* Pigment formation in the feathers depends on the presence of both dominants, C and O. Races A and B are white because each lacks one of the dominants. The F_1 hybrid between them has colored feathers because it receives a different dominant from each parent. *B.* The specific color of the feathers depends on several pairs of alleles, among them B (black) and b (white splashed). Since there is lack of dominance, the hybrid (Bb) is blue. However, for the pigment alleles to express themselves, the dominant alleles C and O must be present; otherwise, a bird will be white, regardless of the genetic factors present at the pigment locus.

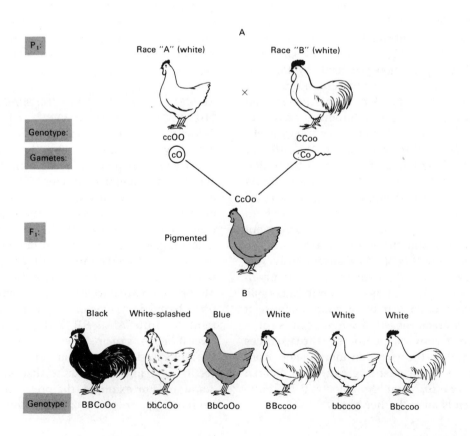

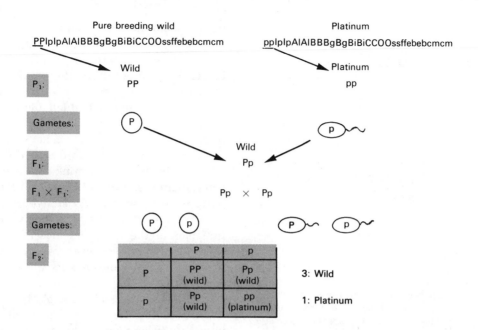

Pure breeding wild
PPIpIpAlAlBBBgBgBiBiCCOOssffebebcmcm

Platinum
ppIpIpAlAlBBBgBgBiBiCCOOssffebebcmcm

	Wild	Platinum
P_1:	PP	pp

Gametes: P p

Wild
Pp

F_1:

$F_1 \times F_1$: Pp × Pp

Gametes: P p P p

F_2:

	P	p	
P	PP (wild)	Pp (wild)	3: Wild
p	Pp (wild)	pp (platinum)	1: Platinum

FIG. 4-2. Cross of wild and platinum-coated minks. Several pairs of alleles interact to produce a given coat color. However, in a simple monohybrid cross, only one locus is under consideration. There is no need to represent the others, because the rest of the genotype is identical in the P_1 parents and does not vary in any of the individuals.

other genes because they are identical in both wild (PP) and the platinum (pp) parents.

Now suppose that a mating is made between a platinum mink and one with brown fur. In this case, two different genes at two different loci are being followed, the p locus and the b locus (Fig. 4-3). We must consider both of these, but we can forget the other 10 because they do not differ. The mating is nothing more than a simple dihybrid cross. It is important at this point to appreciate fully from Figure 4-3 that this cross of a platinum and a brown animal involves two loci and that *both loci* and the gene forms present at them must be represented in the genotype of *each parent*. This is essential because the platinum parent is carrying the wild allele (B) in the homozygous condition at the "brown" locus. In a similar way, the brown parent is carrying the wild gene form (P) in the homozygous condition at the "platinum" locus. As a result of this, the F_1 are all wild. It would be a serious error to represent a cross between platinum and brown as: pp × bb. This fails to take into consideration that *two* separate genes and loci concern us in each parent. It results in a meaningless representation of the cross.

Besides showing us the interaction of many genes, the example of coat color in mink also illustrates at least two other important points. The term "wild type" is commonly used to mean "normal" or "standard," the form which is typically encountered in nature (more details on naming of loci are found at the end of this chapter). We see from the case of coat color that there is *not just one* gene or one form of a gene responsible for wild. There are as many so-called wild or normal alleles for a specific characteristic as there are separate loci which affect that character. In Chapter 1, the point was made that we can study a character only if variations in it are available to us. If everything remained standard, we would never learn the basis of inheritance of any character. But because a gene may mutate, alternative gene forms arise and produce inheritable changes, which result from alterations which take place in the deoxyribonucleic acid (DNA) composing the gene. These may arise spontaneously or be induced by certain environmental factors. All alternative gene forms stem from mutations which have taken place at genetic loci. When a gene mutates and gives rise to an alternative form, we may then recognize the wild form of the gene and its mutant allele. The mutant gene form is a departure from the wild or standard and can result in a phenotypic trait which varies from the typical or normal form of the characteristic. It is these variations from the standard or wild which provide us with contrasting alternative traits of characters and which allow us to study the genetic factors which interact to produce a normal phenotype. So if we have 12 different variations from the standard at 12 different loci, as we do in the case of mink coat color, we are consequently aware of 12 different kinds of departure from the wild. By studying the inheritance pattern of each pair of traits, we learn that fur color in the mink depends on at least 12 separate allelic pairs, each pair representing a wild or normal gene and its allele, a mutant form.

When the dihybrid cross, platinum $\times$ brown is followed through the F_2 generation, another important point comes to light (Fig. 4-3). Among the offspring of the F_1 wild, dihybrid animals,

FIG. 4-3. Cross of platinum and brown-coated minks. In a dihybrid cross, variation is being followed at two loci, and these must be represented in each of the individuals. The other loci which also influence coat color are disregarded, because they are identical in every case.

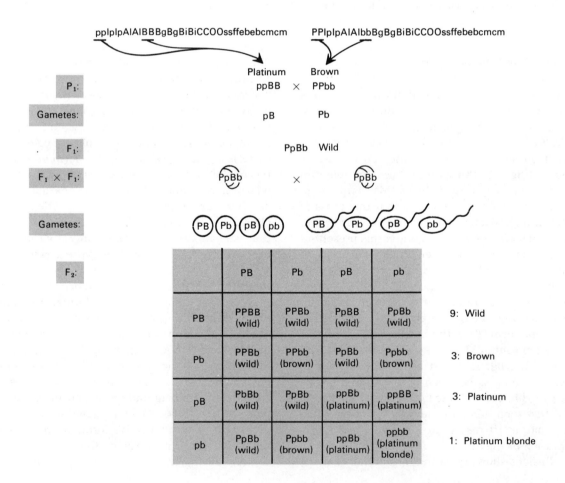

four different phenotypes appear. The chance for a wild is 9:16, for a platinum 3:16, for a brown 3:16, and 1:16 for a completely different type which has been designated "platinum blonde" (ppbb). The cross illustrates beautifully one of the important consequences of the sexual process: independent assortment with its production of new gene combinations. Due to the segregation of two pairs of alleles and their independent transfer to the gametes, various genotypic combinations are possible (3^2 or 9; see p. 27). Since dominance is involved, they fall into four different phenotypes (2^2): three of them are the familiar wild, platinum, and brown, but a new phenotype, platinum blonde, has been derived. Imagine the many genotypes which could result if all 12 pairs were considered (3^{12}) and the variety of possible phenotypes, more than 2^{12}, because there is lack of dominance between certain alleles affecting the coat color. We must keep in mind that the physical basis which generates such variation in sexual species is the meiotic cycle of the chromosomes, a foundation on which evolutionary progress has depended.

GENIC INTERACTION AND MODIFIED RATIOS. Once we understand the concept that many genes interact in the normal development of a character, we will be able to appreciate the basis for various kinds of genetic effects. Let us next take an example from the inheritance of fur color in mice, where more than one pair of alleles is known to influence pigmentation. In addition to the wild grayish color, there are other possibilities, such as black or white. In regard to these three types, gray, black, and white, two pairs of alleles are known to be involved. Wild, or gray, depends for its expression on the presence of a dominant, B; black pigmentation depends on the recessive allele, b. However, if any pigmentation at all, gray or black, is to develop, there must be present at another separate locus the dominant allele, C. The recessive condition, cc, will result in a white animal, regardless of the gene forms present at the first locus. Assume that two dihybrid animals are crossed: BbCc × BbCc (Fig. 4-4). Both are wild in coat color because each possesses the dominants for gray and for pigment production. Among their offspring, the following can be predicted: 9 B– C– (gray): 3 bbC– (black): 3 B– cc (white): and

1 bb cc (white). Notice that the kinds of *genotypes* expected from a dihybrid cross are actually obtained; however, the expression of the genotypes is so altered by genic interaction that the expected phenotypic ratio of 9:3:3:1 is modified to a ratio of 9:3:4. This is so because the genotypes B– cc and bb cc cannot be distinguished from each other. Any pigment in the hair requires the presence of the dominant allele, C. It is as if the c locus can suppress the expression of the pigment factors at the b locus. In one way, this may remind us of dominance, but use of that term is restricted to the interaction between a pair of alleles (B is dominant to b). Another word is needed to describe the suppressive influence of any genetic factor on another which is *not* its allele. Such an effect is called *epistasis*. The example of fur color in mice nicely illustrates "recessive epistasis" because the double recessive condition, cc, is required to mask the expression of genetic factors for pigmentation at the b locus. Epistasis, however, may result from the presence of a dominant allele. Coat color in the dog involves at least two loci. In some varieties, black pigmentation results from the presence of the dominant B; brown pigment depends on the recessive allele, b. But neither color will be expressed if the dominant factor, I (for inhibition of color), is found at another locus on a different chromosome. Figure 4-5 follows the cross of two dihybrid white dogs: Bb Ii × Bb Ii. No pigment is found in their hair because the presence of just one dose of the allele "I" is sufficient to suppress all pigment formation. Among their offspring, the chances become 12:3:1 for white, black, and brown, respectively.

"Epistasis" is a term which was coined before 1910 by Bateson, who discovered that this type of genetic interaction in poultry can modify the classic 9:3:3:1 ratio of Mendel. However, once the basis for the modification is understood, we see that there is no exception to Mendelian inheritance; we must simply remember that no one gene or allelic pair acts alone.

We have already noted that feather color in poultry requires the dominants C and O. Otherwise, no pigmentation will be formed in the feather. We see here an example of two genes (one at the c locus and one at the o locus), either of which can suppress the expression of genetic determinants at a third one, depending on the

FIG. 4-4. Cross of two dihybrid gray mice. The factor B (gray) is dominant to its allele b for black fur. However, no pigment can be produced in the absence of allele C. As a result, two of the four major classes of genotypes cannot be distinguished, and the 9:3:3:1 phenotypic ratio is modified to 9:3:4.

alleles which are present. Birds of the genotypes Bb CC oo or Bb cc OO are white, not blue. Only those birds which have at least one of each dominant allele (C__ O__) can form feather pigment (Fig. 4-1B). Consider a cross of two dihybrids: CcOo × CcOo (Fig. 4-6). Color of some type will be expressed phenotypically by each parent because both dominants are present. Among their offspring, however, pigmented and white will occur in a ratio of 9:7, a modification of the 9:3:3:1 ratio. There are only 9 chances

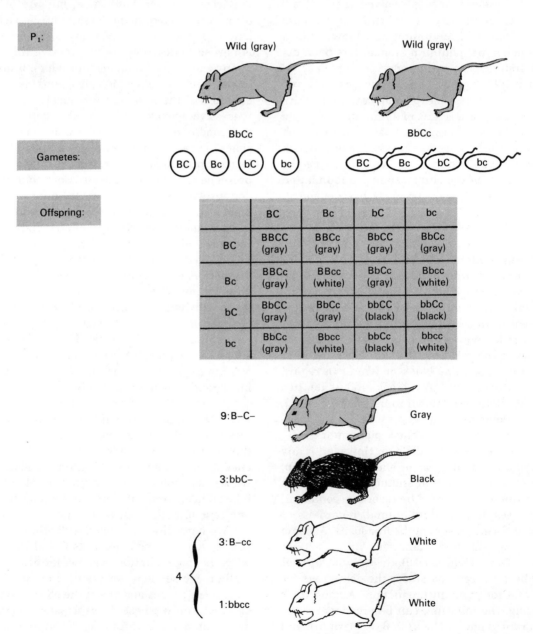

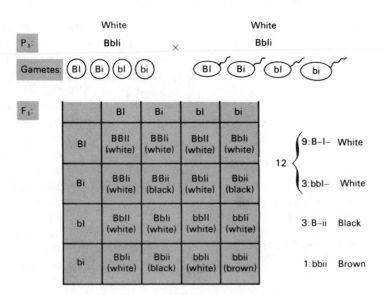

FIG. 4-5. Dominant epistasis. In certain breeds of dogs, the factor for black fur (B) is dominant to its allele for brown (b). However, neither allele can be expressed if the dominant "I" is present. Since two of the four major classes of genotypes cannot be distinguished and are classified as white phenotypes, the 9:3:3:1 ratio is modified to 12:3:1. Compare this dihybrid cross with that shown in Fig. 4-4, an example of recessive epistasis.

out of 16 for any chick to exhibit some type of color because only 9 out of 16 on the average will possess both of the required dominant factors. In cases such as this, we say that "duplicate recessive epistasis" is operating, for we have two genes (one at locus "c" and one at locus "o"), *either of which* independently can alter the effects of a second or a third one (cc can suppress O__ or B__; oo can also suppress C__ or B__).

The genetics of poultry provides many illustrations which demonstrate the complexities of genic interaction. Still other genes are involved in feather color. A gene form "I" at another locus exhibits dominant epistasis and inhibits feather color, just as the allele "I" in dogs prevented any pigmentation of the hair. (Although the letter "I" is used to symbolize both factors, they are obviously not the same allele, because they occur in very different species.) A pure-breeding variety of black birds must have at least the following genotype: BB CC OO ii. With even more genes involved, we see that the picture approaches that of fur color in the mink. We must remember that these examples of many genes influencing the expression of a character reflect the typical situation; they are not unusual or exceptional in the least.

THE BASIS OF GENIC INTERACTION. Additional examples of epistasis could be presented, but this would serve just to reemphasize the important point of genic interaction. But now an important question should be answered, "What is the basis of this interaction; how can a factor at one locus affect the expression of one at another locus?" Today we have a great deal of information from the field of molecular biology to supply us with some answers. In later chapters, evidence will be presented to show that many genes exert their effects on the phenotype through their control of enzyme production. A wealth of research, much of it with microorganisms, tells us that the genetic control of metabolism is largely a consequence of gene control of protein formation. Any factor which governs the protein formation of a cell controls the activities of the cell. Proteins are the most complex of the chemical compounds in the cell. Besides their importance as units in the structural organization of various cell parts, they form the main component of enzymes, the important organic catalysts which enable the cell to carry out most of its chemical steps. We may envision many essential cellular products to be the end result of a series of chemical steps, each of which can take place only if the appropriate enzyme is present in an adequate amount (Fig. 4-7A). Although Figure 4-7A is a definite oversimplification because it eliminates branches in the sequence, it does enable us to picture the average cellular product as the result of a sequence of steps. No step in the sequence can proceed without the product of a previous step. A breakdown anywhere in the chain will prevent the production of the proper end product. Such a breakdown may occur if a step fails to progress due to an enzyme which is defective, absent, or present in insufficient quantities.

We may consider pigment production in the feather of the fowl to depend on a developmental sequence which entails several steps, each controlled by a specific enzyme (Fig. 4-7B). We can see that if a bird should lack the dominant allele

FIG. 4-6. Duplicate recessive epistasis. In poultry, both dominants (C and O) must be present for any pigmentation of the feather. Either double recessive condition (cc or oo) can prevent the expression of genetic factors at other loci which also influence pigment formation. Since three of the four genotypic classes cannot be distinguished phenotypically, the 9:3:3:1 ratio is modified to 9:7.

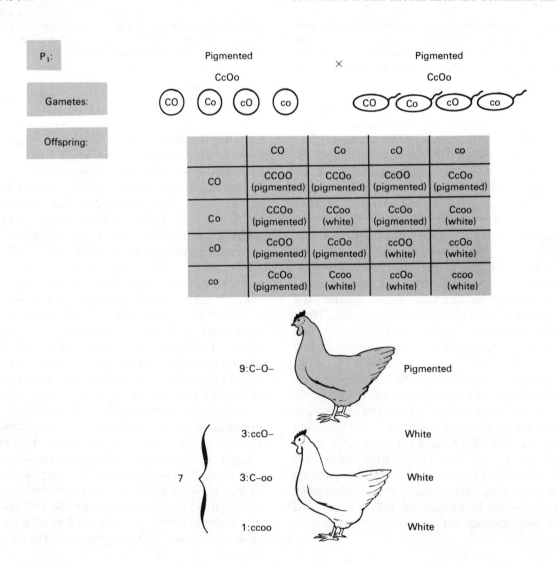

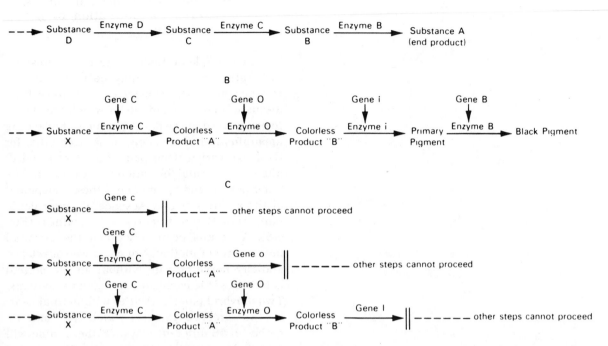

FIG. 4-7. Gene interaction. *A.* Any cellular product is the result of a series of chemical steps, each catalyzed by a specific enzyme. *B.* A simplified version of pigment formation in the feathers of the common fowl. A specific gene governs the formation of an enzyme at each step along the pathway leading to the formation of the end product, the pigment. The type of pigment will depend on the enzymes converting the primary pigment. These, in turn, are controlled by specific genes, such as gene B in this scheme. *C.* Blockage of the pathway leading to pigment formation. The interruption may occur anywhere in the sequence, early (*above*) or later (*below*). Absence of a specific gene form (C, O, or i) and the substitution of its allele (c, o, or I) results in deficiency of the enzyme controlled by the gene. The sequence cannot be completed even though all the other required enzymes are available.

"C" (Fig. 4-7C) or the dominant allele "O," no pigment can form. It does not matter what other genetic factors are present for pigment production if there is a breakdown in a vital step in the chain. If the chain is not interrupted, pigment will form, but the final color will depend on still other genes. Inspection of the sequence of steps will show why a ratio of 9:7 is to be expected after the cross of two dihybrids: CcOo × CcOo. Our knowledge of the classic 9:3:3:1 ratio tells us that on an average, in 9 cases out of 16, both dominants will be present: C__O__. Only these genotypes contain the alleles essential to the completion of the developmental sequence through their control of the needed enzymes. The other seven combinations all lack at least one genetic factor which controls one of the enzymes and hence a step which is essential to a pigment product. If we look at the same chain of reactions (Fig. 4-7C), we see that the presence of allele "I" blocks completion of the sequence because enzyme "i" would be lacking. It does not matter if the factors "C" and "O" are both present. The presence of allele "I," which causes a lack of enzyme "i," interrupts the sequence of steps required for the pigment.

Applying this same reasoning to the case of the dihybrid cross in dogs (see Fig. 4-5), we can see that on the average 12 offspring out of 16 will contain "I" and hence will lack enzyme "i:"

9 I__B__ —white because enzyme "i" is not present

3 I__bb —white because enzyme "i" is not present

3 ii B__ —black because enzyme "i" allows pigment to form and the allele "B" governs the formation of black pigment

1 ii bb —brown because enzyme "i" allows pigment to form and the

recessive condition "bb" governs the formation of brown pigment

An example of duplicate recessive epistasis (9:7 ratio) in humans may be understood in relationship to the concepts which have been discussed. Two pairs of alleles which are known to be involved in the development of the hearing apparatus in humans depend on each other for their expression. One pair of alleles D and d, affects the normal formation of the cochlea. The other pair, E and e, influences the development of the auditory nerve. Absence of either dominant, D or E, will thus result in deafness (Fig. 4-8). A normal cochlea without the essential nerve cannot function. Nor can a well developed auditory nerve operate without all the normal activities of the cochlear canal and its contents. Two dihybrid parents (DdEe) with normal hearing will produce gametes which can come together in 16 different ways. Of these, nine will contain both dominant factors (D__ E__) needed for a complete, functional hearing apparatus. The other combinations lack one or both of the vital genetic determinants: D__ee (3); dd E__ (3) dd ee (1). Although this example does not explain the genic interaction on the cellular level, it does point out that the normal development of a character (the ear in this case) depends on the orderly development of more than one of its parts. Abnormality in one component resulting from a genetic block in development can prevent the normal expression of another portion, even though the latter may have formed properly under the direction of the required genetic factors.

INTERRELATED PATHWAYS IN METABOLISM. Few synthetic pathways are as simple as suggested by Figure 4-7A. Most of them are not independent but are linked to others so that any one product in a sequence may be essential to one additional pathway or more (Fig. 4-9 A and B). In the illustration, a lack of C substance leads to defects in two end products, A and Z, because both pathways leading to them require C for their completion. Since metabolic pathways are commonly interrelated in this way, it is not surprising to learn that a gene, when studied carefully, is often found to have more than one

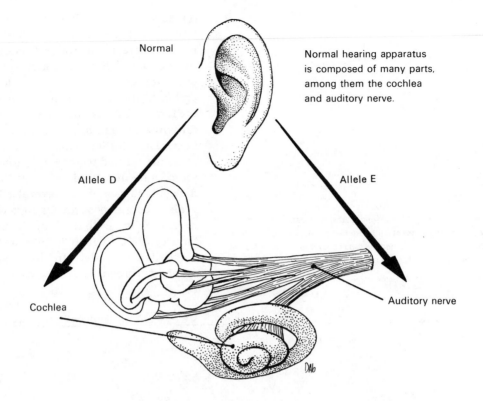

Normal

Normal hearing apparatus is composed of many parts, among them the cochlea and auditory nerve.

Allele D

Allele E

Cochlea

Auditory nerve

Defective hearing apparatus can result from abnormal development of one of its component parts

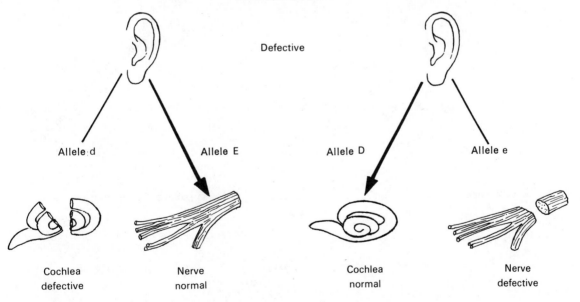

Defective

Allele d

Allele E

Allele D

Allele e

Cochlea defective

Nerve normal

Cochlea normal

Nerve defective

FIG. 4-8. Interaction of genetic factors in development. A normal character may involve the orderly development of many component parts. Abnormality in just one of these due to a genetic block can result in a nonfunctional organ in spite of the presence of the normal components. In the case of the ear, any genotype lacking both dominant alleles (D and E) will be defective as a result of abnormality of the cochlea or auditory nerve.

phenotypic effect. One of these may be more pronounced than the others, but the latter are real and detectable. For example, in the fruit fly, the recessive allele which is responsible for white eye color also influences the color of the testes and even the *shape* of the sperm receptacles in the female. In cats, the genetic factor responsible for white fur and blue eyes also results in deafness. The multiple effects of a single gene or allele are termed *pleiotropy*. Although they may seem to be unrelated, study of the chemical or molecular basis of the several effects often reveals a common basis. One excellent example from humans is that of the disorder phenylketonuria (PKU) which is inherited as a simple Mendelian recessive. In this unfortunate condition, severe mental retardation is a typical

FIG. 4-9. A branched metabolic pathway. Most pathways are interrelated so that a genetic block at one point may have multiple effects. In this example, a genetic block prevents the formation of substance C required for the production of two products, A and Z.

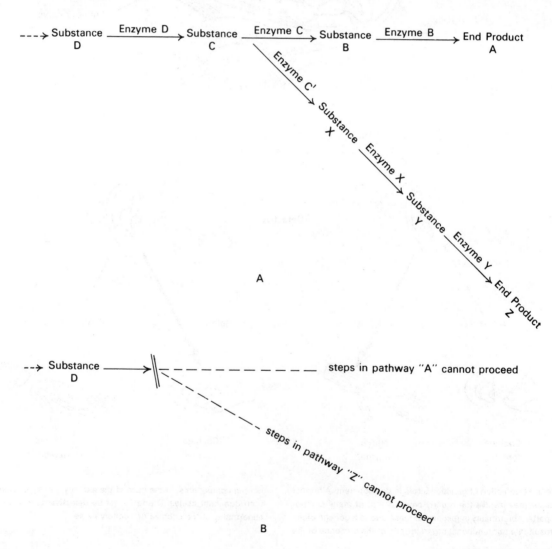

symptom. In addition, however, affected children also tend to have light hair and light skin pigmentation. Blood and urine analyses reveal abnormally high levels of phenylalanine, an amino acid concentrated in the protein of milk, cheese, eggs, and several other common foods. The high levels of phenylalanine are associated with high levels of another chemical, phenylpyruvic acid, a substance not found in the body fluids of normal individuals. Any collection of phenotypic effects such as these which defines a clinical condition is commonly called a *syndrome*. The syndrome which decribes PKU is now known to result from the pleiotropic effects of a recessive allele which causes the block of a single step in a metabolic pathway. As Figure 4-10A indicates, protein from the diet is broken down by the cells into its constituent amino acids. The amino acid phenylalanine is normally converted to another amino acid, tyrosine. This

FIG. 4-10. Interrelated pathways in metabolism of phenylalanine and tyrosine. *A.* Tyrosine normally enters cells through breakdown of dietary protein. In addition, it is formed from phenylalanine by the action of a specific enzyme. The total tyrosine amount then enters other pathways. *B.* In the absence of the enzyme, due to a recessive allele, phenylalanine accumulates, and some is converted to phenylpyruvic acid. The two substances are toxic. The only available tyrosine is preformed and comes from the dietary protein. The excess phenylalanine inhibits an enzyme activity in the pathway leading to the formation of melanin pigment.

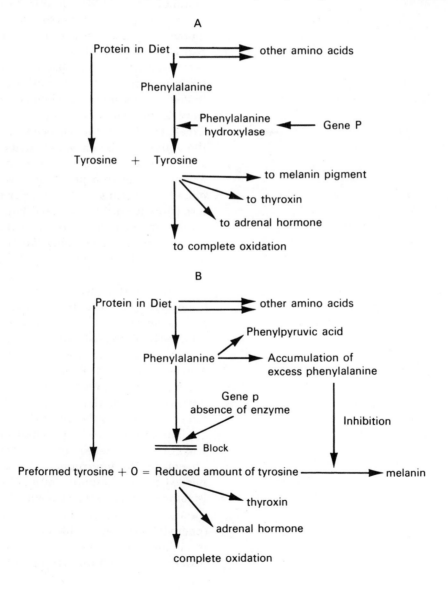

single chemical step involves the substitution of an —OH group for —H and requires a specific enzyme which is manufactured in the liver. The tyrosine which is normally formed is an amino acid which is linked to other major pathways, one leading to the formation of melanin pigment, one to the formation of thyroxin, and another leading to complete oxidation to carbon dioxide and water. Tyrosine may also enter cells as a result of its presence in various dietary proteins; all of it is not formed from the conversion of phenylalanine.

The normal picture of the pathways leading from tyrosine may be altered in the absence of the enzyme which forms tyrosine from phenylalanine (Fig. 4-10B). In this case, the latter substance accumulates in the blood. Appreciable amounts of it become diverted into an alternative pathway which leads to the formation and excessive accumulation of phenylpyruvic acid. The tyrosine required for the pathways which are linked to this amino acid must come entirely from preformed tyrosine in the diet. Moreover, excess phenylalanine has been shown to inhibit the activity of the enzyme which converts tyrosine to another substance needed in the pathway leading to melanin production. We can appreciate why victims of the disorder tend to have less pigment in their skin and hair. The mental retardation is a consequence of the high levels of the accumulated phenylalanine and the phenylpyruvic acid which forms from it. These substances are toxic to the central nervous system and interfere with the metabolism of the brain cells. The unfortunate, irreversible damage to the nervous system can be prevented if afflicted babies are placed on diets low in phenylalanine. This avoids the accumulation of toxic concentrations of this amino acid and its conversion to phenylpyruvic acid. Fortunately, tests are available for the prompt recognition of affected infants shortly after birth.

The example of PKU demonstrates several ways in which a gene may bring about pleiotropic effects: by the interruption of a chemical step needed for the formation of a substance which is common to more than one metabolic pathway; through the accumulation of a metabolite (phenylalanine here) which may be toxic at high levels and which may be converted to another toxic product; and through the inhibitory effect of an accumulated metabolite on enzymes in another metabolic pathway. As in the case of PKU, the assorted phenotypic effects of many genes may be traced to a single chemical block in a sequence of steps. Other examples in humans will be encountered in later discussions which concern genetic phenomena other than pleiotropy.

VARIATION IN GENE EXPRESSION. When we study the phenotypes of several individuals, all of whom are known to possess a particular allele, we often find that the allele is very constant in its expression. For example, the antigens found in human red blood cells are inherited. All individuals possessing an allele for antigen A will show blood protein A in their blood. Individuals having alleles A and B can be easily classed in blood group AB because their red blood cells will contain both antigens, A and B. There are almost no exceptions to this. The alleles for the groups can be expected to express themselves whenever they are present, and they always express themselves in the same way. No person having the allele for blood group A will be more or less "A" than another individual with the same genotype. Both show group A and exhibit it to the same degree.

In contrast to this is a large number of other alleles which do not always express themselves so predictably. A dominant (P) in humans is responsible for the production of extra digits on the hands and feet, an abnormality known as polydactyly. The condition may be present in several members of a family, a parent, and one or more of the children. Occasionally, we find that a phenotypically normal person from such a family marries another normal, unrelated individual but nevertheless produces children with polydactyly. Since the allele for extra digits is a rare one, it would be very unlikely that the unrelated person who marries into the family would also be carrying the allele for the condition. From studying many such cases, it becomes apparent that a person who is heterozygous for the condition (Pp) *may* or *may not* show the trait. The dominant for polydactyly is thus not constant in exerting its phenotypic effect. It does not express itself in all the individuals who actually carry it, and so we say that the allele has *reduced penetrance*. If an allele always

expresses itself, as in the case of the blood antigens, we say that it is 100% penetrant. If 10 people have the same genotype (such as Pp for polydactyly), but only 9 out of the 10 show the dominant effect of the mutant factor whereas the remainder appear normal, the allele is said to have a penetrance of 90%. The degree of penetrance varies greatly for different genetic factors. For example, another dominant in man causes the production of bony projections and has a penetrance of only 60%.

Experiments with laboratory animals have enabled us to demonstrate that reduced penetrance is a phenomenon which must be considered in genetic analysis. In the fruit fly, a certain recessive ("i") may cause an interruption in one of the veins of the wing. When flies with interrupted wing veins are mated to each other, 9 out of 10 of their offspring also show the trait, whereas the remainder have normal wings. The geneticist has an advantage with laboratory organisms such as flies, because he can selectively breed these normal—appearing individuals together (Fig. 4-11). Among their offspring, 9 out of 10 will show the mutant trait. The remaining normal 10% can be bred further and will again produce 90% mutant progeny and 10% normal. We clearly see from such a study that the 10% which are phenotypically normal individuals are in actuality homozygous recessives and that the allele for interrupted wing vein has a reduced penetrance, 90%. It is evident that the true genetic constitution of a certain phenotype (such as the homozygous recessive genotype of the normal appearing flies) may remain undetected if selective breeding experiments cannot be performed. Reduced penetrance is therefore a factor which can complicate genetic investigations and also the interpretation of pedigrees, especially those of humans (Chap. 6).

In addition to the modification of gene expression which can result from reduced penetrance, another type of variation is frequently encountered. In the case of polydactyly, those persons who *do* express the allelle when it is present fall into an assortment of phenotypes. Some of them may have an extra digit on each hand and foot, others on only one. Moreover, the extra digit may range in development from complete all the way down to just a vestige. Obviously, when the allele for polydactyly *is* penetrant, it varies in the kind of phenotypic effect it produces. We say that an allele whose expression varies in degree is of *variable expressivity*. Note the distinction between "penetrance" and "expressivity." The former measures the ability of an allele to express itself at all in any way when it is present in the genotype. Expressivity refers to the kind of phenotype an allele produces when it does express itself (that is, when it is penetrant). One genetic factor may be 100% penetrant but its expressivity may vary greatly. Another may show reduced penetrance but be very constant in its expression. Beside being 100% penetrant, the factors for the blood antigens do not vary at all in their expressivity. Consider in humans still another dominant ("B") which can cause the outer coat of the eye, the sclera, to be bluish. However, only 9 persons out of 10 who carry the allele express it in any way. The sclera of those that do may range from a

FIG. 4-11. Reduced penetrance. When flies with the mutant trait, interrupted wing vein, are mated, 10% of their offspring appear normal. When these are mated to each other, 90% of their offspring are mutant and 10% normal. Further breeding of these latter types will produce the same results. All of the flies, those normal in phenotype as well as those expressing the mutant trait, are homozygous recessives for interrupted wing vein (ii). Therefore, the double recessive genotype is only 90% penetrant.

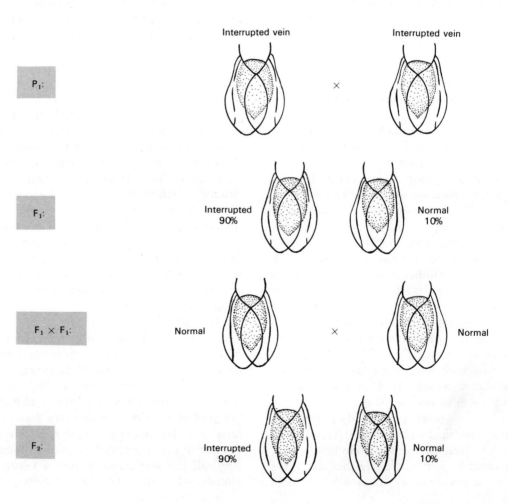

tinge of blue to almost black (Fig. 4-12)! Moreover, the allele causes the bones of the body to be brittle. We have here an example of an allele with reduced penetrance, which is of variable expressivity, and which is also pleiotropic.

BASIS FOR VARIATION IN GENIC EXPRESSION. How can we explain the basis of the phenomena of reduced penetrance and variable expressivity, which can cause marked variations in gene expression? Today we have actual experimental evidence to give us an insight into the problem; however, the correct idea goes all the way back to Johannsen, who stressed that characteristics are not inherited but develop from the interaction of many genes and the environment. We have encountered so far in our discussions several examples of genic interaction in which one gene may alter the expression of another. In the simple example of complete dominance of the allele for tallness in peas, the recessive allele does not express itself at all in the heterozygote (Tt). In a sense, the recessive (t) is not penetrant when its allele (T) is present. Where epistasis is operating, an allele may not be penetrant because of the presence of another genetic factor at another locus (feather color will not be expressed in the genotype CCII). In these examples, we clearly see that the penetrance of an allele may be the direct result of the influence of some other genetic factor, either its own allele (in the case of dominance) or a nonallelic factor (as in epistasis). Therefore, reduced penetrance should not be regarded as any sort of contradiction to the principles of heredity. It is an expected consequence of the role of the rest of the genotype in the expression of any form of one specific gene.

Similarly, variation in the *kind* of expression an allele has when it *is* penetrant often depends on the presence of certain other genes. Many examples relating to spotting have been studied in cattle and rodents. In mice, white spotting may depend on the presence of a recessive (s). However, although spotting may be expressed in the homozygous animals (ss), the size of the white areas can range from tiny points to large spots to a coat that is entirely white. The amount of whiteness depends on a host of genes, each with a small but definite effect, interacting with the allele for spotting (s). The term *modifier* (or modifying gene) is usually reserved for any such genetic factor whose effect is to alter in a small quantitative way the expression of another genetic factor. The cumulative or added influence of modifiers can be very significant. If different modifiers are present in different individuals who possess the same "main allele," the expression of the latter may vary greatly from one individual to the next. The geneticist must consider modifiers when he or she is studying the action of specific genes in experimental animals or plants. The stocks must be kept as similar as possible in regard to modifiers. Failure to do so can lead to erroneous interpretations of inheritance patterns and gene expression.

Some modifiers have such an extreme effect that they completely prevent the expression of some other allele. In the fruit fly, there are several examples of the failure of a mutant genotype to show phenotypically when it is expected. One such case involves the size and shape of the wing. Flies homozygous for the recessive vestigial (vg vg) have reduced and distorted wings. However, the mutant condition may not appear if a certain other allele (called "dimorphos") is present at another locus. We call the latter a *suppressor,* meaning that it is a genetic factor which can prevent the expression of a mutant allele. Some suppressors behave as dominants, others as recessives. Suppressors

FIG. 4-12. Reduced penetrance and variable expressivity. The dominant for blue sclera (B) expresses itself in some way in 90% of those individuals carrying it. The remaining 1 out of 10 is normal. The allele is thus 90% penetrant. Among those showing the trait, there is variation in the shade of the sclera from one person to the next. This variation shows that allele B, when it is penetrant, is of variable expressivity.

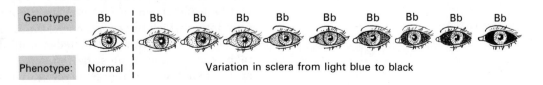

Genotype:	Bb	Bb	Bb	Bb	Bb	Bb	Bb	Bb	Bb	Bb
Phenotype:	Normal				Variation in sclera from light blue to black					

as well as modifiers may confuse the genetic picture if they are not considered in crosses. Moreover, their recognition enables us to appreciate part of the basis for penetrance and expressivity. For modifiers and suppressors, along with other genes composing the total genotype, may interact to produce variation in the expression of any one particular gene.

IMPORTANCE OF THE ENVIRONMENT IN GENE EXPRESSION.   We must not forget that the activities of all the genes are taking place in the cellular enviroment, which in turn may be influenced by the external environment. The geneticist must recognize the importance of environmental conditions, because any characteristic depends on a certain environment as well as on a certain genotype for its typical expression. The environment has been shown to be a critical factor in the degree of penetrance and expressivity of many alleles. In the fruit fly alone, temperature changes are known to be able to change the penetrance of many alleles from 0 to 100%. Similarly, the expressivity of an allele may vary greatly with temperature. The genotype which produces blisters on the wings is much more extreme in its expression at 19° than at 25°. Not just temperature, but all aspects of the environment must be considered in relationship to their influence on gene expression. Some of the variation in the expression of polydactyly is undoubtedly environmental. The fact that any one person showing the trait may have a complete extra digit on one hand and none at all or just a protuberance on the other argues that environmental influences are operating. We are seeing here an example of variable expressivity in *one* individual. Genetic factors for extra digits are also known in animals where there is also evidence for an effect of maternal age on the penetrance of the specific factors.

Nutrition or diet is another critical environmental factor to consider. Recall that the dire effects of phenylketonuria are expressed when there is an accumulation of phenylalanine. A diet which eliminates much of this amino acid will be a factor which will prevent the mental retardation which full expression of the allele for PKU can cause. A frequently cited example from rabbits also demonstrates the influence of diet on the expression of the genotype. Rabbits are herbivorous animals; so their diet would entail the consumption of large amounts of chlorophyll. Associated with chlorophyll in all higher plants is an orange-yellow pigment, xanthophyll, which is usually obscured by the green. In nature, most rabbits possess a dominant, "Y," which governs the production of an enzyme responsible for breaking down the yellow pigments (Fig. 4-13). When the fat of a wild rabbit is noted, it appears white because no xanthophyll pigments have been deposited. On the other hand, some breeds of rabbits are homozygous for the recessive allele y. With this genotype, yy, the essential enzyme is not produced; the yellow pigments therefore accumulate in the fat and give it a yellow appearance. Now suppose we take rabbits of the genotype yy (for yellow fat) and supply them with a source of food lacking green vegetables and the accompanying yellow pigments. The fat of these animals would be white. We could not distinguish them phenotypically from rabbits with the dominant allele.

Another excellent example from rabbits concerns the interaction of temperature with the genotypes which influence fur color. Coat color in rabbits, as in other mammals, involves the interaction of many genes. The formation of the normal dark pigment requires a series of steps and enzymatic reactions. Among the several coat patterns is one known as "himalayan," a phenotype characterized by white body fur and black areas at the tips of the feet, tail, ears, and nose. This pattern results from the fact that himalayan rabbits possess an allele (c^h) which controls the production of a certain enzyme. This is capable of catalyzing a reaction needed for formation of dark pigment. However, at higher temperatures, the enzyme is inactivated. Raised at temperate conditions, rabbits contain this enzyme in its active form *only* in the cooler regions of the body, the extremities such as the feet and tail. That a temperature effect is indeed responsible for the pattern can be shown dramatically by shaving a himalayan rabbit and rearing it under cool conditions. The new fur which grows in will be black. One may also just shave off white areas of an animal's back and keep them cool by application of ice packs. The newly emerging fur in these areas will be black. We see here an example of the environmental

FIG. 4-13. Influence of diet in rabbits on gene expression. Ordinarily, animals of genotype yy can be distinguished from those carrying the dominant Y which governs the formation of an enzyme required to break down yellow pigments in the diet. If the pigments are removed, there is no substrate for the enzyme whether or not it is present. Consequently, all animals fed a diet lacking pigments would possess white fat regardless of their genotypes.

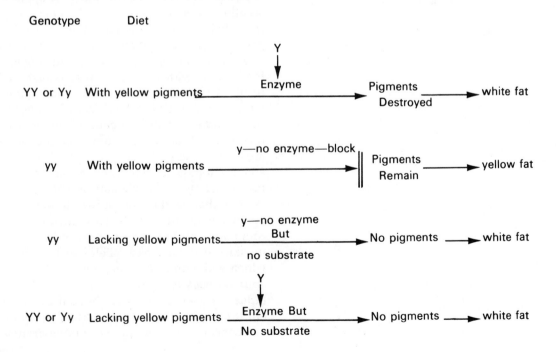

alteration of gene expression. Indeed, the effect of environment may be such that one cannot tell by mere visual inspection whether the phenotype is the result of the presence of certain alleles or whether the environment is operating in such a way that it produces a phenotype which mimics the effect of a specific genotype. For in the case of the rabbit, there is a variety of animal in which black fur depends on the genic constitution. Such animals are black regardless of the temperature under which they are raised. Rabbits with a himalayan genotype may appear identical to these animals if they are raised under cold conditions, even though their genotype is different in relationship to coat pattern (Fig. 4-14A). The term *phenocopy* designates the individual whose phenotype has been altered by the environment in such a way that it imitates the phenotype usually associated with a particular genotype. Environmental production of phenocopies can lead to erroneous interpretations unless their true nature is appreciated.

THE IMPORTANCE OF BOTH ENVIRONMENTAL AND HEREDITARY FACTORS. A very frequently asked question is, "Which is more important, heredity or environment?" We hope it will become clear throughout the many examples which we will continue to encounter that such a question becomes meaningless upon close scrutiny. Does the example of the phenocopy in the rabbit mean that environment is more important? Has it altered the genes? Let us compare a black phenocopy (possessing the allele for himalayan) with an animal that is black as a result of the allele it carries for black pigment formation (Fig. 4-14B). The latter will have black fur regardless of the temperature under which it is reared. It does not matter if we cool or heat areas of its skin; it will still produce black fur, and it will pass down to its offspring the genetic factors for the same phenotypic potential. The animal that is the phenocopy, however, will produce white fur if the environment is altered. And no matter how we raise the animal or cool its skin, it cannot pass alleles to its offspring which will permit them to grow a black coat at ordinary body temperature. The environmental influence does not change the genetic factors, but it can alter the manner of their expression.

Concern with the influence of environment

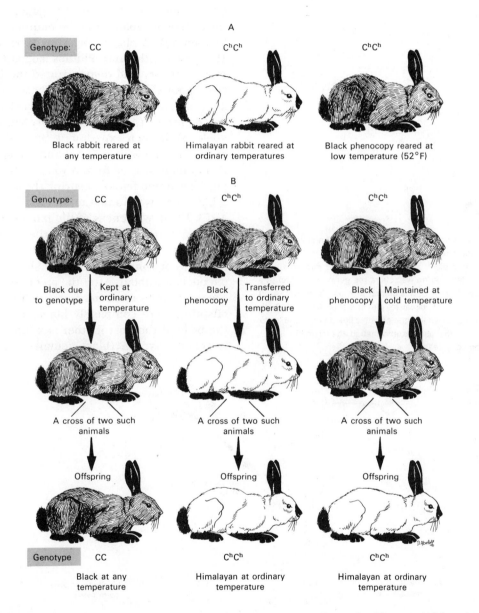

FIG. 4-14. Influence of temperature on expression of genotype. *A.* If a himalayan rabbit is reared at a low temperature from the time of its birth, its fur will be black, and it will appear identical to the black rabbit whose genetic constitution enables it to form pigment at any temperature. *B.* The difference between rabbits of the two genotypes becomes apparent when environment is altered. When a black phenocopy is transferred to a warmer temperature, the new hair will grow in white, except at the extremities of the animal where body temperature is lower. Even if a black phenocopy is kept under cold conditions so that it remains black, its offspring will still be himalayan at ordinary temperatures. The environment did not change the genetic constitution of the himalayan rabbit which is the phenocopy. However, it did alter the expression of the genotype.

on heredity was a matter which troubled many early biologists and has persisted into the present. Weismann's concept of the continuity of the germ plasm (Chap. 1) focused attention on the importance of the gametes as opposed to the body cells. But the idea persisted that perhaps the environment, acting over long periods of time, could slowly change the genes and direct the course of evolution in the manner suggested by Lamarck. A classic experiment performed in 1909 by Castle and Phillips supported Weismann's theory and demonstrated the fallacy of the Lamarckian argument. These geneticists knew that in guinea pigs black coat color depends on a dominant (A) and albino coat on its recessive allele (a). In their experiment, they removed the ovaries from a white animal and transplanted to this animal ovaries from a young, black guinea pig. The albino female, bearing the ovaries from a black one was mated to an albino male (Fig. 4-15). If the whiteness of the female can affect the genes in the gametes of the transplanted ovaries, then we might expect the offspring to be white or some intermediate shade. But this did not come about. After the production of three separate litters resulting from such a cross, the offspring were completely black. Even though the body of the host mother is white, the cross is genetically: AA (black female) × aa (white

FIG. 4-15. Ovarian transplant in the guinea pig. An albino animal lacks the dominant required for black pigment formation. When germ tissue from a black animal is transplanted into an albino, the body of the latter in no way alters the germ cells which are being sheltered. A cross of this animal with an albino will result only in black offspring because the cross is AA × aa. Such experiments demonstrate that the gametes link one generation to the next and that the genetic information they contain is not changed by the body cells.

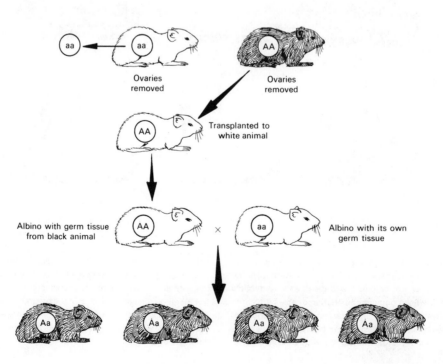

Ovaries removed

Ovaries removed

Transplanted to white animal

Albino with germ tissue from black animal

Albino with its own germ tissue

male). This was an elegant demonstration that the body cells in no way affect the transmission of genetic factors to the next generation. The gametes produced in the transplanted ovary carried the factors to be passed to the next generation. Since the ovary contained only genetic factors for black pigment, it was the allele for black pigment production which was handed down. The gametes or their genetic contents were in no way changed just because they were housed in a white body. We will see (Chap. 15) that environmental agents can induce changes in the germ cells, but these mutations are at random and are in no way directed by the environmental factor which caused them.

These examples from the rabbit and the guinea pig should help clarify the nature of phenotypes and genotypes. They show us that the phenotype is *not* transmitted. The white phenotype of the host mother was not passed to her offspring. Nor did the black rabbit which was really a phenocopy transmit the potential to produce black fur at higher temperatures. What *is* being transmitted in both cases is the genetic constitution of the gametes. At fertilization, two gametes combine to establish a genotype, a combination of genetic factors from both parents. This genotype sets a potential. What *kind* of phenotype is eventually realized will depend on the interaction of all the genes and the environment. And the environmental effect may be the critical one which influences the expression of a certain gene or set of genes and brings about a particular phenotype.

When we are dealing with very complex characteristics, such as body size or intelligence, the interactions of heredity and environment become so complex that they cannot be untangled and precisely measured. But we do know that the expression of the best genotype, let us say one which makes possible the expression of great intelligence, can be suppressed by an unfavorable environment. On the other hand, the most favorable environment will not produce a genius if the hereditary endowment is lacking.

As we learn today of more and more human disorders which have a hereditary basis, we must not despair completely, because the environment can play a major role in alleviating the condition. Recall once more the individuals burdened with the recessive genotype for phenylk-

etonuria. Diet may permit these persons to assume a normal existence so that they appear indistinguishable from other persons. The environmental influence (special diet) has enabled them to escape the damaging effects of the disorder. Nevertheless, the effect of the environment on their bodies has in no way changed the defective allele they carry. Two normal-appearing PKU persons can only produce offspring who are also metabolic defectives. Their survival and chances for a normal life will again be determined by the environment—proper diet in infancy.

When discussing environment, we must not forget that environment is internal as well as external. The cellular environment provides the *milieu* (background) for the activities of the genetic material, and these may be altered by the external environment. However, expression of the genotype may also be changed by factors outside the cell but inside the body itself. One of the most obvious of these internal influences is hormonal. The expression of those genes which affect the secondary sex characteristics depends on the hormonal background (p. 141). Another example is the genotype for early baldness in humans. This genotype is certain to express itself phenotypically in males, whereas young women will not show early baldness at all. It is the internal environment (hormonal) which is interacting with the same genetic factors in the two sexes to give a penetrance of 100% for this genotype in males and 0 in females (Fig. 4-16).

GENES WHICH INFLUENCE SURVIVAL OF THE INDIVIDUAL. In this chapter, we have encountered just a few of the numerous ways in which genes may interact among themselves as well as with their environment. This interaction often results in modifications of the classic Mendelian ratios, such as 9:3:3:1 to 9:3:4 or 9:7, etc. Expected ratios may be altered by still another factor, which can be illustrated by a familiar example in mice. Yellow coat color depends for its expression on a dominant (Y); wild coat results from the homozygous recessive condition (yy). A cross between any yellow mouse and a gray one always produces yellow and gray offspring in a ratio of 1:1. This makes us suspect that all yellow mice are heterozygotes because the ratio of 1:1 is expected when a monohybrid is testcrossed. The truth of this hypothesis is borne out by following crosses between any two yellow mice (Fig. 4-17). The outcome of such a cross is always two yellow to one gray. And these yellow animals always give the typical 1:1 testcross ratio when crossed to grays. Evidently some factor is preventing the birth of homozygous yellow animals. Examination of gravid females reveals that these homozygotes are produced but they they die as embryos while still in the uterus. The dominant for yellow fur (Y) is responsible for killing the individual when it is homozygous. We say that it is a *lethal* allele, meaning that it is responsible for the death of the carrier. Yellow is a *complete lethal*, because it kills the individual before reproductive age. Such lethal alleles are by no means exceptional and must always be considered in populations of plants and animals. Indeed, when a gene mutation arises, it is much more apt to produce an allele which has a harmful or a lethal effect than one which is neutral or conveys a benefit. In the fruit fly, invisible lethal mutations outnumber those which produce a detectable phenotypic effect by 10 to one! The allele for yellow coat in mice is pleiotropic; it has a dominant effect on fur color but a recessive one on viability. Therefore, we can look at the factor for yellow in another way and call it a recessive lethal. Many lethals produce no pronounced effect at all on the phenotype, but they may make their presence known by a decrease in the life span or the very early elimination of the carrier.

How can it be that an allele may have a killing action? The answer is obvious if we remember that metabolism is the result of many interlocked biochemical pathways (see Fig. 4-9). A defect in just one can upset several others. We can appreciate from what has been said about PKU that just one defective step can alter the entire chemistry of the body. A lethal, by blocking a critical reaction, can interfere with normal embryological development of an essential organ such as the heart. The death of the embryo may then follow. Lethals can decrease the chances of survival by causing various kinds of abnormalities in development and physiology. Different lethals eliminate individuals at different stages of the life cycle. The complete lethal removes the carrier before reproductive age so that the affected leaves no offspring (the allele

FIG. 4-16. Expression of early baldness. Allow E to represent the allele for onset of hair loss at a young age. A man will express the condition, but a young woman who carries the same allele will not. The difference in the penetrance of the genotype is due to the internal (hormonal) environment.

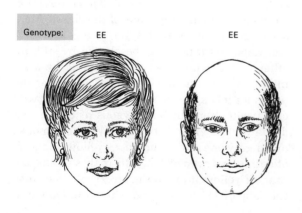

Genotype: EE   EE

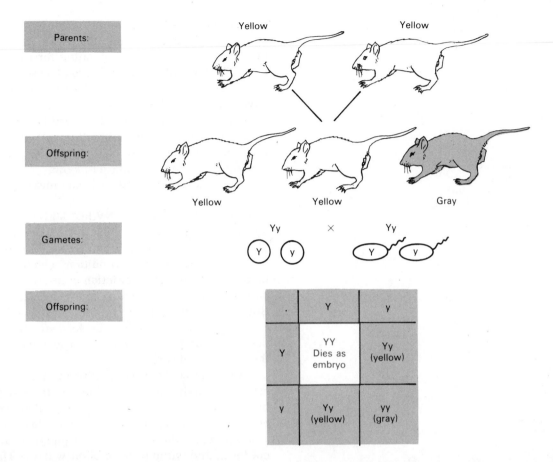

Parents: Yellow Yellow

Offspring: Yellow Yellow Gray

Gametes: Yy × Yy

Y y Y y

Offspring:

	Y	y
Y	YY Dies as embryo	Yy (yellow)
y	Yy (yellow)	yy (gray)

FIG. 4-17. Recessive lethal in mice. When two yellow mice are crossed, the offspring occur in a ratio of 2 yellow:1 gray (*above*). The explanation is that all yellow animals are heterozygotes. Although the yellow phenotype is inherited as a dominant, the allele which is responsible behaves as a recessive lethal. Therefore, no homozygotes for the allele for yellow fur survive.

for yellow fur in mice; in humans the recessive factor for Tay-Sachs disease which kills in infancy). Some lethals exert their killing action later in the life cycle, after the carrier may have left some offspring. In humans, the dominant factor for Huntington's disease, a fatal deterioration of the nervous system, does not usually express itself before the age of 30. Such genetic determinants, which can result in death but which permit the carrier to live to reproductive age, are often grouped as *sublethals*. There are actually no sharp boundaries between lethals as to the stage of the life cycle at which they act to decrease the chances of survival.

Some geneticists calculate that each human on the average carries the equivalent of four lethal alleles. Their high frequency makes us wonder why most new changes or gene mutations should tend to be harmful. Consider the analogy made by H. J. Muller and his watch. The wristwatch may benefit by certain changes in its inner mechanism which would permit it to keep better time. A repairman may make such a beneficial alteration by using his skill and knowledge. But it is also conceivable that a good change could come about by banging the watch on the table. We wouldn't know what we were doing to the insides, but it is possible that the parts of the watch could rearrange by chance in such a way that its operation would be more efficient. However, the most likely result of the banging action would be a complete breakdown because any unplanned change in a complicated mechanism is more likely to be bad than good or neutral. And mutations are unplanned. They bring about modifications which are random. We must always bear in mind that all living things which we see around us today, no matter how simple some of them may seem to be, are really complicated systems upon which evolutionary forces have been operating for eons of time. The force of natural selection has selected those forms which are best fitted to the particular environment in which the species lives. Certain changes may bring about even further advantages in a living system, but the chances are great that the unplanned change, the mutation, will upset it. And so, the damaging or lethal effect of most gene mutations is not unexpected.

LETHALS AND THE ENVIRONMENT. Like any other kind of genetic factor, the lethal may also interact with the environment. In humans, a victim of diabetes, a condition which may involve the complex interaction of many genes, possesses a genotype which may cause death. But today, insulin intake can prevent this lethal effect so that a diabetic may lead an almost normal life and avoid the fatal consequences of the metabolic disorder.

An example from plants illustrates another aspect of environmental influence. In barley, many gene mutations are known which can prevent the development of the chloroplasts. As a consequence, the essential green pigment cannot form, and without it the plant will die. The double recessive white genotype (ww) by preventing chloroplast development causes the death of the seedling once the reserve food stored in the seed is consumed. A cross of two green heterozygotes (Ww × Ww) produces an average of three green seedlings to one white. The latter type is seen *only* in its seedling stage. If we simply observe the adult plants, we may be unaware of the presence of the lethal. Now suppose we take the offspring from the above cross and raise them all in the dark (Fig. 4-18). This will prevent the normal development of the plastid which requires light for its proper formation. In the dark, all the plants will die when the food reserves have been used up. We can keep them all alive, even in the absence of light, by supplying them with organic nutrients. We see in this case that the environment is producing phenocopies; the plants which have the genotypes to permit normal chlorophyll development cannot be distinguished from the mutants. These individuals (WW and Ww) cannot be distinguished from those which possess a

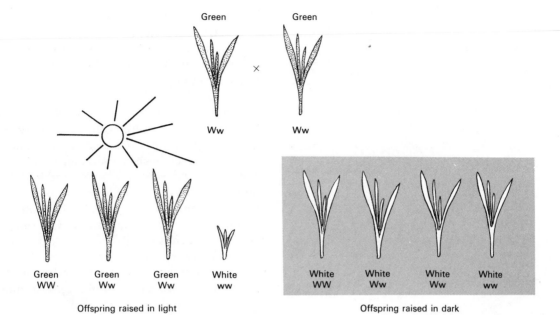

FIG. 4-18. Environment and gene expression. The recessive, w, in barley prevents normal development of a green chloroplast. When two monohybrids are crossed and the offspring grown under natural conditions, the white offspring are seen only as seedlings, because they die when food reserves are used up. If the offspring from such a cross are all grown in the dark, all of the plants will be white and must be supplied with nutrients if they are to survive. The plants whose genotypes permit normal chloroplast development cannot be distinguished from the recessives whose genetic factors normally result in their elimination as seedlings.

genotype (ww) which proves lethal under ordinary environmental conditions. This is just one of numerous illustrations which demonstrate that the best genotype does not ensure normal development; the genetic potential may not be realized in a restrictive environment. However, the most conducive environment will not generate a superior phenotype if the required genetic factors are not there to guide it.

THE NAMING OF GENES AND LOCI. Before proceeding further in our genetic discussions, we must pause to examine the conventions which are used to assign names to genes and the loci on the chromosomes, as well as to the stocks of organisms used in genetic research. Since *Drosophila* is such a widely used genetic tool, we will refer to it in the following discussion, although the same reasoning applies to other organisms as well.

Work with the fruit fly requires familiarity with different types of pure-breeding stocks, with special attention paid to the "wild" or normal. As we noted earlier in this chapter, the designation "wild" implies the standard form which is typically found in nature. Any inherited departure from the wild may result from a mutation at a certain gene site, a locus on a chromosome. The name given to any mutant stock usually suggests the kind of phenotypic variation from the wild type attributed to the genetic alteration. For example, red eye color in *Drosophila* represents wild type. Gene mutation at one of the loci on the X chromosome may result in the absence of eye pigment. Stocks which are homozygous for this change are designated "white," and the position or point on the chromosome where the alteration has taken place is named the "white locus."

The wild gene form for red eye color and its mutant form for white represent a pair of alleles, and either one of the two members may be present at the "white locus" on a particular X chromosome. In elementary genetic discussions such as ours so far, a pair of alleles is usually symbolized by a capital letter for the dominant and a small one for the recessive allele. The symbols W and w are frequently encountered for red eye and white eye, respectively. However, the geneticist does not usually follow this system. Instead, he uses the plus sign (+) to stand for *any* wild trait. The factor for red eye is thus simply designated "+" (or w^+ for extra clarification), and its allele for white eye remains w. The name given to any genetic locus, in this example "white," describes the first mutation or variation which was detected at that site. If there had never been a mutation to alter the color of the eye, we would know nothing about the inheritance of eye color and would be completely unaware of any genes or loci affecting that character. We have noted the fact that the geneticist depends on mutations to learn how a normal character is inherited.

There are many recessive mutations in *Drosophila* which are known to affect an assortment of other traits besides eye color, and they are all named according to this scheme. The recessive "ebony" can result in a fly with a shiny black body instead of the normal gray. The pair of alleles involved is represented by "+" (or e^+) for wild and e for ebony. The recessive forked, f, can cause shortened, split bristles in contrast to the longer, unsplit ones determined by the wild gene form, "+" or f^+. Further examples could be given for a long list of recessive alleles. In this chapter, we have learned that there are *many* wild alleles, not just one. There is a wild allele corresponding to each mutant one, and every one of the wild alleles is represented as "+," no matter what the effect of that particular locus on the phenotype.

In these examples, the first letter of the name of the mutation was used to represent the mutant: w, e, and f for white, ebony, and forked. Since we would soon encounter many names beginning with the same letter, we often must use more than one letter to represent a locus. As e has been used for ebony, it cannot be selected to describe the recessive allele "eyeless," which results in a reduction of eye size. More than just the first letter is required to stand for "eyeless," and the locus is symbolized by "ey." The normal eye condition is still, of course, "+" (or ey^+). And similarly for a long list of others such as dp (for "dumpy," a gene form giving reduced wings), ss (for a recessive causing reduction in bristle size), and many more.

The mutations discussed so far have all been recessive to their wild type alleles (+). Gene alterations which produce a dominant effect, although not as numerous as recessive changes, are responsible for several phenotypic departures from the wild. The naming of dominant gene mutations is identical to that for the recessives. The wild allele, although now the recessive form in such cases, it still represented by "+." The mutant allele, however, is symbolized by a capital letter (or an abbreviation beginning with a capital). A frequently encountered dominant genetic effect in the fruit fly is the "Bar eyed" condition in which eye size is narrowed. It is represented by the capital letter B, and the wild condition as "+" or B^+. There is no reason for confusion if one simply remembers that "+" always designates *any* wild type allele, group of alleles, or wild condition; any mutant is represented by a letter or abbreviation. We can tell at a glance that a mutant gene form is recessive if it is designated by a small letter (as w for white eye). A designation beginning with a capital tells us that the gene form is dominant to wild as in B (Bar), Cy (curly wings), and Pm (plum, a brownish eye color). The wild alleles are respectively B^+, Cy^+, and Pm^+ (or simply just a "+" in each case).

It was mentioned that a locus is named after the first detected mutation at a gene site. This implies that there is more than one kind of alteration possible in a particular gene, and this is indeed the case as illustrated by any multiple-allelic series (Chap. 14). The white eye locus is an excellent example. At this particular site on the X chromosome, there has been more than one change in the genetic material. In addition to the one which results in complete absence of eye pigment, there are several others which have caused different degrees of reduction in the pigment, producing various shades from red to white. We thus find "apricot," "blood," "coral," and several others. These are all changes at the same locus and are consequently all allelic to one another. Because the change to white was the first one described, the other mutations are all represented by the same base letter, w, indicating that the alteration is at the "white locus." To this base letter, superscripts are added giving w^a for apricot, w^{bl} for blood, w^{co} for coral, and so on. The pair of alleles, w^+ and w^a, implies the wild type gene for eye color and a recessive allele at the white locus resulting in apricot (pinkish) color.

There are many benefits in the system which designates wild by a (+). These will become very evident in the discussions related to multiple alleles, linkage, and crossing over. At certain times, we may still prefer to use the older scheme of capital vs. small letters for dominant and recessive when discussing simple monohybrid or dihybrid crosses. The choice often may be just a matter of convenience.

REFERENCES

Beadle, G. W., and E. L. Tatum, Genetic control of biochemical reactions in *Neurospora. Proc. Natl. Acad. Sci.*, 27:499, 1941. (Reprinted in *Selected Papers on Molecular Genetics*. J. H. Taylor (ed.), Academic Press, New York, 1965.)

Eaton, G. J. and M. M. Green, Implantation and lethality of the yellow mouse. *Genetica*, 33:106, 1962.

Knox, W. E., Phenylketonuria. In *The Metabolic Basis of Inherited Diseases*, 3rd ed., J. B. Stanburg, J.

B. Wyngaarden, and D. S. Frederickson (eds.), pp 266–295. McGraw-Hill, New York, 1972.

Rothwell, N. V. *Human Genetics*. Prentice-Hall, Englewood Cliffs, N. J., 1977.

Stern, C., *Principles of Human Genetics*, 3rd ed. W. H. Freeman, San Francisco, 1973.

Sutton, H. E. *An Introduction to Human Genetics*, 2nd ed. Holt, Rinehart and Winston, New York, 1975.

REVIEW QUESTIONS

1. A rare allele for extra digits in the human behaves as a dominant (E) to the normal (e). A man with an extra toe on each foot marries an unrelated woman with the normal condition. The first child has an extra finger; the second child is phenotypically normal. In time, the normal offspring marries a normal, unrelated woman and produces a child with an almost complete extra toe on one foot and the normal number of toes on the other. Offer an explanation

and give the probable genotypes of the persons mentioned.

2. In the human, study of family histories indicates that four out of ten persons who carry a certain rare dominant allele for a skeletal defect do not express it. What would be the penetrance of this allele?

3. In the human, a certain allele results in severe anemia plus damage to the kidneys and the central nervous system. What genetic phenomenon is illustrated by the expression of this allele?

4. In sweet peas, the two allelic pairs C, c and P, p are known to affect pigment formation in the flowers. The dominants, C and P, are both necessary for colored flowers. Absence of either results in white. A dihybrid plant with colored flowers is crossed to a white one which is heterozygous at the "c" locus.

    A. What are the genotypes of these two plants?
    B. What kinds of flowers, colored or white, are to be expected from the cross, and in what ratio?

5. Assume that another allelic pair in sweet peas also affects pigment formation in addition to the genes mentioned in Question 4. The presence of the dominant, R, is required for red flowers. Its recessive allele, r, produces yellow flowers. What would be the phenotypes of the following plants in relation to flower color?

    A. CcPpRr
    B. CcppRR
    C. CcPPrr
    D. ccPPRR

6. In a certain breed of dogs, the dominant, B, is required for black fur; its recessive allele, b, produces brown fur. However, the dominant, I, is epistatic to the color locus and can inhibit pigment formation. The recessive allele, i, on the other hand, permits pigment deposition in the fur. What would be the phenotypes of the following sets of parents, and what would be the results of their mating?

A. bbii × BbIi

B. bbIi × Bbii

C. bbIi × BBIi

7. In the human, the dominants, D and E, are both required for normal development of the cochlea and the auditory nerve, respectively. The recessives, d and e, can result in deafness due to impairment of these essential parts of the ear. Give the phenotypes of each of the following sets of parents and the chance of a deaf child being born as the first offspring.

A. DDee × ddEE

B. DdEE × DDEe

C. DdEE × DdEe

D. DdEe × DDEe

8. In poultry, the shape of the comb varies greatly and involves at least two pairs of alleles. The allele, R, can result in rose comb, and the allele P can result in pea-shaped comb. If both of these dominants are present together, genic interaction produces a walnut comb. When a bird is carrying both recessives, r and p, in the homozygous condition, single comb type results. Give the type of comb of each of the following pairs of birds and the phenotypic ratio to be expected among their offspring as to comb type.

A. rrPP × RRpp

B. RrPp × RrPp

C. RrPp × rrpp

9. Using the information on comb shape given in Question 8, deduce the genotypes of the parents in each of the following crosses.

A. A walnut bird crossed with one with a single comb. These parents produce three walnut, four pea, two rose, and three single.

B. A rose-combed bird crossed with a pea. This parental combination gives rise to rose and walnut in a ratio of 1:1 after several matings.

C. A rose-combed bird crossed with a walnut. These parents produce three walnut, five rose, two pea, and one single.

10. In poultry, the allele, B, can result in black feathers. Its allele, b, can produce white

splashed. The heterozygote is blue. A bird with black feathers and pure breeding for rose comb is crossed with a bird which is white splashed and pure breeding for pea comb. (Use information in Question 8 for comb shapes.)

A. Diagram the cross. What would the F_1 be like?

B. What kinds of birds are to be expected from crossing F_1's with birds that are black and have single combs?

11. In poultry, the factors, C and O, are both required for any color at all in the feathers. Homozygous recessiveness at either locus will result in white. In each of the following 3 cases give the genotypes of the parental birds and their offspring. (Use information on color in Question 10.)

A. Two completely white birds which produce an F_1 which is black.

B. Two black birds which produce 54 black birds and 42 white ones.

C. Two blue birds which produce 16 black, 47 blue, and 14 white splashed.

12. Two pairs of alleles are involved in cyanide production in white clover. Some strains have a high cyanide content, others a low one. When low Strain A was crossed to low Strain B, the F_1 all had a high cyanide content. When the F_1 plants were crossed among themselves, the following F_2 resulted:

high cyanide content 450; low cyanide content 350. Offer an explanation for these observations.

13. Two different highly inbred strains of chickens both have feathers on the upper portion of their legs. When birds from the two different strains are crossed, the F_1 all have feathered legs. In a typical case, crossing the F_1's to each other gave 323. Of these 301 had feathered legs, the remainder being unfeathered. When the unfeathered ones were crossed among themselves, only birds with unfeathered legs resulted. Give a likely explanation.

14. Very little hair is found on a Mexican hairless dog. A cross between a Mexican hairless

and a dog with a typical coat of hair usually produces litters of pups in which half of the animals are hairless and the other half have hair. On the other hand, a cross between two Mexican hairless dogs tends to produce litters in which two-thirds of the pups are hairless and one-third have hair. However, in addition to these surviving puppies, usually some are born dead. These appear hairless and occur in about the same frequency as the pups with hair. Offer an explanation and represent the genotypes of the different kinds of animals using any symbols you choose.

15. Several loci affect hair development in dogs. The allele for wire hair (W) is dominant to its allele for straight hair (w). Allow "H" to represent hairlessness in the Mexican hairless dog and its recessive allele "h" to represent the gene for typical hair growth in other breeds. The Mexican hairless dog is homozygous for straight hair, but, due to the epistatic effect of "H," alleles for hair formation cannot be expressed. Write genotypes of the following animals with respect to the two loci mentioned.

    A.   A Mexican hairless dog.
    B.   A dog with straight hair.
    C.   A dog from a strain which is pure breeding for wire hair.

16. Using information in Question 15, give the results of a cross between:

    A.   A Mexican hairless dog and a dog from a strain which is pure breeding for wire hair.
    B.   A cross between two animals of the first generation resulting from the cross just given. (Cross only the animals which have different genotypes.)

17. Assume that in a disease-resistant variety of plant the recessive lethal "w" occurs with a high frequency, and a considerable proportion of the plants die since they lack the dominant allele "W" and cannot produce chlorophyll. A heterozygote with one dose of the dominant allele cannot be distinguished from a plant homozygous for the dominant. Suggest a program by which the undesirable recessive might be elim-

inated from the variety, considering the fact that the plants are exclusively cross-pollinated.

18. In guinea pigs, black fur, A, is dominant to the albino condition, a, which gives white fur. A black female from a pure-breeding strain carries transplanted ovaries from an albino female. The albino female received in turn the ovaries from the homozygous black animal.

A. What are the expected results among the offspring when the black female is crossed to a black male which had a white parent?

B. What would be expected from the above cross if the black female had also had a white parent?

C. What results are to be expected if the albino female is crossed to this same male?

In each of the following two multiple-choice questions, select any correct answer or answers. More than one answer may be possible, or there may be no correct answer at all.

19. In phenylketonuria (PKU):

A. Phenylpyruvic acid accumulates in the blood.
B. Tyrosine accumulates in the blood.
C. Phenylalanine accumulates in the blood.
D. There is a genetic block in a metabolic pathway.
E. The gene for pigment production is changed so that victims tend to be fair in complexion.

20. The environment:

A. Can reduce the penetrance of some genes.
B. Can affect the expressivity of some genes.
C. Can cause an effect in a following generation due to its influence on the body cells of the parents.
D. Cannot be altered to prevent the expression of any alleles which are harmful.
E. Is usually more important than heredity in the development of an individual.

5

SEX AND INHERITANCE

SEX CHROMOSOMES. Cytogenetic studies related to sex differences have played a major role in proving the soundness of the chromosome theory of heredity. Several early geneticists, including Mendel, realized that sex seems to behave as a Mendelian character. Male and female offspring occur in a 1:1 ratio, the classic monohybrid testcross ratio which is expected from crossing a heterozygote to a homozygous recessive.

In 1902, McClung, working with a species of grasshopper, noted that one chromosome in the male seemed to be unpaired and lacked a corresponding mate. As a consequence, meiosis would produce two kinds of sperms: one-half with the unpaired element and one-half lacking it. McClung connected this unpaired chromosome with the determination of sex and inspired other investigations relating chromosomes to sex. Before 1910, Stevens and Wilson had worked out the cytological picture in many insects. They recognized that the chromosomes occur in corresponding or homologous pairs, but they realized that the two members of a pair are not always identical. The sexes often differ in regard to one of the chromosome pairs. Stevens showed that in *Drosophila melanogaster*, the common fruit fly, the number of chromosomes is eight and that these exist as four pairs (Fig. 5-1A). Cells from females contain one pair of rod-shaped chromosomes (Pair I which is acrocentric), two pairs of V-shaped chromosomes (Pairs II and III with median centromeres), and a very small pair which has been designated the "dot chromosomes" (Pair IV with centromere near one end). In the female, the two members of each pair are identical in appearance, but this is not so in the male. The difference relates to Pair I. Males show only one rod-shaped chromosome instead of two, but they carry a hook-shaped or "J" chromosome which is absent in the female. Similar distinctions between males and females were found in other species as well. The two members of the chromosome complement which differ between the sexes were named the "sex chromosomes." The one found singly in the male and paired in the female was called the "X." The one which is confined to the male was called "Y." As noted in Chapter 2, all the chromosomes of the complement other than the sex chromosomes were designated the *autosomes. Drosophila,* then, possesses three pairs of autosomes and one pair of sex chromosomes, the latter represented as two X's in the female and an X and a Y in the male. Such a chromosome picture is known as an "X-Y" condition and implies that the male has dissimilar sex chromosomes and will produce gametes which differ in their sex chromosome content. For example, male fruit flies will form sperms with three autosomes, but one-half of the gametes will carry an X and the other one-half a Y. Since his gametes fall into two distinct classes, we say that the male is the *heterogametic sex.* The female is termed *homogametic;* all her eggs will be alike insofar as the kinds of sex chromosomes they contain (Fig. 5-1A).

It was realized, however, that this X-Y condition is not universal. Cytological examination had shown that cells of one sex may contain one chromosome more than the other. In such species, the sex chromosome is completely unpaired. Since a Y is not present at all, the X has no mate or pairing partner. We followed the behavior of an X in this so-called "X-O" condition during our discussion of spermatogenesis in Chapter 3. It was the X-O situation which McClung had observed in his study of the grass-

hopper. However, he believed that the male had one more chromosome than the female, when actually the reverse is true. The X-O picture is found in many insects and various other groups as well. In species like the grasshopper, the male is again the heterogametic sex, because the sperms are of two kinds in relationship to their sex chromosome content: one-half will contain one more chromosome than the other. In some organisms with the X-O mechanism, it is the female sex which is heterogametic (X-O) and the male which is homogametic (XX). Indeed, the females of certain species may contain an X and a Y, whereas the males are XX. Such an arrangement is found in moths and butterflies and is typical of birds. It is designated "W-Z" to imply that the female forms two classes of sex cells which differ in their sex chromosome content and that the female is thus the heterogametic sex. This is the reverse of the more common and familiar X-Y arrangement found in *Drosophila* as well as in the human. In our species, we find 22 pairs of autosomes and one pair of sex chromosomes (two X's in the female; one X and one Y in the male).

Geneticists recognized that the presence of sex chromosomes could account for the determination of equal numbers of male and female offspring. It could thus explain what appears to be a monohybrid 1:1 testcross ratio. In an animal with the X-Y mechanism, the eggs are all X bearing and will be fertilized by either X-bearing sperms or Y-bearing ones. Since the male produces these two different classes of gametes in equal amounts, it would be a matter of chance whether any one egg were fertilized by an X-containing sperm or by one with a Y. The outcome is male and female offspring in equal proportion (Fig. 5-1B).

GENES ON THE X CHROMOSOME: SEX LINKAGE. Although such a mechanism seemed to account for sex determination in many organisms, it was not known whether the sex chromosomes contain genes for characteristics *other than* sex or whether the autosomes may also influence sex in some way. The work of T. H. Morgan and his associates at Columbia University was begun in 1909 with the fruit fly and contributed a wealth of information which is basic to an understanding of sex determination.

The investigations of this outstanding group also uncovered other fundamental principles of inheritance which established genetics firmly as a scientific discipline.

Morgan was particularly interested in mutations and their role in evolution. One of the first mutations found under laboratory conditions affected the eye color of *Drosophila,* changing it from the red color found in nature to unpigmented or white. The white-eyed fly which arose was a male, and it was crossed to a normal red-eyed female. All of the offspring were red eyed, suggesting that the genetic determinant for white eyes was recessive to the one for red. A cross between males and females of the first generation produced red-eyed and white-eyed flies in the familiar Mendelian ratio of 3:1. The results indicated that a pair of alleles was involved, W for the dominant red condition and w for the recessive white. However, there was something unusual about this 3:1 ratio. All of the flies with white eyes were males (Fig. 5-2*A*)! One-half of the males had red eyes, but *all* of the females were red eyed. Morgan realized that such results could be explained by assuming that the locus determining red vs. white eye color is on the X chromosome (Fig. 5-2*B*). The females used in these crosses would have two X chromosomes, each bearing the dominant W. They would thus be homozygous for red eye color and can be represented simply as WW. The white-eyed male, however, has but one X chromosome and would therefore have the eye color locus represented only once. The Y chromosome would not carry a locus for any eye color trait. The white-eyed male would thus have an X

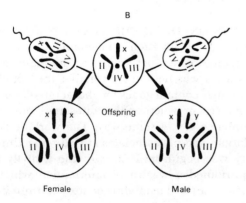

Male A Female

Body cells / Body Cells

Sperms / Eggs

Two types: x bearing
y bearing
(Heterogametic)

All one type : x bearing
(Homogametic)

B

Offspring

Female / Male

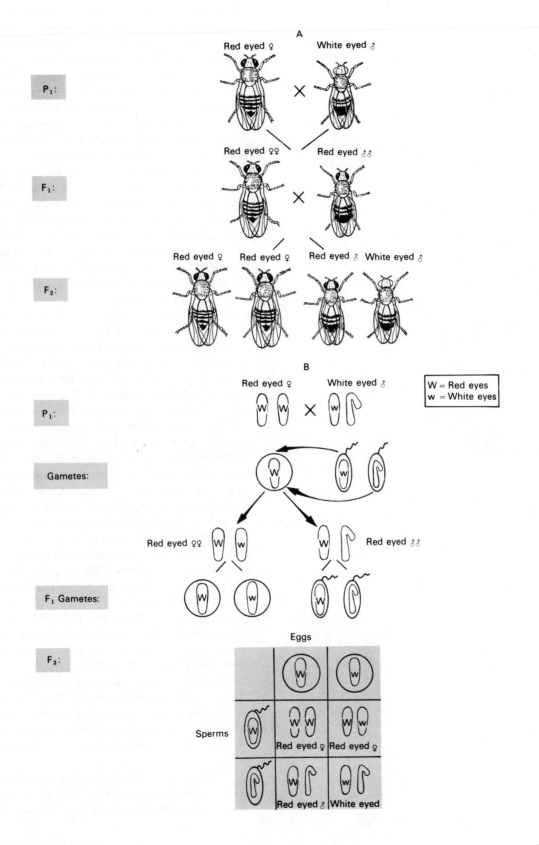

A

P₁:

Red eyed ♀   White eyed ♂

F₁:

Red eyed ♀♀   Red eyed ♂♂

F₂:

Red eyed ♀   Red eyed ♀   Red eyed ♂   White eyed ♂

B

P₁:

Red eyed ♀   White eyed ♂

W = Red eyes
w = White eyes

Gametes:

Red eyed ♀♀   Red eyed ♂♂

F₁ Gametes:

F₂:

Eggs

Sperms

Red eyed ♀ | Red eyed ♀

Red eyed ♂ | White eyed

chromosome with the recessive allele w and a Y chromosome with no corresponding eye color locus. The genotype of such a fly can be represented as wY. The Y chromosome must be shown when the genotype is written to indicate that the male is heterogametic and only carries one factor for white eye color. We cannot say that the male is either homozygous or heterozygous for eye color, because by definition, both of these terms imply that a particular gene is represented twice for a particular character. Since only one dose of a gene can be present in the case of an X-linked gene in the heterogametic sex, we say that the individual is *hemizygous*. Figure 5-2*B* shows why all the F_2 white eyed flies must be males after a cross between a homozygous red-eyed female and a hemizygous white-eyed male.

Morgan realized that if his hypothesis were correct, it should be possible to obtain females with white eyes. It can be seen from the cross in Figure 5-2*B* that all of the red-eyed females of the first generation must be heterozygotes, Ww. Mating them with white-eyed males should produce red-eyed and white-eyed flies of both sexes (Fig. 5-3). When matings of this type were performed, white eyed females were acutally obtained. These white-eyed females were then crossed to red-eyed males. This is the reciprocal of the original cross (red-eyed female × white-eyed male). As Figure 5-4*A* shows, the results of this reciprocal cross were strikingly different from the original mating and strongly supported Morgan's hypothesis. In the first generation, we see that all of the females were red eyed like the male parent, and all of the males were white eyed like their mothers. This is exactly what is to be expected if the gene for eye color is on the X chromosome (Fig. 5-4*B*). We say that a *crisscross* pattern of inheritance has occurred. This simply means that a phenotype present in the female parent appears among all the sons, whereas the contrasting phenotype of the male parent shows in all the daughters. Note from Figure 5-4*B* that the cross of the F_1 red-eyed females with their white-eyed brothers is really the same as the cross diagrammed in Figure 5-3 which first yielded the white-eyed female offspring.

And so a trait other than sex was definitely shown to be associated with a sex chromosome. Morgan then found a short-wing mutant which

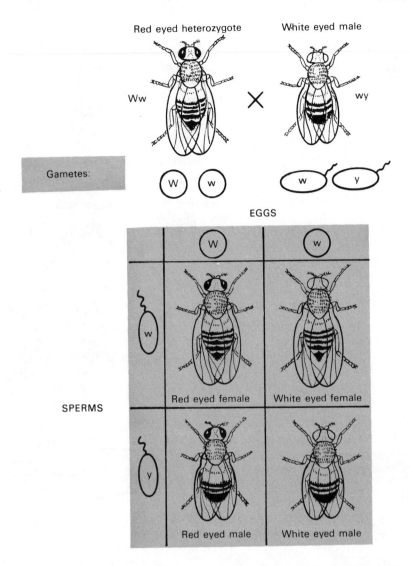

Red eyed heterozygote   White eyed male

Ww   X   wy

Gametes:

W   w   w   y

EGGS

W   w

SPERMS

w

Red eyed female   White eyed female

y

Red eyed male   White eyed male

FIG. 5-3. Derivation of white-eyed females. Red-eyed heterozygotes are obtained from a cross such as that depicted in Fig. 5-2. When these F_1 females are crossed with white-eyed males, one-half of the female as well as the male offspring will be white eyed.

he called "rudimentary." It, too, gave the same pattern of inheritance as eye color. Such genes were called *sex linked,* genes whose loci are found only on the X and not on the Y. It was soon realized that sex linkage occurs in species other than *Drosophila,* and a sex-linked recessive pattern was recognized about this time in man for red-green color blindness and hemophilia.

Actually, the first case of sex linkage had been found in 1906 by Doncaster who was following inheritance of body color in the moth *Abraxis.* The genetic results suggested that the female is the heterogametic sex and the male homogametic. This was later verified cytologically. Knowing now that the female is indeed heterogametic in this species, we can diagram sex-linked crosses in the moth as shown in Figure 5-5. Notice that the reciprocal crosses are different and that a crisscross pattern occurs when dark-bodied females are mated with light-bodied males. Such results strongly indicate sex linkage. The fact that the sex chromosome arrangement is opposite to the one found in *Drosophila* is no contradiction to the concept of sex linkage. However, this feature of the moth made it difficult at

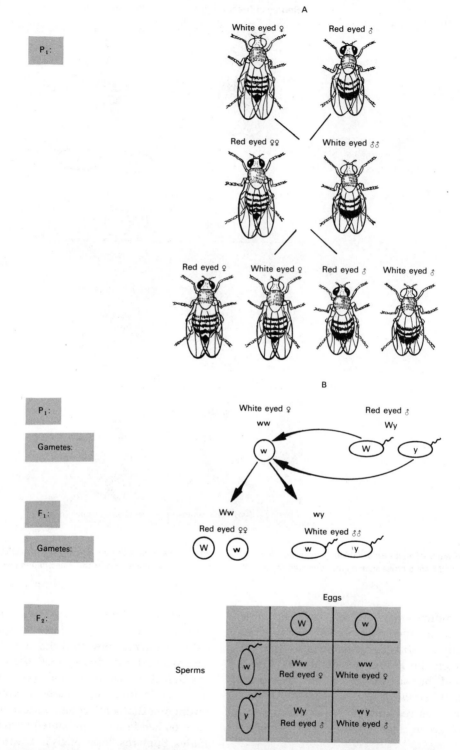

FIG. 5-4. White-eyed female *Drosophila* × red-eyed male. *A.* The results are very different from the reciprocal cross (Fig. 5.2). In the F₁, a crisscross pattern occurs. The trait seen in the female parent occurs in the male offspring; the trait expressed by the male parent is found in the female offspring. *B.* The diagram of the cross shows that in the case of a sex-linked allele, the males will express the trait which the gene governs, regardless of dominance. This is so because males receive the Y chromosome from their fathers and thus have no homologous region to mask any factors on the X contributed by their mothers.

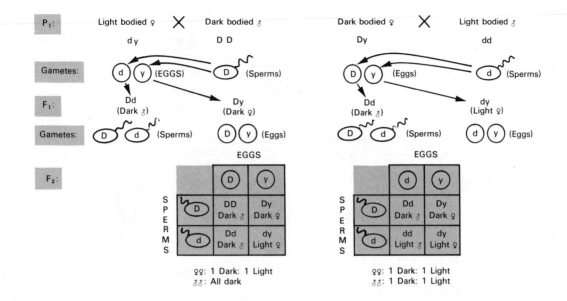

P₁: Light bodied ♀ ✕ Dark bodied ♂
          d y   D D

Gametes: (d) (y) (EGGS)   (D) (Sperms)

F₁: Dd   Dy
        (Dark ♂)   (Dark ♀)

Gametes: (D) (d) (Sperms) (D) (y) (Eggs)

F₂: EGGS
                   (D)   (y)
         S   (D)   DD   Dy
         P   Dark ♂   Dark ♀
         E
         R   (d)   Dd   dy
         M   Dark ♂   Light ♀
         S

              ♀♀: 1 Dark: 1 Light
              ♂♂: All dark

P₁: Dark bodied ♀ ✕ Light bodied ♂
         D y   d d

Gametes: (D) (y) (Eggs) (d) (Sperms)

F₁: Dd   dy
        (Dark ♂)   (Light ♀)

Gametes: (D) (d) (Sperms) (d) (y) (Eggs)

F₂: EGGS
                   (d)   (y)
         S   (D)   Dd   Dy
         P   Dark ♂   Dark ♀
         E
         R   (d)   dd   dy
         M   Light ♂ Light ♀
         S

              ♀♀: 1 Dark: 1 Light
              ♂♂: 1 Dark: 1 Light

FIG. 5-5. Sex-linked inheritance in the moth *Abraxis*. A pair of alleles, D (dark body) and d (light body), is associated with the X chromosome. The reciprocal crosses diagrammed here produce very different results. The crisscross pattern (*on the right*) indicates sex linkage. In moths, the female is the heterogametic sex. However, the principles of sex linkage remain exactly the same as in the more familiar situation where the male is heterogametic. Compare this figure with the crosses in Figs. 5-2, 5-3, and 5-4.

the time to compare it with well known organisms having the X-Y arrangement and in which the complications of sex linkage were being worked out. As noted, we now know that this type of pattern seen in *Abraxis*, where the female is heterogametic, is the typical one in birds and certain other groups.

SEX LINKAGE AND THE CHROMOSOME THEORY OF INHERITANCE. Although the early studies on sex linkage in fruit flies demonstrated that the X carried loci affecting many different characteristics, they did not indicate whether chromosomes other than the X and Y were involved in sex determination. The studies of Bridges settled this point and gave insights into several other basic genetic phenomena. His investigations made use of "exceptional" white eyed females which had been isolated by Morgan. These flies were "exceptional" because they produced offspring of unexpected phenotype. Remember that when an ordinary white-eyed female is crossed to a red-eyed male, a crisscross results (Fig. 5-4). However, the exceptional white-eyed females when mated to red-eyed males produced

the expected white-eyed sons and red-eyed daughters, but some red-eyed sons and red-eyed daughters always appeared among the offspring (Fig. 5-6A). According to the concept of sex linkage, these phenotypes should not arise from the cross.

To account for these unexpected offspring, Bridges proposed that somehow a mishap was taking place at meiosis. He reasoned that perhaps the two X chromosomes in the exceptional females had a tendency to stick together for some reason during gamete formation. This would mean that in addition to eggs of normal constitution, abnormal ones would also be produced (Fig. 5-6B). In a mating of an exceptional female to a red-eyed male, both the normal and the exceptional eggs would be fertilized by the two kinds of sperms. The offspring of the genotypes Ww and wY would represent the expected red-eyed females and white eyed males. But what would the other four combinations be phenotypically, and how could they be identified? Bridges took the unexpected white-eyed females and red-eyed males arising from the cross and studied their chromosome complements. Cells

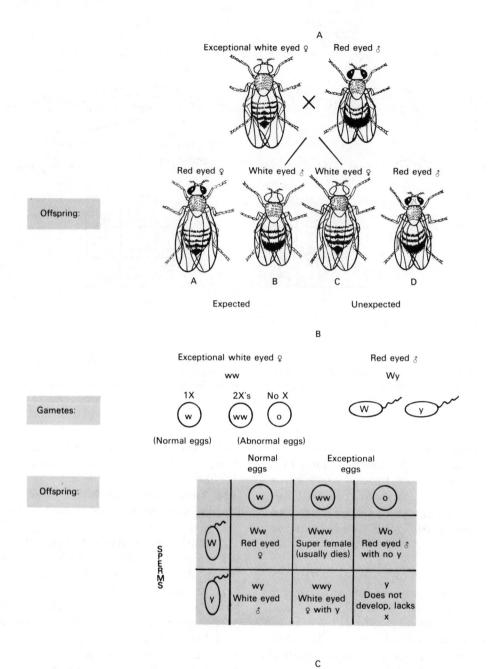

A

Exceptional white eyed ♀   Red eyed ♂

X

Offspring:

Red eyed ♀   White eyed ♂   White eyed ♀   Red eyed ♂

A   B   C   D

Expected   Unexpected

B

Exceptional white eyed ♀   Red eyed ♂

ww   Wy

Gametes:

1X   2X's   No X

w   ww   o

(Normal eggs)   (Abnormal eggs)

Offspring:

S P E R M S		Normal eggs	Exceptional eggs	
		w	ww	o
	W	Ww Red eyed ♀	Www Super female (usually dies)	Wo Red eyed ♂ with no y
	y	wy White eyed ♂	wwy White eyed ♀ with y	y Does not develop, lacks x

C

Red eyed ♀   White eyed ♂   White eyed ♀   Red eyed ♂

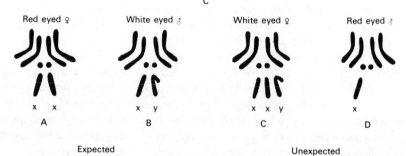

x x   x y   x x y   x

A   B   C   D

Expected   Unexpected

Fig. 5-6.

FIG. 5-6. **FIG. 5-6.** Unexpected results in *Drosophila*. *A.* When a certain exceptional white-eyed female was crossed with an ordinary red-eyed male, the results yielded the expected red-eyed females (A) and white-eyed males (B). However, a small number of unexpected white-eyed females (C) and red-eyed males (D) occurred. *B.* Bridges explained the unusual results by assuming that in a certain percentage of oocytes in the exceptional female the two X chromosomes would not separate at meiosis but would move to the same pole. Thus, in addition to the normal eggs, two exceptional types would arise: one with two X's and the other lacking an X entirely. When fertilized by normal sperms, these exceptional eggs would give rise to exceptional zygotes. *C.* When Bridges examined the offspring cytologically, he found that the expected flies (A and B) possessed the normal chromosome complement. The unexpected females (C) carried a Y; the red-eyed males (D) lacked a Y and were sterile. No flies were found with three X's. These were discovered later and are unusual females (superfemales) which are weak and tend to die. Flies lacking an X are never found, as at least one X is required for normal development.

of the exceptional white-eyed daughters carried the normal number of autosomes as well as two X's. But in addition, a Y was present! The unexpected red-eyed males showed the normal autosome number and one X. But no Y was present at all (Fig. 5-6C)! These males proved to be sterile due to immobility of their sperms. Bridges found no flies with three X chromosomes nor any with just the Y and no X. Because one X at least is necessary for life, the latter never

arise. Flies with three X's were discovered in subsequent work and will be referred to shortly.

The cytological observations completely supported Bridges's hypothesis on the origin of the unexpected offspring. He coined the term *nondisjunction* to describe the failure of the homologous chromosomes to separate at anaphase of meiosis. We now know that nondisjunction may also occur at mitosis when both chromatids composing a chromosome move to the same pole at anaphase. The expression is generally used to describe the failure of chromatid separation at either mitosis or meiosis.

Bridges tested his hypothesis further. He took some of the exceptional white-eyed daughters containing a Y chromosome and crossed them to ordinary red-eyed males (Fig. 5-7). In the unusual females, the two X chromosomes will pair in some of the meiotic cells. At anaphase, the X's will therefore pass to opposite poles, whereas the Y chromosome may migrate to either pole. The result is the formation of two classes of gametes: those of the constitution w and those which are wY. In other meiotic cells, the Y chromosome may pair with one of the X's. In such a cell, this X and the Y will pass to opposite poles. The unpaired X is free to migrate to either pole. This makes possible the formation of four classes of gametes: ww and Y (the unpaired X moved to the same pole as the X which

FIG. 5-7. Exceptional female × ordinary male. When Bridges mated females which carry a Y chromosome, unusual genetic results followed. A small number of white-eyed females and red-eyed males occurred. Cytological examination showed that flies with exceptional chromosome complements were also being produced at the same time. Extra Y's were found in one-half of the males and females. Along with the results depicted in Fig. 5-6, we see a perfect correlation between genetics and cytology.

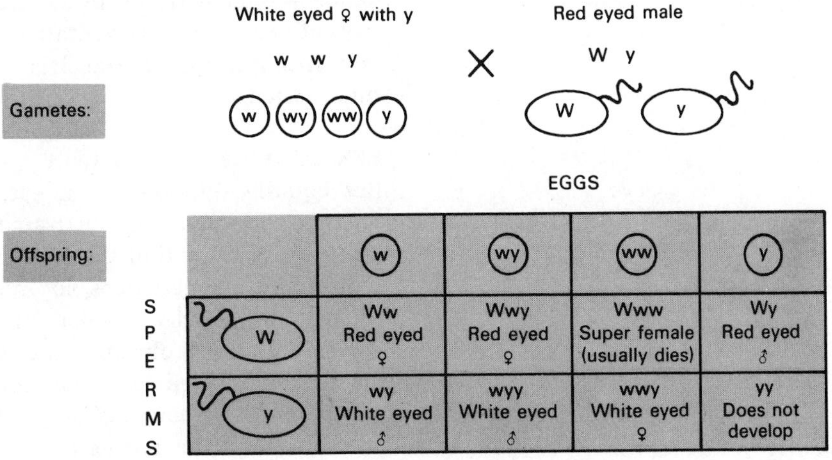

	EGGS			
	w	wy	ww	y
S P E R M S — W	Ww Red eyed ♀	Wwy Red eyed ♀	Www Super female (usually dies)	Wy Red eyed ♂
S P E R M S — y	wy White eyed ♂	wyy White eyed ♂	wwy White eyed ♀	yy Does not develop

Gametes: White eyed ♀ with y: w w y → w, wy, ww, y   X   Red eyed male: W y → W, y

paired with the Y) and w and wY (the unpaired X moved to the same pole as the Y chromosome). Therefore, an overall total of four different kinds of gametes can be formed from these pairing possibilities. However, the four types of eggs were produced in very unequal amounts by the wwY females: 92% were w and wY, whereas 8% were ww and Y. This indicates that the two X's tend to pair more frequently than an X and a Y. Perhaps this is to be expected, because the two X's are completely homologous. In Figure 5-7, the four kinds of eggs are combined in all possible ways with the normal sperm types, W and Y, to produce an assortment of zygotic combinations. How were these different types represented among the offspring? Cytological examinations were performed, and several of them were identified. No flies were found which were XXX, because this combination usually dies. YY flies were never produced at all. The results clearly showed that when exceptional females with a Y chromosome are bred, exceptional results follow. Their white-eyed daughters have a Y chromosome. Although their red-eyed sons have a normal complement of one X and one Y, the Y chromosomes which they carry must come from their exceptional female parents! In addition, extra Y's are present in one-half of the white males and one-half of the red females. This story is an example of perfect correlation between genetic and cytological results. The work is of great significance for many reasons, not the least is its demonstration that the genes are in the chromosomes. For when the chromosome constitution is not determined by the normal process, atypical results arise. This evidence was of extreme importance in the early days of genetic experimentation because it gave very strong support to the chromosome theory of inheritance.

SEX AND GENIC BALANCE. A more subtle but equally important concept, that of genic balance, was also demonstrated by this work. The results show that the presence of a Y chromosome in a zygote does not mean that a male fly will necessarily develop. In *Drosophila*, it seems that the Y chromosome is not needed at all for life or even for maleness, because a fly can be male without a Y or can be female with one. Although the Y is essential for fertility and is thus critical to the survival of the species, it is not required for the life or the sex determination of any one individual. Sex in *Drosophila* seems to depend not just on the presence or absence of a single chromosome but rather on the balance of the autosomes to the sex chromosomes. When the balance is two X chromosomes to two sets of autosomes, a female develops. A balance of one X to two sets of autosomes determines a male.

That this concept of genic balance in sex determination is correct was confirmed by the later work of Bridges. Flies with entire extra sets of chromosomes were produced in the laboratory. Individuals with three or four chromosome sets instead of the normal two are *triploids* and *tetraploids,* respectively. Inasmuch as three or four homologous chromosomes are present in the cells of such flies, the meiotic behavior is irregular (see Chap. 10 for details). As a consequence, when these flies are bred, their offspring may be atypical in chromosome number, having different combinations of autosomes and X chromosomes. Normal males and females, being diploid, possess two sets of autosomes. As shown in Figure 5-1*B*, one set of autosomes in *Drosophila* includes two different V-shaped chromosomes (Chromosomes II and III) and a small, dot chromosome (Chromosome IV). Therefore, in the common fruit fly, one haploid set of autosomes is composed of three chromosomes (Chromosomes II, III, and IV). Normally, two of these sets of autosomes plus two X's result in a female. Two sets of autosomes plus one X produce a male. Unusual flies which developed from mating the triploids and tetraploids did not all have this normal relationship of autosomes to X chromosomes. Studies of their cytology indicated that genes for maleness and femaleness were distributed throughout all the autosomes and the X. As the summary of results in Figure 5-8 indicates, the X would contain genes with a strong female determining tendency. *One set* of autosomes would lean in the male direction as a result of the many genes for maleness scattered throughout. However, the strength of one X toward the female side is greater than the strength of one set of autosomes in the male direction. They are not 1:1. Rather, it seemed that the femaleness of an X was about 1.5 to the maleness of *a set* of autosomes (1.5:1). This

should not be construed to mean that there is any kind of fluid or substance which has male or female determining properties. The ratio 1.5:1 only symbolizes the comparative strengths of the two major factors (one X vs. one set of autosomes) in sex determination.

The observations indicate that the sex of a fly is determined by a balance between many genes distributed throughout the autosomes and those found on the X. The absolute number of chromosomes is not the sole factor, because *any* fly will be a female when the ratio of femaleness to maleness is 3:2 (which in turn results from a ratio of one X to one set of autosomes). It doesn't matter if four sets of autosomes and four X's are present or if there is only one set of autosomes and one X. The ratio is still the same (1:1 between X and autosomes). Any departure from a certain balance, however, will produce a fly which is in some way abnormal. The presence of just one extra X tips the balance heavily in the female direction (4.5:2), giving a fly with exaggerated secondary sex characteristics, a superfemale or metafemale. Such a fly is weak and usually dies. It was not found by Bridges in his original study of the exceptional white-eyed females, but its possible viability was recognized at that time.

A comparable story holds for the determination of maleness. A ratio of one X chromosome to two sets of autosomes (a female to male ratio of 1.5:2) will always result in a male regardless of the presence or absence of a Y. A supermale or metamale, also generally weak, results from the extra strength of maleness provided by one extra set of autosomes. When the female:male ratio is equal (3:3), a sterile intersex, completely intermediate in secondary sex characteristics, is produced.

That many genes are involved in sex determination was shown by Dobzhansky studing such intersex flies containing two X's and three sets of autosomes. After crossing normal flies whose chromosomes had been damaged by X-rays, Dobzhansky found offspring which contained three sets of autosomes and two X's plus additional portions of the X (Fig. 5-9). The larger the piece of the X which was present in a 2X:3A intersex type, the more it resembled a female. Clearly, genes determining femaleness must be scattered along the length of the X.

The involvement of genes on the autosomes is very dramatically shown by a rare recessive autosomal allele in *Drosophila* known as transformer (tra). This is so named because a double dose of this recessive causes flies which have two X chromosomes and two sets of autosomes (and which therefore should be normal females) to be phenotypically male (Fig. 5-10)!! Such recessives cannot be distinguished from normal flies by observation; however the transformed flies are sterile. Because the recessive ("tra") is

FIG. 5-8. Genetic balance and sex determination in *Drosophila*. A female results whenever one X chromosome is present for each set of autosomes. A male develops when one X occurs with two sets of autosomes. Any departure from these ratios results in flies which are in some way abnormal in sexual features.

IX = ♀ Determining strength of 1.5

II).(III
IV

1 Set of autosomes = ♂ determining strength of 1.0

Number of x's	Number of sets of autosomes	Ratio femaleness:maleness	Sex
1 (1.5)	2 (1)	1.5: 2	Male
2 (1.5)	2 (1)	3.0: 2	Female
3 (1.5)	3 (1)	3.0: 2 (4.5:3)	Female
4 (1.5)	4 (1)	3.0: 2 (6.0:4)	Female
2 (1.5)	3 (1)	3.0: 3.0	Intersex
3 (1.5)	2 (1)	4.5: 2.0	Superfemale
1 (1.5)	3 (1)	1.5: 3.0	Supermale

x x

Diploid

x x y

Diploid +y

x x x

Triploid

x x x x

Tetraploid

All ♀ since ratio of x to sets of autosomes is the same in each case: 1x to 1 set of autosomes

BUT

Adding a single x tips the balance toward femaleness

x x x

Superfemale

x x

Intersex

Adding a set of autosomes equalizes strength of male and femaleness

In both of these the ratio of x to sets of autosomes departs from the balance of 1x to 1 set of autosomes

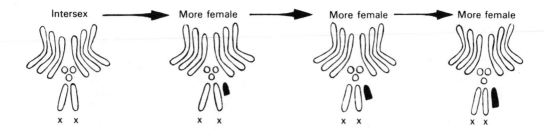

FIG. 5-9. Comparison of intersex types. Flies with two X chromosomes and three sets of autosomes cannot be classified as male or female. Those flies with additional pieces of an X appear more female. The larger the portion of the X, the more female the fly is in appearance. This indicates that genes for femaleness are scattered along the length of the X.

rare in fly populations, most females are Tra Tra XX and most males Tra Tra XY. This is another good example of the interaction of many genes in the determination of a phenotypic characteristic, in this case sex. In the last chapter, emphasis was placed on the fact that any characteristic involves not just one gene or allele but the interaction of many with the environment. Sex is a characteristic whose normal development is no different from that of any other in this respect.

SEX CHROMOSOME ANOMALIES IN HUMANS. Although the concept of genic balance pertains to all living things, not just fruit flies, we must remember that the genetic picture varies in detail from one species to the next. We have already seen that in the grasshopper, the male is normally X-O. No Y chromosome is needed even for sperm development, but a balance of X chromosomes to autosomes is also involved here in normal sex determination. Genic balance certainly plays an important role in humans, but there are distinct differences from the fruit fly story. The Y chromosome in *Drosophila* is unnecessary for life and apparently devoid of genes for male secondary sex characteristics. Thus, an XO individual is male and an XXY is female. However, such is not the case in humans and other mammals, where the Y chromosome is essential for male attributes. Examples of XO and XXY persons are well known. Those of the former constitution have a chromosome number of 45 and are females who display the various abnormalities of Turner syndrome, a condition which affects the development of both sexual and other bodily characteristics. A Turner female, found in about 1 in 3000 female births, is

usually of short stature, possesses a webbed neck, has undeveloped ovaries, and an immature uterus, plus cardiovascular defects and other somatic aberrations (Fig. 5-11). Persons who are of the chromosome constitution XXY are definitely male, but they exhibit an assortment of deviations from normal which describes Klinefelter syndrome. A Klinefelter male arises in about 1 out of 600 male births, a more frequent figure than in the case of Turner syndrome. These males typically show some breast development, small testes, sparse body hair, and some mental deficiency (Fig. 5-12). The Y chromosome in the cat also plays a role in maleness similar to that in humans. Cats which are XXY have been recognized and are the sterile equivalent of the human Klinefelter males.

In order to appreciate the cytology of various clinical conditions in humans, it is important to have an acquaintance with the appearance of the normal human chromosome complement or normal *karyotype* (Chap. 2). Many defects in humans are associated with certain specific chromosome abnormalities, such as variation in number and alteration in chromosome structure. Analysis of the chromosome complement is almost routinely performed today in cases of certain suspected conditions such as Turner and Klinefelter syndromes. Figures 5-11 and 5-12 show karyotypes typical of patients with Turner and Klinefelter syndromes.

However, persons exhibiting Klinefelter syndrome have been identified who have chromosome situations more complex than the XXY. Individuals of the following chromosome constitutions have been found: XXXY, XXXXY, XXYY, and XXXYY. On the average, the XXY male has a lower intelligence than the XY male.

Typical genotypes:

TRA TRA xx

TRA TRA xy

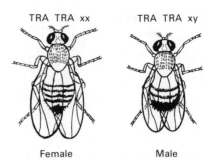

Female

Male

Rare genotypes:

TRA tra xx

tra tra xy

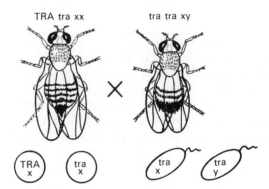

Gametes:

TRA
x

tra
x

tra
x

tra
y

EGGS

Offspring:

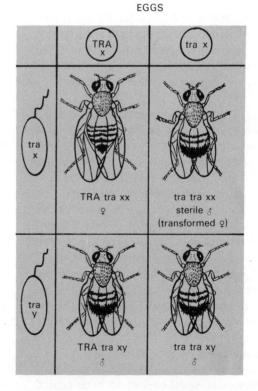

TRA
x

tra x

SPERMS

tra
x

TRA tra xx
♀

tra tra xx
sterile ♂
(transformed ♀)

tra
y

TRA tra xy
♂

tra tra xy
♂

UNDERSTANDING GENETICS

There is evidence that mental deficiency and body abnormalities are still more pronounced in those Klinefelter males with three and four X chromosomes.

Other kinds of atypical sex chromosome conditions have been found among humans. In the general population, males with an XYY constitution occur in about one in 2000 male births. There is the indication that an extra Y tends to increase an individual's height. Cytological studies on inmates of prisons and mental institutions have revealed an unexpected number of men with extra Y's. This has led to the suggestion that aggressive and antisocial behavior may be associated with extra Y's. However, men with additional Y chromosomes are known who display no abnormal behavior traits. Since the role of the environment as the trigger for the expression of antisocial behavior has not as yet been properly evaluated, it is premature to attribute such behavior directly to a surplus of Y-linked genes.

It should be appreciated at this point that the XO, XXY, and other conditions can be explained on the basis of nondisjunction in one of the parents. At zygonema of meiosis in the male, the X and Y chromosomes may form a pairing association referred to as the *sex bivalent.* The pairing does not affect the entire length of the chromosomes but appears to be confined only to short segments in one arm of each of them (Fig. 5-13). This may reflect some homology in short regions of the X and the Y, but it is believed that the pairing is not sufficiently intimate to permit crossing over. Nevertheless, a sex bivalent forms in a primary spermatocyte. The two chromosomes normally separate at first anaphase to produce two cells: one with the X, the other

containing the Y (recall spermatogenesis, Chap. 3). Each of these secondary spermatocytes in turn undergoes the second meiotic division. The final outcome is X and Y containing sperms in equal numbers. Consider the effects of nondisjunction of the X and Y at first meiotic division (Fig. 5-14A). Sperms of the constitutions XY and "O" can be produced. If normal, X-bearing eggs are fertilized by such gametes, the result is XXY (Klinefelter) and XO (Turner). Nondisjunction can also involve a female parent and give rise to abnormal eggs of the constitutions XX and "O" (Fig. 5-14B). This is reminiscent of the situation first discovered by Bridges in *Drosophila.* Fertilization of the exceptional eggs by normal sperms can also produce zygotes which are unbalanced, among them XXY and XO types as well as XXX. The latter constitution as well as XXXX has been found among abnormal females suffering mental abnormalities.

SEX MOSAICS. The fact that individuals with only one Y chromosome but two, three, and even four X's are still males, shows the importance of the Y in the determination of the secondary sex characteristics in humans. This is quite a contrast to the situation in *Drosophila* and is even more impressively illustrated by comparing certain other chromosome anomalies in the fruit fly and the human. In the fly, certain mishaps which affect the sex chromosomes occasionally take place during embryological development. At rare times during the mitotic divisions in some female embryos, one of the X chromosomes may lag and fail to reach the pole at anaphase (Fig. 5-15A). It may be left out in the cytoplasm where it disintegrates. The result is one cell with two sets of autosomes plus two X's and another cell which has two sets of autosomes plus one X. In *Drosophila,* this means that some cells will be genetically female and others male. Flies such as these actually arise and are called *gynandromorphs,* mosaic individuals with discrete female and male body segments. If the loss of the X takes place at the first division of the zygote, the fly will be a bilateral gynandromorph, male on one-half of the body, female on the other (Fig. 5-15B). If the mishap occurs later, smaller patches of male tissue will be present among the female background. Can gynandromorphs arise in humans? Mosaic persons, those who are mix-

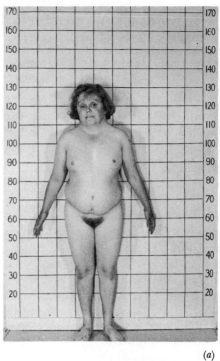

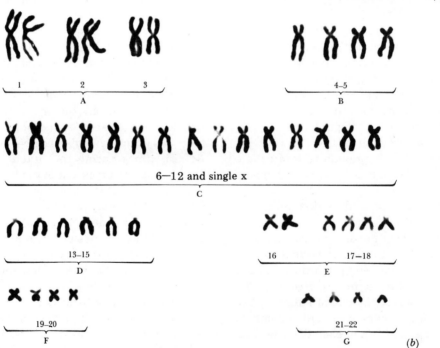

(a)

(b)

FIG. 5-11. The Turner syndrome. *a*. Patient. *b*. Karyotype of patient. Chromosome number, 45. Only one X chromosome is present. The Y is also lacking. (Reprinted with permission from V. A. McKusick, *Medical Genetics 1958–1960*, C. V. Mosby Co., St. Louis, 1961.)

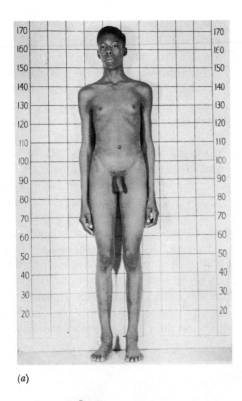

(a)

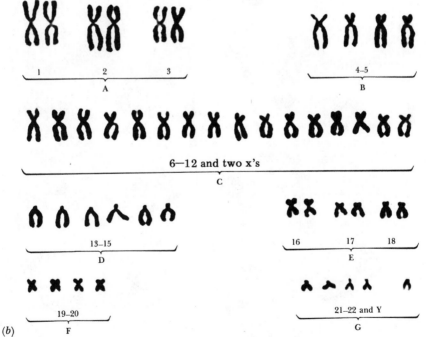

(b)

FIG. 5-12. The Klinefelter syndrome. a. Patient. b. Patient's karyotype. Chromosome number, 47. Two X chromosomes and a Y are present. (Reprinted with permission from V. A. McKusick, *Medical Genetics 1958–1960,* C. V. Mosby Co., St. Louis, 1961.)

tures of different cell lines, are certainly well known. Some of these have groups of cells that are XX and others that are XO. However, such individuals show symptoms of Turner syndrome. No discrete patches of definite male or definite female tissues are seen in any kind of human mosaic. The mosaic may be abnormal phenotypically and may, for example, possess vestiges of the gonads of both sexes. However, hormonal activity in mammals will not allow the development of two distinct classes of cells, those with typical female features intermingled with those which are obviously male.

In cases of chromosome anomalies, such as XXY or XXXY, a testis develops but it is abnormal, probably due to hormonal unbalance caused by the genetic unbalance. Nonetheless, because of the hormonal activity at this early developmental stage, all the body cells are affected so that nothing like a gynandromorph can result.

FIG. 5-13. Meiotic chromosomes of human male at diakinesis. The tiny Y chromosome appears attached terminally to the much larger X. (Reprinted with permission from W. V. Brown, *Textbook of Cytogenetics,* C. V. Mosby Co., St. Louis, 1972.) (Courtesy of Dr. J. Melnyk, City of Hope Medical Center, Duarte, Calif.)

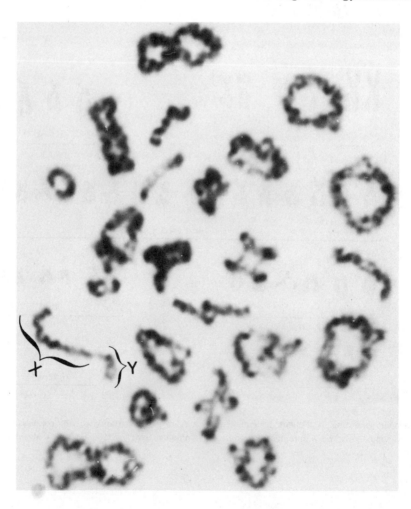

FIG. 5-14. Effects of nondisjunction in humans. *A.* In the male, normal separation of the X and Y chromosomes at anaphase I (*above*) leads to formation of two kinds of haploid gametes (X and Y) which upon fertilization result in balanced female and male zygotes. Nondisjunction (*below*) in which X and Y move to the same pole leads to the origin of unbalanced gametes which can give rise to a Klinefelter male and a Turner female after fertilization. *B.* In the female, all the normal gametes contain one X (*above*). In case of nondisjunction of the X's at meiotic anaphase, unbalanced gametes arise. After fertilization by the X-bearing or a Y-bearing sperm, several types of unbalanced zygotes may form.

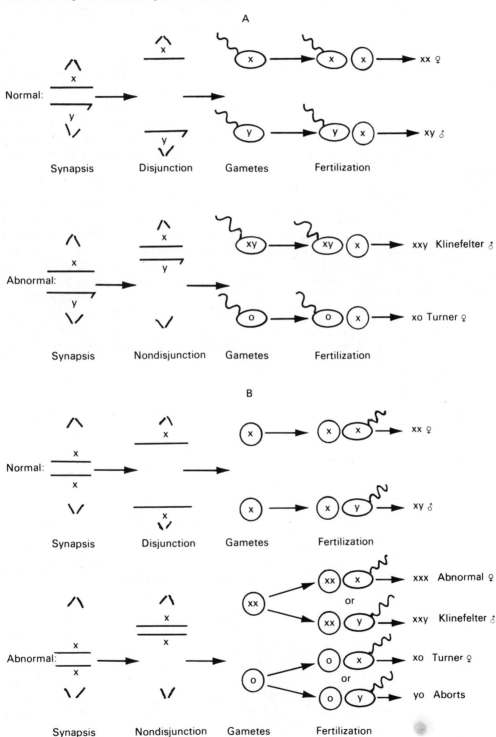

Departures from the normal number of chromosomes due to nondisjunction and other mishaps can affect autosomes as well as sex chromosomes. Moreover, any chromosome may become altered in its structure. These and other chromosome anomalies will be pursued further in Chapter 10.

SEX CHROMOSOMES AND SEX DETERMINATION. In the human embryo, the gonads are first represented by the gonadal ridges which appear about 30 days after the formation of the zygote. These are capable of developing into either testes or ovaries. Normal differentiation is determined by the sex-chromosome constitution of the primitive gonad. The Y chromosome is essential to the development of a testis. In its absence no testes will develop, regardless of the number of X chromosomes which are present. It is clear that the Y chromosome carries a testis-determining factor (or factors). If a Y is present, testes will develop even in persons of unusual karyotypes (XXY, XXXY, XXYY, XXXXY). The differentiation of the testes in the presence of the Y chromosome is established by the sixth week, and once this has occurred, male hormones are

FIG. 5-15. Gynandromorph origin in *Drosophila. A.* A zygote with two X chromosomes plus two sets of autosomes will normally develop into a female. However, at mitotic divisions during embryology, one X chromosome at rare times may lag and be eliminated. The outcome is two cells, one with a nucleus balanced in the female direction and the other with a nucleus balanced in the male direction. If the mishap occurs at the first division of the zygote, a fly which is male on one side of its body and female on the other can arise. *B.* A bilateral gynandromorph which is male on the right side of the body and female on the left. The eye color on the male side is white, because only one X is present and this carries the recessive for white eye color (genotype wO). The X which was lost carried the dominant allele for red color. On the female side, both X's are present, one with the dominant, the other with the recessive (genotype Ww), and so the eye is red. (From E. J. Gardner, *Principles of Genetics,* 4th ed., Wiley, New York 1972.)

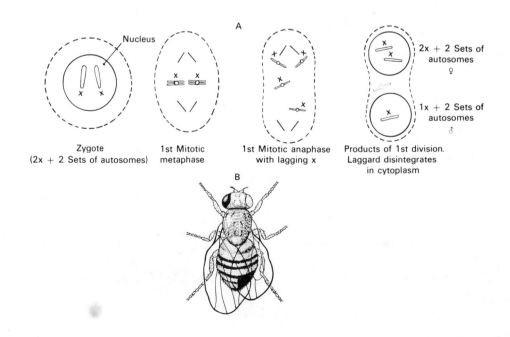

produced by the fetal testes. These then further guide the normal development of other undifferentiated structures into parts of the male reproductive tract.

In the absence of a Y chromosome, as in the XX condition, no testis development can be triggered, and the gonads remain undifferentiated until about the twelfth week when ovarian development proceeds. Completion of normal ovarian development requires two X chromosomes. We see that it is the absence of the Y which enables ovarian differentiation to occur. Moreover, the developing testes produce a substance which acts as an inhibitor to prevent the development of any embryonic structures into certain parts of the female reproductive system, such as uterus and Fallopian tubes. The absence of this inhibitor in the XX condition allows these embryonic structures to develop into the parts of the internal reproductive system of the female.

The X-Y mechanism of sex determination in mammals undoubtedly evolved through the pressure of natural selection, to guarantee equal numbers of males and females. However, the human sex ratio at birth is slightly in favor of males: 106 boys to every 100 girls. The precise explanation for this is still a matter of debate. According to one idea, the Y chromosome is somewhat lighter than the much larger X, and thus a sperm carrying a Y chromosome can travel faster than one carrying an X and has a greater chance of encountering the egg. Another suggestion arrived at from studies of karyotypes of spontaneous abortuses is that the true sex ratio is 1:1, equal numbers of XX and XY zygotes being produced, and that more females are lost *in utero*. Regardless of the correct explanation for the slight excess of males at birth, the ratio of males to females is about 1:1 at the age of sexual maturity.

SEX LINKAGE IN HUMANS.   After the discovery of sex linkage in *Drosophila,* attention was called to similar inheritance patterns for certain human traits. Among the most familiar are red-green color blindness and hemophilia, both of which behave as sex-linked recessives. Approximately 8% of American males have a defect in color recognition, whereas the trait in females is well under 1%. We can easily understand this once we realize that the loci for the major types of color blindness are located on the X chromosome. The ability to perceive color depends not on a single genetic locus. At least four are involved and three of these are X-linked. About 3/4 of all color-blind persons carry a sex-linked recessive which causes the green-sensitive cones in the retina to be defective. Consequently, difficulty is encountered in the recognition of green color and its distinction from red. This is known as the deutan type of color blindness. Less common is the protan type which affects about 2% of males in the population. This sex-linked recessive causes a defect in the cones sensitive to red. In this situation, red color cannot be distinguished, and again red and green are confused. Since red and green color distinction is poor in both deutan and protan types, it was at first thought that only one locus and one sex-linked recessive was responsible, but more refined observations led to the recognition of two separate X-linked loci which affect red-green color perception.

Very uncommon is the tritan form of color blindness in which recognition is poor for all colors. This recessive is found at still another locus on the X and causes defects in cones sensitive to blue as well as to those sensitive to green and red. A very rare autosomal recessive causes total color blindness, since no cones at all develop.

Since the loci for the recessives responsible for red-green color blindness are sex linked, we can understand why this condition is much more common in the male. Any female (other than an X0 Turner female) carries two X chromosomes. For her to express red-green color blindness (let us say of the commoner deutan type), her father must show the trait. The mother must show deutan color blindness herself or at least be a heterozygote. This is so because the daughter must receive a deutan colorblind recessive from each parent in order to express the trait (Fig. 5-16A, B). The same reasoning applies to the protan type of red-green color blindness, as well as to the rare tritan type.

It should be apparent also that a male will always express any gene which is sex linked because he has only one X chromosome (unless he is a Klinefelter male). Being hemizygous, he cannot carry the normal allele and the defective

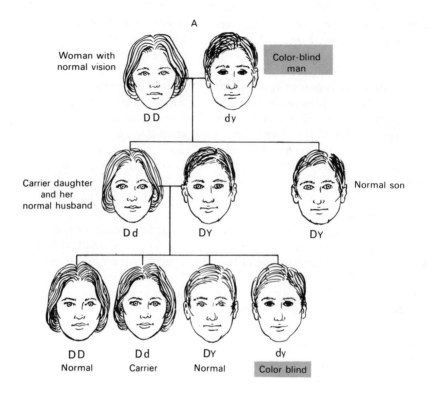

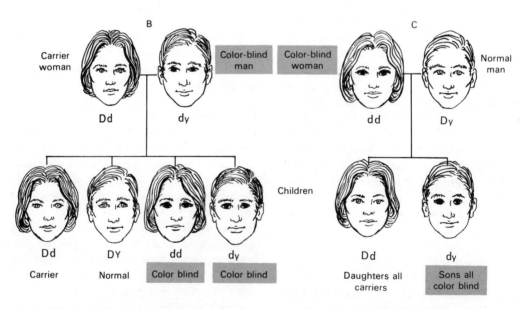

FIG. 5-16. Color-blind inheritance. *A.* The locus for green color perception is on the X chromosome. "D" represents the dominant allele for normal color distinction and "d" its recessive allele for deutan color blindness. The trait passes from a color-blind man to one-half of his grandsons by way of his carrier daughters. Since a male contributes his Y chromosome to his sons and not the X, the sons of a color-blind man cannot receive the recessive from their father. *B.* In order for a color-blind female to arise, her mother must at least be a carrier and her father must be color blind. The color-blind female receives an X from each parent which carries the recessive allele. *C.* All of the sons of a color-blind woman must be color blind because a male receives his X from his mother. The daughters of a color-blind woman will be carriers with normal vision if the male parent has normal vision.

form at the same time. Therefore, any sex-linked allele which expresses itself in a heterogametic male must have come from the female parent. In humans, a son receives the Y from his father; this is a requirement for maleness. Moreover, all the sons of a woman who expresses a sex-linked recessive, such as color blindness, would be affected, even if her mate is normal (Fig. 5-16C). The cross of a color-blind woman and a normal man results in a crisscross pattern and is comparable to a mating in *Drosophila* between a white-eyed female and a red-eyed male.

The sex-linked recessive responsible for the lethal effect of hemophilia, a condition in which the blood fails to clot normally, is of special interest, both for its consequences to the sufferer and its historical import. The affliction, which has been recognized for hundreds of years, received attention when it appeared among European royalty during the reign of Victoria of England. There is little doubt that Queen Victoria was heterozygous for the recessive, "Hh." The probability is great that a mutation from the normal allele (H) to a recessive form (h) arose in either the sperm from her father or in the egg from her mother. At any rate, of the queen's nine children, several definitely received the recessive allele from her. One of her sons was consequently a hemophiliac, and some of her daughters were carriers who were responsible for passage of the hemophilia factor through several royal houses, even down to the present. This is easy to understand from what we know about the inheritance of sex-linked recessives (Fig. 5-17A and B). We would expect from such a mating that one-half of the daughters would be carriers and that one-half of the sons would express the factor for hemophilia.

Today we know a great deal about hemophilia and its molecular basis which bears certain similarities to the story of phenylketonuria (PKU) presented in Chapter 4. We may think of normal blood clotting as the end result of a series of reactions which require various enzymes. The final steps involve the action of the enzyme "thrombin" upon the protein "fibrinogen." The end result is conversion of the soluble fibrinogen to the insoluble fibrin. However, many protein substances are involved in the formation of the thrombin (Fig. 5-18). A defect in any one step governing the formation of any one of these

required proteins can lead to defective blood clotting. Hereditary disorders have been described in which one or more of these substances is deficient in the blood plasma. Therefore, it is not surprising that more than one kind of hemophilia is known. Most cases of hemophilia, designated hemophilia A, result from a deficiency of a protein known as the antihemophilia factor (AHF). However, a completely different sex-linked recessive at another locus can cause a deficiency of another essential product, plasma thromboplastin component (PTC) or Christmas factor. This type of hemophilia, hemophilia B, is also known as Christmas disease, named after the family in which the hemophilia was discovered. Christmas disease is a milder hemophilia making up approximately 20% of all cases. An even less common form of hemophilia results from a rare recessive which is autosomal instead of sex linked and which is needed for the formation of plasma thromboplastin antecedent (PTA), another protein in the reaction chain. We see again an example of the interaction of many genes in the development of a characteristic, here blood clotting. A breakdown in the metabolic chain at any one step due to a defective gene which controls the specific reaction can lead to an abnormal phenotype.

The possibility of female hemophiliacs has been considered for years. It was thought that no such persons could arise, because the homozygous condition "hh" was believed to be lethal to the embryo. We are now aware of a few authentic cases of female bleeders. As we might well expect, they are very rare. In the same way as in the case of color blindness, the female parent of an afflicted girl must at least be heterozygous for the allele and the father must actually express the recessive condition (Fig. 5-16). Since hemophilia is such a serious disorder, most male sufferers do not survive to reproductive age. The few that do manage to have offspring would almost always mate with women who are homozygous normal "HH," because the recessive involved is not common in the population. When a rare mating between a hemophiliac and a carrier does take place an afflicted daughter can be produced.

Well over 100 sex-linked genes have been recognized in humans. Most of them are not as familiar as the one affecting blood clotting and the one for color vision. We should realize that some sex-linked defective alleles have a dominant expression, as does the one responsible for a type of rickets, a skeletal defect. Any trait inherited on the basis of a sex-linked dominant will be more frequent among females because they possess two X's in contrast to the male and therefore have a greater chance of receiving the X-linked allele. Remember that in cases of color blindness and hemophilia, the dominant, normal condition is more frequent among females. For the same reason, an abnormal phenotype will be more common among females if it results from a sex-linked dominant allele.

SEX CHROMATIN. For many years, biologists wondered about the fact that in mammals the female carries a double dose of all the genes on the X as opposed to the single dose in males. This might mean that a female homozygous for the dominant allele for the antihemophilia factor has twice as much of the AHF as does a normal

FIG. 5-17. Inheritance of hemophilia through Queen Victoria. *A.* Hemophilia among European royalty can be traced back to Queen Victoria. The trait was not found among her ancestors, but she was undoubtedly heterozygous for the defective allele. Since hemophilia is sex linked, we would expect her to pass the allele to one-half of her sons and one-half of her daughters. *B.* When the actual pedigree of Victoria and her descendants is studied, the transmission of hemophilia is easily understood on the basis of the expectations summarized in *A.* In the pedigree, circles represent females and squares are males.

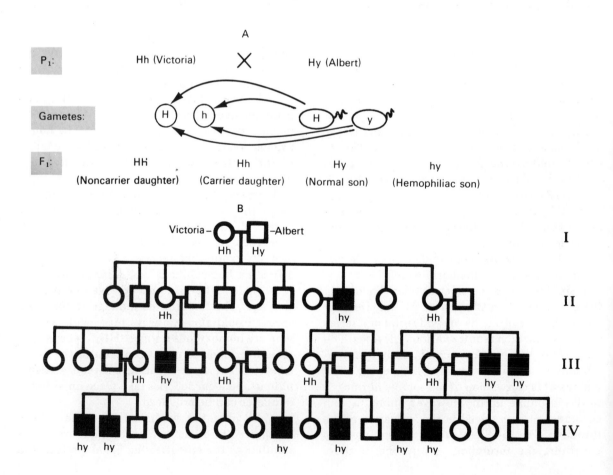

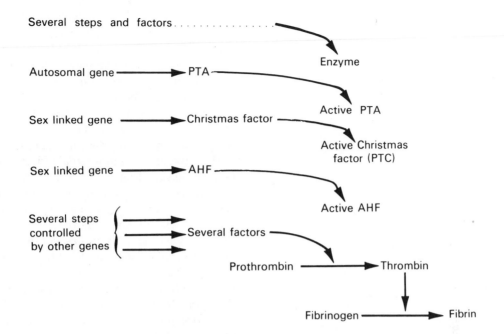

Several steps and factors . . . . → Enzyme

Autosomal gene ————→ PTA ——→ Active PTA

Sex linked gene ————→ Christmas factor ——→ Active Christmas factor (PTC)

Sex linked gene ————→ AHF ——→ Active AHF

Several steps controlled by other genes } ——→ Several factors ——→

Prothrombin ————→ Thrombin

Fibrinogen ————→ Fibrin

FIG. 5-18. Some steps in blood clotting. Normal coagulation involves a series of reactions leading to the formation of insoluble fibrin. Many of the steps in the chain entail the conversion of a gene product to its active enzymatic form by the enzymatic action of the active product of a previous step. If antihemophilia factor (AHF) is not present as a result of the sex-linked recessive, there is nothing for the active product of the previous step to work on. There is also a lack of active AHF to catalyze the following step. The end result is defective clotting.

male. The idea is a disturbing one when considered in relationship to the concept of genic balance. In contrast to the X-linked loci, all autosomal loci are represented twice in both sexes. It seemed to geneticists that something must compensate for the difference in dosage of the sex-linked genes if genic balance were to be preserved. An answer to this problem dates back to 1949 with the discovery by Barr of a nuclear body found in the neurons of the female cat. This small intranuclear entity was completely absent from neurons of the male (Fig. 5-19). Examination of other tissues and other species showed that a constant distinction can be made between nuclei of cells taken from females and from males. A simple way to demonstrate this in the human is by scraping epithelium from the buccal mucosa and staining it with some common dye. The nuclei of most of the cells from a female will show a small rod-shaped structure, more deeply stained than the surrounding chromatin and usually located at the periphery of the nucleus (Fig. 5-20A). Comparable cells from a normal male fail to show such a body, which has

FIG. 5-19. Barr body in neuron of the female cat (× 1050). This body which is very evident in the nucleus of the female is often found in the vicinity of the nucleolus (the large intranuclear body). It is entirely absent from the nucleus of the male.

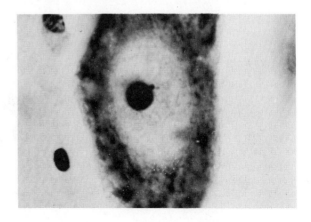

been designated the *Barr body* or the *sex chromatin*. One of the many other tissues which show sex distinctions is the blood, specifically the polymorphonuclear leukocytes. In some of these white blood cells taken from a female, a "drumstick" can be recognized (Fig. 5-20*B*). This is seen as a small appendage attached by a filament to one of the lobes of the nucleus. Although they can be identified in only about 2% of the leukocytes from a female, drumsticks are entirely absent from cells derived from a male (Fig. 5-20*c*). The discovery was then made that drumsticks and Barr bodies are absent from females with Turner syndrome. On the other hand, cells from XXY males contain one Barr body or one drumstick. Males who are XXXY and XXXXY show two and three, respectively. Cases were found of women with multiple X conditions: XXX and XXXX. These persons were found to have two and three sex chromatin bodies, respectively. It became apparent that the number of sex chromatin bodies and drumsticks was one less than the number of X's present in cells of the individual.

THE X CHROMOSOME AND THE LYON HYPOTHESIS. Such observations led to the formulation of a theory which accounts for the relationship of the number of sex chromatin bodies to the number of X's. More important, it suggests a mechanism which compensates for the presence of different dosages of X-linked genes in males and females. Dr. Mary Lyon has presented some of the best arguments for the concept. According to the Lyon hypothesis, all normal females are sex mosaics, composed of different cell lines insofar as the X chromosomes are

FIG. 5-20. Sex chromatin in humans. *A.* Barr body in nucleus of a squamous epithelial cell from the buccal mucosa of a female. *B.* The drumstick in a polymorphonuclear leukocyte of a female. *C.* Polymorphonuclear leucocyte of a male. (Courtesy of Carolina Biological Supply Company.)

A

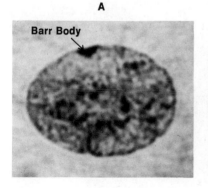

Barr Body

B

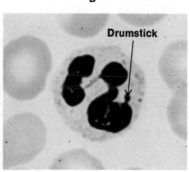

Drumstick

C

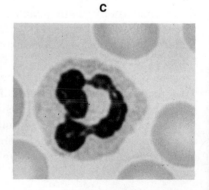

concerned. Only one X is functioning entirely in a given cell; the other is present largely in an inactive state. Which specific X is active in a cell, the one from the maternal parent or the one from the paternal, is a matter of chance. Any female will have the X she received from her mother functioning in some cells and the one from her father acting in others. The choice as to which X will be the active one in a cell is believed to occur during embryological development. There are several lines of evidence to support the Lyon hypothesis. It will be recalled from the discussions of mitosis and meiosis that chromosomes in a compact state, as at the time of nuclear divisions, are actually not active in cellular activities other than those concerned with the division process itself. Evidence from the molecular level has shown that the deoxyribonucleic acid (DNA) in its compact state is relatively inactive in directing the synthesis of other molecules. Cells can be taken from human females and maintained for study in tissue culture. Observations have revealed that at early mitosis one of the two X chromosomes in any cell tends to be more compact and to stain more deeply than those in the rest of the complement. It also tends to undergo synthesis of its DNA later. All in all, the behavior of one of the X's at mitosis has been shown to be distinctly different from the other members of the chromosome set, and some of the features it displays are typical of chromatin which is relatively inactive.

Moreover, sex chromatin is not present in oocytes. During prophase of oogenesis, both X chromosomes appear identical and react the same way to stains. Both are evidently equally active at this time. During first meiotic prophase, they pair and engage in crossing over. They continue to behave in the same fashion even in the earliest divisions of the embryo. However, at approximately 16 days, in the late blastocyst stage, sex chromatin can be recognized. These observations afford strong support for the Lyon hypothesis because they indicate that both X's are equally functional during meiosis and in the cells of the very early embryo. But then at the blastocyst stage, some mechanism becomes operative in the body cells which keeps one of the X's (or a large part of one) in a permanently inactive state by condensing the chromatin. The Barr body is a visual manifestation of this. However, as indicated by oogenesis, both X's remain equally active in the germ line.

In Chapter 3, the heteropyknotic behavior of the X chromosome in the male was discussed. In contrast to the rest of the chromosomes at early meiotic prophase, it is in a compact, deeply staining state. This may be a device which keeps the X in the male inactive during chromosome pairing so that it does not exchange any chromosome segments with the Y as a result of crossing over. Obviously, it would be important to keep any male-determing genes apart on the Y in order to prevent them from being scrambled with genes on the X. This point is discussed a bit more later in this chapter.

Barr bodies and drumsticks show various parallels which indicate that they are both visible manifestations of the same phenomenon, a compacted X which is thus in an inactive state. The fact that not every single cell of a female shows sex chromatin (drumsticks in only about 2% of the polymorphonuclear leukocytes; Barr bodies in 90% of the buccal cells) is considered by many to result from its orientation in a given nucleus. The extreme lobing of the polymorphonuclear leukocyte nucleus could affect visualization of the drumstick.

EVIDENCE FOR THE LYON HYPOTHESIS. Excellent support for the Lyon hypothesis comes from chemical studies of individuals known to be of different genotypes for sex-linked genes, such as the one for hemophilia. For example, normal homozygous women (HH) and normal men (HY) have the *same* amount of antihemophilia protein in the blood, even though the gene dosage is different. The female carrier (Hh) is almost always without symptoms of the disorder, but blood analysis shows the concentration of AHF varies from one carrier to the next. A few of them have about the same amount of the clotting factor as does any normal male or homozygous female, but there are a few in whom the essential substance is very low. Most carriers have about one-half that of the average person. This is precisely what one would expect on the basis of the Lyon hypothesis, which predicts that it is by chance which X chromosome remains inactive in a given cell. Consequently, in a carrier female (Hh), approximately one-half of the cells on an average should have the maternal

X with the normal allele operating (Fig. 5-21). In the rest of the cells, the paternal X with the defective allele is being expressed, and so only one-half the normal amount of the AHF is produced. In a small number of cases, however, by chance the majority of cells will have the maternal chromosome operating. An almost normal level of the antihemophilia factor will be present as a result. Likewise, a few carriers by chance will have the paternal X active in most of the cells and consequently a low level of the factor.

Another sex-linked recessive factor is known to cause a deficiency of the enzyme, glucose-6-phosphate dehydrogenase, which plays a role in the metabolism of glucose. The effect of the deficiency in the red blood cells is not usually of any consequence under ordinary conditions (see Chap. 21). It has been possible to test for presence of the specific enzyme in any given cell. Results of tests on known heterozygotes have shown that in these females, a given cell either has the enzyme activity or lacks it completely. Many other examples could be cited in which two different populations of cells have been demonstrated in females heterozygous for certain other sex-linked recessives. All such examples give very strong support for the idea that the X with the normal allele operates in some cells but that it is in an inactive state in others (Fig. 5-22). Various other lines of evidence have

FIG. 5-21. X inactivation and the AHF. A female carrier of hemophilia may arise if a sperm carrying an X with the recessive unites with an egg carrying an X with the dominant allele responsible for the production of AHF. Both X chromosomes are active in each cell up to the time of the blastocyst when one of them becomes inactive (*arrow A*). The inactive X may be visualized as condensed chromatin, a Barr body, or sex chromatin. It is chance whether the maternally derived X or the paternally derived one remains active. On the average, a carrier will have one-half of her cells with an active X carrying the dominant, and so one-half the normal amount of AHF will be present. In a smaller number of cases (*arrow B*), the maternal X will be active in most cells, and the amount of AHF will approach that of a normal homozygote or of a normal male. Conversely (*arrow C*), in a few carriers the paternal X will be active in most cells, and the amount of AHF will be much lower than the normal.

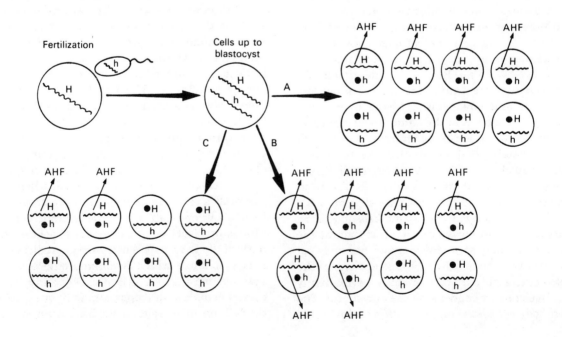

also confirmed the premises of the Lyon hypothesis and lead to the conclusion that it is basically correct. The devices to compensate for the different dosages of X-linked genes in mammalian males and females have probably been perfected through natural selection along with the evolution of the X and Y chromosomes to ensure genic balance in both sexes.

As a result of deactivation of one X chromosome, both sexes would in effect be hemizygous in any one cell for genes on the X. It must be noted, however, that the deactivation of one X in the XX female is probably partial and not entire. This is suggested by the observation that the XX female is quite distinct from the Turner female who is truly hemizygous for genes on the X. Moreover, it is known that two X chromosomes are essential for female fertility. Infertility results in those cases where just a portion of one X chromosome is missing. The observations indicate that at least some segment of both X chromosomes remains active in XX females and that the devices which operate to keep large parts of one X permanently inactive nevertheless permit some genetic material to remain active in the same X chromosome.

INCOMPLETE SEX LINKAGE. In addition to X-linked genes, there are others to be considered in relationship to the sex chromosomes. We have defined sex-linked genes as those which are found only on the X chromosome. We have also noted that the X and the Y chromosomes pair at first meiotic division, which implies that there may be some homologous segments present in each. In the fruit fly, there is a locus named "bobbed" which affects the length of the bristles. It is located at the end of the X chromosome near the centromere. A homologous locus, however, also occurs on the Y. This means that a female fly has the bobbed locus present twice (one locus on each X), and so does the male (one locus on the X, the other on the Y). So a male, like a female, can be heterozygous or homozygous at this locus. The wild allele (+) is needed for long bristles; its recessive allele bb can result in short ones. Alleles such as these are termed *incompletely sex linked,* to mean that they occur on both the X and the Y, in a region of homology between the X and the Y chromosomes. This is in contrast to sex-linked alleles which occur only

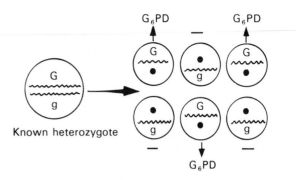

FIG. 5-22. Glucose–6-phosphate dehydrogenase (G_6PD) and red blood cells. Single red blood cells of carriers of the sex-linked recessive (g) for absence of enzyme can be tested. Any one cell either contains the enzyme or lacks it entirely. This indicates that the X which carries the allele for enzyme production (G) is active in some cells but is inactive in others.

on the X. The cross diagrammed in Figure 5-23 brings out some of the unique features of a trait which is inherited on the basis of an incompletely sex-linked gene.

The existence of incomplete sex linkage in humans and other mammals has been debated by students of human cytology and genetics, most of whom consider it unlikely. Instead of extensive areas of homology on the X and Y chromosomes, it would appear to be more advantageous to preserve the genes on the X as a single unit as well as those on the Y as a different unit. This would then help to insure an orderly segregation at meiosis of any genes on the X which influence normal female development and any on the Y which are essential for male development. No advantage would be gained if the X and the Y chromosomes possessed large homologous segments, for this would mean that crossing over could occur freely between the X and the Y at meiosis. As a result of exchange of genetic regions between the two sex chromosomes, genes for maleness and those for femaleness could become distributed on *both* the X and the Y (Fig. 5-24). Even if the number of loci for testis development on the Y is small (possibly only one), free exchange between X and Y would still interfere with clear-cut segregation of sex chromosomes and genes for sex determination. Therefore, there would be value to any mechanism which would prevent this, for it would help to ensure regular sex determination among the offspring. It seems likely that during the evolution of the mammalian sex-determining mechanism, natural selection would favor any changes which would make the X and the Y chromosomes less homologous, that is, more unalike. The end result in this direction would be chromosomes of two distinct types which would guarantee a 1:1 sex ratio. Since there would be few genetic segments which could occur on both the X and the Y, there would consequently be few incompletely sex-linked genes. Any homology which does remain between the X and the Y may permit the pairing of the sex chromosomes which is seen at first meiotic prophase of spermatogenesis. Crossing over between the X and the Y in the male could be prevented by the condensation of the X which appears heteropyknotic at the early stages. The pairing itself may be necessary for the orderly behavior of the sex chromosomes at first metaphase and their even separation at first anaphase. All the features taken together would help preserve the 1:1 sex ratio. Although the possibility of finding incompletely sex-linked genes in mammals cannot be discounted, their number would probably be very low.

HOLANDRIC GENES. There must be a large portion of the Y chromosome in the males of most species which is not homologous to the X. In *Drosophila,* this part of the Y contains genes required for sperm motility. Any gene whose locus is found on the Y chromosome alone is said to be *holandric.* Such a Y-linked gene is passed directly from a male parent to his sons. In mammals, the Y chromosome, as noted previously, is required for development of the testes. There is evidence from studies of individuals with anomalies of the Y chromosome that only a single testis-determining locus is present and that it is located on the short arm of the human Y, close to the centromere. The Y chromosome is also known to carry a locus which is responsible for the production of a certain antigen, the H-Y antigen, in the human, the mouse, rat, and guinea pig. The locus responsible for the H-Y antigen, a histocompatibility antigen, may actually be the same as the locus for testis determination. Except for this locus (or these two loci) there is no good evidence that the Y carries any other genes. In the human, the inheritance pat-

FIG. 5-23. Incomplete sex linkage in *Drosophila.* The locus "bobbed" which affects bristle length is found on both the X and the Y chromosomes. In a cross between a heterozygous male, carrying the dominant wild allele on the Y, and a female homozygous for the recessive allele, the normal trait is passed from father to sons. All of the daughters will show the mutant phenotype. This is to be contrasted with a sex-linked gene. If the locus were on the X chromosome alone and not on the Y, a crisscross pattern would occur.

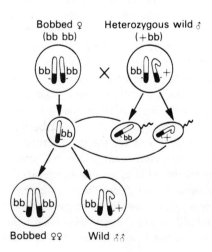

Bobbed ♀   Heterozygous wild ♂
(bb bb)   (+bb)

Bobbed ♀♀   Wild ♂♂

tern of certain traits not associated with the testes or with the H-Y antigen have been considered holandric, but other interpretations now seem more likely.

SEX-INFLUENCED ALLELES. Many species are known in which the manner of expression of certain alleles is affected by the sex of the carrier. Such factors are not necessarily located on the sex chromosomes; most of them are actually autosomal. Among these are various alleles whose expression of dominance depends on the sex of the individual. For example, in certain breeds of spotted cattle, the colored regions on the body may be red or mahogany, a deep reddish brown. When a mahogany female is crossed with a red male (Fig. 5-25A), the results are red-spotted females and mahogany-spotted males. This crisscross pattern suggests a sex-linked recessive in which the female is double recessive and the male hemizygous. However, if the red females are mated to their mahogany brothers, we see that this is not so. Among the male offspring, mahogany and red occur in a ratio of 3:1. The two phenotypes also show in the females but the ratio is the reverse: three red to one mahogany. We can recognize the characteristic monohybrid ratio within the females as a group and within the males as a group. The results suggest that the allele for mahogany (M) is dominant in the male and recessive in the female, whereas its allele for red (m) is dominant in females and recessive in males (Fig. 5-25B). We could just as well have written "M" for red and "m" for mahogany. It makes no difference as long as we keep in mind the distinction in dominance between the sexes. Therefore, the heterozygote, "Mm," will be mahogany if male and red if a female. Alleles whose degree of dominance is determined by the sex of the individual carrying them are called *sex influenced*. Many such examples are known in animals. Another is the presence or absence of horns in some breeds of sheep, where the horned condition behaves as a dominant in males but a recessive in females; the hornless state is dominant in the female sex but recessive in the male.

A trait which is close to most of us seems to be sex influenced; this is pattern baldness in humans. A pair of alleles seems to be involved,

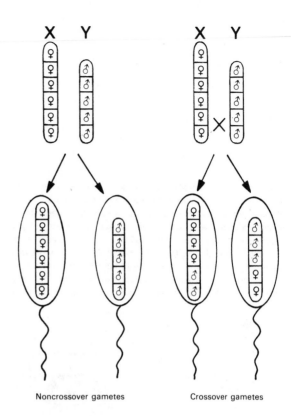

Noncrossover gametes   Crossover gametes

FIG. 5-24. Effect of crossing over between X and Y chromosomes. In mammals, any device to prevent crossing over between the X and Y would be advantageous. If the two sex chromosomes do not exchange segments (left), any genes for maleness and those for femaleness stay together as a unit. If the X amd the Y are free to exchange regions (right), genes for both sexes will become intermingled. This could upset the mechanism for orderly sex determination. Evidence indicates that the Y contains only one locus for testis development.

"B" for bald and "b" for nonbald. Simple monohybrid inheritance occurs, but the allele for baldness (B) behaves as a dominant in males and a recessive in females. Therefore, a person of the genotype "Bb" will be bald if male but nonbald if female. Figure 5-26 shows what is to be expected from a cross of two heterozygotes, a nonbald woman and a bald man. The gene for pattern baldness interacts with other genes which affect the presence of hair on the head. The particular pattern in which the hair is lost, as well as the time of onset, entails the interaction of many genes and the environment. In the last chapter, we encountered several examples of the importance of the environment on **the expression** of the genotype. We might expect

FIG. 5-25. Sex-influenced inheritance in cattle. *A.* A cross of a mahogany-spotted female and a red-spotted male produces an F$_1$ of red-spotted females and mahogany-spotted males. The F$_2$ results indicate the locus involved is not sex linked but that expression of the genes for coat color is influenced by an individual's sex. *B.* The factor for mahogany (M) behaves as a dominant in males and a recessive in females. Its allele for red color is dominant in females and recessive in males. Therefore, heterozygotes (Mm) will differ in color depending on their sex.

the internal environment to exert effects which are as pronounced as those outside the body. And such is difinitely the case. The dominance difference which is typical of sex-influenced alleles is mainly the result of hormonal interaction with the genotype. In the case of baldness, the male hormone is responsible for the expression of the allele B when it is present in just a single dose. The involvement of the hormone is clearly shown in cases where males must take testosterone for some medical reason. This raises the level of the hormone and can result in hair loss in a man who had not yet expressed the trait. Conversely, therapy may prescribe the taking of female hormone by men with a consequent decrease in the amount of male hormone. In these cases, a balding man may start to resume hair growth. However, an increase in breast size

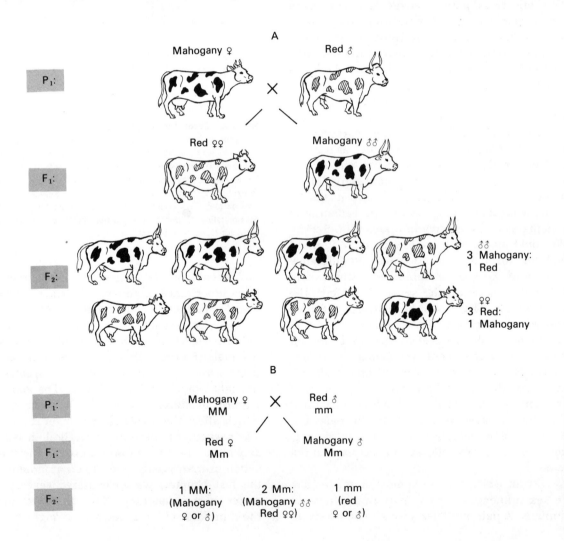

typically ensues, an accompaniment which may discourage thoughts of ingesting female hormone as a device to grow a good crop of hair.

External environmental factors can also cause baldness: scalp infections and contact with radiations or certain chemicals. These agents can mimic the effects of genetic baldness to the extent that the individuals are phenocopies, even though the allele for baldness is absent from the genotype. However, the phenocopy may show good hair growth after therapy or removal of the causative agent. On the other hand, the allele "B" cannot be removed from the man whose baldness stems from his genetic constitution. He will not be able to grow hair on his head as long as his male hormone is there to interact with the genes he has inherited.

SEX-LIMITED GENES. The effects of the internal environment are perhaps most evident in the case of *sex-limited genes*, so called because their expression is confined or limited to just one sex. Among these factors are those responsible for the obvious secondary sex characteristics. Recall that the XY condition can trip the development of the rudimentary gonads of the embryo in the male direction, whereas the XX genotype permits ovaries to develop. Once the gonads, male or female, mature they will produce their respective hormones. These in turn will interact with the genotype, not only to influence the degree of dominance of certain alleles, but also to suppress completely the expression of many others. When we think of secondary sex characteristics, such as breast development in women, it rarely occurs to us that genes for this characteristic have come from both parents, male as well as female. Similarly, the kind of beard which a male will develop is determined equally by inheritance from his mother and his father. This means that both sexes carry genes for breast development and for beard growth. That this is indeed the case is illustrated by many clinical conditions. The bearded lady is one of the most familiar examples. The genetic potential for beard development lies dormant in women and requires the presence of the male hormone to trigger it to expression. The adrenal gland normally produces some male hormone in both sexes. An adrenal tumor can cause an excess production of male hormone in a woman and allow the

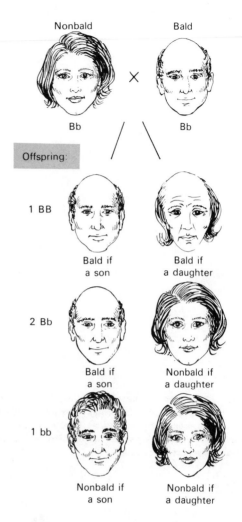

FIG. 5-26. Pattern baldness in humans. The expression of the allele for pattern baldness (B) is influenced by the sex of the carrier. In males, it is expressed in the heterozygote, whereas the heterozygous female expresses the allele for the nonbald condition (b). For a male to remain nonbald, he must be homozygous for the condition. A female will become bald only if she receives the bald-determining allele (B) from both parents, and the onset of hair loss cannot occur until later in her life.

genes she carries for growth of beard to express themselves.

The expression of the sex-influenced trait, pattern baldness which we have just discussed, is affected in still another way by the hormonal environment. Any genetic fctors which determine an early onset of hair loss are limited in expression to males (see Fig. 4-16, last chapter). Therefore, the appearance of hereditary baldness among young women is normally prevented by the limiting effect of the internal environment.

FURTHER ASPECTS OF THE HORMONAL ENVIRONMENT. We have already noted that the taking of estrogen by men can cause breast development, a phenotype usually limited in expression to females. However, it is not simply the presence of a specific hormone which is required for each sex-limited gene to come to expression. The absence of a hormone may permit some genes to be expressed. Although the female hormone is needed for breast enlargement, it is not required for female voice. The latter will develop in the absence of the male hormone. If a male is castrated before puberty, thus removing the main source of male hormone, he will develop neither a deep male voice nor a beard. Both of these attributes require the hormone. Today we read of cases of persons who have undergone "sex changes" after surgery. A man who undergoes such a transformation will always remain genetically male. Having only one X chromosome, such a person can never show a Barr body or a drumstick, because there is no extra X to be deactivated. The individual will be a castrate, but beard growth has been triggered, as has the development of the Adam's apple, also a sex-limited trait. These can be altered only by surgical and cosmetic means. The voice will remain masculine. Breast devel-

FIG. 5-27. Sex-limited inheritance in the fowl. The dominant H causes the male to have hen feathering. Its recessive allele h is responsible for cock feathering in males. Females, however, will not express the allele h. They will be hen feathered regardless of the presence or absence of allele h.

H = Hen feathering in ♂
h = Cock feathering in ♂

Hh
Hen feathered hen

×

hh
Cock feathered cock

Hh
Hen feathered hen

hh
Hen feathered hen

hh
Cock feathered cock

Hh
Hen feathered cock

opment will require continual injections of estrogen. The female-to-male transformation also requires surgical procedures. Male hormones must be administered to achieve beard growth and the expression of other male secondary sex characteristics. However, the XX chromosome constitution remains unaltered and sex chromatin will continue to be present in the body cells.

Some of the most striking examples of sexual dimorphism are found among various groups of birds. The male often appears so different from the female that the novice bird watcher may classify them in separate species. Sex-limited genes are the basis for these striking distinctions. The common fowl is a group which exhibits variation in its plumage pattern. The hen-feathered plumage is quite distinct from the cock-feathered phenotype. Both hen and cock feathering may occur in roosters, but only the hen-feathered phenotype is expressed in females (Fig. 5-27). The allele for hen feathering (H) acts as a dominant in males, whereas its allele for cock feathering (h) behaves as a recessive. But the allele h cannot be expressed at all in females, which are always hen feathered regardless of the genotype.

Experiments with birds have provided a great deal of information on hormonal interaction with the genotype. For example, if the gonads are removed from male or female fowls so that a young bird is deprived of either sex hormone, it is found that the genotypes HH and Hh are expressed phenotypically as cock feathering in both sexes!! This indicates that the factor H prevents cock feathering if either the male or the female sex hormone is present. Its recessive allele, h, when homozygous, allows the cock feathering pattern to develop, but the female hormone will not permit the expression of that phenotype. Therefore, two birds of the genotype hh may show different kinds of plumage. It will be expressed phenotypically as cock feathering in the absence of female hormone but will give hen feathering in the presence of the female hormone. Once such interactions between the genotype and the internal environment are appreciated, certain practices in animal breeding no longer seem strange. For example, it is necessary for a cattle breeder to consider the genes for milk production carried by his bulls when he is interested in producing cows with a high milk yield.

Natural selection has operated throughout countless years in sexual species to perfect a mechanism which ensures the formation of equal numbers of males and females. In mammals, the X and Y chromosomes with their strong potentials for femaleness and maleness, respectively, have become less and less homologous. This is a guarantee that a 1:1 ratio of male and female zygotes will arise. Once the genetic constitution (XX or XY) triggers the development of the specific sex, the appropriate hormonal environment becomes established. This will then

FIG. 5-28. Summary of the interaction of genetic constitution and the hormonal environment in sex determination.

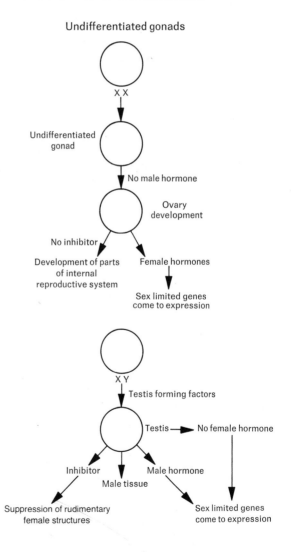

interact with a host of genes throughout the genotype, many of them autosomal, to permit the development of the suitable secondary sex characteristics (Fig. 5-28). We see that the potential set by a normal genotype for the formation of a fertile individual showing typical secondary sex characteristics can be realized only in the setting of a conducive environment, internal as well as external. Sex is but one of the many phenotypic characteristics of an organism which depends for its normal expression both on the entire genotype and the background in which this genotype is allowed to express itself.

REFERENCES

Avers, C. J. *Biology of Sex,* Wiley, New York, 1974.

Cervenka, J., D. E. Jacobson, and R. J. Gorlin, Fluorescing structures of human metaphase chromosomes. Detection of "Y body." *Am. J. Hum. Genet.,* 23: 317, 1971.

Ferguson-Smith, M., Chromosomal abnormalities II: sex chromosome defects. In *Medical Genetics* (pp. 16–26), V. A. McKusick, and R. Claiborne, (eds.). HP Publishing Co., New York, 1973.

Koo, G. C., S. S. Wachtel, P. Saenger, M. I. New, H. Dosik, A. D. Amarose, E. Dorus, and V. Ventruto, H-Y antigen: expression in human subjects with testcular feminization syndrome. *Science, 96:* 655, 1977.

Lyon, M. F., Possible mechanism of X-chromosome inactivation. *Nature New Biol., 232:* 229, 1971.

McKusick, V. A., The royal hemophilia. *Sci. Am., (Aug):* 88, 1965.

McKusick, V. A., Human Genetics.[1] *Ann. Rev. Genet.,* 7: 435, 1973.

Mittwoch, U., *Genetics of Sex Differentiation.* Academic Press, New York, 1973.

Ohno, S. Evolution of sex chromosomes in mammals. *Ann. Rev. Genet., 3:* 495, 1969.

Ohno, S., U. Tettenborn, and R. Dofuku, Molecular biology of sex differentiation. *Hereditas, 69:* 107, 1971.

Rushton, W. A. H. Visual pigments and colorblindness. *Sci. Am. (March):* 64, 1975.

Wachtel, S. S., G. C. Koo, W. R. Breg, H. T. Thaler, G. M. Dillard, I. M. Rosenthal, H. Dosik, S. G. Park, P. Saenger, M. New, E. B. Lieber, and O. J. Miller, Serologic detection of Y-linked gene in XX males and XX true hermaphrodites. *New Eng. J. Med., 295:* 750. 1976.

REVIEW QUESTIONS

1.   In the human, the autosomally linked allele "A" is required for normal skin pigmentation. Its

recessive allele is associated with the albino condition "a." Several pairs of alleles affect color vision. One of the X-linked recessives responsible for red-green color blindness may be represented as p. Its dominant allele P is required for normal color vision. Write as much as possible of the genotypes of the following three persons:

A. A woman with normal vision and normal skin pigmentation whose father is a color-blind albino.
B. A normally pigmented man whose mother is color blind and whose father is albino.
C. A man with normal vision and normal skin pigmentation whose father is a color-blind albino.

2. Using the information in Question 1, show the different kinds of gametes which can be formed by the following three persons.

A. An albino woman who has normal vision but who carries the allele for color blindness.
B. A normally pigmented man who carries the allele for albinism and who is color blind.
C. A normally pigmented woman with normal color vision whose father was a color-blind albino.

3. In the human, a certain rare sex-linked recessive, "i," can result in a cleft in the iris of the eye. Its allele "I" is required for normal iris. What would be the results of the following three crosses?

A. A man with cleft iris and a woman who carries only the normal allele.
B. A woman with cleft iris and a man with normal iris.
C. A woman who is heterozygous for the iris character and a man with cleft iris.

4. The dominant allele, M, is associated with a certain form of migraine headache. Its recessive allele, m, is required in homozygous condition for absence of headache. This pair of alleles is autosomal. Using this information and that given in Question 3 on cleft iris, consider the following. A woman who suffers from no

apparent afflictions takes her daughter to a physician since the young girl is suffering from migraine. The doctor notices that the girl has a cleft iris. From just this information, write what the doctor knows about the genotypes of the girl, her mother, and the girl's father.

5. Using the information given above on cleft iris and migraine, give the expected phenotypic ratio among the children from the following cross: mmIi X MmiY.

6. A rare autosomal recessive, t, results in total color blindness. The dominant allele, T, is required for development of rods and cones. Suppose a red-green color-blind woman (due to the sex-linked recessive, p, noted in Question 1) is homozygous for the dominant autosomal color vision allele. Her husband is totally color blind but carries the dominant sex-linked allele.

A. Diagram the cross and give the expected F_1.
B. Assume a woman and a man having genotypes like the above F_1 have children. What types are to be expected and in what proportion?

7. In the human, a sex-linked dominant allele, R, produces a type of rickets, a skeletal defect. The recessive allele, r, is associated with normal skeletal development. Give the results expected from the following crosses:

A. A man with rickets and a woman who does not have the condition.
B. A woman who has rickets but whose father did not, and a man who does not have the condition.

8. In chickens, the Bar feather pattern is due to a sex-linked dominant allele, B, whereas its recessive allele, b, results in non-Bar. Since the female is the heterogametic sex in birds, what would be the results of the following two crosses? (Give the genotypes and phenotypes of the offspring).

A. A Bar hen and a non-Bar rooster.
B. A non-Bar hen and a Bar rooster whose female parent was non-Bar.

9. A. In mice, as in all mammals, the male is the heterogametic sex. Assume that a sex-linked lethal is present in a strain of animals and that this causes the death of the late embryo. How would this affect the sex ratio?

   B. Answer the same question if a sex-linked lethal were present in a strain of chickens.

10. In the human, a certain sex-linked recessive causes a child to exhibit very abnormal behavior in childhood. An assortment of bodily upsets occurs and the affected child dies. Only males suffer from this genetic disorder. It has never been reported in a girl. Explain.

11. In cats, the factor B (for black fur) is codominant to its allele (b) for yellow fur. The heterozygote is tortoise. The pair of alleles is sex linked. In the clover butterfly, the autosomal factor W (white) is dominant to w (yellow), but the expression of the dominant is limited to females. From a distance, you observe a tortoise kitten and a yellow one playing with a white butterfly. What can you say regarding the sex and the genotypes of the cats and the insect?

12. A noncolor-blind woman is carrying the sex-linked recessive, p, noted in Question 1 which can result in a type of red-green color blindness. Assume that nondisjunction of the X chromosomes occurred at first meiosis in an oocyte of this woman. Give the kinds of offspring which could result if one or the other of the possible exceptional eggs are fertilized by sperms from a noncolor-blind man.

13. Give the possible results if normal eggs resulting from a noncolor-blind carrier female (Pp) are fertilized by unusual sperms resulting from nondisjunction of the X and Y at first meiosis in a noncolor-blind male.

14. An X-linked dominant allele, H, is required for the production of an enzyme, HGPRT, normally present in human cells. Its allele, h, results in lack of the enzyme and the elevation of uric acid in the blood. Suppose 100 fibroblast cells

growing in culture are tested for this enzyme from persons of the genotypes given below. In each case, how many on the average would be expected to show the presence of the enzyme?

(1) HH (3) Hh (5) hh (7) HO
(2) HY (4) hy (6) HhY (8) HYY

15. A certain bull is considered a prizewinner on the basis of his musculature and other fine anatomical points. However, when he is mated with cows, the female offspring sired by him produce much less milk than their mothers. Explain.

16. Consider pattern baldness in the human to result from the expression of the autosomal factor (B) as a dominant in males and a recessive in females. Its allele, b, (nonbald) behaves as a dominant only in females. What would be the results of the following matings?

A. A nonbald man and a nonbald woman who is heterozygous for the pair of alleles.
B. A nonbald man and a woman who became bald later in life.
C. A bald man whose father never became bald and a nonbald woman who is a homozygote.

17. Consider the alleles for baldness in Question 16 along with the sex-linked alleles which affect color vision, P and p (Question 1). Give the results of a mating between a nonbald color blind man and a nonbald woman with normal vision whose father was color-blind and whose mother became bald.

For each of the following, select the correct answer or answers if there are any:

18. Analysis of a buccal smear from a patient reveals the presence of three Barr bodies. The following is indicated:

A. The patient is probably suffering from Turner syndrome.
B. The patient is probably a gynandromorph.
C. The patient is probably an XXY Klinefelter male.
D. The patient will probably show three drumsticks in a blood smear.
E. The patient will probably show four X chromosomes in a karyotype analysis.

19. In the fruit fly, the expression of maleness depends on:

A. Interaction of the X chromosome and the autosomes.
B. Interaction of the X and the Y chromosomes.
C. Interaction of autosomes and Y chromosome.
D. Y chromosome only.
E. Autosomes only.

20. A sex-influenced allele is one which:

A. Is found only on the regions of the X and the Y chromosomes which are homologous.
B. Is found only on the Y chromosome.
C. May be found on autosomes.
D. Will express itself in only one sex.
E. Expresses itself in either sex, but its dominance depends on the hormonal environment.

6

PROBABILITY AND ITS APPLICATION

CERTAINTY VS. PROBABILITY. When the genetic counselor is asked his opinion on a specific problem, he may be unable to give an exact "yes" or "no" reply. The skilled adviser understands that frequently an unqualified response is impossible from the information at hand. However, he can often reply in terms of probability and give the chances between 0 and 100% that a certain event will take place. Starting with extremely simple examples, let us examine the reasoning behind the answers to various questions concerning a few couples and their prospective children.

"When a couple's first child is born, what are the chances that it will be a human being?" Obviously, a clear-cut answer can be given, inasmuch as there is no doubt that the child will be anything other than human. Letting "p" represent the probability for the birth of a human, we can say that the probability here is 1 (representing 100%) or that $p = 1$, because this event is a certainty.

If you had been asked the equally absurd question, "What are the chances that the first offspring of this couple will be a cat?" a precise answer again can be given. The probability of such an occurrence is 0. We can let "q" symbolize

148

the probability of the alternative event, the arrival of a cat, and say that q = 0. Thus, "p" and "q" simply represent alternative probabilities, which in this case have the respective values of 1 (for the occurrence of a human) and 0 (for the arrrival of a cat).

Very often, the probability of some happening lies between 1 and 0 (that is, between 100% and 0). If the couple had asked you the chance of their first child's being a boy, you could not answer with certainty but, undoubtedly, you would not hesitate to offer the correct reply, "1/2." In this instance, p = 1/2 (or 50% or 0.5). It is equally obvious that the chance of a girl is also 1/2. This alternative, the birth of a girl, can be represented by "q." Therefore, we can say that q = 1/2. Note that if there are only two alternatives (p and q) to a particular event, then p + q must equal 1. This is quite obvious in this example, because *either* a male *or* a female must be born at any one birth, and the sum of the separate probabilities, p + q (1/2 + 1/2), equals 1. This important point will be considered in more detail further on.

Let us next suppose that a man and his wife are concerned about the arrival of an albino child among their offspring because each of them has an albino parent. Allowing "A" to symbolize the dominant for normal skin pigmentation and "a" to stand for its recessive allele for albinism, we can represnt these two people as "Aa," for each one must be a heterozygote carrying the recessive allele. We are now in a position to answer questions about the appearance of albinism among the children of this couple in terms of probability.

"What are the chances that the first child born to these two carriers will be albino?" It is quite obvious that in this case we are considering a simple monohybrid cross: Aa × Aa. The genotypes of the offspring would occur in a frequency of 1AA: 2Aa: 1aa, giving an expected phenotypic ratio of three normal to one albino. Therefore, the chance of an albino being born at any birth is equal to 1/4, and, consequently, the probability of a normal child is 3/4. We can let p = 3/4 and q = 1/4. (We could have chosen "p" to stand for the chance of an albino and "q" for the normal. It makes no difference whatsoever as long as we are consistent in any one problem.)

Although these examples may seem trivial,

they are presented to emphasize the fact that we deal in probabilities much of the time without being aware of it. There is a tendency to forget even the simplest answers when a basic question is phrased somewhat formally in terms of probability. We might hesitate if asked, "What is the mathematical probability of an albino being born to two heterozygous parents?" Yet we would probably give the quick response, "1/4," if the question were stated in more familiar terms.

COMBINING PROBABILITIES. Also without knowing it, we frequently combine probabilities, a most important and critical operation in situations dealing with chance. It is often essential to contemplate the probability of one event happening in relationship to a second or third event. In other words, we rarely consider the chance of just one thing happening alone without examining the influence which another event may have upon it. The combining of probabilities is extremely simple, and if the following examples are grasped, little confusion should result in future problems.

Suppose the above couple asked the ridiculous question, "What are the chances that our first child will be *either* a boy *or* a girl?" As mentioned previously, it is obvious at any birth that any one child *must be* one or the other. If it is not a boy, then it must be a girl and vice versa. That is to say, it is *certain* that the child will be *either* a boy *or* a girl. The probability is 100%, and we say that p = 1. Without thinking, we have combined two separate probabilities. In arriving at the answer of one, we have just taken the probability of a boy (p = 1/2) and added it to the probability of a girl (q = 1/2) to get 1. We are simply acknowledging what was said at the beginning concerning two alternatives to an event. In the illustration here, we are dealing with two alternative events which are *mutually exclusive*. The meaning of this expression is apparent in this case, because the birth of a girl would obviously exclude the birth of a boy at any one time. We call two separate events "mutually exclusive" when the occurrence of one of them at a particular time prevents or renders impossible the occurrence of the other. Similarly, if we toss a coin and a head appears, the probability of a tail showing is now 0. Therefore, when dealing with a set of just two mutually exclusive events, "p" and "q," the sum of their separate probabilities must equal one. This also holds for more than two, so that if we have three mutually exclusive events, "p," "q," and "r," then $p + q + r = 1$ and similarly for any number.

"What is the probability that the card you pull from a deck of cards will be the ace of clubs, the ace of hearts, the ace of diamonds, or the ace of spades?" Since the chance of pulling any one of them is 1/52, and because the events are mutually exclusive, the correct reply is "4/52 or 1/13," the sum of the separate probabilities. The chance of pulling any one of the remaining 48 cards in the deck is 48/52. We see again that the sum of the probabilities of all the mutually exclusive events equals one (4/52 + 48/52). To summarize: "The probability that either one or the other of a group of mutually exclusive events will occur at a given time is equal to the sum of the probabilities of the separate events."

Now assume that the couple we are discussing asks about the chance of their having first a boy and then a girl. This is a very different question from the one posed first; for now the events are *not* mutually exclusive but are independent. They are independent because the arrival of a child of one sex at a given birth does not affect the sex of the next or of any other children to be born later. In the same way, the appearance of a head after the toss of a coin does not affect what will show the next time the coin is thrown. So we can say that two (or more) events are independent if the occurrence of one of them at a given time does not affect the occurrence of the second at any other time. To answer the question put to us, we simply multiply $1/2 \times 1/2$ to give a chance of 1/4 that the first child will be a boy and the second one a girl in the precise order. This is so, because the chance that two (or more) independent events will occur together is the product of the chances of each one occurring by itself. Since we know that the probability of a boy alone is 1/2 ($p = 1/2$) and similarly for a girl($q = 1/2$), then the probability of their happening together is the product of their separate chances or probabilities. And it should be evident that this rule for combining probabilities can be extended indefinitely for more than two events. "What are the chances of a boy, a girl, a girl, another girl, and then a boy *in that order?*" The answer is quickly calculated by finding the product of each of the single probabilities of the five independent events in this group: $1/2 \times 1/2 \times 1/2 \times 1/2 \times 1/2 = 1/32$.

Many fail to realize that we combine probabilities whenever we use the Punnett square method in a simple genetics problem. For in the procedure, we do no more than multiply diagrammatically the separate chances of the different kinds of gametes uniting in all possible combinations (Fig. 6-1). With just this basic information in mind we can respond to a large number of questions. A quick response can be given to our couple if these two heterozygous individuals ask, "What are the chances that our first child will be a normally pigmented boy?" Since skin pigmentation does not influence the sex of the individual, we are concerned with two independent events and the chance of their occurring together. We know that the chance of a boy at any birth equals $1/2$ and that the chance of a child with normal skin is $3/4$. So the answer is $1/2 \times 3/4$, which amounts to $3/8$. By using the same logic, the chance of an albino girl is $1/4 \times 1/2$ which is $1/8$.

One very important point concerning independent events must not be overlooked. This is simply the fact that the probabilities remain the same *no matter* how many times any two (or more) may have combined or come together. For example, if this couple has had a boy and then a boy and still another boy, the chance of yet another boy at the next birth remains $1/2$. And the chance of a normally pigmented boy remains $3/8$, regardless of the number of times that this combination between skin pigmentation and sex may have occurred in the family.

What if we were asked the chances of *either* a normal boy *or* an albino girl at the first arrival in the family? Here we are dealing with mutually exclusive events, so that the answer is then $3/8 + 1/8 = 4/8 = 1/2$. It should be obvious that the sum of *all* the probabilities in this group would equal one: $3/8 + 3/8 + 1/8 + 1/8$ (the sum of the separate chances for a normal boy, a normal girl, an albino boy, and an albino girl).

THE SAME COMBINATION MAY ARISE IN DIFFERENT WAYS. Now let us turn to other simple problems which demand no more than a clear understanding of *what* is being asked. Failure to appreciate the meaning can cause an erro-

Gametes:

	A (p)	a (q)
A (p)	AA p^2	Aa pq
a (q)	Aa pq	aa q^2

$= p^2 : 2 \ pq : q^2 = (p + q)^2$
$= AA : 2 \ Aa : aa = 3 \ normal : 1 \ albino$

FIG. 6-1. Monhybrid cross. The Punnett square, which diagrams all the possible combinations of the gametes in a monohybrid cross, is really an expansion of the binomial $(p + q)^2$.

neous reply to the most elementary question. For example, "If a couple has three children, what is the probability that two of them will be boys and one will be a girl?" This is an entirely different question than one which asks, "What is the chance of having first a boy, then another boy, and finally a girl?" In the latter case, we would simply find the product of the separate probabilities ($1/8$). But in the former question, the *order* in which the three children will be born is *not* prescribed. To obtain the answer, a different procedure must be followed which makes use of the basic rules of probability which have just been discussed.

In a family of three, there are several ways that the children could arrive. Remember that the probability for a boy will remain $1/2$ at each birth ($p = 1/2$), as will the probability for a girl at each birth ($q = 1/2$). The different possibilities are shown in Table 6-1. The question asked the chances of two boys and one girl in any order. A glance at the table shows that there is not one but rather *three* set ways in which this can happen. The probability is not $1/8$, as it would be if we were interested in just one of the three possible orders. Instead, it is $3/8$, the sum of the three different ways in which two boys and one girl can arrive, *each one way* with a probability of $1/8$.

Consider the chances of three boys and one girl in any order in a family of four. Again we must acknowledge all the possible manners in

TABLE 6-1 Different ways in which children may be born in a family of three

	ORDER OF ARRIVAL
3 boys	p* p p = 1/8 or p³
2 boys and 1 girl	p p q† = 1/8 ⎫
	p q p = 1/8 ⎬ = 3/8 or 3p²q
	q p p = 1/8 ⎭
2 girls and 1 boy	q q p = 1/8 ⎫
	q p q = 1/8 ⎬ = 3/8 or 3pq²
	p q q = 1/8 ⎭
3 girls	q q q = 1/8 or q³

*p = probability for a boy = 1/2.
†q = probability for a girl = 1/2.

which four children can arrive by chance (Table 6-2). We see that four children may be born to a couple in 16 combinations and that there are four separate possibilities for three boys and one girl. These details demonstrate another fact that is often overlooked. Notice that we are simply combining probabilities in all possible manners while at the same time recognizing the various ways in which the same combination can result. There are six different ways of getting the combination 2 boys and 2 girls; there are four different orders for 1 boy and 3 girls, and so on. Looking back at the families of three and four

TABLE 6-2 Different ways in which children may be born in a family of four

	ORDER OF ARRIVAL
4 boys	p*p p p = 1/16 or p4
3 boys and 1 girl	p p p q† = 1/16 ⎫
	p p q p = 1/16 ⎪ = 4/16 or 4p3q
	p q p p = 1/16 ⎪
	q p p p = 1/16 ⎭
2 boys and 2 girls	p p q q = 1/16 ⎫
	p q p q = 1/16 ⎪
	p q q p = 1/16 ⎪ = 6/16 or 6p2q2
	q q p p = 1/16 ⎪
	q p q p = 1/16 ⎪
	q p p q = 1/16 ⎭
1 boy and 3 girls	p q q q = 1/16 ⎫
	q p q q = 1/16 ⎪ = 4/16 or 4pq3
	q q p q = 1/16 ⎪
	q q q p = 1/16 ⎭
4 girls	q q q q = 1/16 or q4

*p = probability for a boy = 1/2.
†q = probability for a girl = 1/2.

(Tables 6-1 and 6-2), we see that all the possibilities in each example form a binomial distribution. Actually, whenever the Punnett square technique is used to illustrate a monohybrid cross, a diagram of a binomial distribution results. This can be seen in the diagram of the cross between the two heterozygous individuals who are carrying the recessive for albinism (Fig. 6-1).

EXPANSION OF THE BINOMIAL AND ITS APPLICATION. By expanding the binomial, $(p + q)^n$, to a certain power, we can determine the probability of two events occurring in any order without going through the laborious procedure detailed above. It takes little effort to answer the question, "What are the chances of 3 boys and 1 girl occurring in a family of 4?" Inasmuch as this case concerns four individuals, the binomial must be expanded to the 4th power. The power to which the expansion is carried will always be the total number of trials or occurrences in which we are interested, such as five tosses of a coin, $(p + q)^5$, or the arrival of four children, $(p + q)^4$.

The binomial can be quickly expanded with the use of a few simple rules. Consider first just the exponents of the separate terms in the expansion of $(p + q)^4$ (follow Table 6-3 in the expansion):

1. In the first term, p is simply raised to the power of the expansion (n), in this case 4.

2. In the second term, q enters, and the power of p decreases by 1, producing p^3q. In each of the succeeding terms, the power of "p" will continue to decrease by 1, as the power of q increases accordingly to n, and p no longer appears.

3. The number of separate terms in the expansion will always be one more than the power to which the binomial is raised. Because that power here is 4, we consequently have 5 terms.

4. Note that the sum of the exponents of "p" and "q" in each term always equals "n" and is thus 4 in each of the terms in Table 6-3.

We must next derive the coefficients for each of the terms:

1. The coefficient of the first term is always 1, no matter what the power of the expansion; so here it remains simply p^4.

2. In the second term, the figure is just the same as the power of the expansion, and we have here $4p^3q$.

3. The coefficient of each succeeding term is readily determined by examining the term just preceding it. To obtain the value for the third term, we look at the second which has just been completed, $4p^3q$. We then multiply the exponent of "p" (which is 3) by the coefficient of the term (which is 4) giving 12. Next we divide this product by the number of the term, which is 2. This results in a coefficient of 6 for the third term, $6p^2q^2$.

4. Continuing in this manner, the fourth term becomes $4pq^3$ and the last one q^4. Actually, the coefficient of the last term, like the first, will always turn out to be 1.

The expanded binomial can tell us the probability of having 3 boys and 1 girl in a family of 4. We first allow "p" to represent the chance for the birth of a boy and "q" for that of a girl. (Again we could have let "p" represent the chance for a girl and "q" for a boy. It does not matter as long as we adhere to the same symbols in a given problem.) Looking at the expansion

TABLE 6-3 Expansion of $(p + q)4$ (follow text for details)

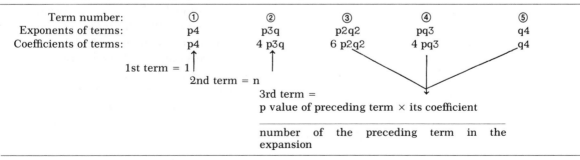

Term number:	①	②	③	④	⑤
Exponents of terms:	p4	p3q	p2q2	pq3	q4
Coefficients of terms:	p4	4 p3q	6 p2q2	4 pq3	q4

1st term = 1

2nd term = n

3rd term =

$$\frac{\text{p value of preceding term} \times \text{its coefficient}}{\text{number of the preceding term in the expansion}}$$

$(p + q)^4$ shown in Table 6-3, we select that term in which the exponents are 3 (for three boys) and 1 (for one girl). Since we have decided to let "p" stand for a boy and "q" for a girl, it is apparent that 3 boys and 1 girl are represented by the term $4p^3q$. Because the probabilities for a boy (p) and for a girl (q) at any one birth are both 1/2, we substitute these for p and q in the term: $4 (1/2)^3 \times 1/2 = 4/16 = 1/4$. This figure gives the chance of 3 boys and 1 girl occurring in any order in a family of 4.

Looking back at the detailed expansion $(p + q)^4$ in Table 6-2 we see that there are indeed four ways out of a total of 16 in which this can take place. By expanding the binomial, we have eliminated the need to construct a laborious diagram. Any question involving just two alternatives can be answered in this manner by expanding the binomial as long as we know the separate probabilities. Take the question, "In a family of 3, where each parent is genotype Aa, what are the chances of finding 2 albino boys and 1 normal girl?" The answer is determined by selecting the term $3p^2q$ from the expansion $(p + q)^3$. Let "p" stand for the probability at any birth of an albino boy ($1/4 \times 1/2 = 1/8$, which is the chance of two independent events, an albino and a boy, occurring together). Similarly, let "q" represent the probability of a normal girl ($3/4 \times 1/2 = 3/8$, the product of the separate chances for a normal individual and for a girl). Substituting in the terms $3p^2q$, we have $3(1/8)^2 \times 3/8 = 9/512$.

MORE THAN TWO ALTERNATIVES. Although we could continue like this for other problems, it would soon become evident that many situations involve more than two alternatives. Suppose the question were asked, "In a family of 3, what is the chance for 1 albino boy, 1 normal boy, and 1 normal girl in any order?" To arrive at this answer, it would be necessary to expand the trinomial $(p + q + r)^3$. This can become cumbersome when even more variables are being followed. Fortunately, we have a formula which enables us to expand binomials, trinomials, etc. in a general way which permits us to consider any number of events or variables:

$$\frac{n!}{s!\,t!} p^s q^t$$

Since this formula is merely a general expression allowing us to arrive at any particular term in an expansion, it can, of course, be used to answer simpler questions, such as the chances of 3 boys and 1 girl in a family of 4. In the formula, "n" stands for the total number of occurrences, in this problem, 4. The familiar symbols "p" and "q" represent the respective probabilities for a boy and a girl. The "s" and the "t" stand for the possibilities for "p" and "q" posed in this question, that is, 3 boys and 1 girl. Therefore s = 3 and t = 1. Substituting we have:

$$\frac{4!}{3!\,1!}(1/2)^3(1/2)^1 = \frac{4 \times 3 \times 2 \times 1}{3 \times 2 \times 1}$$
$$\times\, 1/8 \times 1/2 = 4/16 = 1/4$$

This is the same obtained from selecting the term $4p^3q$ after expanding $(p + q)^4$. Similarly, the chance of 2 albino boys and 1 normal girl in a family of 3 where the parents are both Aa becomes:

$$\frac{3!}{2!\,1!}(1/8)^2(3/8)^1 = \frac{3 \times 2 \times 1}{2 \times 1 \times 1} \times \frac{3}{512} = \frac{9}{512}$$

Since we can find the same answers by using the other method, why should this formula be kept in mind? It is particularly valuable when we are interested in more than two alternatives as in the question, "What is the chance of 1 normal boy, 1 albino boy, and 1 normal girl arriving in any order in the family of 3?" Here we have three possibilities to consider (p, q, and r), and we represent them in the formula letting "p" stand for the probability of a normal boy $(3/4 \times 1/2 = 3/8)$, "q" for an albino boy $(1/2 \times 1/4 = 1/8)$, and "r" for the normal girl $(3/4 \times 1/2 = 3/8)$. The symbols s, t, and u are their respective possibilities (1, 1, and 1). This gives the following:

$$\frac{n!}{s!\,t!\,u!}p^s q^t r^u = \frac{3!}{1! \times 1! \times 1!}(3/8)^1(1/8)^1(3/8)^1$$
$$= 3 \times 2 \times 1 \times \frac{9}{512} = \frac{54}{512} = \frac{27}{256}$$

The formula can thus be adapted for situations involving many variables. Whether you select it to solve a problem containing only two alternatives or whether you prefer to expand the binomial is a matter of choice.

CONSTRUCTING A PEDIGREE. One important application of probability in genetic studies is its use in the analysis of pedigrees. Examination of family histories is an important approach in the field of genetic counseling, a discipline of increasing significance today. Once the molecular basis of an inherited disorder is understood, it then becomes possible using chemical procedures to detect persons who are carriers of the defective gene. Genetic counseling of families having histories of genetic impairments can prevent the birth of afflicted children. At the same time, it can avoid needless worry by family members who do not know that they are actually free of the mutant gene.

The inheritance patterns of several human defects, such as color blindness, are well known. However, it must be borne in mind that certain families may exhibit their own peculiar pattern of inheritance. Because many genes affect a character, a trait inherited as a dominant in one family may follow a recessive pattern in another. Wherever possible, a pedigree should be assembled for the interpretation of a specific case history. A complete and accurate analysis of any lineage depends on a thorough understanding of elementary genetic principles and the laws of mathematical probability. In addition to these basic concepts, little more than common sense is needed to avoid certain common pitfalls often overlooked in hasty interpretations.

When constructing a pedigree, circles and squares are commonly used to represent females and males, respectively (Fig. 6-2). A mating between two individuals is represented by a bar

FIG. 6-2. Construction of a pedigree. See text for details.

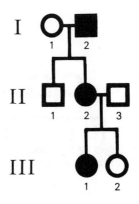

to which their offspring are attached. Roman numerals designate successive generations in the pedigree, whereas arabic numbers denote specific persons. Shading indicates the expression of the trait being followed. In the pedigree in Figure 6-2, persons I-2, II-2, and III-1 exhibit the trait under consideration.

EXAMINING A PEDIGREE. Before examining any pedigree, it is essential to realize that some histories do no provide enough information to permit definitive statements to be made. It is just as important to know that an interpretation is impossible in one case as it is to know that a precise answer can be given in another. These points are illustrated in the following discussion.

When presented with a pedigree, an attempt should be made to determine whether the trait follows a dominant or recessive pattern of inheritance. At the same time, we should ask if the expression of the trait seems to be influenced by the sex of an individual. Suppose we are just handed the pedigree in Figure 6-3A with no additional information and are requested to comment on it. We note that the male parent is expressing some trait and that one-half of the offspring, both sons, also show it. Can we say with any degree of certainty how the condition is inherited? The answer is emphatically, "No!" The reason becomes obvious if we allow "A" to represent any dominant and "a" its recessive allele. Assuming that the trait results from a homozygous recessive condition, then the genotype of the male parent must be "aa" and that of the female parent "Aa". An interpretation is quite possible on this basis. But another conclusion is equally feasible, if we assume the trait is due to a dominant allele and that the male parent is the heterozygote, "Aa"; the female parent would then be homozygous recessive for the normal condition. (Very often the novice will assume that a disorder in a pedigree results from a recessive condition and that the allele for normal expression must be dominant. There is no genetic basis for making such an assumption.)

All three males and none of the females in the pedigree express the trait; therefore, the suggestion might be made that the responsible gene is sex-influenced, sex-limited, or linked to the X or Y chromosomes. Since an argument can

FIG. 6-3. Small family pedigrees. See text for details.

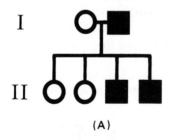

(A)

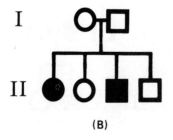

(B)

also be made for each one of these possibilities, it is apparent that nothing about the inheritance of the trait can definitely be stated from this pedigree standing by itself. Those who may argue that the sex of the individual is influencing the expression of the trait should realize that too few people are represented in the pedigree. Elementary probability tells us that theoretical ratios can become distorted due to chance factors when we are working with small samples. The fact that only males in this pedigree show the trait may reflect nothing more than this. Not enough people are involved to indicate anything about the relationship of the trait to sex. Therefore, a valid response to a question about this pedigree is that more information is in order if anything is to be said with any degree of certainty at all.

When studying pedigrees, we are frequently concerned with traits which are not very commonplace in the population. If the above pedigree involves some very rare condition, we might then suggest that the trait is due to a dominant allele. The chance is greater that a dominant rather than a recessive is involved, for if the trait is rare, genes for it would obviously be rare in the population as a whole. Therefore, the mother, *if* she is not related by blood to her husband, would probably *not* be carrying the rare allele. She would probably be homozygous for the normal condition, "aa," and the male parent would be the heterozygote, "Aa." But unless we know whether or not the parents are blood relatives, we still cannot be positive. If they are cousins, for example, the probability is then greater that they will have in common more alleles than any two people just taken at random. It has been reported that traits inherited on the basis of rare recessive alleles are expressed with higher frequency in families where first cousin marriages are common than in other families in the general population. Since most of us harbor in the heterozygous condition at least a few defective recessives, it stands to reason that the chances are greater for two of these factors to come together when the parents have obtained a proportion of their genetic material from a common source, a family pool of genes. So any recessive in a lineage will have a greater chance of coming to expression if the parents are related than if they come from unrelated stocks which contain their own *but different* defective, recessive al-

leles. So if the parents are related, we can make no decision about dominance in a pedigree such as that shown in Fig. 6-3A, a picture where one-half of the offspring expresses a rare trait that is also seen in one of the parents.

But even if we know that the parents are unrelated and that the trait is rare, we still could not be completely certain about a decision made on this pedigree because sex could be exerting an effect in some way. We are therefore forced to conclude that on the basis of a single pedigree such as this, we cannot answer without some qualifying statements. A definite reply may require histories of other families in which the trait is expressed or a pedigree encompassing several generations within the one family.

Can anything more definite be said about the small family represented in Fig. 6-3B? As neither parent is showing the trait, chances are that the allele inherited here is not a dominant; otherwise, we would expect the trait to show in one of the parents. In the case of this simple pedigree, we can feel fairly sure that an autosomal recessive is involved. If only a son and no daughter had expressed the trait, we could not be certain whether the allele is autosomal or sex linked. It should be apparent that the more persons involved in a pedigree, the firmer the basis for reaching a decision.

The phenomenon of penetrance could complicate the picture in Figure 6-3B. In the case of a dominant with incomplete penetrance, a trait may not show even when the factor for it is present in the genotype. In Chapter 4, we learned that the expression of an allele may be suppressed by the presence of certain modifying genes in the rest of the genotype or even by environmental influences. Although reduced penetrance is certainly a factor to consider in family histories, it will be ignored in our discussion in order to illustrate characteristic modes of inheritance of alleles with complete penetrance. The student, however, should appreciate the fact that pedigree analysis demands the recognition of many genetic and environmental factors which can alter a classic pattern of inheritance.

REPRESENTATIVE PEDIGREES. Let us examine a few larger pedigrees to see how much information can be derived from them (Fig. 6-

FIG. 6-4. Pedigrees for analysis. See text for details.

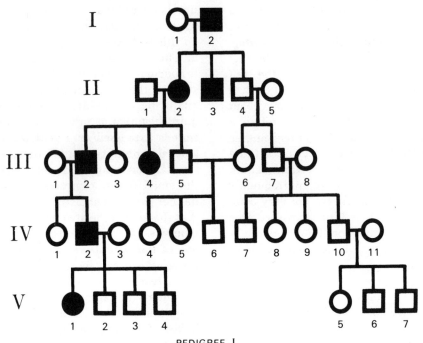

PEDIGREE I

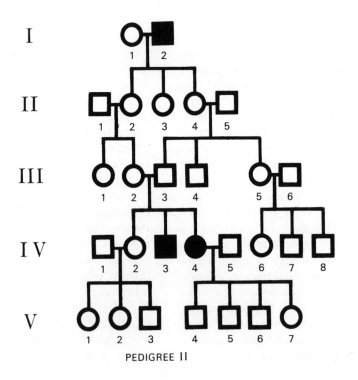

PEDIGREE II

UNDERSTANDING GENETICS

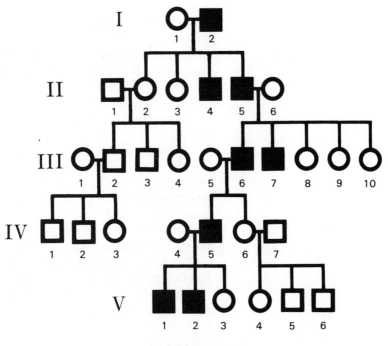

PEDIGREE III

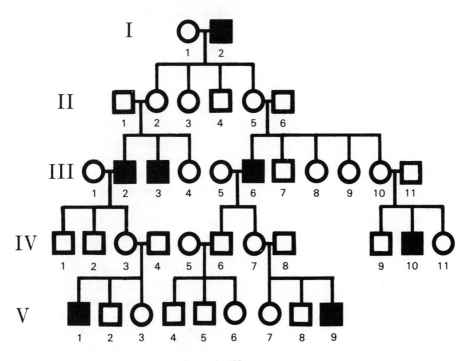

PEDIGREE IV

4). In all of these, we will assume that the trait is an uncommon one in the general population. In Pedigree I, several individuals, both male and female are affected. The major thing that should impress us is that the trait always shows in a family line if one of the parents expressed it. It never skips a generation in this sense. If neither parent shows the trait, then it does not appear among the descendants. This is particularly significant in this pedigree, because the lineage involves a mating between two first cousins (persons III-5 and III-6). An autosomal dominant is strongly indicated by the pattern seen in this family.

Pedigree II is quite different from the first one. We see that although the trait appeared in the first generation male, several generations are free of it. After skipping two generations, the trait reappears. A dominant allele basis seems unlikely. The pattern suggests an autosomal recessive, an interpretation strengthened by the fact that the parents of those affected in the fourth generation are first cousins. When one of the affected mates outside the family, the trait does not show again. This is to be expected, because the outside members probably would not carry the same rare defective allele.

Pedigrees III and IV should be carefully compared. We note in both that the only affected persons are males. Because many individuals are involved, chances are that the sex of the individual *is* playing a significant role in the expression of the allele. However, scrutiny of the two pedigrees will reveal important differences between them. In Pedigree III, *all* the males show the trait if their father expressed it. It is clearly going from father to son, never to daughters. Daughters of affected fathers do not seem to pass it down. The trait cannot be due to a sex-linked recessive, for in that case it would be passed to male offspring by way of a female parent. This is seen in Pedigree IV, which illustrates a sex-linked recessive pattern of inheritance. Here the trait does not go from a father to his sons. When an affected male has offspring, it is not seen among his children but appears among the sons of his daughter. An affected male may have an uncle who showed the trait (persons IV-10 and III-6). Elementary genetics tells us this is exactly the way a sex-linked recessive would be inherited.

PEDIGREES AND PROBABILITY. Now let us try to answer typical questions which could be asked about family members in the four pedigrees. In Pedigree I, suppose person V-4 is concerned about passing the trait down to his offspring and wants to know the chance of doing so. Since it has been determined that the expression of the trait depends on a dominant allele, we can answer safely the the chance is most unlikely. Obviously, he cannot be carrying the dominant because he would be expressing it, even though it is in the family. (We can ignore the highly unlikely complication of a mutation arising in this person's gametes or those of his mate.) If his affected sister (person V-1) asks the same question, again we can give an answer. We know that her genotype must be Aa. She displays the trait and therefore must be carrying the dominant allele, "A." But because her mother did not show it (and is thus genotype aa), she has received a recessive allele for the normal condition from her. If she married a normal man, he too must be aa. The cross is nothing more than a simple testcross: Aa × aa. Basic Mendelian laws tell us that one-half of her gametes will carry the defective dominant (A) and the other one-half the recessive (a). The offspring will consequently be Aa and aa in a ratio of 1:1. Therefore, the chance is 1/2 that the trait will show at the birth of any one child. Suppose this woman is married to a normal person and already has one normal child. What is the chance that the next child will be normal? Elementary probability tells us that it is again 1/2 and that the answer will remain "1/2" no matter how many times the question is asked concerning any one birth. Let us now put the question another way. "If there are two children in the family, what are the chances that both will be normal?" This is the same as asking the chance of first a normal child and then another normal in that order. The answer is the product of the separate probabilities: 1/2 × 1/2 which is 1/4.

Suppose the afflicted person (V-1) is going to marry another similarly afflicted person and wants to know the chance at any birth of a child with the trait. We are now considering the simple monohybrid cross: Aa × Aa. Since the affliction results from a dominant allele, the children will theoretically occur in a ratio of 3 afflicted to 1 normal. So at any one birth, the chance will always be 3 to 1 in favor of having a child with the trait. "What are the chances of 3 normal and 1 afflicted if there are 4 children in the family?" Allowing "a" to represent the probability for a normal child at any birth and "b" the probability for an afflicted, we select the term "$4a^3b$" from the expansion, $(a + b)^4$. Substituting 1/4 for "a" and 3/4 for "b," the answer is: $4 \times (1/4)^3 \times 3/4 = 12/256 = 3/64$.

The same reasoning applies to similar questions about any of the other pedigrees. We must simply remember the manner in which the trait is inherited and apply the laws of probability. An easily answered question but one frequently misunderstood arises in cases involving recessives. Suppose the normal woman in Pedigree II (person IV-2) asks the chance that her first child will be affected if she should next marry a man who shows the same recessive trait as her brother and her sister. To answer such a question, we must first consider the chance that woman IV-2 is a carrier. We do so by referring back to her parents. Neither of them expressed the trait, but each must be a carrier. Letting "A" stand now for the normal dominant allele and "a" for the defective recessive, the parents must both be genotype Aa. A mating between them is again nothing but a Mendelian monohybrid cross. The different genotypes and their frequencies among the offspring would be 1AA: 2Aa: 1aa. We know that the woman asking the question is *not* aa because she does not show the trait, but we cannot tell whether she is AA or Aa. However, we do know the probability of her being a heterozygote. Since the normal offspring occur by chance in a ratio of 1AA: 2Aa, the chance is thus 2 out of 3 or 2/3 that she carries the recessive. To answer her question, we now simply combine probabilities. Since her prospective husband expresses the trait, he is aa. *If* she is a carrier, the cross becomes Aa × aa, and the chance of any child being affected in such a cross is 1/2. So we simply multiply this 1/2 by 2/3 (the probability that she *is* a carrier) and find the answer, 1/3.

In cases like this, we sometimes know the frequency in the general population of normal appearing people who are carriers of a certain defective gene. Suppose we know that the recessive condition depicted in Pedigree II is carried by 1 out of 100 normal-appearing people. Now we can answer the question, "What are the chances that this woman (IV-2) will have a child showing the trait now that she is married to a normal man who is unrelated to her by blood?" Again we simply combine separate probabilities. We have established that the chance is 2/3 that she is a carrier. *If* both of them should prove to be carriers, the cross is Aa × Aa, and the chance of an affected at any birth is 1/4. The answer to the question is easily determined by finding the product of three separate probabilities: $2/3 \times 1/100 \times 1/4 = 2/1200 = 1/600$ (chance that she is a carrier times chance that he is a carrier times chance of double recessive in a monohybrid cross). So it is seen that many times a precise answer in terms of probability is possible from pedigree analysis, even when the exact genotype of a concerned family member is in doubt.

The role of sex as a factor in the expression of a gene does not complicate the picture in any way as long as the particular method of inheritance is kept in mind. In the case of a known holandric trait, answers are very obvious. Any affected man will have a chance of 1/2 of passing it to his offspring, because the chance is 1/2 at any birth that he will have a son (Pedigree III). The chance remains 0 that any daughter will show the trait or pass it down, because she has no Y chromosome.

When a sex-linked gene is involved (Pedigree IV), we can tell any male that he has *no* chance of passing down a recessively inherited trait if he himself does not express it. A sex-linked gene is found on that part of the X with no homologous region on the Y; therefore, the male must express any sex-linked allele he carries, because he is hemizygous for that portion of the X chromosome. What about the normal female (person V-3) in Pedigree IV who has an affected brother? What are the chances that she is a carrier for the sex-linked recessive? Neither of her parents exhibits the trait; so we know that the father carries no allele for it. The mother, however, must be a carrier, because one of her sons expresses it. The genotypes of the mother and father would be, respectively, Aa and AY. The mother will pass the defective allele to one-half of her children, but none of the daughters will express it. They have received a normal dominant from their father, and this will mask the sex-linked recessive. The chance that the woman with the affected brother is a carrier is 1/2, as a simple illustration shows (Fig. 6-5).

What are the chances the woman under discussion will have a child who shows the trait if she marries a normal, unrelated man? The separate probabilities are again determined and multiplied together. If the man does not show the trait, it does not matter whether or not he is related to his wife as far as any sex-linked recessive would be concerned. There is no chance that he is carrying the defective allele. If the woman *is* a carrier, the cross is Aa × AY, exactly as in the case of her parents. Thus there would be again a chance of 1 in 4 that a child will show it. But because there is only a probability of 1/2 that she *is* a carrier, we must take this into consideration and multiply the two separate probabilities: $1/4 \times 1/2 = 1/8$. If the woman had asked the chance of a son's being affected, the answer is 1/4 (the probability is 1/2 of passing it to a son, and this is combined with the 1/2 chance that she *is* a carrier).

Keep in mind that women may also express a sex-linked recessive trait. Such a woman must have a father who shows the trait and a mother who either shows it as well or who is a carrier for it. All of the sons of any affected woman must also be affected, because every son receives an X from his mother and a Y from his father. If the mother is showing the trait and is thus homozygous for it, each X will carry the recessive allele.

From the discussion presented so far, it should be possible to answer the additional Questions 1–12 on the four pedigrees at the end of the chapter.

CHI-SQUARE AND ITS APPLICATION. Once the elementary rules of probability are grasped, we can proceed to a few other basic but essential concepts of chance. The genetic investigator is almost always concerned with probability, and it must be considered in all crossing experiments performed in the laboratory. For example, assume you are following a cross between brown and red-eyed fruit flies. You may suspect that the exercise entails a simple monohybrid cross; so you predict for the second generation a phenotypic ratio of 3 red-eyed flies to 1 brown, assuming that red is dominant. Suppose 400 F_2 flies emerge. Only the most naive would expect 300

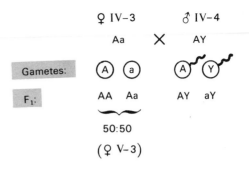

FIG. 6-5. Diagram of cross between persons IV-3 and IV-4 in Pedigree IV.

to be red-eyed and 100 brown. Instinctively, we realize that ideal figures or ratios are seldom obtained. Many factors, which we lump together under the heading "chance," enter the picture to prevent the realization of the exact hypothetical number.

If we scored 400 flies and found 290 to be red eyed and 110 brown, we might not be upset. These figures are close to a true 3:1 ratio, and intuitively we would expect some deviation due to chance. But suppose we had 275:125 or perhaps 250:150. Could these figures also be in agreement with a ratio of 3:1? Have additional chance factors operated to give an even greater distortion from the expected? The answer would be uncertain. We would not know where to draw the line, and confidence in our results would be somewhat weakened. We would feel the need to know with some degree of certainty the probability that the results are still compatible with our hypothesis. If the chances are that we are wrong in our assumption that the data agree with a 3:1 ratio, we would certainly want to know this in order to formulate another hypothesis. Or else we might reexamine the conditions of our experiment to ascertain what outside factors could be influencing our results. Clearly, it is imperative to know if data, although not ideal, are still in agreement with a hypothesis.

Such a dilemma can be avoided through the application of the chi-square (χ^2) formula. This test gives a way to determine the probability that the departures from the expected values which appear in our data are the result of chance factors. We thus have a stronger basis for accepting our original hypothesis or for rejecting it, whatever the case may warrant.

The χ^2 is very easily applied to numerical data; it is indispensable when deciding whether actual figures obtained through experimentation agree with those of an ideal ratio. Returning to the 290 red eyed and 110 brown-eyed flies, how can we determine whether the figures are compatible with a ratio of $3:1$? The values, 290 and 110, represent the observed numerical data, those actually obtained from an experimental procedure. As shown in Table 6-4, they are placed in the proper column, labeled "o" (for "observed")

Next to each observed value, we put the the expected (e). This value can be quickly determined. We have a total of 400 flies and are applying the χ^2 test to an expected ratio of $3:1$. Obviously, there are 4 parts involved (3 parts are red and 1 part brown). Dividing the total of 400 by 4 gives us 100, which represents the "1" of the $3:1$ ratio. Consequently, we would expect $300:100$ in an ideal $3:1$ ratio. We next determine how great the deviation (d) is between our actual results and the ideal. It is -10 and $+10$, respectively, in this example. In order to eliminate the minus sign, each deviation (d) is then squared (d^2). Each of these squared deviations is next divided by the expected value for its particular class (d^2/e). The value of χ^2, as indicated by the formula, is equal to the sum of all these, giving us 1.33 in our example.

What is the meaning of this figure, and how do we put it to use? It is now necessary to turn to a table of probabilities for the different values of χ^2 (Table 6-5). Note that the left column in such a table is labeled "Degrees of Freedom." The number of degrees of freedom is always equal to one less than the number of observed classes. The number of classes here is two (red and brown). Hence, the number of degrees of freedom is one. Some insight into the reason for this can be appreciated if we regard the situation in this way. We have a total of 400, and this number is distributed between the two classes. If we know how many of the total are included in one of the classes, we automatically know how many are in the other class. Similarly, having four classes and knowing the total as well as the amount in three of the classes, we must know what is in the fourth. So we have 3 degrees of freedom when we have 4 classes.

Returning to the table (Table 6-5), next to

TABLE 6-4 Derivation of a χ^2 value (follow text for details)

CLASS	o*	e†	d‡	d^2	d^2/e
Red eyes	290	300	−10	100	100/300 = 0.33
Brown eyes	110	100	+10	100	100/100 = <u>1.00</u>
					$\chi^2 = \Sigma d^2/e = 1.33$

*observed.   †expected.   ‡Deviation.

the degrees of freedom we see numbers corresponding to various values of χ^2. Our value is 1.33, and we see that it falls between the values of 0.455 and 1.64. Reading up, we note that this corresponds to a probability value between 50% and 20%, roughly 25%. It is essential to understand what this implies. A probability value of 25% tells us that in the case of an *exact* 3:1 ratio, chance factors would operate to give a deviation from the expected value at least as large as ours 25% of the time. In other words, when dealing with numbers the size of ours, we should expect a χ^2 as great or greater than the one we calculated in 25 cases out of 100 when we are truly dealing with a ratio of 3:1.

Does this mean we should reject the hypothesis of a 3:1 ratio? On the contrary, a χ^2 value indicating a probability of 25% is *not* considered significant. Statisticians have agreed that only when a probability of 5% or lower is obtained should a χ^2 be considered significant, that is,

large enough to cast doubt on the interpretation of the numerical data. (Return to the χ^2 table and make certain that you see the higher the value of χ^2, the lower the value for the probability.) Of course, the higher the probability that chance is operating to produce any deviation in the data, the more confident we will feel. However, usually there is no reason to reject an interpretation on the basis of χ^2 alone unless the probability is in the vicinity of 5% or lower.

The importance of the χ^2 cannot be overemphasized. However, it is critical to use the information which it provides correctly and to avoid common misconceptions. The following points should be kept in mind. Remember that because χ^2 is a test of numerical data, it must be used only with the actual figures themselves. It is erroneous to apply the method to percentages. Similarly, although we examine ratios by χ^2, we never apply it directly to ratios which have been obtained from the raw data. Wrong concepts

TABLE 6-5 Table of χ^2*

DEGREES OF FREEDOM	P = 0.99	0.95	0.80	0.50	0.20	0.05	0.01
1	0.000157	0.00393	0.0642	0.455	1.642	3.841	6.635
2	0.020	0.103	0.446	1.386	3.219	5.991	9.210
3	0.115	0.352	1.005	2.366	4.642	7.815	11.345
4	0.297	0.711	1.649	3.357	5.989	9.488	13.277
5	0.554	1.145	2.343	4.351	7.289	11.070	15.086
6	0.872	1.635	3.070	5.348	8.558	12.592	16.812
7	1.239	2.167	3.822	6.346	9.803	14.067	18.475
8	1.646	2.733	4.594	7.344	11.030	15.507	20.090
9	2.088	3.325	5.380	8.343	12.242	16.919	21.666
10	2.558	3.940	6.179	9.342	13.422	18.307	23.209
15	5.229	7.261	10.307	14.339	19.311	24.996	30.578
20	8.260	10.851	14.578	19.337	25.038	31.410	37.566
25	11.524	14.611	18.940	24.337	30.675	37.652	44.314
30	14.953	18.493	23.364	29.336	36.250	43.773	50.892

*Reprinted with permission of the Macmillan Publishing Co., Inc., from R. A. Fisher, *Statistical Methods for Research Workers* 14th ed., copyright © 1970 University of Adelaide.

may be avoided simpy by remembering to use the χ^2 test directly on the actual figures collected and not on anything derived from them.

Another note of caution concerns the sizes of the categories of classes being tested. Anyone would realize the absurdity of working with data which include only two items in one class and one in the other. Obviously, more counts are in order, because the larger the totals in each class, the greater the chance of reaching a valid conclusion. To prevent wrong interpretations the χ^2 should never be applied to any set of data where the number of individual counts in any class is less than five. Keep in mind that we do not consider it at all unusual to find human families in which the children are all the same sex. The occurrence of families with a run of 4 boys or of 5 girls doesn't shake our belief in the existence of a ratio of male to female that is close to 1:1. For we realize that the numbers involved in a human family are small and that chance can distort them from falling into an ideal ratio. Only when we score large populations can we hope to approach the expected. Remember that much of Mendel's success resulted from his choice of the garden pea, a species which provides large numbers of individuals for examination through several generations. The fruit fly, whose study has uncovered so many genetic principles, is an ideal genetic tool mainly for the same reason. The rapid advance of molecular genetics in the past decade has stemmed largely from extremely fine analyses of bacteria and viruses whose rapid reproductive rates provide numbers which cannot even be approached in higher eukaryotic forms.

An additional point also entails attention to class size and involves the application of a correction factor. The need for this factor may become clearer after reading the next chapter, but it can be appreciated at least in part at this time. In our genetic experiments, we usually classify flies or other organisms for different traits and assign them to distinct classes, such as red-eyes vs. brown and black body vs. gray, etc. In such cases, we are not involved with any continuous gradation of phenotypes from one extreme to the other, such as dark-red eyes through lighter red, pink, light pink, and finally down to white. We are concerned primarily with *discontinuous* distributions containing groups sharply distinguished from one another. However, a classification of many characteristics in living things does not lend itself to such precise distinctions. If we were to group 50 people into categories based on their weights, they would not fall into just a few sharp classes but rather into many, ranging from heaviest to lightest with most of the people falling somewhere in between the extremes. We could expect a *continuous* distribution. The various classes would not be sharply removed from one another, in contrast to red vs. white or black vs. gray, categories which are separated by big jumps. The relevant point is that the χ^2 formula has been developed from considerations of *continuous* distributions. So when the χ^2 test is used for discontinuous distributions, an error is introduced. This is especially so when dealing with classes of less than 10 figures, and it can also make a difference when working with just one degree of freedom. In borderline cases, failure to apply a correction factor called the "Yates correction" could result in an unwarranted rejection of a hypothesis. The correction is made simply by subtracting a value of 0.5 from each deviation and then proceeding as already described (Table 6-6). We see that the correction in this case made little difference. It can, however, play a role as stated above and should be applied routinely in all analyses involving only one degree of freedom, as well as in those where any one of the class values is close to 10.

One final point about χ^2 must be appreciated. Many fail to understand that the χ^2 does not tell us that a hypothesis is wrong or that it is correct. Rather, it simply indicates that the numerical data we have assembled either are or are not in support of the hypothesis. For example, assume we apply χ^2 to simple counts of the number of male and female flies resulting from an ordinary cross in *Drosophila*. Suppose our hypothesis is that male and female zygotes are formed in about equal frequency, and we expect roughly 200 males to 200 females. However, we actually obtain 250 females to 150 males. A χ^2-analysis will tell us that the data are not in agreement with a ratio of 1:1. The probability is extremely low that this is a distorted 1:1 ratio caused by chance factors. We have good reason to reject the idea of a 1:1 ratio, but the χ^2 *doesn't* tell us that the hypothesis about equal numbers of male and female zygotes is wrong. Instead of rejecting it indiscriminately, we should consider several nonchance factors in the conditions of the experiment which could have upset the results. Are we sure that the stocks of flies were free of sex-linked recessive lethals which could kill off more male than female larvae? Were the expected males mainly mutant types whose viability was decreased at certain temperatures? Was the experiment properly controlled? Perhaps we should not be hasty in rejecting the hypothesis but should repeat the experiment with other stocks of flies while controlling temperature and other environmental conditions more closely. We must not reject basic principles due to a misconception of what χ^2 can tell us.

We also see from this that χ^2 can support an incorrect idea. Suppose the hypothesis had been that sex in the fruit fly is not inherited on the basis of the familiar X-Y mechanism but rather in some fashion which favors an excess of females. Then the χ^2 would support the erroneous hypothesis based on the above results. Intelligent application of the laws of probability and χ^2 further the advance of genetic research by providing a method which permits discrimination between the sound interpretation and one that may mislead the investigator. Misuse of these valuable mathematical tools, on the other hand, can cloud the issue and slow down the progress of valuable efforts. Furthermore, naive ideas about probability can cause unnecessary anxiety when applied to counseling of human families.

TABLE 6-6 Application of correction factor to χ^2 derivation

CLASS	o*	e†	d‡	CORRECTED d	d²	d²/e
Red eyes	290	300	−10	−9.5	90.25	0.30
Brown eyes	110	100	+10	+9.5	90.25	0.90
					$\chi^2 = \Sigma d^2/e = $	120

*Observed.　† Expected.　‡Deviation.

REFERENCES

Bailey, N. T. J. *The Mathematical Approach to Biology and Medicine.* Wiley, New York, 1967.

Dixon, W. J., and F. J. Massey, *Introduction to Statistical Analysis,* 3rd ed. McGraw-Hill, New York, 1969.

Fisher, R. A., and Yates, F. *Statistical Tables for Biological, Agricultural and Medical Research,* 6th ed. Oliver & Boyd, Edinburgh, 1963.

Goldstein, A. *Biostatistics, an Introductory Text.* Macmillan, New York, 1964

REVIEW QUESTIONS

Questions 1–12 comprise an exercise on pedigree analysis based on the pedigrees shown in Figure 6-4. Allow "A" to stand for any dominant allele and "a" its recessive allele.

Pedigree I:

1. Write as much of the genotype as possible for all persons in the pedigree.

2. If person IV-4 marries a normal man from outside the family, what is the chance that at any birth an abnormal child will be born?

3. Answer the same question as the one above for person IV-5, assuming she marries her cousin, person IV-7.

Pedigree II:

4. Write as much of the genotype as possible for all persons in the pedigree.

5. What is the chance that person V-1 carries the recessive allele?

6. What is the chance of having an affected child if person V-1 marries an affected?

Pedigree III:

7. What is the chance that the first child will be affected if person V-1 marries person V-4?

8. Answer the same question as the one above if person V-3 marries person V-5.

9. What is the chance that the first boy will be affected if person V-2 marries a normal woman?

Pedigree IV:

10. Write as much of the genotype as possible for each of the following persons: II-2, II-4, II-5, III-2, III-9, III-10, and V-3.

11. What is the chance of having an affected son at any birth if person V-3 marries person V-9?

12. Answer the same question as above for persons V-6 and IV-10.

In the following questions, consider the normal skin pigmentation trait (A) dominant to albino (a). These traits are autosomal. The sex-linked trait, normal iris (I) is dominant to cleft iris (i). Also consider migraine headache to depend on a dominant allele (M) and the normal condition on its recessive allele (m). These alleles are not sex linked.

13. In a family of six children, give the probability that three girls and three boys will occur in this order: girl, boy, girl, boy, girl, boy.

14. In a family of four boys, what is the chance that the next child will be a boy?

15. If six babies are born in a hospital on January 1, what is the chance that:

  (1) they will all be girls?
  (2) they will be all boys?
  (3) they will all be of the same sex?
  (4) 3 will be girls and 3 will be boys?

16. Assume that a man and wife are each known to be heterozygotes, Aa, carriers of the recessive allele for the albino trait.

  A. What is the chance that the first child will be an albino girl?
  B. What is the chance that the first child will be a normal boy?
  C. What is the chance that the first child will be either an albino girl or a normal boy?

17. A. What is the chance that the couple in Question 16 will have children in this order: a normal girl, an albino girl, an albino boy, a normal boy, and finally, a normal girl?

  B. What is the chance that in a family of five children there will be three normal children and two albinos, in any order?

  C. If these two heterozygous persons already have three normally pigmented children, what will the fourth child be like?

18. You are told that an acquaintance of yours has two children. Your informant knows that one of them is a girl but doesn't know the sex of the other child. What is the chance that the other child is also a girl? Explain.

19. Two parents with normal skin pigmentation carry the recessive autosomal allele (a) for albinism. They have five children, two albinos and three with normal pigmentation.

  A. What is the chance that all the normally pigmented children are homozygous for the dominant allele?
  B. What is the chance that all of the above will be carriers?

20. If a woman who is normally pigmented has an albino brother and is married to an albino man, what is the chance that the first child will be albino? (Her parents are not albino.)

21. Suppose that two parents are genotype Mm and thus suffer from migraine headaches. What is the chance that:

  A. Their first child will be a girl with migraine and their second a boy without the disorder?
  B. In a family of four these parents will have three children free of migraine and one who suffers from it?

22. Suppose a woman with normal iris whose father had a cleft iris is married to a man with normal iris.

  A. What is the probability that a son will have cleft iris?

  B. What is the chance that the first child will be a boy with cleft iris?

  C. What is the chance that the first child will be a daughter who does not carry the recessive allele?

  D. What is the chance that a daughter will carry the recessive?

23. Assume that two married persons are di-hybrids for the skin pigmentation and headache characteristics (AaMm).

  A. What is the chance that they will have children in the following order? normal pigmentation and no migraine, normal pigmentation and migraine, albino with migraine.

  B. What is the chance that the first child will be an albino girl who doesn't suffer from migraine?

24. Considering the two dihybrid parents above in Question 23, what is the chance that if they have four children, two will be completely normal, one will be an albino with migraine, and one will be an albino without migraine?

25. Several pairs of chickens having slightly twisted feathers are crossed. They produce the following kinds of offspring:

Slightly twisted—74

Extremely twisted—33

normal—41

Submit these data to a χ^2 test for compatibility with a ratio of 1:2:1. What is the χ^2 value? What is the P value? Is this a ratio of 1:2:1?

26. Phenotypically normal, dihybrid fruit flies are testcrossed to flies showing both recessive traits, arc wing and black body. Subject the data given below to a χ^2 test on the hypothesis that the genes at the arc and the black loci are undergoing independent assortment. The results are:

$$\begin{aligned}
\text{wild} &\text{---}539 \\
\text{arc} &\text{---}659 \\
\text{black} &\text{---}712 \\
\text{black, arc} &\text{---}490
\end{aligned}$$

7

CONTINUOUS VARIATION
AND ITS ANALYSIS

CONTINUOUS VS. DISCONTINUOUS VARIATION. The basic principles of genetics are based on traits which are very well defined. Mendel's pea plants were either definitely tall or short; flowers were either red or white. Morgan's fruit flies also possessed many clear-cut phenotypic features which could be easily followed from one generation to the next (red vs. white eye color; short wing vs. long wing). Characteristics such as these (height in peas; eye color in the fly, etc.) are said to show *discontinuous variation*, because contrasting traits can be recognized and easily assigned to distinct categories. In contrast are those characteristics which exhibit *continuous variation,* in which the phenotypic differences slowly intergrade. Height, weight, or length and breadth of an organ usually describe a continuous distribution from one extreme to the other. No sharp distinctions are found among the phenotypes to permit the recognition of clear-cut classes. Moreover, extra efforts must be taken to control environmental effects before the genetic basis of such characteristics can be analyzed. And special methods are needed to *describe* the variation which is seen, because the expression of the characteristic intergrades from one individual to another. It is no wonder

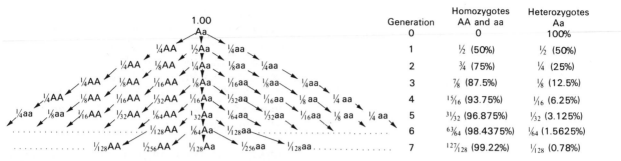

	Generation	Homozygotes AA and aa	Heterozygotes Aa
	0	0	100%
	1	½ (50%)	½ (50%)
	2	¾ (75%)	¼ (25%)
	3	⅞ (87.5%)	⅛ (12.5%)
	4	$^{15}/_{16}$ (93.75%)	$^1/_{16}$ (6.25%)
	5	$^{31}/_{32}$ (96.875%)	$^1/_{32}$ (3.125%)
	6	$^{63}/_{64}$ (98.4375%)	$^1/_{64}$ (1.5625%)
	7	$^{127}/_{128}$ (99.22%)	$^1/_{128}$ (0.78%)

FIG. 7-1. Effects of continued inbreeding. With continued self-fertilization, the percentage of homozygosity increases per generation. Note that the percentage of the heterozygotes decreases by one-half per generation as the percentage of homozygotes increases.

that the nineteenth-century geneticists who selected continuous variation for genetic study encountered great difficulties which obscured the underlying genetic basis. Even after the discovery of Mendel's paper and the demonstration of its relevance to different species, many biologists still believed that Mendelian inheritance did not apply to those characteristics exhibiting continuous variation.

Before 1900, statistical methods were developed to examine complex phenotypic features. The measurement of biological traits and the application of statistics to them forms the basis of the discipline of *biometry*. A mathematical approach is often of great value in the analysis of biological problems, but it may involve many pitfalls. (Recall the discussion on the use of χ^2 in the last chapter.) Using mainly statistical methods and lacking any knowledge of the fundamental principles of genetics, nineteenth-century biometricians often reached erroneous conclusions on the nature of continuous variation. This kind of variation is also referred to as "quantitative variation" and the characteristics which display it are frequently called "quantitative characteristics." These expressions recognize the fact that the variation which is shown can be measured and expressed in mathematical terms.

Johannsen was one of the first biologists whose keen insight helped to elucidate the nature of continuous variation. We have already mentioned (Chap. 4) his importance in distinguishing between "genotype" and "phenotype." Johannsen applied statistical methods to study continuous variation in plants. His work showed that mathematics can be very useful when properly applied to quantitative characteristics.

THE EFFECTS OF INBREEDING. Johannsen worked with beans and the inheritance of length, breadth, and weight of seeds in this plant. His observations and conclusions held far-reaching implications for the science of genetics. Johannsen was able to demonstrate clearly the distinction between the genetic and the environmental components which influence the expression of quantitative characteristics. Before discussing his observations, a few words are in order concerning the effects of inbreeding. The bean is normally a self-pollinated plant. Continued selfing over the course of generations renders the offspring more and more homozygous (Fig. 7-1). Self-fertilization, which can only take place in hermaphroditic organisms, is an extreme case of inbreeding, but it nevertheless demonstrates certain important facts. It is evident that the continued mating of individuals within a line or family increases the chances of bringing together hidden recessives. Hybrid vigor, also called *heterosis*, often accompanies heterozygosity; therefore, it will tend to decrease with continued inbreeding. The basis of the increased vigor associated with heterozygosity will be discussed in more detail in Chapter 16. Figure 7-1 follows only one pair of alleles through several generations of self-fertilization. However, it should be evident that intensive inbreeding will reduce the amount of variation in a population by bringing about a higher percentage of homozygosity for *all* pairs of alleles. Any organism which has sexual mechanisms but which is exclusively self-fertilized, as is true in some plants, goes through the motions of sex without receiving its main advantage: the production of a large assortment of genotypes and phenotypes upon which the force of natural selection can operate.

GENETIC VS. ENVIRONMENTAL VARIATION. Although continued selfing greatly increases the degree of homozygosity, Johannsen showed that a characteristic such as seed length will still display variation when the offspring from a single self-fertilized plant are studied. If one continually selects seeds of longer length *within* a highly inbred line, the same variation continues to be shown (Fig. 7-2*A*); the *average* value for length of seed does not change. Johannsen continued to select the longest and the shortest seeds within a pure line and found it was ineffective. The same average value persisted from one generation to the next. The variation seen within a pure line is thus environmental, because continued self-fertilization will make all the individuals within an inbred line genetically identical. However, different results were obtained when selection was applied to a *mixed* population. Such a population is composed of *many* self-fertilizing individuals, each of which represents a *different* pure line (Fig. 7-2*B*). In the population, there is thus genetic as well as environmental diversity. There are present various pure lines which have larger average seed values than do other lines in the population. This is an expression of genetic diversity among the lines. Choosing only the largest seeds in the *mixed* population will select out those pure lines whose genetic factors enable them to produce larger seeds than the average value for the *population*. This selection can therefore cause a shift in the average size of the seed, but within a line, selection for larger seed will remain ineffective. All the variation within the line is due to the environment. Johannsen's work showed the importance of the environment on complex characteristics, but it also demonstrated that environmental modifications could not be inherited. The concept of Weismann was upheld over the claims of the Lamarckians.

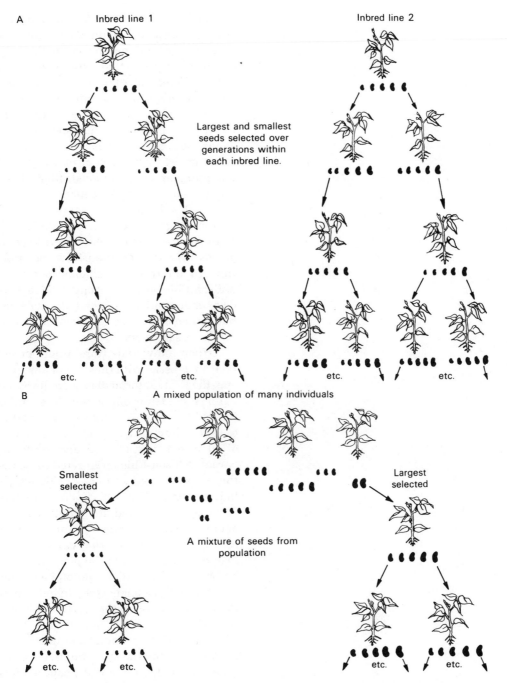

FIG. 7-2. Selection for seed size in beans. *A.* The seeds from a highly inbred, self-pollinated line show a range in size. When two lines are compared, one may have a larger average size value than the other. Selection of the smallest and the largest seeds within a line generation after generation causes no change in the average value for seed length. The same variation continues to be shown, indicating the variation is due to the environment. *B.* If the longest and shortest seeds are taken from a mixture obtained from a whole population of individuals, selection can be effective.

A line may be derived which produces smaller seeds on the average than the population as a whole. Another may be selected which bears seed larger than those found in the population as a whole. The variation in seed size results from both genetic and environmental factors. The population is composed of many genetically diverse lines, and these may be selected out. Within each inbred line, a small-seeded or a large-seeded one, variation which is environmental will continue to be shown.

GENES AND QUANTITATIVE INHERITANCE. Not the least of Johannsen's contributions to genetics was his emphasis on the role of the *entire* genotype in the expression of a characteristic. Too many early geneticists conceived of *one* unit factor for *one* unit character, the erroneous idea that one gene has just one effect and that this is in no way influenced by other genes. Johannsen argued that any character is the result of the interaction of many genes and the environment. The demonstration that quantitative characters were dependent on the interaction of several genes was very important in establishing this idea. Perhaps the first clear-cut example was that of color inheritance in wheat grains, where color intensity varies from white through intermediate shades to red. In 1909, Nillson-Ehle offered strong evidence that the color characteristic in wheat depends not on one but on several pairs of alleles which undergo random assortment.

When individuals from a pure-breeding red race are crossed with those from a white grained one, the F_1 are intermediates. In the F_2, however, various shades occur among the grains, from full red to white. Among the F_2 offspring, 1 out of 64 is white grained and approximately 1 out of 64 is red, like the red and white of the P_1 parent. Nillson-Ehle interpreted these results on the basis of three pairs of alleles which assort independently (Fig. 7-3). Each of the factors for redness, R_1, R_2, and R_3, would contribute an equal amount of pigment to the phenotype. The effects are cumulative; they can simply be added together. The alleles of the pigment genes, r_1, r_2, and r_3, would contribute no pigment and hence no redness to the phenotype; an individual homozygous for all three of these is thus white grained.

Other examples of quantitative inheritance similar to this were found by others and formed the basis for the theory of multiple-factor inheritance. Actually, no new concept is introduced, because the multiple-factor idea explains continuous variation in terms of Mendelian genetics. However, characteristics which exhibit continuous variation typically involve the interaction of more than two pairs of alleles, as well as a pronounced effect by environmental influences. And the effects of the different alleles are quantitative; they may be added up, in contrast to pure dominance or lack of dominance. Multiple-

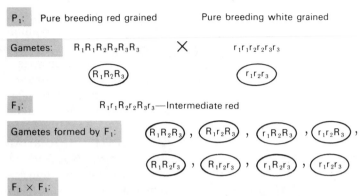

P₁: Pure breeding red grained   Pure breeding white grained

Gametes: $R_1R_1R_2R_2R_3R_3$ ✕ $r_1r_1r_2r_2r_3r_3$

$(R_1R_2R_3)$ $(r_1r_2r_3)$

F₁: $R_1r_1R_2r_2R_3r_3$—Intermediate red

Gametes formed by F₁: $(R_1R_2R_3)$, $(R_1r_2R_3)$, $(r_1R_2R_3)$, $(r_1r_2R_3)$,

$(R_1R_2r_3)$, $(R_1r_2r_3)$, $(r_1R_2r_3)$, $(r_1r_2r_3)$

F₁ × F₁:

	$R_1R_2R_3$	$R_1r_2R_3$	$r_1R_2R_3$	$r_1r_2R_3$	$R_1R_2r_3$	$R_1r_2r_3$	$r_1R_2r_3$	$r_1r_2r_3$
$R_1R_2R_3$	6	5	5	4	5	4	4	3
$R_1r_2R_3$	5	4	4	3	4	3	3	2
$r_1R_2R_3$	5	4	4	3	4	3	3	2
$r_1r_2R_3$	4	3	3	2	3	2	2	1
$R_1R_2r_3$	5	4	4	3	4	3	3	2
$R_1r_2r_3$	4	3	3	2	3	2	2	1
$r_1R_2r_3$	4	3	3	2	3	2	2	1
$r_1r_2r_3$	3	2	2	1	2	1	1	0

FIG. 7-3. Quantitative inheritance in wheat. Kernel color depends on three pairs of alleles. Each genetic factor which contributes to pigment formation (R_1, R_2, and R_3) adds an equal dosage. Their alleles (r_1, r_2, and r_3) contribute nothing to pigment formation The trihybrid carries three alleles for pigment and produces kernels intermediate in color between the parents. The F₁'s form eight classes of gametes. When these combine in all possible combinations, a range in shade is found. The Punnett square shows only the number of effective pigment alleles carried in the offspring. Only 1 out of 64 possesses six pigment factors, and only 1 out of 64 carries none at all. All of the other offspring vary in shade between the original (P₁) parents.

factor inheritance, also known as quantitative or polygenic inheritance, implies that several pairs of alleles are interacting and that each has a similar and measurable effect on a characteristic, which can also be influenced by the environment.

The example of grain color in wheat brings out several important points. Many nineteenth-century biologists might have thought that blending inheritance would explain the results. However, although the F₁ has only one phenotypic class (intermediate to the parents), the F₂ results clearly show that blending has not occurred because the extremes (white and red) appear again unchanged. The range of variation seen in the F₂ generation indicates that genes are segregating and that recombination is taking place. It is also evident that there is not one gene for one character. Grain color depends on the interaction of several genes and their cumulative effect. Moreover, in the F₂, environmental factors may operate to eliminate sharp distinctions among the major color groupings. The cross also reveals that two individuals of the same phenotype may have different genotypes. For example (Fig. 7-3), the genotypes $R_1R_1R_2R_2r_3r_3$ and $R_1r_1R_2r_2R_3R_3$ are in the same general color class because each has four pigment alleles, but the two are quite different genetically.

QUANTITATIVE INHERITANCE AND SKIN PIGMENTATION. Not long after its formulation, the multiple-factor hypothesis was applied to skin color differences in humans. All normal individuals possess skin pigment which is found in the melanocytes (pigment cells) of the live layer of the epidermis. The skin pigment is melanin, a dark brown product which combines with protein to form pigment granules in the melanocytes. Melanin formation depends on the oxidation of the amino acid tyrosine. (Recall the example of phenylketonuria (PKU) in Chapter 4, in which pleiotropy was discussed.) The amount of melanin in the melanocytes is responsible for varying shades of skin color. A certain autosomal recessive gene may block melanin formation and result in albinism, a condition characterized by an abnormally low amount of pigment. This allele is not confined to any one group of humans. Except for those individuals who are albino, the normal range of skin shades is found to vary from the lightest Caucasians to the darkest African peoples. Of course, within so-called "white" or "black" populations, the degree of pigmentation also varies. The earliest studies of inherited pigment differences in human groups were undertaken by Davenport and Davenport. The method of measuring skin color in black and white parents and their F_1 offspring was very limited, but the observations could be explained on the basis of quantitative inheritance in which approximately two pairs of independent alleles are involved. Recently, more accurate methods utilizing spectrophotometry have been employed to estimate the degree of pigmentation. The equivalents of F_1 and F_2 testcross progeny between blacks and whites have been followed. Although still limited, the results suggest that three or four pairs of alleles are operative in skin color differences. Although the present interpretation may still be an oversimplification, there is no doubt that these color variations have a polygenic basis and that only

FIG. 7-4. Inheritance of skin color. A mating between a black and a white person produces offspring intermediate in shade. Two parents of this genotype shown here who are intermediates will have offspring who vary in color from black, through various lighter shades, to white. (Probably three or four pairs of alleles are involved, but the principle is the same as illustrated on the basis of two.)

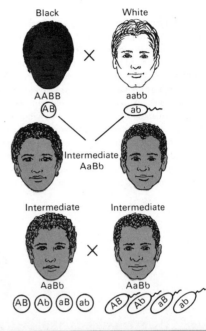

a relatively small number of genes is directly involved.

For the sake of simplicity, we will assume that two pairs of alleles provide the basis for skin shades in humans (Fig. 7-4). An extreme white person would contain no *effective* pigment alleles above a certain basic level. The extreme dark parent would have four pigment alleles in our simplified illustration. The principles demonstrated by the red pigment of the wheat grains also apply here to melanin pigmentation of skin. As in the case of grain color, environment will also influence gradations of shading in humans so that no distinct steps exist between the shades. Again individuals of the same phenotype may have very different genotypes. For example (Fig. 7-4), persons of the following three genotypes would have the same intermediate phenotype, ignoring the environment: (1) AaBb; (2) AAbb; and (3) aaBB. Certain consequences ensue from this. A cross of two persons of genotype 1, the dihydrid AaBb, will produce offspring which can show a range of variation as shown in Figure 7-4. On the other hand, a cross of a person of genotype 2 with one of genotype 3 (Fig. 7-5A) can only give rise to offspring of the same shade as the parents. So we would expect color variation to be great among the offspring in certain families where the parents are intermediate in degree of pigmentation. In other families, little or no variation in shade would be expected from intermediate parents. All of the observations suggest that this is the case. It should also be evident that a mating between a person of any shade with a white individual cannot produce any children who are darker than the more pigmented parent (Fig. 7-5B).

QUANTITATIVE INHERITANCE AND DEEP-SEATED CHARACTERISTICS. Although skin and grain pigmentation represent rather superficial characteristics, a moment's reflection tells us that the most deep-seated characters in any species are ones that show continuous variation and are quantitative in nature—such features as

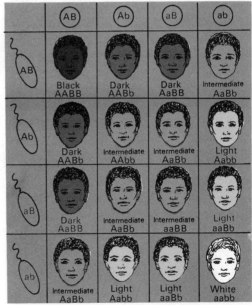

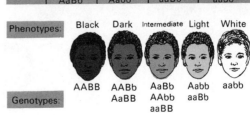

height, weight, length, and breadth of limbs or organs. Although the details of the inheritance of such characteristics are unknown for most species, there is little doubt that they have a polygenic basis. Indeed, polygenic differences undoubtedly provide the distinctions between closely related species. An accumulation of many genetic differences which influence basic physiology may affect the ease with which members of separated groups can mate and so lead to genetic isolation between them, even though both arose from one original population (see Chap. 16 for details).

We still know little about the inheritance of such very complex characteristics as behavior, intelligence, and personality. We actually know

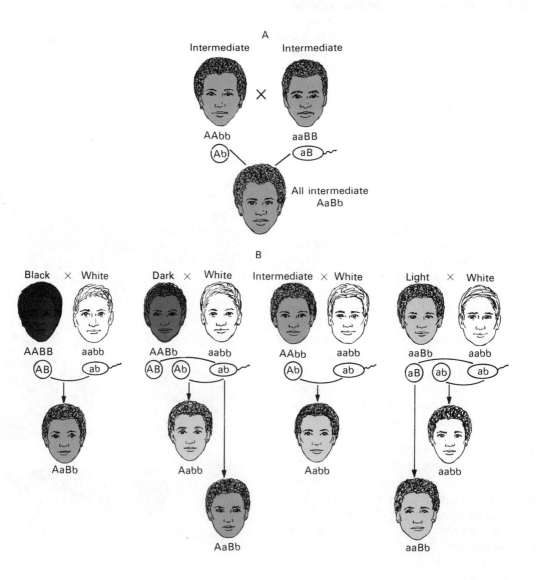

FIG. 7-5. Skin color in certain families. *A.* Some parents who are intermediate in shade give rise to offspring, all of whom are also intermediate. In such cases, each parent is homozygous for pigment alleles at one of the loci. In this example, they happen to be homozygous at different loci, but intermediate offspring would also result if the genotypes of both parents were either AAbb or aaBB. These results should be contrasted with those in Fig. 7-4. *B.* A cross of a person of any shade with a white individual can never give an offspring darker than the nonwhite parent. This is so, because the white contributes no effective pigment alleles which can add to those which may be donated by the darker parent. (Only some of the possible combinations are represented here, but the principle is the same for all of them.)

much more concerning the genetic basis of the more superficial characters of living things than we do about those which are much more fundamental. The reason for this soon becomes obvious. Figure 7-3 shows a Punnett square for the superficial color characteristic in wheat grains in which only three pairs of alleles apparently play a major role. Yet segregation of just these few pairs produces an array of types. Only 1 individual out of 64 has the same genetic combination for pigmentation as does one of the parents, $R_1R_1R_2R_2R_3R_3$ or $r_1r_1r_2r_2r_3r_3$. More complex characteristics must involve many more genes than this. Table 7-1 shows that with just five genes (pairs of alleles) so many new combinations are possible that only a minute number (1/1024) would resemble one of the original parents. This is so, because many pairs of alleles would be segregating and forming an assortment of new combinations. We would certainly expect the genetic basis of a very complex character like intelligence to involve more than 20 different allelic pairs. The number of combinations which are possible becomes staggering. And the expression of the many genotypes will be modified further by the influence of the environment. Moreover, for the sake of simplicity, we have assumed that the effects of each pair of alleles are equal and additive. We now know that this is not so in most cases of polygenic inheritance. Some alleles may contribute more of an effect than others to a quantitative character. We have also ignored the possible involvement of domi-

nance and epistasis. It is no wonder that the genetic analysis of continuous variation has progressed more slowly than that for discontinuous variation. However, we do have approaches to the study of polygenic inheritance. Although certain assumptions must be made when applying them, they can still give some idea regarding the number of pairs of alleles involved and their quantitative contribution to the characteristic.

ESTIMATION OF ALLELIC DIFFERENCES AND THEIR QUANTITATIVE CONTRIBUTION. To approximate the number of pairs of alleles we turn our attention to the fraction of F_2 offspring which expresses a phenotype as extreme as one of the parents (Table 7-1). For example, suppose two races of tobacco plants are crossed (Fig. 7-6A). One race has long flowers in which measurements range from 88–100 mm, with a mean or average of 93 mm. The short flowers of the second race range from 34–42 mm, with a mean of 39 mm. The F_1 offspring are approximately intermediate in flower length, ranging from 55–70 mm, with a mean of 66 mm. When the F_1 are crossed, the 800 F_2 progeny show a much greater range of variation than did the F_1 (from 42–91 mm). Of the 800 F_2 offspring, only 12 were in the size range of the longer parent and 10 in the range of the shorter one. The data from the cross indicate several things and suggest polygenic inheritance. On such a basis, we would expect that two pure-breeding races will produce an intermediate F_1 when they are crossed. The much greater variation among the F_2 than the F_1 offspring results from the segregation of several allelic pairs and the formation of many new genetic combinations. The 12 largest ones (12 out of 800 or approximately 1/66) and the 10 shortest (10 out of 800 or 1/80) suggest that three pairs of alleles are segregating, because three allelic pairs will produce a fraction of approximately 1/64 (Table 7-1) which is as extreme in phenotype as one of the parents.

We can now consider the small-flower P_1 parent (Fig. 7-6) to have a genotype in which the *minimal* effect on flower length, an average of 39 mm, is being expressed by an *unknown* number of genes (Fig. 7-6B). This same minimal effect on length is being exerted also in the long flowered parent. The length in millimeters *above* the minimum (the smaller average of 39 mm)

TABLE 7-1 Fraction of offspring showing one parental extreme*

ALLELIC PAIRS	DIFFERENT KINDS OF GAMETES	FRACTION SHOWING ONE PARENTAL EXTREME
1	$2^1 = 2$	1/4
2	$2^2 = 4$	1/16
3	$2^3 = 8$	1/64
4	$2^4 = 16$	1/256
5	$2^5 = 32$	1/1,024
6	$2^6 = 64$	1/4,096
7	$2^7 = 128$	1/16,384
8	$2^8 = 256$	1/65,536
9	$2^9 = 512$	1/262,144
10	$2^{10} = 1024$	1/1,048,576

* This fraction indicates the number of allelic pairs involved in the inheritance of the characteristic under consideration.

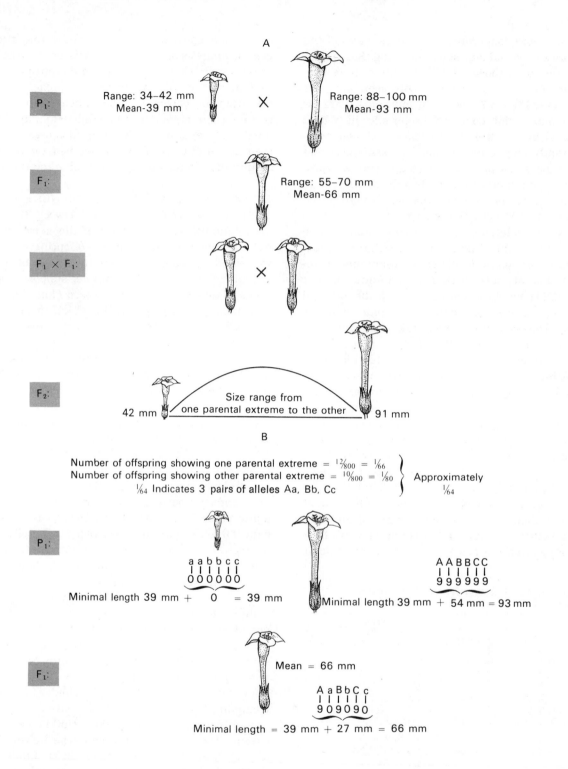

A

P₁: Range: 34–42 mm   ×   Range: 88–100 mm
     Mean-39 mm   Mean-93 mm

F₁: Range: 55–70 mm
     Mean-66 mm

F₁ × F₁: ×

F₂: Size range from
     42 mm   one parental extreme to the other   91 mm

B

Number of offspring showing one parental extreme = ¹²⁄₈₀₀ = ¹⁄₆₆ ⎫
Number of offspring showing other parental extreme = ¹⁰⁄₈₀₀ = ¹⁄₈₀ ⎬ Approximately
¹⁄₆₄ Indicates 3 pairs of alleles Aa, Bb, Cc   ⎭ ¹⁄₆₄

P₁:

a a b b c c   A A B B C C
| | | | | | | | | | | |
0 0 0 0 0 0   9 9 9 9 9 9
Minimal length 39 mm + 0 = 39 mm   Minimal length 39 mm + 54 mm = 93 mm

F₁: Mean = 66 mm

A a B b C c
| | | | | |
9 0 9 0 9 0
Minimal length = 39 mm + 27 mm = 66 mm

FIG. 7-6. Quantitative inheritance in tobacco. *A.* When a short-flowered race is crossed with a long-flowered one, the F_1 has a mean which is about intermediate between the parents. A range in flower size is seen in the P_1 and the F_1. When two F_1 plants are crossed, the F_2 show a range in flower size which greatly exceeds the size range in millimeters as shown in the F_1 or P_1. Some of the F_2 plants have flowers as short as those found in the short parental race and some as long as those in the longer parental race. Most of the F_2's have flowers which range in length between these extremes. The observations indicate that several pairs of alleles which contribute to flower length segregated in the F_1 to produce the variation seen in the F_2. *B.* Calculation of effective allele differences involved in the cross. The genotype of the shorter race can be represented as aabbcc and the longer as AABBCC, because three genes are indicated from the crossing data. The six factors in the longer parent contribute to the difference between the two mean lengths, a difference of 54 mm. Therefore, each allele must add 9 mm.

represents the *effective* genetic contribution of other genetic factors. It is these which differ between the long- and the short-flowered races to cause a difference in flower length. So the lower average length (representing the minimal or basic length) is subtracted from the larger average. The 54 mm which remains is the effective difference between them. Since six genetic factors (two each of alleles A, B, and C) contributed to that amount, each allele donates an average of 9 mm to petal length. The F_1 received three alleles from the larger parent which contribute an effective difference. The small parent gave no alleles which add to the difference. Therefore, the first-generation offspring would have a mean value approximately midway between the two parental means.

In order to study continuous variation in this way, it is apparent that environment must be controlled as closely as possible because the length of any organ can certainly be influenced by environmental factors. Two individuals of the identical genotype could have very different phenotypes simply as a result of these factors. In order to estimate with any degree of accuracy the number of effective alleles which are operating in a specific case, large numbers of individuals must be raised. Where five or more segregating pairs of alleles are involved, it may be impossible in many cases to obtain one of the parental extremes. Moreover, in this example of flower length we assumed equal effects of each contributing allele, and we know this may not

be so. Despite these limitations, however, some knowledge may be gained concerning the genetic basis for a character difference between two different stocks or varieties. It must also be emphasized that the estimation of the number of gene differences in this way *does not mean* that these are the only genes affecting the character. In our example, the smaller P_1 parent certainly has a flower length, and we noted that an unknown number of genes contribute to it, as it does for that *same* amount in the longer parent. The identical idea holds in the case of pigment differences in humans. Normal whites, as noted, possess melanin pigment. Considering the average white genotype to be *minimal*, we try to estimate the number of *effective* allelic differences between white and black populations, just as we approximated that three genes are involved in producing the differences between the large and small-flowered races.

TRANSGRESSIVE VARIATION. Let us next examine a cross which produces results which may at first seem bizarre but which can be easily explained on the basis of polygenic inheritance. Suppose that each of two inbred varieties of corn has a mean value of 64 in. in height. Suppose that the two varieties are crossed in order to obtain greater disease resistance and that height is not being considered. As it happens, the F_1 offspring also average approximately 64 inches in height (Fig. 7-7A). When the F_2 progeny arise, however, a spectrum of sizes is found. Of 2000 F_2 plants, seven are only about 32 in. in height, whereas nine of them reach 96 in. The unexpected results can be readily attributed to the segregation of several pairs of alleles in the F_1. Apparently, the P_1 plants, although homozygous for effective alleles which influence height, were *not extremes*. The possible extremes appeared in the F_2 after segregation and independent assortment in the F_1.

We can now estimate the number of effective pairs of alleles segregating for height, and we can approximate their contribution, again making the assumptions we did previously (Fig. 7-7B). Taking 32 in. as the minimal height resulting from factors *common to both races*, the effective genetic difference becomes 96 in. − 32 in. = 64 in. (The average largest height minus the smallest.) The number of F_2 plants showing

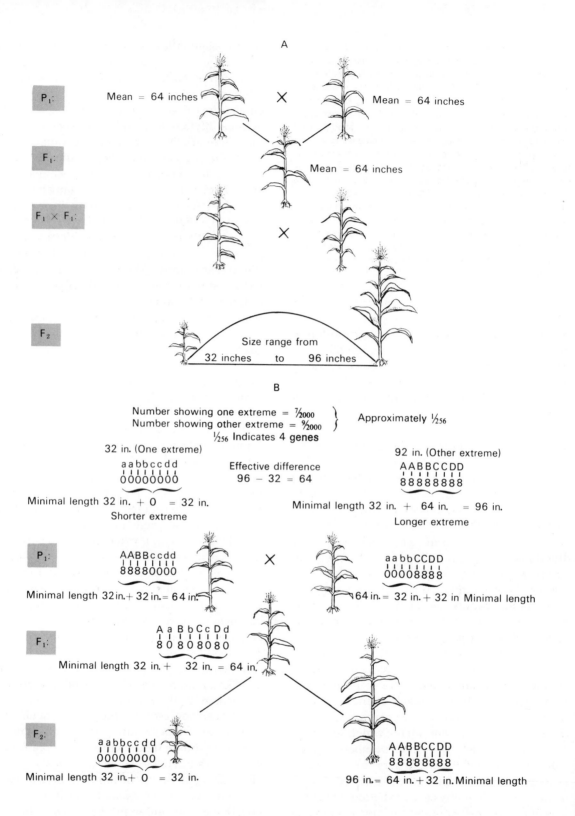

A

P₁: Mean = 64 inches ✕ Mean = 64 inches

F₁: Mean = 64 inches

F₁ × F₁: ✕

F₂: Size range from 32 inches to 96 inches

B

Number showing **one** extreme = $\frac{7}{2000}$
Number showing **other** extreme = $\frac{9}{2000}$ } Approximately $\frac{1}{256}$
$\frac{1}{256}$ Indicates 4 **genes**

32 in. (One extreme)
a a b b c c d d
0 0 0 0 0 0 0 0

Effective difference
96 − 32 = 64

92 in. (Other extreme)
A A B B C C D D
8 8 8 8 8 8 8 8

Minimal length 32 in. + 0 = 32 in.
Shorter extreme

Minimal length 32 in. + 64 in. = 96 in.
Longer extreme

P₁:
A A B B c c d d
8 8 8 8 0 0 0 0

✕

a a b b C C D D
0 0 0 0 8 8 8 8

Minimal length 32 in. + 32 in. = 64 in.

64 in. = 32 in. + 32 in Minimal length

F₁:
A a B b C c D d
8 0 8 0 8 0 8 0

Minimal length 32 in. + 32 in. = 64 in.

F₂:
a a b b c c d d
0 0 0 0 0 0 0 0

Minimal length 32 in. + 0 = 32 in.

A A B B C C D D
8 8 8 8 8 8 8 8

96 in. = 64 in. + 32 in. Minimal length

FIG. 7-7. Transgressive variation in corn. *A.* In certain crosses, two plants of the same height yield an F_1 which is also similar in height to the parental strains. A cross of these F_1's, however, may yield an F_2 which varies greatly and in which some of the individuals exceed the measurements of the original parents. *B.* This transgressive variation results when the P_1's are themselves not the extreme types. The number of pairs of alleles involved and the effective contribution of each can be calculated (see text for details).

---

parental extremes (9/2000 and 7/2000) is close to 1/256 and suggests that four pairs of alleles contribute to this difference of 64 in. meaning that four genes are affecting height. This means that each allele in the largest individuals on the average contributes approximately 8 in. to height *above* the minimal of 32. There is actually no basic difference between this example and the ones preceding it. The only variation is that we did not start out with either possible extreme phenotype in the original parents. We see from this example that offspring can exceed the measurements of the P_1 parents. Such a distribution, in which the offspring exceed the parental measurements, is called *transgressive variation.* It can explain cases in which progeny may be much taller or heavier than their parents, both of which may have more modest measurements.

THE IMPORTANCE OF STATISTICS TO GENETIC PROBLEMS. In several of our examples, we have noted "average" values and acknowledged that a range of variation exists. Fundamental to the analysis of any characteristic showing continuous variation is the need to *describe* it. Phenotypes which exhibit continuous variation are much more difficult to describe than those in which the observed variation can be sharply classified. For example, in the case of weight, we could choose to lump measurements into only three classes: light, medium, and heavy. At first glance, this might seem to satisfy our needs, but it would soon become apparent that an approach like this lacks accuracy. Many of the classifications would be arbitrary, because no sharp boundaries exist among the three categories selected to describe the differences.

Fortunately, a superior approach can be made through the use of elementary statistics.

The full meaning of the word "statistics" may be grasped by a consideration of some very obvious points. Most of us certainly realize that measurements are rarely made on entire populations. In a study designed to compare the weights of adults living in New York and those living in London, it would be absurd to believe that all of the people of each city could be weighed for the purpose. If such an impressive feat could be achieved, we would then have the exact figures to describe weights in the two cities. Such a character value for an entire population is known as a *parameter*. Although we want parameter values in order to describe accurately the characteristics of whole populations, their realization is often either impossible or impractical. Instead we must content ourselves with measurements made on samples taken from a population. Such a sample is considered representative of the entire population. Thus, a measurement of any sample character should reflect the true nature of that character for the entire population. *Statistics* are simply those measurements made on samples rather than on whole populations. We use them to estimate parameters, values which are more difficult or even impossible to obtain.

Various statistics exist to describe any particular characteristic, and it is critical to know the proper one to apply in a certain situation. There are a few basic statistics which have wide, practical application, not only in biological studies but in assorted, unrelated fields where an analysis of figures is required. We will concentrate here on just a few of the commonly used tools which form a basis for the derivation of more intricate formulas. Having a full understanding of their application to simple problems, we may be better prepared in later situations to select other appropriate statistics for the most thorough analysis of a complex set of data.

THE MEAN AND ITS CALCULATION. The best known statistic to describe a set of measurements is the numerical average. Most of us have used it at some time to get an idea of a general tendency, whether it concerns a list of test grades or measurements of height or weight, etc. Suppose we are interested in the effects of a certain drug on weight loss when added to the food of experimental animals. After a 2-month period, one group of 20 is weighed, and loss in grams is recorded in each animal. We might first arrange the figures as shown in Table 7-2A. To obtain an average, we could add up all the individual numbers and then divide by the total of separate values, which is 20. This can, however, be simplified a great deal (Table 7-2B). Starting with the lowest value, we can group together those figures which are the same and note their frequency. In this list, we have five values of 68. We then allow "v" to represent any individual value (or variable), and "f" to indicate the frequency of the particular value. By multiplying $f \times v$, we are merely recognizing the individual values and the number of times that each occurs in the sample. This is obviously simpler than adding together five 68's and similarly for the other numbers which appear more than once. All that remains to be done in order to obtain the mean is to take the total of all the separate values (represented by the symbol Σfv) and divide it by the total number of separate measurements, 20 (n) (Table 7-2C). The choice of symbols to represent different statistical concepts varies from one authority to another, and this many times results in confusion for the novice. Here we will try to use those symbols which are the least cumbersome but which are also frequently encountered.

Once we have the mean for the group of 20 animals, we possess some idea about the tendency toward weight loss in that sample. To obtain additional information from these data, we might reexamine the raw figures to note whether there is any one weight loss value which occurs more often than the others. Inspection immediately tells us that "68" is just such a figure. We have now noted the *mode*, which is the most repeated value in a list of measurements. The mode may give us an additional way of considering the central tendency.

THE STANDARD DEVIATION AND ITS IMPORTANCE. Although it is valuable to have some concept of an average, this knowledge by itself is quite limited and may conceal important aspects of a sample. Such a limitation becomes apparent when we give thought to the following possibility. Suppose we wish to calculate the mean weight loss for another group of 20 animals from a different strain. Conceivably, the two

TABLE 7-2 Recording data and calculation of mean on a set of 20 values. A, Preliminary recording of data and calculation of average. B, Grouping of data and simpler method of summing the values. C, General way to calculate the mean value.

							v*	f*	fv
(A)	35	28	58	16	35	(B)	16	1	16
					28		28	1	28
	63	63	58	75	58		35	1	35
					16		58	2	116
	68	68	63	75	63		63	3	189
					63		68	5	340
	85	68	78	68	58		69	1	69
					75		75	2	150
	93	69	81	68	68		78	1	78
					68		81	1	81
					63		85	1	85
					75		93	1	93
					85	Totals	n* = 20	1280 = Σ*fv	
					68				
					78				
					68				
					93				
					69				
					81				
					68				
					1280				

$$\frac{1280}{20} = 64$$

(C) ($\overline{X}$ = sample mean) $\overline{X} = \dfrac{\Sigma fv}{n} = \dfrac{1280}{20} = 64$

* v, Value or variable; f, frequency; n, total number of values; Σ, sum of.

strains could react differently to the presence of the chemical. A tabulation of the weight losses and a calculation of the mean are made just as for the first set of animals (Table 7-3). The mean for this second sample again gives a value of 64. If we were simply given the means for the two strains of animals, we might erroneously conclude that both reacted identically insofar as weight loss was concerned. However, a glance at the two sets of raw data shows that this is not so (Tables 7-2 and 7-3). The tabulations show that the *range* of weight loss is much greater in the first group. A few animals lost little (16 and 28 g) in the course of the experiment, whereas some lost an appreciable amount (81, 85, and 93 g). On the other hand, much less variation is present in the second group, where the loss in weight of all animals is comparable. With just a knowledge of the means, we would lose sight of the fact that in the first set several individuals

TABLE 7-3 Calculation of mean on a second set of 20 values

v*	f*	fv	
59	1	59	
61	1	61	
62	3	186	
63	5	315	
64	4	256	
65	3	195	
68	1	68	
70	2	140	
Totals	n* = 20	1280 = Σ*fv	$\overline{X} = \Sigma = \dfrac{fv}{n} = \dfrac{1280}{20} = 64$

* v, Individual value or variable, f, frequency of a value; n, total number of values; Σ, sum of.

were not close to the mean at all, whereas in the second, all clustered around the mean. Fortunately, there is another statistic which provides this information clearly and concisely. This is the *standard deviation* (S.D.), also represented by the Greek letter σ. It describes for a sample the amount of variation on either side of the mean. Let us calculate the S.D. for the second group of animals in order to appreciate the valuable information which this statistic can give (follow the steps by referring to Table 7-4).

We must first record the deviation (d) of each value (v) from the mean, which is 64. Next, we square each deviation, eliminating minus signs. We must not fail to recognize the number of individuals which deviate from the mean by each set amount. So we multiply the squared deviations by the appropriate frequency (f) for each respective deviation. The S.D. is then found by applying the formula given in Table 7-4. The reason for dividing by "n − 1" can be appreciated if we remember that although we are dealing with samples, we are actually interested in a whole population and certain of its parameters. However, in taking a sample, there is the danger of underestimating the variation present in the *whole* population. To correct for this, "1" is subtracted from the total size of the sample. There may be times when we are interested only in the sample itself and not in the whole population. In such a case, the correction is unnecessary, and we divide by "n" itself, which in this case would be 20.

TABLE 7-4   Calculation of S.D. from data in Table 7-3

v*	d*	d²	f*	fd²
59	59-64 = −5	25	1	25
61	61-64 = −3	9	1	9
62	62-64 = −2	4	3	12
63	63-64 = −1	1	5	5
64	64-64 = 0	0	4	0
65	65-64 = +1	1	3	3
68	68-64 = +4	16	1	16
70	70-64 = +6	36	2	72
	Totals	n*=	20	142 = Σ*fd²

$$\text{S.D.} = \sqrt{\frac{\Sigma fd^2}{n-1}} = \sqrt{\frac{142}{19}} = \sqrt{7.47} = 2.73$$

* v, Value; d, deviation; f, frequency; n, total number of values; Σ, sum of.

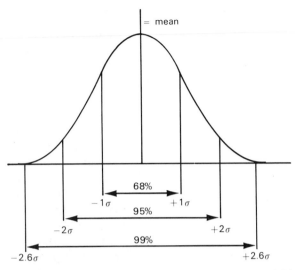

Data from TABLES VII-3 and VII-4 give a mean weight loss in grams of 64 and a Standard Deviation of 2.73. Therefore 99% of the animals in the sample can be expected to show a weight loss value from:

64.0 (mean)	to:	64.0 (mean)
+7.1 (2.6 × 2.73)		−7.1 (2.6 × 2.73)
71.1		56.9

FIG. 7-8. Relationship of standard deviation (S.D.) to a bell-shaped distribution.

Now let us proceed to examine the significance of the value, 2.73, which has been calculated as the S.D. of the second group of animals. To have a basis of comparison, calculate the S.D. for the first sample, following the illustration given in the table as a guide. When this is completed, a much larger value for the S.D. will be obtained. This is to be expected, because S.D. is a measure of the deviation from the mean, and the first group of numbers shows much more variation from the mean than does the second.

When a large group of individuals is studied quantitatively for a certain attribute, such as height, weight, length, etc., usually most of the values will cluster around the mean, and fewer will be found at the extremes on either side of the mean. If the sample is large enough, a normal or bell-shaped distribution could be expected, where the peak of the curve represents the mean as well as the mode. Typically, any large group of measurements tends to conform to a normal or bell-shaped pattern. The S.D. is applicable to such distributions, and it indicates how many individuals in a sample are found with certain values from the mean. One S.D. indicates that 68% of the individuals measured will fall within + or − 1 S.D. on either side of the mean (Fig. 7-8). For our calculated S.D. of 2.73, this indicates that approximately 68% of the individuals in the sample would be expected to show a weight loss value in grams from 61.27–66.73 (64 − 2.73-64 + 2.73). Two S.D.'s tell us that approximately 95% of the individuals will give values between + and −2 S.D. from the mean. So for our group, knowing the mean (64) and the S.D. (2.73), we know that 95% of the measurements fall within the range of 58.54 g to 69.46 g. As we often want to know where 99% of the individuals in a distribution lie, it is convenient to remember that the mean + and −2.6 S.D. will include the values for 99% of the individuals.

Clearly, the larger the S.D., the greater is the departure of the individuals in the sample from the mean. The S.D. for the first of the two groups of animals followed above is much larger than the second and turns out to be approximately 18.52. A bit of reflection on the two means and S.D.'s (64 ± 18.52 vs. 64 ± 2.73) tells us that 68% includes a much greater *departure* from the mean in the first group than in the second. In the latter, 68% of the individuals are relatively close to the mean. This can also be seen diagrammatically in an examination of bell-shaped curves of different heights (Fig. 7-9). In the curve on the right, 68% of the measurements (from + to −1 S.D.) are closer to the mean than in the curve on the left, where a much larger departure is evident for 68% of the values. A good appreciation of the amount of variation present in a sample can be gained by knowing the S.D. as well as the mean. Two samples may be compared for their amount of variability by noting their means and S.D. The higher the S.D., the more variable is the sample. It must be emphasized that S.D.'s may be compared *only* when they are expressed in the same units. It is erroneous to contrast two characteristics (height and weight, for example) for their relative amounts of variability when they are expressed in different units (centimeters vs. grams).

VARIANCE AS A MEASURE OF VARIABILITY. Another frequently encountered statistic is the *variance*, which also measures the amount of variability in a sample. Since it is nothing more than the square of the S.D., it is even more simple to derive than the S.D. Referring back to the calculations of the S.D. for the two sets of animals, we see that the variance of the second group is 7.47 (Table 7-4), the figure we obtained in our calculation before extracting the square root. The variance of the first group is 342.99, the square of 18.52. Immediately, we can tell from the larger value of the variance which sample is the more variable. This statistic is commonly used in this way to compare variability, and it also has various other applications in genetic analysis. Suppose we are studying the

FIG. 7-9. Comparison of curves of different heights. The curve on the left has the larger S.D. In the curve on the right, with its smaller S.D., 68% of the individual measurements included in the range, + and − 1 S.D., are closer to the mean than in the curve on the left.

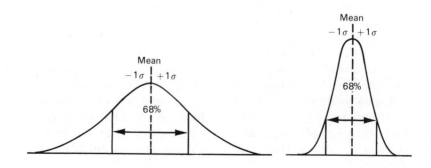

variability of a trait, such as length of ears in rabbits. We observe the total variance in a sample and are now interested in how much of that variation can be attributed to the genotype of the individuals, how much to diet, how much to the temperature under which they were raised, and so on for any number of separate factors which could contribute to the entire amount of variation displayed. After finding the effect of each separate factor on the total, we can look back and see the contribution of each to the entire variance. In other words, we can break up the total variation into different parts. We can add up the amount of variance produced by each part to estimate the percentage of the total that can be described to each single factor. This cannot be done using the S.D. Analyses employing variance may have great value in breeding and population studies; so we should be aware of some of its uses and benefits. We have noted that the S.D. can be related to the normal-shaped curve; however, this cannot be done with the variance. Since the latter is the square of the S.D., it must be expressed in square units. This is an awkward feature of the variance which is eliminated by taking its square root, giving us the S.D.

THE STANDARD ERROR OF THE MEAN AND ITS IMPORTANCE. The S.D. has at least one other very important use in all cases where means are estimated. We have stressed the fact that in most situations, we deal with samples and not with entire populations. Therefore, the mean which we calculate in a particular instance is the mean of the sample; it is not the value of the parameter. So if we get a mean on a sample of 20, 50, or 100 animals, we obviously are not examining all the possible individuals composing the population. If we work with a second, third, and fourth group of animals, all from the same population, and calculate the mean for each sample we would instinctively expect all the individual means to vary from one another. That is, the different sample means will vary among themselves. Each sample mean is just an estimate of the *true mean,* which we will be able to derive only in the rare situation. If we estimate means for a large number of samples and plot them, we will find that these separately calculated means also tend to fall into a normal or bell-shaped distribution. We can take advantage of this to

estimate how close the mean we have calculated for our sample is to the true mean, a value which might be impossible to determine for the whole population.

What are the chances that the means we calculated for weight loss in the two groups of animals (Tables 7-2 and 7-3) reflect accurately the true means, which we can learn only from measuring the effects of the chemical on *all* animals composing the two strains? The answer is provided through the application of another statistic, the standard error of the mean (S.E.), which is quickly computed from the following formula:

$$S.E. = \frac{S.D.}{\sqrt{n}}$$

A glance at the formula shows that the S.E. takes into consideration both the S.D. and the size of the sample (n). The larger the sample, the smaller the S.E. will tend to be. The S.E. is really nothing more than the standard deviation of all the individual sample means which could conceivably be taken and which would tend to follow a normal distribution. Consequently, knowing the mean and the S.E., we have a way to estimate how close our calculated mean is to the true mean. In the case of the second group of animals, the S.E. is quickly found to be:

$$\frac{S.D.}{\sqrt{n}} = \frac{2.73}{\sqrt{20}} = \frac{2.73}{4.47} = 0.61$$

A similar calculation for the first sample gives a S.E. of 4.14. We can now express the two means as 64 ± 4.14 and 64 ± 0.61 for the first and second groups, respectively. This information indicates that in the first case, the chance is 68% that the *true* mean lies within the range of 64 + 1 S.E. and 64 − 1 S.E. (Fig. 7-10). So the probability is 68% that the true mean lies somewhere between 59.86 (64 − 4.14) and 68.14 (64 + 4.14). The probability is 95% that it is between 55.72 and 72.28 (64 − 2 S.E. and 64 + 2 S.E.). There is a 99% chance that the true mean lies within 64 − (2.6 × 4.14) and 64 + (2.6 × 4.14). Obviously, we are using the identical reasoning applied to the S.D. because both statistics, the S.D. and the S.E., are based on bell-shaped distributions. It is apparent that the smaller the

S.E., the greater the *probability* that the sample mean is close to the true mean. In the second sample of animals, there is only a 1% chance that the true mean lies outside the range of 62.42-65.58 (minus and plus 2.6 × 0.61 from 64). Compare this with the first sample, where the range is much greater because its S.D., and hence its variability, is larger than that of the second sample. Sample means are commonly reported along with S.E. so that their reliability can be established. Means reported alone without their S.E.'s may arouse skepticism on the part of the reader. If a S.E. is very high, this suggests the sample may be either too small or very variable, and confidence in the sample mean as a reflection of the true mean may be shaken; additional measurements may have to be taken. Calculation of the S.E. recognizes both variability and the importance of sample size.

THE t TEST AND THE DIFFERENCE BETWEEN MEANS. There is another commonly used statistic which is of further value in the analysis of quantitative data and which employs several of the concepts which have been presented. Its

FIG. 7-10. The standard error of the mean (S.E.). If many separate means (x̄) are taken on samples from a population, the means will tend to fall into a normal distribution. The S.E. is the S.D. of the individual sample means. Approximately 68% of the means will lie within 1 S.E. on either side of the true mean of the whole population. Approximately 95% of them will lie within 2 S.E.'s on either side of the true mean, and 99% will be 2.6 S.E.'s on either side.

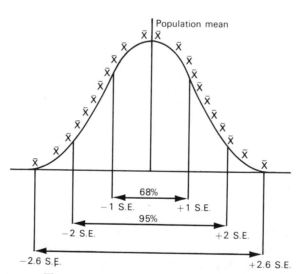

With a sample mean (X̄) of 64 and a Standard Error (S.E.) of 4.4, the probability is: 68% that the true mean lies between: 64.00 (sample mean)   64.00 (sample mean)
                          −4.14 (1 S.E.)   and +4.14 (1 S.E.)
                          59.86   68.14

The probability is 95% that the true mean lies between:
                          64.00 (sample mean)   64.00 (sample mean)
                          −8.28 (2 S.E.)   and +8.28 (2 S.E.)
                          55.72   72.28

significance becomes apparent in the following way. Once we have calculated means and S.E. on two samples, we may ask ourselves the question, "Are these two means significantly different?" Although the numerical value of two means will rarely be identical, we often cannot be certain whether the difference between them reflects a true difference between the samples. Perhaps the two samples are basically the same, and chance has caused the two calculated sample means to assume different values. Suppose, without knowing it, we actually have just one population, and we have taken two or more samples from this same source. The means calculated on the samples could appear to differ appreciably on the surface, and we might as a result erroneously conclude from them that the samples represent two very dissimilar groups. Errors of this type can be averted by application of the *t test,* a procedure that enables us to estimate the probability that two means with different values are truly different. All we need are the sample means along with their S.E.'s in order to calculate *t* from the following formula:

$$t = \frac{\bar{a} - \bar{b}}{\sqrt{(S.E.\ a)^2 + (S.E.\ b)^2}}$$

where $\bar{a}$ = the larger of the two means, $\bar{b}$ the smaller, and S.E. a and S.E. b their corresponding S.E.'s.

To illustrate the application of the *t* test, assume we calculated means and S.E.'s on two other groups of animals whose weight losses are under study, and we obtained the following values: 64 g ± 5.0 and 54 g ± 4.0. From these figures, can we say that the means we have calculated reflect a true difference between the two groups in response to the drug? Perhaps the samples really represent one population of similar organisms; the difference between the means may reflect nothing more than the operation of chance factors. Substituting the figures in the formula, *t* can be derived to help answer the question:

$$\frac{64 - 54}{\sqrt{(5)^2 + (4)^2}} = \frac{10}{\sqrt{25 + 16}} = \frac{10}{\sqrt{41}} = \frac{10}{6.4} = 1.5$$

To determine whether the difference between the two means is significant, we must turn to a table of probability (Table 7-5). The procedure is similar to that already described for the X^2 test. Our number of degrees of freedom is one less than the number of items. Assuming our total number of animals was 40, we see from the table that the probability for *t* with approximately 40 degrees of freedom is not significant. The *t* value is much too low for the two means to be considered significantly different. The table shows that in most cases a *t* value of 3 (and any higher) *would be* significant at the probability level of 1%. So if we arrive at any *t* value of 3 or more, where our number of degrees of freedom is over 14, we can feel confident that the difference between our calculated means is significant at the level of 1% and that the means reflect true differences.

STATISTICS AND THE INHERITANCE OF INTELLIGENCE. The need for statistical methods in the description and analysis of continuous variation is quite apparent. It must be kept in mind, however, that by themselves, statistics may give no insight into the genetic basis of a characteristic. The nineteenth-century biologists were frequently led astray in their evaluations of genetic patterns which they based entirely on statistical methods. Johannsen was well aware of the dangers as well as the values of mathematics in the analysis of quantitative inheritance, and he used statistical tools wisely. It is critical to know which statistic to apply in a given biological problem. We must never forget that living systems are involved and that mathematical tools may not reveal the subtleties which actually exist. The limitations of any statistic must be appreciated to avoid serious misinterpretations.

Perhaps one of the best examples today concerns the application of statistics to the analysis of intelligence in humans. To begin with, it would be impossible to arrive at a satisfactory definition of intelligence which takes into consideration peoples in all different cultures and in the completely different environments of the world. Most students of the problem agree that intelligence involves the ability to handle and interpret abstract symbols in a variety of relationships. The well-publicized intelligence quotient (IQ) is believed by many psychologists to reflect this ability and hence intelligence. The IQ is determined by obtaining the score achieved by an individual on a test which entails handling abstractions. His score is then related to the average score for his age group. The ratio between the former and the latter multiplied by 100 equals the IQ. Therefore, an individual whose score is average for the population age group to which he belongs has an IQ of 100. If he scores 120, whereas the average score is only 100, then his intelligence quotient would be $120/100 = 1.2 \times 100 = 120$.

The average IQ score for the population, the population mean, has thus been established as 100. The scores for the whole population describe a normal curve with 100 as its peak. The S.D. has been estimated to be 15. Thus, 68% of individuals in the population should have IQ's in the range of 85 (−1 S.D.) and 115 (+1 S.D.); 95% of the population would have IQ values between 70 and 130 (±2 S.D.'s).

Performance on an IQ test is believed to have some hereditary component. Obviously, environment would also be expected to play a decided role in the final expression of a characteristic as complex as intelligence. The question arises, "How much of intelligence, as indicated by IQ performance, reflects the expression of genes, and how much results from a host of environmental influences?" To help answer this, studies of IQ performance have been made on identical twins, some reared apart, others raised together. These have been analyzed in relationship to scores achieved by closely related, distantly related, and unrelated persons. From the assemblage of data, it has been estimated that environment may account for only 20% of the *variance* observed in the population on IQ achievement. In other words, 80% of the variance is due to genetic differences among members of the population. It will be recalled (p. 190) that variance is the square of the S.D. and that it can be broken up into different components. Since the value of the S.D. has been estimated to be 15 points for IQ scores, the variance must therefore be 225 (15^2). That portion of the variance which can be attributed to genetic factors is known as the *heritability* of the character. Since the heritability of IQ is estimated to be 80%, this means that 20% of the variance results from environmental factors. So of the total variance of 225, a component of 180 would be due to genes and a component of 45 due to environment.

TABLE 7-5   Critical values of "t".*

df	.20	.10	.05	.02	.01	.001
1	3.078	6.314	12.706	31.821	63.657	636.619
2	1.886	2.920	4.303	6.965	9.925	31.598
3	1.638	2.353	3.182	4.541	5.841	12.941
4	1.533	2.132	2.776	3.747	4.604	8.610
5	1.476	2.015	2.571	3.365	4.032	6.859
6	1.440	1.943	2.447	3.143	3.707	5.959
7	1.415	1.895	2.365	2.998	3.499	5.405
8	1.397	1.860	2.306	2.896	3.355	5.041
9	1.383	1.833	2.262	2.821	3.250	4.781
10	1.372	1.812	2.228	2.764	3.169	4.587
11	1.363	1.796	2.201	2.718	3.106	4.437
12	1.356	1.782	2.179	2.681	3.055	4.318
13	1.350	1.771	2.160	2.650	3.012	4.221
14	1.345	1.761	2.145	2.624	2.977	4.140
15	1.341	1.753	2.131	2.602	2.947	4.073
16	1.337	1.746	2.120	2.583	2.921	4.015
17	1.333	1.740	2.110	2.567	2.898	3.965
18	1.330	1.734	2.101	2.552	2.878	3.922
19	1.328	1.729	2.093	2.539	2.861	3.883
20	1.325	1.725	2.086	2.528	2.845	3.850
21	1.323	1.721	2.080	2.518	2.831	3.819
22	1.321	1.717	2.074	2.508	2.819	3.792
23	1.319	1.714	2.069	2.500	2.807	3.767
24	1.318	1.711	2.064	2.492	2.797	3.745
25	1.316	1.708	2.060	2.485	2.787	3.725
26	1.315	1.706	2.056	2.479	2.779	3.707
27	1.314	1.703	2.052	2.473	2.771	3.690
28	1.313	1.701	2.048	2.467	2.763	3.674
29	1.311	1.699	2.045	2.462	2.756	3.659
30	1.310	1.697	2.042	2.457	2.750	3.646
40	1.303	1.684	2.021	2.423	2.704	3.551
60	1.296	1.671	2.000	2.390	2.660	3.460
120	1.289	1.658	1.980	2.358	2.617	3.373
∞	1.282	1.645	1.960	2.326	2.576	3.291

* Table is taken from table of R. A. Fisher and F. Yates, *Statistical Tables for Biological, Agricultural and Medical Research*, Longman Group Ltd., London, 1973 (previously published by Oliver & Boyd, Edinburgh, 1963). Reprinted by permission of the authors and publishers.

Application of some of these statistical concepts to whites and blacks has led to some questionable conclusions. According to Jensen, the mean IQ score for whites is 1 S.D. above that for blacks. How much of this indicates a genetic, as opposed to an environmental, effect? Jensen reasons that since 20% of the variance is an environmental component, the IQ achievement could be raised proportionately by eliminating all those factors which serve to prevent higher performance on the tests. Reducing the variance value of 225 by 20% gives a new variance of 180. Since the S.D. is the square root of the variance, the new S.D. becomes 13.4. However, this new value is very close to the original one of 15. If whites at present score 1 S.D. higher than blacks, then even under the best environmental conditions mean IQ scores for blacks could not be raised 15 points but only 1.6 (15, the original S.D. minus 13.4, the new S.D.). And so the argument has been proposed that the mean IQ value of whites is truly higher than that of blacks and that this IQ value reflects primarily a difference in genetic factors which contribute to intelligence, not mainly unfavorable environmental forces.

Such conclusions are unfortunate from any viewpoint. Besides contributing to naive emotional prejudices and all their ensuing damage, these ill-founded arguments can cause doubt to be cast on any conclusions concerning intelligence which are actually based on firm scientific reasoning. Many of the assertions about the inheritance of intelligence as reflected by IQ scores are based on shaky foundations. Not only is there no agreement about what the tests truly measure, but also the heritability value of precisely 0.8 (80%) rests on so many questionable assumptions that it cannot be considered reliable. The value for the heritability of any character can vary from a population in one environment to a comparable population in another environment. The data on heritability of IQ have been obtained from studies of white populations in different settings (Europe and the United States), but black populations, even in the United States, have not been studied in a comparable manner. In no way have the studies been able to eliminate the role of the environment. There are no procedures at the moment which allow us to test *both* black and white Americans under the greatest number of different chance environments. Without such information, we cannot hope to suggest a reliable value for heritability. Many other good arguments can be presented against the conclusions reached by those accepting a heritability value of 0.8, but these are far too numerous to present here.

The IQ controversy serves again to caution us in our application of statistical concepts to biological problems. These valuable methods, when used naively, may obscure the true basis of a phenotypic feature. We could become more confused about the actual nature of differences in intelligence among individuals than were the nineteenth-century biometricians about the inheritance of the simplest polygenic characters.

REFERENCES

(Also see references in Chapter 6)

Bodmer, W. F., and L. L. Cavalli-Sforza, Intelligence and race. *Sci. Am. (Oct): 19,* 1970.

Carter, C. O., Multifactorial genetic disease. In *Medical Genetics,* V. A. McKusick and R. Claiborne (eds.), pp. 199–208. HP Publishing Co., New York, 1973.

Eichenwald, H. F., and P. C. Fry, Nutrition and learning. *Science, 163:* 644, 1969.

Erlenmeyer-Kimling, L., and L. F. Jarvik, Genetics and intelligence: a review. *Science, 142:* 1477, 1963.

Eysenck, H. J., *The IQ Argument.* Library Press, New York, 1971.

Falconer, D. S., *Introduction to Quantitative Genetics.* Oliver & Boyd, Edinburgh, 1960.

Harrison, G. A., and J. J. T. Owen, Studies on the inheritance of human skin color. *Ann. Hum. Genet., 28:* 27, 1964.

Jensen, A. R. How much can we boost I.Q. and scholastic achievement? *Harvard Educ. Rev., 39:* 1, 1969.

Jensen, A. R. Race and the genetics of intelligence: a reply to Lewontin. *Bull. Atomic Scientists, (May)* 17, 1970.

King, J. C. *The Biology of Race.* Harcourt Brace Jovanovich, New York, 1971.

Lewontin, R. C. Race and intelligence. *Bull. Atomic Scientists, (March)* 2, 1970.

Stern, C. Model estimates of the number of gene pairs involved in pigmentation variability of the Negro-American. *Hum. Hered., 20:* 165, 1970.

REVIEW QUESTIONS

In the Questions (1–4), assume that two pairs of alleles, A a and B b, form the basis of skin pigmentation differences in humans and are responsible for the recognition of five classes: black, dark, intermediate, light, and white.

1. Give the possible genotypes of each of the following:

   A. A dark-skinned person.
   B. An intermediate who had a white parent.
   C. A white person who had one intermediate parent.

2. A white person and one who is intermediate in skin color produce nine children, all of whom are light skinned. What are likely genotypes of the two parents?

3. Give the phenotypes and genotypes possible from a cross of an intermediate person, who is heterozygous at both loci, and a white person.

4. Two persons of intermediate skin color produce seven children, all of whom are about the same in skin color as the parents. However, two other persons of intermediate skin color produce five children who vary greatly from one another, ranging from white through dark. Give the most likely genotypes of the two sets of parents.

5. Assume that in a particular variety of wheat the color of the grains varies from deep red through intermediate shades to white. Six pairs of alleles are involved: R^1r^1, R^2r^2, etc. Each pigment-contributing allele, R^1, R^2, etc., adds an equal dosage whereas the alleles r^1, r^2, etc. contribute nothing to pigmentation. Plants from a pure-breeding strain which has deep-red colored grains are crossed to a strain with white grains. The F_1's are intermediate in color.

   A. When F_1's are crossed to white-grained plants, how many shades of grain are possible among the offspring?
   B. When the F_1's are crossed among themselves, how many of the offspring can be expected to be white grained, and how many deep red?

6. A sample of plants was measured in inches. The following gives the heights:

Height in inches:	7	8	9	10	11	12
Number of plants:	1	1	4	8	3	3

   Calculate: A. The mean.
   B. The variance.
   C. The standard deviation.
   D. The standard error of the mean.

7. Suppose height measurements are taken on a random sample of 100 men from a population

and that the mean height is found to be 70.50 in. The standard deviation is calculated and found to be 1.50. What would be the range in height that would include:

    A. 68% of the men in the sample?
    B. 95% of the men in the sample?

8. A. What would be the standard error of the mean in the above, Question 7?
    B. What does this value tell us about the mean?

9. A group of adults is given an IQ test with the following results:

Test score:
90 93 94 97 101 108 111 115 119 125
Number of persons:
1 2 2 1 4 7 1 1 3 3

Calculate: A. The mean test score value to the nearest whole number.
          B. The variance.
          C. The standard deviation.

10. A second group gives the following results on an IQ test:

Test score:
99 103 110 111 112 118 119 120 126
Number of persons:
6 2 3 4 3 1 2 2 2

Calculate: A. The mean test score to the nearest whole number.
          B. The variance.
          C. The standard deviation.

11. Calculate the standard errors of each of the two groups in Questions 9 and 10. Determine the "t" value. Does the difference between the two means represent a significant one at the 1% probability level?

12. In one variety of rabbit the ear length of the animals averages about 4 in. In a second variety, the average is 2 in. The ear length of hybrid animals obtained after crossing the two varieties averages 3 in. The hybrids, when crossed among themselves, produce offspring whose ear lengths vary much more than that of the F_1 hybrid parents. Of 496 F_2 animals, two have ears which measure 4 in. and two have

ears about 2 in. long. How many pairs of alleles would you say seem to be involved? How much does each effective allele contribute to length of ear?

13. In one variety of oats, the weight yield of grain per plant is 10 g, whereas in another variety it is only 4 g. Hybrids between the varieties give an average yield of about 7 g per plant. When the hybrids are crossed to each other, the yield varies greatly. About 4 plants in 253 give a yield of 10 g each and about the same number give a yield of about 4 g each. How many pairs of alleles appear to be involved in weight yield per plant? What is the contribution to weight yield by each effective allele?

14. A race of inbred plants has a mean petal length of 12 mm. A second race from another location has the same mean petal length of 12 mm. When the two races are crossed, the F_1's also have a mean petal length of 12 mm. However, the F_2 generation derived from crossing the F_1's to one another shows a very wide spread in petal length. About 3 out of 770 have a length as small as 8 mm, and about 3 out of 777 have a length as long as 16 mm. How many allelic pairs seem to be involved in the difference in petal length? How much does each effective allele contribute to length of the petal?

15. Assuming that height difference in the human depends on four pairs of alleles and that environmental factors are constant, offer an explanation of two parents of average height who produce children much taller than they.

GENES ON THE SAME CHROMOSOME

Once the chromosomes were accepted as the physical basis of Mendelian factors, it became necessary to follow the inheritance patterns of two or more genes located on the same chromosome. The early twentieth-century geneticists realized that certain organisms possess only a small number of chromosome pairs. Does this mean that *Drosophila,* with just four pairs, contains genes associated into four groups and that the genes in each group must always be inherited together? It was suggested by De Vries and several others that the genes in each chromosome group must undergo some kind of exchange, that is, the genes on homologous chromosomes must somehow be able to switch from one homologue to the other. Mendel did not encounter this problem in his work with peas. All seven pairs of alleles exhibited independent assortment, and we know today that each gene he followed was associated with a different chromosome. Therefore, no factor which Mendel followed was tied or linked to any other.

Shortly after Morgan undertook his classic studies with the fruit fly, he showed that two pairs of alleles, and thus two genes, could be associated with the same chromosome. He found this to be so in the case of a gene affecting eye color and one affecting the size of the wing. Recessive alleles of each gene, w (white eye color) and r (rudimentary wing), were both found to be sex linked and hence both located on the X chromosome. When Morgan followed both recessive alleles of the genes at the same time, he found that they did indeed undergo some sort of exchange in relationship to each other (Fig. 8-1A). Then an allele of a third gene, the recessive allele for yellow body color, y, was found to be sex linked and it too showed new combinations in respect to the other two, w and r (Fig. 8-1B). Morgan clearly demonstrated from such observations that two or more genes which are linked (whose loci are on the same chromosome) can in some way separate from each other and form new combinations in respect to their allelic forms. To grasp the full significance of this, let us first examine certain crosses in *Drosophila* which illustrate the nature of linkage and its genetic implications.

INDEPENDENT ASSORTMENT VS. COMPLETE LINKAGE. Suppose we make a cross between two mutant stocks of flies: "eyeless" (having a reduced eye size) and "black" (black body color instead of gray). Both of the responsible alleles are autosomal and are inherited as recessives to the wild. The F_1 offspring are found, as expected, to have the wild phenotype, large eyes, and gray body (Fig. 8-2). A testcross of these F_1 wild flies results in the typical dihybrid ratio, 1:1:1:1. This is exactly what should occur on the basis of independent assortment. The testcross shows the kinds of gametes produced by an organism, as well as the proportion in which they are formed. So the results in this case tell us that the eyeless locus and the black are not in any way linked or tied to each other but are free to segregate independently at gamete formation.

Now let us follow the allele "black" (b) along with another recessive, "purple" (p, for dark eye color as opposed to red). Treating the cross between black-bodied and purple-eyed flies exactly as we did in the above case, we might again expect all wild flies in the F_1 and a testcross ratio of 1:1:1:1. And we will find after making such a cross that the F_1 flies are all wild. However, the testcross will give results which are quite different from our expectations. Let us first testcross F_1 males with double recessive, purple-

eyed, black-bodied females (Fig. 8-3). Instead of four kinds of offspring in equal proportions, we will find only two types in a 1:1 ratio. One type is purple and the other is black; they are phenotypically identical to the original P_1 parental types. Something different must be involved in this cross because independent assortment has not occurred. The reason for this is that the loci "p" and "b" are not found on different chromosomes. They both happen to be located on Chromosome II and thus are linked. As Figure 8-4 shows, we must diagram a cross a bit differently when linkage is involved. It is essential to indicate clearly *which* specific allelic combination is together on a particular chromosome. This is necessary, because those alleles will tend to stay together and thus will not assort independently. The cross shows that each P_1 parent has on each chromosome a recessive in combination with a dominant and that the combination stays together throughout the testcross. The bar or line represents two chromosomes. The combination of alleles above the line will tend to remain together, as will the combination below it. We see that p b$^+$ and p $^+$b entered the cross together through the P_1 parents. The combination did not come apart but remained together among the testcross progeny. The testcross clearly tells us that the two genes are completely linked, because only the old P_1 combinations are formed (purple eyes with wild body color and red eyes with black body).

Now let us make the cross in a slightly different way. In Figure 8-5, the P_1 parents are purple, black (double recessives), and homozygous red eyed, gray bodied (wild eye, wild body). Again we see that independent assortment does not occur. The combination of alleles in the original P_1 parents stayed together and remained completely linked through the testcross. A very significant difference between this mating and the previous one (Fig. 8-4) is seen in the combination of the alleles. Note that in both cases the F_1 hybrids are wild in phenotype, but the allelic arrangement is very different. In the first case (Fig. 8-4), one chromosome has the combination of a recessive with a wild (p b$^+$). The homologous chromosome has the reciprocal of this, wild with recessive (p $^+$b). This arrangement in the F_1 dihybrid, p $^+$b/p b$^+$, is called the *trans arrangement* and results from a cross

A

P₁: White eyed ♀ with long (wild) wings

Red eyed ♂ (wild) with rudimentary wings

F₁: Wild ♀♀ Red eyed-long winged

♂♂ White eyed-long winged

F₂:

White eyed-long winged (males and females)

Red eyed-long winged (males and females)

Red eyed-rudimentary winged (only males)

White eyed-rudimentary winged (only males)

B

P₁: ♀ Yellow body, white eye, rudimentary wing

♂ Wild Gray body, red eye, long wing

F₁: Wild ♀♀

♂♂ Yellow body, white eye, rudimentary wing

F₂:

Wild

Yellow body, white eye, rudimentary wing

Gray body, white eye, rudimentary wing

Yellow body, red eye long wing

Gray body, red eye, rudimentary wing

Yellow body, white eye, long wing

Gray body, white eye, long wing

Yellow body, red eye, rudimentary wing

FIG. 8-1. Genes on the same chromosome undergo some kind of exchange. *A.* White eye color and rudimentary wings are both inherited as sex-linked recessives. When followed through the F_2, a cross of a white-eyed female and a male with rudimentary wings produces flies of four different phenotypes which occur in unequal amounts. The flies with red eyes and long wings and those with white eyes and rudimentary wings represent new combinations which were not found in the P_1 parents. Alleles at the two loci, white and rudimentary, must undergo some sort of exchange to form the new combinations. *B.* When three sex-linked recessives are followed at the same time, new combinations of traits also appear. In a cross such as this one, eight phenotypes arise among the F_2. The fact that new combinations are formed indicates that somehow genes on the same chromosome (the X here) can form new combinations. The various phenotypes shown here in the F_2 do not occur with the same frequency.

---

between two individuals carrying different recessives on the same chromosome. This *trans* arrangement is quite different from the *cis* seen in Figure 8-5. In this case, the dihybrid contains both recessives on one chromosome and both dominants on the other: $p^+b^+/p\,b$. This results from the cross of any individual that is double recessive at two linked loci with an individual homozygous for the corresponding dominant alleles.

Figures 8-4 and 8-5 clearly show that although both dihybrids, the *cis* and the *trans,* will be identical phenotypically, they represent very different arangements of alleles and produce dissimilar results in the testcross. In each case, however, the testcross offspring are like those of the P_1 parents in phenotype. At times, therefore, it may be very important to know whether two dihybrids showing the same phenotype have the same allelic arrangement, *cis* or *trans*, because the two types differ greatly when they are bred. The testcross is the most direct way to determine which of the two arrangements is present in a dihybrid.

CROSSING OVER AND ITS CALCULATION. In both of the testcrosses we have been discussing (Figs. 8-4 and 8-5), dihybrid males were selected and crossed with double-recessive females. We will now return to these same crosses, but this time, F_1 dihybrid females will be mated with double recessive males. As Figures 8-6 and 8-7 show, testcrosses of the F_1 females give not just two types of offspring as in the case of the F_1 males but four kinds! This means that four

different kinds of gametes were formed by the F_1 females; however, the numbers show that independent assortment did not occur. It is the old (P_1) combinations which are the most frequent; the new combinations are in the minority. So although there was a definite tendency for the parental allelic combinations to remain intact, a reassortment did occur in a smaller percentage of the gametes. This separation of linked genes which brings about new combinations is called *crossing over*. In this example, the amount of crossing over is 6%, and consequently, the strength of linkage is 94%. Crossing over does not occur at all in the male fruit fly, where the strength of linkage is normally 100%. But it must be kept in mind that crossing over is typically characteristic of meiosis in *both* sexes in most species. The lack of crossing over in the male fruit fly is an exception. This peculiarity, however, is a valuable one when working with *Drosophila*, for it permits the detection of linkage with relative ease. In most organisms, however, crossing over and the recombination which it generates is a normal and integral feature of gamete formation in both the male and the female.

Figures 8-6 and 8-7 also contrast the results of testcrossing F_1 females with the alleles in the *trans* arrangement and F_1 females with the alleles in the *cis* arrangement. Note that the old or parental combinations among the offspring of the *trans* females [$(p\,b^+/p\,b)$ and $(p^+b/p\,b)$] (Fig. 8-6) are the new combinations among the offspring of the *cis* females (Fig. 8-7). Likewise, the old combinations among the *cis* females [$(p^+b^+/p\,b)$ and $(p\,b/p\,b)$] are the recombinants or new combinations among the progeny of the *trans* females. This is exactly what is to be expected as a consequence of the fact that the gene affecting body color and the one affecting eye color are linked together on the same chromosome. It simply means that they will tend to remain together in the original P_1 arrangement and thus result in a larger number of parental types among the testcross offspring.

The following example illustrates how the amount of crossing over between two linked alleles can be easily calculated from an inspection of raw data obtained in a crossing experiment. A recessive allele of another gene is known to be linked to the one responsible for black body in *Drosophila*. Its effect is to produce a bent wing in the homozygous recessive condition. The allelic pair is "a^+" (for normal wing) and "a" for arc (bent wing). A cross between arc females and black males produces wild-type progeny (Fig. 8-8). Note that they have the alleles in the *trans* arrangement. When the F_1 females are testcrossed, four phenotypic classes can be recognized. To determine the amount of crossing over, the data are inspected to distinguish the parental types from the recombinants, those with new allelic arrangements. The latter are quickly recognized by their phenotypes (wild type flies and those which are both black and

FIG. 8-2. Dihybrid testcross. The eyeless and the black traits are inherited as autosomal recessives. When the F_1 dihybrid is testcrossed to an eyeless, black fly, the offspring occur in the expected ratio of 1:1:1:1.

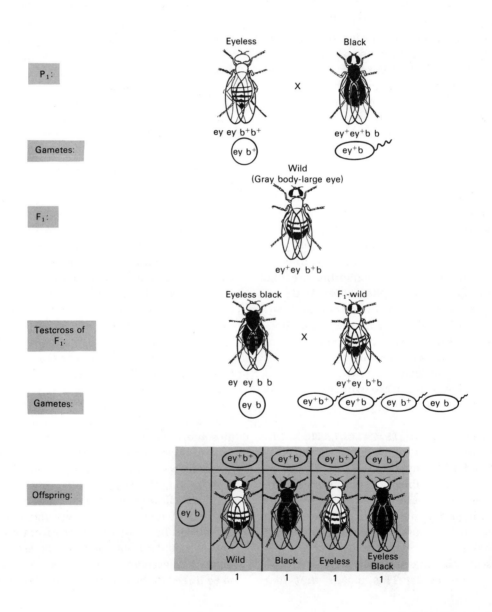

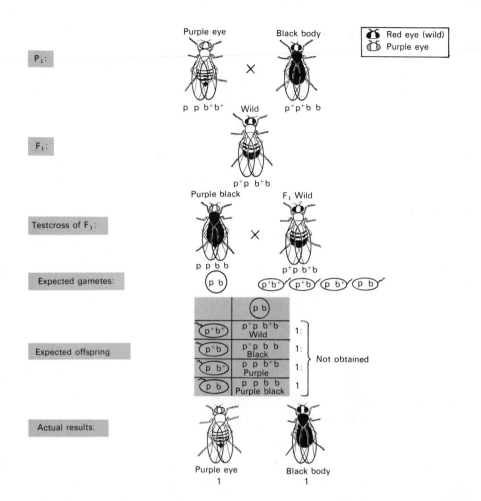

FIG. 8-3. Unexpected testcross results. An F_1 male, dihybrid for purple eyes and black body, is testcrossed. Instead of the expected ratio, only two classes of offspring appear: purple eyed and black bodied. These are exactly like the P_1 parents. No new combinations were formed in this case as seen by the 1:1 ratio instead of the expected 1:1:1:1.

arc) and are the less numerous types. The amount of crossing over is estimated simply by calculating the percentage of these crossover offspring among the total number of progeny (1020/2392 = 0.426 = 42.6%). This example may be used as a guide to estimate the crossover value between any two linked autosomal genes.

In the fruit fly, however, we must remember that the F_1 males will give a very different result in a testcross due to the lack of crossing over (Fig. 8-9). Because linkage strength is 100%, the genes "a" for wing shape and "b" for body color will not separate. Therefore one-half of the flies will show the "arc" phenotype, and the remaining one-half will be black. These phenotypes will be about equally distributed among the progeny.

It should be apparent that the F_1 males cannot be used in *Drosophila* in a testcross to estimate the amount of crossing over, because a ratio of 1:1 will always result regardless of the genes being followed. As mentioned, this lack of crossing over in the male fruit fly is very valuable in showing that two genes are linked and has provided one of the many advantages of *Drosophila* as a tool in pioneer genetic studies.

LINKAGE AND THE F_2 GENERATION. The discussion so far has stressed the use of the testcross in the detection of linkage and the amount of crossing over. Its advantage is clearly seen when we consider a cross which follows two linked genes through the F_2 generation. Suppose we are

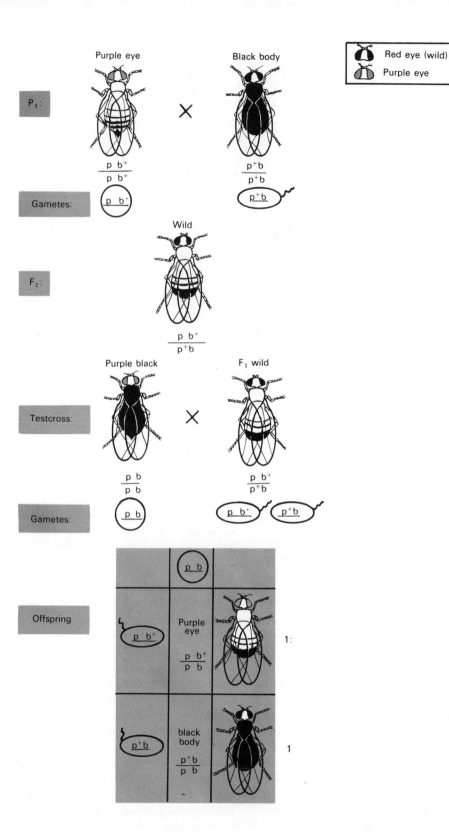

FIG. 8-4. Linkage in *Drosophila*. The gene for eye color and the gene for body color are on the same autosome. Therefore the combination of alleles found in the P₁ parents will tend to remain together. When a wild, dihybrid male is testcrossed, only two types of offspring result, and these are identical to the P₁ parents. The F₁ male forms only two classes of gametes because the two genes are completely linked.

---

working with corn plants and suspect that two particular loci are linked. The alleles of one of the genes affect the leaves (cr^+, the dominant for normal and cr, its recessive allele for crinkly leaves); the alleles of the other influence height (d^+ for normal and d for dwarf). Assume that the actual amount of crossing over which can take place between these genes is 20% but that we are unaware of this at the outset. Instead of making a testcross, we decide to follow the plants through the F_2 generation and to estimate the crossover percentage from the F_2 data. Figure 8-10A represents the cross of a crinkly plant with a dwarf. Since we are not dealing with *Drosophila*, each parent (instead of just the female) produces crossover gametes as well as the parental types. When combining the gametes in all possible combinations, we may use the familiar Punnett square method shown in Figure 8-10A, but we must remember that the gametes are not being formed in equal numbers, as is true in independent assortment. Therefore, we must indicate the frequency of each type of gamete. Each frequency, in turn, indicates the probability of the occurrence of that kind of gamete. When we combine two gametes, we must therefore multiply the two separate probabilities to indicate the chance of the two coming together at any one fertilization. This is no different from what we have been doing all along in problems dealing with independent assortment, where the frequencies of each type of gamete have been equal.

We can see from the diagram of the cross (Fig. 8-10A) that on the basis of 100%, the four different phenotypes would occur in a frequency of 24:24:51:1. If we depended on such an F_2 ratio to tell us the amount of crossing over, we could often be misled by fluctuations due to chance or due to such factors as lethals and genic interactions, etc. But even at its simplest,

the F_2 ratio would usually require intricate calculations to estimate the amount of crossing over. Obviously, the testcross is the more direct route (Fig. 8-10B).

GENES LINKED ON THE X CHROMOSOME. There is a qualification to the statement that use of the F_2 is not the best way to estimate the amount of crossing over. This pertains to F_2 ratios obtained from genes linked together on the X chromosome. A cross between two stocks, mutant for different sex-linked alleles, produces an F_2 generation which is particularly valuable, because it can be used to determine the amount of crossing over. The F_2 generation in a sex-linked cross may be used for this purpose, because *at least* one-half of the F_2 in a sex-linked cross are always testcross progeny. This statement can be appreciated if we recall that sex-linked genes are carried on that part of the X which has no homologous region on the Y chromosome. Consequently, all XY males must express the X-linked alleles and thus will be testcross offspring for any sex-linked genes. This is so, because the Y will not mask any alleles on the X contributed by the female parent. We may, in this sense, consider the Y chromosome devoid of genes being followed on the X. Figure 8-11 illustrates linkage between two recessives on the X of *Drosophila*: "w" (white eyes) and "ct" (cut wings). Note that the alleles in the F_1 females are in the *trans* arrangement. When allowed to mate with the F_1 males, these females produce crossover as well as parental gametes. The males, on the other hand, will contribute only a Y chromosome and an X with the parental combination. The figure shows that the F_2 female progeny will fall into only two phenotypic classes, wild and white. However, four different classes can be recognized among the F_2 males, because their fathers contributed the Y chromosome to them. They are thus testcross offspring, having received from their male parents nothing to mask any recessives on the X from their mothers. The F_2 females clearly are *not* testcross progeny; they received an X chromosome from their fathers which carried a dominant (ct^+) which *did* mask the presence of the recessive. We can ignore these female offspring and concentrate only on the different types of males to estimate the crossover frequency. Suppose the mating

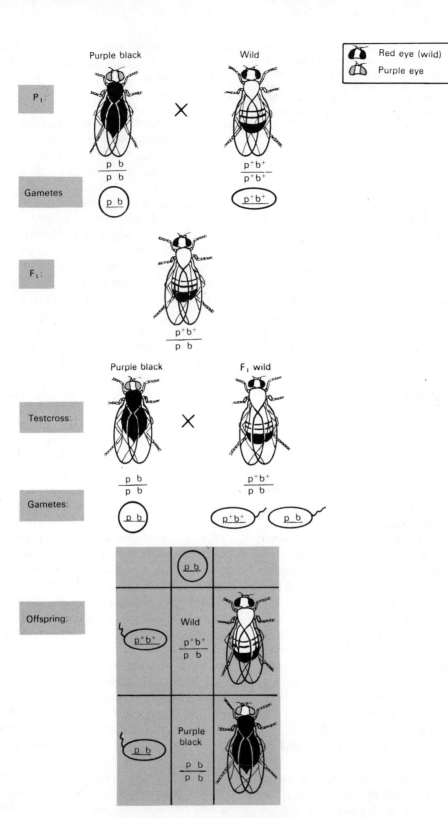

FIG. 8-5. Linkage in *Drosophila*. The principle in this cross is identical to that shown in Fig. 8-4. However, the F_1 dihybrid males in the two crosses must be compared. They are both wild in phenotype, but their allelic arrangements are different. In this case, the alleles are in the *cis* arrangement as opposed to the *trans* in the other cross. The testcrosses give very different results, but in each case, the testcross offspring are identical phenotypically to the P_1 parents. No new combinations of alleles are formed.

FIG. 8-6. Testcross of F_1 female, *trans* arrangement. When a dihybrid female is testcrossed, the linked genes separate or engage in crossing over. The old combinations are more frequent than the new ones. Four types of testcross progeny result, but they occur in unequal proportions which reflect the amount of crossing over (6% here). This testcross is to be contrasted with the testcross of the F_1 male shown in Fig. 8-4.

between "white" females and "cut" males which is diagrammed in Figure 8-11 produced 1000 flies in the F_2 generation, distributed as follows:

Females			Males	
wild	258		white	191
white	250		cut	198
	508		white, cut	45
			wild	38
				472

The F_1 and the F_2 data tell us that the genes are sex linked, because the expression of the genes is seen to depend on the sex of the offspring; only the males are mutant in the first generation; only the males show the "cut" trait in the F_2. To calculate the crossover frequency, we pay atten-

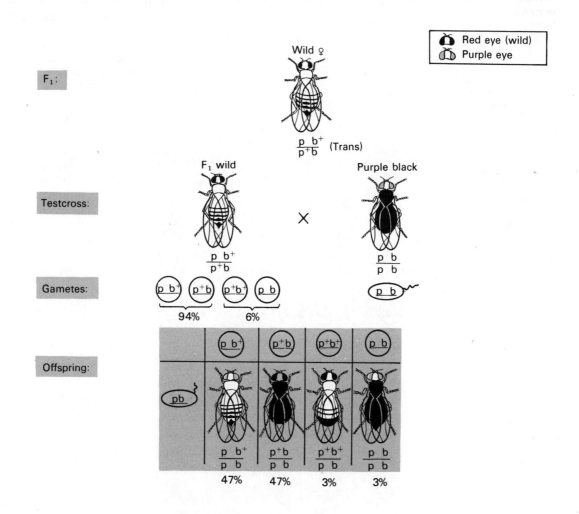

tion only to the F_2 males, the testcross progeny. The parental and crossover combinations among them are recognized and their numbers recorded:

Parentals		Crossovers	
White (w ct⁺)	191	Wild (w⁺ ct⁺)	45
Cut (w⁺ ct)	198	White, cut (w ct)	38
	389		83

The percentage of crossing over is then obtained by dividing the total number of *male crossover types* by the total number of *male* offspring: crossover frequency = 83/472 = 17.6%.

We must remember that when experimental crosses are performed, they are routinely done in reciprocal in order to reveal any differences which might be related to sex. The student should diagram the reciprocal of the cross followed here: "cut" females × "white" males. Al-

FIG. 8-7. Testcross of F_1 female, *cis* arrangement. Because crossing over occurs, four types of gametes are formed, but the old (parental) combinations are the most frequent. These results are just the reverse of those shown in Fig. 8-6, but in both cases, it is the parental types which are more frequent. This cross should be contrasted with the one in Fig. 8-5 in which the corresponding F_1 male is testcrossed.

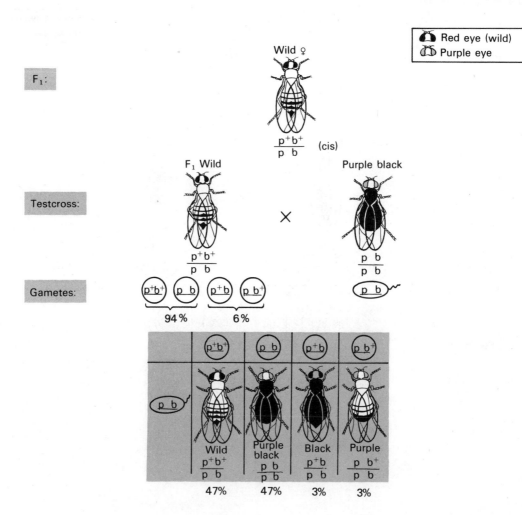

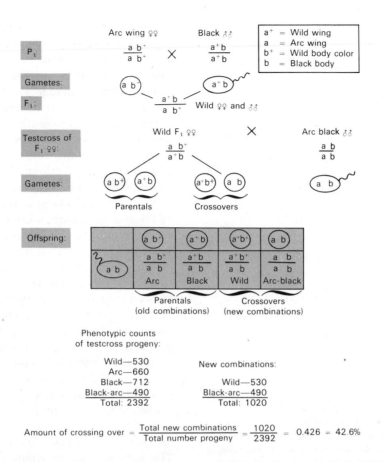

Phenotypic counts
of testcross progeny:

		New combinations:
Wild—530		
Arc—660		
Black—712		Wild—530
Black-arc—490		Black-arc—490
Total: 2392		Total: 1020

Amount of crossing over $= \dfrac{\text{Total new combinations}}{\text{Total number progeny}} = \dfrac{1020}{2392} = 0.426 = 42.6\%$

FIG. 8-8. Calculation of the amount of crossing over. To calculate the amount of crossing over between two autosomally linked genes, the F_1 dihybrid is testcrossed. The crossover types and the parentals are easily distinguished phenotypically. Crossover percentage is calculated by dividing the total number of testcross offspring into the number of new combinations.

though the F_1 male offspring will differ phenotypically from those F_1 males diagrammed in Figure 8-11, the F_2 results among the males will be the same. This is so, because the F_1 females are identical genotypically in both crosses. In actuality, comparable figures will be obtained if the reciprocal matings are performed under similar conditions.

Although males are always testcross progeny in the case of sex-linked genes, females *may* also be testcross offspring, depending on the male parent. Consider the cross shown in Figure 8-12. In the F_2 generation, *all* of the offspring, female as well as male, are testcross progeny. This is so, because the X chromosome of the F_1 males contains both recessives. Consequently, these F_1 males contribute no sex-linked dominants to any

of the offspring. Therefore, there are no factors to mask the effects of any sex-linked alleles coming from the F_1 females. The diagram in Figure 8-12 shows that all of the possible phenotypes are represented equally among the females and the males. Since they are all testcross offspring, every one of the progeny, male and female, can be counted to determine the crossover percentage. In contrast, the reciprocal mating (homozygous wild P_1 females × white, cut males) will give very different results. The student should diagram this cross to convince himself that in this case *only* the F_2 males can be used in the counts to determine the crossover frequency. This is so, because the F_1 males will contribute an X to their daughters which contains both of the dominant wild type alleles.

LINKAGE GROUPS AND CHROMOSOMES. After Morgan's discovery of linkage of genes on the X chromosome of *Drosophila*, several autosomal genes in the fruit fly were also found to be associated into linkage groups. It soon became evident that the nonsex-linked genes were associated into one or the other of two large groups. This made sense, because there are two other large chromosomes present in the fruit fly in addition to the X (Chap. 5). The fourth chromosome is a very small one. Finally, all four groups were established, and the relative sizes of the linkage groups were found to correspond to the sizes of the chromosomes. This is exactly what is to be expected if the genes are located on the chromosomes. Linkage was then found in species other than *Drosophila*, and in these as well, the haploid chromosome number and chromosome sizes corresponded to the number and sizes of the linkage groups. This parallel provided even more support for the chromosome theory of heredity. If the physical basis of heredity is in the chromosomes, there must be many more genes than there are chromosomes, because the latter are found in relatively low numbers. Therefore, the genes would be associated on specific chromosomes, the linkage groups. The haploid number would thus correspond to the number of linkage groups. Large chromosomes would most probably carry more genes

FIG. 8-9. Testcross of dihybrid males in *Drosophila*. The testcross of a male which is dihybrid for two autosomally linked genes yields only the parental combinations in a 1:1 ratio. Linkage strength is 100% in the male fruit fly. Thus, the amount of crossing over cannot be determined by using male *Drosophila* in a testcross. Compare this with the testcross of F₁ females in Fig. 8-8.

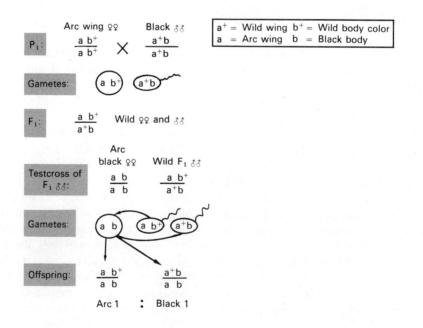

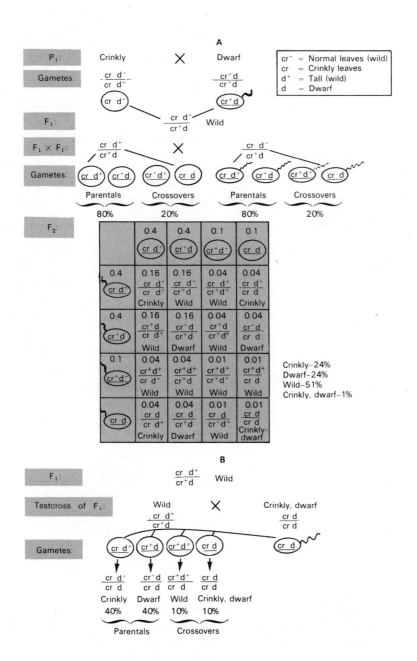

FIG. 8-10. Linkage and the F₂ generation in corn. *A*. There is 20% crossing over between cr and d. Each parent forms crossover as well as parental types of gametes. If the amount of crossing over were unknown to begin with, its calculation on the basis of the F₂ data would be difficult. *B*. A cross of any F₁ hybrid to a double recessive reveals directly the number of new combinations (wild and crinkly, dwarf), and the percentage of crossing over is readily determined.

than small ones, and so there should be the parallel size relationship.

Genes which are linked should therefore tend to stay together. But the fact of crossing over shows that linked genes can separate and enter into new combinations. If it were not for crossing over, all allelic combinations on one chromosome would be forced to remain together (say hypothetically the alleles for blond hair and blue eyes on one chromosome, the alleles for dark hair and brown eyes on the homologue). The only way a new combination (blond hair and brown eyes or dark hair and blue eyes) could be formed would be through the slow chance process of gene mutation. Meiotic crossing over, however, brings about the new combination quickly. Although alleles of linked genes may not assort independently, crossing over extends recombination down to the level of the chromosome. Without crossing over much of the value of the sexual process, the formation of new gene combinations, would be lost. Evolution would be greatly slowed down in those species lacking it

FIG. 8-11. Genes linked on the X chromosome. When genes on the X are being followed, the F_2 generation may be used to estimate the amount of crossing over. This is so, because the Y chromosome in the males masks no factors on the maternal X. Therefore, all of the F_2 males are testcross offspring, and they can always be used in such crosses to estimate the frequency of crossing over. In this cross shown here, *only* the F_2 male offspring can be used because the F_2 females received a dominant allele from the F_1 male parent. This masks recessives on the maternal X chromosome.

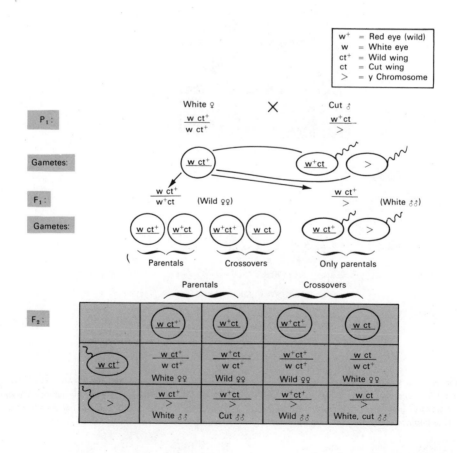

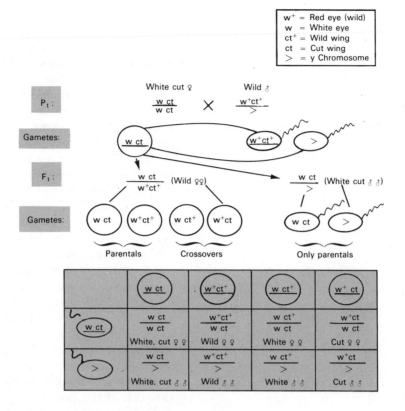

	$w^+ = $ Red eye (wild)
	$w = $ White eye
	$ct^+ = $ Wild wing
	$ct = $ Cut wing
	$> = $ y Chromosome

FIG. 8-12. F_2 females and males as testcross progeny. In this example, the F_2 females receive only recessives from the F_1 male parent. They are thus testcross offspring in the case of the X-linked genes. All of the offspring in this cross may be used to estimate the amount of crossing over. Contrast with Fig. 8-11.

because there would be fewer variants available for the operation of natural selection. Crossing over probably arose long before independent assortment—when only one pair of chromosomes was found in living things. It would have provided the first mechanism to ensure new gene combinations among the earliest cells.

MECHANISM OF CROSSING OVER. Biologists have long recognized the importance of crossing over in sexual species, and speculations on the mechanism whereby it occurs have fascinated cytogeneticists for decades. In 1909, Janssens pointed out that the paired chromosomes at meiotic prophase often assume configurations which appear as cross figures. He interpreted each cross, a chiasma, to mean that a breakage and reunion had taken place followed by a reciprocal exchange between two chromatids in the bivalent, one of them paternal and one maternal. This then was the physical basis of a

crossover event and the chiasma its visible manifestation. Janssens's chiasma theory, while offering an explanation for the mechanism of crossing over, could not be tested, and throughout the years controversies raged on the subject. Several models other than breakage and reunion were proposed. There is no need to review these here, for now recombination studies with microorganisms have provided strong support for a breakage and reunion concept (Chap. 17).

The consensus is that the events involved in crossing over may be much as Janssens proposed. We should therefore be acquainted with several main points in Janssens's theory, also known as the *chiasmatype* theory and which has been expanded by Darlington, one of its leading proponents. According to the chiasmatype theory, the time of crossing over is early meiotic prophase, perhaps at the pachynema stage. Any one crossover involves only two of the four chromatids which are present in the bivalent, which in turn is composed of two intimately paired homologous chromosomes. Breakage of the two involved chromatids, one maternal and one paternal, is the first event to take place. The ends of the broken strands then undergo reunion reciprocally with their homologues (Fig. 8-13). As meiosis proceeds and enters the diplonema and metaphase I stages, the point of the cross-over event can be detected cytologically as a chiasma. Therefore, each chiasma seen actually represents one crossover event. Consequently, counting the number of chiasmata in a bivalent at diplonema would directly indicate the number of crossover events which had taken place earlier. A main point in the chiasmatype theory is that a breakage precedes a chiasma.

A variation of the breakage idea (the so-called classical theory of crossing over) proposes that the crossing of the two chromatids, the chiasma, occurs first. Breakage of the chromatids may follow. If it does, then a crossover takes place. However, breakage need not occur according to this theory, and so each chiasma does not necessarily represent a crossover event. This variation of the breakage and reunion concept is the less favored version. It is difficult to reconcile the classical theory with certain chromosome configurations which can be observed at meiosis in triploids, organisms with three of each kind of chromosome. In triploids, trivalents (associations of 3 homologous chromosomes) occur at meiotic prophase, and in some of these trivalents, chiasmata can be seen holding together the three homologous chromosomes. The

FIG. 8-13. Theories of crossing over. According to the chiasmatype theory (above), breakage of chromatids occurs first and chiasma formation follows, each chiasma representing one crossover event which took place. The classical theory (below) states that the crossing of chromatids and hence the chiasma occurs prior to breakage. A chiasma does not always represent a crossover event, since breakage does not always occur. The outcome is the same according to each theory for every crossover which does occur. Most lines of evidence, however, support the chiasmatype theory.

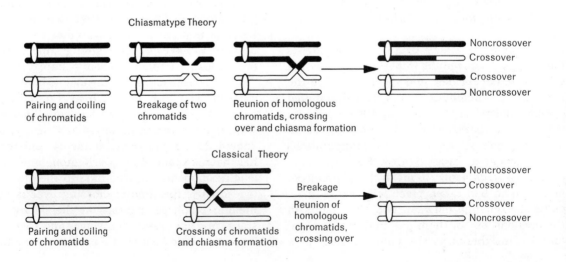

classical theory cannot satisfactorily explain the formation of such chiasmata, whereas the chiasmatype theory provides an acceptable explanation.

In addition, cytogenetic analyses have continued to show a good correspondence between the amount of crossing over detected genetically and the number of chiasmata observed with the microscope. While the relationship has not been reported as 1:1 in all cases (that is 1 genetic crossover detected for each chiasma observed), a higher chiasma frequency as seen at meiotic prophase is associated with a higher frequency of genetic crossing over. Moreover, in those species with large chromosomes, the number of chiasmata present in the bivalents can be determined more accurately than in species with tiny meiotic chromosomes, such as *Drosophila*. In corn, for example, which has large chromosomes, a very good correlation exists between the amount of genetic recombination and the chiasma frequency. The chiasmatype theory, proposed shortly after the turn of the century, still offers the best concept of the mechanism of crossing over, although many questions remain unanswered.

CHROMOSOME MAPPING. The data on crossing over assembled in Morgan's lab from experiments with *Drosophila* were compatible with the hypothesis that the units of heredity are located in linear order on the chromosome. It had been conceivable that the genes are assembled into some complex arrangement with one on top of the other or even scattered haphazardly in the chromosome. But the results on linkage and crossing over showed that this could not be so. For example, in one of his early experiments, H. J. Muller, a student of Morgan, followed 10 mutant alleles linked on the X chromosome. As diagrammed in Figure 8-14, when multiply heterozygous females were testcrossed, the results showed that the great majority of offspring with new combinations had one whole block of genes from one of the X chromosomes up to a certain point; the rest of the genes represent a block from the remaining part of the homologous X. The blocks might be of any length, indicating that an exchange could occur at any point along the length of the chromosome. Nearly always, there was one block up to a certain point and then a block of genes from the other chromosome. But in a few cases, there were arrangements in which a block was present from one chromosome, then a block from the homologue, and then a block from the first one again. These would represent chromosomes in which not one but two breaks had occurred between the two chromosomes, giving a double crossover. Results such as these, as well as others which will be discussed shortly, could only be explained on the basis of a genic arrangement on the chromosome in which the genes were side by side, much as numbers are linearly arranged on a ruler.

It was reasoned that if this idea were correct, one should be able to determine the relative positions of genes on the chromosomes by studying strengths of linkage and the amount of crossing over. When a particular kind of cross is repeated under the same experimental conditions, the amount of crossing over between two genes is found to be constant. The frequency of crossing over between a particular pair of genetic loci is apparently not haphazard. Crossover values for different pairs of linked genes, however, may be quite dissimilar. Consider the two hypothetical crosses in Figure 8-15. There would be a crossover value of 10% for the amount of crossing over between "p" and "l" and a different constant figure, 20%, for "l" and "c." We can interpret such results to reflect distances among these three genes, assuming a linear order. For if crossing over *does* entail some kind of breakage and reunion of chromosome threads, it is obvious that the closer together two genes, the less the chance that a random break will occur between them. So if "l" is closer to "p" than it is to "c," there will be fewer chance breaks between "p" and "l." Thus, they will tend to stay together more than will "l" and "c" when a single break takes place.

Tentatively, let us arrange the three genetic loci in the following order: p   l   c. The reasoning here is that the number of new combinations observed in a testcross stems directly from the amount of crossing over and that this in turns reflects the distance between the genes. If these new combinations can be used to indicate distance, what sort of units should be used as a measure? H. J. Muller suggested taking the percentage of new combinations and using them directly as a kind of measure of distance simply by converting the percentage into *crossover units*. So we would say that genes "p" and "l" are 10 units apart, whereas "l" and "c" are separated by 20 units. A moment's consideration of our tentative arrangement of the three genes (p l c) will tell us that the gene order is not justified from the information at hand so far. It is equally possible that the correct order is c p l. Provisionally, we can only say that "l" and "c" are twice as far apart as "p" and "l." However, "c" could be on either side of "p." Either of the two map arrangements shown in Figure 8-15 is feasible.

Obviously, we must find the number of new combinations between "p" and "c." We cannot place any three genes (1, 2, 3) in the proper relative positions if we only find the crossover values between genes 1 and 2 and between genes 1 and 3. We must also determine the number of new combinations between genes 2 and 3 in order to arrange the loci on the chromosome with any degree of accuracy. Suppose a cross between "p" and "c" gives a value of approximately 30%. This would indicate that the first supposition was correct (left arrangement, Fig.

FIG. 8-14. Genes in a linear order. When a group of genes known to be X-linked is followed, the formation of the recombinant types among the offspring is found to be associated with an exchange of blocks of genes. In this schematic representation, females which are multiply heterozygous for 10 X-linked genes are test-crossed. Among the female and male offspring, new combinations may be detected. These recombinant types possess new arrangements of traits associated with blocks of genes. The size of a block which enters into a new combination can vary from one recombinant to the next. Some recombinants (*lower*) possess a block from one of the X's, then a block from the other, then a block from the first again. These represent double exchanges.

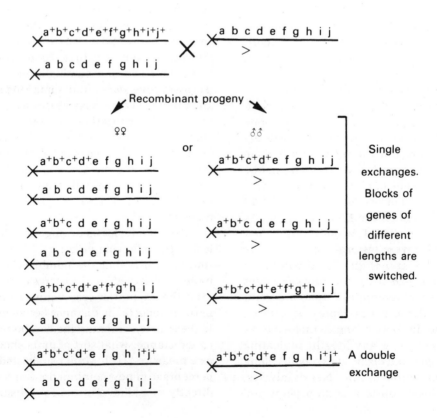

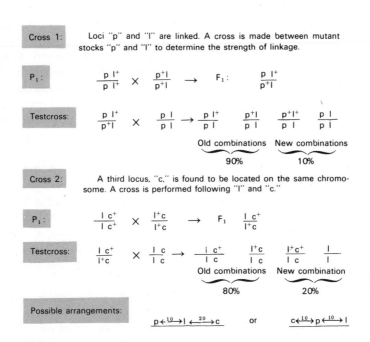

Cross 1: Loci "p" and "l" are linked. A cross is made between mutant stocks "p" and "l" to determine the strength of linkage.

P_1:
$$\frac{p\ l^+}{p\ l^+} \times \frac{p^+l}{p^+l} \rightarrow F_1: \quad \frac{p\ l^+}{p^+l}$$

Testcross:
$$\frac{p\ l^+}{p^+l} \times \frac{p\ l}{p\ l} \rightarrow \frac{p\ l^+}{p\ l} \quad \frac{p^+l}{p\ l} \quad \frac{p^+l^+}{p\ l} \quad \frac{p\ l}{p\ l}$$

Old combinations   New combinations
90%   10%

Cross 2: A third locus, "c," is found to be located on the same chromosome. A cross is performed following "l" and "c."

P_1:
$$\frac{l\ c^+}{l\ c^+} \times \frac{l^+c}{l^+c} \rightarrow F_1 \quad \frac{l\ c^+}{l^+c}$$

Testcross:
$$\frac{l\ c^+}{l^+c} \times \frac{l\ c}{l\ c} \rightarrow \frac{l\ c^+}{l\ c} \quad \frac{l^+c}{l\ c} \quad \frac{l^+c^+}{l\ c} \quad \frac{l}{l}$$

Old combinations   New combination
80%   20%

Possible arrangements:

$$p \xleftarrow{10} l \xleftarrow{\ 20\ } c \quad\quad \text{or} \quad\quad c \xleftarrow{10} p \xleftarrow{10} l$$

FIG. 8-15. Determining gene arrangement. In cross 1, the value of 10% for the new combination indicates the amount of crossing over, the frequency with which the loci "p" and "l" separate. Similarly, in cross 2, the value of 20% indicates the amount of crossing over between "l" and "c." The crossover frequencies reflect the distance between the loci. Thus, "l" and "p" with 10% crossing over (90% linkage) must be closer to each other than "l" and "c" with 20% crossing over (80% linkage). From the data, two map arrangements are possible. The correct order can be established only by making a cross between stocks "p" and "c" to determine the amount of crossing over between them and hence their map distance.

8-15). A value close to 10%, on the other hand, would indicate that "c" is to the left of "l" (right arrangement, Fig. 8-15).

Assuming that the correct order is found to be p l   c, we can appreciate the physical basis for the different crossover values from the illustrations in Figure 8-16. Notice that in all the illustrations of crossing over, each chromosome is represented as a double structure. Chromosomes are depicted in this way because genetic analyses of such organisms as *Neurospora* and *Drosophila* indicate that at the time of crossing over, each chromosome thread has replicated (more details next chapter). Each chromosome would thus be composed of two chromatids, and crossing over would take place in the four-strand stage, which simply means that four chromatids are present when it occurs. Any *one* crossover, however, involves only two of the four strands, and nothing dictates that it must always be the same two. In the illustrations, strands 2 and 3 were selected as the crossover chromatids only for the sake of diagrammatic clarity. This should not be construed to mean that only these two can participate. Strands 1 and 3, 1 and 4, or 2 and 4 could have engaged equally as well.

Although the reasoning discussed in Figure 8-16 seemed to hold, it was soon realized that certain complications could enter the picture. For example, if the gene order actually is p l   c, and "p" and "c" *are* truly 30 units apart, it might be possible for crossing over to occur simultaneously at two points between "p" and "c," that is, to get double crossing over between them. Recall that Muller, studying blocks of genes, found evidence for the occurrence of double crossing over. Let us examine the effects of such an event on the relationship between these two genes and compare them with the results of a single crossover. If just genes "p" and "c" are being followed, *any* double crossover between them would produce the results in Figure 8-17A. Compare these with the results of the single crossover in Figure 8-17B. We see that the single crossovers produce detectable new combinations between "p" and "c." In the case of the doubles, however, "p" and "c" are not placed in a new relationship to each other but still occur together on the same chromatid. Therefore, there would be no way to recognize the fact that *two* crossovers, not just one, had occurred between them. The doubles would be classified along with the parentals, the types with the old combinations. Since we would be losing sight of two single crossovers for every double that was not detected, "p" and "c" would appear to us to remain together in the parental combination more often than is actually so. Since the chromosome map is based on the number of *new* combinations which we can observe, the genes "p" and "c" would appear to be more strongly linked and hence closer together than they really are.

TRIHYBRID CROSSES AND MAPPING. How can we overcome this tendency to arrive at an underestimation of the true amount of crossing over? The problem is quite easily solved by making a trihybrid or 3-point testcross. Instead of following just two genes at a time in the testcross, we consider three. Trihybrid crosses are especially valuable in chromosome mapping and are routinely performed to ascertain gene locations. The same basic procedures are followed when working with microorganisms or with higher species. Because chromosome mapping is so critical to genetic analysis and is assuming an important role in human genetics, the steps and the reasoning behind them are presented here in considerable detail.

If we have established the gene order as p l c and know that "l" is between the other two, we can use this knowledge to detect any double crossovers between the more distant genes, "p" and "c." Now consider a double

FIG. 8-16. Physical basis of crossing over. At the time of crossing over, each chromosome is composed of two chromatids. A single crossover involves only two chromatids of the four which are present in the bivalent. To obtain a crossover between loci "p" and "c," all that is needed is a break and reunion anywhere in the interval between "p" and "c." In the top diagram, the point of crossing over resulted in a new combination between "l" and "c" *as well as* between "p" and "c." To obtain a crossover between "p" and "l," a much shorter distance is involved. Imagine that for a crossover to occur within this shorter interval that the threads must bend more sharply. The lower diagram shows a crossover between "p" and "l." Note that in this case and also in the one shown above, there was a crossover between "p" and "c." This is true simply because *any* one break between "p" and "l" *must* also be a break between "p" and "c." So as a result of their greater distance apart, we would expect more crossing over between "p" and "c" than between "p" and "l."

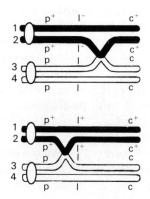

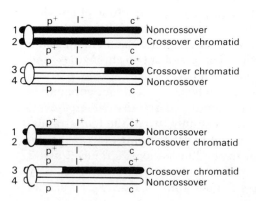

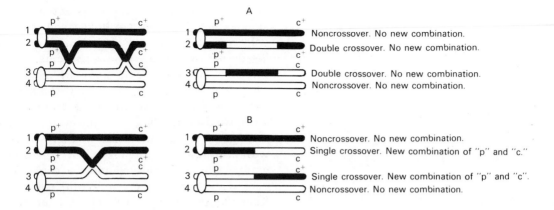

FIG. 8-17. Results of a double and single crossover between two genes. *A*. A double crossover between p and c yields no new genetic combinations. Therefore, if only these two genes are being followed in a cross, we would lose sight of two single crossovers each time a double took place. *B*. A single crossover between p and c can be readily detected by the new combination among the offspring.

crossover when a trihybrid is being followed (Figure 8-18). We see that on either side of gene "l," there are two chromosome regions, region 1 and region 2, where single crossovers can occur. If two singles occur together to give a double, then no new combinations between "p" and "c" are produced. However, we now have a way to detect these doubles. For when a double occurs, the gene which is more central, gene "l" in this case, enters into a new combination with the genes on either side of it. Note the genic parental arrangements in the trihybrid parent. On the one chromosome, the three wild alleles of the three genes are together: p⁺ l⁺ c⁺. On the other, the three mutant alleles are linked: p l c. As a result of the double crossover, we now have p⁺ l c⁺ and p l⁺ c. It appears to us that "l," the gene in the middle, has shifted its position.

The apparent shift has resulted from the occurrence of two crossovers, one on either side of "l." With the trihybrid testcross, we can now recognize double crossovers which would otherwise be lost, and we add these to the number of singles to give the true amount of crossing over between "p" and "c." These two genes on either side of "l" will consequently appear farther apart than they otherwise would. We see from this example that the number of new combinations is *not* the sole criterion for an estimation of the amount of crossing over between two distant genes. Whenever possible, three closely linked loci should be followed to arrive at a more accurate estimation of map distance.

Let us now use a trihybrid testcross to map three hypothetical genes which are autosomally linked in *Drosophila*. We will call the three genes

FIG. 8-18. A double crossover in a trihybrid. A double crossover involves two singles. There are two regions therefore where each crossover can take place. A double crossover in a trihybrid can be detected, because the "middle" gene will assume a new relationship with those on either side of it. It will appear to have switched its position. The alleles on either side of it will remain in the old combination.

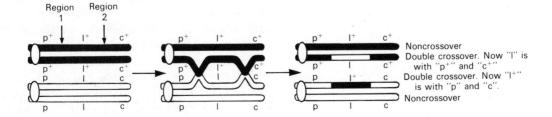

b, n, and t. Nothing is known about these three sites except that they are linked and that the wild type alleles, b^+, n^+, and t^+ are dominant. We do not even know the precise order of the three loci. Since we do not know which gene is the one in between, we cannot say which is the correct way to illustrate the gene arrangement when we diagram the cross. However, if we are consistent, it does not really matter. To start with, we may select the order b n t. This may or may not be accurate, and we could have chosen another arrangement to work with. Data from the actual cross will tell us later which arrangement is actually the correct one.

Suppose a cross is made between homozygous wild females and triple mutant males. Having decided to use the order b n t, we can depict the trihybrid cross and the testcross as shown in Table 8-1. Eight different categories of offspring can be recognized phenotypically on the basis of the combinations of the wild and mutant traits. Eight phenotypic classes are to be expected, because a trihybrid forms eight classes of gametes ($2^3 = 8$). Remembering the meaning of a testcross, we know that these eight classes actually represent the eight different kinds of gametes produced by the trihybrid and that the data reflect the relative frequency in which they were formed. Therefore, we can use the testcross data to map the three genes, for now we have a way to estimate how often the linked genes separated and entered into new combinations during gamete formation. The steps and reasoning go as follows (follow Table 8-1):

1. Since linked genes tend to stay together, we would expect the noncrossover types to be the most common. The phenotypic classes which are the most numerous among the progeny should represent the parental-type gametes, those in which no crossing over has been involved. The data show us these types, and we record them as the parentals: $b^+n^+t^+$ (298) and b n t (289).

2. Single crossovers would be less frequent than parentals, but doubles would be even fewer and can be easily identified. So we record the least numerous classes: b n^+t^+ (15) and b^+n t (21).

3. Once both the parental and double crossover classes have been recognized the correct order of the three genes is readily determined.

TABLE 8-1 Trihybrid testcross data and construction of a map segment

$$P_1 \quad \frac{b^+ \quad n^+ \quad t^+}{b^+ \quad n^+ \quad t^+} \times \frac{b \quad n \quad t}{b \quad n \quad t} \longrightarrow F_1 \quad \frac{b^+ \quad n^+ \quad t^+}{b \quad n \quad t}$$

Testcross $\quad \dfrac{b^+ \quad n^+ \quad t^+}{b \quad n \quad t} \times \dfrac{b \quad n \quad t}{b \quad n \quad t}$

Testcross progeny				
b^+	n	t	....	21
b	n^+	t	....	120
b	n	t	....	289
b^+	n^+	t	....	69
b	n	t^+	....	71
b^+	n^+	t^+	....	298
b^+	n	t^+	....	109
b	n^+	t^+	....	15
		Total		992

Construction of Map Segment from Testcross Data

Parentals
b^+	n^+	t^+	298
b	n	t	289
Total			587

				Doubles
b	n^+	t^+	15	
b^+	n	t	21	
Total			36	

Alleles at "b" appear to have switched position. The "b" locus must be between "n" and "t." One correct way to represent the trihybrid parent is:

$$\frac{n^+ \quad b^+ \quad t^+}{n \quad b \quad t}$$

Crossing over may occur in the segment between "n" and "b"; it may also occur in the segment between "b" and "t." These regions can be called, respectively, region 1 and region 2 and the trihybrid written as:

$$\begin{array}{ccc} 1 & & 2 \\ n^+ & b^+ & t^+ \\ \hline n & b & t \end{array}$$

Single crossover region 1

$$\frac{n^+ \quad \times \quad b^+ \quad t^+}{n \quad b \quad t} \downarrow$$

n^+	b	t	120
n	b^+	t^+	109
Region 1 crossovers			229

Crossover region 1 and 2

$$\frac{n^+ \quad \times \quad b^+ \quad \times \quad t^+}{n \quad b \quad t} \quad \downarrow \quad \downarrow$$

n^+	b	t^+	15
n	b^+	t	21
Doubles total			36

Single crossover region 2

$$\frac{n^+ \quad b^+ \quad \times \quad t^+}{n \quad b \quad t} \quad \downarrow$$

n^+	b^+	t	69
n	b	t^+	71
Region 2 crossovers			140

Total region 1 $\quad \dfrac{36}{265}$

Total region 2 $\quad \dfrac{36}{176}$

$$\text{Amount of region 1 crossover} = \frac{265}{992} = 26.7\% \qquad \text{Amount of region 2 crossover} = \frac{176}{992} = 17.7\%$$

Map segment $\quad \dfrac{n \quad 26.7 \qquad b \quad 17.7 \quad t}{}$

We have seen that when a double crossover takes place, the gene between the outside ones appears to have switched its position in relationship to the two on either side of it. A comparison of the two classes, the parentals and the doubles, indicates that b is the gene which is in between. We see that the three wild alleles went into the cross together on one chromosome and the three recessives on the other. In one group of doubles (b n^+t^+), two of the wild types are still together, and in the other group (b^+n t), two of the recessives are still in the parental arrangement. It is alleles of gene b which do not appear in the original combination, and we conclude that gene "b" must be between the other two.

4. We next rewrite the trihybrid parent in order to show the proper gene arrangement. It makes no difference whether we write $n^+b^+t^+$/ n b t or $t^+b^+n^+$/t b n as long as the proper gene is in the inside position. If this seems confusing, remember that we are primarily concerned with distances. As long as the relationship among the genes is preserved, the sequence selected is of no consequence, just as it would not matter if a road map were turned upside down when it was read. We will choose arrangement n $^+b^+t^+$/n b t to stand for the trihybrid. To convince yourself that the other possibility is equally valid, you should later redo this cross using the alternative order. The same results will be obtained.

5. Looking at the correctly written trihybrid parent, we can see that there are two places where crossing over can occur to rearrange the three linked genes: one between n and b and the other between b and t. For reference, these regions are designated region 1 and region 2, respectively.

6. A break in region 1 will produce the single crossovers: n^+ b t and n b^+t^+. The data show that they amount to 120 and 109 respectively, giving a total of 229 crossovers in that region. The singles in region 2 are n^+b^+t (69) and n b t^+ (71), totaling 140. It is critical to understand that these single crossovers which can be recognized from the data *do not* represent all the crossovers in regions 1 and 2. Double crossing over has occurrred; this means that there were additional breaks in both regions. These cannot be detected by considering the singles alone. Since one double masks the effects of two single

crossovers, the genes n and t will appear closer together than they should on the final map if we fail to take the doubles into consideration. However, we expressly made a triple testcross to detect these doubles, and we identified (step 2) 36 of them among the progeny. A glance at the trihybrid (Table 8-1) shows that to get the classes n^+b t^+ and n b^+t, we must have crossovers in *both* regions 1 and 2. Since this happened 36 times in region 1, we must add this figure to the total of singles which we have already identified in this region. Instead of a total of 229 crossovers, there actually were 229 + 36 or 265. Exactly the same reasoning applies to region 2. Again 36 must be added, this time to 140, giving the correct value of 176 for the amount of crossing over in this region. The value of the triple testcross should now be quite apparent. It enables us to detect in two regions of the chromosome those single crossovers which otherwise would go unobserved. We must not forget to add the doubles identified in any 3-point testcross to each set of singles, those in region 1 and those in region 2 as well.

7. All that remains to be done to obtain crossover values among the three genes is to take the total crossover figure for each single region and divide it by the entire number of offspring. The respective percentages, 26.7% and 17.7%, are simply translated into map units.

SIGNIFICANCE AND USE OF A CHROMOSOME MAP. Once we have a segment of a chromosome map, it is essential to know exactly what it represents and how it can be put to use. Although we may make certain analogies between a chromosome map and a road map, such comparisons should not be taken literally. Although a road map reflects precise distances between points, this is not true of a chromosome map. Although the number of map units between two genes does indeed depend in part on distance, we must not think of the units in terms of accurate spatial measurements. There are several reasons for this. For one thing, the ease with which crossing over occurs along a chromosome is not uniform; there is more crossing over in some parts of the chromosome than in others. For example, the region around the centromere engages in less crossing over than most of the rest of the chromosome. As a result of its

influence, genes in the vicinity of the centromere tend to appear more clustered than genes elsewhere. That this is indeed a centromeric effect has been demonstrated through experiments in which centromere position has been changed as a result of shifts in chromosome segments. Those areas now placed away from the centromere undergo more crossing over. When a map is constructed, the genes in such regions appear farther apart than they previously did when in the vicinity of the centromere (Fig. 8-19).

In addition to the centromere, there are certain other chromosome regions in which crossing over may be cut down and others in which it seems to take place with more ease. Reasons for these local differences in the chromosomes of living things are not known. We must also realize that crossing over, like many other genetic phenomena, may be influenced by environment, both external (temperature, radiations and nutrients, etc.) and internal (hormonal). This latter effect is very clearly demonstrated by the absence of crossing over in the male fruit fly. Therefore, when constructing accurate maps and comparing amounts of crossing over, such factors must be controlled. Since the amount of crossing over can be influenced in several ways, the map cannot be considered a true reflection of distance. However, the amount of crossing over is constant for a given species under a prescribed set of circumstances. Thus, the map can be used to make predictions when a cross is carried out under a defined set of conditions.

If the map can be used in this way to tell us the expected amount of crossing over in different chromosome regions, we should be able to predict from it the number of double crossovers. The map gives us the probabilities of the single events expressed in map units. Since a double crossover is the result of two singles happening simultaneously, we should be able to estimate the probability of a double in a certain chromosome region simply by multiplying together the separate chances for the singles. In our example, the probability is 26.7% that a crossover will occur alone in region 1 and 17.7% that one will take place in region 2. Recalling that the chance for two independent events to happen together is the product of their separate probabilities, we simply multiply the two to find the chance of the double event $(0.267 \times 0.177 = 0.047 = 4.7\%)$. The map as a table of probabilities gives the percentage of doubles to be expected in that particular chromosome region. But let us see if this is borne out by the results of the cross.

We obtained 992 offspring (refer to Table 8-1). Of these 4.7% should be double crossover progeny by our calculations. Therefore, approximately 47 individuals should be distributed between the types n $^+$b t$^+$ and n b$^+$t (992×0.047). The data show that actually only 36 of these types arose from this cross. We might conclude that this lower figure is just the result of chance in this particular instance. However, if we repeated the mating many times, we would continue to find fewer double crossover types than the number predicted from the map. Finding a number of doubles which is smaller than the amount indicated by the map is typical rather than exceptional. This suggests something about the single crossovers themselves. They must *not* be independent events, for if they were, their product should be a reliable prediction of the actual number of doubles which will be obtained in a cross. We are forced to conclude that the occurrence of one single crossover *can* influence the occurrence of another single in the same chromosome. This is particularly so within short regions of the chromosome, where the influence of one crossover on another may be quite pronounced. The term *interference* describes the effect which a crossover in one chromosome

FIG. 8-19. Effect of the centromere. Genes in the vicinity of the centromere generally engage in less crossing over than genes a greater distance away. In this hypothetical map (*above*), note that loci e through m appear clustered, meaning that they are close together in map units. This is due to an effect of the centromere which decreases the frequency of crossing over in its vicinity. The centromere may become shifted (*middle*) due to a chromosome aberration, such as an inversion in which a segment is turned around. When a map is then constructed (*bottom*) on individuals with the new gene order, those genes which had been near the centromere (such as e through i) now appear farther apart in map units. They are engaging in more crossing over. Other genes now placed in the vicinity of the centromere (p, q, and r) map closer together as they now engage in less crossing over.

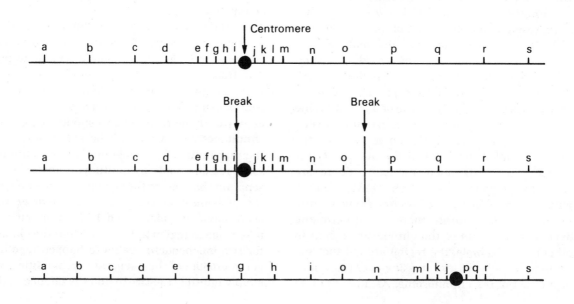

region exerts on the probability of a crossover in an adjacent region.

This interference varies from one part of the chromosome to another. For a more accurate estimate of the amount of double crossovers which will arise in a cross, interference must be taken into consideration. This factor is measured by the *coefficient of coincidence*. This is nothing more than an expression of the ratio of double crossovers actually obtained to those predicted from the map (Table 8-2). We expected approximately 47 doubles but found only 36. The coefficient of coincidence equals the actual number of doubles over the expected number, which in this example is: 36/47 = 0.76. This coefficient tells us that we must modify the number of double crossovers predicted from the map. We expected 47, but knowing the coefficient of coincidence, we multiply this figure by 0.76. This in turn indicates that we should expect only 36 doubles, which is what we actually obtained from the cross. The coefficient of coincidence varies between zero for very short distances (meaning that no doubles at all are to be anticipated) and one (where the expected amount is simply the product of the two singles, and no interference is operating). It is typical for the value of the coefficient to be between the two extremes.

Now let us see how we can use the map to make further predictions. Referring back to the map segment in Table 8-1, we see that we can expect 26.7% crossing over between genes n and b. Similarly, for region 2, we can predict 17.7% crossing over between b and t. Therefore, if 1000 offspring result from a mating, 177 of them (17.7% × 1000) should represent crossovers between "b" and "t." However, remember that the amount of crossing over is *not* equal to the number of *new* combinations, since any one double crossover between two genes will regenerate the old combinations. But knowledge from our map segment can aid us in this regard. We have just calculated that 36 double crossovers are to be expected, taking into consideration the coefficient of coincidence. This figure, 36, was *added* to each set of singles when the genes were being mapped in order to give a more accurate estimate of the true amount of crossing over. Now, however, we must *subtract* the 36 from our expected new combinations, since we will get 36 fewer new combinations in each region (1 and 2) than the distance indicates in crossover units. On the basis of 1000 offspring, therefore, we would expect in region 1 231 *new* combinations (26.7% × 1000 − 36) and in region 2 141 *new* combinations (17.7% × 1000 − 36). What we are doing is using the map to make predictions in terms of probabilities, and that is basically how a chromosome map serves us.

MAPPING THREE GENES ON THE X CHROMOSOME. Once linkage relationships and their major complications are understood, mapping of genes on the sex chromosome should not prove difficult. The procedure is the same as that described above for mapping genes on an autosome and requires no more than an appreciation of what has already been discussed for the determination of crossover values on the sex chromosome. Suppose we are following three sex-linked recessive alleles in *Drosophila*: cross-

---

TABLE 8-2   Calculation of coefficient of coincidence based on data in Table 8-1

Map segment	Number of doubles obtained in actual cross = 36

n   26.7   b   17.7   t

Amount of double crossovers expected from map segment = crossover probability in region 1 × crossover probability in region 2:

$$0.267 \times 0.177 = 0.047 = 4.7\%.$$

Number of doubles to be expected on basis of 992 offspring:

$$992 \times 0.047 = 47$$

$$\text{Coefficient of coincidence} = \frac{\text{no. of doubles actually obtained}}{\text{no. of doubles predicted from map}} = \frac{36}{47} = 0.76$$

---

veinless (missing wing vein), echinus (defective eye), and cut (defective wing). At first, all we know is that they are on the X chromosome, and again we know nothing about their arrangement. Suppose we cross females which are expressing the recessive "crossveinless" with males showing the other two recessive traits, echinus and cut. We may tentatively summarize the cross as shown in Table 8-3. To construct the map segment, we follow the identical steps just given for linked autosomal genes. The only additional thing to remember is that males are always testcross progeny in the case of genes on the X chromosome. The F_2 females, on the other hand, may or may not be. Females are testcross offspring *only* if their fathers were expressing all the recessives. In this example, we see that this is not so. The F_1 male parents are showing the dominant traits wild eye (ec^+) and normal wing (ct^+). Consequently, their daughters will not reveal the results of all the crossing over that occurred in the F_1 females. The F_2 males, however, having received a Y from their fathers, will represent all eight types of gametes (two parental and six crossovers) formed by their mothers. They also indicate the proportion in which those eggs were produced. So, we simply disregard the females and direct our attention solely to the F_2 males for a construction of the map (Table 8-3).

We pick out the parentals which are the most numerous phenotypes: cv^+ ec ct (1099) and cv ec^+ ct^+ (1051). The doubles are clearly cv^+ ec^+ ct^+ (5) and cv ec ct (4). A comparison of the doubles and parentals shows the "cv" appears to have shifted its position and thus must be the more central gene. So the trihybrid female must be rewritten in the proper way: ec^+ cv ct^+/ec cv^+ ct. It is most important at this point to note the genic composition of the parental chromosomes. One chromosome has two wild alleles and one mutant (ec^+ ct^+ cv), whereas the other has the two mutants and one wild (ec ct cv^+). The beginner often assumes that in the parents all three wild alleles must occur together on one chromosome and all three mutants on the homologue. Although this was the case in the example given for the autosomal genes, there is nothing which dictates that this must always be so. A moment's thought tells us that a trihybrid individual can arise in various ways, depending on the way the cross was made.

TABLE 8-3 Trihybrid cross and F_2 data for three X-linked genes*

$$P_1: \quad \frac{cv \; ec^+ \; ct^+}{cv \; ec^+ \; ct^+} \; (\text{♀♀ crossveinless}) \quad \times \quad \frac{cv^+ \; ec \; ct}{Y} \; (\text{♂♂ echinus, cut}) \quad \longrightarrow$$

$$F_1: \quad \frac{cv \; ec^+ \; ct^+}{cv^+ \; ec \; ct} \; (\text{wild females})$$

and

$$\frac{cv \; ec^+ \; ct^+}{Y} \; (\text{crossveinless males})$$

$$F \times F_1: \quad \frac{cv \; ec^+ \; ct^+}{cv^+ \; ec \; ct} \; (\text{wild}) \quad \times \quad \frac{cv \; ec^+ \; ct^+}{Y} \; (\text{crossveinless})$$

F_2:

Females				
Wild 1340				
Crossveinless 1308				
2648				

Males				
Echinus, cut	cv^+	ec	ct	1099
Wild	cv^+	ec^+	ct^+	5
Echinus	cv^+	ec	ct^+	111
Cut	cv^+	ec^+	ct	132
Crossveinless, cut	cv	ec^+	ct	109
Crossveinless, echinus	cv	ec	ct^+	135
Crossveinless	cv	ec^+	ct^+	1051
Crossveinless, echinus, cut	cv	ec	ct	4
			Total	2646

* The order of the loci shown here is only tentative. Consult text for details.

A trihybrid is any individual heterozygous at three loci which are under consideration in a cross. So it would be incorrect to believe that the F_1 trihybrid *must* be: $ec^+ \; cv^+ \; ct^+/ec \; cv \; ct$.

This is just one possibility; it could have resulted if the cross had been made differently (by crossing homozygous wild females with triple mutant males or vice versa). So when given a list of progeny from a 3-point testcross, we must peruse it for the most numerous classes in order to identify the parentals. It is fallacious to assume that the parental types must occur in one set arrangement.

Returning to the correctly written trihybrid, we pick out the crossovers in region 1: $ec^+ \; cv^+ \; ct$ (132) and $ec \; cv \; ct^+$ (135) = 267. Region 2 crossovers are: $ec^+ \; cv \; ct$ (109) and $ec \; cv^+ \; ct^+$ (111) = 220. The doubles total nine and must be added to each of these sets of singles. The final arithmetic becomes: 276/2646 for the crossover distance between ec and cv and 229/2646 for

that between cv and ct. The map for this small chromosome segment is thus: ec 10.4 cv 8.7 ct. The coefficient of coincidence is calculated in the usual manner and gives a value of 0.37, as simple arithmetic will show.

MAP LENGTH, MULTIPLE CROSSOVERS, AND THE AMOUNT OF RECOMBINATION. Chromosome maps have been constructed in considerable detail for certain organisms, particularly *Drosophila, Neurospora,* the mouse, and corn. Figure 8-20 shows the location of some of the better known loci in the fruit fly. When detailed chromosome maps are made, they are usually built up by determining the crossover values between genes that are closely linked. Trihybrid testcrosses are used to establish gene order, but if genes are a considerable distance apart, multiple crossovers (doubles, triples, and even higher degrees) may occur between them to obscure the true crossover value. Therefore,

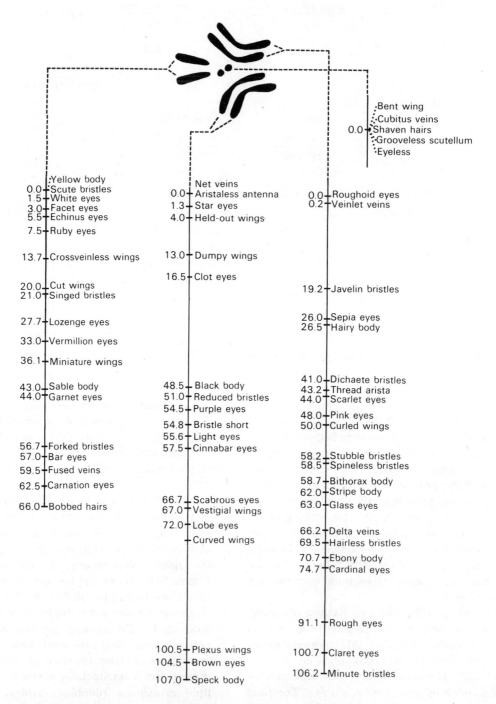

FIG. 8-20. Genetic or linkage map of the four chromosomes of *Drosophila melanogaster*. (Reprinted with permission from E. W. Sinnott, L. C. Dunn, and T. Dobzhansky, *Principles of Genetics*. McGraw-Hill, New York, 1958.)

more accurate map units are determined by studying recombination values between genes close enough to minimize the effects of these factors.

Suppose a gene order such as the following has been established by trihybrid testcrosses: f g h i j k l. The map would then be more precisely built up in map units by finding the crossover value between h and i, then h and g, g and f, and so on. Obviously, this would be more accurate than just relying on recombination values between f and l to reflect the number of map units between these two loci, because multiple crossovers could occur in the f-l interval. To have a more reliable estimate of the distance between f and l in map units, we would add up the small crossover values between the genes in the f-l interval.

It soon becomes apparent from glancing at chromosome maps that they may include more than 50 map units, even more than 100 (note the two large chromosomes of *Drosophila*, Fig. 8-20). We have noted that a good correspondance has been found between the number of chiasmata and the amount of genetic crossing over. The relationship between chiasmata and crossover events is often taken to be 1:1 when estimating map lengths in various species. If the correspondence is close to 1:1, then an average of one chiasma in a bivalent would reflect a map length of 50 units. This would be so, since genetic recombination between the pairs of genes located at the opposite ends of the chromosome would take place 50% of the time. An average of two chiasmata, or three or four per bivalent would give respectively for the specific chromosomes map lengths of 100, 150, and 200 units.

What does this mean in terms of the amount of crossing over to be expected between very distant genes? Does this mean that 60% or even higher amounts of recombination will occur? The answer is that the *number of recombinants cannot exceed 50%*. A moment's reflection tells us that 50% crossing over is equivalent to independent assortment. If one crossover between two linked genes, "a" and "b" (Fig. 8-21A), occurs in every meiotic cell, that will mean that equal numbers of parentals and crossover gametes will form. Indeed, if genes are far enough apart on a chromosome, this will tend to occur, and we may not be able to tell at

first that the two genes are linked; they would be assorting independently, as if they were on separate chromosomes (Fig. 8-21B).

The same principles of map construction used in lower forms can be applied to higher groups. Well over 200 genetic loci have been assigned to specific chromosomes in the human, and the information is being used in a very practical way (see Chap. 9). An estimate has been made of the map length of the human X chromosome and the *total* map length of all the autosomes. This has been reasoned from studying the frequency of chiasmata in meiotic cells of testicular tissue. Recall that a single chiasma is probably equivalent to a crossover value of 50%. Counts of chiasmata average about 52 for all the autosomal bivalents in a meiotic cell. This gives a total map length for all the autosomes as 2600 units. Since the human X chromosome is about 6% the length of one total haploid set of autosomes, its genetic length has been calculated to be at least 160 map units. The figure cannot be considered precise, however, since chiasmata counts were made in male meiotic cells and on autosomes. There is apparently less crossing over in the human male, as reflected in less recombination than in the female. Therefore, the true value of the X chromosome in map units may be greater, closer to 200 units. Nevertheless, the length of the X in map units is seen to be sufficient to account for the fact that certain genes, known to be X-linked still appear to be undergoing independent assortment. Students of human genetics employ the term *synteny* to describe the situation in which two loci have been assigned to the same chromosome, but still they may be separated by such a large distance in map units that genetic linkage has not been demonstrated. The failure to show linkage would result from the fact that crossing over usually separates the two distant loci.

The complete lack of crossing over in the male fruit fly gives a great advantage in showing that even widely separated genes are located on the same chromosome. Remember that only two of the four chromatids participate at any one crossover. Therefore, even when a crossover occurs in every meiotic cell, there will still be only 50% recombination. One might believe that multiple crossovers could increase the recombination value above 50%. A consideration of the different kinds of double crossovers (Fig. 8-22) may clarify this. It was stated in our discussion that a double crossover can involve any of the four strands. We have considered for the sake of clarity only *two-strand doubles,* those double crossovers in which the same two strands participate in each of the two crossover events. Figure 8-22 shows that *three-* and *four-strand doubles* may also take place. Because they are independent events, the doubles occur in a random frequency of 1:2:1 (one two strand double: two three strand doubles: one four strand double). If the numbers of parental and recombinants are added up, it is seen that they total 50: 50. Again the recombination value does not exceed 50%, even on this basis.

What about even higher numbers of crossovers between two genes? The overall effect of multiple crossing over is to place alleles of genes back in parental combinations and thus to reduce the frequency of new combinations to *less than* 50%. This is generally so, even for widely separated genes. Such genes usually produce less than 50% recombinants. In *Drosophila*, the loci "arc" and "black" are over 50 map units apart, but the amount of recombination which is found is less than 50% for all the reasons given here. Therefore, one cannot always look at a map and translate the distance indicated by map units

FIG. 8-21. Effects of a high frequency of single crossovers. *A.* If a crossover occurs between two genes in a meiotic cell, four different kinds of gametes are formed. Only two of the four chromatids participate in any one crossover. If one crossover between a and b occurs in every meiotic cell, it will produce recombinants in a frequency of 50%, and it will appear as if the loci are assorting independently. *B.* The diagram shows independent assortment, assuming a and b are on separate chromosomes. The two possible anaphase separations occur with equal frequency. The result is four kinds of gametes in equal numbers. It can be seen between a comparison of diagrams *(A)* and *(B)* that a single crossover in every cell at meiosis will have the same effect as independent assortment.

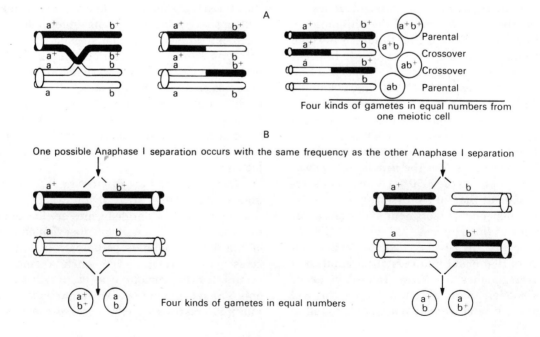

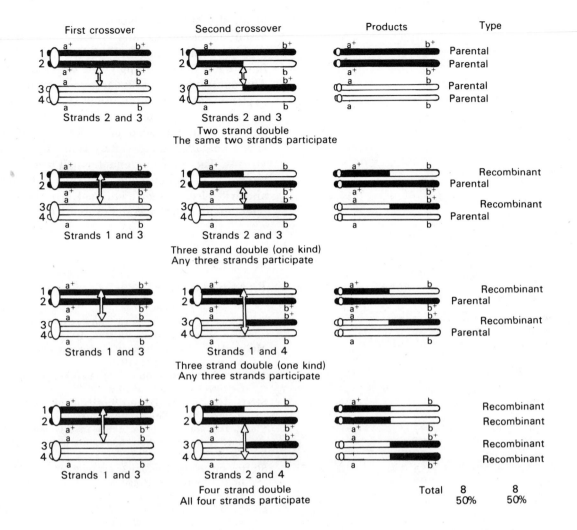

First crossover	Second crossover	Products	Type

FIG. 8-22. Types of double crossovers. Double crossing over can involve any of the four chromatids and may take place in a variety of ways. This figure illustrates just one of these. There are always two different manners in which a three-strand double can occur. Since double crossing over is at random, the two-, three-, and four-strand doubles occur in a frequency of 1:2:1, respectively, as indicated here. The number of recombinant types, however, cannot exceed 50%. (The second crossover usually does not occur at exactly the same position as the first. The diagram here is shown to clarify events between loci a and b.)

into the amount of recombination to be expected. The number of map units corresponds closely to the amount of recombination only for those genes which are relatively closely linked, usually in the vicinity of 20 map units or less.

LINKAGE AND A LINEAR ORDER OF GENES. Linkage and crossover values with all their complications (interference and multiple crossing over, etc.) can be explained only on the basis of a linear arrangement of genes. No sense can be made of the data on the other assumptions. As Muller showed, exchanges occur in blocks (see Fig. 8-14). Janssen's theory of breakage and reunion can explain this. There is even cytological evidence for a linear arrangement. When a piece of a chromosome is accidentally lost as a result of two breaks in a chromosome (details in Chap. 10), the missing region at times may be detected with the microscope. Genetic studies of flies with known cytological deletions have revealed that a gene or a group of genes is missing. We will see in Chapter 11 that the concept of a linear order agrees well with what is known about the molecular biology of the gene. The essentials of linkage and crossing over, which were worked out in much higher species, also have been found to apply to microorganisms. The same idea of a linear order therefore is applicable to these lower groups. Later chapters describe rapid advances in molecular biology and the modern concept of the gene, made possible by detailed linkage analysis in bacteria and viruses. The following chapter will discuss the relevance of a knowledge of linkage to humans and will also present additional aspects of the phenomena of linkage and crossing over.

REFERENCES

Hotchkiss, R. D. Models of genetic recombination. *Ann. Rev. Microbiol:* 28, 445, 1974.

Muller, H. J. The mechanism of crossing over. II. *Am. Naturalist, 50:* 284, 1916.

Nicklas, R. B. Chromosome segregation mechanisms. *Genetics, 78:* 205, 1974.

Stern, H., and Y. Hotta. Biochemical controls of meiosis. *Ann. Rev. Genet, 7:* 37, 1973.

Stern, H., and Y. Hotta. DNA metabolism during pachytene in relation to crossing over. *Genetics, 78:* 227, 1974.

Sturtevant, A. H. The linear arrangement of six sex-linked factors in *Drosophila,* as shown by their mode of association. *J. Exp. Zool., 14:*43, 1913.

Swanson, C. P., T. Merz, and W. J. Young. *Cytogenetics.* Prentice-Hall, Englewood Cliffs, N. J., 1967.

D. von Wettstein. The synaptinemal complex and four-strand crossing over. *Proc. Natl. Acad. Sci., 68:* 851, 1971.

Whitehouse, H. L. K. *Toward an Understanding of the Mechanism of Heredity.* Edward Arnold, London, 1969.

Whitehouse, H. L. K. The mechanism of genetic recombination. *Biol. Rev., 45:* 265, 1970.

REVIEW QUESTIONS

1. Assume that the two allelic pairs, r^+, r and s^+, s are so closely linked that crossing over can be ignored. What kinds of gametes can be formed by:

A. A double heterozygote with the alleles in the *trans* arrangement?

B. A double heterozygote with alleles in the *cis* arrangement?

2. Assume that the two allelic pairs, t^+, t and u^+, u are linked and that the frequency of crossing over between the two loci is 30%.

A. What kinds of gametes will be formed by a dihybrid with the genes in the *trans* arrangement, and what would be the expected frequency of each type considering crossing over?

B. Answer the same question for a dihybrid with the alleles in the *cis* arrangement.

3. Assume that a double crossover takes place in a meiotic cell of the dihybrid in Question 2B. What would be the combination of the alleles resulting from the double crossover between "t" and "u"?

4. In the fruit fly, the allele for purple eye color (pr) is recessive to its allele for red (pr^+). The allele for vestigial wings (vg) is recessive to the wild allele for normal wings (vg^+). The two genes are autosomally linked. Females from a purple stock are crossed to males from a vestigial stock. The F_1 flies are all wild (red eyes and normal wings). F_1 females are testcrossed with the following results:

purple—210

wild— 40

vestigial—215

purple, vestigial— 35

A. Diagram the testcross.
B. Calculate the amount of crossing over.

5. Give the results to be predicted if an F_1 dihybrid male from the cross described in Question 4 is testcrossed, assuming 500 offspring.

6. In tomatoes, round fruit (o^+) is dominant to long (o). Simple flowering shoot (s^+) is dominant

to branching flowering shoot (s). Plants from two different pure-breeding varieties are crossed. One variety bears round fruit and has branched flowering shoots. The other variety has long fruits and simple flowering shoots. F_1 plants were testcrossed, and the following progeny were obtained:

round, simple—23

long, simple—83

round, branched—85

long, branched—19

A. Diagram the testcross.

B. Calculate the amount of crossing over.

7. Using information from Question 6, assume that two dihybrids having the alleles in the *cis* arrangement are crossed to each other.

A. Give the kinds of gametes and the frequency of each expected from each parent.

B. Give the phenotypes and the expected frequencies from a cross of two such hybrids with the *cis* arrangement.

C. Assuming 1000 plants are obtained in this cross, how many of them would you expect to have the combination: long fruits and simple flowering shoots?

8. In chickens, the dominant allele, I, prevents pigment formation so that the feathers are white. The recessive allele (I^+) permits the feathers to be colored. This pair of alleles is autosomally linked to a pair which influences the development of the feather. The allele F^+ results in normal feathers, but the allele, F, causes brittle feathers when in the homozygous condition. However, there is a lack of dominance so that the heterozygote has mildly brittle feathers.

A hen with colored, brittle feathers is mated several times to a rooster with white, normal feathers. Over three dozen eggs are produced which yield chicks with white, mildly brittle feathers. Diagram the cross, giving the genotypes of the parents and the F_1.

9. F_1 hens from the cross in Question 8 are crossed to roosters having colored, normal feathers. The following chicks are obtained:

> white, mildly brittle—17
> colored, mildly brittle—64
> white, normal—61
> colored, normal—15

A. Diagram this cross.

B. Give the genotypes of all the kinds of birds.

C. Calculate the amount of crossing over.

10. In the fruit fly, the allele w^+ for red eye color is dominant to its allele for white eye, w. This pair of alleles is sex linked, as is another pair which influences body color: the dominant for gray body (y^+) and the recessive for yellow (y). Females from a pure-breeding, white-eyed, gray-bodied stock are crossed to males from a stock pure breeding for red eye and yellow body. Give the genotypes of the P_1 and the genotypes and phenotypes of the F_1.

11. F_1 females from the cross described in Question 10 are crossed to white-eyed, gray-bodied males. From the results presented below, calculate the amount of crossing over between alleles at the white (w) locus and alleles at the yellow (y) locus.

Females: red-eyed, gray-bodied—340
white-eyed, gray-bodied—350

Males: white-eyed, gray-bodied—346
red-eyed, yellow-bodied—324
red-eyed, gray-bodied— 9
white-eyed, yellow-bodied— 6

12. In the human, two pairs of alleles d^+, d and p^+, p are associated with color perception. The recessives, d and p, can result in color blindness. Both loci, d and p, are found on the X chromosome. Linkage is so close that the amount of crossing over is very low.

A. Ignore crossing over, and diagram a

cross between a color-blind woman who is homozygous for the recessive d but who is homozygous for the dominant allele at the p locus and a man who is color blind due to the recessive he carries at the p locus. Show the expected offspring.

B. Assume a dihybrid female with the color perception alleles in the *trans* arrangement marries a man who carries neither recessive. Diagram the cross and show the offspring, ignoring crossing over.

C. In the last case, show the expected results if crossing over occurs.

13. The dominant allele o^+ is required for pigment deposition in the iris of the human eye. Its recessive allele, o, is responsible for ocular albinism. The locus is found on the X chromosome, as is the d locus associated with color blindness which was mentioned in Question 12.

A. Assuming no crossing over, what are the possible results of a cross between a woman with ocular albinism who is homozygous for the allele for normal color vision and a man who has a pigmented iris but who is color blind due to the recessive d.

B. Assuming no crossing over, give the results of a cross between a woman who is dihybrid at the albinism and color vision loci (*cis* arrangement) and a man with both normal traits.

14. The amount of crossing over between the linked genes e and s is found to be 11%. The gene r is found to be linked to e and s. Alleles of the r gene engage in crossing over with those of the s gene with a frequency of 7% and with those of the e gene with a frequency of 4%. What is the order of these genes on the chromosome and the distance between them in map units?

15. Given the following portion of a chromosome map, answer the questions below:

d   14   m   18   p

A. The amount of double crossing over in this region is found to be 1.5%. What is the coefficient of coincidence?

B Alleles of the s gene are found to be linked to alleles of the genes shown on the map. A crossover value of 13% is found between the s and the p genes. Where should the s gene be placed on the map?

16. In corn, the genes v, b, and l are linked. The data given below summarize the results of a trihybrid testcross. From the data:

A. Give the correct genotype of the trihybrid parent showing the correct sequence of genes.

B. Construct a map segment showing the distances among the genes.

C. Calculate the coefficient of coincidence.

v^+	b^+	l	304
v^+	b	l	119
v	b	l	18
v	b^+	l	70
v^+	b	l^+	64
v^+	b^+	l^+	22
v	b^+	l^+	108
v	b	l^+	295

17. In the Oriental primrose, short style (l^+) is dominant to long style (l). Magenta flower color (r^+) is dominant to red (r), and green stigma (g^+) is dominant to red (g). A trihybrid testcross is performed. From the data presented below, construct a map giving the positions of the three genes and the distances involved.

l^+	r^+	g	292
l^+	r^+	g^+	153
l	r	g^+	286
l	r	g	139
l^+	r	g^+	36
l^+	r	g	22
l	r^+	g	40
l	r^+	g^+	18

18. In *Drosophila*, assume the three genes, x,

y, and z, are linked and that the alleles x^+, y^+, and z^+ are dominant to x, y, and z. Crosses of trihybrid females and males showing the three dominant traits, x^+, y^+, and z^+, give the results tabulated below. From the data:

A. Give the genotypes of the P_1 males and females, showing the correct sequence of the genes.

B. Construct a map giving the locations and distances.

C. What can you say about the interference in this map region on the basis of the data?

Females:	$x^+y^+z^+$	1012
Males:	$x^+y^+z^+$	30
	x^+y^+z	32
	$x^+y\ z^+$	441
	$x^+y\ z$	2
	$x\ y^+z^+$	3
	$x\ y^+z$	430
	$x\ y\ z^+$	27
	$x\ y\ z$	38

19. The following is a map segment showing the location of loci, r, s, and t:

$$\underline{r \qquad 12 \qquad s \qquad 10 \qquad t}$$

The coefficient of coincidence in the region is equal to 0.5. The following cross is made in the fruit fly:

$$\frac{r^+s^+t}{r\,s\,t^+} \times \frac{r^+s^+t^+}{Y\,\text{chromosome}}$$

Answer the following on the expectation of 4000 offspring:

A. How many will be wild-type females ($r^+s^+t^+$)?

B. How many males will be of the constitution $r^+s\ t$?

C. How many males will be r^+s^+t and $r\ s\ t^+$?

20. In the fruit fly, the allele for sepia eye color (se) is recessive to the allele for red (se$^+$). Straight wings (cu$^+$) are dominant to curled (cu). The two genes, se and cu, are linked on an autosome, Chromosome III. The amount of crossing over between them is 24%. A cross is made between a female from a sepia-eyed stock and a male from a curled stock. The F$_1$'s are wild (red-eyed with straight wings). They are crossed to one another and produce an F$_2$.

A. What kinds of offspring will appear in the F$_2$, and in what ratio?

B. The se gene is also linked to the e gene (for ebony body). The allele e is recessive to e$^+$ for gray body. Diagram a cross between a female from a sepia stock and a male from an ebony stock. Can you tell the amount of crossing over between the two genes on the basis of the F$_2$ results? Explain.

FURTHER ASPECTS OF LINKAGE AND CROSSING OVER

ASSIGNMENT OF LOCI IN THE HUMAN. The subject of assignment of genes to chromosomes in the human raises the question of the actual gene number present in the human chromosome complement. The exact figure is certainly not known, but the amount of DNA can be estimated, and from this value some suggestion can be made. Sufficient DNA exists for as many as 5 million different kinds of genes which could determine the amino acid sequence of polypeptide chains of proteins, so called structural genes which are transmitted according to classical Mendelian principles (See Chap. 19). However, the true gene number is believed to be well under this estimate, perhaps more on the order of 50,000. There are several reasons why the lower number is probably closer to the truth. For example, much of the DNA in the human is known to exist in more than one copy, meaning that a lot of gene repetition exists (Chap. 17). Moreover, some of the DNA of the chromosomes is not associated with genes which govern the formation of protein products. Some DNA such as that composing the centromere region, is involved with the very structure of the chromosome itself. A considerable portion is also concerned with control mechanisms as well as with the formation of transfer RNA and ribosomal RNA. Much of this kind of DNA is not associated with specific protein products and would not be recognized as genes which are inherited in a Mendelian fashion.

In the human, we are aware of well over 1200 specific genetic loci, over 1100 on the autosomes and over 100 on the X chromosome. Of this number, about 900 are associated with disorders. However, probably no more than 1 out of 50 of the genes of the human have actually been identified.

Until 1967, the study of family pedigrees was just about the only method available for the recognition of human linkage groups. Linkage of genes to the X chromosome can be established much more readily than the linkage of two or more genes on an autosome. Consequently, the first advances in assignment of loci were made with the X. The reason that linkage of genes to the X chromosome is more easily detected stems from the fact that pedigree analysis may show more affected males or a criss-cross pattern of inheritance. As noted, over 100 human genes have been found to be sex linked. The actual map distances have been calculated for some of these through analyses of many pedigrees and the application of the fundamental principles of sex linkage and genetic recombination. For example, knowing that the locus for deutan color blindness and the one for the production of the enzyme glucose-6-phosphate dehydrogenase (G_6PD) are sex linked, one can proceed to study family pedigrees which include doubly heterozygous mothers and their sons. The pedigree must be complete enough to establish whether the alleles in any heterozygous woman are in the cis or the trans arrangement. Such information can be obtained if the pedigree also includes information on the woman's father. If the latter, for example, is color blind and cannot produce G_6PD, whereas the woman expresses both dominant traits, then we know that her genotype is: $\dfrac{gd^+ d^+}{gd^- d^-}$. We can deduce this since her father can only give her an X chromosome carrying both recessives.

Similarly, we might reason that another woman carries these alleles in the trans arrangement. The pedigrees are then examined with respect to the sons of such women whose genotypes are known as to the cis or the trans arrangement. When this was actually done, it was established that 1 son in 20 of doubly heterozygous mothers showed new combinations of the alleles, meaning that crossing over must have occurred in 5% of the cases (Fig. 9-1). The two loci can then be considered 5 map units apart. Studies of this type have established the map distances for several X-linked genes. The loci for hemophilia A and protan color blindness are so closely linked to those for deutan color blindness and G_6PD production that the precise order of the loci in the linkage group is still unknown.

There are genes known to be located on the X chromosome which, however, seem to be assorting independently from other genes whose loci are also known to be X-linked. The reason for this is undoubtedly due to the length of the X (Chap. 8), so that two loci which are appreciably far apart may be separated frequently by crossover events and so appear to be assorting independently. Many students of human genetics would say that two such genes are syntenic (are on the same chromosome) but that linkage has not been established between them. In our discussions, we will continue to use the term "linkage" in the classical sense, two genes being linked if they are on the same chromosome.

In contrast to the X chromosome, it is a much more difficult task to establish that two or more loci are linked on an autosome, when working with family histories. Even when this is accomplished, pedigree analysis does not allow assignment of the linkage group to a specific autosome. Those pedigrees most useful would be those in which one parent is doubly heterozygous and the other doubly recessive (a testcross). Again knowledge of whether the double heterozygote has the alleles under consideration in the cis or trans arrangement is necessary for precision. Those loci at which dominant or codominant allelic forms reside are valuable in pedigree studies of autosomal genes. A good example of such a valuable locus is that for the ABO blood grouping. At least three gene forms are possible at this locus rather than two: the allele for A blood antigen, the allele for B antigen, and the one for O type (no A or B antigen present). We call this a multiple allelic series, meaning that more than two allelic forms exist and that any one member of the group can be found at that locus on a given chromosome. The allele for A antigen is dominant to the O allele. However, A and B alleles behave as codominants to each other so that A and B antigens will both be present when both A and B alleles are present, as in genotype AB. This makes the codominants A and B (and others like them) valuable as marker genes when studying linkage groups, since we know with certainty when A or B alleles are present in a genotype. Moreover, when more than two alleles are possible at a locus, as in a multiple allelic series, more variety is possible among persons in the population and the chances are increased that any one person is heterozygous at that locus. All of these features make it easier to follow an allele at such a locus along with some rare dominant allele. We can be more certain that the rare dominant is in a given combination with an allele at that locus and can study that rare dominant in several families in combination with the different possible alleles at the locus. This accelerates the possibility of detecting whether the dominant tends to be inherited along with the different alleles at that

FIG. 9-1. Use of family pedigrees in linkage analysis. To ascertain the amount of crossing over between the locus for deutan color blindness and the locus for G_6PD production, pedigrees must be complete enough to establish whether the doubly heterozygous mothers of sons being studied carry the genes at these loci in the *cis* or in the *trans* arrangement. This requires information on the maternal grandfathers. Once the *cis* or *trans* arrangement is determined (only one possible *trans* arrangement is shown here), a count of the sons is made to determine how many are recombinants and how many nonrecombinants. In this case, 5% of the sons of doubly heterozygous mothers were found to be recombinants, giving a map distance of 5 units between the two loci.

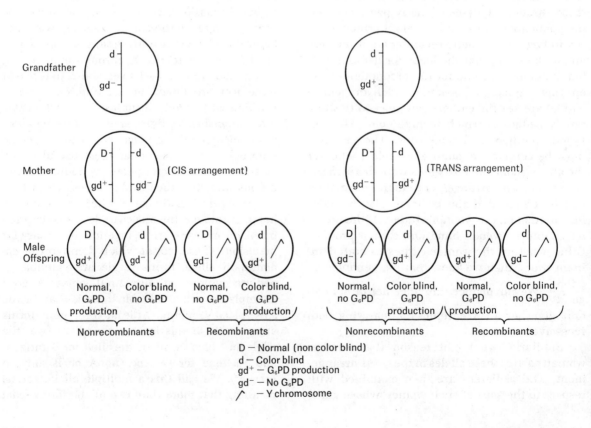

D — Normal (non color blind)
d — Color blind
gd^+ — G_6PD production
gd^- — No G_6PD
⌃ — Y chromosome

locus. Thus we can establish linkage between the two loci, if it exists, much more rapidly than if less variety were possible and there was less chance of knowing when a given gene was present in the genotype.

Completely recessive autosomal alleles provide the most difficult problems of all, and linkage cannot be determined as precisely for them from family pedigrees. Frequently mathematical procedures are used in such cases to estimate the chances that two loci are autosomally linked.

Early in the 1950's, the locus for the secretor trait (described later in this chapter) was shown to be linked on an autosome with the locus for the Lutheran blood group (a locus associated with certain red blood cell antigens) by analyzing family histories. We still do not know the precise chromosome on which these occur. A few years later, similar studies showed that the ABO locus is linked to the locus responsible for a disorder known as the nail-patella syndrome in which the fingernails are abnormal as well as the knee caps, which may actually be missing. Also established was linkage between the locus for the Rh blood types and one associated with elliptocytosis, an abnormality of the red blood cells. Today we know that the former two loci are on Chromosome 9, the latter on Chromosome 1.

Today, family pedigrees continue to be used in the assignment of loci to linkage groups and in the mapping of chromosomes, but complementing this is the important method of somatic cell genetics which has greatly accelerated the assignment of genetic loci in the human.

SOMATIC CELL HYBRIDIZATION Mapping by somatic cell hybridization entails the fusion of somatic cells of different species. It was found that certain inactivated viruses (such as the Sendai viruses) can so alter the properties of cell membranes that cells of unrelated organisms may fuse, forming a hybrid cell containing chromosomes derived from both groups. The human cell types employed in such a procedure are often the fibroblast, obtained from bits of skin growing in tissue culture, and the white blood cell. The cells selected would be chosen for their ability to carry out a specific biochemical reaction or to produce some easily detected product. Suppose that the human cells carry a gene form responsible for the formation of an enzyme (enzyme A)

and that cells of a certain mouse strain do not. After fusion between the cells of the two species, the hybrid cell divides. However, there is a pronounced tendency for human chromosomes to be lost at each cell division (Fig. 9-2) until the cells eventually contain only a few of the human chromosomes. Hybrid cell lines with a small number of human chromosomes are then established and their chemical properties are studied. If enzyme A is found to be present *only* in those lines which contain a certain chromosome (say 17), and if it is always absent when that chromosome (17) is absent, then the gene locus controlling the formation of enzyme A can be assigned specifically to that particular chromosome. In this way, more than one gene may be associated with a single chromosome, and consequently, two or more genes would fall into a linkage group. The cell hybridization method also permits the assignment of loci in humans which do not control the production of a biochemical product. For example, loci associated with susceptibility of cells to viruses and certain drugs can be assigned by this procedure.

Genes linked to the X chromosome as well as to autosomes can be identified by this method as well as by pedigree analysis. The procedure has provided a great impetus to the assignment of genes to specific autosomes, a most difficult task on the basis of family histories alone. The cell fusion technique depends on the ability to select out the hybrid cells which have actually formed following the mixture of two cell lines. The number of hybrid cells formed constitutes a very small proportion of the cells in the mixed culture, and these must be isolated from the bulk of the unfused ones. In order to screen these hybrids, each of the two parental cell lines used must be deficient in some way. For example, say line I carries a mutant gene which prevents its forming enzyme I, whereas line II cannot produce enzyme II as a result of a different genetic defect. As a consequence of these defects, each parental line, if it is to grow in a culture medium, must be supplied with a specific product, let us say a purine for line I and a pyrimidine for line II. If the cell fusions are carried out on a growth medium deficient in both the purine and the pyrimidine, neither parental type will grow. Only hybrid cells will survive, since line I supplies enzyme II and line II contributes enzyme I. Once the hybrid cells are thus selected, they can then be followed as outlined above in respect to the assignment of a specific locus to a specific chromosome.

The cell hybridization method has greatly accelerated human chromosome mapping. However, the family method continues to be very productive as well. Pedigree analyses are still required to measure map distances between two loci. The cell hybridization method does not reveal distances. The two methods actually complement each other. For example, if the family method shows that two or more loci are autosomally linked and a certain distance apart in map units, the cell hybridization method may establish the exact chromosome to which one of the members of the linkage group belongs due to the fact that the one gene produces a product (say an enzyme) which can be followed in somatic cell studies as already discussed. In this way, the ABO blood locus was assigned to Chromosome 9, since a genetic locus to which family studies showed it is linked controls the produc-

FIG. 9-2. Cell fusion and human linkage groups. Hybrid cells may be formed from fusion of human and mouse cells. However, in the divisions of the hybrid cells, the human chromosomes tend to become lost from the cell nuclei. Cells can eventually be isolated which contain just a few human chromosomes in proportion to those of the mouse. Such cells may give rise to various cell lines, carrying different combinations of specific human chromosomes. If an enzyme is present in the original human cells and not in those of the mouse, the cell lines derived from the hybrid cells can be tested for the presence of the enzyme. If the enzyme is present only in those cells carrying a certain chromosome, and if it is always absent when the chromosome is not present, then the gene responsible for the production of the enzyme can be assigned to that chromosome. In this simplified example, it can be seen that Chromosome 17 is associated with the hypothetical enzyme, A.

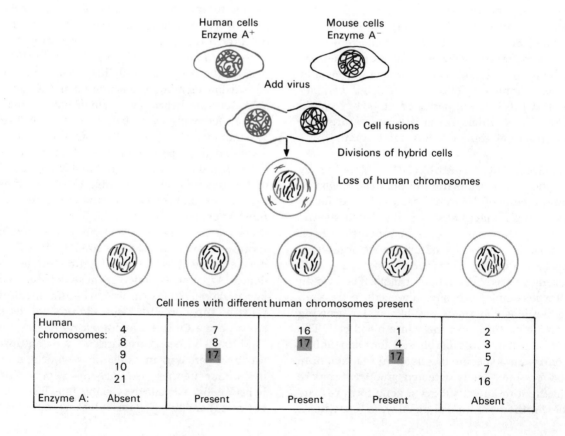

Human chromosomes:				
4	7	16	1	2
8	8	17	4	3
9	17		17	5
10				7
21				16
Enzyme A: Absent	Present	Present	Present	Absent

Cell lines with different human chromosomes present

tion of an enzyme which is detectable in cell cultures.

There are a few other methods used to assign genes to chromosomes, but the family and the cell hybridization methods have provided most of the information. At least one genetic locus has been assigned to each of the 22 human autosomes, in addition to the X and the Y chromosomes. We know over 20 specific loci which are located on Chromosome 1. A whole array of traits is now associated with genes on specific chromosomes: production of various kinds of specific enzymes; clinical disorders such as the nail-patella syndrome, cataract, deficiency of clotting factors; factors involved in resistance of cells to infection; production of certain hormones; and more. In all, approximately 210 genetic loci have been assigned to specific chromosomes, about 110 of these being assigned to autosomes and 100 to the X chromosome.

Precise assignment of genetic loci to specific arms or parts of chromosome arms has also been achieved for some genes. This has been made possible through studying human cells in which chromosome breakage and rearrangement of some kind have occurred (Chap. 10). For example, a piece of one chromosome may be shifted or translocated to another chromosome. The piece may be identified by studying the pattern of chromosome bands (Chap. 2). If it is found that a band or bands are missing from one chromosome and now have been inserted into another, cell hybridization may show that a certain gene product, usually associated with a given chromosome, is now associated with another chromosome. The knowledge that the band is missing from a specific location in the first chromosome and is now associated with a new position in another chromosome, tells us that the locus associated with the product is normally found in the region of that band on the normal chromosome.

Deletions, chromosome aberrations in which segments of chromosomes are actually missing, have similarly been used to assign genetic loci to specific locations on given chromosomes. A given chromosome band or a portion of a chromosome arm may be found missing in some cell lines. If absence of the band can be associated with absence of a product known to be controlled by a locus assigned to that chromosome and if presence of the band is associated with presence of the product, then the particular genetic locus can be related precisely to that portion of the chromosome. For example, a person who was heterozygous for Rh blood antigen (Rh^+ is dominant to rh^-) was found to be producing both Rh^+ red blood cells and rh^- ones. The person was then found to have a cell line with a deletion in Chromosome 1 in the short arm, away from the centromere. Thus the Rh locus, known to be on Chromosome 1, was assigned to that specific region. Further significance of chromosome mapping in the human will be discussed in Chapter 10 in relation to structural chromosome changes and chromosome evolution.

THE APPLICATION OF LINKAGE INFORMATION IN THE HUMAN. The close linkage known to exist between the hemophilia A and G_6PD loci can be put to use in prenatal diagnosis. Suppose it has been established (Fig. 9-3) that a woman is a carrier for hemophilia. She received from her hemophile father the defective hemophilia allele as well as the linked allele enabling her to produce G_6PD. Her mother donated an X carrying the normal allele for blood clotting and the recessive for G_6PD production. The alleles are in the *trans* arrangement. Because the two genes involved are so closely linked, there is just a small chance for crossing over to occur. This woman will, of course, donate one X chromosome to her male offspring, who receive a Y from the male parent. There is, therefore, a 50% chance that a son will receive the defective allele for hemophilia. In this case, the allele is carried along on the same X with the normal allele for enzyme production; so any hemophile son will probably produce G_6PD. A nonhemophile son would lack the ability to produce the enzyme. Amniotic cells may be withdrawn during pregnancy. If they indicate a male on the basis of the lack of a Barr body and the presence of a fluorescent Y body and if the enzyme is shown to be present, the probability is very high that the woman is carrying a hemophile male embryo. In a case such as this, the enzyme locus is a marker; the presence of the product (the enzyme) associated with the locus, G_6PD, tells us that the defective allele is most likely present too. The mapping of human chromosomes thus is seen to have a decided value; it can increase the number of human disorders susceptible to prenatal diagnosis.

The disorder myotonic dystrophy is the first serious human affliction to be assigned to a gene which is a member of an autosomal linkage group. The dystrophy locus, as noted, has been found to be linked to two others, one which is responsible for the blood antigens of the Lutheran grouping and the other which determines the secretor trait, the secretion of A and B antigens into various body fluids such as the saliva. When the secretor allele (S) is carried by a fetus, it can determine the secretor phenotype of the amniotic fluid (whether A or B antigens are present). This fact, plus that of close linkage with the dystrophy locus, has a very valuable application because the dystrophy allele has reduced penetrance and delayed onset of expression. Moreover, it produces no product in the body fluids and thus can remain completely undetected until it strikes a victim. But the knowledge of linkage can guide us in some cases. Suppose, as illustrated in Figure 9-4, that a male parent is known to carry the dominant secretor allele on the same chromosome as the dominant allele for the dystrophy. The man's wife is found to be a nonsecretor. Since she is phenotypically normal and is not related to her husband by blood, it is almost certain that she is not carrying

FIG. 9-3. Detection of a defective sex-linked allele. In this pedigree, the P_1 female parent is known to have received an X from her father which carries the recessive hemophilia allele and the dominant for the production of G_6PD. The X she received from her mother carries the contrasting gene forms. This information, plus the fact that the two loci are tightly linked on the X, can be put to practical use in prenatal diagnosis. Any male offspring receives one X from his mother, making the chance 1/2 that any son of the P_1 woman will be a hemophile. The enzyme locus can be used as a marker to check for the presence *in utero* of the defective allele. Because the two loci are closely linked, the crossover eggs will be much rarer than the noncrossover types. Therefore, if the amniotic cells indicate a male offspring and if the enzyme is present, the chances are very high that the unborn son has received the parental X carrying the recessive for hemophilia.

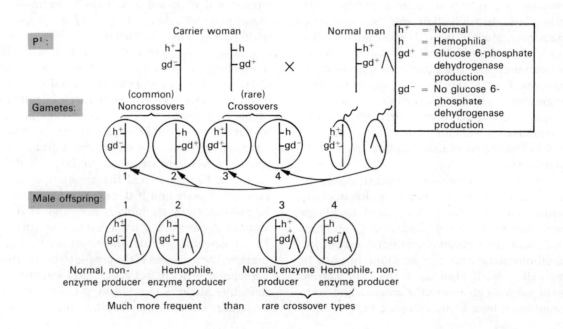

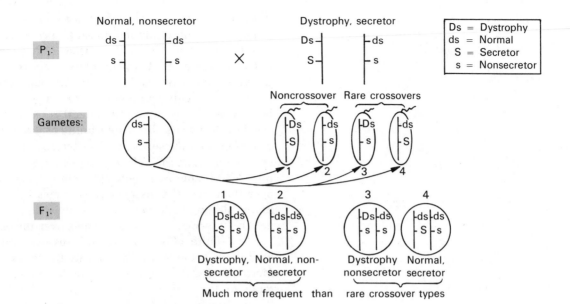

FIG. 9-4. Detection of a defective autosomal allele. The locus associated with myotonic dystrophy and the secretor locus, which determines the ability to secrete antigens into body fluids, are autosomally linked. In this case, the male parent is known to carry the dominant factors for dystrophy and for the secretor ability on one chromosome; the recessive alleles are on the other. His wife is homozygous recessive. Since the two loci are tightly linked, the crossover gametes produced by the male would be much less frequent than the noncrossovers. If the amniotic fluid reveals antigens of the A or B type, this means the secretor allele is present and that most probably the embryo carries the dominant for dystrophy.

an unexpressed, dominant dystrophy allele. If she is blood group AB and the man is group A, the offspring may be blood type A, B, or AB (see Chap. 14 for details). Antigen A, B, or both A and B will be secreted into the amniotic fluid if the fetus is a secretor. If a couple such as this is concerned about the possibility of their unborn child being afflicted with dystrophy, amniotic fluid can be withdrawn. If it is found to give a positive reaction for the presence of A or B antigens, this would mean that the secretor allele from the male parent is present. Since it is closely linked to the dystrophy locus, there is only a small chance for crossing over to take place. Therefore, the probability is very high (actually over 90%) that the fetus carries the harmful allele. Abortion thus can be seriously considered. If the fluid should test negative for A and B antigen, it would mean that the offspring received from the father the chromosome which carries the nonsecretor allele. Since this is closely linked to the normal allele, the chances are over 90% that the dystrophy allele is not present. The value of the knowledge of linkage is quite obvious

in this example. Even though the exact chromosome on which the alleles are located is still unknown, the knowledge that they are linked is all that is needed for the prenatal diagnosis. If this can be extended to include loci for other serious disorders, prenatal genetic counseling will have a much wider application and will provide very accurate predictions.

FACTORS INFLUENCING CROSSOVER FREQUENCY. Because crossing over is of such universal importance, biologists have tried to gain a precise understanding of its exact nature and the factors which may influence it. Although answers are still not known to many of the questions concerning the mechanism of crossing over, several important facts have been established. Various factors which influence the frequency of crossing over have been recognized throughout the course of genetic investigations. The effect of the centromere in decreasing the amount of crossing over has already been briefly discussed (see Fig. 8-19). The ability of the internal or hormonal environment to alter the crossover frequency is clearly seen in *Drosophila,* where normally no crossing over occurs in the male. This effect of sex is seen in a few other species as well. In the female silkworm, crossing over is normally absent. Although the occurrence of equal amounts of crossing over in both sexes seems to be the general rule for all species, it appears that *when* a difference does occur between the sexes, it is the *heterogametic* sex which shows the reduced frequency. Female mice show approximately 25% more crossing over than males. Crossover frequencies in humans, although still not firmly established, suggest a 40%–50% excess in females over males.

A definite effect of age on crossover frequency was found by Bridges in *Drosophila.* More crossing over occurs in young females and tends to decrease in frequency with age of the fly. In humans, the effect of age on crossing over has not been studied sufficiently to give a completely accurate appraisal. However, counts of the number of chiasmata in the meiotic cells of males show an increase with age, whereas in females, there is a reduction. The few chiasmata observed in older oocytes seem to be associated with an increase in the number of unpaired chromosomes. This in turn could reflect the effect of maternal age on the frequency of nondisjunction.

Temperature must be controlled when precise measures of crossing over are desired. Departures from the normal temperature range, that is, extremes of heat and cold, tend to raise the frequency of crossing over. Radiation has a pronounced effect on the amount of crossing over and definitely acts to stimulate it in the fruit fly. Both the temperature and the radiation effects seem to encourage more crossing over in regions where it is normally reduced. For example, both of these environmental factors increase the amount of crossing over in the region of the centromere. They may even induce it in the male fruit fly and stimulate its occurrence in body cells, where it is normally inhibited.

In the 1950's, R. P. Levine showed that certain divalent cations such as Ca^{++} and Mg^{++} could alter the amount of crossing over. Flies raised on a diet with an excess of calcium showed less crossing over after a 4-day period than those on normal food medium. The decrease persisted for over a week. When the metallic ions were reduced below that of normal (by adding the chelating agent, Versene), the crossover frequency increased by a significant amount.

Alterations in chromosome structure (Chap. 10) may decrease the amount of crossing over. The presence of a deletion, for example, in one chromosome may interfere with crossing over in the chromosome parts which *do* pair properly (see Fig. 10-16). Inversions, reciprocal translocations, and duplications all seem to cause a decrease in the crossover frequency, as if the alteration in chromosome structure were somehow causing an interference with the normal forces of attraction between the homologous chromosomes.

CYTOLOGICAL PROOF OF CROSSING OVER. The model of crossing over, as suggested by Janssens, implies that a definite physical exchange actually occurs between homologous chromosomes. An experimental test of this idea encounters serious difficulty if one considers the following point. If the idea is correct and if an exchange of chromosome segments does occur between maternal and paternal chromosomes, one could not normally detect this exchange cytologically. The reason is that homologous chromosomes are alike in size and shape. Consequently, a reciprocal exchange between them at a particular point will result in two chromosomes that are identical to the originals (Fig. 9-5). This difficulty was surmounted independently and in the same year (1931) by two separate investigations, that of Stern with *Drosophila* and that of Creighton and McClintock working with corn. In both studies, the investigators appreciated the need to obtain for cytogenetic analysis a pair of homologous chromosomes which did *not* appear identical when observed with the microscope. Both experiments are equally elegant in their design and represent a landmark in the correlation between genetics and cytology. However, only the work of Stern will be presented here, because the two procedures have many features in common and rest on the same reasoning.

Making use of accidental chromosome aberrations, Stern was able to develop female flies in which the two X chromosomes differed conspicuously in appearance (Fig. 9-6). One of the X's was broken into two pieces. The one portion

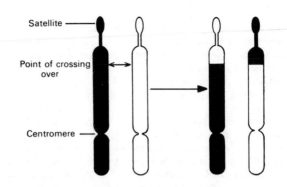

FIG. 9-5. Cytological effect of crossing over. According to the popular model, crossing over is reciprocal and involves the exchange of homologous segments. Therefore, no matter what the morphology of the chromosomes, after crossing over they will be exactly the same morphologically as before.

remained attached to the centromere of the X. The second piece became associated with a centromere derived from the small Chromosome IV. The other was also altered in morphology, as a portion of a Y chromosome joined it at one end, giving the impression of a tail. The "broken X" carried two mutant alleles in the part associated with the X centromere. These were the recessive "car" (carnation or pinkish eye color) and the dominant B (bar or reduced eye size). The X with the piece of the Y contained the wild type alleles. Stern crossed females with the morphologically different X's to males which had carnation eye color (refer to Fig. 9-6). He reasoned that if crossing over *does* entail some kind of physical exchange of chromosome material, then a crossover between the carnation and the bar loci should produce two chromosomes different in appearance from the two original ones. A chromosome of normal size and shape would result and also a broken one with a tail. Since he knew that "car" and B were present on the original "broken" chromosome and the "+" alleles were on the X with the tail, Stern had the chromosomes marked genetically as well as cytologically. He was actually performing a testcross. Since the male parent had contributed both recessives, "car" and B^+, he was able to recognize recombinant offspring among both the females and the males. Figure 9-6 shows that the recombinant types would be of two distinct phenotypes: carnation and bar. The two parental types would be carnation-bar and wild. When he examined the flies cytologically, Stern found in the carnation

FIG. 9-6. Cytological demonstration of crossing over. Stern utilized females having a pair of X chromosomes which differed in appearance. One of them was broken and carried the recessive "carnation" and the dominant "bar." The other X chromosome appeared to have a tail due to an attached segment of the Y chromosome. This chromosome carried the wild alleles. Females of this constitution have red eye color (wild) but are bar eyed. They were crossed with carnation males whose X was normal in appearance. If an exchange of chromosome segments occurs when crossing over takes place between the carnation and bar loci, this should alter the appearance of the X in the female parent. Two chromosomes of different morphology should arise in the eggs: a normal-shaped X and a broken one with a tail. When Stern examined the F₁ offspring, he found these chromosomes in the crossover offspring. The noncrossovers carried the original maternal X's.

male flies a normal X chromosome. In the bar males, a broken X with a tail could be seen. Since a male gets his X chromosome from the maternal parent, this means that the "new" kinds of X chromosomes came from the female parents which did not possess them in their body cells. The X of normal morphology and the one with the tail must have arisen as a result of crossing over in meiotic cells of the female parents. The "new" chromosomes were also found in the female offspring as predicted, along with the structurally normal one from the male parents. In the nonrecombinant offspring, only parental kinds of X's were present. Thus, we see again a

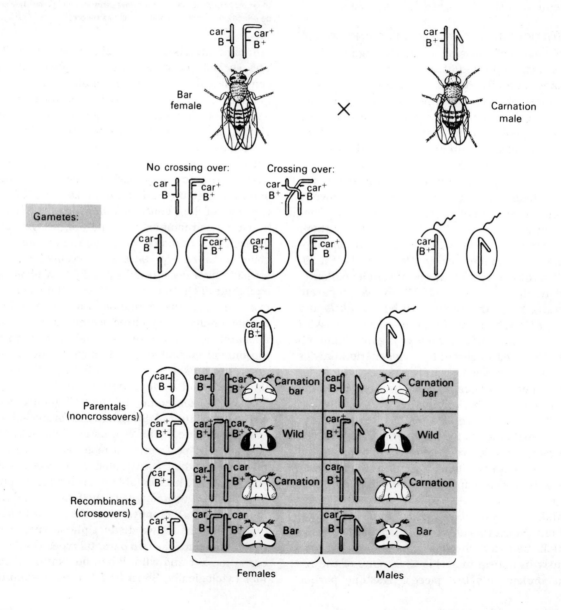

perfect correlation between cytology and genetics (recall the work of Bridges on nondisjunction). When something new arose genetically in this example (the new combination of linked alleles), something new was created cytologically (chromosomes of different morphology from the parental ones).

This work of Stern plus that of Creighton and McClintock left no doubt that after crossing over a new physical arrangement of chromosome segments arises. However, it did not tell exactly how it takes place, a matter argued even today. As mentioned in the last chapter, an actual breakage of chromatids does seem to take place. According to Janssens, only two threads out of four participate in any one crossover event. This idea implies that at the time of crossing over, each chromosome composing a bivalent is already double so that four strands compose the bivalent. The matter cannot be settled by cytological observation, because the exact time of crossing over is unknown and sufficient resolution of the chromatids is not possible at all stages of meiosis. However, conclusive genetic results with *Drosophila* and certain fungi have shown that crossing over takes place when four chromatids are actually present, that is, in the four-strand stage. Since the red bread mold, *Neurospora*, has been such a valuable tool in various kinds of genetic analyses, we will concern ourselves here with the kind of valuable information it provides on crossing over.

NEUROSPORA AS A TOOL IN CROSSOVER ANALYSIS. Many fungi are composed of a mass of threads known as a *mycelium*. One of the threads of a mycelium is a *hypha*. In the "sac fungi," or *Ascomycetes* to which *Neurospora* belongs, the sexual spores are formed in specialized, sacshaped hyphae called *asci*. The life cycles of these molds can be rather complicated. It is sufficient here to understand just a few points. All the nuclei found in the hyphae of a plant are haploid (Fig. 9-7). Plants arising from different spores may be of different mating types. If plants of the proper mating types come into contact, a male nucleus of one plant may enter the female sex organ of the other. However, fertilization does not immediately take place. Instead, the male and female nuclei undergo a series of mitotic divisions without combining.

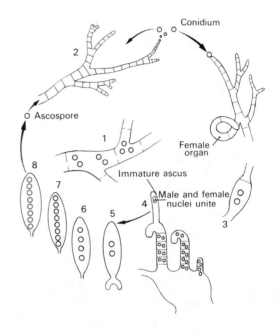

FIG. 9-7. Features of life cycle of *Neurospora*. *1.* A hypha is composed of cells containing haploid nuclei. These may wander from cell to cell through pores in the crosswall. *2.* An ascospore, a haploid cell produced after sexual reproduction, can grow into a new mass of hyphae (a mycelium). A new plant may also arise from a conidium, a special cell which can act like an asexual spore. A conidium may also serve as a male gamete. If it contacts receptive hyphae of the opposite mating type, it can act as a male gamete. Its nucleus is able to migrate to the female nucleus contained in the female sex organ. *3.* Once it enters the female organ, the male nucleus approaches the female nucleus, but the two do not unite. *4.* A number of special hyphae then grow out of the female organ. At the same time, the female-male pair of nuclei undergoes a series of mitotic divisions, still without fusing. Finally, in a cell at the end of each special hypha, male and female nuclei unite. This end cell will become an ascus. Because many special hyphae grow from a female organ, many asci are produced. The fertilization nucleus in each ascus is the same, because each traces back to the original male-female pair. *5.* After fertilization, first meiosis soon takes place. *6.* This is followed by second meiosis, giving four haploid nuclei. *7.* Each haploid nucleus in the ascus undergoes mitosis to give a duplicate of each of the four products of meiosis. *8.* Each one of the eight nuclei becomes the nucleus of an ascospore. Every ascus thus contains the products of one meiotic division in duplicate.

Eventually, saclike hyphae (asci) arise. A pair of nuclei, one from the maternal plant and one from the paternal, enters each ascus. It is only in the ascus that the two different nuclei finally unite to give true fertilization. Immediately after fertilization, meiosis takes place, producing four haploid nuclei in each ascus. The crucial point is that *each* ascus contains a tetrad of nuclei which has resulted from *one single* meiosis. In other words, when an ascus is formed, the products of one meiotic division are held together in the structure. In *Neurospora*, each haploid nucleus divides once more in the ascus, but this division is mitotic. The end result is that the nuclei in the ascus become the nuclei of the sexual spores. These spores are arranged in a linear order. Each ascus preserves the products of one meiosis in duplicate. Therefore, *Neurospora* has an advantage in cytogenetic analysis which most organisms do not. Typically, as in an animal testis for example, the gametes of countless meiotic divisions are all mixed up together. It is impossible to tell which nuclei were formed from the same one original cell as the products of one meiotic division. But in *Neurospora*, we can take an individual ascus, open it, take each spore out in order, and finally test the spore for some trait.

Suppose we are following a single pair of alleles, "al$^+$" (for normal spore color) and "al$^-$" (for white; Fig. 9-8). A cross of two different plants of the proper mating types will give a fertilization nucleus which is heterozygous at the spore color locus, "al$^+$/al$^-$." When we later examine the asci which arise, we find that some of them give a spore arrangement of 4 colored: 4 albino, arranged in the linear order: al$^+$, al$^+$, al$^+$, al$^+$, al$^-$, al$^-$, al$^-$, and al$^-$. Other asci are found which have an arrangement which is 2:2:2:2, for example: al$^+$, al$^+$, al$^-$, al$^-$, al$^+$, al$^+$, al$^-$, and al$^-$. Figure 9-8 gives an explanation for these arrangements. The 4:4 segregation results from the fact that no crossing over has occurred, and the alleles segregate at the first meiotic division. In the case of a crossover, homologous segments are exchanged between two of the four strands composing a bivalent. Note from the figure that the effect of this is to bring about an association of a *pair* of alleles with the *same* chromosome. Consequently, segregation of the alleles is delayed until *second* anaphase, when the centrom-

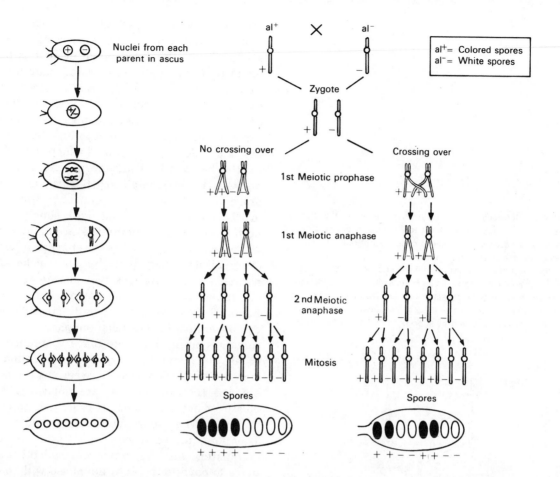

FIG. 9-8. Segregation of ascospores in *Neurospora*. The nuclei in an ascus are held in the same orientation which they had after the meiotic divisions. If no crossing over occurs, a pair of alleles segregates at first anaphase, and the spores, are arranged in a 4: 4 pattern in the ascus. If a crossover takes place, segregation of a pair of alleles is delayed until the second meiotic division, and the final pattern is thus 2:2:2:2.

ere divides and the chromatids are separated. These second-division segregations tell us that crossing over has taken place when four strands are present. If only two were present at the time of crossing over (Fig. 9-9), we could find only 4: 4 segregation in the ascus.

Conclusive proof of crossing over in the four-strand stage in *Neurospora* is seen when two pairs of linked alleles are being followed (Fig. 9-10a and b). The "al" locus is linked to "lys," a locus which affects the synthesis of the amino acid lysine. Suppose we cross a lysine-requiring mutant with an albino. After making the dihybrid cross, we can test each spore in any F_1 ascus for its lysine trait, as well as for its color trait. When this is done, we find whenever we have a

new combination of traits that the parental combinations (al⁻ lys⁺; al⁺ lys⁻) *are* always present along with the recombinant types (al⁺ lys⁺; al⁻ lys⁻). The critical point is that we find the recombinant types of spores and the parental types *in the same ascus*. This would not be possible if crossing over occurs in the two-strand stage, as shown in Figure 9-10c. In this case, we would find only the new combinations in an ascus. Therefore, the coexistence in one ascus of both the parental and the crossover types of spores tells us that crossing over must have taken place in the four-strand stage. We see that *Neurospora* is ideal for studies which relate the segregation of genetic factors to the segregation of specific chromosome segments at meiosis.

SOMATIC CROSSING OVER. We have noted that radiation can stimulate crossing over in somatic cells; this exceptional event has been found to occur in *Drosophila* and many fungi. At first, it may seem unimportant whether or not somatic crossing over takes place. But let us consider the consequences, using a hypothetical example from humans (refer to Fig. 9-11). Suppose a female zygote is formed which is dihybrid for the hemophilia trait and for the ability to produce the enzyme glucose-6-phosphate dehydrogenase. The alleles are carried in the *trans* arrangement. An individual of this genotype will normally express the normal blood condition and will be able to manufacture the enzyme. At birth, every cell of the body should be identical, because normal mitotic divisions will guarantee equal distribution of genetic material to daughter cells. Figure 9-11 shows how somatic crossing over can upset this. In rare cells, the phenomenon of somatic pairing may take place spontaneously or as the result of exposure to radiation. A crossover may then occur between the centromere and the two loci we are following. This crossover event produces two metaphase chromosomes whose chromatids are not identical. Therefore, unlike a normal mitosis, the two chromatids which separate from each other and move to opposite poles at anaphase will be different. One outcome is that two new cell lines can rise from the one original genotype. The individual would be a mosaic, a mixture of wild type cells, of cells homozygous recessive for the hemophilia trait, and cells which lack the ability to synthesize G_6PD. The extent of the mosaicism will depend on how early in development the somatic crossover occurs. The earlier it takes place, the larger will be the number of cells in the two new cell lines, because the recombinant cells will have undergone more mitotic divisions before completion of embryology. We see from this example that somatic crossing over can uncover recessives in the body, leading to patches of mutant cells scattered among normal cells. Obviously, this can produce undesirable effects. Therefore, it would be advantageous for crossing over to be suppressed in body cells. It is logical to assume that in the evolution of living things, crossing over provided the first mechanism of recombination, since the simplest diploid cells must have had only one pair of homologous

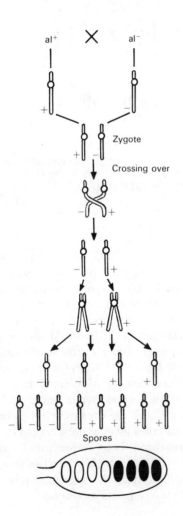

FIG. 9-9. Depiction of crossing over in the two-strand stage. Assume in *Neurospora* that after zygote formation, crossing over occurs when each chromosome is composed of only one chromatid. After the crossover, the alleles on a crossover segment will simply have shifted to a new centromere attachment. When each chromosome duplicates, the outcome in each case is a chromosome composed of two identical chromatids. The only segregation possible would be 4:4. Since 2:2:2:2 segregations also occur, this suggests that crossing over must occur in the four-strand stage.

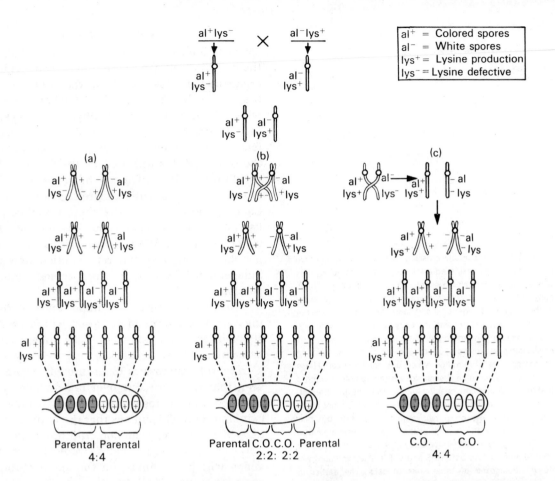

FIG. 9-10. Segregation of allelic forms of two linked genes in *Neurospora*. *a.* If no crossing over occurs, the ascospores will be found in a 4:4 arrangement in the ascus. *b.* If crossing over occurs, the parental types and the crossover types of ascospores will be found in an arrangement of 2:2:2:2. *c.* Assuming that crossing over takes place in the two-strand stage and is then followed by chromatid replication, *only* the recombinant types could be found in an ascus. They could not occur along with the parentals as shown in *b.* However, when recombinant spores are found in an ascus, they always occur along with the parentals. This means that crossing over must take place in the four-strand stage as depicted in *b* and not in the two-strand stage as in *c*.

chromosomes. And so, the most primitive kinds of diploids must have relied on crossing over for recombination. As the diploid stage evolved and the body became differentiated between germ cells and somatic cells, mechanisms must have evolved to allow crossing over in the germ line where it was needed to provide new combinations of linked alleles but to prevent it in body cells where the effects could be harmful due to the production of mosaicism. Certainly there are mechanisms which prevent crossing over at meiosis in one sex of certain species, such as the fruit fly and the silkworm. It would seem likely that comparable genetic controls must have evolved to prevent its occurrence at mitotic divisions in order to ensure a soma composed of cells with the identical genetic information.

IMPORTANCE OF CROSSING OVER TO EVOLUTION. We can aptly conclude this chapter with a consideration of the important role of crossing over in evolutionary progress. An organism which is completely asexual would not have the mechanism of meiosis to bring about new allelic combinations through independent assortment. The only way a new combination can result is through gene mutation. Consider the two diploid members of the same species, with a chromosome number of four, which are represented in Figure 9-12. One of them expresses the phenotype ab⁺, the other the phenotype a⁺b. The offspring will be identical to each parent: all ab⁺ in the one case, all a⁺b in the other. If two diploid organisms of these same genotypes were sexual, they would be able to produce haploid gametes which could unite to form dihybrids. These, in turn, by independent assortment can give rise to four kinds of haploid gametes. A cross between the two dihybrids will produce the familiar dihybrid ratio, 9:3:3:1. We now have *four* different phenotypes (2^2) and nine different genotypes (3^2). Lacking the sexual process, there would be just two phenotypes and two genotypes, one for each of the cell lines. Much more variation is made possible through sexual reproduction and the independent assortment which meiosis makes possible.

Now let us consider the importance of crossing over as a source of variation which greatly supplements that contributed by independent assortment. Referring back to Table 8-1 on which a trihybrid testcross is diagrammed, it can be seen that if no crossing over takes place, only two types of gametes will be formed, just the old combinations. This means that the main advantage of sexual reproduction, the production of new combinations of alleles, would be lost for genes which are linked on the same chromosome. However, if all possible crossovers occur among the three genes, then eight different types

FIG. 9-11. Effects of somatic crossing over. In this hypothetical example, a zygote is considered to be dihybrid at the hemophilia locus and the locus which governs the production of G_6PD. Normally (*left*) mitotic divisions follow which ensure that every cell of the body will be identical and have the potential to make antihemophilia factor (AHF) and the enzyme G_6PD. On very rare occasions, pairing of homologous chromosomes occurs in the body cells. If this is followed by a crossover event, the two resulting chromosomes carry chromatids which are not identical. These chromosomes may arrange themselves on the spindle in such a way that the products of the ensuing mitotic divisions are different. In this case, a cell results which can make AHF but cannot make the enzyme G_6PD. The other cell cannot make AHF but can produce the enzyme. Since most of the mitotic divisions will be normal, the majority of cells in the body in this case will have the potential to manufacture both AHF and G_6PD. The individual will be mosaic, composed of three lines of cells. The earlier in development that the somatic crossover takes place, the larger the number of exceptional cells and the extent of the mosaicism.

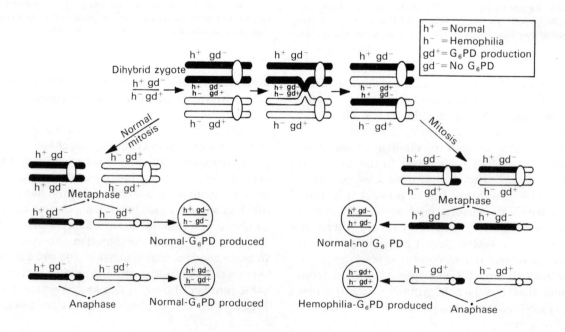

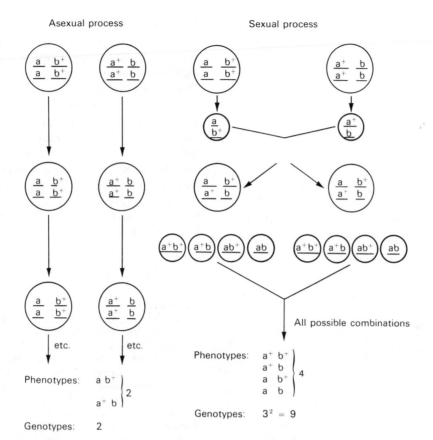

Phenotypes: a b$^+$ } 2
            a$^+$ b

Genotypes: 2

All possible combinations

Phenotypes: a$^+$ b$^+$
               a$^+$ b   } 4
               a   b$^+$
               a   b

Genotypes: $3^2 = 9$

FIG. 9-12. Asexual and sexual reproduction compared. If members of a species are completely asexual, they continue to reproduce their own type indefinitely (*left*). With two pairs of alleles and assuming dominance, only two phenotypes and two genotypes are possible in the population. This will remain so until rare spontaneous mutation causes one of the genes to mutate. In contrast (*right*), the sexual process generates new combinations swiftly. The genetic material of the two types is brought together. With just two pairs of alleles, four phenotypes and nine genotypes are possible. These different combinations may vary in the advantages which they give. Natural selection will favor those which are more beneficial. Because the sexual process brings about more variation than the asexual in a given length of time, those lines which are sexual have evolved more quickly than those which are exclusively asexual. Crossing over speeds the process up by contributing even more to the variation effected by independent assortment.

of gametes are formed, the two parentals plus the six crossover types. It is evident that crossing over, along with independent assortment, contributes greatly to the variation generated by the sexual process. With crossing over, recombination is extended down to the level of the chromosome. Inasmuch as crossing over typically occurs at every meiosis, new gene combinations are being continuously formed. Crossing over, therefore, speeds up the production of new combinations of genetic material. Its contribution to the pool of variations in humans can also be appreciated by referring back to Chapter 1. When discussing new combinations provided through independent assortment, we noted that 2^{23} types of gametes, would be a minimal number for a human. It should now be evident to us that each of the 23 pairs contains thousands of genes and that much heterozygosity is present, because the human is highly hybrid due to outbreeding. Since crossing over can occur at any point along a chromosome, the number of new combinations which can be generated becomes staggering. Each crossover point doubles the number of types of gametes. The new allelic combinations make possible a vast assortment of phenotypes. These are the variations upon which the evolutionary force of natural selection can operate.

The greater the number of variations, the greater the chances for types better adapted to different sets of environmental conditions. The larger the number of different types, the greater is the chance for the occurrence of superior forms in a given period of time and, thus, the faster the rate of evolutionary change. Therefore, it is not surprising that sexual organisms have provided the main branches leading to the evolution of the diverse kinds of living species. The contribution of crossing over to the pool of variation, by extending recombination to linked alleles, has been a major factor in the evolution of higher forms of life and cannot be overestimated.

REFERENCES

Creighton, H.S., and B. McClintock. A correlation of cytological and genetical crossing over in *Zea mays. Proc. Natl. Acad. Sci., 17:* 492, 1931.

Emerson, S. Meiotic recombination in fungi with special reference to tetrad analysis. In *Methodology in Basic Genetics,* W. J. Burdette (ed.), pp 167–206. Holden-Day, San Francisco, 1963.

Ephrussi, B., and M. C. Weiss, Hybrid somatic cells. *Sci. Am. (April): 26,* 1969.

Harper, P. S., M. L. Rivas, and W. B. Bias, *et al.* Genetic linkage confirmed between the locus for myotonic dystrophy and the ABH-secretion and Lutheran blood group loci. *Am. J. Hum. Genet., 24:* 310, 1972.

Mayo, O., The use of linkage in genetic counseling. *Human. Hered., 20:* 473, 1970.

McKusick, V. A. Human genetics. *Ann. Rev. Genet., 4:* 1, 1970.

McKusick, V. A. The mapping of human chromosomes. *Sci. Am. (April): 104,* 1971.

McKusick, V. A., and F. H. Ruddle, The status of the gene map of the human chromosomes. *Science, 196:* 390, 1977.

Ruddle, F. H. Linkage analysis in man by somatic cell genetics. *Nature, 242:* 165, 1973.

Ruddle, F. H., and R. S. Kucherlapati, Hybrid cells and human genes. *Sci. Am. (July): 36,* 1974.

Stern, C. Somatic crossing over and segregation in *Drosophila melanogaster. Genetics, 21:* 625, 1936.

Whittaker, D. L., D. L. Copeland, and J. B. Graham. Linkage of colorblindness to hemophilias A and B. *Am. J. Hum. Genet., 14:* 149, 1962.

REVIEW QUESTIONS

1. In the human, the allele for normal blood clotting (h^+) is dominant to the recessive allele

(h) which results in hemophilia. The allele for the production of the enzyme glucose-6-phosphate dehydrogenase (gd⁺) is dominant to its allele (gd) for the absence of the enzyme. Both allelic pairs are X-linked.

A man with hemophilia has a daughter who is concerned about the chance of her bearing a child with the disorder. It is found that the hemophiliac man does not produce the enzyme but that his wife, the woman's mother, does. The concerned daughter also produces the enzyme. She is married to a man without hemophilia who is an enzyme producer also. Amniocentesis and chromosome analysis show that the daughter is carrying a male fetus which is an enzyme producer. What is the chance that the baby has hemophilia, considering the fact that a chromosome map shows the h and the gd loci to be two map units apart on the X chromosome?

2. Assume that in the human a dominant allele (A) is responsible for a fatal nervous deterioration which usually has its onset after the age of twenty. The normal condition depends on the recessive allele (a). The presence of the dominant allele is associated with no detectable product before the onset of the disease. Assume that the locus associated with the disorder is on an autosome and is found to be linked to a locus associated with a specific enzyme which can be detected *in utero*. The allele for enzyme production (E) is dominant to the recessive allele for absence of enzyme (e). Pedigree analyses indicate that the two loci, A and e, are ten map units apart.

A man with the nervous disorder is also found to be an enzyme producer. Family history reveals he is a dihybrid with the alleles in the *trans* arrangment. The man's normal wife is not an enzyme producer. During her pregnancy, amniocentesis shows that the offspring is not an enzyme producer. What is the chance that the fetus carries the gene for the nervous disorder?

3. The ability to produce the enzyme G_6PD depends on a sex-linked dominant allele (see Question 1). Also sex linked is the dominant allele, P, which is necessary for normal color vision. The recessive allele, p, produces protan color blindness. The loci for the enzyme and for color vision are 6 map units apart. A dihybrid woman with alleles in the *cis* arrangement marries a man with normal vision who is an enzyme producer. Diagram the cross, and show the possible offspring and their expected frequencies.

4. In *Neurospora,* the loci "pro" and "leu" are associated with the production of the amino acids proline and leucine. The two loci are linked. The centromere is to the "left" of pro, and pro is to the "left" of leu. A strain of *Neurospora* which can produce neither amino acid (genotype: pro leu) is crossed to a strain which can produce both of them (genotype: pro⁺ leu⁺). What will be the sequential arrangement of the ascospores in an ascus derived from a dihybrid nucleus in which each of the following has taken place?

A. A single crossover between pro and leu involving the two middle strands.
B. A double crossover between pro and leu involving the same two strands (the middle ones).
C. A four-strand double in which each of the crossover events between pro and leu involves a different pair of chromatids.
D. A single crossover between the centromere and pro, involving the two middle strands.

5. The "ser" locus for the production of serine is found to map between "pro" and "leu." Following a cross between a strain which can produce all three amino acids and one which cannot produce any of them, what will be the sequential arrangement of the ascospores in an ascus derived from a trihybrid nucleus in which the following is true?

A. No crossover has occurred.
B. A single crossover occurred between pro and ser involving the two middle strands.
C. A single crossover occurred between ser and leu involving the two middle strands.
D. A double crossover took place involving the same two middle strands. One crossover is to the "right" of ser and the other one to the "left" of ser.

6. In *Drosophila*, the loci for yellow body (y) and distorted or singed bristles (sn) are sex linked. The alleles for gray body (y⁺) and normal bristles (sn⁺) are dominant to their recessive alleles, y and sn. The arrangement on the chromosome is:

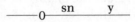

A dihybrid zygote is formed with the alleles in the *trans* arrangement. Assume that somatic synapsis takes place and that it is followed by crossing over between "sn" and the centromere in one of the cells of the developing fly. What will be the consequences?

7. In the fruit fly, the locus for dumpy wings (dy) is on chromosome No. II. The locus for sepia eyes (se) is on chromosome No. III, and the locus for the eyeless trait (ey) is on chromosome No. IV. The traits dumpy, sepia, and eyeless are recessive to the wild type or normal traits. Suppose a mutation arises in a wild-type laboratory stock of flies and causes the development of extra hairs on the body. The genetic factor causing hairiness (h) is found to be recessive and not sex linked. The following three crosses are made and followed through the second generation:

1. hairy × dumpy 2. hairy × sepia 3. hairy × eyeless. The results of the F₂ are: Cross 1— 9 wild: 3 hairy: 3 dumpy: 1 hairy, dumpy. Cross 2— 2 wild: 1 hairy: 1 sepia. Cross 3— 9 wild: 3 hairy: 3 eyeless: 1 hairy, eyeless.

What conclusions can be reached on the assignment of the "h" locus to a chromosome? Explain.

8. Following fusion of human cells with those of the mouse, an attempt is made to associate a gene for the production of enzyme D with a specific chromosome. The human cells used in the fusion are producers of enzyme D, whereas the mouse cells are not. Several stable lines are established following cell fusions. From the results below, what chromosome appears to be associated with enzyme D production?

Five cell lines with the indicated human chromosomes present				
3	3	2	3	3
4	4	4	5	4
7	10	7	8	7
8	21	8	12	11
10	X	12	18	14
11		21	X	15
21		X		X
X				

Enzyme present (+)   +   −   +   +   −
or absent (−)

9. Assume that production of a certain enzyme in the human has been associated with Chromosome 1 through somatic cell hybridization. However, studies of cells from one person indicate that the enzyme is associated with Chromosome 5. Offer an explanation, and explain how one might get evidence for it.

Select the correct answer or answers, if any, for each of the following:

10. Crossing over may be influenced by:

   A. Age.
   B. Sex.
   C. Temperature.
   D. Alteration in chromosome structure.
   E. Radiation.

11. Crossing Over:

   A. Produces only crossover spores in any one ascus of *Neurospora* following a crossover in that ascus.
   B. Is an abnormal occurrence in the males of most species.
   C. Produces new gene combinations which would not be possible by independent assortment alone.
   D. Generally produces beneficial effects when it occurs in body cells.
   E. Does not entail any sort of physical exchange in the chromosomes.

12. The recessives a, b, and c are sex linked in *Drosophila*. A cross of two parental strains results in the following F_1: $\dfrac{a^+b^+c^+}{a\,b\,c}$ and $\dfrac{a\,b\,c}{Y}$. If these two F_1's are crossed, the following would be true:

   A. The amount of crossing over cannot be estimated from the resulting F_2.
   B. The amount of crossing over can be determined from the F_2 from the male offspring alone.
   C. The amount of crossing over can be determined by considering all of the resulting offspring.
   D. The alleles "a" and "b" are in the *trans* arrangement in the female F_1's.
   E. Crossing over can only be determined from the F_2 females.

10

CHANGES IN CHROMOSOME STRUCTURE AND NUMBER

POINT MUTATIONS VS. CHROMOSOME ANOMALIES. Throughout the preceding chapters, we have frequently referred to sudden inheritable changes or gene mutations. They are also called "point mutations," because they represent modifications at specific points along the deoxyribonucleic acid (DNA). These are the alterations which lead to the origin of alleles, alternative forms of a gene, and they will be the major topic of Chapter 15. They are not, however, the only kind of modification which can affect the hereditary material and bring about inheritable variation. Some changes alter the actual morphology of the chromosome by modifying its size or shape. Other changes involve the addition or subtraction of one or more entire chromosomes. Although this kind of genetic variation can be technically classified as "mutation," the term is usually reserved for point mutations. The more gross modifications, which are the subject of this chapter, are generally referred to as "chromosome aberrations" or "anomalies."

Variations in chromosome structure or number were recognized in early genetic studies (Chap. 5). Throughout the years, they have been found in a large number of different plant and animal groups. Their occurrence in humans is a matter of special concern today to the medical practitioner, the genetic counselor, and to certain family groups. The most common cause of congenital defects in humans can be attributed to chromosome anomalies. Their alarming frequency is indicated by the fact that more than 90% of abnormal embryos are lost through spontaneous abortion, and among these, chromosome anomalies are very frequent. An understanding of the nature of chromosome aberrations and their genetic consequences in a variety of plant and animal species can lead us to approach intelligently the problems which they present for human society.

THE INVERSION AND ITS CONSEQUENCES. On rare occasions, a chromosome may break spontaneously into one segment or more. Certain environmental agents, such as radiations or certain chemicals, may actually induce breakage and greatly increase the number of resulting aberrations. One of the main types of chromosome alteration is the *inversion*. As the term itself implies, a stretch of the chromosome is turned around (Fig. 10-1A). This requires two breaks in the chromosome, followed by a healing of the breakage points. It is important to note from Figure 10-1A that when an inversion arises, the interstitial segment (cdef) does not attach to either of the original chromosome ends (a or g). The reason is that the ends or end genes of a chromosome appear to be polarized and do not attach to other segments. The two extremities of each chromosome have been designated *the telomeres* and are often described as being "nonsticky." Any other gene segments between the telomeres are considered "sticky," meaning that they can attach to one another in various new arrangements, but *not* to telomeres.

An inversion can arise in a somatic cell or at some time during gamete formation in either sex. Assume that a sperm contains a chromosome with an inverted segment and that it fertilizes an egg in which the corresponding chromosome has the standard sequence of genes (Fig. 10-1B). The zygote would consequently have a normal chromosome and one carrying the inversion. Any individual with a normal chromosome and one in which the structure has been altered in any way at all is called a *structural heterozygote*. The inversion heterozygote in this example is genically balanced, because no genetic loci are missing or modified in any detectable way. Gene order has been somewhat changed, but there is no change in the kind or number of genes as a result of the inversion. Consequently, we would expect such an individual to be normal phenotypically, and this is

FIG. 10-1. The inversion. *A*. Origin of an inversion after two breaks in a chromosome. *B*. A sperm carrying an inversion fertilizes an egg in which the homologous chromosome has the normal gene sequence. The result is a zygote which is an inversion heterozygote. (The centromere is represented by an open circle.)

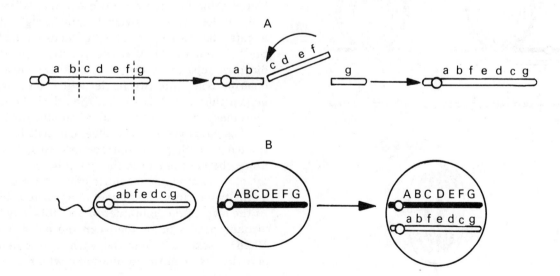

FIG. 10-2. Pairing and crossing over in an inversion heterozygote. *A.* Loop formation is shown followed by a crossover. A crossover between loci d and e (or any other two loci in the inverted segment) results in the origin of a dicentric chromosome and one which is acentric. Both are genically unbalanced. For the sake of clarity, each chromosome is represented as being single. Actually, at the time of crossing over, each is double, but only two chromatids engage in any one crossover event. *B.* An inversion may be detected with the microscope, because anaphase I cells will show a bridge and a fragment. These represent the dicentric being pulled to both poles and the acentric which cannot move at all.

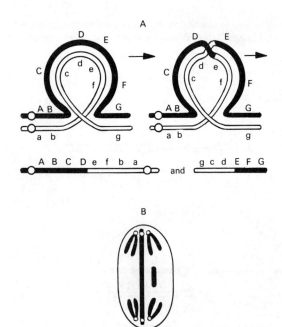

generally so. Some position effect (p. 289) is possible, but it can be ignored for the moment. The inversion heterozygote can usually live to maturity without any ill effects from the anomaly. Why, then, is this condition not more common among living things, since there is no upset in genic balance in the heterozygote? The answer comes when the meiotic picture is considered. It will be recalled that homologous chromosome segments undergo synapsis at zygonema of meiotic prophase. Two homologous chromosomes will encounter pairing difficulties if one member contains a segment that is reversed. However, the forces of synapsis are powerful enough to ensure pairing of homologous regions if at all possible, even if the chromosomes must undergo contortions. In the inversion heterozygote, the pairing is accomplished through the formation of a loop in one of the chromosomes, either the normal or the inverted one (Fig. 10-2*A*). The two chromosomes may separate normally at anaphase I, and there would be no complications. However, if the inversion is large enough, a crossover is certain to occur in some cells. In Figure 10-2*A*, a crossover is indicated in the loop between loci "d" and "e," and its consequences are seen. As a result of a crossover anywhere within the inverted segment, two chromosomes will be formed which are very abnormal in structure. One of them will possess two centromeres (a dicentric chromosome); the other will lack a centromere (an acentric). Both are also genically unbalanced; each has certain loci represented more than once, whereas other loci are missing (the dicentric has the "a" and the "b" loci present twice but lacks "g"; the acentric has two doses of the "g" locus but lacks the "a" and the "b"). If the acentric and the dicentric separate at first anaphase and move to opposite poles, the two nuclei which form will be genically unbalanced. But the affected chromosomes are apt to be involved in still further cytological problems. The dicentric will be attracted to both poles, the acentric to neither. Since the centromere is the dynamic center of the chromosome, the dicentric may be stretched as the two centromeres travel to opposite poles. At the level of the light microscope, this may be visualized as a bridge between the poles (Fig. 10-2*B*). Next to it may be seen a fragment, actually the acentric chromosome which is unable to move at all and which will disintegrate in the cytoplasm after being left out of a nucleus. The stretched chromosome may break anywhere along its length. No matter, it is evident that cells will arise which are genetically unbalanced. Since these events would be taking place at meiosis, the consequence would be a certain proportion of unbalanced sperms or eggs. If the gametes survive, any offspring resulting from them would be abnormal in some way or would fail to survive at all due to the imbalance. In general, therefore, inversion heterozygotes suffer a certain decrease in fertility as a result of the formation of acentrics and dicentrics formed after crossing over in the inverted segment. Since these structural heterozygotes would be at a reproductive disadvantage, leaving fewer offspring in the long run than the normal individuals, natural selection would tend to weed the aberration out of the population. If an inversion is tiny, it has a greater chance of persisting in a population, because the probability of crossing over within it is less. Although the small inversions often tend to persist, large ones are not usually found to be widespread in most plant or animal populations.

INVERSIONS IN *DROSOPHILA*. The outstanding exception to a low level of inversions in the wild is found in the genus *Drosophila*. In many fruit fly populations, inversions of appreciable size are very common and may acctually be a normal part of the hereditary composition of the population. How can these persist without causing a reduction in fertility? We have noted that the fruit fly is unusual in that there is normally no crossing over in the male. Therefore, the complications discussed above would not be encountered, and no reduction in male fertility would ensue from the inversion. In the female, however, crossing over is the rule at meiotic prophase. Still, female fertility is not reduced by the presence of the inversion. Remember that crossing over takes place in the four-strand stage. Only two of the chromatids of the bivalent participate in any one crossover. In the case of an inversion, two chromatids usually will not be involved in a crossover in the inverted segment and will be normal in their genic balance. The abnormal, dicentric chromosome is held back as a result of the chromatid bridge, and the acentric

usually will not move at all. Consequently, one of the noncrossover chromatids (either the one with the inversion or the one with the normal order) completes meiosis and enters the egg (Fig. 10-3A). The overall effect of the chromatid bridge in the females of certain animal species, such as those of *Drosophila*, is that the defective chromatids are shunted into the polar bodies. This preferential inclusion of a normal chromatid in the egg cell of an inversion heterozygote is termed "selective orientation." The balanced chromatid carrying the inversion has just as much chance as the standard one to enter the egg, and there is no reduction in fertility. So due mainly to a lack of crossing over in the male and to selective orientation in the female, inversions can accumulate in *Drosophila* populations without reducing the fertility.

But because inversions are so common in *Drosophila*, do they carry some sort of advantage? The answer seems to be "Yes." The reason for the value can be appreciated from a reconsideration of the consequences of a single inversion within the inverted segment (Fig. 10-3A). A single crossover brings about the recombination of linked alleles, but in the case of the inverted region, the recombination is associated with abnormal chromatids, and these are cast off into the polar bodies. Only the old combinations (the chromatids with the standard arrangement and the one with the original inversion) are preserved to enter the egg. Therefore, the effect of the inversion is a decrease in the amount of recombination among genes in the inverted region. We have learned that crossing over and the recombination it generates are highly desirable aspects of the sexual process and have enhanced the variation available for natural selection. There is, however, a reverse side to the coin (Fig. 10-3B). For it is quite possible for a group of certain closely linked alleles to constitute a very desirable combination. The combination and the advantage it gives can be destroyed by recombination. The presence of an inversion ensures that the combination will stay together from one generation to the next. It also cuts down the chances that all the genes in the affected segment will become homozygous for one of their allelic forms in any one individual. Populations of *Drosophila* have been studied for the possible adaptive value of their inversions.

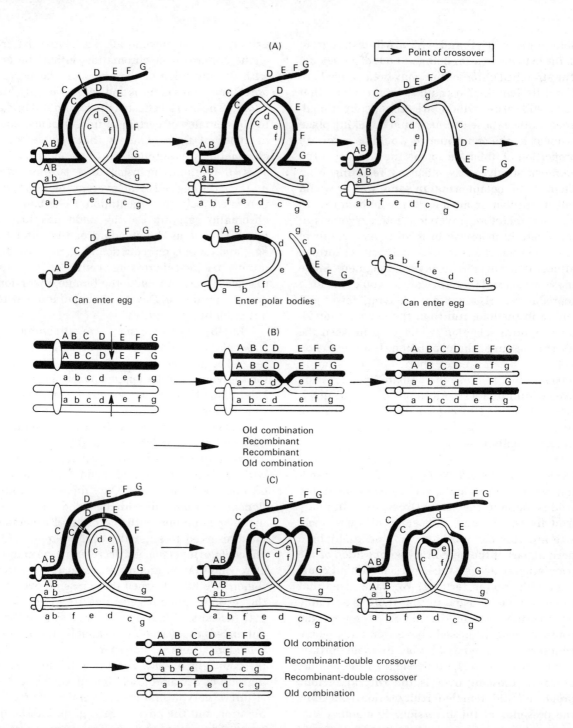

FIG. 10-3. Bridge formation at oogenesis. *A.* Only two of the four chromosome strands participate in any single crossover. As a result of the formation of the bridge, the defective dicentric strand is held back at the end of the first meiosis. The strands of normal morphology (those with just one centromere) become directed toward the "outside" of the cell. Upon completion of second meiosis, one of the "outside" nuclei becomes the nucleus of the egg. The defective dicentric and acentric chromosomes are never oriented toward the outside and always enter polar body nuclei.

Note that the egg will therefore contain one of the parental gene combinations, CDEF (the noninverted arrangement) or fedc (the inversion). Since only the parental combinations are present, the chance remains good that many individuals will be heterozygous for the inverted region and also heterozygous for the genes in the region. *B.* Suppose the gene combinations CDEF and cdef (in any order) bring an advantage when present as a unit on a chromosome. If no inversion is present, any single crossover in the "cdef" region can break up the combination. The inversion, therefore, as

And it has been found that the frequency of certain inversions seems to be correlated with natural climatic conditions. Strong support for this concept has been provided by laboratory experiments in which fluctuations in temperature have been associated with characteristic fluctuations in the frequency of specific inversions (see Chap. 16 for more details). It is suspected that small inversions in many animal and plant groups may provide a mechanism for preserving combinations of alleles favorable under a variety of environmental conditions.

If an inversion is sufficiently large, double crossing over may take place within it. Fig. 10-3C shows that this will produce two balanced chromatids, each normal in structure and with a new combination of genes. Double crossing over, then, will tend to break up old combinations of alleles. However, double crossing over would be much less frequent than singles and would be quite rare within a small inversion.

DIFFERENT TYPES OF INVERSIONS. Since chromosome breaks may occur anywhere along the length of a chromosome, this makes possible countless types of inversions. Two major categories may be recognized. In our discussion so far the two breaks have both been in one chromosome arm, to one side of the centromere. Such inversions are called *paracentric*. These are in contrast to *pericentric inversions*, those in which the two breaks are on either side of the centromere (Fig. 10-4A). The story of synapsis and loop formation is the same for the pericentric heterozygote as for the paracentric. A big difference is seen, however, when the products of a crossover are examined (Fig. 10-4B). No dicentrics or acentrics are formed. The two resulting crossover chromatids are normal morphologically in that they possess one centromere each, but each is genically unbalanced. Since there is no dicentric chromatid, no chromatid bridge will

be formed at anaphase. The two genically unbalanced chromatids are not held back at meiosis and have as much chance to enter the egg as do the balanced ones. The consequence is some reduction in fertility in the female. Therefore, we find the pericentric inversion a much less common feature of populations, even *Drosophila*. Again however, small ones have a chance to accumulate, and there is evidence that they play some role in changing chromosome morphology. From Figure 10-4C, it can be seen that if the two breaks on either side of the centromere are *not* equidistant, the centromere position will shift. Small shifts of the centromere due to pericentric inversions have almost certainly operated in *Drosophila* during the evolution of the various species to alter chromosome morphology. It accounts for the presence of "J-shaped" chromosomes in some groups which are related to those having rods or "V's" (see p. 292) for more details). Each "J" probably has been derived from a rod- or "V"-shaped chromosome through pericentric inversions which shifted the centromere. In the X chromosome of a stock of *Drosophila melanogaster,* Muller was able to shift centromere position experimentally over a period of time by using radiations to induce breakage.

Once an inversion has occurred in a chromosome, there is no reason why additional ones cannot occur later. And each of these would usually involve different points of breakage. However, a second inversion in a chromosome may have one point of breakage in a segment which has already been reversed in a previous inversion. Consequently, one inversion would overlap the other (Fig. 10-5A). If a structural heterozygote possesses the standard chromosome and the one in which the two inversions have taken place, a complex loop, rather than a simple one, is seen at meiosis (Fig. 10-5B). Overlapping inversions have occurred in the evolution of various *Drosophila* populations. Within a species, certain races of flies differ by two or more inversions. These changes can be detected by studying the complexities of the loops seen in the salivary gland chromosomes which are in a permanently synapsed state (Fig. 10-5C).

In humans, pericentric inversions have been identified in a few families. Thirteen children

shown in *A* tends to preserve it. Without the inversion, there is also a greater chance for genes in the region to become homozygous. *C.* A double crossover within an inverted segment produces chromosomes of normal morphology and also brings about recombination. Since no bridge is present, the recombinant chromosomes can enter an egg nucleus. The advantageous combinations CDEF and cdef will thus be broken up.

have been detected who have inversions in Chromosome 9, and all of them are normal. A pericentric inversion in humans, as well as in other species, may be recognized by a change in the position of one of the centromeres; this is reflected as a change in the length of the two chromosome arms. Crossing over within the inverted segment in a pericentric heterozygote results in the production of unbalanced chromosomes which will differ, depending on the point of crossing over. An individual heterozygous for a pericentric inversion could potentially produce offspring with different types of abnormalities. This may explain the case in which several abnormal children were born to a phenotypically normal woman who was found upon karyotype analysis to have a pericentric inversion in Chromosome 10. The possibility of inversions in humans as a cause of some complex defects must be considered when karyotypes are examined in families with a history of congenital disorders.

THE TRANSLOCATION. Another well-known chromosome aberration is the translocation, an alteration in which chromosome segments, or even whole arms, are transported to new locations. A simple type of translocation, commonly called a *shift*, involves three breakage points. All three may be in one chromosome (Fig. 10-6A). One consequence of this type of shift is discussed later in this chapter. Other simple

FIG. 10-4. Pericentric inversions. *A*. Note that each break is on either side of the centromere. Contrast this with the paracentric inversion (Fig. 10-1A) where both breaks are on the same side of the centromere. *B*. A crossover anywhere in the inverted region in the case of a pericentric inversion will produce two chromatids, each normal morphologically (with one centromere each) but each genetically unbalanced. Note that one chromatid has loci a and b represented twice and lacks g. The other has a double dose of g but lacks a and b. Only two chromatids are shown for the sake of clarity. The other two would not participate in this crossover and would be balanced parental types. However, since there is no bridge to hold back the unbalanced ones, the egg nucleus may receive a defective chromosome, and so a decrease in fertility ensues. *C*. If the two breaks are not equidistant from the centromere, the centromere position will be shifted. In this case it is now closer to one end, giving a J shape to what was originally a chromosome with a median centromere.

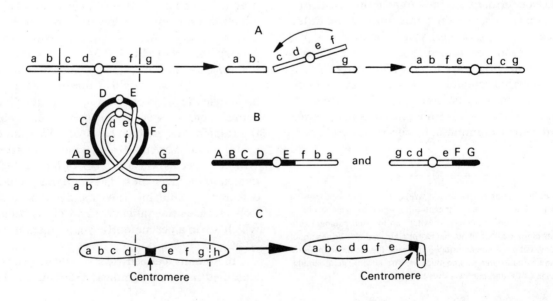

shifts may involve two breaks in one chromosome and a single in a nonhomologue (Fig. 10-6B). The segment from Chromosome 1 is then inserted into Chromosome 2.

A shifting of chromosome segments is termed a *reciprocal translocation* when there is a mutual exchange between two nonhomologous chromosomes. Reciprocal translocations have been studied extensively in many plant and animal species. Figure 10-7A illustrates a reciprocal translocation in which whole arms are exchanged. Assume that the alteration occurs during spermatogenesis, so that a sperm with the chromosome alteration fertilizes a normal egg. The result is a translocation heterozygote which, as in the case of the inversion, would be genically balanced and most probably normal phenotypically. Like the inversion heterozygote, the individual with the translocated chromosomes would encounter problems during synapsis at first meiotic prophase. The forces of attraction may achieve pairing by the formation of atypical chromosome associations. When whole arm translocations are present, a characteristic cross figure forms which may be easily seen at pachynema in certain favorable material (Fig. 10-7B). All the homologous segments are paired, and crossing over may occur. The chief problem in the translocation heterozygote does not concern crossing over but rather disjunction, the manner in which the chromosomes will separate at anaphase. The association of four chromosomes at first meiotic division is termed a *tetravalent* to distinguish it from the normal meiotic bivalent. At diplonema and diakinesis, the cross figure will open out into a circle as the forces of repulsion become operative. By first metaphase, the circle of four will be at the equatorial plate. The manner in which the chromosomes of the tetravalent disjoin will determine the genic balance of the resulting gametes. Inspection of Figure 10-7C shows that if any adjacent chromosomes in the circle move to the same pole, unbalanced gametes will arise. These will contain extra doses of some genes and will lack others entirely. However, if alternate chromosomes in the circle move to the same pole, balanced gametes will be formed. Of those arising from a single meiotic cell, one-half will carry the normal chromosome arrangement, and one-half will have the translocated chromosomes.

The end effect, then, of the reciprocal translocation is to produce a reduction in fertility in the structural heterozygote. Usually, the unbalanced gametes will not survive or else they will give rise to abnormal offspring.

In some cases, more than two homologous chromosomes may engage in reciprocal translocations. This increases the size of the circle or even the number of circles seen at meiosis, depending on how the arms have been distributed. For example, in the cultivated plant, *Rhoeo discolor*, all the chromosome arms in the complement have been shifted. The result is a multivalent association which includes all 12 chromosomes (Fig. 10-8). At times a chain rather than a circle forms in some meiotic cells. The overall effect of the large number of reciprocal translocations in this plant is to produce a certain number of unbalanced gametes due to irregular chromosome distribution at anaphase.

Decreases in fertility due to the presence of reciprocal translocations have been recognized in corn and *Drosophila*, to mention just two familiar species. As in the case of the inversion, reciprocal translocations do not tend to accumulate to any great extent in natural populations, but again there is at least one striking exception to the rule. This is seen in the plant genus *Oenothera*, the evening primrose.

IMPORTANCE OF TRANSLOCATIONS IN THE EVENING PRIMROSE. When meiotic cells of *Oenothera* are examined, it is found that the 14 chromosomes do not usually form seven pairs as would be expected. Instead, cytological examination reveals a large circle composed of an end-to-end association of 14 chromosomes, resembling that shown in Figure 10-8 for the chromosomes of *Rhoeo*. The work of Cleland clarified the situation and showed that during their evolution, the chromosomes of the evening primroses had undergone a series of translocations. Whole arms of chromosomes were shifted around to such an extent that *no* one chromosome is completely homologous to any other in the whole chromosome complement. The only way the homologous segments can associate is by the formation of a large circle (Fig. 10-9). As we noted above, adjacent disjunction will lead to unbalanced gametes; so it might be expected that such a large circle would give rise

A

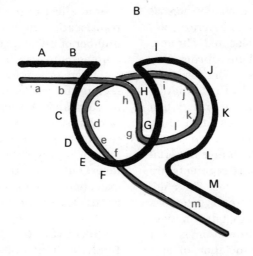

		1st					1st					
a	b	c	d	e	f	g	h	i	j	k	l	m

Standard order

				2nd								2nd
a	b	h	g	f	e	d	c	i	j	k	l	m

Result of first inversion

a	b	h	g	l	k	j	i	c	d	e	f	m

Result of second inversion
which overlaps the first

B

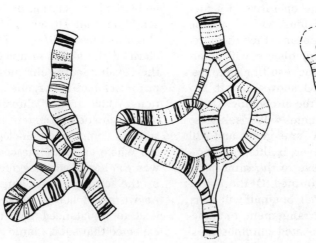

C

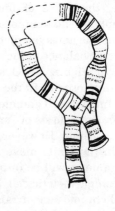

FIG. 10-5. Simple and complex inversions. *A.* Origin of an overlapping inversion. Assume 1 is the standard order and that the first breakage points result in an inversion. When another inversion occurs, the second points of breakage can involve loci which were inverted in the first rearrangement. *B.* An individual heterozygous for just one inversion in a pair of chromosomes will show a simple loop at meiosis. However, an individual carrying a chromosome containing two inversions which overlap will show a compound loop. *C.* Simple loop (*left*) and compound ones seen in the salivary gland chromosomes of *Drosophila azteca.* If two races are crossed and a simple loop forms in a chromosome pair, we know that the races differ by a single inversion. If a compound loop forms, we know there are overlapping inversions. By studying the loops and their complexities, the cytogeneticist can tell by how many inversions two races may differ. Relationships among races may be worked out by making use of this reasoning. (Reprinted with permission from J. Hered., 30: 3–19, 1939.)

to an appreciable amount of adjacent disjunction and, consequently, many unbalanced gametes. But such is not the case in *Oenothera,* where disjunction is usually alternate. If alternate disjunction should always occur, all of the gametes will be genically balanced. But how can alternate disjunction take place almost exclusively? When the chromosomes of *Oenothera* are studied, it is seen that they are all approximately the same size and shape. The arms of any one chromosome are about the same as those of any other. Moreover, the translocations which have taken place have been mainly *whole arm* translocations in which entire arms, not just small chromosome segments have been exchanged. The result is

that all the chromosomes tend to retain their original size and shape. No matter how the arms are shifted about, the chromosomes all remain quite similar. When such chromosomes separate at anaphase, the forces of segregation are balanced in the circle so that one chromosome is repelled equally by the chromosomes on either side of it. This means that all the chromosomes which alternate with each other will stay together, because they will always move to the same pole (Fig. 10-9). The alternate chromosomes thus compose a set which will tend to remain intact generation after generation. For example, if set 1 comes from an egg and set 2 from the pollen parent, the same sets again will be reconstituted after alternate disjunction. Therefore, *Oenothera,* although it has 14 chromosomes and should consequently have seven linkage groups, behaves as if it has just two chromosomes and one linkage group! A moment's thought tells us that if set 1 is in a sperm nucleus, it could fertilize an egg which also bears set 1 (and similarly for set 2). This would produce plants homozygous for set 1 (or set 2), and we would expect the offspring to have 14 chromosomes which would form seven pairs. However, *Oenothera* has another very unusual feature which makes it impossible for any individual to be homozygous for any one kind of chromosome set. For each set carries a specific kind of recessive lethal allele. The lethal of set 1 is different from the lethal of set 2. As Figure

FIG. 10-6. Shifts. *A.* A shift involving three breaks in one chromosome. The genes c and d become switched to a new location in the same chromosome. *B.* A shift involving breaks in two nonhomologous chromosomes. The two genes c and d are now switched to a new location in a different chromosome.

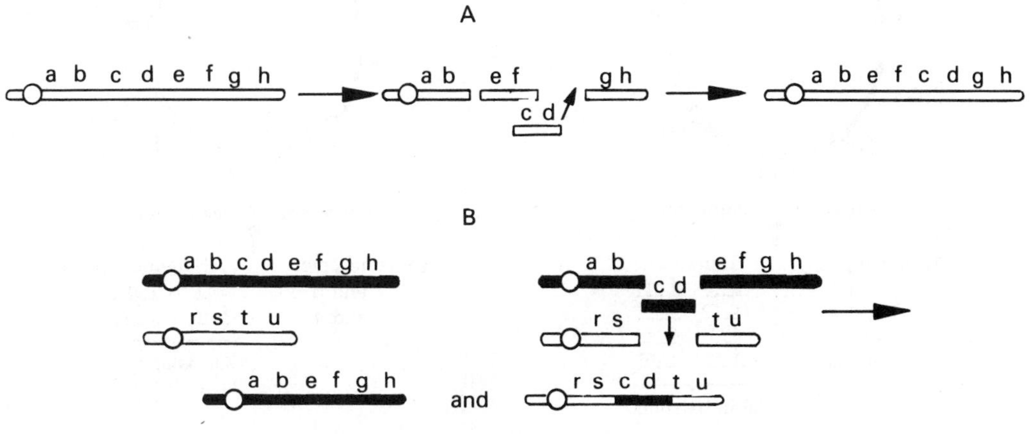

A

B

FIG. 10-7. Reciprocal translocation and its consequences. *A.* Two pairs of nonhomologous chromosomes engage in a reciprocal translocation. This follows two breaks, one in each of two non-homologous chromosomes, shown here as occurring next to the centromere. If a sperm carrying the translocation fertilizes a normal egg, the result is a translocation heterozygote. (Only the end genes are labeled in each arm for sake of clarity.) *B.* Since arms have been shifted between nonhomologous chromosomes, the pairing of homologous arms is achieved by formation of an association of four chromosomes, a tetravalent, seen as a cross in early meiosis and then as a circle by late diplonema. *C.* Adjacent disjunction (*left*) may occur at anaphase. If Chromosomes 1 and 2, for example, move to the same pole, the gametes will be unbalanced. They will have some genes represented twice (those in arm f) and will lack others completely. The other adjacent combinations are also seen to be unbalanced. If alternate disjunction occurs (*right*), the gametes will be balanced. One-half of them will carry the original arrangement and the other one-half the translocation.

10-10 shows, this mechanism prevents the development of any individual with two identical sets. One-half of the offspring will die due to the homozygous condition of the recessive lethals. Only the hybrids will survive; they contain two different lethals and two different sets. The circle of 14 is thus perpetuated. The situation is just as if two chromosomes instead of 14 are present, because the seven chromosomes of any set tend to remain together as if they were linked.

It might seem that the loss of so many offspring as a result of the lethals would be a disadvantage. However, *Oenothera* is normally self-pollinated. This ensures the maximal amount of pollination and fertilization which can give rise to a large number of viable seeds. And the plants are highly vigorous due to valuable gene combinations which are held together in each set from one generation to the next. We see here a shining example of two usually disadvantageous features (translocations and lethals) combined in such a way as to produce a

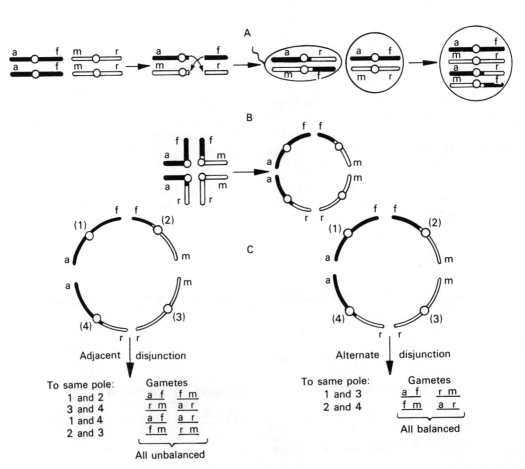

highly advantageous mechanism which ensures hybrid vigor (see p. 463 for a further discussion of hybrid vigor). The evening primroses compose a group of plants which is widespread throughout North America, and its success can be attributed mainly to its unusual genetic system.

Oenothera is also of historical interest, because it was this group of plants which led DeVries to formulate his mutation theory. However, the "new" plants which arose suddenly in his gardens were not the consequences of gene mutation as DeVries thought. At the time, De Vries was completely unaware of the presence of both the translocations and the unusual chromosome segregation. He crossed evening primroses from many localities in various combinations. The "new" types or "sports" which arose were actually offspring derived from hybrids in which the circle of 14 did not always form as described above. And so, these hybrids did not behave as if they had only two chromosomes. The new types were not gene mutants at all but rather recombinants with new gene combinations. The irony is that DeVries focused the attention of geneticists on gene or point mutations which *do* occur. But he formulated the

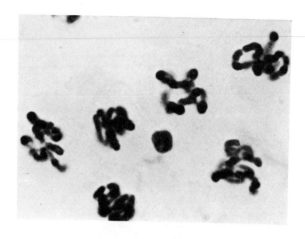

FIG. 10-8. Multivalent associations in *Rhoeo discolor* as a result of translocations.

FIG. 10-9. Chromosomes of *Oenothera*. The arms of all 14 chromosomes have been shifted by translocations so that no one chromosome is completely homologous to any other. Pairing of homologous arms results in a circle of 14 chromosomes. Alternate disjunction always occurs and produces balanced gametes. The same seven chromosomes remain together, constituting a set.

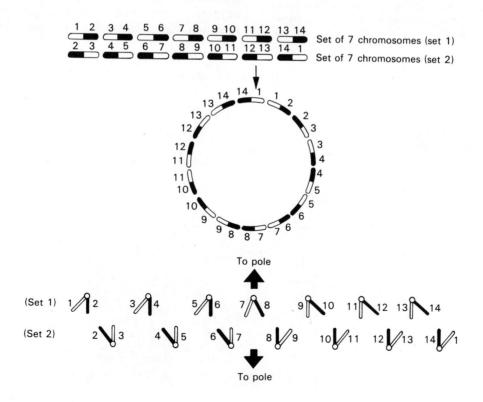

concept of sudden gene changes on results which were not associated with point mutations at all.

SIGNIFICANCE OF TRANSLOCATIONS IN HUMANS. Translocations are definitely known to occur in humans, and their consequences are of great concern to the medical geneticist. As will be discussed in more detail later in this chapter, Down's syndrome (Fig. 10-11) more often results from the presence of one extra chromosome, Chromosome 21, so that an afflicted individual has 47 chromosomes. This, however, is not the only picture in these unfortunate victims, for some of them show the normal number, 46. But karyotype analysis reveals that a structural chromosome change has altered the appearance of one of the chromosomes in group D, believed to be Chromosome 15. Two normal number-21 chromosomes are present, but a change is evident in pair 15. One of the members is normal, but the other is conspicuously larger and does not match its homologue. Therefore, the normal 15 remains unpaired. The change is explained as the result of a translocation between Chromosomes 21 and 15 (Fig. 10-12A). It can be seen from the figure that the new chromosome at the left contains most of the material from the two original chromosomes, 21 and 15. The tiny chromosome which forms may get lost. There is

FIG. 10-10. Lethals in *Oenothera*. Any plant is composed of two sets of seven chromosomes each. The meiotic behavior shown in Fig. 10-9 keeps each set intact. Each carries a different recessive lethal which prevents homozygosity for lethal 1 or 2 and consequently for set 1 or 2. Only heterozygotes survive carrying the two original sets with the lethals.

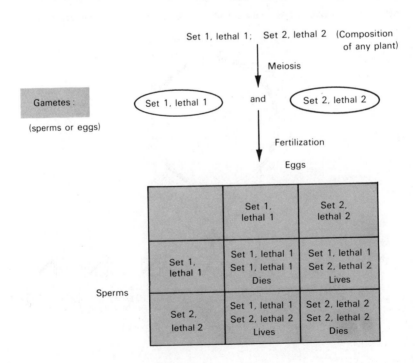

evidence, however, that the translocations studied in humans have not been reciprocal. This would mean that most of the genetic material of Chromosome 21 would be shifted into chromosome 15. The tiny chromosome might not be formed at all. Nevertheless, the large one containing most of the genetic material of Chromosomes 21 and 15 does arise, and it can be passed down through a gamete. Let us say that an individual inherits it through a sperm (Fig. 10-12B). He or she would receive from the female parent a normal 15 and a normal 21. However, the person would have a chromosome total of 45 instead of the normal 46 and can be easily identified by karyotype analysis. The reduced number is present since this individual does not possess a *pair* of 21 and a *pair* of 15. Instead, there is a normal 15, a normal 21, and the altered chromosome, 15/21. Since the essential genetic material of Chromosomes 15 and 21 is present, the person would exhibit no ill effects and might be considered completely normal until a karyotype analysis revealed the aberrant chromosome number.

The meiosis in such a person would be quite irregular, because the translocated chromosome, 15/21, can synapse with both the normal 21 and the normal 15 (Fig. 10-12C). A trivalent association of three chromosomes may be seen at meiosis. The segregation of these three will not be regular and can result in the formation of gametes of different constitutions. Of these, some will be normal, containing one Chromosome 21 and one number 15. But an appreciable amount will contain number 21 and the translocated chromosome, 15/21. When a gamete of this sort combines with a normal one, Chromosome 21 is in effect present three times in the zygote (Fig. 10-12D). The chromosome number of such an individual would be normal (46), but he or she would be indistinguishable from those sufferers of Down's syndrome who have 47 chromosomes. This is so, because both classes of victims have most of Chromosome 21 in three doses. Down's syndrome arises with an unfortunate frequency (approximately 1 in 500 newborn); it therefore accounts for a large number of mentally defective children. Most of them possess 47 chromosomes and have arisen as a result of nondisjunction in the female parent, but a percentage arises, as discussed here,

through a carrier parent. Identification of carriers of the translocation by karyotype analysis is important where the occurrence of Down's syndrome appears to run in a family line, since the heterozygous carrier of the 15-21 translocation can continue both to transmit gametes which may give rise to Down's children and to transmit the carrier condition to other children who may in turn do the same (Fig. 10-12C). Identification and proper genetic counseling of carriers may help reduce the incidence of afflicted children. Those cases of Down's syndrome resulting from the presence of one extra Chromosome 21 rather than a translocation (Fig. 10-11) are not familial, since the aberration arises in meiotic cells of the immediate parent of the Down's individual, who will almost always fail to reproduce.

Translocations have been reported between Chromosome 21 and chromosomes other than 15, and these, too, are associated with Down's syndrome. Moreover, some cases of translocations in humans do not involve Chromosomes 15 or 21 at all. These have been associated with mental retardation and a high probability of death before or shortly after birth.

As noted in Chapter 9, studies of cells with translocations are especially valuable in the assignment of a gene to a localized region of a chromosome.

THE DETECTION OF DELETIONS. Another type of structural aberration, which has now been described in humans, as well as in other species, is the deletion. Actually, it was the very first type of structural change to be identified and was recognized in *Drosophila* in 1915. As its name implies, a deletion entails the actual loss of a portion of the chromosome so that a block of genes is missing. A deletion can follow a double break in a chromosome as shown in Figure 10-13. The affected chromosome is actually smaller as a result of the loss, although a size change is not detectable when the eliminated piece is minute. It should be obvious that the deletion will certainly cause a degree of genic unbalance, because certain loci will be completely missing from one of the chromosomes. The possible effects of such unbalance are demonstrated by two early cases in *Drosophila*. The very first one, described by Bridges, involved a

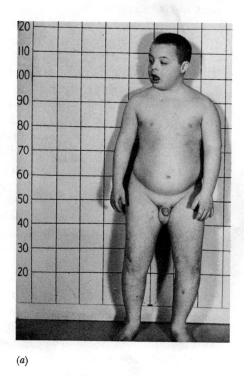

(a)

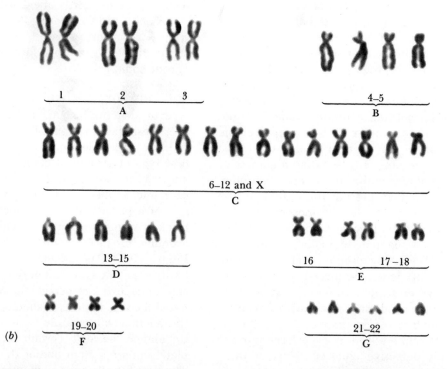

(b)

FIG. 10-11. Down's syndrome. *a*. Child with characteristic features. *b*. Typical karyotype showing 47 chromosomes. The extra chromosome is in the G group. (Reprinted with permission from V. A. McKusick, *Medical Genetics 1958–1960,* C. V. Mosby Co., St. Louis, 1961.)

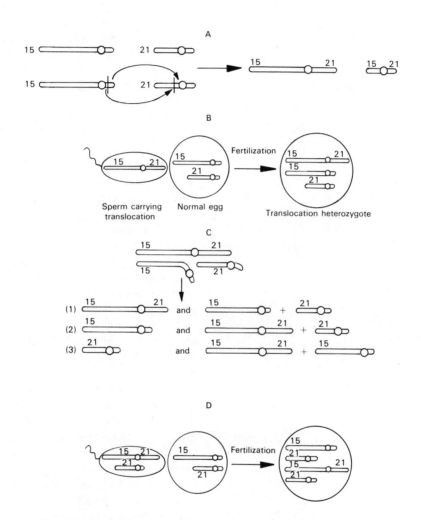

FIG. 10-12. Translocation in humans with the origin of trisomy 21. *A.* Arms are exchanged between Chromosomes 15 and 21. The small chromosome which may arise would contain little genetic material. The larger chromosome which arises contains most of the original Chromosomes 15 and 21. *B.* A sperm carrying the translocated chromosome fertilizes a normal egg. The zygote which results will be essentially balanced, because two Chromosomes 15 and two Chromosomes 21 are in effect present. The chromosome number is one less than normal due to the fact that 15 and 21 are joined as a result of the translocation. *C.* A trivalent forms at meiosis in the translocation heterozygote. This brings about pairing of homologous regions, but it does not guarantee balanced segregation. Because any one of the three chromosomes

may move by itself to one pole, several possibilities follow. Segregation 1 will produce gametes which will pass down the translocation; it will also produce gametes which are normal. Segregation 2 will produce a gamete lacking Chromosome 21 (a lethal condition) and a gamete which can result in Down's syndrome as shown in *D.* The third segregation produces gametes which are very unbalanced. One type lacks Chromosome 15. The other has it present twice. The unbalance proves lethal. *D.* A sperm formed in a translocation heterozygote as a result of segregation of type 2 (*C above*) fertilizes a normal egg. The zygote has three doses of Chromosome 21 and will develop into a victim of Down's syndrome. The chromosome number, however, will be normal.

deletion in the "Bar" region of the X chromsome. The "Bar" effect, in which the eye is narrowed, is inherited as a sex-linked dominant. As illustrated in Figure 10-14A, a cross between a normal female and a Bar male should produce Bar females and normal males. However, in a certain cross, Bridges mated a normal female with a Bar male and obtained a normal female among the offspring. At the same time, a lethal seemed to appear. After a series of appropriate crosses, it became clear that a piece of the chromosome, including the Bar locus, had been deleted and that this deletion could be carried in the female but not in the male. Figure 10-14B shows that the deletion causes the elimination of the dominant Bar, and so the wild trait, which is recessive, can be expressed. When the female with the deletion is crossed, one-half of the male offspring die as a result of the missing chromosome region (Fig. 10-14C). Evidently, the deleted segment contains genetic information essential to life, because its complete absence in the hemizygous male prevents development. Further crosses showed that in addition to Bar, at least one other locus was missing. This proved to be the locus "forked," which has a pronounced effect on the bristles.

A different deletion arose in the X of *Drosophila*; this affected the development of the wings. A female heterozygous for this deletion has wings which contain a nick or notch at the tip. This notch condition has arisen independently several times in the fruit fly. Each notch fly has been shown to contain a deletion which always includes the locus "facet," a genetic region which affects the texture of the eye. The extent of these different notch deletions varies and may include the "white" locus in addition to facet (Fig. 10-15). All notch flies must be females because the deletion is a complete lethal in the male, for the same reason as given above for Bar.

DELETIONS AND MAPPING IN DROSOPHILA. There have been many cases in the fruit fly in which the presence of suspected deletions has

FIG. 10-13. Formation of a deletion. Two breaks may arise in a chromosome, followed by loss of the interstitial piece. The portions with the centromere and telomeres rejoin.

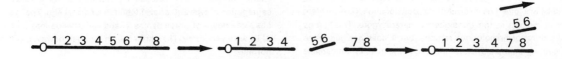

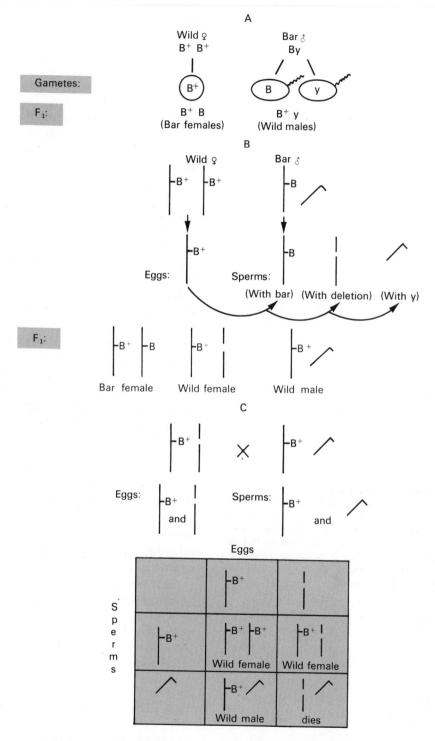

FIG. 10-14. Effects of deletion in Bar region. *A.* Ordinary cross of wild with Bar. (B^+ = wild; B = bar). *B.* A deletion arises in some of the sperms, eliminating the dominant Bar (B). Unexpectedly, some of the females appear wild. *C.* When the wild female carrying the deletion is crossed with a wild male, all of the female offspring are wild, but one-half carry the deletion. Since one Bar region is necessary for life, it acts as a sex-linked recessive lethal, and one-half of the F_1 males fail to develop.

been verified cytologically. If a deletion is present, there will be a region on the normal chromosome which is unable to pair at the time of synapsis because the homologous region has been lost. Consequently, the normal chromosome may form a buckle or loop whose size depends on the extent of the deletion (Fig. 10-16A). The loci absent from the deleted chromosome will be present in the loop formed by the normal one. This fact has enabled the cytogeneticist to ascertain the location of specific genes and frequently to associate them with definite bands in the salivary chromosomes. For example, if the genetic results indicate that a specific locus has been deleted, one may search for a loop in the specific chromosome to which the gene is linked (Fig. 10-16B). One would look at the X in the salivary nuclei in the case of notch or white, because these loci map on the X. If a loop is found, one may compare the bands in the loop with those in the other chromosome. Any bands absent from the latter would thus represent the deletion. In this way, definite loci can be associated with definite chromosome sites. The locations of genes, as determined in this way, correspond well with their positions as determined by genetic mapping. However, although the linear sequence of the genes is the same, the map may show regions where the loci appear closer together than they do when a map is made cytologically from the salivaries (Fig. 10-17). The reason, of course, is due to the fact that the genetic map depends on the frequency of crossing over. In some chromosome regions, genes engage in less crossing over than those in other regions. We have noted that loci near the centromere, for example, appear clustered due to its inhibiting effect on crossing over. The genetic map tells us the *probability* of crossing over, not the exact spatial distances among genes. The latter are more precisely resolved by the cytological map, built up through the study of deletions.

EFFECTS OF DELETIONS IN VARIOUS SPECIES. The examples of the deletions illustrate several other important points. When a deletion is present, atypical genetic effects usually follow. We see that a recessive may unexpectedly express itself; the recessive "wild" was expressed when the dominant "Bar" was assumed to be

FIG. 10-15. The notch phenotype always involves a deletion of the locus "facet," which influences eye development. Any deletion may include more than "facet," in this case, the "white" locus. All notch flies must be females because the facet deletion acts as a sex-linked recessive lethal. In the cross shown here, the recessive "white" expresses itself in one-half of the females because the white locus is included in the deletion. Only one-half of the males will develop due to the deleted genes.

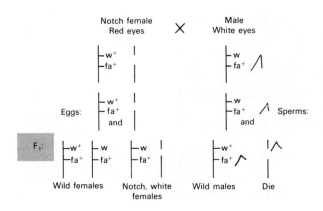

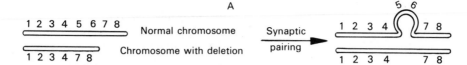

A

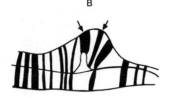

B

FIG. 10-16. Cytological detection of deletions. *A.* At the time of meiotic pairing, a buckle may be observed in a bivalent. This results from the fact that the normal chromosome is longer than the homologue with the deletion. The genetic region present in the normal chromosome has no corresponding part with which to pair. The buckling makes possible synapsis of all those homologous regions which are present. *B.* The giant salivary chromosomes are permanently synapsed and contain conspicuous bands. Any regions in the buckle (indicated by the arrows) must be absent from the homologue. Bands in the buckle can be seen to be deleted from the other chromosome. If known loci are absent, as determined genetically, they may be associated with bands in the buckle. In this diagram, the bands indicated are absent from the other chromosome. If certain genes are known to be missing, their chromosome location may thus be associated with these bands.

present. This is often referred to as "pseudodominance." In the case of a "notch" deletion which encompasses the "white" locus, the recessive white could suddenly appear (see Fig. 10-15). The mutant recessive, white, has superficially acted as if it were dominant, because it was expressed when the wild gene was supposedly present. Actually, it has remained recessive; its pseudodominance is entirely the result of the deletion.

The case of the "notch" deletion also demonstrates that the deletion itself may produce a characteristic phenotypic effect which acts as a dominant. This is very often true of deletions in a variety of organisms. The "notch" deletion, moreover, behaves as a recessive lethal (see Fig. 10-15). No males can develop because they have only one X chromosome, and no females homozygous for the notch deletion can arise. Again, this is often the case with small deletions, autosomal as well as sex linked. If their effect on genic balance is not too severe, small deletions may accumulate in a population to some extent. But when they become homozygous, they usually prevent development and act as if a recessive lethal were present. Of course, if a deletion is of sufficient size or includes certain critical genes, it may also act as a dominant lethal and kill the heterozygote.

These observations also tell us something else which reinforces what has been said in Chapter 4 and elsewhere concerning the interaction of genes. The "notch" deletion shows that the elimination of a locus, the facet locus which has a pronounced effect on the eye, may cause a defect in another organ, here the wing. This is another good example that any gene may have more than one effect (pleiotropy) and that the normal development of any character depends on the interaction of the entire genotype. The very viability of the organism can also depend on the presence of certain loci, at least in a single dose. Obviously, the gene products of "Bar," "facet," and others are essential to the development of the embryo in the fruit fly.

Work with mice has shown that orderly development may depend on products governed by genes. A certain product may be needed at a critical step or time in the developmental process. If it is absent, a crisis results which can terminate development and prove lethal. Detailed studies of different-size deletions in the X of *Drosophila* have also illustrated the crucial role of various loci at specific times in differentiation. It is very important to note at this point that a deletion of any size can come to exist in a heterozygous state only if it can be transmitted through haploid cells. Deletions frequently prove lethal to the gametes of an animal, the mouse for example. Moreover, the haploid (gametophyte) generation of plants is especially vulnerable to deletions. The genes essential to the formation of viable gametes would appear to be widespread throughout the chromosomes in these species which, unlike *Drosophila*, cannot tolerate deletions of any substantial size in their sex cells or cells in the haploid generation.

From these observations of the dire effects of deletions in experimental organisms, it comes as no surprise that a deletion in a human can have serious consequences. Unfortunately, this is borne out by the effects associated with a deletion in the shorter arm of Chromosome 5. Afflicted persons exhibit an assortment of effects known as the cri du chat syndrome. The name refers to the catlike cry which such individuals produce in infancy and sometimes even beyond. In addition to various characteristic facial features (Fig. 10-18), the carriers of the deletion suffer growth defects and mental retardation. The products of genes in that particular segment of Chromosome 5 are evidently necessary for normal development.

FIG. 10-17. Comparison of the genetic and cytological maps of the third (III), second (II), and X chromosomes (X) of *Drosophila melanogaster.* C is the cytological map and G the genetic map. Figures indicate the genetic distances in map units. The lines connecting the cytological with the genetic maps indicate the microscopically observed and the genetically determined positions of certain chromosome breakages. (Reprinted with permission from E. W. Sinnott, L. C. Dunn, and T. Dobzhansky, *Principles of Genetics,* 5th ed., McGraw-Hill, 1958.)

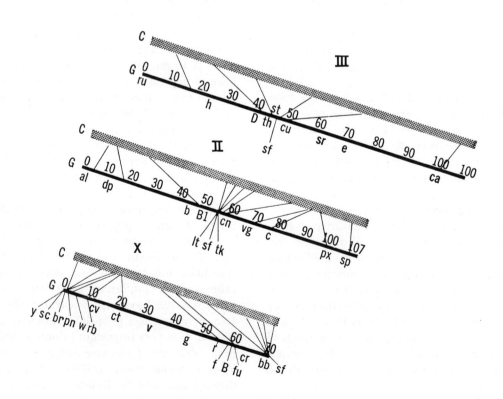

Especially interesting is the association of a particular chromosome aberration with one form of leukemia, chronic myeloid leukemia. The leucocytes of affected individuals typically show a portion missing from one of the small chromosomes in the G group (see Table 2-1). The chromosome with the deletion was designated the "Philadelphia chromosome." It was uncertain at first whether Chromosome 21 or 22 was the one involved, but the consensus was that it is the same chromosome as that associated with Down's syndrome, number 21. The reasoning was that a victim of Down's is more apt to develop leukemia than is the average person. However, improvements in staining procedures with fluorescent dyes have resolved differences between Chromosomes 21 and 22 in the G group and have positively shown that the Philadelphia chromosome is not the same as the one implicated with Down's but that it represents Chromosome 22. It has been a matter of debate whether the presence of the deletion in the Philadelphia chromosome, found only in the white blood cells, somehow causes the leukemia or whether it is the result of the abnormal condition of the bone marrow. To complicate the matter a bit more, the cytological banding picture shows that the Philadelphia chromosome actually represents the effects of a translocation and that the missing portion has been shifted from Chromosome 2 and inserted onto Chromosome 9! What had been considered a deletion represents instead the association of a translocation with the disorder. It could still be argued that some tiny piece of the chromosome was lost during the chromsome alteration, but there is no evidence for this. Nevertheless, the association of a chromosome alteration with a specific malignancy remains highly provocative, even though the full significance of the relationship is as yet unknown.

The importance of deletions in human cells in the assignment of specific genes to definite regions of a given chromosome was noted in Chapter 9. Deletion mapping in the human, which involves techniques of chromosome banding and somatic cell hybridization, along with pedigree analysis, will probably play a very significant role in the localization and ordering of genes on a given chromosome.

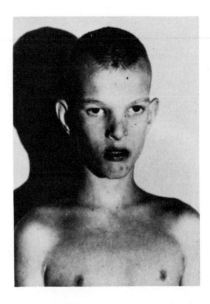

FIG. 10-18. Child showing features of the cri-du-chat syndrome. (Reprinted with permission from *Birth Defects: Atlas and Compendium*, ed., D. Bergsma, The National Foundation—March of Dimes, White Plains, New York with permission of the editor and contributor.)

DUPLICATIONS: THEIR ORIGIN AND THEIR EFFECTS. There is still another kind of structural chromosome alteration which may produce unexpected genetic results. This is the *duplication,* a condition in which a portion of a chromosome is present in excess in somatic cells. If a region is present three or more times instead of the normal two, it is said to be "duplicated." One way in which a duplication can arise is shown in Figure 10-19A and B. A simple translocation or shift may place a chromosome segment in a new location in the chromosome. The chromosome containing the shift would still be balanced; therefore, there probably would be no phenotypic effect in the individual carrying the altered chromosome along with the one with the normal sequence of genes. The figure shows that pairing problems would again pertain at meiosis, and two loops might form in order to permit synapsis of most of the homologous regions. A crossover anywhere between the loops, however, will lead to two altered chromosomes, one with a deletion and one with a duplication. The consequences of the former have just been discussed. The chromosome with the duplication can also cause phenotypic effects due to an upset in genic balance. If it enters a gamete which then combines with a normal gamete, the zygote will contain some loci in three doses instead of two (the loci FG in Fig. 10-19C). The outcome will depend on the size of the duplication and the specific loci involved. In some instances, a dose of two recessive alleles along with the dominant allele results in the expression of the recessives. For example, a female *Drosophila* carrying two doses of the sex-linked recessive, v, (vermilion eye color) and one of the wild allele for red eye, v^+, has vermilion eyes. The genotypes v^+v v and v v are the same phenotypically. In contrast to this, many examples could be given in which one dose of the dominant will express itself over two doses of the recessive. In the fruit fly, the wild allele sp^+, is dominant over two doses of its recessive allele, sp, the gene "speck," which is responsible for a certain body abnormality. Therefore, sp^+sp sp is wild. On the other hand, the recessive "pl" (for abnormal wing venation), when present in an extra dose with wild pl^+ produces a weak mutant phenotype; the fly which is pl^+ pl pl can be distinguished from both the typical wild (pl^+ pl^+ or pl^+ pl) and the homozygous recessive (pl pl). In still other cases, intermediate conditions of some type are produced. Obvously, if a duplication is unsuspected, its presence in a particular mating can lead to puzzling results.

Duplications are considered to have played a very important role in evolution as a major factor in increasing the number of genes in the chromosome complement. Once a genetic region is present in additional amounts, the stage is set for the originally identical genes to diverge. This is so, since each of the duplicated regions can undergo its own history of independent mutation. As different mutations and hence different gene changes continue to arise and as the force of natural selection operates, what were once identical gene forms may eventually be recognized as two very different genes, each controlling the formation of a somewhat different product.

There is good evidence for this in the fruit fly, mouse, and the human. In these species, there are various cases of very close linkage among genes which determine the amino acid composition of very similar protein chains. If the two protein chains controlled by two separate genes are vey much alike in amino acid se-

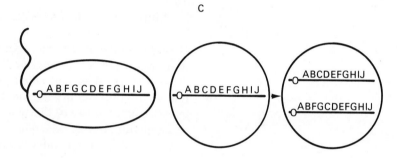

FIG. 10-19. Origin of a duplication. *A.* A shift of a genetic region takes place after three breaks in one chromosome. *B.* At meiosis, the chromosome with the shift cannot pair exactly throughout its length with its homologue. A buckle forms in each chromosome to bring together most of the homologous regions. A crossover anywhere between the loops results in a chromatid with a deletion (FG) and another with a duplication. *C.* A gamete with the duplication fertilizes a normal gamete. The zygote has one genetic region present three times. The region in excess (FG) is said to be duplicated.

quence, this means that the two genes involved must also be very similar, since their information content is similar. This is seen in the human in the case of the beta and the delta polypeptide chains of hemoglobin (Chap. 16). Not only are the two genes tightly linked (actually adjacent to each other), each protein chain is strikingly similar to the other, 146 amino acids long and differing only in 10 of them! It is reasoned that the beta and delta genes represent duplications of one original gene and that the two genetic regions have diverged as a result of independent gene mutations.

DUPLICATION OF THE BAR LOCUS AND POSITION EFFECT. The unexpected genetic effects of a duplication are well illustrated by the story of the Bar "mutation" in *Drosophila.* As already noted, the effect of Bar is to decrease the number of facets in the eye so that it resembles a band or bar. The condition is inherited as a sex-linked dominant (actually an incomplete dominant) and was discovered in 1914 (see Fig. 10-14*A*). It was then found that homozygous Bar stocks, when used in crosses, occasionally give rise to wild flies. This should not be so, because the Bar "allele" is a dominant. The frequency of the reversal to normal is approximately 1/1500, a figure much too high to result from mutation of the mutant Bar gene back to normal (B → B⁺). It was also found that when these unexpected normals arose, another phenotype also appeared, and with the same frequency of 1/1500. This new type was a narrower, more extreme Bar than the typical one, and so it was designated "ultrabar." In a series of crosses, Morgan and Sturtevant mated male and female Bar flies separately by crossing them with normals. Their results showed that the normal and the ultrabar arose only when a female was used as the Bar parent, never when a Bar male was crossed with a wild female. For example, the cross Bar female × wild male (B B × B⁺Y) can result in some unexpected offspring with wild eyes and some which are ultrabar. The reciprocal cross, wild female × Bar male (B⁺ B⁺ × B Y) gives only the expected results as diagrammed in Figure 10-14A.

Sturtevant made a series of crosses in which he followed not only Bar but also genes very closely linked to it. He was able to demonstrate that whenever the normal or the ultrabar appeared, a crossover had taken place in the Bar female in the vicinity of the Bar locus. The question arose as to why this phenomenon should be happening in this region of the chromosome. Why weren't similar results obtained when other loci were studied? Muller suggested that the enigma could be explained on the assumption that the dominant Bar effect is *not* the result of a point mutation but rather of a duplication. Under this idea, a homozygous Bar female would have two Bar loci represented on each X chromosome and would be B B / B B (Fig. 10-20A). The normal female would thus be "B/B," with just one dose of Bar on each X.

As a result of the presence of duplicated genetic material in the homozygous Bar female, difficulties could be expected to arise at synapsis. Since four identical regions are present, the pairing need not always be two by two. A certain amount of synaptic confusion might lead to the pairing of the "right-hand Bar" on the one chromosome with the left of the other (or vice versa). Crossing over in the area could then produce unequal products: "B" and "BBB" (Fig. 10-20B). The former, according to Muller's hypothesis, would be the normal, nonbar condition; that is, the presence of a single Bar on the X of a male or a single Bar on each X in the female produces the normal-eye phenotype. The latter chromosome, BBB, with three adjacent Bar loci would result in ultrabar. Muller suggested a cytological examination of the banding pattern of the X chromosome in the salivaries. In the region of the X where Bar would be present, as suggested by genetic maps, a certain segment was found to be represented twice in the Bar fly, three times in the ultrabar, and just once in the wild (Fig. 10-21)!! Again we see a perfect correlation between genetic results and cytology. Thus, it was shown conclusively that the Bar effect is the result of a tandem duplication (two identical segments next to each other). The more segments which are adjacent, the greater the mu-

tant Bar effect and the narrower the eye. When flies with the several genotypes shown in Figure 10-22 are compared phenotypically, a rather unusual fact becomes evident. For example, a comparison of flies 3 and 4 reveals that they differ phenotypically. Fly 3 has a much wider eye than fly 4. However, when the number of Bar loci is counted, it turns out to be the same; both flies have four Bar genes (BB/BB is Bar; yet BBB/B is ultrabar). This was the first indication that the dosage of a gene is not the only factor which might be involved in the expression of a phenotype. Also to be considered is the arrangement of the genes: three Bars on one chromosome and one Bar on the homologue is different from two Bars on each chromosome. This type of phenotypic effect resulting from the

FIG. 10-20. The Bar region. *A.* The normal wild fly carries one Bar locus on any X. Because it is sex linked, the wild female carries two X chromosomes, each with one Bar locus. A Bar-eyed fly results when the Bar locus is duplicated on an X, one Bar adjacent to the other. *B.* Crossing over may occur within the Bar region. Typically, pairing will be normal (*upper diagram*) and the two crossover chromatids will be identical to the original. However,

forces of pairing may at times lead to some confusion, because four homologous loci are present (*lower*). A crossover in the Bar region then leads to production of two different chromatids, a wild type and one with three adjacent Bar loci which can produce an ultrabar effect. The figure depicts a male which has arisen after fertilization of an egg carrying the X with Bar present three times.

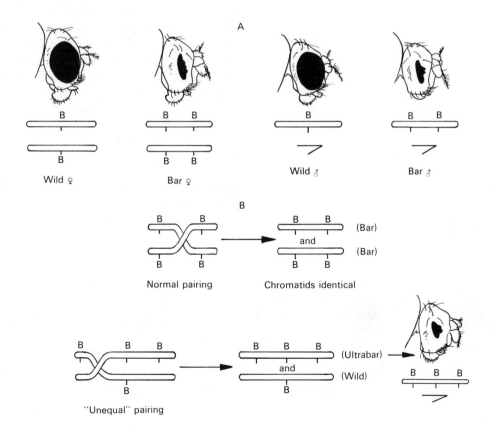

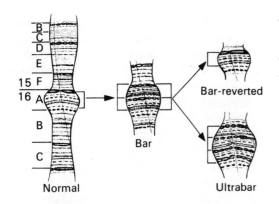

FIG. 10-21. The Bar region in the salivary chromosomes. Certain bands appear duplicated in the X chromosome of flies with the Bar phenotype. In those with the ultrabar condition, the bands are present in triplicate. The normal as well as the ultrabar condition can arise from a Bar female in which "unequal pairing" occurs, followed by a crossover in the Bar region. (Reprinted with permission from C. B. Bridges, *Science*, 83:210–211, 1936.)

FIG. 10-22. Comparison of female flies with different numbers and arrangements of the Bar locus.

position of one gene in respect to another was named the *position effect*. Position effects have been studied in detail in the fruit fly and have now been demonstrated in other organisms as well.

Additional observations with the Bar locus clearly verified that the position of a gene can be very important. A case was found in which breaks had occurred on either side of Bar. This was followed by a translocation of Bar from its normal location in the X to the small fourth chromosome (Fig. 10-23A). Various comparisons could now be made. For example, a male having an X with no Bar locus but having a fourth chromosome carrying Bar was found to possess a very narrow eye, a condition called "Stone's Bar." Figure 10-23B illustrates that a Bar fly with just one Bar locus can have the same reduced number of eye facets as a fly with two or even three doses of Bar, depending on the location of the Bar region. Size of eye decreases with the number of Bar loci which are in the fourth chromosome. Comparison of male and female flies carrying Bar in different doses left no doubt that the location of the Bar locus is as important as its dosage. When Bar is taken out of its normal location and placed in another chromosome or placed next to another Bar, it can cause a reduction in eye size. So we must be aware of the possibility of position effects when studying any structural heterozygote. In our discussion of inversions and translocations, we stated that the heterozygote is "usually normal." This is so, but a qualification must be made, because position effects may possibly arise.

HETEROCHROMATIN AND POSITION EFFECT. Not all genes show position effects when they are placed in new locations. Studies with the fruit

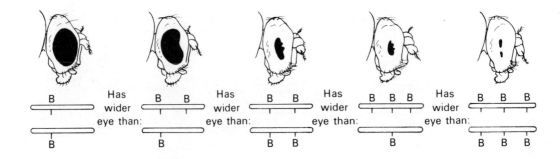

fly indicate that it depends on the particular gene involved, as well as on the new location. Very interesting position effects are produced when certain loci are shifted to the vicinity of a type of chromatin called "heterochromatin." When the behavior of sex chromosomes at meiosis was discussed (Chap. 3), we noted that the X chromosome in the male grasshopper reacts differently than the other chromosomes to staining. It tended to be condensed and to stain deeply at times when the others were more diffuse. We also saw in Chapter 5 that one X chromosome in the mammalian female is repressed and that we can detect this inactive X as condensed, deep-staining, sex chromatin (the Barr body or drumstick). In addition to these points, we also find when chromatin is studied in detail in the nondividing nucleus that some of it always tends to remain condensed, in contrast to the chromatin which is stretched out. We use the term "heterochromatin" in all these cases to refer to that chromatin which seems "out of step" with the rest. Some portions of the chromatin may behave as heterochromatin only at certain times. In the case of mammalian females, the X chromosome derived from the paternal parent may be condensed as heterochromatin in some cells, whereas it can be extended and active in others (recall the Lyon hypothesis). However, there are other regions of the chromatin which always tend to act differently from the rest and are thus typically heterochromatic. *Euchromatin* refers to the chromosome material which undergoes the familiar cycle of condensation and despiralization as the nucleus divides and returns to the metabolic state. Most of the genes well known to us are located in euchromatic regions. Relevant to our present discussion is the peculiar type of position effect which can be exerted by heterochromatin on the euchromatin. Suppose that the dominant sex-linked allele for red eye color (w+), normally found in euchromatin, is placed next to heterochromatin as the result of a translocation (Fig. 10-24A). If a female fly contains the dominant w+ adjacent to heterochromatin on one X chromosome and the recessive allele w in its normal location on the other one, an unexpected phenotype may result. Normally, a fly which is w+/w should have red eyes. But if such a fly has the w+ allele next to heterochromatin, it may have a variegated eye, one which has patches of red and patches of white. Many such examples demonstrate that dominant alleles from the euchromatin, placed next to heterochromatin, tend to lose their dominant effect. It is as if the heterochromatin were rendering them inactive in some cells, so that the recessive phenotype is expressed. The degree to which the recessive condition appears depends both on the amount of heterochromatin and the particular kind involved. If exposed to enough heterochromatin of a certain kind, a dominant allele may be unable to express itself at all. There are several studies in the fruit fly which demonstrate that the dominant allele has not been altered or in any way changed in structure when the variegated effect is produced. In several cases, the rearranged dominant allele has been restored to its normal location in the euchromatin. It then acts again as a normal dominant, illustrating that it was the neighboring heterochromatin which had been responsible for the variegation and not a mutation in the gene.

Not only may one allele at a time show the variegation effect. If a transposed euchromatic segment contains two or more alleles, more than one of them may be influenced (Fig. 10-24B). The allele closest to the heterochromatin would show the greatest effect. The influence of the heterochromatin decreases with its distance from a particular euchromatic gene. If a more distant allele in a group is affected, those between it and the heterochromatin are certain to be affected also. This illustrates the *spreading effect,* the tendency of more than one allele to be repressed by the presence of heterochromatin. The closer the allele to the heterochromatin, the greater its chance of being influenced.

The assortment of effects which can arise from duplications and from simple changes in gene position underline the complications which must be considered in cytogenetic analysis. It is important to keep in mind the consequences which such alterations in chromosome structure may prove to have for development in humans.

CHROMOSOME ALTERATION AND EVOLUTION. There is no question that structural changes have altered the size, shape, and number of chromosomes during the evolution of many plant and animal groups. In the genus *Droso-*

FIG. 10-23. Origin and effects of Stone's Bar. *A.* Stone's Bar arises when the Bar locus is shifted from the X to the tiny fourth chromosome. The Stone's Bar phenotype is due to a position effect, the insertion of the Bar region into a new location, Chromosome IV. *B.* A normal male (*above*) has one X with Bar as well as two Chromosome IV's. A normal female has two X's, each with one Bar, as well as two fourth chromosomes. The male and female both have the same eye size. Apparently, Bar in its normal position does not reduce the number of facets in the eye. A comparison of flies differing in the number and position of Bar loci (*middle and below*) shows that the absolute number of Bar loci is not the sole determinant of eye width; rather it is the position which is important. Eye size tends to decrease when Bar is taken out of its normal location and placed elsewhere.

phila it is very evident that paracentric inversions and whole-arm translocations have been involved in the differentiation of species within the genus. When chromosomes of the several fruit fly species are compared, one finds a very striking picture (Fig. 10-25A). The chromosomes are found mainly in the shape of rods and various V-shaped configurations. The chromosome number may differ among the species, but the number of arms tends to remain the same from one to another. The X chromosome may be a rod in some (*melanogaster* and *virilis,* for example) or it may be a V (*pseudoobscura, miranda*). When maps are made of the different species, it is found that the same types of loci tend to be found linked together in an arm (pink and ebony are always in the same arm; so are facet and white).

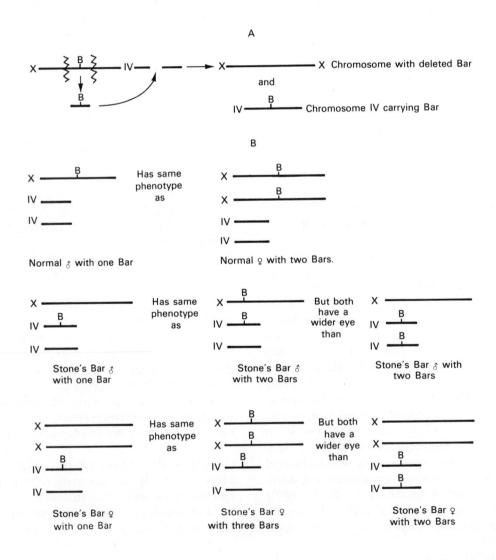

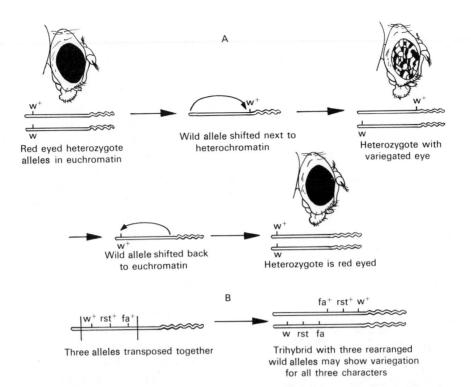

A

Red eyed heterozygote
alleles in euchromatin

Wild allele shifted next to
heterochromatin

Heterozygote with
variegated eye

Wild allele shifted back
to euchromatin

Heterozygote is red eyed

B

w^+ rst^+ fa^+

Three alleles transposed together

fa^+ rst^+ w^+

w rst fa

Trihybrid with three rearranged
wild alleles may show variegation
for all three characters

FIG. 10-24. Heterochromatic position effect (wavy line indicates heterochromatin). *A.* Wild type allele placed next to heterochromatin in the same or in a different chromosome after a chromosome alteration may result in failure of the wild allele to express itself in some cells. When returned to a normal euchromatic region after a second chromosome alteration, the wild allele again expresses itself in every cell. *B.* Spreading effect is seen when more than one gene is shifted to a region of heterochromatin. If the gene farthest away is influenced ("facet" in this case), all those intervening up to the heterochromatin will also show the position effect.

This implies that the arms in the several species are basically the same and have been derived from an ancestral type. In those species with V-shaped chromosomes, the two arms associated to form the V may be different. For example, the X chromosome of *Drosophila americana* is composed of two arms, but one of these is very different from one of the arms of the X of *pseudoobscura*. The other arm in the X of both species corresponds to each other as well as to the X in *D. melanogaster,* which has but a single arm. It seems as if genes tend to stay together in an arm but that whole arms have been shifted about by translocations. The order of the genes, however, may not be the same in a corresponding arm of the different species. The evidence suggests that inversions within corresponding arms have altered the gene order (Fig. 10-25B).

Pericentric inversions have undoubtedly played a role in the shifting of the centromere, with a consequent change in the chromosome morphology. Studies of the banding patterns between closely related groups indicate that the J-shaped chromosomes present in some have been derived from rod- or V-shaped chromosomes as a result of inversions in which the two breaks are not equidistant from the centromere.

In *Drosophila,* it is possible to cross some of the species and to study the association of the arms in the salivaries of the hybrid. The observations support the hypothesis that arms are homologous in the various species but that translocations have changed their attachments and inversions have rearranged the gene sequences.

The evolution of human chromosomes can now be studied in a more precise way. The techniques of chromosome banding are exceptionally valuable in revealing the details of chromosome morphology and thus permitting comparisons of chromosomes between the human

(A)

(B)

and other species. The chromosomes of the human and some other species may be compared as to number, size, shape, and banding pattern. In this way, karyotype analyses can give some insight into possible relationships among different groups and any chromosome changes which have arisen as they diverged. For example, the chimpanzee is most probably the closest relative of the human, as strongly suggested by the similarities in their DNA and many of their proteins. The chromosome number of the chimpanzee is 48. One more pair of autosomes is present than in the human. However, this difference in number can be attributed to a translocation involving two acrocentric chromosomes. In the translocation, the two larger arms of two nonhomologous chromosomes would have become associated, much as seen in Figure 10-12A. The banding pattern suggests that human Chromosome 2 has arisen in this way from two nonhomologous ancestral chromosomes, each having one very short arm. Many other differences seen between the chromosome complements of the human and the chimpanzee can be attributed to pericentric and paracentric inversions.

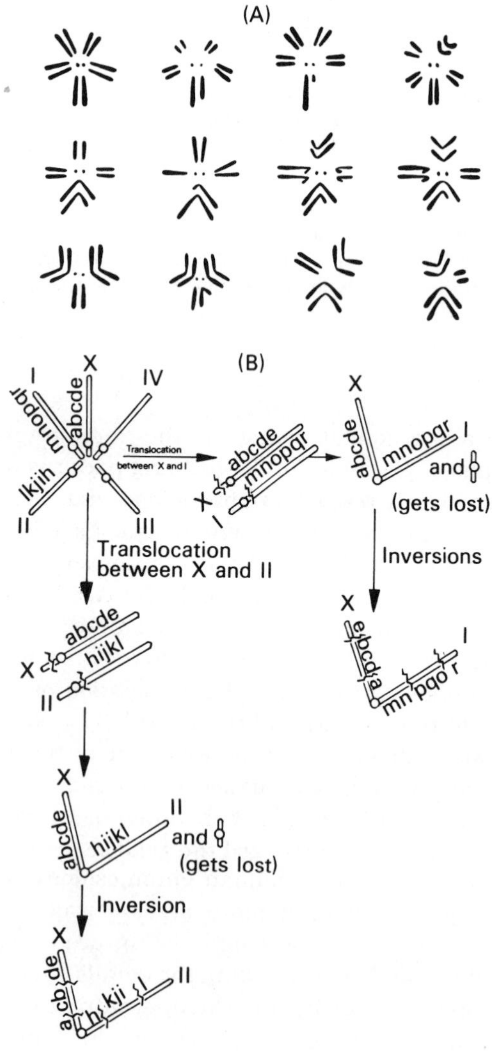

FIG. 10-25. A. Chromosome complements of some species of *Drosophila*. The karyotypes are of males, and the X and Y chromosomes are the two lower ones in each case. The different arrangements of rods, V's, and J's, can be explained on the basis of shifting of chromosome arms and segments during evolution of the species by translocations and inversions. The species shown are: (1) *Drosophila virilis*; (2) *Drosophila funebris*; (3) *Drosophila repleta*; (4) *Drosphila montana*; (5) *Drosophila pseudoobscura*; (6) *Drosophila miranda*; (7) *Drosophila azteca*; (8) *Drosophila affinis*; (9) *Drosophila putrida*; (10) *Drosophia melanogaster*; (11) *Drosophila willistoni*; and (12) *Drosophila prosaltans*. B. Alterations in karyotype. In *Drosophila*, chromosome arms may have been shifted about in a fashion similar to this. From an ancestral type having five rod-shaped chromosomes, V-shaped ones can arise. The X is shown here undergoing a reciprocal translocation with Chromosome I in one case and with Chromosome II in another. The tiny chromosome may be dispensable, because it contains large amounts of heterochromatin, which is found around the centromere. Eventually two species may be derived in which certain genes are sex linked in one but not in the other. However, a group of genes would be similar and sex linked in both species. These genes would be found in the same arm. This arm would correspond to the single arm in a species having just a rod-shaped X, because this arm traces back to the same ancestral X. The order of genes in the arms may differ due to different inversions as the species evolved separately.

UNDERSTANDING GENETICS

Similar information from other animal and plant groups raises the question of the role which chromosome rearrangements play in species formation. There is no strong evidence to support the idea that they are ever the decisive factor in setting up reproductive isolation between two related populations, which then go on to evolve independently as separate species. It seems more likely that the influence of the chromosome alterations in speciation is exerted at a later stage, when they may enhance the genetic isolation already established by gene mutations. Gene mutations appear to be the major genetic changes which establish reproductive isolation, and they do so by causing inharmonious allelic combinations when a hybrid is produced. Gene mutations occur much more frequently than structural rearrangements. Moreover, the *same* gene mutation may recur in a population, so that a mutant allele may become quite frequent. If it has adaptive value, it will spread through the population and replace the original form of the gene. However, each structural rearrangement is usually unique. When an inversion occurs, the chance of another occurring with exactly the same breakage points is very unlikely. Consequently, specific inversions would not tend to spread as fast as gene mutations, even if they did entail an adaptive value. Gene mutation is by far the most important genetic change that provides the source of variation for the operation of natural selection.

VARIATION IN CHROMOSOME NUMBER. In addition to the various aberrations which change the structure of the individual chromosomes, other anomalies alter the actual chromosome number of an organism. We have seen that normal development depends on genic balance. Therefore, a departure from that chromosome number which is characteristic of a species can be expected to upset this balance and lead to adverse effects. Any variaiton from the normal number of chromosomes is termed *heteroploidy*. A heteroploid organism can differ from a typical member of its species by one or more entire sets of chromosomes. Such a condition is called *euploidy*. For example, if a diploid number of 10 chromosomes, composed of two haploid sets of five, is the normal number for a species, then the presence of an entire extra set would result in a euploid with 15 chromosomes. Euploidy will be discussed more fully in the latter part of this chapter.

In contrast to a euploid, an individual may have one or a few extra chromosomes or perhaps even one or a few chromosomes less than normal. This situation is termed *aneuploidy,* a condition in which the chromosome number differs from the normal diploid by less than an entire set $(2n - 1; 2n + 1; 2n - 2,$ etc).

ANEUPLOIDY IN HUMANS. We would expect aneuploidy to lead to severe genic unbalance, and we have already encountered several examples of this so far in humans. Down's syndrome (see Fig. 10-11), resulting from an extra chromosome 21, is a good example which illustrates the serious consequences of the unbalance. Individuals with one extra chromosome are called *trisomics,* meaning that their body cells contain three doses of a particular chromosome. Trisomics, such as the victims of Down's syndrome, may result from nondisjunction at meiosis in one of the parents. (Fig. 10-26). If two chromosomes fail to separate at anaphase I of meiosis, one-half of the gametes will contain an extra chromosome and one-half will have one chromosome less than normal. A gamete of the former type, when it combines with a normal gamete, will produce a zygote which is trisomic. The zygote receiving only one member of a chromosome pair is designated a *monosomic.*

The extra chromosome present in cases of Down's syndrome has been shown by fluorescent staining techniques to be the smaller of the two autosomes in group G. Since the chromosomes in any group are numbered according to decreasing size, this means that the Down's chromosome is technically number 22, not 21 as has been thought for so many years!! However, inasmuch as Down's syndrome has been designated "trisomy 21" for a considerable period of time and the terminology has become well established, geneticists have agreed to keep the designation "21" for the Down's chromosome in order to avoid confusion. The Philadelphia chromosome, therefore (earlier this chapter), which is actually the larger of the two in the G group, has been arbitrarily numbered "22."

The trisomy which produces Down's syndrome is definitely correlated with the age of the female parent. Women over 40 years of age have a much higher probability of bearing a defective child than do younger mothers. The age of the father does not seem to be a factor. There is reason to believe that chromosomes of oocytes become much more subject to nondisjunction the longer they remain in first meiotic prophase. It will be recalled that the first meiotic division in the human oocyte is not completed until the follicle matures. This means that an oocyte has been lying in a sustained meiotic prophase for years, perhaps decades, by the time meiosis is resumed. The spermatogonia, on the other hand, provide a supply of spermatocytes which are continually undergoing spermatogenesis throughout adult life.

It is interesting to note that trisomy 21 occurs in the chimpanzee and that the condition results in a syndrome closely resembling Down's syndrome in the human.

Trisomy in humans is also known for Chromosomes 13 and 18. Infants with the extra chromosome suffer severe abnormalities, such as gross facial defects, as well as skeletal and brain abnormalities. The unbalance is so severe that early death ensues, undoubtedly due to the excess genetic formation present in the extra chromosome.

Those aneuploid persons with an XXXY or XXXXY condition (Chap. 5) are trisomic and tetrasomic, respectively, for the X chromosome. Although, they have the abnormalities of Klinefelter syndrome, they nevertheless survive and are in no way as defective as those individuals trisomic for one of the autosomes. This ability to tolerate extra doses of the X may be

FIG. 10-26. Nondisjunction of an autosome at meiosis. Two homologous chromosomes (assume number 21 in the human) fail to separate at first anaphase. They move to the same pole and enter the same nucleus. A gamete derived from this will contain two chromosomes which are number 21, instead of just the normal one. A normal gamete from the other parent will contribute one Chromosome 21. The resulting zygote is thus trisomic, carrying three instead of two. A gamete derived from the other nucleus at meiosis (bottom one in figure) will lack chromosome 21 entirely. After fertilization, the zygote will be monosomic, having just a single chromosome 21 which was contributed by the balanced gamete.

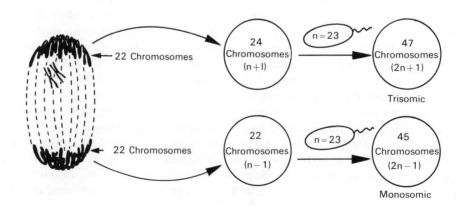

possible as a consequence of some deactivation of those X's in excess of one. Remember that one, two, and three Barr bodies are present, respectively, in persons who are XXY, XXXY, and XXXXY.

Human aneuploids monosomic for an autosome are extremely rare. Most monosomic zygotes or early embryos apparently fail to continue development. Autosomal monosomy is apparently extremely lethal in the human.

However, monosomics for the X chromosome are certainly well known. These individuals (XO) show the assortment of characteristics which describe Turner syndrome (see Fig. 5-11). Again this could arise from nondisjunction during parental gamete formation. If the two X chromosomes of the female move to the same pole during oogenesis, two sorts of eggs could result: XX + 1 set of autosomes and 0 + 1 set of autosomes. Fertilized by a Y-bearing sperm, the former could be responsible for an XXY Klinefelter male. The latter, fertilized by an X-bearing sperm, could produce a victim of Turner syndrome. It was mentioned that Turner births are less frequent than Klinefelter. The suggestion has been made that monosomy in humans may cause a greater degree of unbalance than polysomy, the presence of extra chromosomes. Only those monosomics who have a missing X (Turner females) can apparently survive to birth, but a large proportion of such XO embryos evidently do not reach full term.

ANEUPLOIDY IN PLANT SPECIES. Aneuploidy has been widely studied in plants which seem to tolerate aneuploid conditions much more readily than do animals. A variety of plant species has provided valuable information concerning the genetic and cytological consequences of aneuploidy, as well as its phenotypic effects. One of the most thoroughly investigated species is the thorn apple or Jimson weed, *Datura stramonium,* a very poisonous plant whose leaves are the source of certain narcotic drugs, among them atropine. Blakeslee undertook work with *Datura* in 1913, and over a period of years, he and his colleagues were able to isolate a variety of different polysomics. *Datura* has a haploid chromosome number of 12, which means that 12 kinds of trisomics are possible, one kind for each of the 12 chromosomes in a haploid set. Thus, any trisomic plant would have 25 chromosomes, two haploid sets plus one extra chromosome. This trisomic can be compared phenotypically to others which carry different, extra chromosomes. All 12 possibilities were obtained. Comparisons among them showed the effects of the genic unbalance on such organs as the leaf and the seed capsule. Each of the different extra chromosomes was found to produce characteristic phenotypic departures from the normal (Fig. 10-27).

When the trisomics were examined cytologically, their meiotic cells showed the presence of the extra chromosome. Since a trisomic has three doses of a certain chromosome, meiotic irregularities are to be expected. At synapsis, the three homologues may form an association of three chromosomes, a trivalent. When trivalents (or any multivalents for that matter) are formed, there is never an association of all three chromosomes throughout their lengths. Instead, the pairing is always two-by-two at certain points, so that a chain of three may be seen (Fig. 10-28). However, the three may not necessarily enter into trivalent formation. Instead, a bivalent and an unpaired univalent may result. Because the trisomics in *Datura* proved to be fertile, they could be crossed in various combinations, either with other trisomics or with normal diploids. This permitted the derivation of other types, such as tetrasomics (2n + 2) which could be studied further for the effects of two extra identical chromosomes on genic balance. Crosses between trisomics also enabled Blakeslee and his associates to study the inheritance patterns and also the modifications of typical Mendelian ratios resulting from the segregation of three homologous loci instead of the normal two. As might be suspected, there is a greater chance in trisomics for the recessive allele to be masked (one dose of wild is often dominant to two recessives). The usual 3:1 ratio is therefore not obtained but is replaced by other ratios in which the homozygous recessive type is less frequent.

In addition to these polysomics, Blakeslee and Belling obtained other plants whose extra chromosomes had undergone translocations and other kinds of structural alterations. All of this enabled Blakeslee to observe more closely the consequences of extra doses of specific genes on the phenotype. The observations left no doubt that adding extra genes to the diploid resulted in plants whose phenotypes departed in some way from the typical. Increasing the dosage of one specific kind of gene increased the phenotypic change in a certain direction. The extensive work with *Datura* has illustrated the important role of genic balance in the development of the normal characteristics of an organism.

Several plant groups are known in which a range of chromosome numbers exists within a genus or species. One of the most extensive aneuploid series occurs in *Claytonia virginica*, the common spring beauty, a single species in which 20 different diploid chromosome numbers have been reported ranging from 12 to 200. This species is quite unusual in its apparent ability to tolerate fluctuations in chromosome number with no loss of viability or fertility.

AUTOPOLYPLOIDY: ITS ORIGIN AND EFFECTS. Instead of containing just one or two extra chromosomes, some euploid organisms possess entire extra *sets* in addition to the typical diploid number. These individuals are called *polyploids*. Polyploidy is a very characteristic feature of flowering plants and has played an important role in the evolution of many plant groups, including a number of our valuable crop plants, such as wheat and cotton. When we examine a list of the chromosome numbers of species within a plant genus, we often find that the numbers are multiples of a basic one. For example, in different wheats, 2n = 14, 28, 42 (multiples of 7); in *Chrysanthemum*, 2n = 18, 36, 54, 72, 90 (multiples of 9). We can explain these observations on the basis of polyploidy; the groups with the higher numbers have been derived from those with lower ones through the multiplication of whole chromosome sets.

Before explaining how this multiplication can come about, let us first examine one of the main types of polyploidy. This is *autopolyploidy*, a condition in which the extra chromosome sets have originated from within the species. A plant species with a diploid number of 10 would have two haploid sets: AA. If one or two extra sets are present we would have an autotriploid (AAA) and an autotetraploid (AAAA). The origin of the extra sets could have come about through some failure at meiosis. Accidental disruption of the spindle might suppress anaphase movement,

FIG. 10-27. Capsule types in *Datura*. The capsule of the normal plant with 24 chromosomes differs from the various trisomics. Each trisomic has 25 chromosomes, but the extra one in each represents a different chromosome. The extra chromosome in each case is designated here as A, B, C, etc. (Redrawn with permission from *J. Hered.*, 25: 86–108, 1934.)

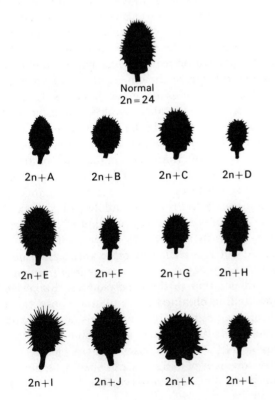

Normal
2n = 24

2n + A   2n + B   2n + C   2n + D

2n + E   2n + F   2n + G   2n + H

2n + I   2n + J   2n + K   2n + L

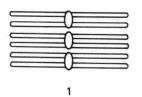 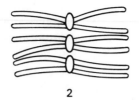 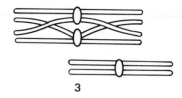

| 1 | 2 | 3 |

FIG. 10-28. Multivalent associations. When three (or more) homologous chromosomes are present, the meiotic picture is never one in which all three associate throughout their lengths as shown in (1). Rather, the pairing is always two by two in a trivalent (2). However, not all cells will contain a trivalent, because the three homologues may become distributed as a bivalent and unpaired univalent (3).

giving rise to unreduced gametes (AA). Union of such a gamete with a normal one (A) would produce the autotriploid (AAA; Fig. 10-29). A fusion of two diploid gametes would produce an autotetraploid. The accident responsible for autopolyploidy could occur in somatic tissue as well. Nuclei of two body cells might fuse; or anaphase of mitosis could fail. The result would be a tetraploid cell (AAAA). If this somatic cell goes on to divide, it can give rise to an entire plant which is tetraploid. Any flowers on a tetraploid branch would produce unreduced gametes (AA). By self-fertilization, tetraploid offspring would result. Although tetraploids arise spontaneously, they may also be induced artificially by a variety of treatments. The most extensively used agent is the drug colchicine. This is a plant alkaloid derived from the European autumn crocus and is used medicinally to alleviate the pain of gout. The primary action of this drug is on the spindle apparatus. Colchicine upsets the organization of the microtubules comprising the spindle and so prevents anaphase movement. All of the chromosomes become incorporated into one nucleus, and a tetraploid cell (4n) is formed. Colchicine does not interfere with the duplication of the chromosomes, and so its presence during a series of chromosome replications can lead to high levels of ploidy. Solutions of the drug may be sprayed on buds; seeds or roots may be dipped or soaked for several hours. Because the drug is very toxic, the proper concentration and exposure time must be rigidly controlled.

Polyploids are often produced artificially, because a polyploid at times may encompass some features which make it more desirable than the diploid from which it is derived. Some polyploid cereals, potatoes, and fruits possess desirable commercial qualities. Some polyploids are larger and more vigorous than the diploid. This has unfortunately given rise to the impression that all polyploids are large or "gigas" types. It is true that polyploid cells tend to be larger than those which are diploid, but this is not always so. Even the presence of larger cells does not ensure larger plants, as the cell number may be decreased in the polyploid. One generalization which can be made is that the polyploid tends to grow more slowly than the diploid. Consequently, it may flower later and over a longer period of time. Another feature of polyploids is some reduction in fertility as compared to the diploid. Some autopolyploids are very sterile, others quite fertile. The safest generalization is that the effects of the ploidy depend on the particular species involved. Genic interaction and the ability to tolerate extra sets of chromosomes may be quite different from one species to another.

Cytologically, most autopolyploids do carry some meiotic irregularities due to the presence of extra homologous chromosomes. This is well demonstrated by the triploid, which is usually highly sterile. Figure 10-30 shows why this is so. As was the case in the trisomic (Fig. 10-28), any three homologous chromosomes may enter into a trivalent association, or they may form a bivalent and a univalent. It is even possible that no pairing will be established and that three univalents will result at first meiotic prophase. So in triploid meiotic cells, one may encounter trivalents, bivalents, and univalents. It should be obvious that there is no mechanism to ensure an anaphase separation in which two chromosomes of each kind will move to one pole and one of each kind to the other pole. Such a separation would produce balanced diploid gametes and balanced haploid ones. However, there is usually random segregation of the members of each group of three homologues so that only a few gametes come to have a balanced combination of chromosomes. This in turn means that the gene dosage is unbalanced in most of the cells after meiosis. Consequently, the pollen of triploids is often very sterile, and seed set is very low. The same sort of meiotic irregularity accounts for the reduced fertility characteristic of polyploids with high but odd chromosome numbers (5n, 7n, etc.). This sterility has commercial value in plants such as the watermelon, where the almost seedless fruit is more desirable than that of the fertile diploid. To maintain sterile triploids, the plant breeder must propagate them asexually by means of leaf or root cuttings; he may also obtain seeds which will grow into triploids by crossing a tetraploid with a diploid.

In the autotetraploid, there is a greater chance than in the triploid for the formation of balanced gametes. Since four chromosomes of each kind are present, multivalents composed of an association of four are seen (tetravalents). Trivalents and univalents are also possible. But there is a higher probability of bivalent formation in the tetraploid (or any polyploid with an even chromosome number). Therefore, anaphase segregation results in a much larger proportion of cells which have balanced gene combinations. In certain species, such as the Jimson weed, *Datura,* the autotetraploids suffer little decrease in fertility.

ALLOPOLYPLOIDY AND PLANT SPECIATION. Although an autopolyploid may exhibit some phenotypic differences from the diploid, autopolyploidy does not contribute any new genetic information to a species, nor does it bring about

FIG. 10-29. Diploidy and autopolyploidy. The extra set or sets of chromosomes in an autopolyploid are derived from within the one race or group so that more than two completely homologous sets are present.

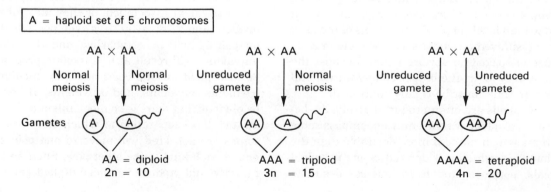

UNDERSTANDING GENETICS

any new combinations of alleles. All the alleles of the autopolyploid originate from within the species or the race; any different phenotypic effects are primarily a consequence of the interaction of extra gene doses. The second main type of polyploidy, however, does make possible new combinations of alleles from different sources. This is *allopolyploidy,* a polyploid condition in which the chromosome sets are derived from different races or species. In order for an allopolyploid to arise, there must be hybridization between two genetically different groups. One criterion which distinguishes two groups as separate species is their inability to cross and produce fertile offspring. The horse and donkey are certainly separate species. They can be readily crossed, but their offspring, the mule, is sterile as a result of meiotic incompatibility between horse and donkey chromosomes. This leads to genic unbalance and defective gametes. The very fact that two distinct groups can cross at all is evidence of a relationship between them as would be expected from an understanding of evolutionary processes. If two originally identical groups are isolated long enough, different mutations in each group may produce enough incompatibility between them so that they will be unable to cross (exchange genes) when eventually brought together again. Evidence tells us that two separate species may arise in just such a manner when one population is separated into two groups which then become spatially isolated (Fig. 10-31A). The forces of evolution may proceed so far that the two can no longer cross at all to produce any offspring when brought together again. In other cases, the two may be similar enough to cross but incompatibilities in their genetic material may result in sterile progeny (Fig. 10-31B). Other groups may have diverged somewhat, but still not far enough to prevent some gene exchange through crossing (Fig. 10-31C). In a case such as this, we would recognize subspecies, races, or varieties instead of distinct species. If left alone to cross freely, two varieties or races will eventually become one similar, freely interbreeding group. In the following discussion of allopolyploidy, the two groups involved must still possess sufficient genetic similarities between them to permit some gene exchange to take place. Some may be true species in every sense of the word, so that pro-

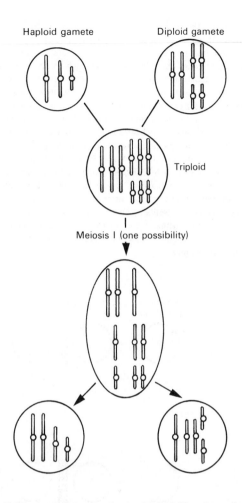

FIG. 10-30. Irregular segregation in a triploid. Since three of each kind of chromosome are present, trivalents and univalents may occur at meiosis, and segregation will tend to be irregular. Balanced gametes would contain two of each kind of chromosome or one of each kind. Most cells, however, will contain unbalanced combinations. For example, the gamete at the left has one extra "long" chromosome. The gamete at the right has two extra chromosomes, one of "medium" size and a "small" one; if it possessed another "long" one it would be a balanced diploid gamete.

duction of offspring between them on the diploid level never or rarely occurs. Others may not be this completely isolated genetically and may actually be subspecies. In other words, there is no sharp distinction between autopolyploidy and allopolyploidy, because the distinction between species is frequently blurred.

One of the best examples of allopolyploidy is the one which was the first to be reported, the experimental production of a hybrid between radish and cabbage. Although both of these are

FIG. 10-31. Degrees of compatibility between related groups. In each of these three cases, an original population becomes separated into two subpopulations which are prevented from exchanging genes due to some barrier, such as spatial isolation. In *A*, the accumulation of different kinds of changes in A and B has been sufficient to result in complete inability of members of the two groups to cross when they are brought together. Speciation is complete, and A and B are valid species. In *B*, the accumulation of genetic differences has been sufficient to prevent the formation of a fertile hybrid. Speciation has not progressed as far as in *A* but since all hybrids are sterile, A and B remain distinct and are recognized as species. In *C*, the two groups, although recognizably different, have not diverged sufficiently to prevent gene exchange. They are not two distinct species but rather subspecies or races of one species. The fertile hybrids can cross with each other and with the parental types. If free interbreeding continues, the two groups will give rise to one new population, and the two subspecies will not be recognizable.

members of the mustard family, they are certainly genetically isolated and differ enough to be placed in separate genera by taxonomists. They would rarely, if ever, produce offspring under natural conditions. The radish, *Raphanus*, and the cabbage, *Brassica*, both contain a diploid chromosome number of 18. In Figure 10-32, a haploid set of radish and a haploid set of cabbage are represented by 9R and 9B, respectively. The Russian cytologist, Karpechenko, succeeded in producing an F_1 between them. This F_1 hybrid was a diploid and contained 9R and 9B chromosomes. It was also quite unfertile due to meiotic irregularities. The two parental species have been reproductively iso-

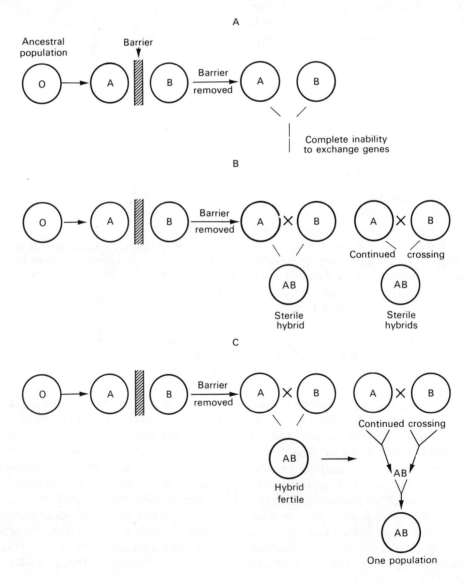

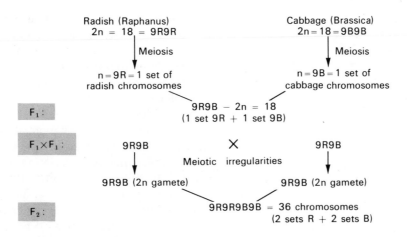

FIG. 10-32. Origin of an allotetraploid. One haploid set of radish chromosomes is represented by 9R and one haploid set of cabbage by 9B. The irregular meiotic behavior in the highly infertile F₁ on rare occasions produces a gamete which is diploid. If two of these unite, an allotetraploid arises which is fertile, because each parental set (9R and 9B) is represented in the diploid condition and can form bivalents.

lated long enough to allow the accumulation of a significant number of changes between their chromosome complements. These changes have produced incompatibilities which prevent synapsis between a set of R (radish) and a set of B (cabbage) chromosomes. Since no bivalents form, irregular chromosome segregation leads to unbalanced combinations and nonfunctional gametes. Among the abnormal meiotic segregations would be some in which *all* of the chromosomes would become incorporated into a single cell. These would give rise to diploid gametes. Such occurrences would be rare, but nevertheless, two such diploid gametes were produced by the F₁ hybrid, and these combined to form a plant with four sets of chromosomes, 2R and 2B. Since this individual has twice the chromosome number of either original parent, it is a tetraploid. It cannot be called an autotetraploid, because the extra chromosome sets have come from different species. And this allotetraploid, unlike the F₁ hybrid which gave rise to it, is very fertile. This is quite understandable, because there are now two sets of radish chromosomes which can form harmonious pairs at meiosis, and similarly two sets of cabbage chromosomes which can also form bivalents. As a result of the normal synapsis and orderly chromosome segregation, the gametes are balanced, each containing one set of radish (R) and one set of cabbage (B) chromosomes.

Unlike the autopolyploid, the allopolyploid contains very new combinations of alleles which would not arise under ordinary circumstances. The allopolyploid in our example was given a new name, *Raphanobrassica*, because it combined the traits of both parental species. Unfortunately, the combination did not produce a plant with economically valuable attributes. However, some experimentally produced allopolyploids do contain a combination of features making them more desirable than the diploids, as seen in the case of several hybrid wheats. Allopolyploids may be produced experimentally with the aid of colchicine, which is applied to the diploids in order to obtain diploid gametes. The polyploid at times possesses an advantageous assortment of characteristics which enables it to thrive under conditions which would be unfavorable for either one of the diploid parents. The polyploid may actually be more robust than either diploid and may actually replace them. Allopolyploidy definitely occurs in nature, and the process, more so than autopolyploidy, has operated in the history of the major plant groups. It should be appreciated from a reappraisal of the radish-cabbage cross that the allotetraploid is in every sense of the word a species. It is morphologically distinct from either diploid parent. But more important to the definition of a species is its reproductive isolation from either diploid parent. As Figure 10-33 shows, a cross of the allotetraploid to either one of the parents yields triploid combinations. One set of chromosomes in either kind of triploid would be unpaired at meiosis, with the ensuing formation of unbalanced gametes. Therefore, allopolyploidy provides one mechanism for the evolution of new species. Unlike the process requiring spatial isolation, it occurs when two genetically different groups become associated in the same locality. Moreover, it occurs abruptly, not as the result of the accumulation of point mutations over a long period of time.

The process of allopolyploidy does operate in nature and has produced tetraploid derivatives from diploids which still exist today. This has been experimentally demonstrated by the actual synthesis of certain known species from crossing known diploids. An excellent and early demonstration was made by Muntzing, working with European herbs of the genus *Galeopsis*. Muntzing crossed two different species, *Galeopsis pubescens* and *Galeopsis speciosa* (Fig. 10-34). Each has a haploid chromosome number of 8. He managed to obtain an F_1 hybrid, but this proved to be quite sterile. However, from this F_1, he finally succeeded in obtaining an F_2 offspring. This plant proved to have 24 chromosomes and was actually a triploid. Muntzing then backcrossed this plant to *G. pubescens* and obtained one seed. This in turn grew into a perfectly fertile plant which was exactly the same as the naturally occurring species, *Galeopsis tetrahit*. When he crossed his experimentally synthesized plant to a natural *G. tetrahit,* he obtained perfectly fertile *G. tetrahit* offspring. This proved that the natural plant and the synthesized one were genetically identical and that natural *G. tetrahit* was an allopolyploid which undoubtedly arose in nature through hybridization between *G. pubescens* and *G. speciosa*. Many similar syntheses have now been performed with other groups and have established the allopolyploid origin of several natural species, among them many commercially valuable wheats and cereal grasses. Indeed, cytotaxonomic studies leave no doubt that some of our most economically important plants (tobacco, cotton, and potatoes, et al.) arose through allopolyploidy.

POLYPLOIDY IN ANIMALS. Since polyploidy is so characteristic of the plant kingdom, we might

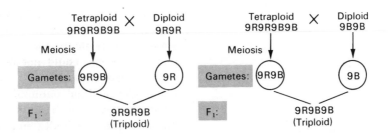

FIG. 10-33. Reproductive isolation of the allotetraploid. If an allopolyploid such as *Raphanobrassica* is backcrossed to either parent species (radish or cabbage), the F₁ is a sterile triploid. The backcross to the radish parent (*left*) leaves the nine cabbage chromosomes unpaired. Similarly, the cross to the cabbage parent leaves nine unpaired radish chromosomes. Irregular segregation of the univalents causes sterility. The allotetraploid can be recognized as a distinct species on the basis of reproductive isolation.

expect it to have been important in animal evolution as well. Polyploid animals have been produced experimentally in some species, such as *Drosophila* and the silkworm; however, when animal groups are surveyed, it becomes quite evident that the number of species with naturally occurring polyploidy is very low. Good reasons for its scarcity among animals were first presented by H. J. Muller. In the first place, most animals are dioecious, an individual being either male or female. Higher plants, on the other hand, are commonly hermaphroditic. Therefore, if a polyploid flower should arise, diploid eggs and sperms would be formed on the same individual, and the chances of union between two diploid gametes would be good. However, in the animal, where the sexes are separate, any diploid gamete which arises would usually combine with a haploid one. This would be so, because there is an extremely low probability that a diploid egg and a diploid sperm would arise at the critical time in two separate individuals which chanced to mate. The combination of a haploid and diploid gamete would produce a triploid, and we have seen that such individuals are sterile. So autopolyploidy would encounter difficulties in just getting started. But suppose that the rare combination of events takes place and that a polyploid female and a polyploid male are formed. In *Drosophila*, the former would have four sets of autosomes and four X chromosomes, AAAAXXXX; a corresponding male would be AAAAXXYY. At spermatogenesis, the X's of the male would usually pair with each other, as would the Y's (Fig. 10-35). This would produce gametes which are AAXY. These would combine with eggs which are AAXX, resulting

FIG. 10-34. Synthesis of a natural allopolyploid. Muntzing crossed two species and obtained a highly sterile F₁, but continued inbreeding of this F₁ yielded one offspring. This was a triploid, containing two sets of *speciosa* but only one set of *pubescens* chromosomes. Backcrossing to the *pubescens* parent finally produced a fertile plant. This arose as a result of a combination of an unreduced gamete from the triploid parent and a normal haploid *pubescens* gamete. The plant is fertile because two of each parental set are present. It was identical to a naturally occurring species, *Galeopsis tetrahit,* and crossed easily with the natural species. The latter undoubtedly arose in nature as an allotetraploid between *Galeopsis pubescens* and *Galeopsis speciosa.*

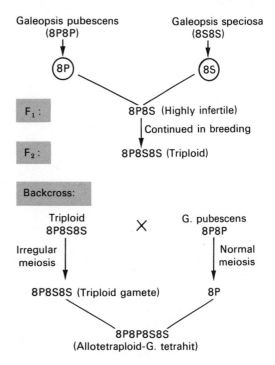

in offspring that are AAAAXXXY. This combination of four sets of autosomes and three X's would be a sterile intersex (Chap. 5), and so the tetraploid races could not be perpetuated and would disappear. Although the situation would not be identical for mammals, it is obvious from what has been said about sex determination in humans that the XXXY individual would be a sterile Klinefelter type.

And so we would not expect to find any appreciable amount of polyploidy in those species having separate sexes. An exception to this is found in the plant genus, *Melandrium,* which has an XY sex-determining mechanism similar to that in animals. Polyploidy occurs in this dioecious species, because the presence of just one Y is sufficient to produce a fertile male, even in the presence of three X's (XXXY is male, not a sterile intersex). In most animals however, this is not possible due to the interaction of many sex-determining genes on the X, the Y, and even the autosomes in many species.

It should be added that allopolyploidy, which has been so important in plant evolution, would be most unlikely in higher animals, because it requires interspecific hybridization. Mating behavior would place a barrier in the way of this, because any animal is not apt to respond to the mating signals or devices of a member of another species. In animals, one of the mechanisms for reproductive isolation of species is provided by mating preference. This is obviously not the case in plants.

Actually, in man and other mammals, polyploidy exerts a lethal effect, and the consequences of tetraploidy are apparently more severe than those of triploidy. The frequency of observed spontaneously aborted embryos which are tetraploid is quite low, indicating that the zygotes do not usually develop very far; no such fully developed fetuses have been recovered. However, nearly 4% of all spontaneous human abortions show triploid cells. And these embryos have usually developed further than those with tetraploid cells and may approach that of a full-term fetus. A very small number of triploids have actually lived for a few hours, but they showed severe bodily defects. In contrast, examples of tetraploid cells in surviving humans have been extremely rare, and these cases involved somatic mosaicism (see below for more details).

CHROMOSOME ANOMALIES AND THE HU- MAN. Man and many other mammals seem unable to tolerate the unbalance resulting from both aneuploid and polyploid conditions. The significance of this becomes more apparent when we realize that chromosome anomalies are the most common cause of congenital defects in humans. They would cause even more suffering if it were not for the fact that over 90% of all abnormal embryos are spontaneously aborted. Among these, chromosome aberrations are very frequent. The most common single chromosome disorder is the XO anomaly, or Turner condition. However, it seems to have lethal consequences, because not more than 2% of the zygotes which are XO develop into full-term fetuses. And those that do often have severe birth defects, particularly in the cardiovascular system.

The consequences of an extra autosome are extremely severe, and few infants aneuploid for one of the larger autosomes survive. The most common aneuploidy for an extra autosome in humans is trisomy for the small Chromosome 21, whose unfortunate effects have been discussed earlier in this chapter. It was mentioned that the incidence of Down's syndrome increases with the age of the mother. Although women over 40 produce only about 4% of the babies born, 40% of the Down's infants are born to women over 40. The correlation between maternal age and chromosome anomalies of many kinds is very striking, and there is no doubt that the frequency of spontaneous abortions increases with maternal age. In the aborted embryos and fetuses from women over 40, one-third are found to have unbalanced chromosome conditions. However, among the aborted offspring of very young mothers, 17 or under, 40% were also heteroploid. It seems possible that at either extreme of maternal age, there may be a higher probability for the occurrence of chromosome disorders. No correlation has been found between maternal age and the occurrence of Turner syndrome or polyploidy.

In Chapter 15, we will see that radiations, in addition to their induction of gene mutations, are also potent agents in the production of chromosome anomalies. Exposure of cells to high-energy radiations greatly increases the frequency of all types of structural changes, as well as nondisjunction. These same aberrations may

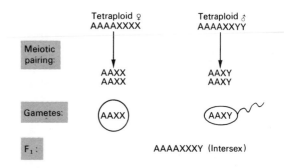

FIG. 10-35. Cross of tetraploid fruit flies. A = one set of autosomes. Tetraploid males and females are genetically balanced; the ratio of X chromosomes to autosomes is normal (2:2 in the female and 1:2 in the male). However, the balanced tetraploid condition cannot be perpetuated. Pairing and segregation in the tetraploid male produces a preponderance of gametes which contain one X and one Y. Union of these with a diploid egg results in offspring which are sterile intersexes, because the balance of X to autosomes departs from 2:2 or 1:2.

also be induced by certain chemicals. The list of substances which can alter chromosome structure is ever multiplying and is becoming a matter of growing concern in today's world where liberal use of food additives and drugs poses a threat of unknown genetic consequences. Maternal ingestion of certain drugs during pregnancy has been suggested as the cause of the production of some tetraploid cells in the leukocytes of infants. Patients receiving chemical therapy and radiation for cancer treatment may also show an increase in tetraploid leukocytes.

In Chapter 5, mosaic individuals were mentioned whose bodies consist of cell populations which differ in their chromosome constitutions. The origin of genetically diverse lines of cells in an organism results from mitotic mishaps during embryological development and may involve either the autosomes or the sex chromosomes. The outcome is an individual composed of a mixture of cell types. The earlier the mitotic error, such as nondisjunction, the greater the chance for a larger number of cells to have the aberration. The mishap may occur in a cell of a specific organ or tissue, so that only a restricted part of the body will show the mixture. Individuals with a variety of types and degrees of mosaicism have been described. The effects of the aberrations also vary accordingly. A number of mosaics for the sex chromosomes has been identified. Klinefelter individuals are known who are mixtures of cells that are XX and others that are XXY (written as XX/XXY). Turner syndrome has been found in persons who are XO/XXX and even XO/XYY.

Although complete tetraploidy has never been reported in a live infant, a 6-year-old mentally retarded child with an assortment of bodily abnormalities was found to be a mosaic of normal cells, tetraploid cells, and those with an extra Chromosome 18. A married man who was fertile and normal was also shown to be a mosaic, a mixture of cell lines. Among the genetically diverse cells the most common were XXY, but 15% of the fibroblasts were tetraploid. This man had five abnormal children out of 10, of whom two suffered from Down's syndrome. Another appeared to be an XX/XO Turner's mosaic. The case suggests a family tendency for nondisjunction. This example, plus those which follow, illustrate the problem which mosaicism presents to the genetic counselor.

Two other phenotypically normal fathers produced children with Down's syndrome. After cells were taken from them from various tissues and then grown in culture, both men were found to be mosaics for Chromosomes 21. In one man, the extra Chromosome 21 was found in skin and testicular fibroblasts but *not* in leukocytes. In the other man, mosaicism for the chromosome was found among the leukocytes. In these cases, the Down's aberration was undoubtedly coming through the male parent. All of the persons involved were under 30 years of age. Such examples indicate the need to consider both parents in cases of suspected inherited anomalies and to look for mosaicism in different tissues.

The possibility of chromosome unbalance as a causative factor in abnormal human behavior is suggested by studies of XYY males. Much of the data come from prison populations, and they tend to indicate a correlation between the Y chromosome anomaly and tallness. Several of the clinically observed cases describe aggressive behavior among males with extra Y's; however, others do not. The correct interpretation is made difficult by the possibility that those persons with violent or extreme types of behavior come more often to the attention of the investigator. Confirmation depends on studies of the general population in order to determine the overall frequency of XYY males and their adjustment to society.

Facilities are available today for proper genetic counseling of prospective parents who are concerned about the birth of children defective as a result of chromosome anomalies. Techniques are at hand not only for karyotype analysis of infants and adults but also for the detection of chromosomal and chemical defects in the unborn. These procedures, plus a basic knowledge of genic balance, can prevent some of the unfortunate consequences which follow upsets in chromosome structure and number.

REFERENCES

Arrighi, F. E., and T. C. Hsu, Localization of heterochromatin in human chromosomes. *Cytogenetics, 10*: 81–86, 1971.

Astaurov, B. L. Experimental polyploidy in animals. *Ann. Rev. Genet., 3*: 99–126, 1969.

Atnip, R. L., and R. L. Summitt, Tetraploidy and 18-trisomy in a six-year-old triple mosaic boy. *Cytogenetics, 10*: 305–317, 1971.

Carr, D. H. Genetic basis of abortion. *Ann. Rev. Genet., 5*: 65–80, 1971.

Caspersson, T., G. Gahrton, and J. Lindsten, *et al.* Identification of the Philadelphia chromosome as a number 22 by quinacrine mustard fluorescence. *Exp. Cell Res., 63*: 238–240, 1970.

Francke, U. Quinacrine mustard fluorescence of human chromosomes: characterization of unusual translocations. *Am. J. Hum. Genet., 24*: 189–213, 1972.

Hamerton, J. L., *Human Cytogenetics*, vols. I and II. Academic Press, New York, 1971.

Hirschhorn, K. Chromosomal abnormalities I: autosomal defects. In *Medical Genetics*, V. A. McKusick, and R. Claiborne (eds.), pp. 39–50. HP Publishing Co., New York, 1973.

Hook, E. B., Behavioral implications of the human XYY genotype. *Science, 179*: 139, 1973.

Hsu, L. F., M. Gertner, and E. Leiter, *et al.* Paternal trisomy 21 mosaicism and Down's syndrome. *Am. J. Hum. Genet., 23*: 592–601, 1971.

O'Riordan, M. L., J. A. Robinson, and K. E. Buckton, *et al.* Distinguishing between the chromosomes involved in Down's syndrome (trisomy 21) and chronic myeloid leukemia Ph' by fluorescence. *Nature, 230*: 167–168, 1971.

Rothwell, N. V., and J. G. Kump, Chromosome numbers in populations of *Claytonia virginica* from the New York metropolitan area. *Am. J. Bot., 52*: 403–407, 1965.

Stebbins, G. L. *Variation and Evolution in Plants.* Columbia University Press, New York, 1950.

Stebbins, G. L. *Processes of Organic Evolution,* 2nd ed. Prentice-Hall, Englewood Cliffs, N.J., 1971.

Wahrman, J., J. Atidia, and R. Goitein, *et al.* Pericentric inversion of chromosome 9 in two families. *Cytogenetics, 11*: 132–144, 1972.

White, M. J. D. Chromosome rearrangements and speciation in animals. *Ann. Rev. Genet., 3*: 75–98, 1969.

Wilson, M. G., J. W. Touner, and G. S. Coffin, *et al.* Inherited pericentric inversion of chromosome no. 4. *Am. J. Hum. Genet., 22*: 679–690, 1970.

REVIEW QUESTIONS

1. Let the following represent a pair of homologous chromosomes in an inversion heterozygote. (The 0 represents the centromere.)

—0 A B C D E F G   —0 a b f e d c g

Assume that a crossover takes place between the c and the d loci at meiosis. Give the two products which will result from the crossover event.

2. In the inversion heterozygote in Question 1, what will be the two products assuming a double crossover occurs between the same two strands, the first one between c and d, and the second crossover between e and f?

3. Let the following represent a pair of homologous chromosomes in another inversion heterozygote. (The 0 represents the centromere.)

$$\underline{\text{H I J}_{0}\text{K L M N}}\qquad\underline{\text{h l k}_{0}\text{j i m n}}$$

Assume that a crossover takes place between the loci "k" and "l." Give the two products resulting from the crossover event.

4. In the inversion heterozygote in Question 3, what will be the two products, assuming a double crossover between the same two strands, one crossover betwee "i" and "j," the other one between "k" and "l"?

5. Assume that an individual has one chromosome with the gene order as follows:

$$\underline{_{0}\text{R S T U V W X Y Z}}$$

In the homologue, a shift resulted in the insertion of the segment *T U V* between "X" and "Y." Show how this situation can lead to the origin of a chromosome with a duplication and one with a deficiency.

6. Below are two pairs of homologous chromosomes with the centromeres in a median position. Assume a reciprocal translocation occurs so that whole arms are shifted between two of the nonhomologous chromosomes following the breaks whose positions are indicated:

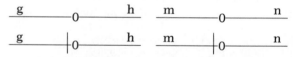

A. What will be the composition of a translocation heterozygote carrying the two pairs of chromosomes?

B. If adjacent disjunction occurs, give the two kinds of gametes possible for each of the two possibilities.

C. Give the kinds of gametes possible from alternate disjunction.

7. A certain race of *Oenothera* carries two complexes of seven chromosomes each. One complex of seven is called alpha; the other set of seven is beta. Following self-fertilization, what will be the composition of the zygotes? Indicate which ones will survive.

8. Assume that a deletion arises in the X chromosome of the mouse and removes a gene which

is required in at least single dose for the survival of the embryo.

  A. What would be the result of crossing a female carrying the deletion and an ordinary male?

  B. Assume that the deletion of an autosomal gene required in single dose accumulates in a population. What will be the outcome when carriers of the deletion mate with each other?

9. Allow "B" to represent one dose of the Bar locus on the X chromosome. A Bar-eyed female of the constitution $\frac{B\,B}{B\,B}$ is mated to a Bar-eyed male of the constitution $\frac{B\,B}{Y}$. Assume that "unequal" pairing occurs in some of the meiotic cells of the female. What kinds of offspring can result from the fertilization of both the exceptional kinds of eggs and the eggs in which pairing is exact?

10. Assume that a certain animal species has a haploid chromosome complement which consists of the five chromosomes designated I, II, III, IV, and V. Give the chromosome constitution of a body cell of an individual which happens to be.

  A. Trisomic for II.
  B. Triploid.
  C. Monosomic for IV.
  D. Tetrapolid.

11. Species A has a diploid chromosome number of 24; so does another plant, Species B. After many attempts a highly infertile F_1 is produced. From this F_1, a fertile allotetraploid is eventually derived.

  A. Give the chromosome number of the F_1 plant.
  B. Give the chromosome number of the fertile derivative.

12. Assume plant species No. 1 has a diploid chromosome number of 16. Give the following chromosome numbers:

  A. An autotriploid derivative of species No. 1.

  B. A trisomic member of species No. 1.
  C. An allotetraploid derived from a cross with species No. 2, which has a haploid chromosome number of 8.
  D. A monosomic member of species No. 1.

13. Assume that three separate populations of plants A, B, and C have many similar phenotypic features. However, many combinations of traits make each population distinct from the other. When plants from Population A are crossed with those of Population B, the F_1's are intermediate in phenotype, but they are completely sterile. When plants from Population B are crossed with members of Population C, very fertile intermediate F_1's arise. Plants from Population A crossed with those from C give F_1's, which are intermediate between A and C, but they are sterile. Give an explanation of these observations.

14. Give at least one way to account for the origin of humans with the following chromosome constitutions:

  A. XXY.
  B. XO.
  C. XYY.
  D. A person with only one normal metacentric chromosome No. 1 but with an unusual chromosome the size of No. 1 but which is acrocentric.
  E. Three complete sets of chromosomes.
  F. A person with only one chromosome No. 21 and only one No. 13 plus one unusually large chromosome.
  G. XXX.
  H. XX/XO mosaic.
  I. XXY/XX mosaic.
  J. XY/XYY/XO mosaic.

15. Match the disorder in column A with the letter of an appropriate chromosome anomaly in column B (there may be more than one possible choice in a given case and an item in column B may be used more than once):

A	B
_____ Turner syndrome	1. polyploidy
_____ Klinefelter syndrome	2. aneuploidy
	3. monosomy
_____ Down's syndrome	4. triploidy

_____ *cri-du-chat*
syndrome
_____ Philadelphia
chromosome
_____ XYY male

5. inversion
6. translocation
7. deletion
8. nondisjunction

16. Demonstrate or explain how a metacentric chromosome can be derived from two acrocentric chromosomes with centromeres very near the end of a chromosome arm.

17. Occasionally males with Klinefelter syndrome are found to have an XX sex chromosome constitution. No Y chromosome is present! Offer an explanation using information on the sex chromosomes in Chapter 5.

Select the correct answer or answers, if any, which best complete each of the following (Questions 18–24):

18. The dominant "P" in *Drosophila* results in red eye color. Its recessive allele "p" is associated with purple eyes. The locus of this pair of alleles is in the euchromatin. A shift places the "P" allele next to heterochromatin. A fly arises with the following genotype:

$$\frac{P\,(\text{heterochromatin})}{p\,(\text{euchromatin})}$$

A very possible result would be that:

A. Purple eye color must be expressed, since red eye color cannot be expressed at all.
B. The fly must have red eyes, since purple cannot be expressed due to the chromosome change.
C. The fly could have eyes with patches of red and patches of purple.
D. The fly would die as a result of genic unbalance.

19. Allow "A" to represent a haploid chromosome complement in a plant species. Suppose that an individual is AAAA. The following can be said:

A. The individual is certain to be larger than the diploid.
B. The individual may possibly grow more slowly than the diploid.

310

C. This is an example of euploidy.

D. There may be some reduction in fertility due to irregular chromosome segregation.

E. This is an example of tetrasomy.

20. A diploid gamete (AA) fertilizes a normal haploid gamete (A) of the same species. The following can be said about the resulting offspring:

A. The offspring is an allotriploid.

B. The offspring will be completely fertile.

C. The offspring will be quite infertile due to the random segregation of trivalents.

D. The offspring will be quite infertile due to the lack of chromosome homology.

E. The offspring is an autotetraploid.

21. Tetraploid races are rare among animals because:

A. Diploid gametes cannot arise in animals.

B. The sexes are usually separate in animals.

C. Polyploidy is lethal in all animals.

D. Tetraploidy produces individuals which are not normal in the balance of sex chromosomes to autosomes.

22. An individual is heterozygous for a pericentric inversion. Which of the following would be true:

A. The individual would probably be abnormal in some way, since certain genes are deleted

B. The inversion can be recognized cytologically by the formation of cross figures or circles during meiotic prophase.

C. A chromosome with two centromeres will result if crossing over occurs within the inverted segment.

D. A complex loop will be seen cytologically, since one inversion overlaps the other.

E. The position of the centromere will be shifted if breaks on either side of it are not equidistant.

23. Reciprocal translocations:

A. Will be widespread in nature if there is no crossing over in the male.

B. Result in unbalanced gametes if alternate chromosomes in the multivalent move to the same pole.

C. Result in the formation of acentric fragments at first meiotic prophase.

D. Do not occur in animal cells.

E. Can result in shifts of chromosome arms and bring about new linkage arrangements.

24. In *Oenothera*:

A. Typically no bivalents form at meiotic prophase.

B. The pure-breeding individuals are homozygotes with very favorable gene combinations.

C. At least two lethals are present, and yet the plants are vigorous.

D. All the chromosomes from the male parent go to the same pole at meiosis, and all from the female parent go to the opposite pole.

E. No chromosome in the complement is completely homologous ot any other one.

11

CHEMISTRY OF THE GENE

Throughout the preceding pages, we have encountered many fundamental genetic principles and analyzed various inheritance patterns. We can continue to examine further variations on these themes, but before we proceed, let us try to answer the following questions. "Precisely what is being inherited when we talk about the inheritance of genes? Of what are they composed, and how do they manage to exert control over the cell and bring about their different effects?" These thoughts occurred to the earliest geneticists. As more and more information was assembled from studies of inheritance patterns in various species, investigators realized that this type of approach provided no insight into the nature of the gene itself and of its action on the molecular level. A clarification of these points demands some knowledge of the chemistry of the chromosomes, because they are the sites of the genetic loci. By 1940, it had been demonstrated that the chromosomes of eukaryotes contain protein associated with nucleic acids, deoxyribonucleic acid (DNA) and ribonucleic acid (RNA). The gene itself was generally considered to be composed of protein, since proteins are the most complex of all the chemical substances in the cell. It was reasoned that the gene must be complicated, because it exerts a profound influence on the cell. In higher organisms, there must be thousands of different genes involved in a multitude of chemical interactions. Only proteins seemed complex enough to afford the diversity needed to build up so many different kinds of genes, each with an important and specific function.

THE *PNEUMOCOCCUS* TRANSFORMATION.

A landmark in biology was reached in 1944 with the report of an investigation which implicated DNA and not protein as the genetic material. The history of the announcement goes back many years before (1928) to the English bacteriologist, Griffith, who was working with the pneumonia organism, *Diplococcus pneumoniae*. This is a spherical cell which typically occurs in pairs surrounded by a sizable common capsule. Such encapsulated cells are virulent and can kill experimental animals. On agar plates, they form colonies which are glistening and mucoid and which have been described as "smooth." On occasion, cells from smooth colonies (or S cells) may change and lose the capsule. Such a mutation renders the cells harmless, because they can now be attacked by the white blood cells of the body. Moreover, their colonies on agar appear granular in contrast to the smooth, encapsulated forms and have been termed "rough" or R.

The S cells from smooth colonies, when injected into an animal (such as a rabbit) can elicit the formation of specific types of antibodies. This is because the capsule, a polysaccharide, acts as an antigen. When different strains of S cells are injected into animals, it is found that the antibodies formed may be of different types. Therefore, there are really different kinds of virulent cells which can be designated types I, II, III, etc. Their distinctions are detected by the fact that the antibodies against the several types are specific, so that cells of type I react only with type I antibody, not with that of type II. The specific types are genetic characteristics which are inherited from one cell to another. When a virulent cell of type I mutates to a rough form, the R cells on rare occasions may later revert (back mutate) to the S or virulent form. When they do so, they always mutate back to the S type from which they were derived. For example,

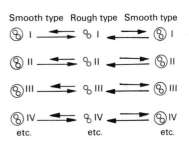

FIG. 11-1. Spontaneous mutation in *Pneumococcus.* Cells may mutate spontaneously but when they do so, they remain true to their antigenic type which is a genetic characteristic. The back mutation from rough to smooth occurs at a much lower frequency than the mutation smooth to rough. A smooth or rough of any type never changes spontaneously to the rough or smooth of a different type.

descendants of a rough cell which arose from type I would revert only to S type I, never to type III or any other (Fig. 11-1).

A series of experiments by Griffith produced some very unexpected results. He injected virulent cells from smooth colonies into mice. As expected in this case, the animals died, and autopsy revealed the presence of living, smooth cells. When he injected animals with rough cells, they survived (Fig. 11-2). Again this was to be expected, because such cells are not virulent. Griffith destroyed virulent S cells by heat treatment. When those cells were injected into mice, the animals lived since the killed bacteria cannot multiply to produce toxin. A most unusual outcome occurred when the experiment was performed in another way. Griffith injected mice with two kinds of cells: a small amount of living R cells derived from type II and also heat-destroyed S cells of type III. The predicted result would be that the animals would live, because neither the living R cells nor the killed S ones can cause disease by itself. However, many of the animals died, and when examined were found to contain living cells which produced smooth colonies and which were of type III. Such cells can kill, but how did they arise? Griffith said that the living R forms had somehow been able to take on the particular capsule type of the killed virulent cells. They had been "transformed," and the phenomenon became known as the "pneumococcus transformation." In this process, R cells derived from any type of S could acquire the capsule of any type S cell. Apparently, then, the R cells could somehow be changed genetically through incubation with nonliving S cells. Something was causing the genetic change, and it was termed "TP" for transforming principle. Others confirmed the work of Griffith and demonstrated that the transformation could be accomplished in the test tube and did not require an animal for incubation. R cells of one type could be transformed *in vitro* by heat killed-S cells of any type.

Avery, MacLeod, and McCarty of Rockefeller Institute, employing refined chemical procedures, were able to obtain small amounts of highly purified TP extracted from type III S cells. This purified TP was very active in transforming RII cells into SIII. Avery and his associates then subjected the purified TP to an assortment of exacting chemical tests in order to ascertain its chemical nature. Their extraction procedures removed most of the protein and fat. Tests of the final preparation showed the TP to be rich in DNA and to contain some RNA. The transforming activity of the TP preparation was not affected by treatment with protein-digesting enzymes. Since only traces of protein could be present in the TP after the extractions, it seemed highly unlikely that the transforming ability was due to any protein molecules. When the enzyme RNase was added to destroy any RNA which was present, the activity of the preparation was still not affected. However, treatment with the enzyme DNase caused transforming power to be lost. Thus, there was little doubt that the transforming power resided in DNA. This in turn implied that pure DNA by itself could bring about a genetic change. Therefore, the gene must be composed of this substance, at least the gene for capsule type in the pneumonia bacterium. It was not a complex protein as had been believed.

ADDITIONAL EVIDENCE FOR DNA AS THE GENETIC MATERIAL. Other lines of evidence were assembled in the 1940's and early 1950's which also implicated DNA as the genetic material. The Feulgen staining reaction, which had been developed in the 1920's was shown to be highly specific for DNA. When cells were stained by this method, it was found that the chromosomes or nuclear contents stained and not the cytoplasm. Since the genetic material is in the

314

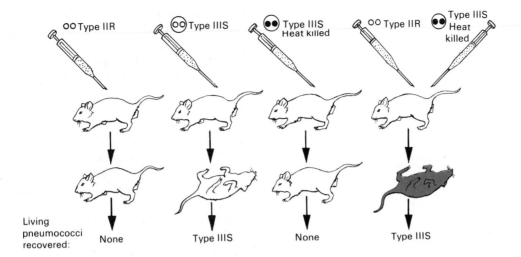

FIG. 11-2. Pneumococcus transformation. Mice injected with rough cells alone or heat-killed cells alone survive as expected, because the rough are not virulent and dead cells should not kill. However, a combination of living rough and heat-killed smooth cells can cause the death of an animal. Living smooth cells can be recovered after autopsy, and they are always found to be of the same bacterial type as the heat-killed smooth.

chromosomes, this is exactly where we would expect to detect it by staining procedures. Stains for substances such as protein or carbohydrate showed them to be located throughout the cell. Moreover, the amount of DNA in a single nucleus could be measured quantitatively. This was made possible by the development of a photometric method by Pollister. Cells can be stained for DNA by the Feulgen reaction and then placed on the stage of a microscope. A beam of light can be directed to a single nucleus. The amount of light passing through will depend on how much dye is complexed to the DNA. The more dye bound, the more light absorbed, and so there would be less light to pass through the mciroscope. The amount actually passing through can be measured in a precise way by using sensitive photo cells. Use of this Feulgen photometric procedure on a variety of plants and animals demonstrated that the nuclei of gametes contained exactly one-half of the amount of DNA present in body cells or diploid cells. This is exactly what we would expect of the genetic material, the haploid amount in the reproductive cells, and the diploid amount in somatic lines. Cells known to have three or four sets of chromosomes (triploid and tetraploid cells) have exactly three and four times the amount of the DNA in the haploid gametes. Thus, there was found an exact parallel between the amount of

DNA and the level of the chromosome number, haploid, diploid, and polyploid. Cell components which stained for the presence of proteins and other substances did not give these sharp results.

Further observations showed that DNA, unlike protein, fat, or carbohydrate, does not undergo continual breakdown and resynthesis in the cell. It tends to remain the same in the various cells of an organism, whereas other substances are continually metabolized. Again this feature of DNA is what we would expect of the hereditary material. It would certainly seem unlikely that the genetic material would be undergoing continual change in the cells. If it is needed to supply information, we would expect it to maintain its integrity. Only the DNA appears to do this.

Many biologists were still reluctant to give up the idea of protein as the substance of heredity. However, other kinds of observations argued strongly against protein as the candidate. The DNA of the chromosomes is associated with various kinds of proteins, among them the histones. These are relatively simple, basic proteins which are distributed over the entire length of the DNA and which are present in about the same amount as the DNA on a weight basis. The apparent constant association of the histones with the chromosomes of eukaryotes might suggest that they contain the genetic information.

CHEMISTRY OF THE GENE 315

However, there are a few exceptions to the typical association between histones and DNA. The main one is found in the sperms of many animals (some birds, fish), where the bulk of associated protein is not histone but protamine, a similar but detectably different type. When the sperm fertilizes the egg, the protamine of the chromosomes from the male is replaced by histones during early divisions of the embryo. If histones compose the gene, we would not expect them to be absent from sperm cells. Nor would it seem logical to assume that genes are usually histones but that in certain cells they can undergo change into another substance. The DNA, however is constant and does not change in kind during gamete formation.

Today we know much more about the non-histone proteins which are also associated with chromatin. These proteins are quite variable both quantitatively and qualitatively from one cell type to another within a species. They also vary markedly during different stages of the cell cycle. Their role, as well as that of histones, in genetic regulation will be explored more fully in Chapter 19.

A classic experiment performed in 1952 by Hershey and Chase argued strongly against protein as the substance of the gene. It also includes other important observations needed for our discussions on microorganisms in Chapters 17 and 18. Hershey and Chase selected the bacterium *Escherichia coli* and a virus which attacks it as their objects of study. They did so because the T/2 virus or phage has a relatively simple morphology. From electron microscope studies, they knew it was composed of a protein envelope in which resides a core of DNA (Fig. 11-3). This structure is characteristic of this and the other T-even phages which can reproduce only in *E. coli* cells. After infecting the bacteria, the viruses somehow command the cell machinery to manufacture more virus and consequently the cell's activities become involved with the manufacture of phage particles. As a result, the cell eventually bursts. The viruses released are in all ways like the original infecting ones. Therefore, the viruses must contain genetic information which is continuous from one generation to the next. Somehow, a virus which attacks a bacterial cell must take over the cell and provide it with the information needed to make more virus. Hershey

FIG. 11-3. Structure of T-even viruses. Diagram of some features of the virus. A contractile sheath surrounds the core. Except for DNA in the head, all of the other parts are protein.

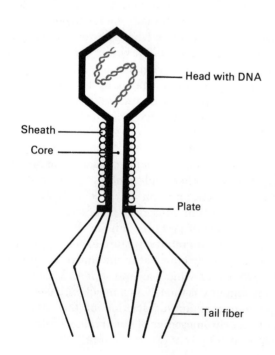

Head with DNA

Sheath

Core

Plate

Tail fiber

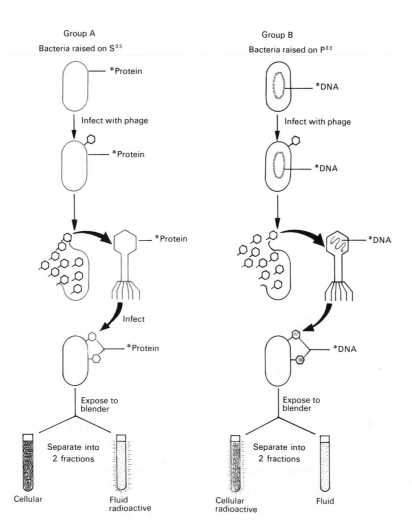

*Protein

Infect with phage

*Protein

*Protein

Infect

*Protein

Expose to
blender

Separate into
2 fractions

Cellular

Fluid
radioactive

*DNA

Infect with phage

*DNA

*DNA

*DNA

Expose to
blender

Separate into
2 fractions

Cellular
radioactive

Fluid

FIG. 11-4. The Hershey-Chase experiment. Bacteriophage with radioactive (*) protein coats were obtained by infecting bacteria whose protein was all radioactive. Phages with radioactive (*) DNA were obtained by infecting another group of bacteria whose DNA was all radioactive. The phages whose protein was tagged (group A) and those whose DNA was tagged (group B) were allowed to infect ordinary bacteria. Shortly after infection, the cultures were separated into a cellular fraction and a fluid one. The radioactivity in group A was associated wtih the fluid, meaning that the labeled protein of the infecting phage did not enter the cells. In group·B, the cellular fraction was labeled, indicating that the labeled DNA did enter.

and Chase set out to determine exactly what is getting into the cell with the information for virus construction. They knew that the phage seemed to attach to the bacterial cell wall, as if it were injecting something into the cell. Was this protein, DNA, or both which was entering the cell with genetic information of the phage? Using a very elegant experimental design, Hershey and Chase obtained striking results (Fig. 11-4). They took a group of bacteria (group A) and raised it on radioactive sulfur, S³⁵. Sulfur is a component of protein, and so all the bacterial protein would be labeled. Viruses were then allowed to infect these bacteria. The new viruses released from these cells would all contain radioactive protein envelopes, because only labeled protein was available for their coats which were assembled in the labeled bacterial cells.

A second group of bacteria (group B) was raised on a medium supplied with radioactive phosphorus, P³². The phosphorus would be incorporated not into the protein but into the nucleic acids of the cell, as phosphorus is one of the critical ingredients of DNA and RNA. The group-B phages were then allowed to infect the labeled cells. The released phage particles would have radioactive phosphorus incorporated into their DNA, which would consequently be tagged or labeled. Hershey and Chase now had two groups of viruses, group A with protein coats tagged with radioactive sulfur and group B with DNA labeled with radioactive phosphorus. The group A phages were then allowed to infect ordinary *E. coli*. A few minutes after infection, the bacterial cells were placed in a Waring blender. The tiny cells cannot be injured by the blades, but any particles adhering to their walls can be torn away. After exposure to the blender,

FIG. 11-5. Sugars and bases in the nucleic acids. *A.* Structure of purine and pyrimidine, the parent compounds of the nitrogen bases in DNA and RNA. *B.* Structure of sugars, purines, and pyrimidines in DNA and RNA. Note that although both types of nucleic acids contain similar building units, an important difference exists in the sugar component and in respect to two pyrimidine bases.

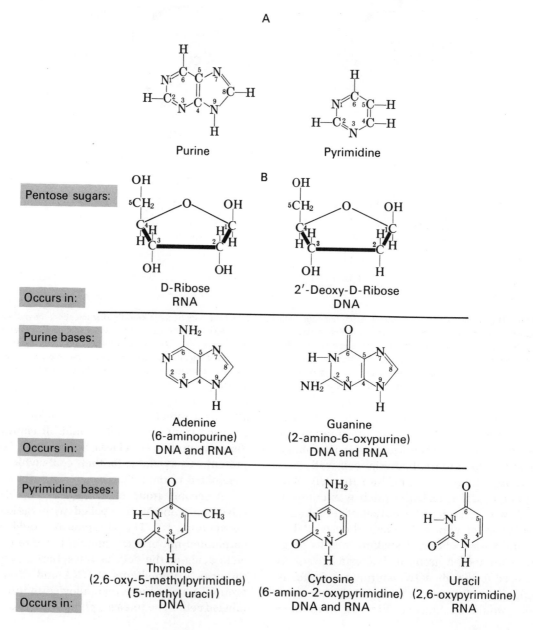

A

Purine

Pyrimidine

B

Pentose sugars:

Occurs in:

D-Ribose
RNA

2'-Deoxy-D-Ribose
DNA

Purine bases:

Occurs in:

Adenine
(6-aminopurine)
DNA and RNA

Guanine
(2-amino-6-oxypurine)
DNA and RNA

Pyrimidine bases:

Occurs in:

Thymine
(2,6-oxy-5-methylpyrimidine)
(5-methyl uracil)
DNA

Cytosine
(6-amino-2-oxypyrimidine)
DNA and RNA

Uracil
(2,6-oxypyrimidine)
RNA

the cells were separated from the fluid medium by centrifugation to give a cellular fraction and a fluid fraction. Both fractions were then tested for radioactivity of S^{35}. It was found that almost all of the activity was in the fluid fraction. This meant that only a little of the sulfur was associated with the cells.

Group B phages with tagged DNA were also allowed to infect ordinary *E. coli*. The identical procedure was followed to give a cellular fraction and a fluid fraction. Both were then tested for the labeled DNA. It was found that almost all of the radioactivity was with the cells, in contrast to the results with group A. In other words, most of the DNA of the phages entered the bacteria, whereas most of the protein remained outside in the fluid. It was also shown that the small amount of protein which was found with the cells did not appear in the *new* viruses which arose if the cells were allowed to burst. The protein tended to disappear and not enter into the construction of new viruses from the infected cells. These results give strong support in favor of DNA as the genetic material of the phage which bears instructions for the manufacture of new virus particles.

CHEMISTRY OF NUCLEIC ACIDS. Still other lines of evidence implicated DNA as the genetic material, and we will encounter some of these throughout later discussions. It is now necessary to review what was known about nucleic acids in the early 1950's. The first studies go all the way back to the 1860's, those of Miescher in Basel. Studying pus cells and sperm, he demonstrated that nuclei contain a protein-phosphorus compound. His work was followed by others in Europe and the United States. The best source of nucleic acid from animals proved to be the calf thymus gland. Hydrolysis of the nucleic acid showed that it was composed of: (1) phosphoric acid, H_3PO_4; (2) a 5-carbon sugar, later identified as $D(-)$2-deoxyribose; (3) the purine bases, adenine and guanine; and (4) the pyrimidine bases, thymine and cytosine (Fig. 11-5A). Both kinds of bases contain significant amounts of nitrogen and are referred to as "nitrogen bases."

Another rich source of nucleic acid is yeast. Actually, it is very abundant in RNA, and it was this nucleic acid which was analyzed. Unfortunately, this gave rise to the erroneous notion that RNA was plant nucleic acid and DNA the animal kind. Both types, of course, are essential ingredients of all animal and plant cells. Analysis of the RNA yielded the same kinds of products as the DNA upon hydrolysis. One big distinction between the DNA and RNA lies in the sugar component; a second one lies in one of the pyrimidine bases (Fig. 11-5B). The sugar of RNA, D-ribose, is slightly different from the deoxyribose sugar. The main difference is at the carbon 2 position; the deoxyribose has one less oxygen at carbon 2 than does the ribose. In RNA, the pyrimidine base, thymine, does not occur. Instead, thymine is replaced by the pyrimidine, uracil, which in turn is not found in DNA. Another distinction between the two kinds of nucleic acid also became evident, and this was in their cellular distribution. The DNA is confined almost exclusively to the nucleus, whereas the RNA has a wider distribution, being found in the nucleus and throughout the cytoplasm.

Once DNA was implicated as the genetic material, extra attention was directed toward it. For, if the gene is composed of DNA, a knowledge of how the gene operates on the molecular level depends on an understanding of how the DNA itself is put together. Chemical analyses showed the DNA (and the RNA also) to be constructed of repeating units called *nucleotides*. Each DNA nucleotide is a combination of a phosphoric acid, a deoxyribose sugar, and one of the nitrogen bases (either a purine or a pyrimidine). Therefore, there are four different kinds of nucleotides possible in DNA, depending on the nitrogen base (Fig. 11-6A): adenylic acid, guanylic acid, thymidylic acid, and cytidylic acid. The DNA is thus a polynucleotide. The individual nucleotides are connected through chemical bonds between the sugar and phosphoric acid components (Fig. 11-6B). The nitrogen bases in a strand of DNA are not joined to each other; they are linked to the sugar. As the figure indicates, the bases provide the only distinction among the nucleotides. In any stretch of DNA, a nucleotide of one kind need not be followed by any particular other nucleotide. As DNA from various sources was analyzed, no restriction seemed to be imposed on the sequence of nuclotides within the DNA molecule.

Note (Figs. 11-5 and 11-6) that the sugar component of the nucleic acids exists in the furanose form, in a ring of five instead of six, as is found in the pyranose form. The sugar may occur in the cell in association with one of the bases *without* being linked to a phosphoric acid. Such a combination of a pentose sugar and a purine or pyrimidine base is called a *nucleoside*. Since a nucleoside is equivalent to a nucleotide devoid of the phosphate component, there are four different kinds which correspond to the four nucleotides of DNA: adenosine, guanosine, thymidine, and cytidine (Fig. 11-7A). Uridine is the nucleoside in RNA which corresponds to the RNA nucleotide, uridylic acid (Fig. 11-7B). Chemically, the nucleotides are phosphoric esters of the nucleosides. It should be noted that nucleotide structure is found not only in nucleic

FIG. 11-6. Nucleotides in DNA. *A.* Four different nucleotides occur in DNA. Each is a combination of deoxyribose sugar joined to a phosphate and to one of the four bases: adenine (A), guanine (G), cytosine (C), or thymine (T). *B.* DNA is a polynucleotide. The nucleotides are attached to each other through the sugar and phosphate components. Note that the bases are not connected to the phosphate but are linked only to the sugar.

A

Deoxyadenylic acid

Deoxyguanylic acid

Deoxycytidylic acid

Deoxythymidylic acid

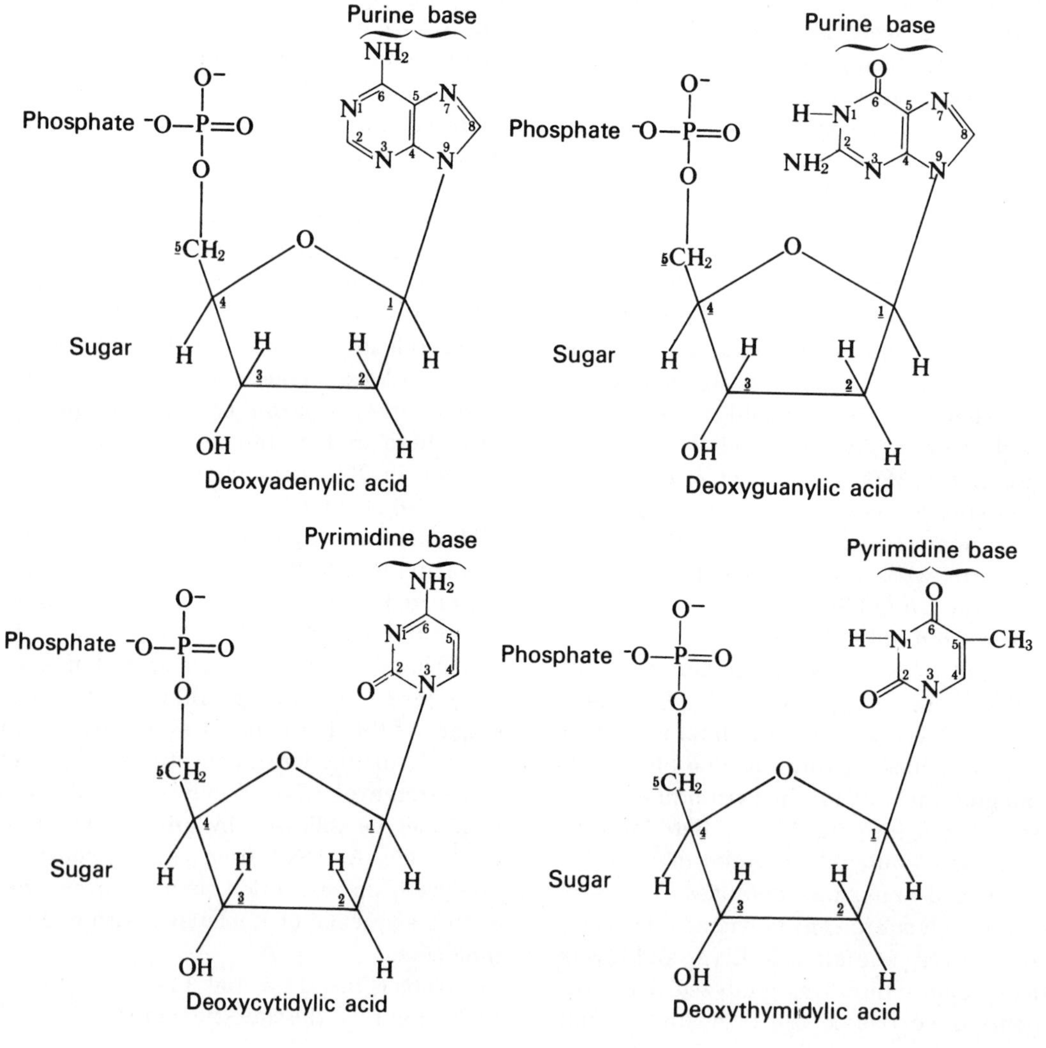

acids but also in several biologically important compounds, such as certain coenzymes and the universal energy source, adenosine triphosphate (ATP).

The 1940's yielded a great deal of information on the way the nucleotides are assembled in DNA. One idea had been that the DNA is composed of repeating tetranucleotides, each tetranucleotide being a sequence of four nucleotides in which the bases adenine (A), guanine (G), cytosine (C), and thymine (T) are always represented. In other words, the four kinds of bases were supposed to occur in equal amounts. Chargaff and his associates showed that this is not the case. The DNA does not contain equal amounts of the four different bases. There is nothing which restricts the sequence of the nucelotides. Moreover, the proportion of bases varied greatly with the different sources of the DNA. But one very important fact came to light. It was found that the amount of A was always equal to the amount of T. Similarly, the amount of G equalled that of C. Therefore, the A:T ratio is 1 and so is the G:C ratio. However, the amount of

TABLE 11-1 Base composition of DNA from various organisms*

	ADE-NINE	THY-MINE	GUA-NINE	CYTO-SINE
Man (sperm)	31.0	31.5	19.1	18.4
Chicken	28.8	29.2	20.5	21.5
Salmon	29.7	29.1	20.8	20.4
Locust	29.3	29.3	20.5	20.7
Sea urchin	32.8	32.1	17.7	17.7
Yeast	31.7	32.6	18.8	17.4
Tuberculosis bacillus	15.1	14.6	34.9	35.4
Escherichia coli	26.1	23.9	24.9	25.1
Vaccinia virus	29.5	29.9	20.6	20.3
E. coli bacterio-phage T₂	32.6	32.6	18.2	16.6*

* Note that the amount of A/T tends to be equal and so does the amount of G/C. However, the proportions of A/T:G/C vary from one species to another. Some have low amounts of G/C (sea urchin); some have a very high amount (tuberculosis bacillus). (Reprinted with permission from Herskowitz, I. H. *Principles of Genetics*, Macmillan Publishing Co. Inc., New York, 1973.)

A:T is not necessarily equal to the amount of G:C. Indeed, wide differences were found in these proportions. Some types of DNA have much more A:T than G:C, whereas an abundance of G:C over A:T is found in the DNA from other species (Table 11-1). Thus, the makeup of the DNA was shown to vary from one species to another in its base proportions, but the different tissues of any one species showed the same kind of DNA. This, of course, made sense. We would expect the hereditary material to differ among the diverse forms of life but not from one cell type to another within an organism or within a species.

Thus the DNA of each species has its characteristic $A + T/G + C$ ratio. This ratio is commonly called the "percent GC content" for short. A striking finding is that the percent GC content of lower forms (prokaryotes, for example) shows great variability from one species to the next, perhaps 25% in one and 75% in another. With evolutionary progression, the percent GC appears to vary less and less within the divisions of plants and animals and to be a more fixed value. In mammals, the overall average value is about 40%, one species of mammal showing a similar value to the next one. The reason for this stability of $G + C$ value in higher life forms and its evolutionary significance are unknown.

More detailed chemistry showed exactly how the nucleotides are linked to one another through the sugar and phosphate (Fig. 11-8). The number 1 carbon in the sugar is attached to the nitrogen base. The second carbon is the one lacking the oxygen. Carbon 4 is completely in the ring. Carbons 3 and 5 are involved in the internucleotide linkages. The —OH group of carbon 3 in the sugar component of one nucleotide forms an ester linkage with the phosphoric acid. The phosphate in turn forms another ester linkage with the —OH group of carbon 5 in the adjacent nucleotide. Thus, the phosphoric acid engages in the formation of a double ester. It is these diester linkages which hold together the sugar-phosphate backbone. Because positions 3 and 5 of the sugar are involved, we say the nucleotides are joined by 3'-5' phosphodiester linkages.

PHYSICAL STRUCTURE OF DNA AND THE WATSON-CRICK MODEL. By the early 1950's a great deal was known about the chemistry of the DNA, but this told little about the architecture

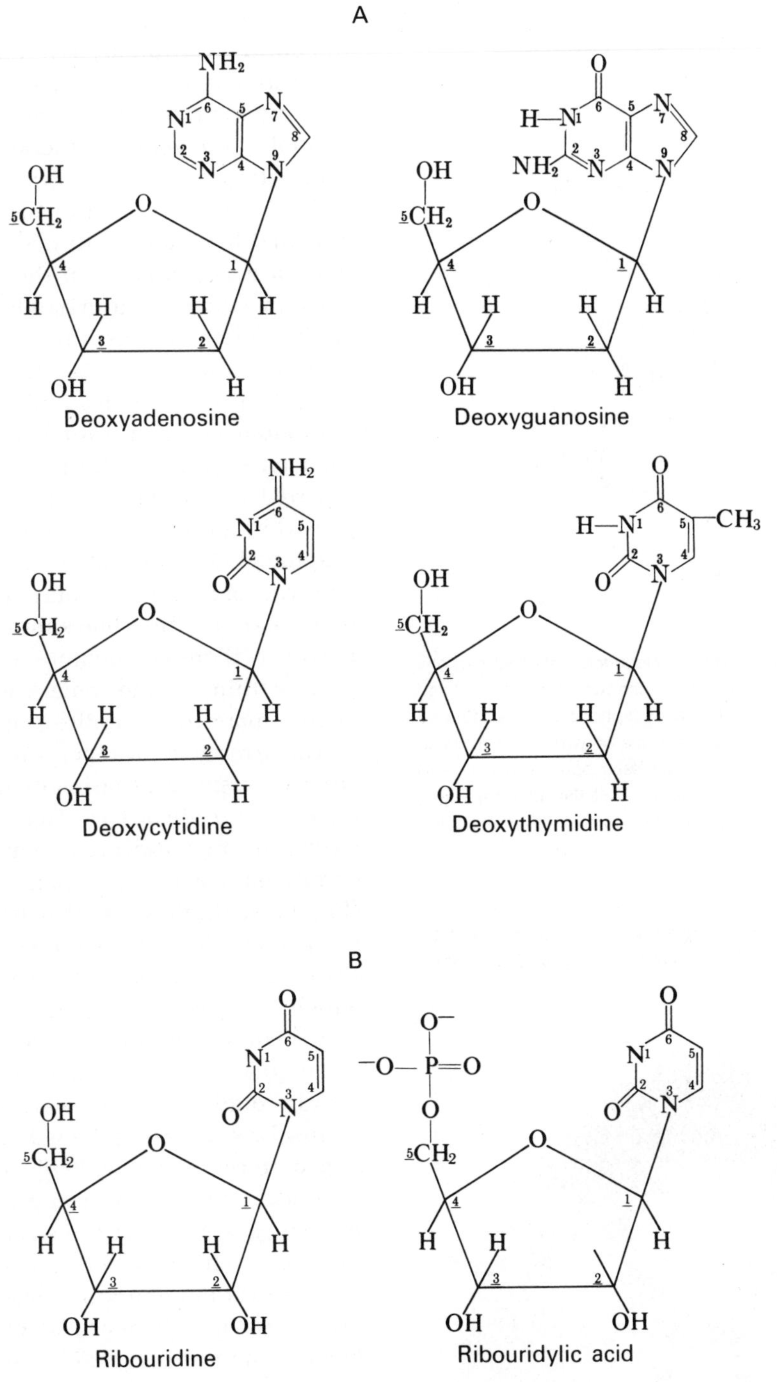

FIG. 11-7. Nucleosides in nucleic acid. *A.* A nucleoside is a combination of a purine or pyrimidine base with a pentose sugar. The four kinds in DNA are deoxyribonucleosides and correspond to the nucleotides (Fig. 11-6) which contain a phosphate group linked to the sugar. *B.* Nucleosides and nucleotides of RNA correspond closely to those of DNA. One major difference is in the sugar. The ribose of RNA contains an oxygen at position 2 in the sugar ring which is absent from that position in DNA. Another distinction is that in RNA the nucleoside, ribouridine, and nucleotide, ribouridylic acid, occur in place of thymidine and thymidylic acid.

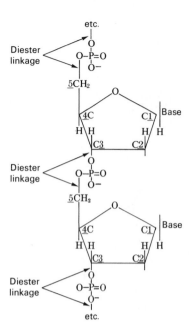

FIG. 11-8. The linkage between nucleotides. Note that carbons 3 and 5 of the sugar are involved in the formation of a linkage with the phosphate component. A phosphoric acid has formed one ester bond with the —OH associated with the carbon in position 3 of one sugar and another ester bond with the —OH associated with the carbon in position 5 of the adjacent sugar. (Refer also to Fig. 11-6B, which does not show the C atoms in the sugar).

FIG. 11-9. X-ray diffraction. (Copied with permission from J. D. Watson, and F. H. C. Crick, *Cold Spring Harbor Symp. Quant. Biol.*, *18*: 123–131, 1953.)

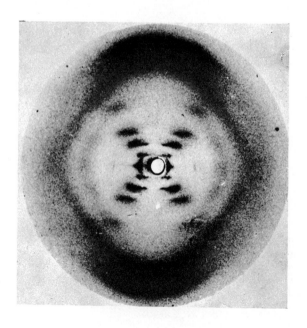

of the molecule. Were these polynucleotides in the form of straight, unbranched chains? Or were they twisted or folded in some way? Was the organization of the nucleotides somehow related to the genetic function of the DNA molecule? Important data on the physical structure of DNA were obtained in the laboratory of Wilkins and his group at King's College. The information was primarily in the form of X-ray diffraction studies and crystallographic analyses. In diffraction studies, the material is subjected to a beam of X-rays. Assuming the material is composed of molecules and atoms arranged in a random or haphazard manner, the rays, in their passage through the material, will not be diverted in any characteristic fashion. If they pass through the material and fall on a photographic plate, they will not describe any definite pattern, and when the plate is processed, little more than a diffuse blackening may be seen. If, however, there is a definite arrangement of the parts within the molecule, the rays will be deflected more in some directions than in others. As the arrangement repeats itself, the photosensitive emulsion may be bombarded over and over again in certain areas. This will heighten the darkening in these areas, with the result that certain dark spots may repeat and form a pattern (Fig. 11-9). Studies of such patterns can therefore tell a great deal about the spatial organization within a molecule and the repetition of the units which compose it. Diffraction studies by Wilkins's group showed that the DNA molecule is not haphazard in arrangement but possesses a definite organization in which the parts repeat themselves. Utilizing the valuable X-ray diffraction patterns along with the information from chemical analyses, Watson and Crick were able to deduce a model of the structure of the DNA molecule. The Watson-Crick model is perhaps the greatest landmark in modern biology. It has given rise to the new discipline of molecular biology and has provided a deep insight into the very nature of gene action.

The diffraction patterns suggested that the DNA molecule is composed not of just a single long strand of nucleotides but rather two strands bonded together in some way. Moreover, the two polynucleotide chains are not simply stretched out but are twisted in a spiral, much as are the railings of a spiral staircase. The DNA molecule

FIG. 11-10. The Watson-Crick double helix. According to the model, the DNA is composed of two polynucleotide chains twisted in a spiral. The sugar and phosphate components provide the backbone of the chains. These in turn are held together by pairing between the nitrogen bases (see Figs. 11-11 and 11-12). The bases would be stacked inside the double helix, much like stairs in a spiral staircase. Note that a large groove and a small one are present. This is due to the orientation of the sugar in the chains. (Reprinted with permission from J. D. Watson, and F. H. C. Crick, *Cold Spring Harbor Symp. Quant. Biol., 18:* 123–131, 1953).

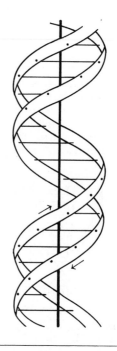

is in the form of a double helix (Fig. 11-10) having a diameter of about 20 Å. One complete turn of each chain is 34 Å, and this includes 10 nucleotides. Therefore, the distance between one nitrogen base and the next one on a chain is 3.4 Å. The chemical information told Watson and Crick that the nitrogen bases in DNA are in the ratios: A:T = 1 and G:C = 1. Added to what was known about the physical structure of the DNA, this information suggested to them that preferential base pairing occurs in the molecule and that this pairing is responsible for holding the two chains together. The purine adenine would thus pair preferentially with the pyrimidine thymine, whereas the purine guanine would pair with the pyrimidine cytosine (Figs. 11-11 and 11-12). The two bases composing each base pair would be held together by hydrogen bond formation. Hydrogen bonds are much weaker attractive forces than covalent bonds, which entail the sharing of electrons between atoms. Knowledge of the molecular structures of the four bases indicated that two hydrogen bonds could hold the A-T pair together, whereas three would be formed in the G-C pair. As a consequence of this hydrogen bond formation between the base pairs all along the molecule, the two polynucleotide chains would in turn be held together. The base pairs extend from the sugar-phosphate backbone and would be stacked up inside the double helix resembling the stairs in the analogy of the spiral staircase. The two chains of the molecule would run in opposite directions with respect to the

FIG. 11-11. Preferential base pairs in DNA. Adenine on either one of the two chains pairs with thymine; cytosine pairs with guanine. Note that two hydrogen bonds hold the A-T pair together, whereas three are involved in the G-C pair.

Adenine   Thymine

Thymine   Adenine

Cytosine   Guanine

Guanine   Cytosine

CHEMISTRY OF THE GENE

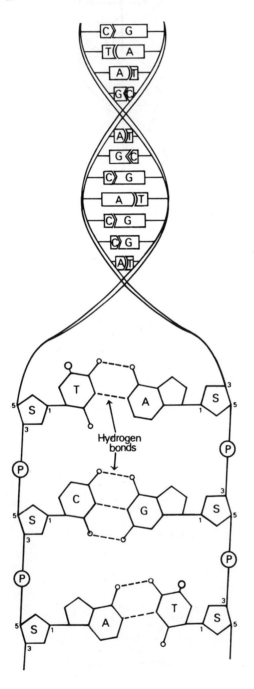

FIG. 11-12. Diagram of DNA double helix. Note the difference in orientation of the two chains: one runs in the order 3'-5' and the other in the order 5'-3'. It can also be seen that the two chains are complementary, not identical. (Reprinted with permission from Bonner and Mills, *Heredity*, 2nd ed., Foundations of Modern Biology Series. Prentice-Hall, Inc., Englewood Cliffs, N. J., 1964.)

attachment between sugar and phosphate groups. Recall that 3'-5' phosphodiester linkages hold the nucleotide units together in a chain. If we start at the "top" of a DNA molecule, the orientation of one chain would be 3'-5'-3'-5'- etc. The other chain would be in reverse order and would run 5'-3'--5'-3'-- etc. As a result of this orientation, the molecule looks the same from the "top" as from the "bottom": a 3' chain and a 5' chain terminate at each end (Fig. 11-12). Thus each end of the molecule will have one 5' or phosphate terminating chain and one 3' or —OH terminating chain. The sugars in the chain are so arranged that they cause the double helix to contain two unequal grooves, one conspicuously larger than the other (Fig. 11-10).

The structure of the DNA molecule as proposed by Watson and Crick was not based on mere assumptions. The model was constructed through careful, painstaking correlation between chemical and crystallographic data. All of the suggestions in the model were in complete agreement with established chemical and physical principles. One critical feature of the model is that the two polynucleotide chains are not identical due to the preferential base pairing. Not only do the chains run in the opposite direction, but most important, they are complementary: the A and C in one strand corresponding to a T and G in the other. Since the model imposes no restrictions on the sequence of the bases in any one chain, A could be followed by T, G, C, or another A. However, if one knows the sequence of bases along a length of one of the chains (say, A-T-C-C-A-G), one can be certain of the sequence of bases along the complementary chain (T-A-G-G-T-C).

IMPLICATIONS OF THE WATSON-CRICK MODEL. A burst of excitement followed the announcement of the Watson-Crick model. One of its many contributions was to provide a key to the interpretation of several familiar genetic phenomena about which little or no insight had been provided. Although everyone knew that the genetic material undergoes self-duplication, there was no well-founded suggestion as to how this might take place. The Watson-Crick model offered a valid basis for this. According to the model, at the time of replication, the hydrogen bonds between the bases of the two complementary chains would be dissolved (Fig. 11-13). The two chains would then unwind and separate. Each strand maintains its integrity in the process and does not break down in any way. Instead, each one acts as a template or pattern for the assembly of another strand, one which is complementary to it. These complementary strands would be constructed from their building blocks, the nucleotides present in the cell. As the sugar-phosphate backbone is assembled, each nitrogen base in the original strands attracts the complementary one. The purine adenine (A) in one chain attracts the pyrimidine thymine (T) in the cellular environment. Likewise, T in the other original chain attracts nucleotides of A in the cell. This complementary attraction by all the bases in each of the two original strands, along with the assembly of the sugar-phosphate backbone, results in the construction of two new chains, each complementary to the original "old" ones. The overall effect is two double helical DNA molecules, each identical to the one original. This fits in perfectly with what we would expect if the genetic material is to remain identical from one cell generation to the next. For the genetic material must contain a store of information. The Watson-Crick model has important implications on this extremely critical point. Since the model assumes no restriction on the sequence of base pairs, the genetic information could somehow be stored in the form of a code. The code cannot reside in the sugar or phosphate, because these are the same from one nucleotide to the next; only the bases may differ. The sequence of bases then could provide the basis of a code. If the sequence is A-T-T along one stretch of the DNA, this could mean something different than the sequence G-T-C. A gene, a segment of double-stranded DNA, would then contain information coded in the sequence of its base pairs. And one gene or one allele would differ from the next due to the difference in the sequence. Since a gene would most likely be composed of many nucelotides, it would include many base pairs. The unrestricted sequence allows a limitless number of differences and could account for the endless number of different genes which are found in living species.

But it is known that the genetic material may suddenly change as the result of a spontaneous gene mutation. This implies a change in

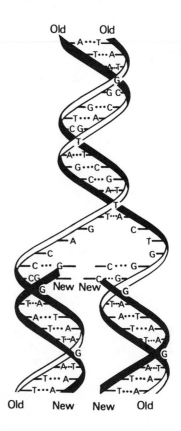

FIG. 11-13. Self-duplication of DNA. During replication, the hydrogen bonds between the strands are dissolved. Each original strand remains intact and acts as a pattern for the assembly of a complementary chain. (Reprinted with permission from James D. Watson, *Molecular Biology of the Gene,* 2nd ed., copyright 1970 by J. D. Watson; W. A. Benjamin, Inc., Menlo Park, Calif.)

the coded information. How could this come about? Chemical analysis of the DNA showed that the molecule is tautomeric. This means that the molecule undergoes occasional shifts in the positions of certain protons within it. One form of the molecule would be the most common, but at any one moment, a hydrogen may change position, so that a molecule could be in its rarer form or state (Fig. 11-14A). This rare state might occur at the time the gene is replicating or building more of itself. Since the tautomeric changes can affect hydrogen atoms in the purine and pyrimidine bases, the change could alter the pairing preferences of the bases (Fig. 11-14B). Let us assume that in a stretch of DNA a shift occurs in the position of a hydrogen in one of the pyrimidine bases, thymine. As a consequence of the shift, it may not be able to pair with its

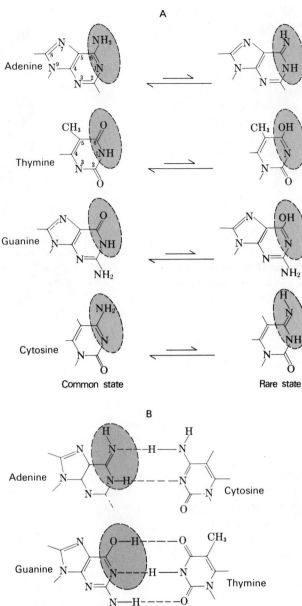

A

Adenine

Thymine

Guanine

Cytosine

Common state

Rare state

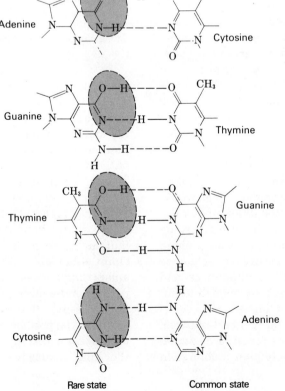

B

Adenine — Cytosine

Guanine — Thymine

Thymine — Guanine

Cytosine — Adenine

Rare state

Common state

FIG. 11-14. Tautomerism and formation of exceptional base pairs. *A.* Each of the four bases in DNA typically exists in a common state. However, hydrogen atoms may at times shift position so that a rare state may exist at a particular time. For example, adenine bears a —NH₂ group. At rare times, one of the H atoms shifts so that the —NH₂ becomes —NH, and the H is in a new position. *B.* Such tautomerism can change pairing preference. For example, the shift which can occur in adenine now allows it to bond with cytosine. Shifts in the other bases cause similar changes in pairing preference.

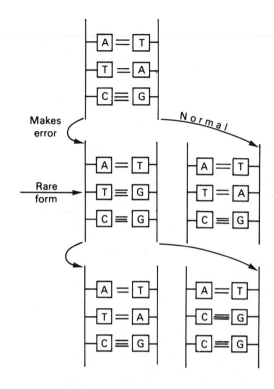

FIG. 11-15. Tautomerism and mutation. A sequence of normal nucleotide pairs (*above*) is typically found in a stretch of DNA. If a base should exist in its rare form (*middle*) in one of the strands at the time of replication, an exceptional base pair may form. At the next replication (*below*), the bases in each strand would usually behave normally and attract their complementary bases. However, because a substitution was made at the previous replication, a new base pair now substitutes for the original.

normal partner, adenine, but instead, pairs with guanine (Fig. 11-15). The next time the molecule undergoes replication, there would be no reason to expect a tautomeric shift, because this would be a rare event in any case and would not have to affect the same bases. Therefore, normal bonding preferences would be expected at the next replication. However, because G has been inserted in the one strand (the right one in the figure) as a result of the first error, C will now be placed in the corresponding position on the new complementary strand. Instead of the original base pair sequence A-T, T-A, C-G, we now have A-T, C-G, C-G. This T-A pair has been replaced by a C-G. If the base pairs do provide the basis for a code of stored information, the code would now be changed. The new sequence of base pairs could mean something very different from the original message. This model could also explain the dire effects of a deletion in which a part of the chromosome is lost. If a sequence of base pairs is omitted, then part of the message would be missing. Any change in the structure of the chromosome could alter the base sequence and so upset the genetic message.

If the Watson-Crick model is correct, this means that the coded information in the DNA must somehow be transferred to other parts of the cell. Obviously, a code is of no use if its information cannot be communicated. The model triggered a wealth of investigations on the mechanism of information transfer from nucleus to cytoplasm, a most important topic which is the subject of Chapter 12. Before many of the investigations on information transfer, gene replication, and gene mutation could be undertaken, it was essential to test the soundness of the Watson-Crick model. One very admirable feature of the model was that its for-

mulation did permit tests of its validity. If the model were sound, then experiments could be designed whose results would be predictable. Throughout the years, the Watson-Crick model has passed the scrutiny of many investigations the results of which support all the essential features of the model. Several of these studies will be considered later as they pertain to other specific topics; two of them are given below, because the techniques employed are fundamental to many of our following discussions.

EXPERIMENTAL TESTING OF THE WATSON-CRICK MODEL. In 1958, Meselson and Stahl presented strong support for the concept of replication as pictured by Watson and Crick. According to the model, DNA replication is semiconservative. This means that the entire molecule does not remain intact when new DNA

FIG. 11-16. Density gradient ultracentrifugation. Substances which are very similar in density may be separated by this sensitive method because each will tend to settle where its density matches that of a solute. Different types of DNA can be separated, such as the three kinds indicated here: one type composed of two light chains containing N^{14}, another composed of two heavy chains with N^{15}, and an intermediate DNA made of one light and one heavy. (See text for details, and compare with Fig. 11-17.) (Reprinted with permission from James D. Watson, *Molecular Biology of the Gene,* 2nd ed., copyright 1970 by J. D. Watson; W. A. Benjamin, Inc., Menlo Park, Calif.)

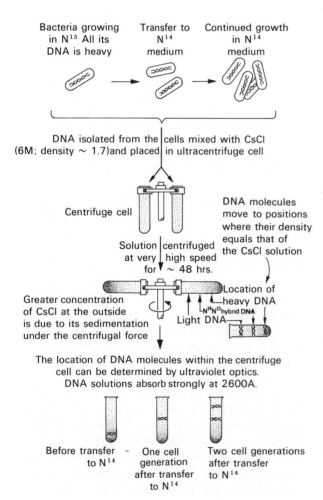

is being formed; instead, the two strands come apart. However, each strand is preserved intact, as shown in Figure 11-13; it does not dissociate in any way. Besides this semiconservative concept of replication, there are two other possibilities which can be imagined: conservative (the two strands stay together, and a new double stranded molecule is built up next to them) and dispersive (the two original strands break down and entirely new strands are constructed from these and other precursors in the cell). To gain information on this critical point, Meselson and Stahl employed a procedure now routine in laboratories today, density gradient ultracentrifugation. The technique is based on the knowledge that a solute (sugar, salt, etc.), when spun at a high speed in a centrifuge, will become distributed from one end of the tube to the other. It will eventually reach an equilibrium, and at this time, the molecules will be most dense at the centrifugal end and will gradually decrease in density toward the centripetal end. Such an equilibrium density gradient has practical laboratory application. If one has a mixture of substances which are very similar in density, they may be separated by this procedure. For when a mixture is spun in the appropriate solution, a density gradient is established in the centrifuge tube. The substances composing the mixture tend to separate out at characteristic levels in the tube. This dissociation occurs, because a substance settles at a location where the density of the solute molecules matches *its own* density. The technique is so sensitive that materials of very similar densities may be separated by using the proper solute. For example, a mixture of DNA molecules of different densities may be separated in a gradient of cesium chloride (Fig. 11-16). Under laboratory conditions, a given kind of DNA can be made heavier than normal by supplying organisms with a medium containing only heavy nitrogen, N^{15}, an isotope of the common N^{14}. If "heavy" DNA (heavy, because its bases would contain N^{15}) and light DNA are mixed, they may be dissociated in a cesium density gradient. This is possible, since they will come to equilibrium and form distinct bands at slightly different locations in the centrifuge tube.

In their experiment (follow Fig. 11-17), Meselson and Stahl employed the bacterium *E. coli.* The organism was grown on a source of labeled, heavy N^{15} for several generations until all of its DNA was heavy. When extracted from the cells, this heavy DNA settled out in the density gradient at a characteristic position. Meselson and Stahl removed samples of bacteria from the "heavy" cultures and placed them on a medium containing ordinary N^{14}. They tested the DNA of some of these immediately, without allowing any cell division (0 generations). Some samples were allowed to divide once (1 generation), others two and three times (2 and 3 generations) before their DNA was extracted. As Fig. 11-17 shows, all of the DNA from the "0 generation bacteria" was heavy as expected. It was reasoned that if the Watson-Crick model was correct, each of these heavy strands would separate during the first round of cell division. Each could only direct next to it the construction of a light strand, because only light nitrogen would now be present. The DNA of the first generation cells would thus contain DNA which is hybrid, one which is intermediate in density between double-stranded heavy and double stranded light molecules. It should therefore settle out at a location in the centrifuge tube which is distinct from that of the other two. And this is what was found, a single band at a position characteristic of neither N^{14} nor N^{15}. The DNA was then tested from the bacteria which had undergone two divisions. According to the Watson-Crick model, each strand will maintain its integrity. But under the conditions of the experiment, each strand can only order the construction of light strands. The original heavy ones build up light ones and so do the light ones which were made during the first division. The result is equal numbers of light and intermediate (hybrid) strands. A perfect correlation was found between the prediction and the experimental results, for now *two* bands of *equal* size were present in the density gradient: one at the level of the hybrid DNA and the other at a less dense position, that typical of light DNA. The third generation DNA again gave exactly what would be expected from the Watson-Crick model. Hybrid and light strands are again both present, because the original heavy ones would continue to act as templates. However, they would be fewer in proportion to the light ones, since only light strands can be constructed. Observations clearly showed that the light band in the density gradient was con-

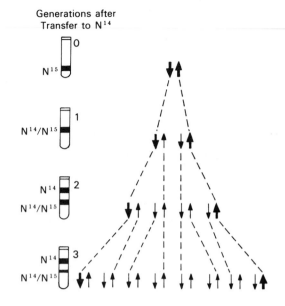

Generations after
Transfer to N^{14}

FIG. 11-17. Meselson and Stahl experiment. In this highly schematic diagram, a pair of reversed arrows represents a double helix. It can be seen that if the cells are grown on medium with heavy (N^{15}) nitrogen, all of the cellular DNA is heavy if no division is permitted (O generation). As the number of cell divisions increases, following transfer to N^{14} medium, the number of light strands increases. Note that the strands maintain their integrity. Any strand formed in one generation acts as the template for a new light strand in the next generation. In the third generation, the hybrid (N^{14}/N^{15}) band is narrower than the N^{14}, because only light strands can be made. Still, the heavy strands continue to act as templates.

spicuously thicker than the slightly heavier hybrid one. If replication were conservative or dispersive, such results could not be explained. Replication as proposed by Watson and Crick offers the only answer.

According to the model, as replication takes place, the two chains separate from each other while each base attracts the complementary one. It is conceivable that chain separation is unnecessary for the construction of two new chains. It is also possible that both strands must separate completely along their lengths and exist entirely as single chains at the time of replication. The work of Cairns with *E. coli* DNA provided details on these aspects of the duplication process. Cairns was a pioneer in the study of the "chromosomes" of bacteria. His superb isolations of bacterial DNA showed that the DNA of *E. coli* is in the form of a circle (Fig. 11-18). Bacteria and other prokaryotes do not possess chromosomes in the true structural sense of the word as used for eukaryotes. The genetic material of these simple organisms contains naked strands of DNA which are not complexed with histone protein. The DNA is much smaller in amount than in the "true" chromosomes of eukaryotes. When extracted from a cell, the prokaryotic DNA can often be followed through its entirety. The overall simplicity of this DNA provides a system for detailed study which is extremely difficult in eukaryotes. To follow the replication of DNA in *E. coli,* Cairns used the technique of autoradiography. When cells of any kind are grown in a medium containing thymidine, this nucleoside is incorporated only into the DNA. The thymidine can be rendered radioactive if it contains tritium (the heavy isotope of hydrogen). As it disintegrates, tritium emits beta particles (electrons). If these fall upon a photographic emulsion, the film will contain darkened spots when it is processed. Cairns grew *E. coli* cultures in a medium of tritiated thymidine for different periods of time. He extracted the DNA and covered it with an emulsion which was eventually developed in a fashion similar to that of any photographic film. The dark spots on the resulting autoradiograph indicate the breakdown of radioactive atoms in the DNA. The more such atoms in one location, the more spots, due to the greater number of beta rays affecting the emulsion. Thus, the number of spots or their density would be proportional to the number of radioactive nucleotide chains in the DNA molecule. If two radioactive chains are present at a certain location and only one at a second location, the former will give off twice as many emanations as the latter. The darkening will thus be twice as dense.

Cairns compared autoradiographs from bacteria which had grown in labeled thymidine for different lengths of time. Particularly significant were autoradiographs from those cultures of cells which had completed one round of DNA replication in the labeled medium and were now undergoing a second replication in the presence of the label. On each of these, Cairns was able to identify a region which appeared as a fork (Fig. 11-19). This single fork appeared at different positions depending on the length of time the cells had grown in the presence of the radioactive

FIG. 11-18. Diagram of an autoradiograph of the *E. coli* chromosome. The chromosome is actually in the act of duplicating. The DNA is from cells which have been grown in the presence of radioactive thymidine for two generations (see text for details). (Drawn after J. Cairns, *Cold Spring Harbor Symp. Quant. Biol., 28:* 43–46, 1963.)

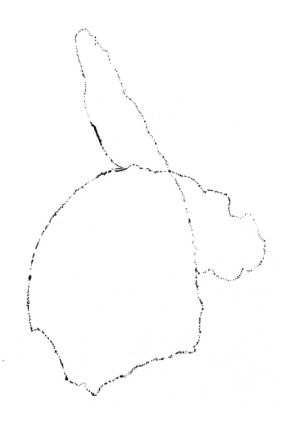

thymidine. The observations can be readily interpreted according to the Watson-Crick model. The fork marks the point in the DNA where replication is taking place. If one round of replication is completed, the DNA is composed of one labeled and one nonlabeled strand. As a second round starts, the two strands unwind, and each proceeds to direct the formation of a radioactive partner. However, one strand will have two labeled chains, the other only one. Consequently, a "Y" figure is produced on the emulsion at this point due to the fact that one arm of the "Y" has a density of dots twice that of the base of the "Y" and the other arm. Cairns was able to time the movement of a single replicating site from the initiation point around the circular chromosome by comparing autoradiographs from cultures grown for different lengths of time in the labeled thymidine. A fork was estimated to travel approximately 20–30 μ/min. It was later established that replication of the DNA in the *E. coli* chromosome is *initiated* at a single site, as indicated by Cairns's original observations, but that two growing points are established at that site. Replication is then *bidirectional*; these two replicating sites move in opposite directions around the circular bacterial DNA until they eventually converge and fuse. The observations clearly indicate that chain separation goes hand in hand with replication of the DNA strands in much the way predicted by the Watson-Crick model.

The DNA of higher organisms including the human has been studied in a similar manner by refined autoradiography. In eukaryotes, however, there are many initiation points and thus many replicating forks along the length of the DNA. As in *E. coli*, replication proceeds bidirectionally from any one point. The rate of movement of a single fork, however, is slower, only about 1/50 the rate of that in *E. coli*, approximately 0.5 μ/min. This might be expected, because the eukaryotic chromosome is in a highly compact condition and probably needs to unfold when it is replicating. Since the prokaryotic chromosome is extended and free of proteins, the replication site would be able to travel at a higher rate.

The work of Cairns and that of Meselson and Stahl are only two of the many experimental

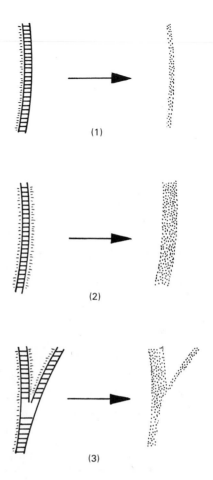

FIG. 11-19. Point of replication in DNA. *1.* If one chain of the double helix is labeled with tritiated thymidine, dark spots will appear on the autoradiograph due to the emission of electrons as the label decays. *2.* If both chains are fully labeled, the density of dots on the autoradiograph will be twice as great as in the case of the one-half labeled DNA. *3.* One-half labeled DNA which is undergoing a second round of replication in labeled medium will show a region which appears like a fork with one arm having a density of dots twice that of the other. In the denser arm, two radioactive chains are present (the original and the new one being constituted). In the other arm, only the new chain will be radioactive, because it is being formed in a labeled medium on a nonlabeled template.

approaches which have tested and supported implications of the Watson-Crick model. Others will be encountered in our later discussions of gene action at the molecular level.

OTHER ASPECTS OF SEMICONSERVATIVE REPLICATION. Many details concerning DNA replication at the molecular level still await clarification. However, many important facts have been established since the publication of the Watson-Crick model, and a few of these must be grasped for a basic understanding of the replication process. It is clear that as a DNA chain grows, each nucleotide which is being added is present in its triphosphate form (a nucleoside triphosphate). Two phosphates are eliminated as each nucleoside triphosphate is added to the chain. Phophodiester bonds are formed between the adjacent nucleotides composing a DNA chain, and this requires enzymatic activity. An enzyme, DNA polymerase, was discovered which can catalyze the joining of nucleotides and thus build up a chain complementary to the template chain. At least three forms of the enzyme DNA polymerase have now been identified in cells. Apparently one of these plays the major role in chain elongation, although all forms can catalyze the formation of phosphodiester bonds. We will therefore simply continue to refer to "DNA polymerase" as if it were a single entity.

The mode of action of DNA polymerase has also made it clear that the already formed DNA chain does indeed act as a template. The enzyme is able only to recognize the sugar-phosphate portions of the nucleotides. Preformed DNA must be present to establish the sequence of the four kinds of nucleotides, as depicted by the Watson-Crick model. Another feature of DNA polymerase is that it can increase the length of a growing DNA chain only in the 5' to 3' direction. This means that as a chain grows, the 5' end of the nucleoside triphosphate being added is joined to the 3' end (the —OH end) of the nucleotide already in the chain (Fig. 11-20). In other words, a nucleotide just inserted would be found at the 3' end of the growing chain. The "first nucleotide" in the growing chain would have a free 5' or phosphate end. Recall that the two complementary chains in DNA show opposite polarity, one running 5' to 3', the other 3' to 5' along the length of the DNA molecule (see Fig. 11-12). Nevertheless, as the double helix unwinds, replication in both newly growing strands is 5' to 3'.

With these basic facts, let us see in the next chapter how a knowledge of DNA structure can improve our understanding of the gene and our insight into certain fundamentals of genetics.

FIG. 11-20. Growth of DNA chain in the 5' to 3' direction. Each nucleotide added to a growing DNA chain is in its triphosphate form and is inserted by the union of its 5' (PO₃) end with the 3' (OH) end of the nucleotide already in the chain. A pyrophosphate is produced and is hydrolyzed to two phosphates in the elongation process, which is catalyzed by DNA polymerase, working only in the 5' to 3' direction.

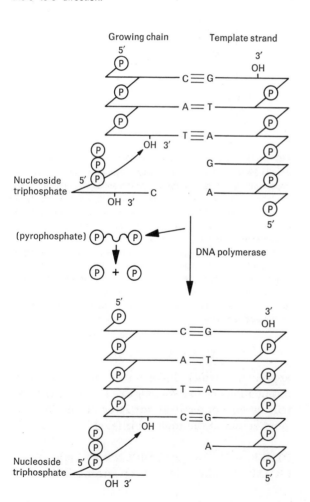

REFERENCES

Avery, O. T., C. M. Macleod, and M. McCarty. Studies on the chemical nature of the substance inducing transformation of pneumococcal types. *J. Exp. Med., 79:* 137–158, 1944. Reprinted in *The Biological Perspective, Introductory Readings.* W. M. Laetsch (ed.), pp. 105–125. Little, Brown, Boston, 1969.

Cairns, J. The chromosome of *Escherichia coli. Cold Spring Harbor Symp. Quant. Biol., 28:* 43–45, 1963.

Dickerson, R. E. The structure and history of an ancient protein. *Sci. Am. (April):* 58, 1972.

Hershey, A. D., and M. Chase. Independent functions of viral protein and nucleic acid in growth of bacteriophage. *J. Gen. Physiol., 36:* 39–54, 1952. Reprinted in *Papers on Bacterial Viruses,* G. S. Stent (ed.), 2nd ed., pp. 87–104. Little, Brown, Boston, 1965.

Kornberg, A. *DNA Synthesis.* W. H. Freeman, San Francisco, 1974.

Masters, M., and P. Broda, Evidence for the bidirectional replication of the *Escherichia coli* chromosome. *Nature (New Biol.), 232:* 137–140, 1971.

Meselson, M., and F. W. Stahl. The replication of DNA in *Escherichia coli. Proc. Natl. Acad. Sci., 44:* 671–682, 1958. Reprinted in *Papers in Biochemical Genetics,* G. L. Zubay (ed.), pp. 397–402. Little, Brown, Boston, 1966.

Mirsky, A. E. The discovery of DNA. *Sci. Am. (June):* 78, 1968.

Watson, J. D., and F. H. C. Crick. Molecular structure of nucleic acids. A structure for deoxyribose nucleic acid. *Nature (Lond.), 171:* 737–738, 1953. Reprinted in *Classic Papers in Genetics,* J. A. Peters (ed.), pp. 241–243. Prentice-Hll, Englewood Cliffs, N.J., 1959.

Watson, J. D., and F. H. C. Crick. Genetical implications of the structure of deoxyribonucleic acid. *Nature (Lond.), 171:* 964–969, 1953. Reprinted in *Papers on Bacterial Genetics,* E. A. Adelberg (ed.), 2nd ed., pp. 127–131. Little, Brown, Boston, 1966.

REVIEW QUESTIONS

1. In each of the following, DNA preparations from a donor bacterial strain are incubated with living cells of a recipient. What would be the nature of any transformed cells with respect to each of the specific traits being studied?

    A. Donor: capsulated, type III. Recipient: rough, type V.

    B. Donor: streptomycin sensitive. Recipient: streptomycin resistant.

    C. Donor: penicillin resistant. Recipient: pencillin sensitive.

2. Bacteria were infected with phages labeled with S^{35} and then whirled in a food blender. Cells were then separated from the fluid medium. Cells and fluid were next tested for radioactivity. Where will most of it be found, and why?

3. Of what building units is each of the following composed?

    A. A nucleotide.

    B. A nucleoside.

4. How does DNA differ from RNA with respect to:

    A. Its base content.

    B. The chemistry of its sugar component.

    C. Its cellular distribution?

5. A DNA sample taken from an invertebrate reveals the following on analysis of its base composition: thymine 33% of the nitrogen bases present. What can you predict as likely amounts of the other bases?

6. Given below is a 5′ end of a segment of DNA in which the base sequence is represented. What would be the orientation and the base sequence in the complementary strand?

    5′ T-A-T-T-A-G-A-T-C-C-A-T

7. Consider the nucleotide sequence, T-A-G. Assume at the time of DNA replication that "A" in this strand cannot pair with its normal partner but pairs with "C" instead. At the replication following this rare event, what will be the resulting sequence in the two new DNA molecules?

8. Consider four separate nucleoside triphosphates, one with the base adenine, one with cytosine, one with guanine, and one with thymine. Each becomes inserted into a growing DNA chain next to a template through the action

of DNA polymerase. The nucleoside triphosphate with thymine becomes the first residue at the beginning end of the growing chain. The one with adenine becomes the second, that with cytosine the third. The one with guanine becomes the very last residue in the chain. Answer the following:

A. In the new chain formed, which base will be in the nucleotide which has a free —OH end?

B. In the new chain, which base will be in a nucleotide with a free phosphate?

C. Which end of the second nucleotide will be joined to that of the third one which is added to the growing chain?

D. In the template strand, what will be the base in the nucleotide with the free —OH end.

9. Suppose that DNA did not replicate semiconservatively as proposed by Watson and Crick. Assume the double helix stays intact and a double helix composed of two new strands is formed at replication. This would be a conservative rather than a semiconservative method of replication. If the DNA replication were conservative, what would have been the observations in the density gradient experiment of Meselson and Stahl?

Select the correct answer or answers, if any, in each of the following:

10. The following would appear to be true about DNA:

A. It undergoes continual breakdown and resynthesis during cellular metabolism.

B. An exact parallel is found between the amount of DNA and the number of chromosomes.

C. On rare occasions, hydrogen atoms in the molecule may shift position.

D. Its X-ray diffraction patterns indicate no regular arrangement of its parts.

E. Its replication is accompanied by separation of the two chains composing the molecule.

11. In DNA:

A. The sugar components of one strand are linked directly to the sugar components of the complementary strand by hydrogen bonds.

B. The sugar component exists as a 5-membered ring.

C. The purine and pyrimidine bases are attached to the sugar and not directly to the phosphates.

D. Carbon number 3 and carbon number 5 of the sugar are involved in internucleotide linkages.

E. Guanine and cytosine are held together by two hydrogen bonds.

12. In *E. coli:*

A. There is no histone associated with the DNA.

B. The rate of movement of a replication fork in the DNA is slower than in a eukaryote.

C. A nuclear envelope surrounds the genetic material.

D. Several chromosomes are present per nuclear region.

E. The chromosome is circular.

12

INFORMATION TRANSFER

GENE CONTROL OF PROTEIN SYNTHESIS. One important implication of the Watson-Crick model is the concept that information which is coded in the deoxyribonucleic acid (DNA) molecule is somehow transferred to the rest of the cell. A gene localized in the nucleus would exert its influence on the cell by sending out specific instructions. Such an idea raises several questions. Two of the most important are: "What sort of influence does the gene have?" and "How can the gene exert any effect on other parts of the cell since it is confined to the chromosome?" Many lines of evidence from an assortment of investigations were offered to provide a valid answer to the first question. These indicated that the gene probably exerts control on the organism through its influence on protein formation. Genetic research conducted in the 1940's and 1950's with microorganisms (mainly bacteria and *Neurospora*) contributed an abundance of information concerning the effect of gene mutation on the chemistry of the cell. Numerous examples accumulated which showed that a particular gene mutation could cause the block of a chemical step through its effect on a specific enzyme. As more and more such cases were found, it seemed that one important way in which a gene can influence a cell is through its control over an enzyme. This led to the formation of the "one gene, one enzyme" theory which states that each enzyme is under the control of a specific gene. Because enzymes are protein in nature, it seemed possible that the gene might control the formation of proteins in general. In support of this idea, gene mutations were found in humans which alter the structure of proteins. The first such case was that of the effect of the recessive allele for sickle-cell anemia on the hemoglobin molecule. In 1952, Ingram demonstrated that the hemoglobin of victims of sickle-cell anemia differs from the normal, but that this difference is surprisingly small. The normal hemoglobin molecule contains two kinds of protein chains composed of over 140 amino acids each. The only way in which the defective sickle-cell hemoglobin departs from it is in the replacement of just a single amino acid (glutamic acid in normal hemoglobin) by another (valine in sickle-cell hemoglobin) at one position in one of the chains. Comparable alterations at just one position in the hemoglobin molecule also became known for other kinds of hereditary anemias.

Investigations with bacteria and the viruses which attack them (phages) showed that one of the first detectable changes in a bacterium after the entry of the DNA of the virus is the appearance in the cell of proteins which are specific to the virus, *not to* the bacterial host. This indicated that the DNA of the phage exerts it influence on the cell by directing the formation of new kinds of protein. Another argument for a relationship between DNA and protein can be made on the basis of the physical structure of the two types of molecules. Although DNA is a double helix, it is essentially a linear molecule (see Fig. 11-12). Each of the two strands is a polynucleotide composed of nucleotide units which follow one another in a linear order along the length of the molecule. Classical data on linkage and crossing over established that genetic maps are also linear. This concept of one gene following another along the length of the chromosome fits in well with the molecular picture of the gene, a segment of a linear molecule. So the linear genetic map corresponds to the linear sequence of nucleotides composing the DNA. Furthermore, protein mol-

ecules are also essentially linear. Protein molecules are composed of amino acids, approximately 20 different kinds. The backbone structure of all proteins is provided by linkages between the amino acids, in which one amino acid is joined to another in a linear order.

Although the types of amino acids differ in their complexity (Fig. 12-1), they all possess certain features in common. All have a carboxyl group (—COOH) and an amino group (—NH$_2$), except for proline in which the latter is represented by an —NH group which is in a ring. The R portion attached to the carbon atom adjacent to the carboxyl group is what distinguishes one amino acid from the next. The R may represent nothing more than a hydrogen atom, as in glycine, the simplest of all the amino acids. It may, however, be a more complex side grouping, as in tyrosine. When the amino acids unite in the formation of a protein, it is their carboxyl and amino groups which play the most important role. For the separate amino acids are joined by the reaction of the carboxyl group of one with the amino group of another (Fig. 12-2). A molecule of water is evolved, and a bond is formed known as the "peptide linkage." As a result of these linkages, chains form which are composed of a sequence of amino acids in a linear order. A relatively short chain of amino acids is commonly called a *polypeptide*. A protein is a longer or more complex polypeptide chain; it may also be an aggregate of separate polypeptide chains. Proteins can thus become very complicated and diverse and are the most complex of all the chemical substances in the cell. Twenty kinds of amino acids arranged in different ways in chains of varying lengths permit countless varieties. In addition, a complex R grouping on one amino acid may react with the R grouping on another amino acid in the same chain or in an adjacent one. These interactions can produce complex foldings of the separate polypeptide chains. All of this makes possible a diversity of protein molecules which differ not only in their amino acid content but in their three-dimensional architecture as well. But no matter how complex the protein, it is still a linear molecule in the topographical sense. For if we unfold a protein, we see that it is fundamentally a series of amino acids joined together by peptide linkages. We see, therefore, a correspondence from the linear genetic map to the linear DNA molecule to the linear protein molecule.

From such considerations, it seemed quite likely that the specific linear order of the amino acids in a protein might be determined by the specific linear order of the nucleotides in a gene. A gene mutation (a change in a nucleotide) might bring about a change in a specific amino acid at a specific location in a protein molecule. This could explain such cases as the difference between sickle-cell hemoglobin and the normal protein. It could also provide an explanation for the relationship between the gene and the enzyme, as stated in the one gene, one enzyme theory. For a mutant gene might result in an altered enzyme which is unable to catalyze an essential step in a biochemical pathway. If the control of the gene on cell activities *is* actually

FIG. 12-1. The 20 common amino acids. Amino acids possess a carboxyl group, an amino group, and an R group (outlined in the figure). It is the R group which is the distinguishing feature among the amino acids. Its complexity varies from a single H, as in glycine, to the complex groups seen in the aromatic amino acids. Figure 12-1 continues on pages 341 and 342.

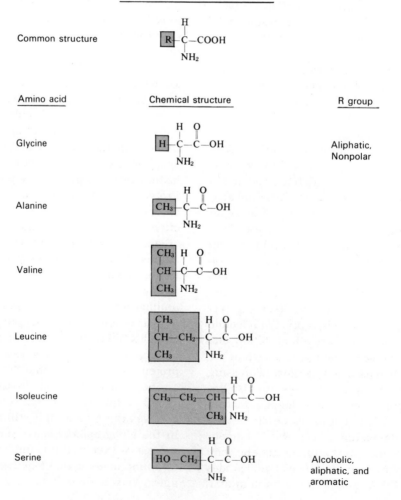

through its control of protein formation, it then becomes necessary to answer the second question posed at the beginning of the chapter, the mechanism whereby information is transferred. Cell biologists had established that protein formation takes place mainly in the cytoplasm, on the submicroscopic bodies, the ribosomes (refer to Fig. 2-2). If information on protein structure is coded in the DNA, then it becomes necessary to explain how this information is transferred to the sites of protein assembly. It was conceivable that the ribosomes themselves contained some information which determined amino acid sequence in a protein. The composition of the ribosomes was shown to be a complex of protein and ribonucleic acid (RNA). As a matter of fact, most of the RNA of the cell is located in the ribosome fraction. It was suspected, therefore,

that RNA molecules might somehow play a significant role in the problem of information transfer. Although RNA was known to resemble DNA in many respects, details of its structure were not clear in the 1950's. It was evident that both kinds of nucleic acid, RNA and DNA, are polynucleotides. Both contain the same purine and pyrimidine bases, with one exception. (Review Figs. 11-5 and 11-7.) The pyrimidine base, thymine, is not found in RNA. In its place, uracil, a different pyrimidine, occurs and this is absent from DNA. Another difference between the two nucleic acids was found in the sugar component. The ribose of RNA and the deoxyribose of DNA are both pentose sugars which are identical except at position 2 in the five-membered ring. The deoxyribose lacks the one oxygen present at the 2 position in ribose.

FIG. 12-1 (continued).

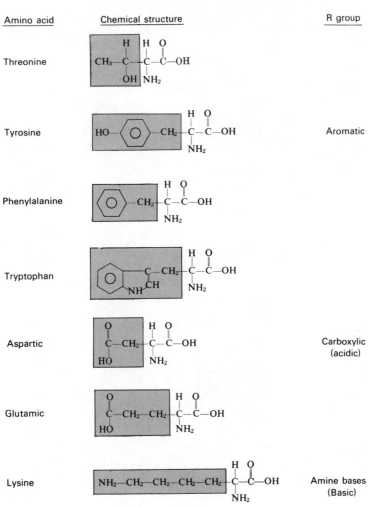

TRANSFER RNA AND SOME OF ITS INTERAC-
TIONS. The close similarities between the two
macromolecules suggested that RNA could
somehow carry information coded in the DNA.
A key to the solution of the problem of information transfer was provided through research
conducted at Harvard University by Hoagland
and his associates. The important discovery was
made that each amino acid, *before* engaging in
protein synthesis, attaches to a special kind of
RNA which is *not* the RNA of the ribosome. This
special RNA makes up a small portion of the
soluble part of the cytoplasm. The ribosomes are
not in the soluble fraction; so this soluble kind

FIG. 12-1 (continued).

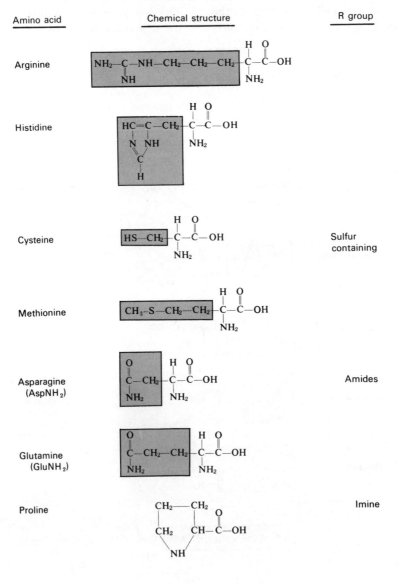

Amino acid	Chemical structure	R group
Arginine		
Histidine		
Cysteine		Sulfur containing
Methionine		
Asparagine (AspNH₂)		Amides
Glutamine (GluNH₂)		
Proline		Imine

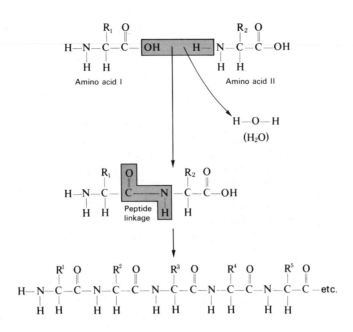

FIG. 12-2. The peptide linkage. The backbone of the protein molecule consists of amino acids joined together through peptide linkages. Reactions may take place between various R groups of the many amino acids so that the molecule may come to assume a complex architecture. Still, when unfolded, the protein is seen to consist of amino acid units in a linear order joined by peptide linkages (*below*).

of RNA was named sRNA to distinguish it from the rRNA of the ribosome. The attachment of amino acids to sRNA was demonstrated not only in the cell but in the test tube as well. For example, if isolated ribosomes are supplied with amino acids and certain other essential substances, they may manufacture polypeptides *in vitro,* even though intact cells are not present. Test-tube studies demonstrated that in these cell-free systems the amino acids bind to the sRNA independently of the ribosomes. Once each amino acid is attached to an sRNA, it is then transported to the ribosome where it may then participate in protein formation. Because the sRNA transfers the amino acid to the ribosome, it was also designated "transfer RNA" or "tRNA." (We will continue to refer to the soluble RNA by the more descriptive term, transfer RNA.)

The tRNA was then shown to be more than a mere vehicle which carries amino acids. Since there are 20 types of amino acids which take part in protein synthesis, there are at least 20 different kinds of tRNA, one for each type of amino acid. The structure of certain tRNA molecules was worked out in the late 1960's. It was shown to be a single-stranded RNA and to have a "cloverleaf" shape (details in Chap. 13). An amino acid in the cell, once it is joined to its specific tRNA, receives an identity. After attachment to its tRNA, the amino acid will then be placed in a specific position in a growing polypeptide chain. Without being identified in a way such as this, the amino acids could not be assembled into any specific sequence. Therefore, the binding of each amino acid to its identifying tRNA is a critical event in protein synthesis, which was shown to entail a series of steps.

Before any free amino acid in the cell may be bound, it must be activated so that it can react with tRNA and also have the energy needed to form peptide linkages with other amino acids (Fig. 12-3A). The energy required to activate an amino acid is supplied by adenosime triphosphate (ATP), the universal energy donor of the cell. The transfer of energy from ATP to an amino acid requires a specific enzyme which can recognize both the amino acid and the ATP. Since there are 20 different kinds of amino acids, there are 20 varieties of the enzyme, one for each specific amino acid. These essential enzymes are called *aminoacyl-tRNA synthetases.*

FIG. 12-3. Activation of an amino acid and its attachment to tRNA. *A.* An amino acid reacts with the high-energy compound ATP. This is brought about by an enzyme specific for the particular amino acid. The enzyme forms a complex with the amino acid in which a high-energy bond is transferred to the amino acid from the ATP. Two phosphates are released. *B.* The activated amino acid-enzyme complex possesses sufficient energy to react with a tRNA and to transfer the amino acid component to the tRNA. Each type of amino acid has a specific type of tRNA with which it reacts. After the reaction, the enzyme and adenosine are released. The tRNA, carrying its specific amino acid, can now move to the ribosome where the amino acid will be placed in a specific site in a growing polypeptide chain. The tRNA is then free to recycle and to react again with its specific kind of amino acid. Until an amino acid reacts with its kind of tRNA, it has no identity an cannot react with the ribosome.

Not only are they necessary for the activation of the amino acids, they are also responsible for joining the energized amino acids to their appropriate tRNAs (Fig. 12-3*B*). Each synthetase thus has two critical roles to perform: (1) making the specific amino acid active by effecting the transfer of energy from an ATP molecule; and (2) joining the activated amino acid to its proper tRNA. Once it is coupled to its tRNA, the amino acid is carried to the ribosome where it may now form peptide linkages with other amino acids which have undergone the same sequence of events. The tRNA will drop away and can recycle to pick up another specific amino acid.

TRANSCRIPTION AND MESSENGER RNA. Although this information supplied many details on steps in protein synthesis, it also raised many additional questions. Among these were, "How

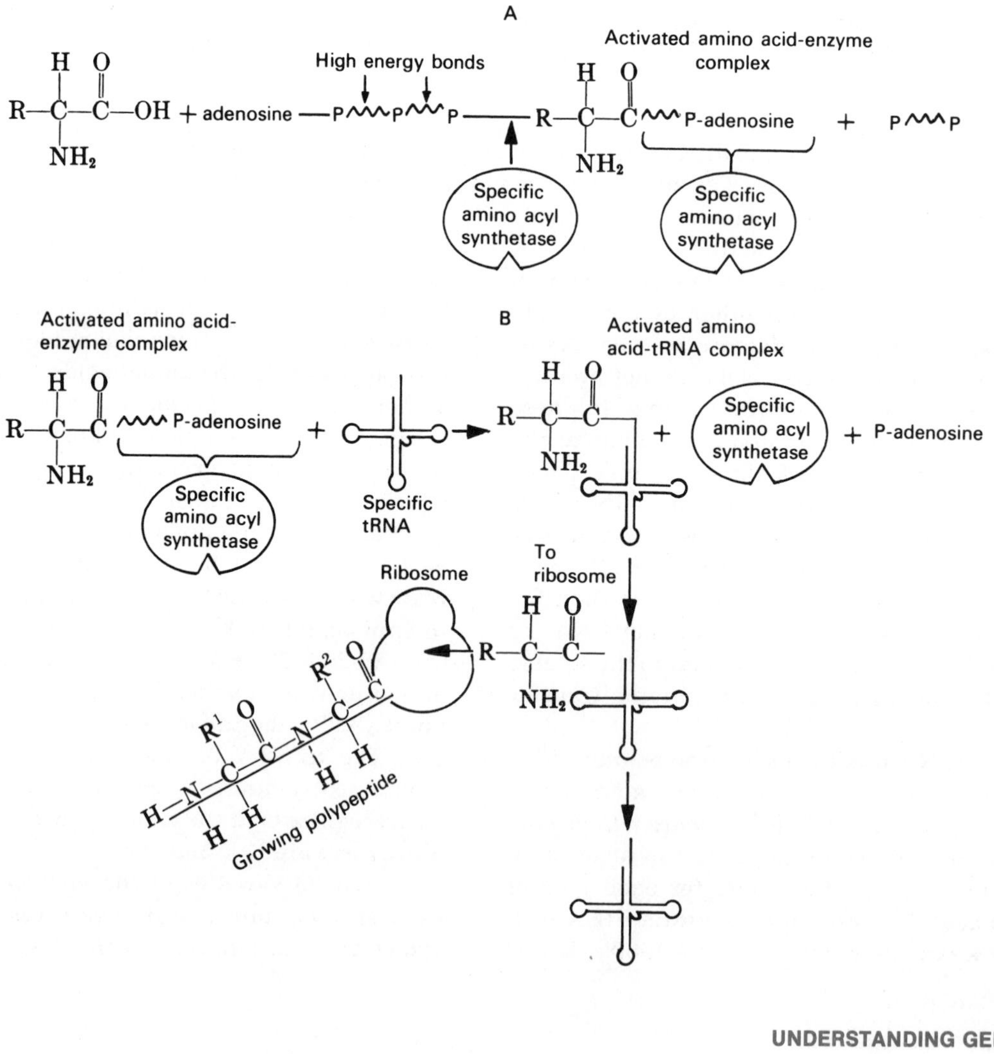

A

B

does a specific tRNA recognize the proper amino acid? Is a code involved? If so, where is it located, and what is it?" Since the ribosomes are the sites where peptide linkages are formed, do they contain a set of instructions which directs the kind of protein to be made?" A series of brilliant experiments performed by different teams of investigators provided some answers to these important questions. One line of research pursued the synthesis of RNA from its building units, the four different kinds of ribonucleotides. This kind of study was advanced by the isolation of a certain enzyme, RNA polymerase, from the bacterium *Escherichia coli*. It was found that this catalyst can stimulate the synthesis *in vitro* of RNA. Whole cells were not necessary for the synthetic RNA to be made, but certain requirements had to be met (Fig. 12-4). Among these was the need for the ribonucleosides to be present as triphosphates, their activated or energized form. If any RNA is to be made in the test tube, it is also essential for some DNA to be present. The reason for this is that the enzyme RNA polymerase cannot link the nucleotides together on its own. The DNA is required as a blueprint or template; RNA formation by the polymerase depends on DNA. The RNA which is synthesized resembles the DNA which is present and which serves as a guide or template in the RNA formation. This latter point is clearly seen in the following way. The DNA used as the template in the cell-free system can be derived from different sources. For each type of organism, the DNA has a characteristic ratio of the paired bases (refer to Table 11-1). For example, in *E. coli* the amount of adenine-thymine pairs (A-T) in proportion to the guanine-cytosine pairs (G-C) is equal to 1 (A + T: G + C = 1). DNA from calf thymus gives another value, 1.3, and so on for any number of organisms. When DNA from *E. coli* was used as the template in the synthesis of RNA, it was found that the base pairs in the new RNA occurred in the same ratio as they did in the template DNA. This means that the ratio of A + U (because uracil replaces thymine in RNA) to G + C = 1. If calf thymus were used, the A + U: G + C ratio would be 1.3.

Additional investigations of this type left no doubt that the DNA can direct the manufacture of RNA which resembles it closely. On the basis of the DNA model, it seemed feasible that the

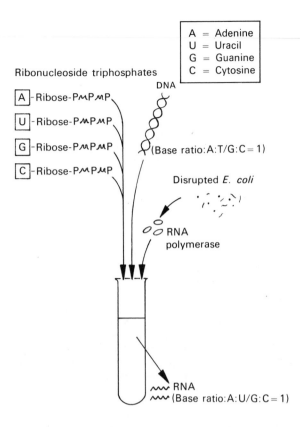

FIG. 12-4. *In vitro* synthesis of RNA. Cell-free synthesis of RNA can be achieved in the presence of several basic ingredients. Among these are: the building units of RNA in their energized triphosphate form, DNA to act as a template in their assembly, and RNA polymerase needed as the catalyst. The RNA formed will possess the same base ratio as the DNA which served as the template.

double helix can unwind and engage in at least two different essential processes. The first, self-replication or the synthesis of more DNA, has already been discussed (Chap. 11). But obviously, if the DNA did no more than direct its own synthesis, there would be no possibility for the origin of anything other than DNA. If evolution is to proceed, genetic material must not only be able to replicate itself precisely (except for rare mutations) but also to transfer stored information. Through both of these processes, replication and information transfer, genetic material can direct cellular activity and make possible the construction of other complex cellular products. At the same time, transmission of the stored information would be ensured from one cell generation to the next.

If DNA in the cell does direct the formation of other kinds of molecules besides DNA, one of these certainly seemed to be RNA. It has now been established that the double helix unwinds, not always to engage in replication, but to permit the formation of RNA which is complementary to it (Fig. 12-5). For example, along one strand the enzyme RNA polymerase, using the DNA as a guide, assembles the RNA precursors (the ribonucleoside triphosphates) into an RNA strand complementary to the one of DNA. The bases along a stretch of DNA (CAATG, for example) pair off with the complementary ones of the ribonucleosides (GUUAC). This process is called *transcription,* the formation of RNA complementary to, and under the direction of, a segment of DNA. The RNA formed after transcription closely resembles the DNA stretch which acted as a template. The sequence of bases in the DNA is reflected in the sequence of the complementary bases in the RNA.

It must be noted here that RNA as well as DNA has a definite orientation, a 5′ and a 3′ end

FIG. 12-5. Self-replication and transcription. Self-replication (*left*) is necessary to ensure continuity of the genetic information from one cell generation to the next. However, this information must be transmitted to the cell if it is to be of any consequence. A critical step in this transfer is transcription (*right*). A portion of the double helix unwinds and acts as a template for the formation of a stretch of RNA. The RNA resembles the DNA closely, but one major difference is the replacement of T by U in the RNA. Also, the polarity of the RNA is opposite that of the DNA template.

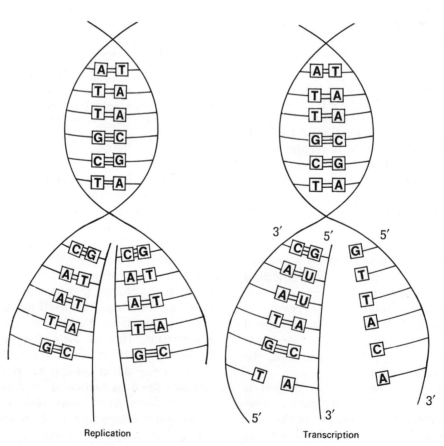

Replication

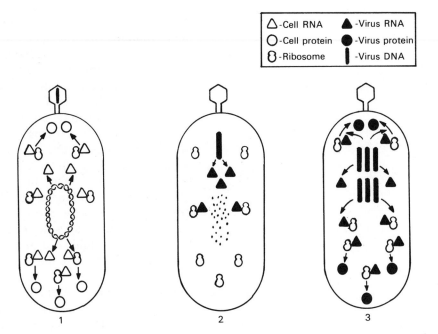

FIG. 12-6. Phage infection of a bacterial cell. *1.* Before the phage injects its DNA into the bacterial cell, the DNA of the cell is engaged in transcription, forming RNA which resembles the DNA of the cell. This RNA associates with the ribosomes of the cell to form necessary cell proteins. Shortly after the viral DNA enters the cell (*2*), a new type of RNA appears. This is a transcript of the viral DNA and it associates with the ribosomes of the cell. The cellular DNA decomposes; RNA and protein synthesis typical of the cell cease to take place. Shortly thereafter (*3*), more viral DNA and viral RNA appear in the cell. Viral-specific protein also is synthesized. No new ribosomes are made; the viral mRNA associates with the ribosomes which were originally present.

due to the ester linkages between the phosphate and the sugar. This was discussed in the case of DNA along with the fact that the assembly of a DNA strand occurs in the 5′ to 3′ direction (Chap. 11). Similarly, the nucleotides in RNA are joined in the 5′ to 3′ direction; the 5′ end of the nucleotide being added is linked to the 3′ end of the nucleotide in the growing chain. Therefore, the polarity of RNA formed in transcription is just the opposite of that of the DNA template on which it was assembled. For example, the DNA sequence reading 3′ *C A A T G T* 5′ would form a transcript with the sequence 5′ *G U U A C A* 3′ (Fig. 12-5).

If the base order in DNA corresponds to coded information, then the RNA contains information which is stored in the DNA. The RNA formed in this way could therefore be involved in the transfer of information. But exactly what kind of role does this RNA play in the cell? Is this the RNA of the ribosomes, or is it still another kind of RNA which is distinct from the rRNA and tRNA?

Answers came from research using *E. coli* and certain phages which attack it (T/2 and T/4). It had been established that once the DNA of the virus enters the bacterium, the host is directed to make protein which is not normally found in the cell. Clearly, the virus can instruct the host cell to make products specific to the virus. Therefore, it must have some mechanism which enables it to convey the necessary information to the host for the synthesis of the viral necessities. Studies of virus-infected bacteria revealed that immediately after entry of the virus DNA, an RNA appears (Fig. 12-6). This RNA was found to resemble the DNA of the virus, *not* the DNA of the host. Another very important fact also emerged from studies of infected bacterial cells. It was demonstrated after viral infection that *no* new ribosomes are formed in the cell; those present in the cell *before* infection are retained. The new RNA formed under the direction of the virus DNA was shown to associate with the ribosomes of the cell. Newly formed protein, protein specific to the virus, can then be detected in association with the ribosomes. No doubt remains that the bacterial ribosomes can somehow synthesize viral protein by assembling amino acids in the proper order. To do this, the ribosomes must require direction in the form of a set of instructions. This is apparently provided by RNA, an RNA different from the ribosomal and the transfer RNAs. This special variety was named "messenger RNA" (mRNA) because it contains a message which directs the ribosomes to assemble amino acids in an order which is specific for a given kind of protein. Without the mRNA, the ribosomes can do nothing. The existence of mRNA, as well as tRNA and rRNA, has now been established in all groups of organisms from bacteria to mammals.

EVIDENCE FOR A TRIPLET CODE. The process of transcription, the formation of coded information in RNA from a DNA template, is essential for the orderly transfer of genetic instructions. In transcription, there is a passage of information from one kind of nucleic acid to another, from DNA to RNA. To be meaningful, this information must now undergo *translation*; the information coded in the RNA must be read and finally transferred to protein. Before the mechanism involved in translation can be understood, it is essential to decipher the code which resides in the DNA. Any such code must depend on the sequence of the bases in the DNA molecule. If this is so, then the bases must somehow correspond to the amino acid units which compose protein. It is obvious that one base cannot correspond to one amino acid, because there are only four bases to designate 20 amino acids. Perhaps a sequence of two bases, a doublet such as CA, indicates an amino acid. However, if four different entities of any kind are involved, such as four different bases, and if two of these are taken at a time, only 16 different permutations or arrangements of the four (4^2) are possible. This is four less than the needed 20. If the code is based on triplets, three bases taken at a time, then 64 (4^3) distinct arrangements can be realized. This is many more than necessary. Many ideas were offered on this point; among these was the suggestion that only 20 triplets designate amino acids, one triplet for one amino acid. The remaining triplets would be meaningless or nonsense. Another idea was that the code is *degen-*

erate; most or all of the triplets would represent amino acids. In a degenerate code, two or more triplets could code for the same amino acid.

Before such coding matters could be settled, it was necessary to obtain evidence for or against a triplet code. It was still conceivable that something which had not been considered was at the basis of the coding. One of the first clear lines of evidence to support the concept of a triplet code came from the work of Crick and his colleagues. Crick favored the idea that the simplest explanation might be the correct one. He suggested that the message coded in the gene is based on triplets and that the gene is read three bases at a time from a definite starting point. Only one of the two DNA strands would undergo transcription in the formation of the messenger. The reading of triplets would continue until the end point of the gene is reached (Fig. 12-7). To test these ideas, Crick employed *E. coli* and T/4 phage. The wild type bacterial virus can infect *E. coli* and cause cell lysis. On an agar plate, the lysis is evident as plaques, clear areas where cells have been destroyed. The size of the plaque is characteristic of wild T/4. Several types of mutations may affect the ability of the virus to produce lysis so that mutant plaques, or even no plaques at all, form on an agar plate.

Crick studied a number of different mutations in a gene (the B gene of the virus) which affects plaque formation. The mutations were induced by mutagens, and the separate mutation sites within the gene were mapped from recombination data (crossover frequencies). In the studies, acridine dyes were employed as mutagens. These chemicals, by actually inserting into the DNA, can cause deletions and duplications of one or a few base pairs (Chap. 15). Most other mutagens typically produce their effect by causing base pair substitutions in the DNA. After exposing the T/4 virus to acridines, it was noted that the different mutations within the gene often resulted in complete loss of gene function. The sites of various acridine-induced mutations were mapped. This made it possible to infect bacteria with viruses mutant at two different sites. From such a mixed infection, offspring viruses could be obtained. Among these progeny viruses, recombinants could be formed; some would be double mutant, others wild (Fig. 12-8). The doubly mutant ones could then be studied for their ability to produce plaques on *E. coli*. It was found that certain double mutants were nonfunctional, whereas others retained some function or behaved as if they were wild. It was possible to classify the single mutants into two categories: "+" and "−." A combination of a "+" and a "−" could suppress the mutant effect, whereas two "+" or two "−" mutations in one virus DNA would be completely mutant. Such results can be easily explained according to Crick's hypothesis. As Figure 12-7 shows, the gene would be read from a fixed starting point, three bases at a time, each triplet designating an amino acid. In the absence of mutant sites within the gene, a wild-type protein is formed. Allowing "+" mutations to stand for added bases, it can be seen that the addition of just one nucleotide will upset the reading of the gene (Fig. 12-9A). After the point of insertion, the reading is out of phase, because there is nothing within the gene to set the triplets apart. The

FIG. 12-7. Reading of a gene. According to Crick's concept, the gene would have a starting point and an end point. Reading would proceed from the beginning to the end of the gene three bases at a time. One of the DNA strands of the gene would be transcribed into the messenger RNA whose bases would be complementary to those in the strand which is read.

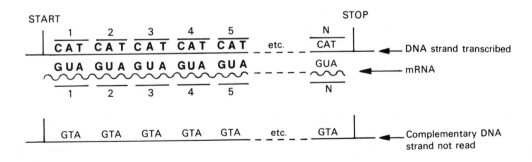

reading of triplets from the point of insertion on will therefore be changed. Many of the triplets will now designate amino acids which should not be placed in particular positions in the protein chain. The result is an altered protein and a mutant effect. The same sort of reasoning applies to a "−" or deletion mutation, where the reading is again out of phase from the point of alteration where the base is deleted (Fig. 12-9B). However, a combination of a "+" and a "−" mutation can bring the reading back into phase again (Fig. 12-9C). Therefore, if the stretch between the "+" and the "−" is not too long, wild type function might be restored. A combination of three "+'s" or three "−'s" would also result in a return of function if enough of the correct message is in phase. However, two "+'s" or two "−'s" together would not restore correct reading to the frame, and gene function would be lost (Fig. 12-9D).

Various combinations of mutants were made, and results were all compatible with Crick's hypothesis. They also demonstrated that at least two different classes of gene mutation can be recognized. There are the familiar kinds of mutations in which a single base pair becomes changed, rather than removed or duplicated (Fig. 12-9E). These can be designated *missense mutations*. Although they may cause a mutant effect, the gene often retains some of its function. Except for one triplet, all the rest of the gene reads correctly. Therefore, only one amino acid will be changed in the protein dictated by this mutant gene. Missense mutation can also mutate back to wild on rare occasions. In contrast to these are the *frame shift mutations*, illustrated by the "+" and "−" alterations. In this kind of mutation, a base is added or deleted, with the consequence that the reading of the gene may be shifted over a long stretch of nucleotides. Any protein associated with this type

FIG. 12-8. Intragenic recombination. After exposure to a mutagen such as an acridine, mutations may arise in the particular gene under study. (1.) The location of each separate mutation can be mapped. A bacterium may be infected with two types of virus (2). A process akin to crossing over can take place between the virus particles (3). Among the progeny viruses, wild types and double mutants can be recognized (4).

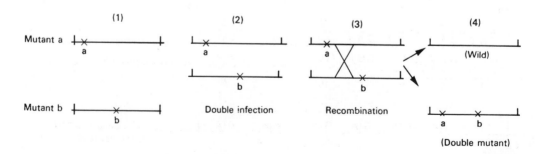

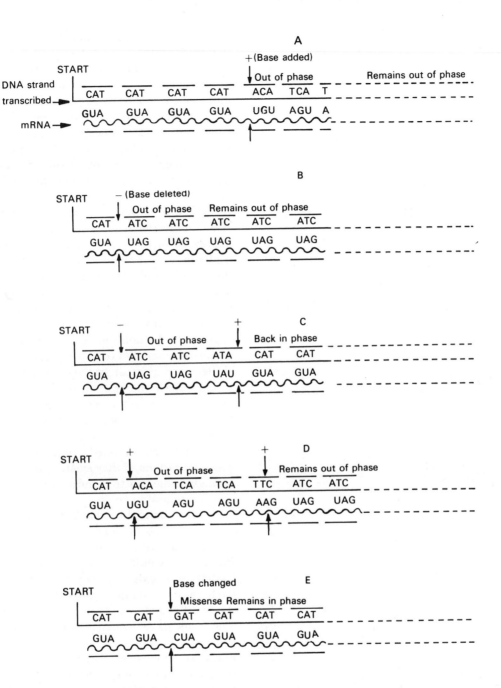

FIG. 12-9. Types of mutations and their effects on the reading of the gene. The addition or deletion of a base (*A* and *B*) will cause the reading frame to go out of phase, because the bases are read in triplets starting from a fixed point within the gene. Note that the mRNA which is transcribed from the DNA will reflect the change and will also have stretches which are out of phase. A combination of a "+" and a "−" mutation (*C*) can bring the reading frame back into phase. If the stretch which is out of phase is not too extensive, the gene may show some function. A combination of two "+" mutations (*D*) cannot bring the frame back into phase. This is also true of a combination of two "−" mutations. Most spontaneous mutations are probably the result of a base substitution (*E*), rather than the addition or deletion of a nucleotide. Except for one triplet which codes for one "incorrect" amino acid, all the other bases are in phase and would be read correctly. The gene would retain some function. These missense mutations can revert spontaneously to normal if a later mutation causes the original base to be inserted for the substituted one. A "+" or a "−" mutation, on the other hand (*A-D*), would not be expected to revert spontaneously with any ease, because a whole nucleotide must be either eliminated or added, rather than just a base substituted.

of genetic alteration would thus possess a long stretch of amino acids different from the normal and would very likely be nonfunctional. Spontaneous revertants of the "+" and "−" frame shift mutations would not be expected, and this seems to be the case. All the experimental evidence supported Crick's concept of a gene which is read in groups of three nucleotides from a fixed starting point. Normally, no sort of punctuation would occur within the gene to halt the reading in any way.

The work also gave good evidence that most of the nucleotide triplets designate amino acids. If only 20 of them represent amino acids, one triplet for each kind of amino acid, then most of them (44) would be nonsense; they would not denote any amino acid at all. If this were so, in the cases of combinations between the "+" and "−" mutations, most recombinants would be expected to be nonfunctional instead of wild or partly functional. For if the reading frame is shifted and *if* most triplets are nonsense, we would expect the shift to produce a high incidence of nonsense triplets which would bring the reading to an end. Since most "+" and "−" combinations (as well as three "+'s" and three "−'s") are usually partly functional, the nonsense mutations, if they exist, must be few. This strongly indicates that most of the triplets correspond to amino acids.

According to this model, the gene is pictured as "commaless;" it contains no sort of signals or punctuation within it to indicate a stop or start. However, it would seem that some kind of punctuation must exist *between* separate genes. For each gene should be a complete unit read from its own fixed starting point. The work with the T/4 virus gave evidence for the existence of some sort of intergenic punctuation as well as nonsense triplets. As noted above (Fig. 12-9E), missense mutations, which may arise spontaneously or which can be induced by a variety of mutagens, do not upset the reading of most of the gene, because only one triplet is changed. However, a suspected missense mutation arose which completely eliminated the function of the gene. It was reasoned that perhaps the mutant change had resulted in the formation of a triplet which did not represent an amino acid but which brought the reading of the mRNA to a premature halt. Consequently, an incomplete protein would be formed. If this idea is correct, then it should be possible to eliminate the responsible triplet by combining it with a "+" and a "−" frame shift mutation (Fig. 12-10) on either side of it. This combination of three mutations actually resulted in a partial restoration of gene function. A frame shift mutation before the nonsense triplet and another one after it shifts the reading frame, eliminates the nonsense triplet, and brings the reading back into phase again.

Additional evidence for the existence of some DNA-coded signals or punctuation was provided by additional genetic analysis in the T/4 virus. A genetic alteration was found which affected not only the "B" gene which was being studied in detail but also gene "A" adjacent to it. The particular mutant change proved to be a deletion in which the end of gene "A" was removed as well as the beginning of gene "B" (Fig. 12-11A). As a result of this change, gene "A" could no longer function but gene "B" retained some of its activity. The important point was that *if* punctuation does normally exist between two genes, it would be removed by the deletion covering part of the "A" and "B" genes. To test this, the deletion mutant was combined with an acridine mutation produced in gene "A." Normally, such a frame shift mutation in gene "A" will not affect gene "B," because the two are separate units. However (Fig. 12-11B), when a frame shift mutation is present in "A" along with the deletion, activity of gene "B" is completely eliminated. This is readily explained by assuming that the deletion has removed some code which normally signals the end of gene "A" and the start of gene "B." After its removal, genes "A" and "B" become joined. Instead of being read as

FIG. 12-10. Nonsense triplet. *A*. Assume the normal nucleotide sequence is as shown and results in the formation of a completely functional mRNA which leads to the formation of a normal polypeptide chain or protein. *B*. If a mutation occurs which alters just one base in a triplet, it may cause the formation of a nonsense triplet (shown here as ACT) within the gene. Since the triplet codes for no amino acid, the mRNA which forms carries a nonsense triplet (UGA) complementary to the one in the DNA. This will cause premature termination of the protein chain. The gene thus will not be able to perform its function. *C*. A "+" and a "−" mutation on either side of the nonsense mutation causes a shift in the reading frame. Although the gene is out of phase in the one area, it nevertheless eliminates the nonsense codon. A protein may form which has a few incorrect amino acids but which still has some function. We see here that the presence of three mutations can lead to a protein with some function, whereas the presence of just the nonsense mutation causes formation of an incomplete, nonfunctional protein segment.

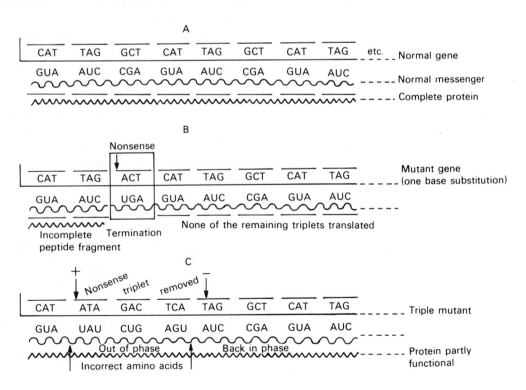

separate genes, the two are read as one starting from the beginning of gene "A." From the point of the added frame shift mutation in gene "A," the rest of the message in both "A" and "B" becomes distorted, and no functional protein can be formed.

CODONS AND THEIR ASSIGNMENTS. As supporting evidence for the triplet nature of the code was assembled, important investigations were conducted to elucidate the very meaning of the 64 triplets. Which triplets stand for which amino acids, and which of them are nonsense? The pioneer studies in deciphering the genetic code were carried out by Nirenberg and Matthei, who reported a breakthrough in 1961. The logic behind the work was to provide the protein-making machinery of the cell with a *known* genetic message. Any protein made would then reflect the known information which had been supplied by the investigator. Nirenberg and Matthei employed cell-free test-tube systems containing components from ruptured bacteria plus mixtures of amino acids. If any polypeptide is to be made *in vitro,* certain essentials are required. Besides the amino acids, all of the parts of the system needed to assemble them into protein must be present (Fig. 12-12). This includes an energy source, ribosomes, the various tRNAs along with the corresponding synthetases, and a coded mRNA. The ability to manufacture a synthetic messenger RNA with a known sequence of nucleotides provided the critical key to deciphering the genetic code. We noted that the transcription of DNA into mRNA by the enzyme RNA polymerase requires a DNA template. Fortunately, in the mid 1950's Ochoa and Grunberg-Mangano had isolated from *E. coli* still another enzyme, polynucleotide phosphorylase, which is capable of synthesizing RNA. Although it has another role in the intact cell, it can be used *in vitro* to assemble RNA nucleotides in a *random* fashion without following a DNA template. This meant that one could provide the enzyme with certain specific ribonucleotide units and obtain an RNA of known composition. Using the phosphorylase, Nirenberg and Matthei prepared a synthetic messenger which consisted only of repeating units containing uracil (polyuridylic acid). Such RNA could then be used as

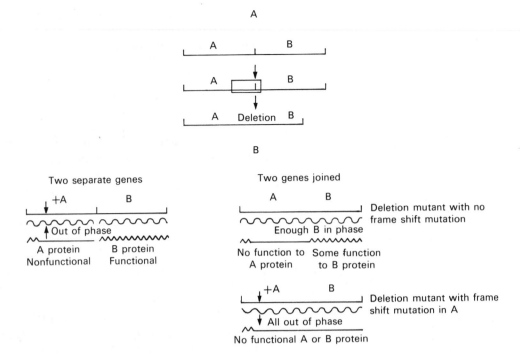

FIG. 12-11. Elimination of punctuation between two genes. *A.* Normally "A" and "B" behave as separate functional units, each governing the formation of a specific protein. A deletion arose removing the end of gene "A" and the beginning of gene "B." Gene "B" still retained some function, although "A" function was eliminated. *B.* Normally (*left*) a frame shift mutation ("+" or "−") in gene "A" will render the "A" protein nonfunctional but will have no effect on "B" protein. This is to be expected if there is some signal between genes which indicates the end of one gene and the beginning of the next. If this signal is removed (*right*), the two genes will become joined and their triplets will be read as if they were all part of one gene. Therefore, a frame shift mutation in a part of the "A" gene can upset the reading in the "B" gene and destroy any residual function which may have existed.

a messenger and supplied to the protein-making system described above (Fig 12-12).

In the experiments, separate test-tube mixtures of the essential factors were followed (Fig. 12-13*A*). In each mixture, one of the 20 different kinds of amino acids was labeled with C^{14}, the remaining 19 being nonradioactive. This made it easy to tell which specific amino acid was being incorporated into any insoluble polypeptide which formed in response to the messenger RNA which had been supplied. For if a polypeptide formed and if it was labeled, then it must contain the particular labeled amino acid in that mixture.

When the polyuridylic acid (UUU) was used as mRNA in the cell-free system, a polypeptide was formed. Chemical analysis showed it to be polyphenylalanine, a polypeptide composed of repeating units of the amino acid, phenylalanine (Fig. 12-13*B*). The messenger contained only repeating units of nucleotides with uracil. Therefore, if the code *is* based on triplets, the RNA triplet UUU must indicate phenylalanine during translation. Since the RNA triplet is normally a transcription of DNA, the DNA triplet AAA in a cell must correspond to phenylalanine (Fig. 12-14). A sequence of three nucleotides designating an amino acid or other information was termed a *codon*. Thus, there are DNA codons and RNA codons, such as AAA and UUU, respectively, for phenylalanine.

In the same way, polymers composed of other single kinds of ribonucleotides were prepared and supplied as messengers. Polycytidylic acid (repeating nucleotides containing cytosine) was found to induce the formation of a polypeptide composed of repeating units of the amino acid proline. Thus, the RNA codon CCC and the complementary DNA codon GGG represent proline. Polymers were then made containing two and three different bases. For example, suppose the enzyme polynucleotide phosphorylase is supplied with equal amounts of two kinds of nucleotides, one-half of them containing cytosine and the other one-half uracil. At random, the enzyme will assemble an RNA containing eight different triplets in equal amounts: UUU, UUC, CUU, UCU, UCC, CCU, CUC, and CCC. When used as messenger, this polyribonucleotide directed the incorporation of phenylalanine and proline, because the codons UUU and CCC are present. But in addition, the polypeptide which forms also contains leucine and serine. This means that these two amino acids must be coded by RNA codons which contain both uracil and cytosine. However, the specific triplets for leucine and serine (or for any amino acid coded by more than one base) cannot be determined by this procedure. This is so because the exact order (UUC, UCU, or CUU, etc.) cannot be established. Later techniques devised by Nirenberg permitted the precise assignments of codons and clarified other features of the genetic code.

In these procedures, simple trinucleotides are synthesized which are composed of just three ribonucleotides in a known sequence. It was found that each simple trinucleotide was able to act as mRNA by binding to both a ribosome and to tRNA carrying its specific amino acid. For example (Fig. 12-15), the trinucleotide CCC will form a complex with a ribosome and a transfer RNA charged with the amino acid proline. Because the small trinucleotide is obviously too short to act as a regular messenger, protein synthesis does not take place. However, ribosome-transfer RNA-trinucleotide complexes can be isolated on filters along with free ribosomes. As in the other method, any one mixture contains all 20 kinds of amino acids with only one kind labeled. The ribosome fraction of the system is isolated on the filters. This also contains the ribosomes which are complexed with the charged tRNA and the trinucleotides. By determining whether the fraction is labeled, a specific trinucleotide sequence can be directely related to a specific amino acid.

FEATURES OF THE GENETIC CODE. The procedure just described gives conclusive proof of the concept of a triplet code. But in addition, it has also shown that most of the triplets represent

FIG. 12-12. Cell-free system for polypeptide synthesis. Simple polypeptides can be made *in vitro* without the presence of intact cells. However, several basic cellular ingredients are required as shown here.

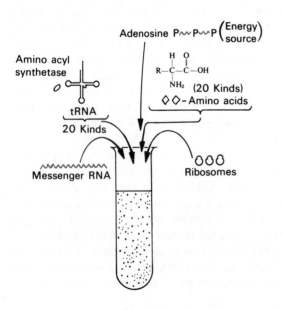

amino acids. As we will discuss later in more depth, only 3 of the 64 codons do not correspond to amino acids. This means that the code is "degenerate"; more than one triplet can designate a specific amino acid. This is the case for almost all of the amino acids (Table 12-1). At first, it might seem surprising that the code is degenerate, but this in no way means that the code is haphazard. For any one amino acid is designated only by specific codons. In addition, the degeneracy of the code provides an actual advantage to living things. As can be seen by reference to Table 12-1, it is usually the third letter of the codon which differs when two or more triplets signify the same amino acid. Consider the case of arginine, which is specified by six RNA codons. Of these, the first two letters are identical in four of them. In the remaining two codons, the first two letters are the same. An advantage becomes apparent when we consider the codons more closely. Suppose the CGU sequence which designates arginine is altered by mutation to CGA (Fig. 12-16). This codon also designates arginine. Actually, in four of the codons, any base change at the third position will still code for arginine. A change in the CGA triplet at the first position to "A" produces AGA, which again corresponds to arginine. We see

FIG. 12-13. *In vitro* synthesis of polypeptide using known mRNA. *A.* To determine whether a specific codon is related to an amino acid, a mRNA of known composition is supplied to cell-free systems. Twenty separate systems must be followed. In each, one of the 20 amino acids is labeled. If the label is recovered in the resulting polypeptide manufactured by the system, it is then known that the messenger which was used contained a codon for that amino acid. *B.* When a synthetic mRNA composed of repeating units of uridylic acid (UUU) is supplied to each of 20 separate systems, only the system containing labeled phenylalanine gives rise to a labeled polyphenylalanine chain. Since none of the other 10 yielded a labeled polypeptide, it could be concluded that UUU is the triplet which codes for phenylalanine.

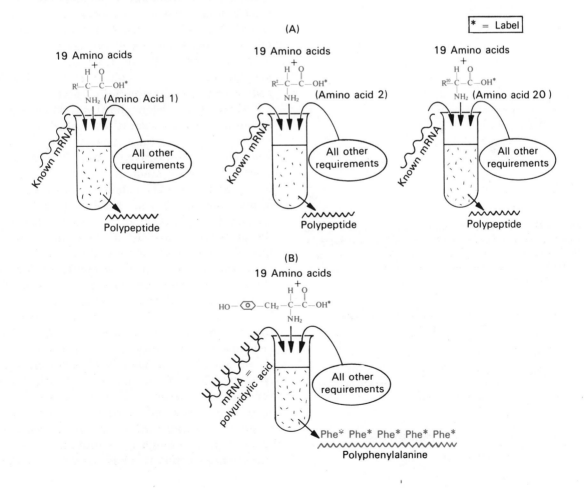

that the degeneracy of the code actually acts as a buffer against the effects of mutation. If only one triplet coded for one amino acid and the remaining 44 were nonsense, then a mutation would always result in an altered protein, either by causing an amino acid substitution or by bringing about premature termination of the protein chain. Degeneracy on the other hand results in a number of "silent mutations," codon changes which give rise to different codons designating the *same* amino acid. Since no amino acid substitution would follow a silent mutation, the protein product would be normal and the very fact of the mutation would remain unnoticed. Examination of the code also shows that amino acids with similar chemical properties have similar codons which differ by only one letter. This means that if the AGA codon for arginine were changed by mutation to AAA, lysine would be substituted. Inasmuch as both are basic amino acids, the mutation might not cause too great a change in the properties of the protein and could also possibly remain undetected. Natural selection has undoubtedly operated during evolution of the genetic code to reduce the effects of gene mutation as much as possible through the establishment of a genetic code with the greatest buffering power.

Three of the codons, however, do not stand for amino acids. These are the so-called nonsense codons, UAA, UAG (also referred to as "ochre" and "amber," respectively), and UGA. As suggested by Crick's studies with the T/4 virus, these codons have been shown to provide a type of punctuation. They are better termed "chain-terminating codons" than "nonsense," because the presence of any one of them on a mRNA indicates the end of the message and hence the end of translation by the ribosome. The existence of these RNA chain-terminating codons also tells us that the complementary DNA codons must be ATT, ATC, and ACT (Fig. 12-17A). The identification of the chain-terminating codons has been made possible through genetic studies with *E. coli* and phage, as well as through work with synthetic RNAs of known nucleotide sequence. For example, presence of the triplet UGA will bring about termination of the reading of a synthetic messenger RNA by a ribosome. If it arises *in vivo* through gene mutation, we say that a nonsense mutation has arisen; translation

FIG. 12-14. RNA and DNA codons. If the mRNA codon UUU specifies phenylalanine, then the DNA triplet AAA must designate UUU, because the mRNA is a transcript of DNA. Therefore, each amino acid has a DNA as well as an RNA triplet or codon which is specific for it.

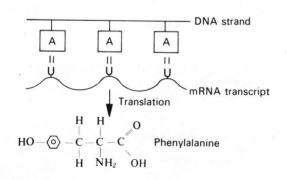

TABLE 12-1 List of RNA and DNA codons and the specific amino acids which they designate

RNA	DNA		RNA	DNA		RNA	DNA		RNA	DNA	
UUU	AAA } phe		UCU	AGA ⎫		UAU	ATA } tyr		UGU	ACA } cys	
UUC	AAG		UCC	AGG ⎮ ser		UAC	ATG		UGC	ACG	
UUA	AAT } leu		UCA	AGT ⎮		UAA	ATT } "nonsense" or chain terminating		UGA	ACT } "nonsense"	
UUG	AAC		UCG	AGC ⎭		UAG	ATC		UGG	ACC } try	
CUU	GAA ⎫		CCU	GGA ⎫		CAU	GTA } his		CGU	GCA ⎫	
CUC	GAG ⎮ leu		CCC	GGG ⎮ pro		CAC	GTG		CGC	GCG ⎮ arg	
CUA	GAT ⎮		CCA	GGT ⎮		CAA	GTT } gln		CGA	GCT ⎮	
CUG	GAC ⎭		CCG	GGC ⎭		CAG	GTC		CGG	GCC ⎭	
AUU	TAA ⎫		ACU	TGA ⎫		AAU	TTA } asn		AGU	TCA } ser	
AUC	TAG } ileu		ACC	TGG ⎮ thr		AAC	TTG		AGC	TCG	
AUA	TAT ⎭		ACA	TGT ⎮		AAA	TTT } lys		AGA	TCT } arg	
AUG	TAC met		ACG	TGC ⎭		AAG	TTC		AGG	TCC	
GUU	CAA ⎫		GCU	CGA ⎫		GAU	CTA } asp		GGU	CCA ⎫	
GUC	CAG ⎮ val		GCC	CGG ⎮ ala		GAC	CTG		GGC	CCG ⎮ gly	
GUA	CAT ⎮		GCA	CGT ⎮		GAA	CTT } glu		GGA	CCT ⎮	
GUG	CAC ⎭		GCG	CGC ⎭		GAG	CTC		GGG	CCC ⎭	

FIG. 12-15. System for precise codon assignments. A short known sequence of three bases, a trinucleotide, can be synthesized. When added to the in vitro system, it attaches to a ribosome and also binds to a specific tRNA carrying its specific amino acid. In the procedure, only one amino acid of 20 is labeled in any one test-tube system. In this figure, it is proline, and the synthetic trinucleotide is CCC. The complex, composed of the ribosome, trinucleotide, and charged tRNA, can be isolated on filters. In this case, the ribosome fraction will be labeled because the proline is labeled. The presence of a label in the ribosome fraction of any one test tube indicates the presence of a labeled complex. This, in turn, indicates that the particular labeled amino acid is coded by the known trinucleotide which was supplied. Such procedures have shown that more than one kind of trinucleotide can code for the same amino acid, meaning that the code is degenerate.

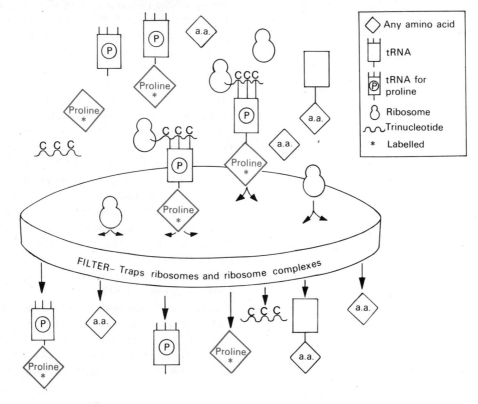

comes to a premature halt with the production of an incomplete protein or polypeptide fragment (Fig. 12-17B).

We see from all of these studies that on the molecular level, mutations may represent very different alterations of the codons. A *missense mutation* (see Fig. 12-9E), we have seen, is a change from one codon which is specific for a certain amino acid to another codon specific for a different amino acid. An altered protein results, and the magnitude of the effect will depend on which amino acid has substituted for the original. A *nonsense mutation,* on the other hand, occurs when a codon specific for an amino acid is changed to one of the chain-terminating codons and thus brings translation to an end. An incomplete protein or fragment forms. A *silent mutation* (Fig. 12-16) represents a change from a codon specific for an amino acid to another codon specific for the same amino acid. No effect is produced on the protein product because no amino acid substitution has occurred. Figure 12-18 summarizes these classes of mutation. A deleted base or an added one can bring about a *frame shift mutation* in which a portion of the coded information in the gene will be out of phase (Fig. 12-9). These can result in very defective protein products with stretches of incorrect amino acids. A frame shift mutation may even cause the origin of a chain-terminating codon with the consequent formation of an incomplete protein.

CHOICE OF DNA STRAND READ IN TRANSCRIPTION. One very important decision which must be made is the choice of which of the two DNA strands of a gene is to be read at the time of transcription. A moment's thought tells us that the RNA polymerase probably does not read both DNA strands when it forms mRNA. One reason why we would not expect this is due to the fact that the genetic code is degenerate. If both strands were read, then two different kinds of protein would form after the translation of the two different kinds of messenger. This does not seem to be the case, for only one kind of protein or polypeptide chain is associated with a specific gene, an observation which led to the formulation of the "one gene-one enzyme" theory. We will encounter more detail on this subject in Chapter 17. Work with certain microorga-

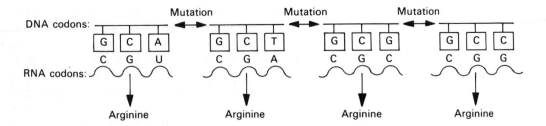

FIG. 12-16. Silent mutations. Shown here are some of the possible silent mutations which involve codons for the amino acid arginine. If any base substitution occurs at the third position in any one of these four DNA codons which are shown, the result is still an RNA codon which designates arginine. Therefore, certain mutations may take place and remain undetected, because they bring about no amino acid substitutions. The buffering effect of a degenerate code is apparent; if a change in a codon always caused a change to a codon standing for another amino acid or for termination of the protein chain, every mutation would have a detectable effect, most of them harmful.

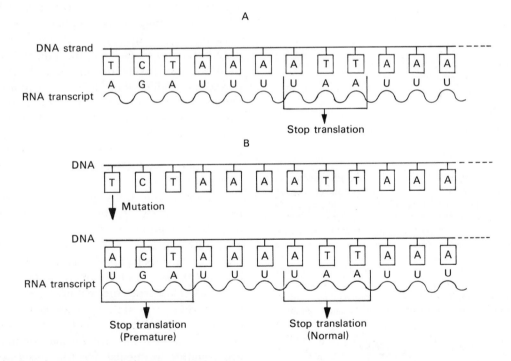

FIG. 12-17. Chain-terminating codons. *A.* The RNA codons UAA, UAG, and UGA establish the end of translation of a messenger RNA. The complementary codons on the DNA would be ATT, ATC, and ACT. When the ribosome reads the messenger, the chain-terminating codons indicate the end of a polypeptide and the chain is terminated. *B.* If a nonsense mutation arises in a gene (shown here as ACT), a chain-terminating codon will be transcribed onto the mRNA in a position ahead of normal one. A premature stop signal will thus be carried on the messenger and the ribosome will not translate the message past that point. Thus, that particular polypeptide chain will be incomplete.

nisms has clearly shown that only one strand of the DNA is read. For example, in a particular bacterial virus, the two strands which make up the "chromosome" can be quite easily separated. Only one of the two strands forms hybrids with the mRNA which was formed from the corresponding *double stranded* DNA inside an infected bacterium (see Fig. 18-13). Similar evidence from other microorganisms has also demonstrated that within a gene only one of the two strands of the double helical DNA of the gene undergoes transcription.

It is not known exactly how the decision is reached as to which strand shall be read, but the enzyme RNA polymerase is certainly involved. In a living cell, the RNA polymerase must be able to recognize the beginning of a gene. If the enzyme could start transcription at any point in a gene, it would copy only portions of the DNA into messenger RNA. A definite starting region for the attachment of the enzyme is now known to exist.

Such a region, known as a *promoter*, contains start signals which can be recognized by the enzyme RNA polymerase. However, the enzyme does not begin transcription from the site which it first recognizes. Instead, it attaches loosely to the DNA and travels further along the promoter to another region where it becomes more firmly bound. Transcription begins a few nucleotides past this point. It is the DNA itself which contains information coded in certain base-pair sequences in the promoter region which permits recognition by the RNA polymerase for the start of transcription.

Actually the RNA polymerase has been found to be far from simple. While DNA polymerase is composed of a single polypeptide chain, RNA polymerase is composed of five different kinds of polypeptide chains. One of these is called the *sigma chain* or *sigma factor* and is easily removed from the rest of the enzyme, the core. Actually, the sigma portion is not involved in catalyzing the formation of RNA. This can be accomplished by the core without sigma. However, sigma is needed for the recognition of signals in the promoter region (Fig. 12-19A). Without sigma, as seen in test tube systems, incorrect signals are read, and transcription may begin at random anywhere along the DNA. But before the sigma of the normal RNA polymerase

FIG. 12-18. Three types of mutations. If the RNA codon for arginine, CGA, is changed by mutation to AGA, the altered codon still codes for arginine. No change will occur in the polypeptide governed by the gene carrying this mutation. However, if CGA is changed to UGA, the protein will be incomplete. If the codon AGA for arginine is changed to AAA, a missense mutation will have occurred, because another amino acid will be placed in the polypeptide instead of the correct one. The protein may or may not be greatly altered, depending on the particular substitution. In this case, the substituted lysine would probably not cause a pronounced effect, because its properties are similar to arginine.

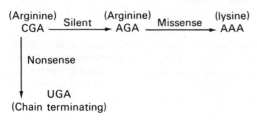

UNDERSTANDING GENETICS

can even bind to the promoter, two other factors may be required in the transcription of certain genes. One is a simple molecule, cyclic adenosine monophosphate (cyclic AMP), which is formed from cellular adenosine triphosphate (ATP). This small molecule has been designated the "second messenger," because of its interactions with hormones in their control of cell activity. But cyclic AMP has also been shown to stimulate transcription of certain genes in bacteria. However, it does not do this directly. It must form a complex with a specific protein which also occurs in the cell, CAP (catabolite gene activator protein) (Fig. 12-19B). When complexed with cyclic AMP (cAMP), CAP binds to the DNA. Only then can the RNA polymerase attach to the promoter and start transcription.

It has been demonstrated in the bacterium *E. coli* that CAP binds to a region in the promoter which is distinct from the one which interacts with RNA polymerase. Thus, the promoter contains a CAP site and an RNA polymerase interaction site and is now known to be much more complex than suspected at first. The length of the promoter for the lac gene (Chap. 19) has been shown to be 80 base pairs long, half of which are involved in the CAP site. The exact sequence of base pairs is known for the promoter in this case as well as in a few others. However, the base sequences recognized by the CAP-cAMP complex are not known.

Once transcription begins, the RNA polymerase travels along the DNA template to form RNA which is complementary to the DNA seg-

FIG. 12-19. Factors involved in transcription. *A.* Transcription of a gene depends on a definite starting point in the gene, the promoter. This region is recognized by the sigma factor, a portion of the enzyme RNA polymerase. One strand of the double helix is selected for reading. As mRNA formation proceeds, the sigma factor may drop away to recycle, because it is the remaining portion of the enzyme which performs the catalytic action. *B.* At least two other factors are known which are required before the sigma factor can bind to the promoter. These are cyclic AMP and CAP. The two form a complex which must first bind to the proper DNA strand before the events shown above in *A* can take place. *C.* A protein factor, rho, is somehow able to recognize the end of certain genes. The combination of rho with the DNA would prevent the enzyme RNA polymerase from continuing transcription in an indiscriminate fashion.

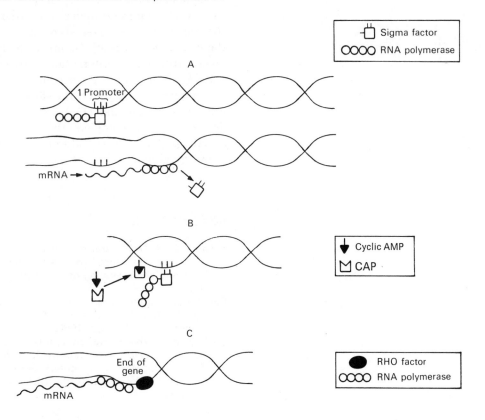

ment. This continues until the end of the gene is reached. Stop signals must be built into the DNA (recall the work with T/4 phage in which the punctuation between the "A" and "B" genes was removed). One type of signal which brings a halt to transcription appears to be a stretch of DNA nucleotides all containing the base, adenine. When the RNA polymerase encounters this stretch on the DNA strand being read, it produces a complementary stretch of RNA containing the base uracil. This is attached to the end of the mRNA transcript. Transcription then terminates. Also involved in the termination of transcription is a protein factor. It has been designated "rho" and is required if transcription is to take place in the normal way (Fig. 12-19C). Without it, the RNA polymerase continues to transcribe one gene after another along the DNA, instead of stopping when one or a few genes have been read. It is not yet known if there are different kinds of rho factors or whether they must actually attach to the DNA in order to perform their function in termination. However, it is believed that rho can recognize certain base sequences which indicate where transcription is to come to a halt for some genes.

We see that the transfer of information from the gene to its messenger RNA transcript entails the interaction of many factors at the molecular level. Still other interactions involving the mRNA, tRNA, and the ribosome in translation are required to achieve the final transfer of information from the DNA codons within the double helix to complete the formation of a specific portein. These will form the basis of the following chapter.

REFERENCES

Brenner, S., A. O. W. Stretton, and S. Kaplan. Genetic code: the "nonsense" triplets for chain termination and their suppression. *Nature, 206:* 994, 1965.

Cold Spring Harbor Symposium on Quantitative Biology. *The Genetic Code,* vol. 31. Cold Spring Harbor Laboratory, New York, 1966.

Cold Spring Harbor Symposium on Quantitative Biology. *Transcription of Genetic Material,* vol. 35. Cold Spring Harbor Laboratory, New York, 1970.

Crick, F. H. C. The genetic code. III. *Sci. Am. (Oct):* 55, 1966.

Kornberg, A. *DNA Synthesis.* W. H. Freeman, San Francisco, 1974.

Lake, J. A. Amino-acyl-tRNA binding at the recognition site is the first step of the elongation cycle of protein synthesis. *Proc. Natl. Acad. Sci.,* 74: 1903, 1977.

Pastan, I. Cyclic AMP. *Sci. Am. (Aug):* 97, 1972.

Roberts, J. W. Termination factor for RNA synthesis. *Nature,* 224: 1168, 1969.

Roth, J. R. Frameshift mutations. *Ann. Rev. Genet.* 8: 317, 1974.

Suguira, M., Okamoto, T., and Takanami, M. RNA polymerase ρ factor and the selection of initiation site. *Nature,* 225: 598, 1970.

Travers, A. A. Positive control of transcription by a bacteriophage sigma factor. *Nature,* 225: 1009, 1970.

Watson, J. D. *Molecular Biology of the Gene* (3d ed.), W. A. Benjamin, Menlo Park, Calif., 1976.

Zubay, G., D. Schwartz, and J. Beckwith, Mechanism of activation of catabolite-sensitive genes: a positive control system. *Proc. Natl. Acad. Sci.,* 66: 104, 1970.

REVIEW QUESTIONS

1. Assume a segment of a DNA strand has the following nucleotide sequence:

3′ T A C A A A T C T C A T T G T A T A G G A   5′

    A. Give the base sequence and polarity in the complementary DNA strand.

    B. Give the base sequence and polarity of a mRNA strand formed by transcription from the first-mentioned 3′ strand.

    C. How many amino acids would this segment represent?

2. A. Assume that exposure to a mutagen brings about a base substitution so that the first nucleotide in the third codon from the 3′ end of the DNA strand is changed to A. What would be the ultimate effect? (Consult the coding dictionary, Table 12-1).

    B. What do we call this class of mutations?

3. A. Assume exposure to a mutagen brings about a base substitution so that the third nucleotide of the second codon from the 3′ end is changed to G. What would be the ultimate effect?

    B. Name this class of mutations.

    C. Assume the second nucleotide of the second codon from the 3′ end was changed to C. What would be the effect?

    D. Name this class of mutations.

4. A. Suppose exposure to an acridine causes elimination of the third nucleotide in codon 1 and elimination of the first nucleotide in codon 3. What will this do to the reading of the gene, and what will be a likely outcome?

    B. Suppose the second nucleotide in codon 4 were removed in addition to the other two mentioned above. Now what would be the effect?

    C. Suppose the first elimination, removal of the third nucleotide in codon 1, is followed by the addition of a nucleotide containing T between the first and the second nucleotides of codon 3. What would happen to the reading of the gene, and what would be a likely outcome?

    D. What do we call the class of mutations described here?

5. Of the two amino acids, tryptophan and arginine, which is more likely to be substituted in a polypeptide chain by another amino acid as a result of a mutation? Explain why after consulting a coding dictionary.

6. Give three features which the various kinds of amino acids have in common.

7. In addition to the 20 kinds of amino acids, the appropriate buffers and inorganic substances, give five major essentials which must be supplied in a cell-free *in vitro* system for polypeptide synthesis.

13

MOLECULAR INTERACTIONS IN TRANSCRIPTION AND TRANSLATION

THE RIBOSOME AND SOME OF ITS INTERACTIONS. The burst of genetic research in the late 1950's and early 1960's contributed valuable insights into the mechanism of translation. The process had been shown to encompass interactions among the ribosomes, the mRNA, and the tRNAs, each kind charged with its specific amino acid. Although ribosomes themselves cannot assemble amino acids into protein, it soon became evident that they are much more than passive entities which somehow simply form peptide linkages. Ribosomes from different kinds of cells and creatures are surprisingly uniform in their physical and chemical characteristics. Their size and weight may be measured by the rate at which they sediment or settle out in a centrifugal field. The rate of sedimentation is expressed in "S" units (Svedberg). On this basis, two size classes of ribosomes can be recognized, those from prokaryotes such as *E. coli* with an S value of 70 and those from the cytoplasm of eukaryotes with an S value of 80. Ribosomes are also found in the chloroplasts and mitochondria of eukaryotic cells, and these are similar to the 70S ribosomes of prokaryotes. The significance of these will be discussed more fully in Chapter 20.

Any ribosome is in turn composed of two subunits which can be separated by lowering the magnesium concentration in the environment (Fig. 13-1). The smaller of the two subunits sits like a cap on the larger. The subunits also have characteristic S values, 50S and 30S, respectively, for the larger and smaller ones of prokaryotes and 60S and 40S for the corresponding subunits in eukaryotes. The subunits may reassociate to form a complete ribosome if the magnesium concentration is again raised. Each of the subunits is composed of RNA and protein. The RNA found in each ribosomal subunit has a characteristic S value also. In prokaryotes, the smaller subunit contains a 16S RNA molecule. The 50S subunit contains two kinds of RNA molecules, a larger one with a sedimentation value of 23S and a smaller one with a value of 5S. The smaller (40S) subunit of eukaryotes contains an RNA molecule with a value of 18S. The 60S subunit has one large RNA component whose sedimentation value differs with the species grouping (Table 13-1). For multicellular animals, the S value of the large RNA molecule is 28S; for protozoans, fungi, and plants, the S value is 25–26. Two smaller kinds of RNA molecules are also present in the 60S subunit of all eukaryotes. One is a 5S molecules, and the other is a different type with an S value of 5.5. This latter species of RNA molecule, not found in prokar-

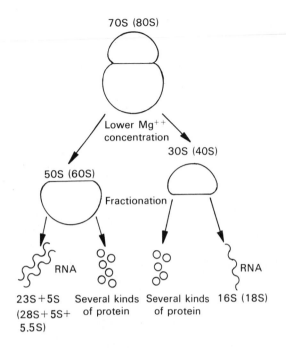

FIG. 13-1. Ribosomal components of prokaryotes and multicellular animals. A ribosome consists of two subunits which may dissociate and associate depending on the Mg++ concentration. Each subunit may be fractionated into its RNA and protein components. The sedimentation value is characteristic for the ribosomes and their parts. In the figure, the S values are given for prokaryotes with the values for multicellular animals in parenthesis.

TABLE 13-1 Characteristic S values of cytoplasmic ribosomes and their components

SOURCE	INTACT RIBOSOME	RIBOSOME SUBUNITS	rRNA IN SUBUNIT
Prokaryotes	70S	30S	16S
		50S	23S, 5S
Eukaryotes	80S	40S	
		60S	
Protozoa		40S	18S
		60S	25–26S, 5S, 5.5S
Fungi		40S	18S
		60S	25–26S, 5S, 5.5S
Plants		40S	18S
		60S	25–26S, 5S, 5.5S
Animals		40S	18S
		60S	28S, 5S, 5.5S

yotes, is very closely associated in the larger subunit with the larger RNA molecule.

The proteins of the larger subunits are very complex and include approximately 50 different kinds. Some of them have been shown to be essential for the formation of peptide bonds and the binding of tRNA molecules. The complexity of the ribosome is undoubtedly related to the active role it plays in protein synthesis. The larger subunit must be able to recognize mRNA and tRNA; it must also join amino acids by peptide linkages and move the mRNA along so that the coded information in the message can be completely translated. Important roles of the smaller subunit include binding to the first codon of the mRNA, which indicates that translation is to begin, and assuring that the mRNA retains a single-stranded configuration so that all the codons are exposed for reading. If the mRNA looped upon itself, certain codons might not be available for translation. Also binding to the smaller subunit is the tRNA bearing the amino acid which will be the first one in the protein which is to be constructed (more details follow). Recent evidence indicates that the two subunits of the ribosome actually exist in the cell as separate parts and do not associate until they engage in protein synthesis (Fig. 13-2). Once a ribosome reads a messenger RNA strand, the two subunits again dissociate and are free to recycle. This means that the "top" of a ribosome may associate with the "bottom" of another if they engage again in translation. As Figure 13-2 also shows, single ribosomes do not read a messenger strand. Instead, groups of them associate with a single mRNA, an association termed a *polyribosome* or *polysome*. It must be emphasized that in the polyribosome, each single ribosome is forming a separate protein molecule independently. One ribosome does not cooperate with another to form a protein. Therefore, at any moment as a message is being read, protein molecules of the same kind in different stages of completion will be associated with different ribosomes along the length of the mRNA. Once a ribosome reaches the end of the messenger, the completed protein molecule is released. The ribosome leaves the mRNA and dissociates into subunits which may again participate in the translation of this same message or of a different one.

THE MOLECULAR INTERACTIONS IN TRANSLATION.   In order for the genetic information to be read, a mRNA must be singled stranded. The messenger strand is composed of ribonucleotide

FIG. 13-2.   Polysome and ribosome. The ribosomal subunits exist as separate entities in the cell. A 30S and 50S subunit associate at the beginning of the mRNA strand (*right*). Any one mRNA contains a group of ribosomes, each in the process of forming a polypeptide. The entire association is termed a "polysome." As any one ribosome moves toward the end of the messenger, its polypeptide chain increases in length. At the end (*left*), the finished polypeptide is released and the two subunits dissociate to recycle and enter the pool of subunits in the cell. (The diagram does not show tRNA and other details.)

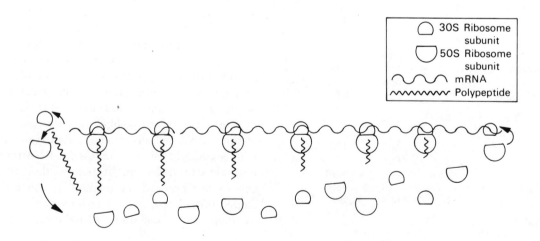

	30S Ribosome subunit
	50S Ribosome subunit
	mRNA
	Polypeptide

units, and these are complementary to the DNA segment from which they were transcribed. Since DNA segments represent specific genes and therefore contain different kinds of genetic information, the lengths of genes must vary. Consequently, the mRNAs in a cell would be different in size. Any small protein includes at least 100 amino acids. Since it takes three nucleotides to designate an amino acid, even the smallest messenger strand would contain 300 nucleotides. Since proteins occur in an assortment of sizes, so do the corresponding mRNAs. Unlike the RNA of the ribosomes, the mRNA of a cell, as expected, is far from uniform. Nor is the mRNA as simple to characterize as the tRNA. The latter is a relatively small molecule composed only of about 80 nucleotides and with an S value of only 7.

Since each kind of tRNA gives identity to a specific amino acid, transfer RNA molecules cannot all be identical. They must contain different kinds of coded information. The structure and exact nucleotide sequence is known for several tRNA molecules, and this prediction has been upheld. The tRNAs have been found to be single stranded, but X-ray diffraction studies indicate a double-helical structure in certain portions of the molecule. The reason for this apparent contradiction is that the single strand of the tRNA loops to form folds, somewhat in the shape of a cloverleaf (Fig. 13-3). Four loops can be recognized: a large one near the 3' end followed by a small one which varies in size, and two other large loops as one moves toward the 5' end. The molecule is held in this shape due to hydrogen bonds between complementary base pairs which lie opposite each other within a fold.

Chemical analysis of the various tRNAs has shown that the 5' end usually begins with a nucleotide containing guanine. The 3' end always terminates in the sequence of bases: cytosine, cytosine, and adenine. It is this adenine-terminating end (the 3' end) which joins to the activated amino acid. It will be recalled that this takes place when the tRNA and amino acid are linked by a specific enzyme, an aminoacyl-tRNA synthetase. In addition to the four bases typically found in RNA (A, U, G, and C), the transfer RNAs have been found to contain exceptional bases, those which differ slightly from the usual ones. It is believed that these unusual

bases are derived from the more familiar ones once the latter have been inserted into the RNA strand. Since they do not engage in base pairing, they may serve to prevent pairing in the loop region and thus permit these regions to interact with certain specific molecules and cell components. One of these odd bases, pseudouradine (ψ) always occurs in the sequence G C ψ T in the loop near the 3' end (Fig. 13-3). It is thought that this sequence may be involved in attaching the tRNA to the ribosome. The large loop near the 5' end contains the unusual base dihydrouridine, and this loop is thought to be the portion of the molecule which binds to an aminoacyl synthetase.

One important difference among the kinds of tRNA is found in a sequence of three nucleotides in the lower loop of the molecule (Fig. 13-3). This triplet is critical in placing the amino acid in the correct position in the growing protein. It accomplishes this by recognizing the complementary triplet in the messenger. Because it is complementary to the codon in the mRNA, it is called the *anticodon*. The anticodon in the transfer RNA can base pair with the codon of the messenger.

It is important to note here some points relating to the polarity of codons and anticodons. Remember that a DNA strand and its RNA transcript possess opposite polarities (refer to Fig. 12-5). The RNA codons given in Table 12-1 are written as they would occur in mRNA reading in a 5' to 3' direction. The corresponding DNA codons are written in the reverse order, 3' to 5'. Thus, the RNA codon 5' *AUG* 3' (for methionine) corresponds to the DNA codon 3' *TAC* 5'. The anticodon in the tRNA, while complementary to the codon in the mRNA, possess the opposite polarity, and in this case would be 3' *UAC* 5'. The ability of the anticodon to base pair with a codon in the mRNA enables the tRNA to position its amino acid properly in a growing polypeptide chain according to the scheme in Fig. 13-4.

The smaller unit of the ribosome attaches to the messenger RNA and to the first transfer RNA. This first tRNA contains an anticodon which pairs with the first codon in the messenger. It is the 5' end of the messenger strand which becomes attached to the ribosome and this first tRNA. The mRNA is read (translated) in the 5′ to 3′ direction, the same direction in which it was synthesized.

After the complex is formed composed of the smaller ribosomal subunit, the mRNA, and the tRNA, the larger subunit of the ribosome associates with the smaller to form a complete ribosome. The larger subunit contains two adjacent sites which can bind tRNA molecules. One of these sites, the *amino site* (or "A" site), is the one which accommodates all of the tRNA molecules arriving *after* the first one. For example, the second tRNA, carrying its specific amino

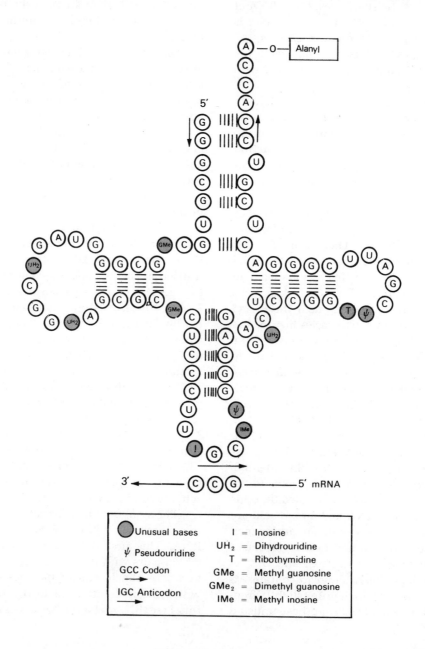

FIG. 13-3. Structure of tRNA. Complete structure of yeast tRNA specific for alanine. (Reprinted with permission from James D. Watson, *Molecular Biology of the Gene*, 2nd ed., copyright 1970 by J. D. Watson; W. A. Benjamin, Inc., Menlo Park, Calif.)

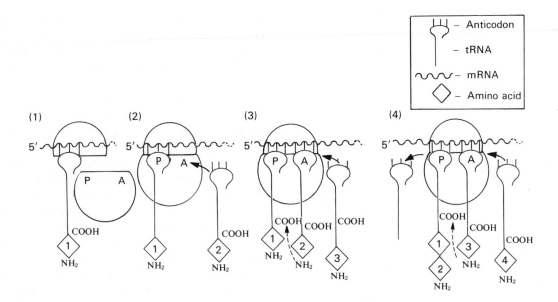

FIG. 13-4. Beginning of translation. *1.* A tRNA, carrying the first amino acid coded by the first codon in the messenger, attaches to the smaller subunit. *2.* The larger subunit then associates with the smaller one. A second tRNA with its amino acid moves toward the "A" site of the larger subunit. *3.* The first tRNA is now at the "P" site and the second one is at the "A" site. A peptide bond forms between the NH_2 group of amino acid 2 and the COOH of

amino acid 1. *4.* The first tRNA is ejected from the ribosome and the second one, now carrying both amino acids, moves to the "P" site. Meanwhile, a third charged tRNA has moved to the "A" site. The process repeats until a chain terminating codon is reached. Note that it is the 5' end of the mRNA which is read first by the ribosome. The ribosome translates the messenger from the 5' to the 3' end.

acid, moves into the amino site of the larger subunit. The other site, which is bound to the first charged tRNA, is the *peptide* or "P" site. After entry of the second transfer RNA molecule to the "A" site, a peptide bond is formed between the carboxyl group of amino acid 1 at the "P" site and the amino group of amino acid 2 at the "A" site. This bond formation requires a specific enzyme, peptide synthetase, found in the larger subunit of the ribosome. Transfer RNA 1 is now ejected from the ribosome. It can now recycle and pick up another specific amino acid. The second tRNA moves from the "A" to the "P" site, carrying the two linked amino acids. As this takes place, the mRNA moves so that the next codon (the third in our example) is exposed in the proper position. A third charged tRNA now moves to the "A" site. Again this involves the interaction of the codon (the third one here) in the mRNA with the complementary anticodon of the tRNA. The amino group of the third amino acid, now in the "A" position, forms a peptide bond with the carboxyl group of amino acid 2 at the "P" site. The second tRNA is now ejected;

tRNA 3, holding a tripeptide, moves to the "P" site, while the mRNA moves again to expose the next codon. A fourth charged tRNA enters, and the process repeats over and over until the polypeptide chain is completed and finally released from the ribosome. It is evident that the first amino acid in the polypeptide chain has a free amino group and thus an "amino end." Each amino acid which is added to the growing polypeptide chain is linked by its amino group to the carboxyl group of the preceding one. The last amino acid will therefore have a free carboxyl end. Once the linear polypeptide or protein chain is completed, it may then assume other configurations due to the interaction of amino acid residues along its length. The complex foldings typical of many proteins result from such interactions of amino acid units along one polypeptide chain or between two polypeptide chains in those cases where the protein is many stranded. The folding, therefore, takes place after polypeptide chains are formed; the ribosome does not guide the three-dimensional architecture of a protein. The messenger RNA in the cell may or may not be used over and over again, depending on the cell type and its stage in development. If no longer needed, it may be degraded by enzymatic action, probably by polynucleotide phosphorylase.

INITIATION OF TRANSLATION. Once formed, we see that the mRNA becomes translated by forming a complex with the ribosome and the charged tRNA. Being a transcript of the entire gene, a messenger RNA strand includes transcriptions of many DNA codons, each designating an amino acid. As we will learn in Chapter 19, one messenger RNA may even carry the transcriptions of two or more genes which control steps in the metabolism of the same substance. Therefore, a messenger strand must contain some kind of signal to indicate where translation into protein by the ribosome is to begin. This is apparently achieved by the RNA codon "AUG." The coding dictionary (see Table 12-1) shows that this triplet designates the amino acid, methionine, and that this is the only RNA codon assigned to methionine. Chemical analyses of proteins from microorganisms demonstrate that this amino acid is the one most frequently found at the beginning (the amino end) of the protein chain. Studies using mRNA from certain phages have shown that the protein made *in vitro* frequently contains as its "first amino acid" n-formyl methionine, a form of methionine with a formyl group in place of the amino group (Fig. 13-5). When compared with the same protein made by the virus *in vivo* in the bacterial cell, the n-formyl methionine is not present. Such observations have suggested that this special form of methionine is normally placed in the first position in a polypeptide chain. Reference to its structure shows that it has a blocked amino group. Therefore, it cannot join to the carboxyl group of any amino acid. However, it has a carboxyl group of its own with which any amino acid can form a peptide linkage. This feature of the blocked amino group would make n-formyl methionine well suited to serve as the first unit starting a protein chain. Since it is not found in normal protein which is formed in the living cell, it would seem that it is later changed to a typical methionine residue containing an amino group and lacking the formyl grouping. It may also be cleaved from the chain so that the second amino

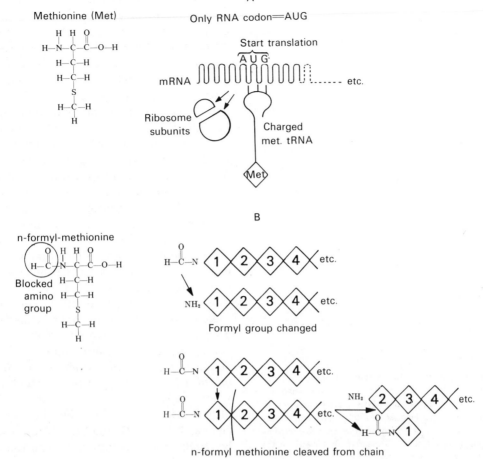

FIG. 13-5. Start of translation. *A.* The amino acid methionine is represented by only one RNA codon. Methionine is the most frequently occurring amino acid at the beginning of a polypeptide chain as shown by analyses of protein from cells of microorganisms. It appears that the RNA codon, AUG, signals the start of translation at that point in the messenger. The specific tRNA carrying methionine would thus be the first to complex with the messenger and the ribosome. *B.* In the test tube, however, n-formyl methionine, which has a blocked amino group, is frequently found at the first position. Evidently, this form of methionine is placed first in the chain. Its formyl group is then later changed to an amino group. It may also be cleaved from the polypeptide chain so that the "second" amino acid becomes the leading one.

acid residue in the chain later becomes the leading one with the free −NH₂ end (Fig. 13-5*B*).

Analyses on the molecular level have revealed a special kind of transfer RNA in the cell. It is called "initiating tRNA," or tRNA$^{f\,met}$, because it is complexed with n-formyl methionine. Actually, the formyl group is added by enzyme action *after* ordinary methionine has become attached to the tRNA$^{f\,met}$. This special transfer RNA thus makes the enzyme reaction possible, whereas the ordinary methionine tRNA (tRNA^{met}) cannot, and so any methionine attached to it remains unchanged. The tRNA$^{f\,met}$

can also recognize the triplet AUG when it is at the beginning of the messenger RNA. If AUG is found elsewhere along the messenger strand, "ordinary" methionine is inserted by the "ordinary" tRNA^{met} (Fig. 13-6). Thus, we see that a codon may mean different things, depending on its position in the mRNA chain. AUG at the beginning indicates that the protein is to be started from that point, and it is recognized by the special tRNA charged with *n*-formyl methionine. Located elsewhere, the triplet AUG is recognized by the usual tRNA (tRNA^{met}) carrying methionine.

Surprisingly, both tRNA^{met} and tRNA$^{f\ met}$ have been found to have the same anticodon, 3′ UAC 5′, and therefore are coded by the same triplet, AUG. One of the main distinctions between them appears to depend on their ability to react with certain other protein factors in the cell. Only the initiating tRNA carrying n-formyl methionine can bind to certain initiating factors required for binding of the tRNA to the 30S subunit of the ribosome and to the mRNA marking the beginning of translation (Fig. 13-4). On the other hand, it cannot bind to another kind of factor called an elongation factor. Ordinary tRNA^{met} as well as the other kinds of tRNAs can react with this factor and enter the "A" site of the ribosome as a polypeptide chain grows during translation.

We will see shortly that in many cases two or more genes are normally transcribed on the same messenger RNA strand. There must, therefore, be stop signals along the messenger to indicate the end of translation. Otherwise, two or more normally separate proteins would become joined as one large abnormal unit. Such signals are just as important as the stop signals on the DNA and the rho factor which indicates when transcription is to terminate. We have already encountered the three-chain terminating codons, UAA, UAG, and UGA. These codons can be recognized by certain other specific proteins, the so-called "release factors." Somehow, a release factor acts to remove the "last" tRNA from the ribosome and bring translation to a stop. We now see that the three-chain terminating codons make up only one part of the complex of interacting factors necessary for orderly information transfer.

THE ANTICODON AND THE WOBBLE HYPOTHESIS. It has been well established that the genetic code is degenerate and that 61 of the 64 codons designate amino acids. It seems logical to expect to find 61 different kinds of tRNA molecules, one kind with a specific anticodon for one kind of codon. And indeed, more than one kind of tRNA is known for certain specific amino acids. However, the 1 to 1 correspondence does not seem to hold. Moreover, it has been demonstrated that one given kind of tRNA (and hence one given anticodon) can recognize more than one RNA codon! In addition, some anticodons contain the unusual base, inosine (see Fig. 13-3). This base

FIG. 13-6. Initiating transfer RNA. A special kind of tRNA can recognize n-formyl methionine and the triplet AUG when this triplet is the first on the messenger to designate an amino acid. The triplet AUG when located at any other position on the messenger designates ordinary methionine which is identified by the ordinary methionine tRNA. This diagram shows AUG at the first and the third positions and the codon for phenylalanine at the second. Phenylalanine will thus be placed at position 2 and ordinary methionine at position 3 in the polypeptide chain as the ribosome reads the messenger.

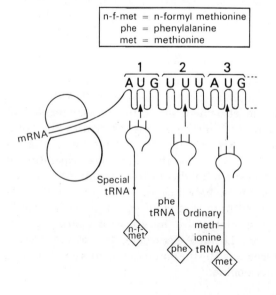

n-f-met	= n-formyl methionine
phe	= phenylalanine
met	= methionine

is derived by enzymatic action from adenine, which is originally in that position in the tRNA loop.

Crick has proposed the wobble hypothesis which helps to explain these several points. Remember that the RNA codon in the mRNA is written in a 5' to 3' direction and that the anticodon which base pairs with it has the opposite polarity, 3' to 5'. According to the wobble hypothesis, when the codon and anticodon interact, the first base pairing occurs between the bases at the 5' end of the codon and the one at the 3' end of the anticodon. The middle bases pair next, and finally the two at the third position, the base at the 3' end of the codon and the 5' end of the anticodon. The first two pairings are specific ones. The base at the 5' end of the anticodon, however, can wobble in the sense that it is freer to move about spatially than the other two and in some cases may form hydrogen bonds with more than one kind of base at the 3' end of the RNA codon. Consider an anticodon for arginine: 3' UCU 5'. The base uracil is at the 5' or wobble position. At this position, uracil can pair with either the base adenine or guanine in a codon, hence with either the codons 5' AGA 3' or 5' AGG 3'. Both these RNA codons are for arginine.

The unusual base inosine (I on Fig. 13-3) at the wobble (5') position may pair with A, U, or C at the 3' position in an RNA codon. G at the wobble position may pair with U or C in the codon. There are restrictions, however; C in the wobble position may pair only with G (as in the case of the codon-anticodon interaction for methionine). Similarly, A at the 5' position can base pair only with U. No single kind of tRNA can recognize four different codons.

The wobble hypothesis explains why fewer than 61 different tRNAs may exist and also the fact that one purified kind of tRNA has been shown to interact with more than one kind of RNA codon. It also can account for the presence of inosine in certain anticodons. Moreover, the hypothesis has been strengthened by the fact that certain predictions made from it have been verified. It correctly predicted, for example, the minimum number of tRNAs required for the amino acids serine and leucine.

In the process of translation, we see the active role of the ribosome in reading the message, forming the peptide linkages, and moving along the mRNA. We also see that tRNA is not just a passive carrier of an amino acid. Through its anticodon, it actively places its amino acid in the proper position, following the instructions coded in the messenger strand. The interactions of the mRNA, tRNA, and ribosome in translation achieve the final transfer of information from the DNA codons within the double helix—the formation of a specific protein.

PROCESSING OF RIBOSOMES. Cells which are actively growing and synthesizing protein at a high level are very rich in ribosome content. In an *Escherichia coli* cell, it may account for as much as 30% of the cell mass. A decrease in growth rate is followed by a corresponding drop in the number of ribosomes. Ribosome production is apparently very closely coordinated with other parts of the cell in the many aspects of normal metabolism. In prokaryotes, the ribonucleic acid (rRNA) of the ribosomes is coded by specific regions of the deoxyribonucleic acid (DNA), the ribosomal genes, or ribosomal DNA (rDNA). The ribosomal RNA which is transcribed from this DNA is released into the cell where it is clothed in protein. In eukaryotic cells, however, the story is a bit more complex. Cytochemical work had shown that the cytoplasm of a metabolically active cell stains deeply with basic dyes due to the presence of a large number of ribosomes. Such a cell typically contains a nucleolus which is larger than that in a less active cell (Fig. 13-7). Conclusive evidence that the nucleolus controls ribosome formation and is therefore intimately involved in protein activity was presented in the mid-1960's from investigations with the African clawed toad, *Xenopus laevis*. In this organism, the normal diploid cell contains two nucleoli. A mutation (actually a deletion of the organizer) was discovered which can affect the formation of the nucleoli. An individual that is heterozygous for this change has cells which contain only one nucleolus. The homozygous mutant dies in the embryonic stage, and its cells contain no nucleoli at all. It was found that the cells of these mutants are incapable of forming ribosomal RNA. The embryo reaches the tail bud stage of development only because of the ribosomes which came from the egg. Once these are depleted, the embryo dies.

Thus, the nucleolus was shown to be essential for ribosome formation.

Additional work with *Xenopus* gave more information on the genesis of the ribosomes. DNA was isolated from the nuclei of normal animals (those with two nucleoli per cell) and then subjected to density-gradient ultracentrifugation. This procedure revealed the presence of two kinds of DNA which differ in their densities. Most of the DNA separated into a major band, but a small amount, less than one-half of 1% could be isolated as a small or satellite band (Fig. 13-8). This band has not been demonstrated in the DNA taken from mutants which lack nucleoli. Experimental evidence indicates that this satellite DNA is the DNA of the nucleolar-organizing region of the chromosome (Chap. 2). Not only is this region required for the formation of nucleoli, it also contains the genes necessary for the formation of ribosomal RNA.

How can we demonstrate that a certain DNA is responsible for the formation (transcription) of a certain kind of RNA? This can be achieved through the technique of DNA-RNA hybridization. In this procedure, the double-stranded DNA must first be transformed to a single-stranded state. This can be achieved by raising the temperature above the melting point of the DNA, a point at which the two strands of the double helix separate (Fig. 13-9A). If the melted DNA is quickly cooled, the two complementary strands will not have the opportunity to get together to reform double strands. RNA may then be added to these single-stranded DNA preparations. The mixture of DNA and RNA may then be incubated at a temperature which is too low to permit double stranded DNA to form but high enough to allow RNA to combine with any complementary DNA segments. The formation of hybrid double helixes, composed of a DNA chain and an RNA strand, indicates the specific DNA segments and the RNA transcribed from them.

Ribosomal RNA of wild-type *Xenopus* was made radioactive by labeling with tritium. When such labeled RNA was used to form RNA-DNA hybrid molecules, it was found that it was the DNA in the satellite band which was forming hybrids with the rRNA. This could be easily seen, because it was mainly the satellite band in the density gradient which was labeled, due to the presence of the labeled RNA (Fig. 13-9B).

FIG. 13-7. Epidermal cells of a grass. A cell forming a root hair is much more active metabolically than a cell which fails to produce one. Note the large nucleolus in the nucleus of the hair (*above*) in contrast to the smaller nucleolus in the nucleus of the hairless cell (*below*).

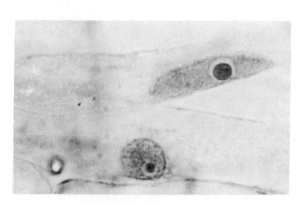

Almost no labeling was associated with the major band. Studies of this type have clearly shown that most of the genes needed for ribosomal RNA formation are in the nucleolar organizer. Other procedures have demonstrated that not just a few genes but many hundreds are normally involved. So a good percentage of the DNA (approximately 0.1% in *Xenopus*) is concerned just with the production of ribosomes. The use of the DNA-RNA hybridization procedure in the actual isolation of ribosomal genes is discussed in Chapter 21.

The formation of ribosomal RNA tells us something about information transfer: information coded in nucleic acid is not necessarily translated into protein. We see that the DNA concerned with the production of ribosomal RNA (the rDNA) undergoes transcription to RNA which then becomes assembled into the larger and smaller subunits of the ribosome. This takes place without translation of the RNA. Another such example is the transfer RNA which is formed from nuclear DNA and which participates directly in cellular activities without being translated.

The RNA of the ribosomes becomes complexed with protein. Formation of this ribosomal protein takes place on the ribosomes of the

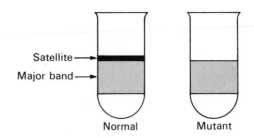

FIG. 13-8. DNA from *Xenopus*, DNA from a normal animal whose cells contain nucleoli separates into a major band and a small satellite band when it is subjected to density gradient ultracentrifugation. The DNA from a mutant with no nucleoli shows no satellite band.

FIG. 13-9. DNA-RNA hybridization. *A.* Single DNA strands will hybridize with those RNA strands which contain stretches of complementary nucleotides and which are transcripts of the DNA. *B.* Labeled RNA of the ribosomes can be used to hybridize with the total DNA from cells of a normal *Xenopus*. When the DNA is then subjected to a density gradient, only the satellite DNA is labeled. This indicates that the RNA of the ribosomes contains stretches complementary to the satellite DNA and that the rRNA is a transcript of it.

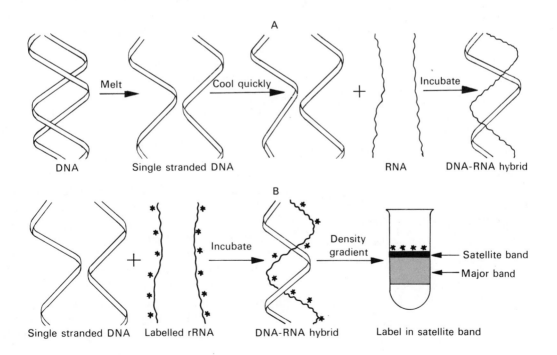

cytoplasm. In eukaryotes, some of it moves into the nucleolus where it unites with RNA to complete the larger ribosomal subunit. It appears that the RNA of the smaller subunit moves out to the cytoplasm, where it is combined with protein (Fig. 13-10).

While the genes for the 18S and 28S ribosomal RNA are located in the nucleolar-organizing region of eukaryotes, the genes for the small 5S RNA of the larger subunit are located elsewhere in the chromosomal complement. In the human, genes for the 18S and 28S rRNA are localized to the short arms of the satellited chromosomes, Chromosomes 13, 14, 15, 21, and 22. On the other hand, genes for the 5S RNA are found in the long arm of Chromosome 1, which bears no satellite and has no nucleolar-organizing function. In the toad *Xenopus* the 18S and 28S rRNA genes are confined to just one chromosome site associated with the nucleolus, whereas the genes for 5S rRNA are located at the ends of the majority of the chromosomes. Somehow the 5S rRNA, found on a completely different molecule than the 18S and the 28S rRNA, becomes packaged into the larger ribosomal sununit along with the 28S rRNA, a process which must entail more complex molecular interactions than that indicated in Figure 13-10.

ONE GENE-ONE POLYPEPTIDE CHAIN. The relationship between gene mutation and enzyme activity was very evident long before any insight had been gained into the molecular basis of information transfer. So many examples became known of the effect of a point mutation on an enzymatic step that the concept of "one gene-one enzyme" became widely accepted. We have already noted that the idea implies that each enzyme or protein is controlled by a specific gene. Essentially, the theory has been proven correct, but today we know a great deal more about enzyme and protein structure than we did 20 years ago. In many cases, an enzyme is composed of one kind of long polypeptide chain, and it *is* under the control of one gene. An example is the enzyme tryptophan synthetase in the red bread mold, *Neurospora*. This important catalyst is needed for the final step in the manufacture of the amino acid, tryptophan. A segment of DNA, a gene, controls the formation of this specific protein or long polypeptide chain

FIG. 13-10. Biogenesis of eukaryotic ribosomes. The rDNA genes of the nucleolar organizing region undergo transcription to form the rRNA. This RNA is processed in the nucleolus into the 18S and 28S RNA of the ribosomal subunits. In the processing, about 20% of the RNA transcript is discarded. The 18S RNA moves out of the nucleolus and becomes clothed in protein made by ribosomes in the cytoplasm to form the 40S subunits. Some cytoplasmic protein (*lower*) moves into the nucleolus where it combines with the 28S RNA to form the 60S subunits. These then pass out of the nucleolus into the cytoplasm.

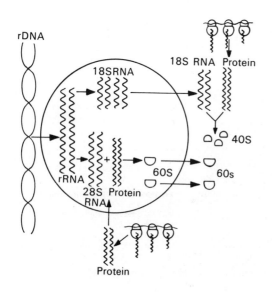

UNDERSTANDING GENETICS

which makes up the entire enzyme, tryptophan synthetase. We can think of any gene as a functional unit, meaning that it has a specific function to perform. The function of the gene in this example is to guide the formation of a particular kind of polypeptide, tryptophan synthetase (or "t'ase"), which has a specific enzymatic action (Fig. 13-11A). The one gene-one enzyme concept is very clear-cut in this case; however this is not always so. In *E. coli*, the enzyme t'ase is under the control of two distinct genes (Fig. 13-11B). Two different kinds of polypeptide chains make up this bacterial enzyme. The function of each one of the genes is to transfer information for the formation of a specific polypeptide. The two kinds of polypeptides then assemble to form tryptophan synthetase, essential to the final step in the formation of tryptophan. Since the completed enzyme in this case is composed of two kinds of polypeptides, each specified by a different gene, the relationship of one gene-one enzyme is not clear-cut. However, because any protein is really one or more long polypeptide chains, we can modify the original statement to say "one gene-one polypeptide chain." This statement is basically sound, for every kind of polypeptide chain is specified by a particular gene. On the other hand, many proteins are composed of several separate chains which become associated. Indeed, the tryptophan synthetase of *E. coli* is actually composed of four, but only two different kinds are present, each one of them represented twice. In Chapter 14, we will reexamine the gene as a unit of function in its relation to a specific polypeptide. We will see how the information we now have on the molecular level enhances our understanding of the nature of allelism.

COLINEARITY. We have noted that both DNA and protein are essentially linear molecules. And we have seen that a codon (a sequence of three nucleotides in a DNA segment and in the complementary RNA triplet transcribed from it) is typically specific for a certain amino acid type and determines the position of that kind of amino acid in a polypeptide chain. It is tempting to speculate that perhaps the linear sequence of the nucleotides in the DNA and RNA corresponds to the linear sequence of the amino acids along a stretch of a protein molecule. Actually, this

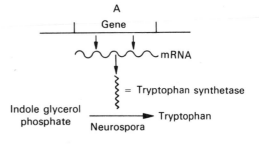

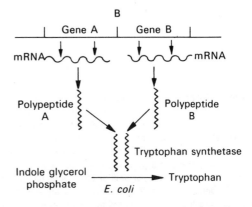

FIG. 13-11. One gene-one polypeptide. *A.* In *Neurospora*, the enzyme tryptophan synthetase is composed of one kind of polypeptide chain which is under the control of one gene. In this case, the relationship one gene-one enzyme is direct. *B.* In *E. coli*, the enzyme t'ase is composed of two different kinds of polypeptides. Each is controlled by a separate gene. In this case, therefore, the concept of one gene-one enzyme does not strictly hold. However, in both *Neurospora* (above) and *E. coli* (below), the function of each gene is to govern the formation of a specific type of polypeptide, and the relationship one gene-one polypeptide is clear.

has been demonstrated quite clearly in microorganisms. For example, Yanofsky and his colleagues studied in detail one of the two genes whose function is to guide the formation of the bacterial enzyme, tryptophan synthetase. We have just discussed that the *E. coli* enzyme contains two different kinds of polypeptide or protein, each controlled by a separate gene. Several separate mutations in one of these two genes, gene "A," were mapped very precisely (Fig. 13-12A). The defective protein associated with each of the mutations was analyzed in detail. Each defective protein was found to have a single amino acid substitution. The substitution was at a different position in the protein chain from one mutation to the other. When the position of the amino acid substitution was related to the site of mutation in the gene, it became evident that a mutation mapping at one end of a gene produced an amino acid change at one end of the protein, let us say the amino end. A second mutation might map at the other end of the gene. The corresponding protein was shown to have an amino acid alteration at the other end of the protein, the carboxyl end. A third mutation might map between the first two. The amino acid substitution associated with it would then fall at a position in the chain between the other two amino acid changes. In other words, the position of a mutation within a gene corresponds to the position of an amino acid substitution in a protein chain. The relative position of a triplet in the DNA and in the complementary RNA transcribed from it is related directly to the position of a specific amino acid in a protein. The nucleic acid molecules and the protein are *colinear* (Fig.13-12*B*). A change at a site in the DNA is reflected at a corresponding site in the messenger RNA. This in turn is reflected at a corresponding position in the polypeptide after translation of the messenger.

FINGERPRINTING AND MUTANT HEMO-GLOBINS. Although most of our knowledge of molecular interactions has come from elegant experiments with microorganisms, the findings apply to higher forms and are especially valuable in clarifying the molecular nature of certain genetic diseases. Surprisingly, it was a human

FIG. 13-12. Colinearity. *A.* Mutations occur at different sites within gene "A" of *E. coli* which determines the amino acid sequence of one of the polypeptide chains of tryptophan synthetase. The relative positions of the separate mutations can be mapped. This schematic representation shows that if a mutation maps at the beginning of a gene, an amino acid substitution is made at the beginning of the polypeptide chain. A mutation at the end of the gene (mutation 2) is reflected by an amino acid substitution near the other end of the protein, the carboxyl end. Mutations, such as mutation 3, which map between ends of the gene are associated with amino acid substitutions between ends of the polypeptide. The gene and the polypeptide chain it governs are thus colinear. *B.* Messenger RNA, carrying the information for amino acid sequence, is a transcript of the DNA. Therefore, alterations in the nucleotides of the DNA are reflected at corresponding positions in the mRNA. These in turn are responsible for the amino acid substitutions at positions in the protein. These amino acid changes, therefore, correspond to altered nucleotide sites in the mRNA and in the DNA. The gene, its RNA transcript, and the polypeptide are thus colinear.

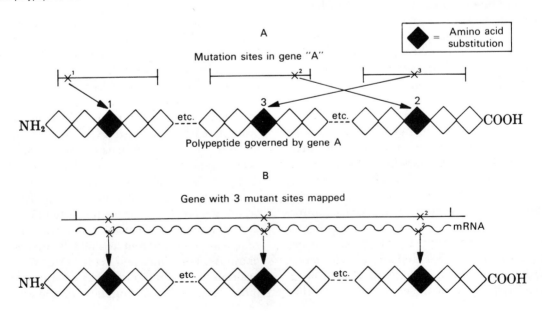

disorder, sickle-cell anemia, which provided the first experimental demonstration that the amino acid sequence in a protein is under the control of a gene. The unfortunate effects of the disease had been shown to stem from some sort of change in the hemoglobin molecule, but it was not known whether the change was an alteration in the amino acid content or in the folding of the molecule. An answer to the question demands the detailed analysis of the amino acid sequence in hemoglobin. Such a task is seen to be an immense one when it is realized that each hemoglobin molecule in an adult is not just a single polpeptide chain but consists of four chains, two of one kind and two of another. Each kind of chain is controlled by a different gene at a different genetic locus. In the adult, most of the hemoglobin in a red blood cell is hemoglobin A, composed of 2 alpha chains combined with 2 beta chains. About 2% of the hemoglobin present is a type called hemoglobin A_2 in which 2 alpha chains are combined with 2 delta chains. Each alpha chain contains 141 amino acids and is quite distinct from the beta and the delta chains. However, each beta and each delta chain is composed of 146 amino acids, and the two resemble each other very closely. As mentioned in Chapter 10, the beta and delta loci, which control the formation of these two chains, are adjacent to each other on the chromosome and are believed to have arisen as a result of a duplication.

In sickle-cell anemia, normal hemoglobin A is not present in the red blood cell, since a change has occurred in the beta chain. Inasmuch as hemoglobin A_2 is not involved, it will not be considered further in the discussion.

The analysis of hemoglobin was greatly facilitated by a shortcut procedure known as "fingerprinting," which was used by Ingram in his pioneer determination of amino acid sequence in the protein. In this method, a protein which is to be analyzed is broken up into short fragments by cleaving it with a protein-digesting enzyme such as trypsin (Fig. 13-13). The resulting fragments, the peptides, then must be separated from one another. This is accomplished by using the technique of paper chromatography coupled with that of electrophoresis. The products of the enzyme action are placed near one edge of a piece of filter paper. The paper is then exposed to a specific solvent which flows across

it in one direction. The peptide fragments will travel along in the direction of the solvent, but they will not all move at the same rate in a particular solvent and so can be partially separated. To obtain a more complete separation of the fragments, the paper is then turned 90 degrees and exposed to an electric field. Due to differences in their net electric charges, the fragments move at different rates to the "+" or "−" pole. The overall result is a good separation of peptide fragments, which then may be stained by methods which color proteins. The technique is called "fingerprinting," because a particular protein gives a characteristic picture. When the fingerprints of normal and sickle-cell hemo-

FIG. 13-13. Fingerprinting technique. The figure shows a fingerprint of the enzyme ribonuclease from the sheep pancreas. The enzyme was exposed to trypsin. A portion of the digested material was then applied to the small spot at the left. It was next subjected to paper chromatography, as indicated by the arrow below. After allowing the solvent to evaporate, the paper was moistened with buffer solution and subjected to electrophoresis. The sheet was sprayed with ninhydrin solution which stains the areas containing peptides. These areas can be cut out and the peptides washed from the paper. The amino acid composition of each peptide can then be determined by further analysis. (Reprinted with permission from C. B. Anfinsen, *The Molecular Basis of Evolution,* p. 145, Wiley, New York, 1959).

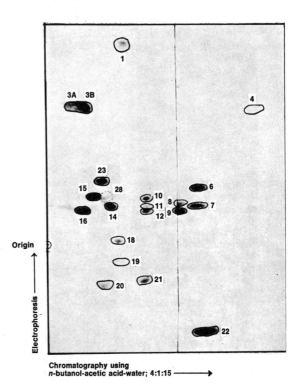

globin were compared, they were found to be identical except for one fragment (Fig. 13-14). This indicated the presence of some difference which alters the rate of migration of the specific fragment in the solvent and electric field. Because the peptide fragments are small (approximately eight amino acids each), their exact amino acid composition can be determined with relative ease.

Recall that the "first" amino acid is considered to be the one with the free amino group and the "last" one that with the free carboxyl group. These two "free" amino acids can be identified by stains which impart distinct colors to them. In this way, the peptides which include the "first" and "last" amino acids can be recognized. The stained regions bearing the fragments are cut out and the peptides are then washed out of the paper. The amino acids composing each peptide can be determined by hydrolysis followed by paper chromatography. A solvent separates the amino acids in one direction. The paper is turned 90 degrees, and a second solvent floods the paper. The different kinds of amino acids become separated due to their different mobilities in the solvents.

Ingram was able to show that the peptide fragment difference between normal and sickle-cell hemoglobin was due to a difference in just a single amino acid (Fig. 13-15). Position 6 in the beta chain of normal adult hemoglobin (numbering from the amino acid with the free amino group) was shown to be occupied by glutamic acid, whereas this position was occupied by valine in sickle-cell hemoglobin. After Ingram's attack on the problem, it eventually became possible to piece together the different peptide fragments and to show the exact sequence of amino acids in hemoglobin. This was possible by using different enzymes, getting different-sized fragments, and recognizing within these fragments the "first" and "last" amino acids. It was finally shown that the only difference between normal adult hemoglobin (hemoglobin A) and sickle-cell hemoglobin (hemoglobin S) is the single amino acid substitution shown by Ingram.

Shortly after this was established, still another abnormal hemoglobin was analyzed, hemoglobin C which also causes an anemia. It results from a recessive mutation, but the cells do not sickle, and the effects are not as severe

FIG. 13-14. Fingerprints of normal and sickle-cell hemoglobin. The fingerprints are identical except for one peptide which has become altered in its mobility in an electric field. The shading and arrows indicate the peptide difference. (Reprinted with permission from V. M. Ingram, *Nature, 180:* 362, 1957.)

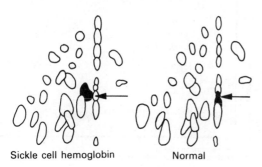

Sickle cell hemoglobin   Normal

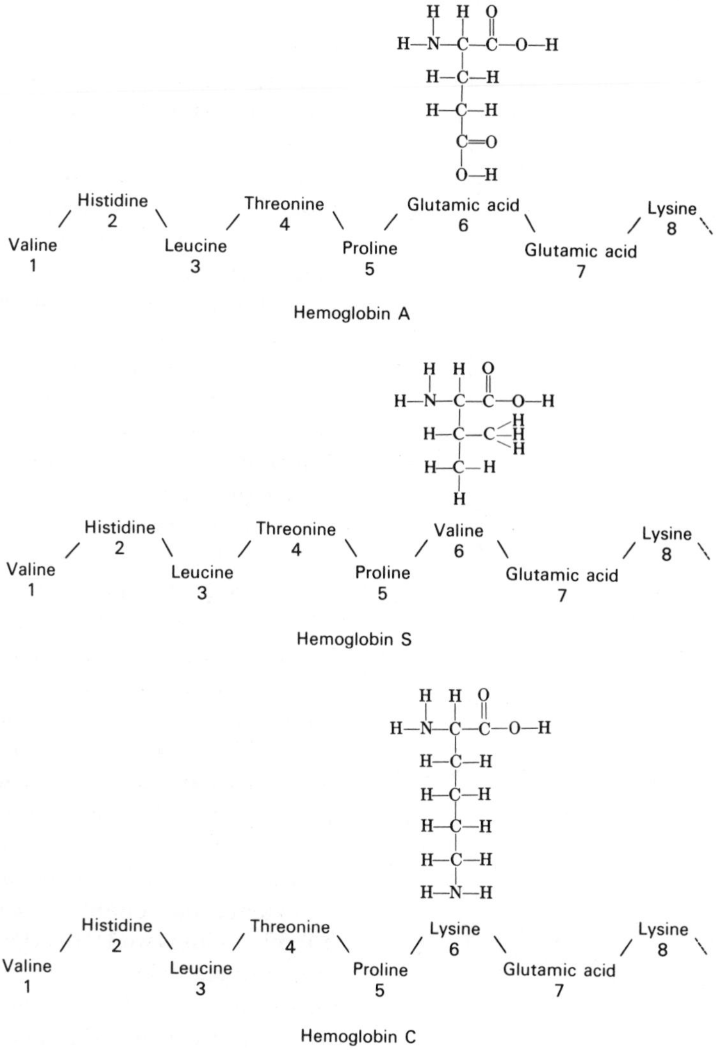

Glutamic acid structure (Hemoglobin A side chain):
H—N—C—C—O—H with H, H, O; H—C—H; H—C—H; C=O; O—H

Hemoglobin A sequence:
Valine 1 — Histidine 2 — Leucine 3 — Threonine 4 — Proline 5 — Glutamic acid 6 — Glutamic acid 7 — Lysine 8

Hemoglobin A

Valine structure (Hemoglobin S side chain):
H—N—C—C—O—H with H, H, O; H—C—C with H, H, H; H—C—H; H

Hemoglobin S sequence:
Valine 1 — Histidine 2 — Leucine 3 — Threonine 4 — Proline 5 — Valine 6 — Glutamic acid 7 — Lysine 8

Hemoglobin S

Lysine structure (Hemoglobin C side chain):
H—N—C—C—O—H with H, H, O; H—C—H; H—C—H; H—C—H; H—C—H; H—N—H

Hemoglobin C sequence:
Valine 1 — Histidine 2 — Leucine 3 — Threonine 4 — Proline 5 — Lysine 6 — Glutamic acid 7 — Lysine 8

Hemoglobin C

FIG. 13-15. The difference between normal hemoglobin and two mutant types. The precise amino acid sequence was determined for normal hemoglobin (hemoglobin A) and for sickle-cell hemoglobin (hemoglobin S). A difference between them was found at one position. When the exact amino acid sequence was established for the entire beta chain of the hemoglobin, it was shown that this was the only difference in the entire chain. At position 6 (near the NH₂ end), the glutamic acid in hemoglobin A is replaced by valine in hemoglobin S. Another mutant hemoglobin, hemoglobin C, also has an alteration at position 6. In this case, lysine is substituted.

as those of sickle-cell anemia. However, hemoglobin C was shown to differ from the normal hemoglobin A at exactly the same position as hemoglobin S (Fig. 13-15). In hemoglobin C, position 6 is occupied by lysine. The two amino acids (valine and lysine) substituting at this position for the glutamic acid in normal hemoglobin have electric properties which differ from each other and from glutamic acid as well. These differences may cause changes in the shape of the beta polypeptide chain. An alteration in the shape of this chain may in turn affect the ar-

chitecture of the entire hemoglobin molecule and so cause a series of metabolic disturbances.

The knowledge of these specific amino acid alterations has other implications. For one thing, we can relate these changes to certain codons which are specific for the amino acids involved. Reference to the dictionary of codons (Table 12-1) shows that two RNA codons designate glutamic acid (GAA and GAG). Four of them represent valine (GUG, GUU, GUC, and GUA), and two stand for lysine (AAA, and AAG). As Figure 13-16 shows, we can in turn relate these RNA codons back to the DNA codons from which they were transcribed. We do not know which of the two codons for glutamic acid has been involved in the change to a codon for valine and to a codon for lysine, but we can narrow down the possible changes somewhat. Our knowledge of mutation on the molecular level tells us that a gene mutation arises from a change in just one base. We can see that the codon CTC, for example, could by one step become changed at the second position to produce CAC, a codon for valine. The CTC codon, by a change in the first position, could produce TTC, a codon for lysine. It would require two changes in the CTC codon to give rise to the other valine and lysine codons. We can consider these changes to be rather unlikely. Similar reasoning applied to the CTT codon suggests that *if* it is the glutamic acid codon involved, the change was to CAT (for valine) and to TTT for lysine.

The sickle-cell story illustrates the progress which has been made in our understanding of the molecular changes which accompany a genetic disease. Such an advance has been made possible through the accumulated knowledge acquired by many research teams investigating information transfer on the molecular level. Such knowledge has enabled us to relate sickle-cell anemia to just two DNA triplets and the possible single nucleotide changes which may have occurred in them. The single alteration involved results in the transcription of a messenger RNA carrying information for a specific polypeptide chain in which an amino acid substitution will be made. The translation of this mRNA produces a polypeptide, an altered beta chain, differing from the standard in just a single amino acid. This alteration in turn affects the interaction of the beta chain with the other kind of chain, the alpha chain, which is governed by a different gene. The final protein, hemoglobin S, departs sufficiently from the normal hemoglobin A to be less efficient as an oxygen carrier. The whole syndrome of serious effects typical of the disease follows as a consequence.

Dozens of other mutations are now known which alter the polypeptide chains of hemoglobin A. When compared, each atypical hemoglobin is found to possess one specific amino acid substitution which may involve either the alpha or beta chain.

All the evidence with the mutant hemoglobins indicates that two separate functional units (genes) govern the production of hemoglobin. The function of one of them is the formation of an alpha chain and that of the other a beta chain. Two alpha and two beta chains then associate to form normal adult hemoglobin. A point mutation (a change in a single nucleotide) at any site in one of the genes can result in an abnormal alpha chain. Similarly, a mutation at a site within the other gene can cause an abnormal beta chain to be produced. The hemoglobin story presents a good example of two units of function associated with a complex protein product made of more than one polypeptide chain. It also illustrates clearly the one gene-one polypeptide concept.

MOLECULAR BASIS OF CERTAIN GENETIC DISEASES. Another serious genetic disorder whose molecular basis has been recently clarified is ganglioside lipidosis. This lethal condition is commonly called Tay-Sachs disease and is expressed in children homozygous for a certain autosomal recessive. The mutant gene is found with a higher incidence among Jews of Northern European origin than in other groups. The severe symptoms (blindness, growth retardation, insanity, and loss of motor coordination) typically end in death by the age of four. The severity of the disorder follows from abnormal accumulation of lipids in cells of the central nervous system. The overcrowding produced by the lipid deposits can actually cause death of nerve cells. We now know that this serious defect is associated with the blockage of a critical step in the breakdown of fats. Victims of Tay-Sachs disease show an abnormality in the enzyme hexosaminidase (Hex). This enzyme is actually composed of two different polypeptide chains, hexosaminidases "A" and "B." Only the latter is found in normal amounts in victims of the disorder. It is a deficiency of Hex A which blocks the normal metabolism of fats in the central nervous system. The parents of affected children are able to handle fats normally, but examination of their blood shows them to have less than the typical amount of Hex A. Any prospective parents showing such a deficiency run a risk of one in four of producing a Tay-Sachs child. It is now possible to detect such a child *in utero*. Cells shed into

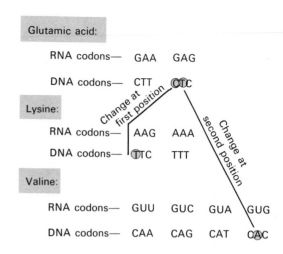

FIG. 13-16. Codons and hemoglobin alterations. A gene mutation arises from a single base change in a triplet. This change can cause an amino acid substitution in a polypeptide chain. Position 6 in the beta chain of hemoglobin is governed by a DNA codon which designates glutamic acid. By a single change, it can give rise to a codon for valine. By a different change, the codon can produce a triplet for lysine. Only certain codons for valine or lysine can be derived from each of the two DNA codons for glutamic acid by single steps. The single-step changes for the codon CTC are illustrated here.

the fluid of the amniotic cavity can be obtained through aminocentesis. If cultures of these cells reveal that Hex A is missing, an abortion may be considered to prevent the birth of a seriously afflicted child. Now that this genetic disease has been related to a defect in a specific polypeptide chain composing an enzyme, chances are good that the precise alteration in the nucleic acid eventually may be identified, as in the case of sickle-cell anemia. Such an approach may be applied to other genetic diseases whose molecular bases are now only partly understood. Deficient proteins which upset steps in biochemical pathways, in disorders such as phenylketonuria and hemophilia, may some day be shown to correspond to point changes at specific sites in the DNA. But can this knowledge ever be used to alleviate the suffering of individuals afflicted with the genetic disorder? Is it conceivable that DNA can ever be made in the test tube and then actually be used to transfer correct genetic information? In Chapter 21, we will learn that artificial, biologically active genes have been synthesized, and we will discuss the potential benefits of such accomplishments in the alleviation of human disorders.

REFERENCES

Avers, C. J. *Cell Biology*. Van Nostrand, New York, 1976.

Clark, B. F. C., and K. A. Marcker. How proteins start. *Sci. Am. (Jan.): 36,* 1968.

Cold Spring Harbor Symposium on Quantitative Biology. *The Mechanism of Protein Synthesis,* vol. 34, Cold Spring Harbor Laboratory, New York.

Crick, F. H. C. Codon-anticodon pairing: the wobble hypothesis. *J. Molec. Biol., 19:* 548, 1966.

Conley, C. L. and S. Charache. Inherited hemoglobinopathies. In *Medical Genetics,* V. A. McKusick, and R. Claiborne (eds.), pp. 53–61. HP Publishing Co., New York, 1973.

Dickson, R. C., J. Abelson, W. M. Barnes, and W. S. Reznikoff. Genetic regulation. The lac control region. *Science, 187:* 27, 1975.

Dudock, B. S., G. Katz, and E. K. Taylor, *et al.* Primary structure of wheat germ phenylalanine transfer RNA. *Proc. Natl. Acad. Sci., 62:* 941, 1969.

Engelman, D. M., and P. B. Moore, Neutron-scattering studies of the ribosome. *Sci. Am. (Oct): 44,* 1976.

Nanninger, N. Structural aspects of ribosomes. *Internat. Rev. Cytol., 35:* 135, 1973.

Nomura, M. Ribosomes. *Sci. Am. (Oct):* 28, 1969.

Nomura, M. (ed.). *Ribosomes.* Cold Spring Harbor, New York, 1973.

O'Brien, J. S. Ganglioside storage diseases. *Adv. Human Genet., 3:* 39, 1972.

Rich, A. R. and S. H. Kim, The three dimensional structure of transfer RNA. *Sci. Am. (Jan):* 52, 1978.

Spiegelman, S. Hybrid nucleic acids. *Sci. Am. (May):* 48, 1964.

Stamatoyannopoulos, G., The molecular basis of hemoglobin disease. *Ann. Rev. Genet., 6:* 47, 1972.

Yanofsky, C. Gene structure and protein structure. *Sci. Am. (May):* 80, 1967.

Yanofsky, C., G. R. Drapeau, and J. R. Guest, *et al.* The complete amino acid sequence of the tryptophan synthetase A protein (a subunit) and its colinear relationship with the genetic map of the A gene. *Proc. Natl. Acad. Sci., 57:* 296, 1967.

REVIEW QUESTIONS

These questions are based on information in both Chapters 12 and 13.

1. Place the number which corresponds to a specific enzyme in the space next to the statement which applies to the enzyme. A number may be used more than once or not at all. Some statements may be associated with more than one enzyme.

A. Can link ribo-
nucleotides to-
gether without a
primer ____

B. Lacking in Tay-
Sachs disorder ____

C. Found in larger
subunit of
ribosome ____

D. Needed for iden-
tification of
specific amino
acids ____

E. The enzyme of
cellular tran-
scription ____

1. polynucleotide
phosphorylase
2. aminoacyl
synthetase
3. tryptophan
synthetase
4. hexoseamino-
dase A
5. RNA
polymerase
6. DNA
polymerase
7. peptide
synthetase
8. phenylalanine
hydroxylase

2. Match the number of the term on the right
with the appropriate description on the left. A
number may be used more than once or not at
all.

A. Region of gene which
contains recognition
signals

B. Needed for normal
termination of tran-
scription of some genes

C. Signal for the start of
translation

D. Is formed only on
initiating tRNA

E. Actively recognizes the
signals in the gene for
the start of transcription

F. Signal to end a
polypeptide

G. Found at position 6 in
the beta chain of
hemoglobin A.

H. Protein needed to stimulate transcription of
certain genes before
binding of RNA
polymerase

1. promoter
2. sigma factor
3. cyclic AMP
4. AUG
5. *n*-formyl
methionine
6. rho factor
7. methionine
8. valine
9. glutamic
acid
10. UUU
11. UGA
12. CTT
13. CAP

3. Name the appropriate kind of nucleic acid
which applies in each of the following: (It is
possible that there may be more than one answer.)

A. RNA that is not translated.

B. Contains the genes whose transcription
is required for ribosome formation.

C. Cannot be formed to play its cytoplasmic
role in an organism lacking nucleoli.

D. Carries information for a polypeptide
chain.

E. Eukaryotic ribosomal nucleic acid
whose genes are not found in the region
of the nucleolus.

4. Suppose the ribosomal genes of some prokaryote, say *E. coli*, are isolated. Ribosomal RNA
is also isolated from cells of the prokaryote and
then hybridized with the DNA strands following
the technique of DNA-RNA hybridization. What
percentage of DNA strands would you expect to
form hybrids with the RNA? Why?

5. Refer to the DNA strand shown in Question
1, Chapter 12, and answer the following:

A. How many individual tRNA molecules
are required to translate this segment?

B. What are possible anticodons which
could be used for translation of this
segment?

6. For each of the following anticodons, give
the corresponding RNA codons and DNA codons
which are possible, as well as their polarities:

A. 3′ UUA 5′

B. 3′ UCU 5′

C. 3′ AAU 5′

D. 3′ CCI 5′

7. Consult the genetic dictionary (Table 12-1),
and give the amino acid designation in each of
the four cases in Question 6.

8. From the genetic dictionary, note the RNA codons for the amino acid, serine. What would be the *minimum* number of tRNAs needed to insert serine into a growing polypeptide chain during translation?

9. Answer the same question stated in Question 8 for the amino acids, leucine and arginine.

10. Consider an mRNA carrying ribonucleotides numbered 1, 2, 3, etc., from the beginning of the message for a specific polypeptide. The normal polypeptide has 300 amino acids.

A. Ribonucleotide No. 14 undergoes a change resulting in a missense mutation. At which position from the NH_2 end in the polypeptide will an amino acid substitution occur?

B. Assume ribonucleotide No. 23 changes and a "nonsense" mutation arises. How many amino acids would you expect in the peptide fragment?

11. A certain polypeptide chain is composed of 100 amino acids. During its synthesis, a methionine molecule was picked up by a specific tRNA and placed in position No. 1 in the growing polypeptide. An asparagine molecule was placed in the second position, a leucine in the third, a histidine in the fourth, and so on and so on, until a trytophan is placed in the last position. Answer the following, identifying these five particular molecules.

A. The tRNA carrying this amino acid was the first to move to the "A" site of the ribosome.

B. This amino acid will have a free carboxyl group.

C. The tRNA carrying this amino acid did not associate with the "A" site of the ribosome.

D. The carboxyl group of this amino acid formed a peptide linkage with the amino group of leucine.

E. This amino acid has a free amino group.

12. Answer the following concerning ribosomes:

A. Give the characteristic S value of the intact eukaryotic ribosome.

B. Give the S value of the subunit in prokaryotes which attaches to the mRNA and to the first tRNA.

C. Give the S value of the subunit of *E. coli* which contains "A" and "P" sites.

D. Give the S value of the eukaryotic subunit which contains the enzyme required for the formation of peptide linkages.

E. Name the ion required for the cohesion of the two subunits.

14

ALLELISM, ANTIGENS, AND ANTIBODIES

DETECTION OF MUTATIONS AT DIFFERENT LOCI. Knowledge of the gene at the molecular level has provided a sound physical basis for some established genetic concepts. This is well demonstrated by a return to the subject of allelism. We have defined a genetic locus as a position on a chromosome occupied by a gene. Any specific locus in an organism is occupied by a gene which has primary effects on a particular characteristic; so we speak of a locus for eye color or a different locus for wing shape, etc. A gene residing at a certain locus may have alternative forms, alleles, which may produce different effects on the phenotypic characteristic. For example, a certain locus on Chromosome II of *Drosophila* is concerned with wing size and shape. Normal wing is long and smooth in outline. A mutation occurred at this locus producing an alternative form of the normal gene which can cause the wing to be shrivelled and nonfunctional. Following the system of naming genes in Chapter 4, the locus was designated the "vestigial locus" or "vg," because the mutation is recessive to the wild, "vg+." Only one of the alternative forms of the gene may be present at that locus on any one chromosome. But the

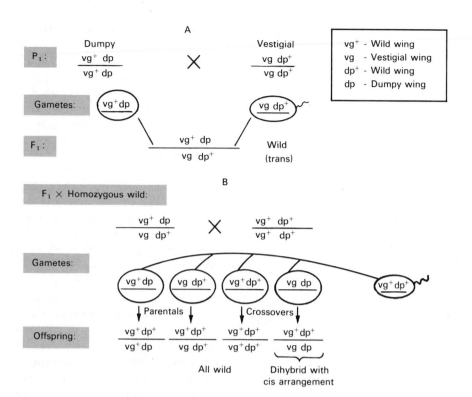

FIG. 14-1. *Cis* and *trans* arrangements in *Drosophila. A.* a cross of two mutants which are homozygous for recessives at two different loci produces an F₁ which is wild and which carries alleles in the *trans* arrangement. The F₁ is wild in phenotype, even if the parents showed a mutant trait for the same characteristic. This is so in this example because the dumpy parent contributes the wild allele of the vestigial, and the vestigial parent contributes the wild of dumpy. We say that such alleles complement each other. *B.* Since vestigial and dumpy are separate loci, crossing over may occur between them. This can give rise to offspring which are dihybrid and which have the *cis* arrangement. These dihybrids are wild, just as those with the *trans* arrangement.

diploid may be one of three possible genotypes: $vg^+ vg^+$; $vg^+ vg$; or $vg\ vg$.

Another recessive mutation occurred which also influences wing shape, causing it to be reduced and inadequate for flying. The locus at which the mutation took place is also on Chromosome II, and it was named "dumpy" or "dp." If a fly with dumpy wings is crossed to one with vestigial wings (Fig. 14-1A), the F₁ flies are all wild. This is not at all surprising. It simply means that the dp and vg loci, although both affecting the wing and both on Chromosome II, are separate loci, and thus the genes at these loci are different ones. Indeed, mapping of Chromosome II indicates that the two are over 50 units apart. When two wing mutants are crossed, each parent contributes a normal "wing" gene form. The F₁ heterozygous flies are therefore normal in phenotype and the alleles are in the *trans* arrangement. The gene forms dy and vg

are not allelic to each other. We see here again an example of the fact that two separate genes at different loci may affect the same characteristic. Since the loci, dp and vg, are so far apart, we can expect crossing over to occur between them. This does indeed take place, so that the dihybrid with the *trans* arrangement can produce chromosomes with both mutant alleles on the same chromosome and both wild alleles on the other (Fig. 14-1B). As a result, dihybrid flies can arise which have the *cis* arrangement. Although these points are simply a review of basic principles, it is important to realize here that when we are dealing with two recessive mutations at different loci, the F₁ heterozygote in *cis* or *trans* will be wild. This is so, because both heterozygotes will contain a normal dominant allele for each of the defective ones. We say that mutations such as dp and vg complement each other. This means that when two mutant forms are crossed, the recessive defect carried by either mutant parent can be "covered up" by the corresponding normal allele contributed by the other mutant parent. The F_1 resulting from a cross between dp and vg is normal, since the dp parent contributes a functional vg^+ allele and the vg parent a normal dp^+ allele. Two completely normal wing alleles are present, and so complementation takes place.

DETECTION OF SEPARATE MUTATIONS AT ONE LOCUS. Now let us turn to a third recessive mutation which also influences wing shape. The mutation was named "antlered," because wing form suggests the antlers of an animal. Tentatively, we might name the recessive "an," and attempt to map it. It is quite possible that there is a third locus which is also able to affect the shape of the wing. A cross between "an" and "dp" (Fig. 14-2A) produces wild F_1 offspring, and the gene maps on Chromosome II. The F_1 heterozygotes with the *trans* arrangement are wild, identical to those with the *cis* arrangement. This is exactly what we would expect for two separate loci. Additional data suggest that the "an" locus is over 50 map units from dp and must be very close to the "vg" locus. A mating is then performed between "vg" and "an" flies in order to map the "an" locus more accurately. We might expect the F_1 to be wild, because we are thinking

---

FIG. 14-2. Complementation and allelism. *A*. When flies showing antlered wing and those with dumpy wing are crossed, the F_1 is wild. This indicates that the two recessives are at separate loci. Complementation occurs because each mutant type contributes the wild type allele which the other parent lacks. Having both dominant alleles, the F_1 dihybrid is wild. *B*. When vestigial and antlered are crossed, we might expect the F_1 to be wild if we interpret the cross as one involving two separate loci. However, the actual results give an F_1 that is mutant, intermediate between the vestigial and antlered. This indicates that the F_1 did not receive a necessary wild type allele from each parent. Complementation did not take place. *C*. The reason for the unexpected results seen above is understandable when the cross is interpreted correctly. Vestigial and antlered are not allelic forms of genes at separate loci. They are actually variant forms of the same gene which affects the wing. Neither vestigial nor antlered possesses a wild-type form of the gene (vg^+) and so the F_1 is mutant. The antlered × dumpy cross actually does involve two separate loci as shown in *A*; however, the antlered effect is the result of a mutation at the vestigial locus, and so the correct symbol for antlered wing, vg^a, should be employed.

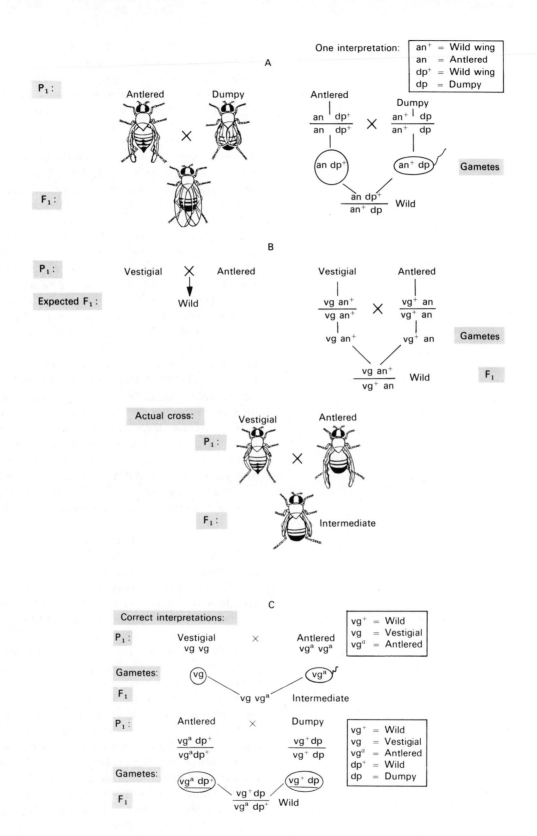

of "an" and "vg" as separate loci. However, the F_1 resulting from the cross are found *not* to be wild (Fig. 14-2B). Rather, their wings are distorted in size and shape and resemble some sort of intermediate between antlered and vestigial. Evidently, "vg" and "an" *did not* complement each other. This means that the F_1 does not have a normal gene form for antlered and a normal one for vestigial. As a result, the F_1 has a mutant phenotype. It would appear, therefore, that the "vg" mutation and the "an" mutation are both defects in the same gene. They are not identical defects, because each produces a slightly different phenotype. Nevertheless, they are alleles, alternative forms of the *same* gene. This should not surprise us, because the definition of "alleles" does not imply that only two forms of a gene can exist. Indeed, one gene may change in many different ways, and our knowledge of the gene at the molecular level makes this very logical. For the gene is a stretch of deoxyribonucleic acid (DNA). It includes many nucleotide pairs, and a change in a base pair may cause a mutant effect. Since there are many base pairs within a gene, there are thus many different points within it at which a change can take place. We might expect the phenotypic effect of many of these to be different. The gene is not an entity which can change in only one way so that only one pair of contrasting forms or alleles is possible at each locus. Rather, the genetic locus is complex; it encompasses a segment of DNA which includes a number of base pairs which are normally in a specific sequence.

Returning to "vestigial" and "antlered," we are therefore forced to recognize them *not* as alleles of separate genes but as different forms of the *same* gene, each with a different defect. This means that antlered and vestigial are alleles, variations of the same gene. We cannot symbolize a cross between antlered and vestigial as we did in Figure 14-2B, for this implies that they are not alleles. Because vestigial was the first mutation found at the locus involved in this case, the locus has been designated by the name of that mutation and represented as vg. But now, a second mutation has been detected at this same genetic region. To designate it, we keep the same base, vg, and add a superscript to give us vg^a. This tells us that we are dealing with another mutation at the "vestigial" locus. Figure

14-2C illustrates the cross between an antlered and a vestigial fly in the correct way, as well as the cross between antlered and dumpy. Following this method, we immediately see that vestigial and antlered are alleles and that the F_1 will be mutant. They cannot complement each other because only the one locus, vg, is involved, and both alleles at that site are defective. At the level of the DNA molecule, we can envision a segment of DNA, a gene, with the vestigial and antlered mutations at different points within the gene, each affecting a different nucleotide pair (Fig. 14-3A).

Many examples of loci with three or more alternative gene forms are nown in a variety of species, from microorganisms to humans. We recognize these genes as members of a series of *multiple alleles*, three or more forms of a gene at a genetic locus.

CROSSING OVER WITHIN A GENE. From the information presented above, we see that the detection of a multiple allelic series is not difficult. It is relatively simple to tell if two recessive mutations are allelic or not. For if we cross two organisms showing mutant phenotypes and obtain mutant offspring instead of wild, we then know that the two mutant genes are not complementary. They must be alleles. However, if they are mutations at separate and distinct sites within the same gene, we should be able to detect crossing over between them (Fig. 14-3B). We know today that *intragenic* crossing over is a frequent occurrence in microorganisms, where

FIG. 14-3. Allelism. *A.* A gene at a locus is a stretch of DNA composed of many nucleotide pairs. Any gene, therefore includes many sites. Each wild-type allele at a locus has specific nucleotide pairs which makes it distinct from genetic factors at other loci. The dumpy locus is distinct from vestigial, and the two are over 50 map units apart. A change may occur at one of the sites within a gene, giving rise to an allele or variant form, such as vestigial. A change does not always involve the same site within a gene, and so more than one variant form of a gene may arise, such as antlered and vestigial. Because the variant forms and the wild allele involve the same locus, they are alleles of the same gene. *B.* An individual may have two variant forms of the same gene and will show the mutant phenotype, as is the case in the F_1 between vestigial and antlered. Rare crossing over within the gene can produce a wild type gene form and one with two defective sites. (The illustration is highly schematic and does not represent the actual number of sites or the positions of the altered sites.)

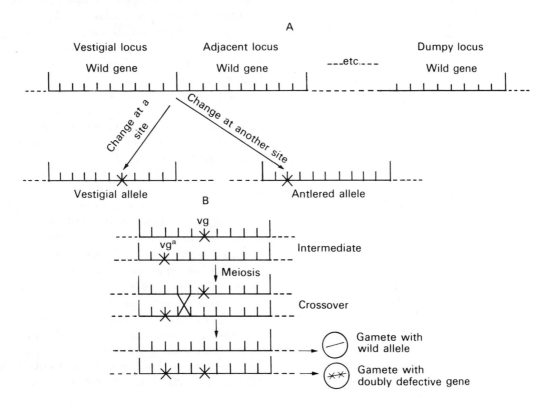

it can be more readily detected than in higher organisms. The reason for this ease of detection of crossing over within a gene should be evident from what we have learned about linkage and chromosome mapping (Chap. 8 and 9). Genes that are closely linked on the DNA of the chromosome would tend to stay together in the parental combination. There would be less crossing over between them than between two genes which are widely separated. And obviously, different sites within the same gene would be very closely linked, usually more so than two sites in two different genes. Therefore, in order to detect crossing over within a gene, we need to raise a very large number of offspring, because the new combinations (the recombinants) will be very few. To detect intragenic crossing over, even in a species as prolific as the fruit fly, requires the breeding and examination of thousands of offspring. With microorganisms (viruses, bacteria, and certain molds), it is possible to raise millions of progeny in a very short time. And techniques are available for the detection and scoring of recombinants. This extremely rapid rate of multiplication is one of the main reasons that crossing over within genes was fully analyzed in microorganisms. Intragenic mapping will be discussed in more detail in Chapter 18.

Cases of crossing over within the gene came to light slowly in *Drosophila*, although their true nature had to await clarification from studies with microorganisms. A well-known example in *Drosophila* concerns a locus which affects the shape of the eye, specifically the number of facets present, and its degree of pigmentation. The locus is located on the X chromosome and was named "lozenge" for the first mutation detected there. Additional independently occurring recessive mutations with a lozenge effect were also recognized. Some of these altered the eye in slightly different ways. When these recessives were crossed among themselves, the F_1 offspring were mutant in respect to their eyes. This indicated that the various recessives with the lozenge effect were all alleles. They fell into three main groups. For the sake of simplicity, we will summarize the situation as if there were three separate sites: lz^1, lz^2, and lz^3 (Fig. 14-4). Any combination of two mutations in the *trans* arrangement produces a mutant lozenge effect. After a detailed study of this region of the X

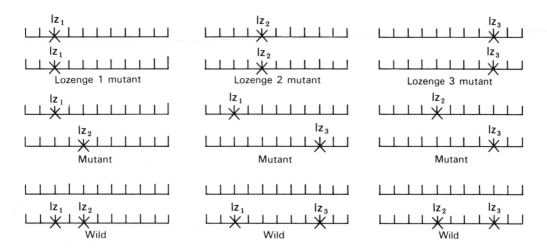

FIG. 14-4. The lozenge locus. Several recessive mutations have arisen independently at the lozenge locus, and some of these are at different sites. Each homozygote for a specific mutation shows a mutant phenotype (*above*). When these different mutants are crossed among themselves (*middle*), the F_1 hybrids are mutant.

They carry the mutant sites in the *trans* arrangement and have no completely normal lozenge gene form. Those hybrids with the *cis* arrangement (*below*) are wild, because one of the alleles is completely normal. The two defects are in the other allele.

chromosome, which entailed raising a very large number of flies, it was found that crossing over could take place among the three lozenge sites. As a consequence of crossing over, two defects could be placed together on the same chromosome. This in turn made possible the production of flies with the *cis* arrangement. Such flies, as opposed to those with the *trans* arrangement, were found to be wild.

The lozenge story revealed the fact that the *cis* and *trans* arrangements could result in different phenotypes, mutant in *trans* and wild in *cis*. This was very annoying to the geneticist of the early 1950's who knew nothing about the molecular biology of the gene and who was reluctant to say that crossing over could take place within a gene, something unheard of in classical genetics. But today, in light of our knowledge of gene structure, the difference between the effect of the *cis* and *trans* arrangements is readily explained as seen in Figure 14-4. The dihybrid with the *trans* condition possesses two defective forms of the same gene and consequently shows a mutant effect. There is no complementation. In the *cis* arrangement, however, both defects are within the *same* stretch of DNA so that the dihybrid has one completely normal gene, and the phenotype is wild. The term *pseudoalleles* was coined to designate cases

such as lozenge, where a difference is apparent between a dihybrid with the *cis* arrangement and one with the *trans*. Such a difference came to be called the *cis-trans* effect. The various cases of pseudoalleles which were detected in the fruit fly were very bothersome to the geneticist who was reluctant to admit that the classical gene is divisible, and that it is composed of parts separable by crossing over. Once research with microorganisms clearly showed that the *cis-trans* effect is quite common when analyzing very large numbers of offspring, the concept became acceptable that the gene includes many separate sites which can undergo independent mutation and crossing over. Actually, it fit in beautifully with the knowledge that the gene is a polynucleotide. Today we still encounter the term "pseudoalleles," but it is a misnomer, because mutations at sites within the one gene are truly "alleles."

Wherever feasible, we may try to map sites within a gene, but since this entails vast numbers of offspring, intragenic maps in higher organisms are few. In the fruit fly, several genes have been mapped for a few mutant sites. In the case of lozenge (Fig. 14-4), three of the mutant sites are separated by much less than 1 map unit. Progress has been made in mapping certain genes in mice which influence the coat color (the yellow gene) and the development of the embryo (the tailless gene). Failure to demonstrate recombination between two alleles does not necessarily mean that crossing over cannot occur between them. It may mean only that we have as yet failed to detect it because the mutant sites are so very close together. There is little wonder that many genes have been mapped in bacteria and viruses in contrast to higher organisms with their longer generation times and smaller number of offspring.

ALLELISM AND THE FUNCTIONAL UNIT AT THE MOLECULAR LEVEL.   Let us now see how allelism and the *cis-trans* effect can be explained in light of our knowledge of molecular interactions. We have considered the gene in relationship to its function at the molecular level, the transfer of information to direct the formation of a specific polypeptide chain. Any gene, because it has a certain function to perform, therefore, may be considered as a "unit of function." Each gene, or unit of function, contains smaller units within it, the nucleotides. Mutation may occur at any one of these sites within a gene, and as we shall see in more detail in Chapter 18, crossing over may take place between nucleotides within the gene. Therefore, the gene, the unit of function, is composed of smaller units of mutation and crossing over, the nucleotides. Viewed in this way, we see that the genes on the chromosomes are not comparable to a group of indivisible beads tied together on a string. Each gene is instead a stretch of DNA with a specific function to perform, and each includes separate sites of mutation and crossing over. The term *cistron* (Chap. 18) has been coined to refer to the gene as a unit of function. The expression should not confuse us if we understand what has been said about the physical basis of the gene and its molecular interactions. The words "gene" and "cistron" mean the same thing insofar as both designate a genetic unit with a specific function. The expression "cistron" implies that the functional unit, the ordinary gene, is divisible. It represents a genetic region within which separate mutations can occur and crossing over can take place. It is equally correct to say "one gene-one polypeptide" or "one cistron-one polypeptide."

In our discussion earlier in this chapter, we

learned that a cross between two recessive types gives normal progeny if the mutations being followed are in separate genes. However, if two separate mutations represent defects at different sites within the *same* gene, mutant offspring are produced. In the latter case, we say that the two mutations do not "complement" each other and are therefore allelic. If no mutant offspring had resulted, we would know that the mutations *do* "complement" each other. The mutations would then be defects in separate genes and would *not*

be allelic. Let us now examine these ideas in terms of gene function. Suppose two separate recessive mutations arise which cause the blocking of a metabolic step controlled by a certain enzyme (Fig. 14-5A). The blocking is shown to result from the production of a defective enzyme by the mutant types. The genetic defects in each mutant are mapped and are shown to lie at closely linked, but nevertheless separate, sites on the chromosome. The question now arises, "Are these defects separate mutations within

FIG. 14-5. Lack of complementation. *A.* Assume enzyme X, which is composed of two kinds of polypeptide chains, is needed to catalyze the step in which Y is changed to Z (*above*). Two separate mutations occur which block formation of enzyme X so that the step does not proceed (*below*). When the sites of the mutations are mapped, they are found to occur at separate, closely linked positions. *B.* When the two mutants are crossed in this example, the combination of the two separate mutations gives no enzyme. This is to be expected if the two defects are in the same gene. In this diagram, both are represented in gene "B." If the normal enzyme structure depends on both polypeptide "A" and polypep-

tide "B," the F_1 hybrid will not be able to form normal enzyme, because no normal polypeptide "B" can arise. This follows from the fact that the F_1 received a defective "B" allele from each parent. The two separate defects or mutations do not complement each other. *C.* An individual with the mutations in the *cis* arrangement will be able to produce normal enzyme. This is so, because at least one completely normal "B" allele is present to form the required polypeptide. Both defects are within the same stretch of DNA, giving a doubly defective allele, but the completely normal one can carry out the essential function.

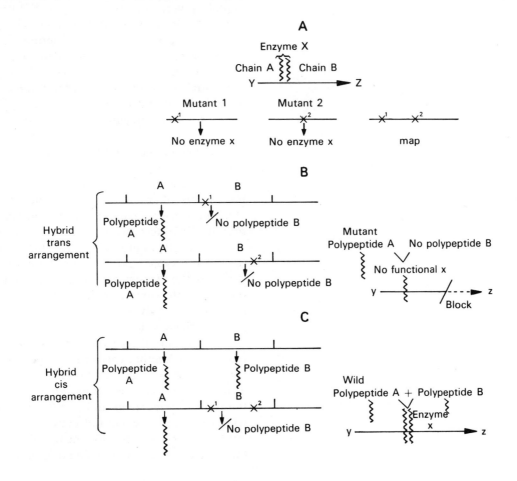

the same gene or are they in different genes?" A cross is made between the two mutant types, each carrying a mutation which maps at a different site. When the offspring are produced, they are also found to be mutant, defective for this same enzyme. What does this mean in terms of the unit of function (Fig. 14-5B)? Since the mutations are not complementary, no normal enzyme can be produced by the F_1 offspring. This is so for the following reasons. One parent contributed a genetic factor with a defect at a certain site, and this results in the production of a defective polypeptide, one incapable of catalyzing the needed metabolic step. The other parent also contributed a defective genetic factor, and it is a form of the same gene that is defective in the first parent, and so, the same polypeptide is affected. The only difference between the two mutants is that they carry defects at different sites within the same gene. The two factors are alleles. Therefore, the *trans* arrangement produces mutant types, because no functional polypeptide governed by that one gene is being produced. No intact enzyme can be formed.

Crossing over in the F_1 offspring with the *trans* arrangement may yield a doubly defective allele and one which is normal (Fig. 14-5C). Any individual carrying both a normal allele and a gene form with two defects within it would produce normal enzyme. This is so, because the individual with the *cis* arrangement possesses one completely normal allele (B in Fig. 14-5C). Contrast this with the F_1 heterozygote with the *trans* arrangement. The latter carries no completely normal unit of function for the formation of the necessary polypeptide and is consequently mutant. Therefore, when crossed, if two recessive mutant types produce an F_1 which is defective for the same characteristic (here a specific enzyme), the two separate mutations are allelic; they affect the same unit of function. The *trans* arrangement gives a mutant phenotype, whereas the *cis* yields wild.

Now let up apply exactly the same kind of reasoning to another case of two separate recessive mutations (Fig. 14-6A). However, in this example, a cross of two mutants defective for the same enzyme gives an F_1 which is wild, one which can produce normal enzyme and carry out the essential step. This means the two mutations can complement each other in the *trans*

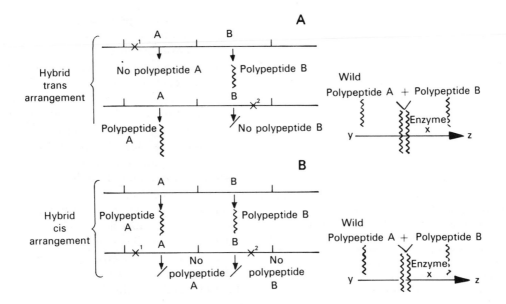

FIG. 14-6. *Cis-trans* arrangements. Complementation. *A.* In this example, an F₁ hybrid is formed between two mutants (A and B), each with a defect in a separate gene. Consequently, one functional "A" allele and one functional "B" allele are present to allow formation of the enzyme. *B.* This *cis* arrangement is wild, because one chromosome contains both the functional "A" and the functional "B" alleles of two different genes.

arrangement. The defects must be in separate units of function, separate cistrons. They complement each other because gene "A" controls a polypeptide chain which is distinct from the polypeptide chain controlled by gene "B." In order for the enzyme to be functional, there must be at least one good chain "A" and one good chain "B." Mutant "A" contributes a defective "A" but a good "B." Mutant "B" contributes a good "A" but a defective "B." As result, the F₁ has some good "A" and some good "B" and can therefore form functional enzyme. The hybrid with the *cis* arrangement will also form functional enzyme (Fig. 14-6*B*). The "*cis-trans*" test is a fundamental one used to determine allelism. The basis for the different results in the *cis* and *trans* arrangements is easily understood on the basis of gene function and molecular interactions.

MULTIPLE ALLELIC SERIES IN ANIMALS AND PLANTS. The number of alleles composing a multiple allelic series has been found in many cases to be quite large, and there is the possibility that the detection of additional alterations at a particular locus will increase the number fur-

ther. Even in higher forms such as the fruit fly, as many as a dozen alleles are known for certain loci. At the sex-linked white-eye locus, for example, some 12 alleles have been recognized. These produce a spectrum of pigmentation from white through shades which gradually deepen to the normal red of the wild type. The mutations have been named to suggest their phenotypic effects such as white, ivory, pearl, apricot, and cherry (w, w^i, w^p, w^a, and w^{ch}). A heterozygous female has an eye color intermediate between that of the homozygotes. Crossing over has been demonstrated among some of these, proving that they are separate sites within the "white" gene. As noted above, failure to demonstrate recombination among all of them does not mean that this is not possible but may only reflect an inability to detect it with the number of flies studied so far.

The occurrence of numerous mutations at a locus also shows us that although many changes can occur at a site and produce differing phenotypic effects, the *primary* effect is still confined to the same characteristic. So we may find many alterations within the "white" gene affecting eye pigment in different ways, but no one of these changes has a pronounced influence on another characteristic, such as wing shape. This again makes sense, because any locus has a primary effect in a certain chemical pathway (Chap. 4) due to its control of a specific metabolic step.

In the rabbit, a well-known multiple allelic series is known which influences pigmentation of the fur (Fig. 14-7). At least four alleles are known in the series: C (wild); c^{ch} (chinchilla); c^h (himalayan); and c (albino). In Chapter 4, we used the interaction of the himalayan allele, c^h, and the temperature of the surroundings as an example of the influence of environment on gene expression. The expression of the other mutant allele, c^{ch}, is similarly affected by body temperature, because the enzyme controlled by the allele is heat sensitive. The amount of pigment in a chinchilla rabbit is therefore less than in

FIG. 14-7. Coat color in rabbits. At least four alleles are known to affect pigment of the fur. In this multiple allelic series, the mutant types are due to changes at different sites within the pigment gene. The wild-type allele is completely dominant to its variant forms. The himalayan allele is dominant to albino. However, the chinchilla allele is not completely dominant to the himalayan and albino alleles. (The representation of the alleles does not intend to represent actual sites within the gene.)

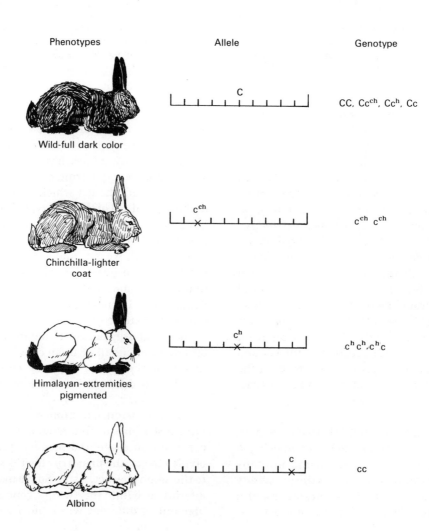

Phenotypes	Allele	Genotype
Wild-full dark color	C	CC, Cc^{ch}, Cc^h, Cc
Chinchilla-lighter coat	c^{ch}	$c^{ch}c^{ch}$
Himalayan-extremities pigmented	c^h	c^hc^h, c^hc
Albino	c	cc

the wild. Any one animal can have any two of the four alleles. What this series illustrates is that in a multiple allelic series, some mutant alleles may be dominant to others; others may show codominance or incomplete dominance. The wild allele, C, is completely dominant to the other three in the rabbit, but different relationships exist among the mutant forms. Himalayan is dominant to albino. However, chinchilla animals are homozygous ($c^{ch} c^{ch}$) because the chinchilla allele is not completely dominant to albino and himalayan. The heterozygotes are light gray. This series in the rabbit contrasts with the eye pigment alleles in *Drosophila*, where the mutant gene forms are all incompletely dominant to each other.

In some plant species, genetic factors which form multiple allelic series have been shown to exert a very important role in the control of crossing ability. The situation was first described by East in the 1920's with *Nicotiana*, but comparable series are known to occur in other species. It is a well-known fact that some plants are normally self-pollinated, as are the pea plants of Mendel. Pollen falling from the anther onto the stigma of the same flower or a flower on the same individual will produce a pollen tube which delivers the male nucleus to the egg. The pollen may also grow when transferred to a stigma on another individual. Many species of plants, however, are normally only cross-pollinated. If the pollen is to function, it must encounter a stigma in a flower on a different plant; otherwise, it will cease growth. There is actually a pronounced value to a system which ensures cross-pollination, for this in turn guarantees outbreeding. Crossing with other genetic types increases the chances of hybrid vigor and decreases the probability that harmful recessives will be expressed. Self-pollination is the height of inbreeding. As mentioned in Chapter 7, species which are exclusively self-pollinated go through the motions of sex (meiosis and gamete formation), but they lose all of its advantages (the production of new combinations). Some outbreeding would seem essential to a species if it is to evolve at any appreciable rate or even survive in a radically changed environment. Exclusive self-fertilization can deprive any species of the flexibility needed to produce a range of new allelic combinations, some of which may have a selective advantage in a changed environment. Plant groups which are only self-pollinated are generally at a disadvantage to those in which a certain amount of outcrossing is ensured. The genetic basis for cross-pollination in tobacco resides in a large number of self-sterility alleles, which form a multiple allelic series: S_1, S_2, S_3, S_4, etc. Any one plant can be hybrid for any two of these, but no plant can be homozygous for a particular sterility allele (Fig. 14-8). This is so, because pollen carrying a specific sterility allele will abort on the stigma of a flower which contains the same allele. If a plant is S_1S_2, none of its pollen can survive on the stigma of any flower on that same plant. Even if two different plants have the same sterility alleles, they are still incompatible, and so we cannot cross two individuals which are both S_1S_2, for example. However, an S_1S_2 can be crossed with S_2S_3, but only one-half of the pollen from each can be successfully exchanged, because they each possess the S_2 allele in common. On the other hand, the S_1S_2 type and the S_3S_4, having no sterility alleles in common, can freely exchange pollen and genetic material reciprocally. This example of multiple alleles in plants is one which illustrates a genetic mechanism to bring about outbreeding and all the advantages to the species which it entails.

MULTIPLE ALLELES AND THE ABO BLOOD GROUPING.   In humans, about 15 major blood groups have been recognized, and most of these are inherited on the basis of multiple allelic series. The ability to classify the blood of any individual into the various groupings depends on the presence or absence of specific antigens in the red blood cells. Antigens are large molecules usually proteinaceous, at least in part. They can elicit the formation of specific antibodies, other large protein molecules (the gamma globulins), with which they react. The reaction can be recognized by a clumping or a lysis of the cells. Certain antibodies occur naturally in the body fluids and are independent of the presence of a corresponding antigen. The ABO grouping illustrates all of these points. It is distinguished by the presence of A or B antigens in the red blood cells and of the a or b antibodies in the blood serum (Table 14-1). The presence of a specific antigen in the red blood cells, A, B, or both A and B, defines the corresponding blood

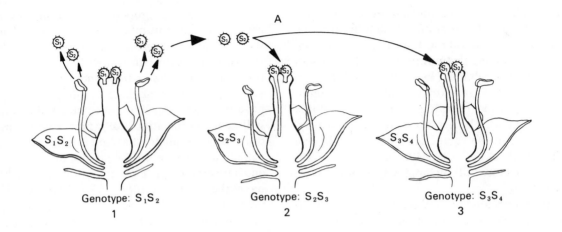

Genotype: S_1S_2
1

Genotype: S_2S_3
2

Genotype: S_3S_4
3

B

Summary of three reciprocal crosses

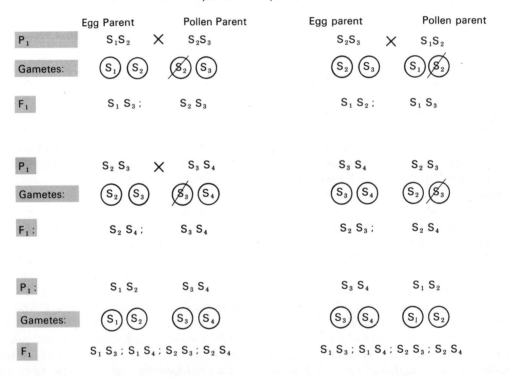

FIG. 14-8. Self-sterility alleles. *A*. In many plants, the inability to self-pollinate is based on sterility alleles which make it impossible for pollen to grow on the stigma of a plant which contains the same allele. All plants must be heterozygous for the sterility alleles. In this example, only one-half of the pollen from plant 1 can grow on plant 2; the other one-half which carries the allele in common aborts. Similarly, only one-half of the pollen from plant 2 will grow on plant 1. However, all of the pollen from plant 1 will grow on plant 3, and vice versa, because they have no sterility alleles in common. Plants 2 and 3 will only be able to exchange one-half of their pollen successfully. As a result of this system, only heterozygous individuals can develop, as the summary of the reciprocal crosses possible with these three genotypes indicates (*B*).

TABLE 14-1 General scheme of antigens and antibodies in the ABO system

BLOOD TYPE	A AND B ANTIGENS IN RED BLOOD CELLS	A AND B ANTIBODIES IN SERUM
A	A	b
B	B	a
AB	A and B	None
O	None	a and b

TABLE 14-2 General scheme of genotypes in the ABO system

BLOOD TYPE	POSSIBLE GENOTYPES	A AND B ANTIGENS IN RED BLOOD CELLS
A	$I^A I^A$, $I^A i$	A
B	$I^B I^B$, $I^B i$	B
AB	$I^A I^B$	A and B
O	ii	None

type within this major grouping. As Table 14-1 shows, if an antigen is absent from the blood cells, the corresponding antibody is present. Since antigen A reacts with antibody a (and likewise B with b), a knowledge of the ABO type is crucial in blood transfusions. Any transfusion is incompatible if the cells of the donor are clumped by those of the recipient. A person of type A contains antibody b in his serum. The millions of cells from a B donor would be clumped upon contact with the b antibody and would clog the smaller blood vessels, resulting in severe damage or death. Only type O can donate blood to the other types with any amount of safety. Although both types of antibodies, a and b, are present in the serum of type O and will react with type A and B cells, the serum of the O donor would become diluted by the A, B, or AB recipient so that a severe effect is avoided.

The A and B antigens are actually identical in structure as far as their protein portion is concerned. The difference between them resides in a sugar component to which they are complexed. The A and B antigens are mucopolysaccharides, combinations of a sugar and a protein. We say that their antigenic specificity resides in the sugar component, because it is that portion, not the protein, which reacts specifically with the antibody. The antigenic specificity is under the genetic control of the ABO locus. At its simplest, three different allelic forms may occur at this locus, I^A, I^B, and i. The alleles I^A and I^B determine the presence of antigens A and B, respectively, and each is dominant to the allele, i, for blood grouping O, the absence of both the A and B antigens (Table 14-2). The I^A and I^B alleles provide an excellent example of codominance, because the heterozygote, $I^A I^B$, produces both the A and B antigens. The number of alleles at the ABO locus is now known to total more

than three as originally described. The A blood types have been shown to include more than one subgrouping: A_1, A_2, and several others. The existence of several kinds of B type is also indicated. The I^{A1} allele appears to act as a dominant to I^{A2}. If we just consider the alleles I^{A1}, I^{A2}, I^B, and i, we see that different subgroupings of type AB are possible, as well as of type A (Table 14-3). Six different types can be recognized, and these are distinguished from one another on the basis of specific antigen-antibody reactions. If we were to consider additional subgroupings of A and B, the number of possible ABO types would increase accordingly. The accurate detection of antigenic differences among the subgroupings has value not only for its medical applications but for its legal aspects as well.

Ignoring the subgroupings for the moment, let us see how knowledge of the inheritance patterns of the blood groups can aid in a case of disputed paternity. Suppose that a man is suing his wife for divorce and accuses another man as the father of the wife's child. Blood typing reveals the following: wife, type B; baby, type AB; husband, type O; and man, type A. Our genetic

TABLE 14-3 Two subgroupings of type A and possible blood groups

	BLOOD GROUP
$I^{A1} I^{A1}$	A_1
$I^{A1} i$	A^1
$I^{A1} I^{A2}$	A^1
$I^{A2} I^{A2}$	A_2
$I^{A2} i$	A_2
$I^B I^B$	B
$I^B i$	B
$I^{A1} B$	$A_1 B$
$I^{A2} B$	$A_2 B$
ii	O

knowledge tells us that the husband has a well founded suspicion (Fig. 14-9). He cannot be the baby's father because he is blood type O and cannot therefore contribute an allele for the production of either A or B antigen. The baby must have received an allele for antigen A from the male parent. In reaching a decision regarding the correct male parent, however, caution must be exercised. Although the accused man possesses blood type A, an appropriate type for the baby's father, this evidence by itself cannot incriminate the man. For many men have blood type A. Moreover, the father could also be type AB, because such a person can contribute either the allele A or the allele B to the offspring. So we see that the genetics of blood types may eliminate a person as a possible parent, if he lacks the allele required for the production of a specific antigen. However, just because a person has the suitable blood grouping does not mean that he *is* the parent in question.

In our example, suppose that subgroupings of A were identified, and the baby was found to be A_1B and the man A_2. Since an A_2 person (Table 14-3) can be only genotype $I^{A2}I^{A2}$ or $I^{A2}i$, he cannot give the offspring the I^{A1} allele. The man would thus be exonerated on this more detailed blood typing. In this way, a person may be eliminated as a possible parent as more detailed blood groupings are examined. Although the probability of parenthood would become greater in a case where more and more subgroupings between child and suspected parent agree, the evidence by itself would still be inconclusive.

The ABO locus has now been assigned to a specific chromosome, Chromosome 9, which also carries the locus associated with the nail-patella syndrome. The value of multiple alleles in the establishment of human linkage groups was discussed in Chapter 9 in relation to the ABO series. The ABO locus is known to interact with certain loci to which it is not linked. We have already noted the importance of the secretor locus in genetic counseling. The secretion of the A or B antigens into various body fluids depends on the presence of the dominant "Se." Approximately 70% of persons with these antigens are also secretors. An AB secretor must be either genotype I^AI^B Sese or genotype I^AI^B SeSe and will secrete both antigens into the body fluids. An O type individual, even if he or she has the "Se"

FIG. 14-9. Blood groups and disputed paternity. In a case such as this, the husband cannot be the father of the child, because he is blood type O and can contribute to his offspring neither allele I^A nor I^B for the corresponding antigens. The child received the I^B allele from the mother and must have received the I^A allele from the father. The other man here could be the father on the basis of his blood group, but the possession of an appropriate grouping is inconclusive by itself.

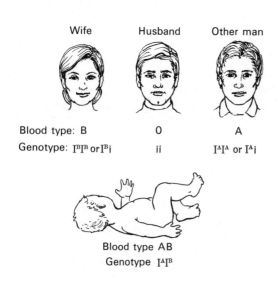

Wife	Husband	Other man
Blood type: B	O	A
Genotype: I^BI^B or I^Bi	ii	I^AI^A or I^Ai

Blood type AB
Genotype I^AI^B

allele, cannot secrete antigen but may pass the secretor ability down to offspring.

Another locus, also not linked to the ABO, has been found to interact with it in an interesting manner. We have just seen that a person of AB blood type cannot have an O parent. However, very rare cases have been described in which an individual carries the alleles for A or B antigen, but neither is expressed. A person of genotype $I^A I^B$ could thus appear to be type O. The ability to produce A or B antigen at all depends not only on the presence of allele I^A or I^B at the ABO locus, but also on at least one other locus, the h locus. The presence of the dominant allele H determines the presence of substance H, which can also be detected by antigen-antibody reactions. This H substance is essential for the formation of either A or B antigen (Fig. 14-10A). Since most persons have

this dominant allele they consequently contain H substance in their blood serum, no matter what their blood group. The rare individual of genotype hh lacks H which is required for the formation of A or B antigen. He or she is said to express the Bombay phenomenon, so called for the city where the first case was reported. Not only is the individual negative for H substance but will appear as type O, regardless of ABO genotype (Fig. 14-10B). Although expression of the Bombay phenomenon is rare, it is an excellent example of an epistatic effect and underscores the caution which must often be exercised in a specific case before reaching a conclusion on the basis of a single pedigree.

THE MN BLOOD GROUPING. Another major blood grouping discovered many years ago is the MN group. Each of us falls into one of three

FIG. 14-10. The "H" substance and the Bombay phenomenon. *A.* The allele "H" is present in most persons and is responsible for the production of a substance required for the development of A and B antigens. In the absence of "H," the recessive allele is unable to produce the substance. A and B antigens do not complete development, and therefore, A and B antigen cannot be detected in the blood. *B.* Normally, all persons are homozygous for "H." In a typical cross of an AB person with one who is type O (*left*), the offspring will be either type A or B. A very rare person (*right*) may be homozygous for the recessive allele "h" and manifests the Bombay phenomenon. Although phenotypically type O, the person has the genetic information to produce incomplete A and B antigens. Since the other parent carries the "H" allele, the offspring will be either type A or B. As seen here, the most unusual situation would arise of two type O parents with offspring who have A or B antigens.

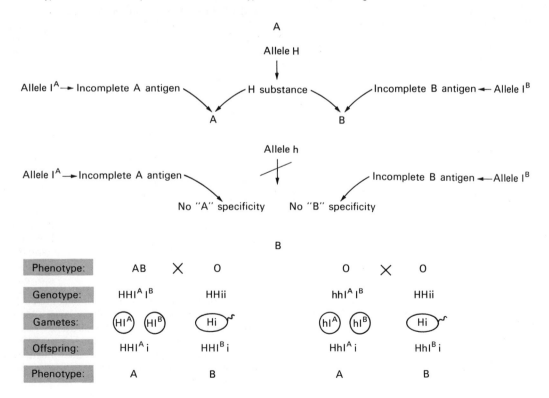

classifications: M, N, or MN, due to the possession of M or N antigen in the red blood cells. The
average person is unaware of his MN type,
because the MN antigens play no role in blood
transfusions. Antibodies to M or N antigens do
not naturally occur in the blood serum. Such
antibodies, however, can be elicited by injecting
human blood into an experimental animal such
as a rabbit. Antibodies to M or N form in the
animal's serum and when mixed with human
blood, can reveal the presence of the antigens in
the human red blood cells, as they cause the
latter to clump. At its simplest, the formation of
M or N antigens depends on a pair of alleles, L^M
and L^N. The two forms exhibit codominance, so
that three possible genotypes and phenotypes
are recognizable: $L^M L^M$ (type M), $L^M L^N$ (type MN),
and $L^N L^N$ (type N). These react specifically with
antibodies M, M and N, and N (Table 14-4A).
The situation at the MN locus has proved to be
more complicated than was first suspected. The
three phenotypes are always associated with the
presence of one or two other antigens, antigen
S or antigen s (Table 14-4B). An M person could
be type MS, Ms, or MSs and likewise for MN and

TABLE 14-4 MN and MNSs blood groupings

PHENO-TYPE	GENOTYPE	ANTIGENS IN RED BLOOD CELLS
A*		
M	$L^M L^M$	M
MN	$L^M L^N$	M and N
N	$L^N L^N$	N
B†		
MS	$L^{MS} L^{MS}$	M and S
Ms	$L^{Ms} L^{Ms}$	M and s
MSs	$L^{MS} L^{Ms}$	M, S, and s
MNS	$L^{MS} L^{NS}$	M, N, and S
MNs	$L^{Ms} L^{Ns}$	M, N, and s
MNSs	$L^{MS} L^{Ns}$ or $L^{Ms} L^{NS}$	M, N, S, and s
NS	$L^{NS} L^{NS}$	N and S
Ns	$L^{Ns} L^{Ns}$	N and s
NSs	$L^{NS} L^{Ns}$	N, S, and s

* At its simplest, the MN blood type is inherited on the basis
of a pair of alleles, each of which is responsible for the
production of a specific antigen. The alleles are codominant.
 † The M and N blood types are always associated with S or
s antigens. One interpretation, the one shown here, is that a
multiple allelic series composed of four alleles is associated
with the MN locus.

UNDERSTANDING GENETICS

N types. Some authors interpret the phenotypes on the basis of a series of multiple alleles at the MN locus in which four allelic forms occur: L^{MS}, L^{Ms}, L^{NS}, and L^{Ns}. More recent evidence favors the idea that two separate but very closely linked loci are involved: the MN locus and the Ss locus. Crossing over would be so rare due to the extremely close linkage that the separate genes involved would appear to form one multiple allelic series. It should be evident that the MN grouping by itself may also aid in cases of disputed paternity. Because the L^M and L^N alleles are codominant, a child of blood type N (genotype $L^N L^N$) whose mother is type MN could not have a father of type M (genotype $L^M L^M$). The latter can only provide an L^M, and this allele will express itself when present in the genotype.

ANTIGENS AND THE RH BLOOD GROUPING. One major blood grouping which has been well publicized is the Rh. Its discovery goes back to 1940. Landsteiner and Wiener injected blood from the Rhesus monkey into the rabbit, which in turn responded by the production of antibodies. These antibodies were then able to agglutinate the monkey's blood when mixed with it. The rabbit antiserum containing the antibodies against the monkey's blood was also mixed with blood from humans. It was found that the blood of 85% of those tested was also clumped by the antiserum. Landsteiner and Wiener had removed from the antiserum all those other antibodies known at the time which could possibly react with human antigens. Since clumping still took place, this meant that the rabbit serum contained an antibody against an antigen present in *both* human and monkey blood and which had been undetected up to that time. The antigen was designated "Rh" to denote its presence in the Rhesus monkey.

The presence of Rh antigen was shown to be under genetic control. Those persons with the Rh factor were designated Rh+. The 15% of the white population whose blood did not react with Rh antibodies were termed Rh− because the antigen or factor was absent from their blood. The inheritance of the Rh type seemed to depend on a single pair of alleles: the dominant R for the presence of the antigen and the recessive r for its absence. Unlike the ABO system, no naturally occurring Rh antibodies of any kind

are formed in the blood. The Rh− person (rr) is not born with antibodies against the Rh antigen, nor is the Rh+ person born with antibodies against Rh− blood. However, the Rh− individual, it was found, could form antibodies against Rh antigen if he or she received an injection of Rh+ blood. These antibodies can clump cells containing Rh antigen once the antibodies are formed by the Rh− person in response to the presence of Rh+ blood in the bloodstream. It soon became evident that the Rh factor is just as important as the ABO type in transfusions. An Rh− person may be able to tolerate one transfusion of positive blood, because he contains no preformed antibodies. However, once sensitized, he will produce a high level of Rh antibodies on a second transfusion, and their reaction with Rh+ antigens in red blood cells can lead to fatal results.

Not only did the Rh type demand consideration in transfusions, it also became implicated in certain cases of blood incompatibility between mother and child. Some babies are born severely jaundiced and seriously anemic. This condition, *erythroblastosis fetalis,* is the consequence of fetal red blood cell destruction which not only produces an oxygen defect but which also causes clogging of the blood vessels of the liver with damaged red blood cells and the entry of bile pigments into the blood. In very severe cases, the child may suffer permanent damage. Death may even occur shortly before or after delivery. The destruction of the baby's red blood cells results from an antigen-antibody reaction in the child's bloodstream. We have noted that a person may become sensitized to Rh antigen after a transfusion. A high antibody level can be reached in an Rh− woman who has been so sensitized. If her husband is Rh+ his genotype may be either RR or Rr. If he is heterozygous there would be a probability of 1/2 that the dominant allele will be transmitted. If the offspring does inherit the dominant allele and is thus Rh+, any antibodies against the Rh antigen will react with the blood of the fetus *if* they pass the placental barrier from mother to child due to some defect in the placenta (see below). So an Rh− mother, already sensitized, can possibly pass antibodies to her offspring *in utero.* Serious consequences can then follow (Fig. 14-11A).

Unfortunately, this is not the only way in which an incompatible Rh reaction can come

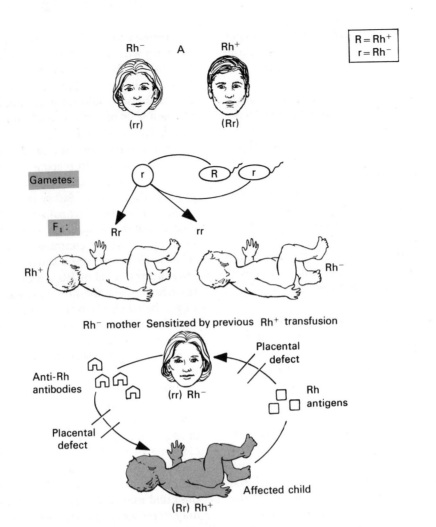

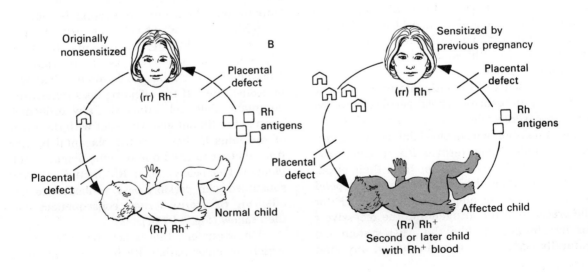

FIG. 14-11. Rh incompatibilities. *A.* An Rh⁻ woman and an Rh⁺ man who is heterozygous have a 50:50 chance of having an Rh⁺ child (*above*). If the child is Rh⁺ and the mother is sensitized due to a previous transfusion, Rh antibody formation may be evoked and antibodies may pass the placenta (*below*). This will occur only if a defect exists in the placental capillaries to permit fetal-maternal blood exchange. Consequently, an Rh⁺ child carried by the mother can suffer from *erythroblastosis. B.* A woman not carrying antibodies against Rh⁺ blood may become sensitized by an Rh⁺ offspring, but only if a placental defect exists (*left*). The mother usually will not form a sufficient number of antibodies to affect the first Rh⁺ child, even if she becomes sensitized by the pregnancy. An Rh⁻ woman already sensitized by a previous pregnancy with an Rh⁺ child can respond to antigen from a second Rh⁺ fetus (*right*). Antibody production can be high enough to affect the second Rh⁺ child. This arises only if a placental defect is present.

about (Fig. 14-11*B*). An Rh⁻ woman who has never been exposed to Rh antigen *may* become sensitized during a pregnancy if she is carrying an Rh⁺ child. It is possible for some blood from the fetus to leak into the circulatory system of the mother. This can result from some defects in certain placental capillaries which may actually break or allow seepage of blood from fetal to maternal circulation. Once this happens the Rh⁻ mother may become sensitized, just as if she had received an injection of Rh⁺ blood. The antibodies formed by the Rh⁻ mother may in turn cross the placental barrier and cause their damaging effects. Although Rh incompatibilities between mother and child are certainly a matter for concern, knowledge of the subject should be used properly. Publicity has caused unnecessary alarm to many parents who differ in their Rh types. We have seen that trouble may ensue if the mother is Rh⁻ and the baby Rh⁺. This follows only from a mating between an Rh⁺ man and an Rh⁻ woman. If the woman is positive and the man negative, there is no problem, because an Rh⁻ baby carried by an Rh⁺ mother cannot cause the mother to form antibodies which will destroy Rh⁻ blood. But *even if* a woman is Rh⁻ and her mate is positive, there may still be no problem at all. If the male is heterozygous (Rr), there is a 50% chance that the child will be Rh⁺. If the child should be Rh⁺, he will most probably not be affected if he is the first Rh⁺ born to an unsensitized mother. Even if the first Rh⁺ child does sensitize the mother, the build up of antibodies in the maternal circulation usually will be insufficient to cause a strong reaction. And it

must also be kept in mind that an Rh⁺ offspring does not necessarily sensitize the mother. The placenta is normally a very effective barrier between mother and child. In most cases, there will be no exchange of blood between mother and child. *Erythroblastosis fetalis* is actually uncommon, because placental defects are exceptional rather than the rule. This means that an Rh⁻ mother may have one or more Rh⁺ children and encounter no difficulties. So we must apply the genetic knowledge intelligently. Techniques are now available to detect the presence of fetal red blood cells or Rh antibodies in an Rh⁻ woman. If such antibodies are present, suppressant drugs may be administered so that the blood of the fetus will not receive a large dose of them. Moreover, a vaccine has been developed which prevents an Rh⁻ woman from producing antibodies against Rh⁺ blood, even if fetal blood should cross the placental barrier. Proper medical counseling can avoid unnecessary alarm, as well as the unfortunate consequences of Rh incompatibility.

Not long after its discovery, the Rh grouping was shown to have a more complex genetic basis than was just described and to involve more than the one pair of alleles, R and r. With the improvement of techniques for the detection of antigenic differences, it became clear that not all Rh⁺ individuals are the same. As shown by antibodies produced against Rh⁺ blood in experimental animals, there are actually many different Rh⁺ types. These various Rh⁺ individuals possess different antigens which may be detected through antigen-antibody reactions. Some Rh⁺ persons have two kinds of Rh antigens in their red blood cells; other persons may possess both of these antigens as well as additional kinds. Indeed, Rh⁻ people were also found to vary. The many Rh variations stem from the fact that several distinct Rh antigens occur in humans, not just one or two. Even Rh⁻ persons possess Rh antigens which can bring about antibody production. But the most commonly occurring antigens in Rh⁻ people are not responsible for the Rh incompatibilities which we have just discussed.

To account for these many Rh types, two different ideas have been proposed which have caused some confusion in terminology. Essentially, one concept, that of Wiener in the United

States, pictures the Rh genetic region as a complex locus on the chromosome (Fig. 14-12). We may think of it as a stretch of DNA. Changes at different points within this DNA segment would result in the production of different gene forms or alleles which would form part of a multiple allelic series. According to the Wiener idea, at least eight different Rh alleles exist. Each allele controls the formation of a certain *combination* of antigens (Table 14-5). An individual can have any two of the eight alleles and would be Rh$^+$ or Rh$^-$, depending on the genes he carries (Table 14-7).

A group of English hematologists, Fisher, Race, and Sanger, interpret the situation somewhat differently. According to their hypothesis, the Rh region on the chromosome includes not just one but three loci (Fig. 14-13). Each locus is the site of a pair of alleles: Cc; Dd; and Ee. Eight different combinations of the six alleles are possible on any one chromosome: CDE, CdE, etc. Each one of the six alleles controls a distinct antigen, and each of these has been detected serologically by antigen-antibody reactions. (Anti-d serum is not easily obtained.) Since eight combinations of alleles are possible, there are eight different combinations of antigens, one for each arrangement of alleles (Table 14-6). The three loci are so closely linked that crossing over in the "cde" region of the chromosome is very rare. Therefore, any one of the allelic combinations (CDE, cDe, etc.) would usually stay linked and thus be inherited together as a unit. In essence, each combination would appear as if it were one allele, even though three are actually present at each Rh region on the chromosome. Based on this idea, any one person would possess six Rh alleles, any two combinations of three each, such as *CDE* on one chromosome and *cdE* on the other. The genotype of such a person could be represented as CDE/cdE (Table 14-7).

No matter how we view the Rh story, either as one locus with a series of multiple alleles or as three tightly linked loci, the various Rh alleles all show codominance. As Tables 14-5, 14-6, and 14-7 indicate, each allele when present is expressed and causes the production of its specific antigen. The antigens are in turn detected by reactions with their corresponding antibodies. The maternal-fetal Rh incompatibility is mainly the result of the presence of D antigen (approx-

FIG. 14-12. The Wiener concept of the Rh region. According to this interpretation, a series of multiple alleles exists at one locus. Each allele controls not one but a combination of antigens. For example, the "r" allele would govern the formation of antigens c, d, and e, which would react with the corresponding antibodies. A total of eight different alleles exists, each controlling a different combination. Only four are depicted here. Actually, the "CDE" designation shown here is employed in the other system (Fig. 14-13) rather than in the Wiener. However, the antigens and antibodies correspond in both, and the one designation is used here for the sake of clarity.

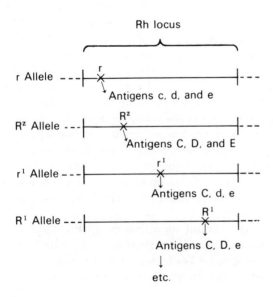

TABLE 14-5 The eight different Rh alleles, according to Wiener*

DIFFERENT ALLELES	ANTIBODIES FORMED AGAINST ANTIGENS GOVERNED BY THE ALLELE				
	Anti C	Anti D	Anti E	Anti c	Anti e
Rh⁻ { r				+	+
r'	+				+
r''			+	+	
rʸ	+		+		
Rh⁺ { R⁰		+		+	+
R¹	+	+			+
R²		+	+	+	
Rᶻ	+	+	+		

* Each allele is responsible for the production of a combination of two or more antigens, and these can cause the formation of corresponding, specific antibodies (anti-d sera not available).

imately 90% of the cases). The C antigen seems to be involved in the remainder. The Rh⁻ condition is often defined as the absence of D antigen, because that antigen is the most important in the Rh incompatibilities. Following this definition, an Rh⁻ person could contain the C antigen. At any rate, two Rh⁻ persons may have different genotypes. They may contain different combinations of antigens (cde/cde or cde/cdE), but these usually play no significant role in cases of blood type differences. Whether the Wiener or the Fisher concept is correct is still unsettled, and there are good arguments in favor of each.

TABLE 14-6 The eight combinations of the Rh loci, according to Fisher*

	ANTIBODIES FORMED AGAINST ANTIGENS GOVERNED BY GENES AT THE LOCI				
	Anti C	Anti D	Anti E	Anti c	Anti e
Rh⁻ { c d e				+	+
C d e	+				+
c d E			+	+	
C d E	+		+		
Rh⁺ { c D e		+		+	+
C D e	+	+			+
c D E		+	+	+	
C D E	+	+	+		

* Each combination of three alleles is strongly linked and behaves as if it were one allele. Each combination of three, according to this idea, corresponds to one of the alleles as pictured by the Wiener concept (Table 14-5).

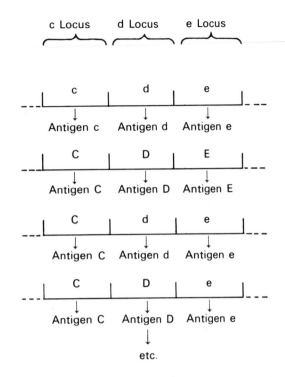

FIG. 14-13. The Fisher concept of the Rh region. According to this idea, the Rh region includes three very closely linked loci which rarely are separated by crossing over. At each locus a pair of alleles exists, and each allele controls the formation of a specific antigen. On any one chromosome, any combination of three of the six alleles may occur. A total of eight different combinations of three is possible. Contrast this figure with Fig. 14-12.

TABLE 14-7 Some common Rh⁻ and Rh⁺ genotypes*

PHENOTYPE	GENOTYPE		ANTIBODIES
	Weiner	Fisher	
Rh⁻	r/r	cde/cde	c, e
Rh⁻	r''/r	cdE/cde	c, E, e
Rh⁺	R⁰/r	cDe/cde	c, D, e
Rh⁺	R¹/R¹	CDe/CDe	C, D, e
Rh⁺	R²/r	cDE/cde	c, D, E, e
Rh⁺	R¹/R²	CDe/cDE	C, c, D, E, e
Rh⁺	R¹/r	CDe/cde	C, c, D, e

* Any person, according to Wiener, can have any combination of the eight alleles, one allele on each chromosome. According to Fisher, any person may have a combination of any three alleles on one chromosome and a combination of any three on the other. No matter how the situation is viewed, the individual forms antibodies against specific antigens. Antigen D is the one most commonly involved in Rh incompatibilities. The Rh positive condition is often defined as "presence of D antigen." (Anti-d serum is not readily available.)

Nevertheless, we know today that the genetic region associated with the Rh blood grouping is located at a specific region of Chromosome 1. To explain a very complex story in the simplest way, we can think of the Rh grouping based primarily on a pair of alleles, D and d, because it is the D antigen which must be primarily considered when Rh types differ.

It should be evident from the assortment of different Rh types which actually can exist among people that the Rh blood groupings can go a long way in settling cases of disputed parentage. Added to the information on the ABO and MN groupings, the specific Rh type provides another basis for establishing the possible innocence of an individual accused in a paternity suit.

FURTHER SIGNIFICANCE OF BLOOD GROUPINGS. Although the Rh system is the one popularly known to be involved in maternal-fetal blood problems, it is not the only one. Indeed, the familiar ABO grouping has also been implicated in such incompatibilities and may account for the failure of as many as 5% of human zygotes to reach full-term delivery. We might expect some antigen-antibody reaction to take place if blood from an O type mother should leak into the circulation of an offspring who is type A, B, or AB. The O mother's serum contains both antibodies "a" and "b," which can clump the red blood cells of the fetus. There is reason to believe that these antibodies can be introduced into the fetal circulation as a result of placental defects in much the same way as the Rh factors.

Except for their importance in blood transfusions, the ABO types have generally been little more than natural curiosities. We now know that they must be acknowledged in maternal-fetal relationships, but in addition, evidence is accumulating that the ABO type may have other associations. A prevalence of type O is found among patients with bleeding peptic ulcers in contrast to healthy controls or patients with nonbleedings ulcers. However, few persons of blood type O are found among victims of certain circulatory disorders in which heart and blood vessels are damaged. More persons of type A seem to suffer from the circulatory disease, *arteriosclerosis obliterans*, which involves hardening of the arteries and the formation of blood clots. The significance of this association of blood groupings and seemingly unrelated traits is unknown. However, it is conceivable that we are for the first time detecting pleiotropic effects of these blood type alleles. The phenotypic effects of the many blood group genes may yet prove to be quite diverse and to depend on an interaction with other genetic loci and the environment.

Our discussion of blood groupings up to this point has by no means exhausted the number of them which has been identified. There are several others which are much less well known but which may be encountered in specific situations. To mention just a few, there are such groupings as the Kell, the Kidd, the Duffy, Lutheran, and Lewis, *et al.* The names usually refer to the family in which the grouping was first detected. For example, the Kell factor (K) denotes an antigen found in most persons. Women who are double recessive (kk) may give birth to erythroblastotic Kell positive children. As in the case of the Rh⁻ mothers, the Kell negative women may produce antibodies (anti-Kell) which can react with the Kell antigen in the fetal circulation. These many additional blood groups can also be of aid in legal matters. They also demonstrate how unique each one of us is. The number of possible combinations of the many blood group loci, some of which are multiple allelic, defies comprehension.

ANTIGENS, GRAFTING, AND TRANSPLANTS. Intimately related to the topic of blood groups is the subject of tissue grafting and organ transplants. We might expect other tissues as well as blood to contain distinct antigens which can react with specific antibodies. And indeed, the rejection of an organ such as the kidney or the failure of a skin graft to take are examples of an immunological response. All cells possess antigens which vary from one individual to the next and which may cause the rejection of transplants. Such factors are called histocompatibility antigens and the genes which control them histocompatibility genes (H genes). The antigens, as well as those associated with the blood types, are present in the cell membrane, where they may possibly play some role in its structure and biological properties. The most thorough studies on the genetics of grafting have been performed with mice. However, the same principles are

believed to apply to other species, including the human. Mice within one family line may be mated for generations to produce strains which are so inbred that they are homozygous for an appreciable number of alleles. Valuable genetic information on tissue compatibility factors and their role in tissue transplants may be gained from studying graft tolerance within and between the inbred strains. Such work gives strong evidence for the following concept: a recipient will accept tissue from a donor if the recipient possesses the same kinds of histocompatibility alleles as the donor. Conversely, if the donor, and hence any tissue from the donor, possesses H alleles which are *not* present in the receiver, the graft will be rejected. Table 14-8 represents hypothetical results from two inbred strains, A and B, in which two pairs of histocompatibility alleles are being followed. As was true for the ABO and Rh alleles, the H alleles are codominant in their expression. It is obvious from the table that within an inbred strain, tissue can be successfully grafted and accepted because there is no difference between donor and recipient in the kinds of H alleles which they possess. When the two inbred strains are crossed, their F_1 offspring contain H alleles from each parent. These hybrids have all the alleles which are typical of each strain. Thus, they can accept grafts from either parent type. However, the F_1 cannot act as donor to either inbred parental strain, because the F_1 has some alleles not found in strain A or B. If the F_1's are interbred to produce F_2 progeny, the latter may have various new combinations

TABLE 14-8 Grafting compatibilities between animals of various genotypes

DONOR	DONOR'S GENO-TYPE	RECIP-IENT	RECIP-IENT'S GENO-TYPE	ABIL-ITY TO ACCEPT GRAFTS
Strain A	$A_1A_1B_1B_1$	Strain A	$A_1A_1B_1B_1$	+
Strain B	$A_2A_2B_2B_2$	Strain B	$A_2A_2B_2B_2$	+
F_1 hybrid	$A_1A_2B_1B_2$	Strain A	$A_1A_1B_1B_1$	−
F_1 hybrid	$A_1A_2B_1B_2$	Strain B	$A_2A_2B_2B_2$	−
Strain A	$A_1A_1B_1B_1$	F_1 hybrid	$A_1A_2B_1B_2$	+
Strain B	$A_2A_2B_2B_2$	F_1 hybrid	$A_1A_2B_1B_2$	+
F_2 hybrid	9 Types	F_1 hybrid	$A_1A_2B_1B_2$	+
F_1 hybrid	$A_1A_2B_1B_2$	F_2 hybrid	9 Types	− in most cases

(3^2) of the two pairs of alleles, such as $A_1A_1B_1B_2$. But no matter what the new combination in the F_2 animals, the F_1's will be able to accept grafts from them. This is so, because the F_1's will contain *all* the types of alleles present in any F_2. The reverse, however, is not necessarily so, because independent assortment will scramble the H alleles into a variety of new combinations (nine in this case). The only F_2 animals capable of accepting grafts from F_1 animals must be dihybrids, just as the F_1's, and they would be in the minority.

This simple example illustrates the principle behind graft tolerance, but it is an oversimplification. In mice, about 30 separate loci are associated with histocompatibility genes. The number of different possible combinations becomes staggering. Moreover, some of these loci exert a greater effect than others. In mice, the genetic region known as the "H_2 complex" is the main one in the sense that the antigens it controls can bring about a very strong immunological response, much more so than other loci. This region has been studied in detail and has been shown to involve several very closely linked loci adjacent to each other. At each locus, a multiple allelic series exists. Each allele determines the structure of a protein, a specific histocompatibility antigen. The specific tissue antigens can be recognized by specific antibodies, and the ensuing reaction accounts for the tissue rejection.

In addition to circulating antibodies which form in the body of a mammal in response to foreign antigens, there is another immune response, and it is the one which brings about the actual rejection of foreign tissue received in a transplant. This is the cellular response, a reaction which entails those particular lymphocytes, the so-called "T lymphocytes" or "T cells," which have become differentiated in the thymus gland. The T cells are able to recognize foreign antigens on the surfaces of cells in tissue grafted to a host. A T cell has antibody molecules embedded in the surface of its cell membrane, and these apparently enable T cells to respond to the presence of antigens on the surfaces of cells in foreign tissue.

As in the mouse, several genetic regions in the human regulate tissue compatibility, and one of them, the "HLA complex," is of primary

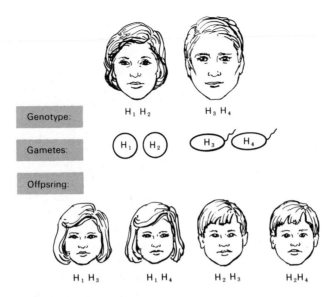

Genotype:

$H_1 H_2$ $H_3 H_4$

Gametes:

H_1 H_2 H_3 H_4

Offpsring:

$H_1 H_3$ $H_1 H_4$ $H_2 H_3$ $H_2 H_4$

FIG. 14-14. Tissue incompatibility at the HLA locus in humans. Any one person is apt to be heterozygous at the HLA locus, at which a large number of alleles exists. The parents in this case, as would probably be true of any two unrelated persons, do not possess any "H" allele in common and would be unable to exchange grafts. Among their children, diverse types occur. These cannot give grafts to or accept them from either parent, because there will be a difference in one "H" allele. For the same reason, children of different genotypes cannot exchange grafts. However, two children can occur who carry the same alleles at this locus and conceivably they might be able to exchange tissue. However, still other loci are involved; differences at these will reduce the chances of free exchange, even if compatibility exists at the HLA locus.

importance. This major histocompatibility complex has been assigned to the short arm of chromosome 6. The HLA complex is believed to include four very closely linked loci, two of which are responsible for antigenic determinants which can react with circulating antibodies. The other two loci control cell antigens to which T lymphocytes can react. Apparently, at each locus in the complex, a multiple allelic series exists, the exact number of possible gene forms at each one being unknown. Since the loci in the HLA complex are so tightly linked, crossing over between any two would be rare. For the sake of simplicity we can think of the complex as a single locus, one which is extremely important in determining those antigens involved in the T cell response. Since many allelic forms are possible, any one person can be thought of as heterozygous at this region for H alleles, let us say H_1/H_2. An unrelated person would almost always be heterozygous for different gene forms, such as H_3/H_4. Considering the HLA complex in this simplified way, we can nevertheless see why most grafts will not be tolerated between individuals (Fig. 14-14). A cross between two persons of the genotypes just given will result in the following combinations of HLA alleles: H_1/H_3; H_1/H_4; H_2/H_3; and H_2/H_4. None of the children will be able to accept a graft from either parent. There is a chance, however, that two offspring will have the same combination of HLA alleles, such as two of genotype H_1/H_4 in our example. This means that two siblings could possibly exchange tissue. Although the chance for this is greater than it is for a successful transplant between any other two persons (except for identical twins), we must not forget that several loci, not just one, are involved. Even if compatibilities should exist at the HLA locus, a graft may eventually be rejected due to incompatibilities at other loci which control tissue antigens. We see again, as in the case of blood types, that each of us possesses a vast number of histocompatibility alleles in a combination which is distinct from that found in most other persons.

There is good reason to believe that the evolution of the great complexity in the immune system has resulted largely from the value it imparts in the recognition of those cells in the body which may suddenly undergo malignant transformations. Any newly arisen cancer cell would have alterations in its cell surface and could be recognized as "foreign." A malignant transformation probably entails a vast assortment of possible alterations, each one quite distinct and requiring a specific cellular response if it is to be eliminated.

Striking illustrations of the profound importance of the immune response are seen in those genetic disorders where there is a decreased efficiency in the ability to respond to foreign antigens. Several types of immune-deficient disorders exist, and these are termed "the agammaglobulinemias." The several types vary in their severity, in their response to treatment, and in their genetic bases. In some families, the disorder is due to a sex-linked recessive; in other families, an autosomal inheritance pattern is seen. In one of the severest forms, both the T lymphocytes and the circulating antibody responses are impaired. Due to the lack of antibodies, such persons are susceptible to every type of infection, viral and bacterial, as well as to malignancies. Such unfortunate persons can readily accept grafts, since no antibodies or T cells can participate in a rejection. Still, they are no different from anyone else in the ability to donate tissue, since their unique antigenic determinants will be present in cells of their tissues.

REFERENCES

Bach, F. H., and J. J. van Rood. The major histocompatibility complex. *New Eng. J. Med., 295:*872, 1976.

Cooper, M. D., and A. R. Lawton. The development of the immune system. *Sci. Am. (Nov):* 58, 1974.

Cunningham, B. A. The structure and function of histocompatibility antigens. *Sci. Am. (Oct):* 96, 1977.

Edelman, G. M. The structure and function of antibodies. *Sci. Am. (Aug):* 34, 1970.

Eisenbarth, G., P. Wilson, F. Ward, and H. E. Liebovitz. HLA type and disease occurrence in familial polyglandular failure. *New Eng. J. Med., 298:* 92, 1978.

Jerne, N. K. The immune system. *Sci. Am. (Jul):* 52, 1973.

McDevitt, H. O. Genetic control of the antibody response. In *Medical Genetics,* V. A. McKusick and R. Claiborne (eds.), pp. 169–182. HP Publishing Co., New York, 1973.

Munro, A., and S. Bright. Products of the major histocompatibility complex and their relationship to the immune response. *Nature*, 264: 145, 1976.

Paul, W. E., and B. Benacerraf. Functional specificity of thymus-dependent lymphocytes. *Science*, 195: 1293, 1977.

Pontecorvo, G. *Trends in Genetic Analysis.* Columbia University Press, New York, 1958.

Race, R. R., and R. Sanger. *Blood Groups in Man*, 6th ed. Scientific, Oxford, 1973.

Reisfield, R. A., and B. D. Kahan, Markers of biological individuality. *Sci. Am. (June)*: 28, 1972.

Rubinstein, P., N. Suciu-Foca, and J. F. Nicholson. Close linkage of HLA and juvenile *diabetes mellitus. New Eng. J. Med.*, 297: 1036, 1977.

Silvers, W. K., and S. S. Wachtel. H-Y antigen: behavior and function. *Science*, 195: 956, 1977.

Vogel, F. ABO blood groups and disease. *Am. J. Hum. Genet.*, 22: 464, 1970.

Weiss, N. S. ABO blood type and arteriosclerosis obliterans. *Am. J. Hum. Genet.*, 24: 64, 1972.

Wiener, A. S. Blood groups and disease. *Am. J. Hum. Genet.*, 22: 476, 1970.

REVIEW QUESTIONS

1. In a certain fungus, several recessive mutations occurred independently, each of which prevents the manufacture of adenine. Normal wild-type fungi of this species have the ability to make adenine. The several adenineless stocks have been named ad_1, ad_2, ad_3, etc., to distinguish them according to their order of occurrence. Various crosses among some of the stocks were made with the following results:

    (1) $ad_1 \times ad_3$ (wild)
    (2) $ad_1 \times ad_8$ (wild)
    (3) $ad_1 \times ad_{16}$ (wild)
    (4) $ad_8 \times ad_{16}$ (adenineless)
    (5) $ad_3 \times ad_8$ (wild)
    (6) $ad_3 \times ad_{16}$ (wild)

How many loci are being followed in these crosses, and which of the adenine mutations seem to be allelic to each other?

2. Assume that a, b, and c are recessive mutations in a certain mold and that the sites of the mutations are very closely linked. The following combinations give the results indicated for each:

$$\frac{a^+\ b^+}{a\ b} = \text{wild} \qquad \frac{a^+\ b}{a\ b^+} = \text{mutant}$$

$$\frac{b^+\ c}{b\ c^+} = \text{wild} \qquad \frac{b^+\ c^+}{b\ c} = \text{wild}$$

Do any of the mutations complement each other? Are any of them allelic?

3. The multiple alleles w (for white eye), w^a (for apricot eye color), w^{ch} (for cherry or light red) are found at the w (white) locus on the X chromosome of *Drosophila*. Heterozygotes for these alleles have intermediate eye colors, since the alleles in the series are incompletely dominant to each other. However, each allele is recessive to the wild-type allele (w^+) for red eye color. Give the results expected from the following crosses:

    A. An apricot female × a white male.
    B. An apricot female × a cherry male.
    C. A cherry female × a red male.
    D. F_1 females from the cross in "C" × an apricot male.

4. S_1, S_2, S_3, etc., are sterility alleles which form a multiple allelic series in tobacco. A plant cannot be homozygous for any one of them. What will be the constitutions of the F_1 plants in regard to the sterility alleles in each of the following crosses?

    A. S_1S_2 egg parent × S_4S_5 pollen parent.
    B. S_3S_4 egg parent × S_4S_5 pollen parent.
    C. S_1S_2 egg parent × S_1S_2 pollen parent.
    D. S_4S_5 egg parent × S_3S_4 pollen parent.

5. In the rabbit, a multiple allelic series affects the color of the fur. The wild-type allele (c^+) for dark fur is dominant to all the other alleles in the series. The allele c^h for Himalayan is dominant to albino (c). However, the allele for chinchilla coat color (c^{ch}) gives light gray color when present with either the allele for Himalayan or albino. Give the genotypes and the phenotypes of the offspring from the following crosses.

    A. A Himalayan animal whose male parent was albino × a chinchilla.
    B. An animal with wild-type fur which had a chinchilla parent × an albino.
    C. A light gray animal which had an albino parent × a light gray which had a Himalayan parent.

6. Give the probable genotypes of the following pairs of parents in respect to the A, B, O blood type alleles.

    A. An A type parent and an O who have one child who is type O.

    B. An AB type parent and an A who have one child who is type B and one who is type A.

    C. A type A parent and a type B who have eight children who are type AB.

7. There are two subgroups of type A blood which can be recognized. The allele A_1 is dominant to A_2 and to O. A_2 is also dominant to O. Give the results of the following crosses:

    A. An A_1 type person who had an O parent × an A_2 who had an O parent.

    B. An A_2 type person who had an O parent × an A_1B.

    C. $A_1B \times A_2B$.

    D. $A_1B \times O$.

8. A woman is blood type B. Man No. 1 is type O. Man No. 2 is A_2. The woman's baby is type A_1B. Which man could be the father? Explain.

9. The secretor trait (Se) which enables A and B antigens to be secreted into the body fluids is dominant to nonsecretor (se). Give the blood type and indicate whether the saliva will contain A or B antigens in the offspring from each of the following parents:

    A. $I^A I^B$ Se Se × ii se se.

    B. $I^A I^B$ Se se × ii Se se.

    C. I^Ai Se se × I^Bi Se se.

10. The allele H is required for the completion of A and B antigens. In its absence, the allele h cannot direct the formation of A and B antigens. The h allele is thus epistatic to the ABO locus and can cause a person to be blood type O regardless of the presence of the A and B alleles. Give the blood types expected among the offspring in each of the following matings, as well as the ratio in which they would occur:

    A. $I^A I^B$ Hh × $I^A I^B$ Hh.

    B. ii Hh × I^Ai Hh.

    C. I^Ai Hh × I^Bi hh.

11. A woman is blood type A MN. Her baby is type O, N. Which of the following two men is the father? Man No. 1: Type A M. Man No. 2: Type O N. Explain.

12. Considering Rh⁺ blood to be associated with the dominant, D, and Rh⁻ with its recessive allele, d, give the possible offspring which can occur from the following matings:

    A. An O, Rh⁻ woman (who has an Rh⁺ parent) and an O Rh⁺ man (who has an Rh⁻ parent).

    B. An A, Rh⁺ woman and a B, Rh⁺ man, both of whom had one Rh⁻ parent.

    C. A woman of type O who suffered *erythroblastosis fetalis* and a man who is AB negative.

13. A woman is blood type O MN Rh⁺. Her baby is O MN Rh⁻. Man No. 1 is A N Rh⁺. Man No. 2 is O MN Rh⁻. Which of the two could be the father?

14. Of the following matings, which run a risk of producing a baby with erythroblastosis fetalis?

    A. Female $\dfrac{C\,D\,E}{C\,D\,E}$ × Male $\dfrac{c\,d\,e}{c\,d\,e}$

    B. Female $\dfrac{c\,d\,E}{c\,d\,e}$ × Male $\dfrac{C\,D\,e}{c\,D\,e}$

    C. Female $\dfrac{C\,d\,e}{c\,d\,e}$ × Male $\dfrac{c\,D\,E}{C\,d\,e}$

    D. Female $\dfrac{C\,d\,E}{C\,d\,e}$ × Male $\dfrac{C\,d\,E}{C\,d\,e}$

15. In the human, the HLA histocompatibility locus is a complex one. According to one concept, four loci actually occupy the HLA region on the chromosome, and these four loci are so closely linked that crossing over can be ignored. The four loci have been designated A, S, B, and D. At each one of them, a large series of multiple alleles exists. Assume that a woman and her husband have the following genotypes, respectively, for the histocompatibility loci:

$$\frac{A^1\,S^2\,B^1\,D^2}{A^2\,S^1\,B^2\,D^1} \quad \text{and} \quad \frac{A^3\,S^4\,B^3\,D^4}{A^4\,S^3\,B^4\,D^3}$$

    A. Give the genotypes possible among the offspring.

    B. Can the children donate to or accept tissue from the parents?

    C. How would you answer the above question if one of the children is a boy with agammaglobulinemia and cannot produce T lymphocytes?

    D. What is the chance that any two children in the family might be compatible in respect to these loci?

16. In the human, there are two different genetic blocks in the pathway leading to melanin pigment formation. One of them, a, interrupts a step which prevents the formation of the immediate precursor of melanin. The second block, b, prevents the conversion of the precursor to melanin pigment. Each of the two types of albinism which results from these blocks is an autosomal recessive.

    A. What kinds of offspring are to be expected if an albino person of type a has children with an albino person of type b?

    B. What is the chance that two persons of the children's genotype will produce an albino at any birth if two such persons should marry?

17. Assume that in a certain biological step substance A is converted to substance B by enzymatic action. The step can be blocked if an individual is homozygous for the mutations a¹, a², a³, or a⁴. These separate mutations map at closely linked but different sites which are separable by crossing over. A series of crosses gives the results below. Offer an explanation on the concept of the cistron, and diagram the crosses.

    (1) a¹ × a¹—step blocked
    (2) a¹ × a⁴—step blocked
    (3) a¹ × a²—step proceeds
    (4) a¹ × a³—step proceeds
    (5) a² × a³—step blocked
    (6) a¹a⁴ double mutant × normal—step proceeds
    (7) a¹a² double mutant × normal—step proceeds
    (8) a²a³ double mutant × normal—step proceeds.

18. Two married persons suffer from an anemic condition which follows an autosomal recessive pattern of inheritance in each family line. Fingerprinting of their hemoglobin A shows that the female has one amino acid substitution in the alpha chain and the male has one amino acid substitution in the beta chain. Their children however, do not suffer from the anemia. Explain.

15

GENE MUTATION

SOMATIC AND GERMINAL POINT MUTATIONS.
Without knowing their physical basis, De Vries
called attention to the importance of sudden
inheritable changes as the ultimate source of
variation in the process of organic evolution. For
if the concept of the continuity of the germ plasm
is correct (Chap. 1), only the hereditary material
can transfer information from one generation to
the next. Any advantageous changes which may
benefit the species must result from variations
in the hereditary substance of the germ cells.
Although most mutations might be harmful,
evolutionary progress would necessarily depend
on the ability of the gene to mutate at times to
another form (an allele) which conveys an advantage. Indeed, all hereditary variation, good
as well as bad, which is seen in living things,
has ultimately stemmed from mutations in the
hereditary material.

The pioneer geneticists, such as T. H. Morgan, realized the importance of mutation in
genetic research. If no variation at all occurred
in a character, nothing at all could be learned
about its mode of inheritance. It was possible to
work out the inheritance of eye pigments in the
fruit fly only because mutations to white and
various shades were uncovered in the laboratory.

Crossing only pure breeding red-eyed flies is meaningless and contributes no information on the genetic loci which interact and influence color of the eye. Although the importance of spontaneous mutation was appreciated, the physical basis of the hereditary alteration was unknown and had to await the discovery of the chemical nature of the gene. Let us first examine certain significant facts about point mutation which were assembled in earlier studies. We will then review them in the light of more recent information acquired through research at the molecular level.

Since point mutations are changes in the DNA, we can expect them to occur in any kind of cell, somatic as well as germ cells. In diploid organisms, a somatic mutation at a specific locus would tend to remain undetected. This is so, because the wild form of the gene at that locus on the homologous chromosome would also be present in the cell, and its effect would most probably be dominant and would offset that of the mutant form (Fig. 15-1A). However, somatic mutations in some plants have given rise to plant varieties of commercial value, such as the navel orange. In each case, a mutation arose in a single somatic cell. By repeated mitotic divisions, cells derived from the original mutant cell produced a branch composed of genetically identical cells, each with the original mutant allele. Since the desirable fruits of these mutant forms are often sterile (no seeds in the navel orange), the varieties must be propagated asexually by cuttings or grafts. Somatic mutations are also known which result in mosaic individuals, those with phenotypically different patches of tissue. Variegated flowers with segments of different color, as seen in certain varieties of roses, are

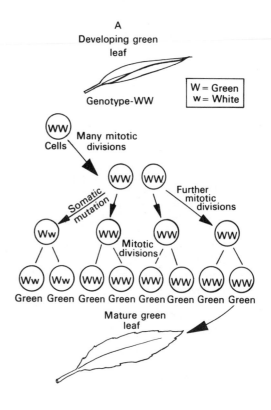

A
Developing green leaf

Genotype-WW

W = Green
w = White

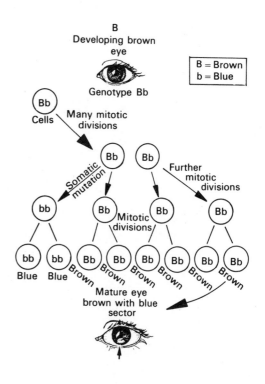

B
Developing brown eye

Genotype Bb

B = Brown
b = Blue

FIG. 15-1. Effects of somatic mutations. *A.* Many somatic mutations remain undetected. The wild allele would usually be present in the cell and would typically be dominant to the mutant form. If a recessive mutation arises as a result of a mutation during development, the dominant will still be expressed in the heterozygous cells. All the cells, heterozygous and homozygous, will show the same phenotype. *B.* If an individual is heterozygous, a somatic mutation may affect the dominant allele and thus generate a cell which is homozygous recessive. Further cell divisions will give rise to two populations of cells which differ phenotypically as well as genotypically.

examples. In humans, cases of a brown-eyed person with a blue segment in one iris (or vice versa) can be explained on the basis of somatic mutation (Fig. 15-1B).

Although somatic mutations may be of consequence to the individual, the germinal mutations are of greater significance to the population, as these can be transmitted from one generation to the next. Because the consequences of gene mutation are so great for all species of living things, it is important to appreciate the kinds or types of mutations which can occur. Those mutations which arise for no apparent reason are called "spontaneous." A spontaneous mutation may occur at any site on a chromosome. If we could follow every locus within a species, we would undoubtedly find that at some time or other a mutation would arise at each locus. However, the frequency of mutation at the different loci would vary greatly. In one of the first precise studies of spontaneous mutation rates, Stadler followed eight loci which affect the phenotype of the kernels on an ear of corn. Since each kernel represents one individual, many hundreds of offspring of one plant can be observed by examining several ears. Because spontaneous mutation of any one gene is such a rare event, it is essential to examine very large numbers of individuals in order to estimate the mutation rate. During the course of his study, Stadler noted (Table 15-1) that mutations occur much more frequently at some loci than at others. Indeed, during the course of the investigation, no mutations at all were detected at the waxy locus ("Wx") in contrast to the high number which arose at "R" and "I." Mutation analysis in microorganisms has shown that certain sites on the DNA are more likely to undergo mutation

TABLE 15-1 Spontaneous mutation rates of eight genes in corn

GENE	NO. OF MUTATIONS PER MILLION GAMETES
R	492
I	106
Pr	11
Su	2.4
Y	2.2
Sh	1.2
Wx	0
C	2.3

than others. Controlled studies of mutation rate are difficult to conduct in higher forms. However, variation in mutation frequency among different loci has also been demonstrated in several species, including *Drosophila*, and is even indicated in humans.

TYPICAL CONSEQUENCES OF SPONTANEOUS MUTATION. When the many spontaneous mutations are studied in an organism, it is found that the majority of them produces just a slight effect, as opposed to a very pronounced change in a character. This should not seem surprising on the basis of our knowledge of gene interactions (Chap. 4). We have learned that many genes influence the expression of a character and that many biochemical pathways are interrelated in cellular metabolism. Some genes control chemical steps which are more important or critical than other steps. Although many genes influence a character, the majority of them influences it in only a slight way (consider the many modifying genes and genes with quantitative effects). Therefore, any mutation, being random, will more likely occur in a gene with a slight effect than in one which can alter the character drastically.

When the effects of these gene mutations on the individual are examined, it is found that the overwhelming majority is harmful in some way. Again, this is to be expected on the basis of the interrelated pathways in metabolism. An unplanned change in a complex system is certainly more apt to produce a harmful effect than one which shows foresight or planning. And spontaneous mutation *is* a random, unplanned event which leads to a genetic alteration without any regard for the consequences. We have already considered this fact as the reason for the high incidence of lethal alleles resulting from gene mutations (Chap. 4). A complete lethal has an effect so drastic that it eliminates the individual before reproductive age. Lethals outnumber by far those mutant alleles which produce some visible effect on a characteristic. Indeed, most mutant alleles which bring about a visible change in the phenotype carry some sort of detriment with them. Since we have just said that gene mutations resulting in the origin of mutant alleles producing slight effects are the most common, we would expect those resulting

in just a slightly harmful or detrimental effect to be more common that the complete lethals. And this is found to be so. Well-controlled studies with the fruit fly have shown that gene mutations giving rise to alleles with a visible effect on a character are in the minority and that the more pronounced the visible effect, the greater the accompanying detriment. But more common than these are the complete lethals with no visible effect but which can completely remove the individual before the age of sexual maturity. Most frequent of all however, are those mutant alleles with no visibly detectable effect but which contribute a slight degree of harm. We may call these *detrimentals*. A detrimental may not kill the individual possessing it, but the allele is adversely affecting him or her in some way and is decreasing the probability of survival. These alleles with their slight, but nonetheless deleterious, effects are of serious consequence to a population and will be discussed more fully in the following chapter.

Once a gene has mutated to an allelic form, this new allele will continue to duplicate itself, unless, of course, it is eliminated at once. It must be remembered, however, that the new gene form, as well as the original one, may itself mutate further, giving us still other allelic forms, as we saw in the case of multiple alleles. However, the mutant allele at times may mutate back to the original or wild form of the gene. This is called "reverse mutation" or "back mutation." (Fig. 15-2). The rates of back mutation have been established for certain genes in several species, particularly in microorganisms. The frequency of mutation back to wild is also characteristic of a particular gene, but usually the back mutation is rarer than the *forward* mutation, the change from wild to mutant.

When a gene mutation arises, it may result in the origin of either a dominant or recessive mutant allele; but by far, the most numerous mutant forms of a gene have recessive effects. This is quite understandable when we consider that natural selection has operated over long periods of time so that today the wild alleles in any species are the products of millions of years of screening. Those alleles which are most conducive to the survival of the species will have been established by natural selection. Any wild allele would thus be one which is beneficial to the individual organism living in its specific environment. Part of the benefit of any wild allele would be its ability to express itself in the presence of mutant forms. In other words, wild alleles have actually been selected for their dominant effects. We can appreciate the value of this when we consider that most spontaneous alterations in a gene are harmful. If the wild form can "cover up" the deleterious effect of the mutant, the heterozygote will not be harmed. More individuals will survive than if the deleterious mutant allele were dominant. If a rare recessive mutation does occur which carries an advantage superior to that of the wild, natural selection will favor the mutant. It will also favor those forms of the mutant gene which later arise to make it more dominant. Eventually the "mutant" allele may supplant the original "wild" and thus become the new wild type.

MUTATION RATE AND SOME FACTORS WHICH INFLUENCE IT. Although spontaneous mutation rates for individual loci are very low (a probability on the average of 1–10 per million gametes), the frequency of spontaneous mutation should not be underestimated, as any higher organism contains thousands of genetic loci. Because there are so many genes present, the chance that *some one* of them will undergo a change becomes quite significant. Studies on the total mutation rate in *Drosophila* indicate that in one generation there is the probability that 5% of the gametes will contain a mutation which arose in that generation time. Calculations of total mutation rate in other organisms give a figure comparable to this. In the human estimates have been made on the mutation frequency of certain genes associated with pathological disorders. Sex-linked recessives and mutant alleles with dominant effects have been the most frequently studied because autosomal recessives demand more involved procedures. If a disorder, known to be inherited typically as a dominant defect, appears in an offspring of two normal parents, it is then considered that a dominant mutation has occurred. Adding up the number of cases of children born with the defect in a given number of births is taken as an indication of the mutation rate. Such a procedure may entail many errors. One obvious source is that the phenotype of the affected individual

FIG. 15-2. Forward and back mutation. A wild allele ("A") may mutate to more than one allelic form. Each allele in turn may mutate further to still another form. The mutation frequencies (as indicated by varying lengths of the arrows) are not necessarily the same but are characteristic for each type of alteration. Mutation back to the allelic form from which a gene arose usually takes place at a much lower rate than the mutation in the "forward" direction.

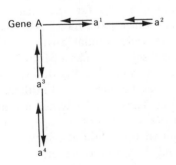

may be environmentally caused and not the result of a gene mutation at all. We have seen how phenocopies (Chap. 4) mimic genetic effects. It is also evident that two similar phenotypes may have two very different genetic bases (Chap. 7). Moreover, penetrance and expressivity may complicate the picture even more. Therefore, many factors can reduce the accuracy of studies of mutation frequency in humans. However, the estimates made for certain individual loci are comparable to those determined for other species, in the vicinity of 1 mutation per 100,000 gametes per generation. So what has been found for other organisms seems applicable to man as well. The figure 5% for an overall mutation frequency per generation in the human would not appear to be an overestimate.

It is important to understand what this 5% value means in the life cycles of different species. Note that the value for the overall mutation frequency is expressed, *not* in relationship to definite time units such as months or years, but in relationship to generation time. Different species obviously have different generation spans, approximately 2 weeks in the fruit fly in contrast to approximately 30 years in the human. But when different species with their different generation times are compared (excluding microorganisms), the overall mutation rates expressed in *generation time*, although not identical, are certainly comparable. Obviously, in terms of weeks or months, this would mean that a species with a short life cycle of just a few weeks would have a mutation rate hundreds of times higher than humans or any other species with a much longer life cycle. The very fact that the figure is similar when expressed in generation time implies that natural selection has somehow geared the spontaneous mutation rate to the life cycle. Evolutionary progress depends on the ability of the gene to change. However, a very high or uncontrolled amount of gene mutation would prove fatal to a species, because most mutations are harmful. So it would seem logical to expect that mutation rate itself is a genetic feature, a characteristic adjusted to the life cycle of the species. If this is so, there must be genes which can influence the mutation frequency of other genes. This brings us to a discussion of factors which can affect the typical mutation frequency. The genetic composition itself has indeed been

shown to play an important role. In the fruit fly, spontaneous mutation rates may vary as much as 10-fold among different stocks. Such marked increases in mutation frequency have been attributed to definite genes, *mutator genes,* which can increase the mutation rate of other genes. Normally, we would not be aware of the presence in the genotype of genes which keep the mutation frequency adjusted to the life cycle. However, if a mutation occurs in such a gene, the adjustment is interrupted. An increase in mutation frequency follows, and a mutator gene can be recognized. Genetic elements which greatly increase the mutation frequency of other genes have also been recognized in corn, where they have contributed much information on the subject of the genetic control of development.

The very first factor found to alter mutation frequency was temperature. Morgan and his students noted that fruit flies raised at 27°C had two to three times the mutation rate per generation than those raised at 17°C. Later studies showed that not only high temperature but also abnormally low ones elevated the mutation frequency. Increased mutation rate also followed sudden temperature shifts back and forth from high to low. This increase was noted even if the shift was from a high to a low temperature within the normal range for the species. So it would appear that departures from temperature conditions to which the organism is normally adapted may increase the mutation rate.

Observations made on some plant and animal groups indicate that aging can increase the mutation rate. An increase is seen after the aging of seed and pollen in plants and the aging of sperms in *Drosophila.* There is even evidence that in humans a similar effect pertains. As stated above, a mutation is said to have occurred in humans if a condition, known to be inherited as a dominant defect, suddenly appears in an offspring of two unaffected parents. Certain dominant disorders, such as achondroplastic dwarfism, have been found to appear with a greater frequency where the male parent is considerably older than the average.

The importance of malnutrition as a causative agent in mutation is indicated by the doubling of the mutation frequency in plants, such as the snapdragon, when they are deprived of certain mineral elements. Such effects as those of malnutrition, aging, and other environmental influences on mutation rate may be related to the failure of repair mechanisms in the cell, a topic discussed more fully later in this chapter.

HIGH-ENERGY RADIATIONS AND THEIR GENERAL EFFECTS ON MATTER. Some of the features of spontaneous mutation which were noted in early studies indicated something about the physical basis of the mutation process. H. J. Muller, one of Morgan's students, was impressed by the fact that the gene, a very stable entity which goes through countless accurate duplications, can change suddenly to another form, which in turn can duplicate itself accurately. These rare sudden changes in a gene were seen to be random ones; a particular environment did not favor a particular kind of mutation over another. And when a rare gene change occurred in a cell, it was always *just one* change in one gene. Even in a diploid cell, if a gene mutates, the same gene on the homologous chromosome is not affected. These observations all indicated to Muller that mutation was some kind of chance disturbance on the molecular level. Although environmental factors might conceivably increase the frequency of such disturbances, they wouldn't favor one kind over another. *Any* point in the hereditary material might be affected at a given moment. A disturbance affecting a particular gene did not have to influence the homologous position in the same cell, because it was a random event. Muller reasoned that if a mutation actually was the result of a chance pointwise excitation, then high-energy radiations should be able to increase the frequency of mutation. This would be expected, because radiations were known to cause chance disturbances as they pass through matter. Any agent which could cause such changes might be considered a potential instrument for raising the mutation rate. In his classic experiment performed in 1927, Muller obtained almost immediate results which showed conclusively that X-rays increase the frequency of both gene mutations and structural chromosome changes.

To appreciate this classic work and its relevance today, let us review a few basic facts concerning radiations. The radiations pertinent to the topic of mutation are those which can travel through space which is empty of matter.

They do not, therefore, include sound waves. Of the two major types of radiation, the electromagnetic ones are probably the most familiar (Table 15-2). In one sense, they may be considered to be waves, whose lengths can be measured from the long radio waves to the extremely short cosmic rays. Note that they form a continuous gradation or spectrum from the longest extreme to the shortest without any sharp boundaries between them. Visible light includes certain wavelengths in part of this continuum. It is, however, those electromagnetic waves whose lengths are shorter than those of light which are effective as mutagenic agents, factors which can increase the rate of mutation. To appreciate part of the reason for this, we must consider the energy which is associated with waves of different lengths. Although these radiations have many features of waves, in certain other respects they resemble particles. For when they are absorbed by matter, they are absorbed in definite units. Similarly, when they are given off by a source, such as an X-ray tube or ultraviolet lamp, they are emitted in definite units. These units are called *photons*, and a photon of a particular wavelength has a certain characteristic amount of energy associated with it, called the *quantum*. As the wavelength decreases, the quantum energy associated with a unit (a photon) of the electromagnetic radiation increases. Thus, *more energy is associated with photons of shorter wavelength*. This energy value is of importance to the material through which the radiation

TABLE 15-2 Spectrum of electromagnetic radiations

WAVE TYPE	WAVELENGTH	Å = 1/10,000 μ
Radio	10^7–0.04 cm	Insufficient energy to excite or to ionize atomic electrons
Infrared	20,000 Å–7,800 Å	
Visible light	7,800 Å–3,800 Å	
Ultraviolet	3,800 Å–150 Å	Energy to excite
X-rays	150 Å–0.15 Å	Sufficient energy to excite and to ionize
χ Rays	0.15 Å–0.005 Å	
Cosmic rays	0.005 Å–0.00008 Å	

passes. If there is sufficient quantum energy associated with the photons, the energy may be sufficient to influence the internal organization of the molecules composing a substance. Visible light and the longer wavelengths do not possess enough energy to disorganize subatomic particles. A photon of X-radiation, however, has sufficient energy to be absorbed by submolecular elements, usually electrons. Thinking of the atom as a planetary system, the electrons would be circling the nucleus in one or more shells or orbits (Fig. 15-3). Normally, these electrons are found at a characteristic distance from the nucleus, as they are attracted and held by the nuclear binding forces. The inner orbital electrons are held more strongly by the attractive forces of the nucleus than those in the outer orbits. If a single electron should absorb excess energy, it may move farther away from the nucleus than it normally would. We say that

FIG. 15-3. An atom in three states. In this diagram (*left*), the atom has three electrons circling a nucleus which contains three protons and four neutrons. The atom is neutral in charge and the electrons circle in orbits which are a characteristic distance from the nucleus. If excess energy is imparted to the atom, a single electron, usually in an outer orbit, may move a greater distance away from the nucleus (*middle*). If the electron receives sufficient energy, it may actually break away, leaving the atom positively charged (*right*).

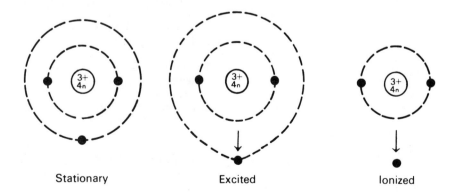

Stationary   Excited   Ionized

such an atom is in an "excited state," meaning that a single particle in the atom now possesses excess kinetic energy. This means that the electron concerned is bound less firmly by the nuclear attractive forces. If sufficient energy is absorbed, it may allow the electron to break away from the nucleus to which it belongs. In such a case of an outright separation of electric charges, we say that an ionization has occurred. It is the electrons in outer orbits which are more likely to be affected, because the inner ones are more firmly held. Once an electron has been knocked out of an atom, it may in turn knock out electrons in other atoms. Eventually, the energy would be dissipated. The free electrons would attach to other atoms, resulting in positively and negatively charged particles. The overall effect of excitations and ionizations is to make the material more reactive. So much energy is associated with gamma and cosmic radiations that inner electrons and atomic nuclei themselves may be influenced. Ionizing radiations cause both excitations and outright ionizations. Ultraviolet light, however, with its relatively long wavelength, does not cause ionizations. Its effects are mainly in the nature of excitations and are discussed more fully later in this chapter.

The other major class of radiations includes the corpuscular radiations (Table 15-3). Unlike the electromagnetic radiations, these possess a definite mass. They are discrete bodies which may behave as streams of particles. They affect matter in the same way as described for the other class. We may think of a stream of protons, neutrons, or electrons actually hitting a suba-

TABLE 15-3   Some of the corpuscular radiations

PARTICLE	WEIGHT (ATOMIC MASS UNITS)	CHARGE
Electron (beta particle)	0.00055	Negative
Positron (positive beta particle)	0.00055	Positive
Proton (nucleus of common isotope of hydrogen)	1.007	Positive
Deuteron (nucleus of heavy isotope of hydrogen)	2.013	Positive
Alpha (nucleus of helium)	4.002	Positive
Neutron	1.009	Neutral

tomic particle and imparting excess kinetic energy to it which is sufficient to cause an ionization. As mentioned above, once an electron is separated from an atom, it may in turn collide with other electrons and ionize a second atom. Similarly, a proton or neutron may pass through matter, knocking out electrons which can then go on to affect still other electrons in other atoms. From the behavior of both classes of radiations, we can thus appreciate what is meant by a chance, pointwise disturbance at the level of the atom and why Muller thought it likely that high-energy radiations could induce mutations.

RADIATIONS AND MUTATION RATE. In order to obtain quantitative data on any possible effects

of X-rays on the mutation rate, Muller designed a method which is ingenious for its simplicity and lack of ambiguity. Instead of selecting for study mutations which produce visible effects, Muller chose sex-linked lethals which occur with a higher frequency than the visibles. Morever, any sex-linked recessives will express themselves sooner than will autosomal recessives. For his investigation, Muller derived stocks of flies in which the females were ideally suited for the detection of new mutations on the X chromosome (Fig. 15-4). In these females, one X is an ordinary wild or standard one. The other, however, possesses two mutant sites: the Bar mutation (B) which behaves as a dominant and a recessive lethal (l). This same X also contains

FIG. 15-4. The C/B method. When a C/B female is mated, one-half of the normal number of male offspring will develop, because one-half of them receive the C/B chromosome which carries a lethal. If the C/B female is mated to a male treated with radiation, the treated X chromosome is transmitted only to the F$_1$ daughters. Only the Bar-eyed daughters are then selected, because it is known that they carry at least one lethal, that linked to the Bar

locus. When these F$_1$ Bar females are mated to any male, it is known that at least one-half of the males will not develop, again due to the lethal linked to Bar. The other one-half receive the treated X, which traces back to the original treated P$_1$ male. If it too carries a lethal due to the treatment, no male offspring at all appear in the F$_2$. The appropriate controls must be run to consider spontaneous mutation (see text for details).

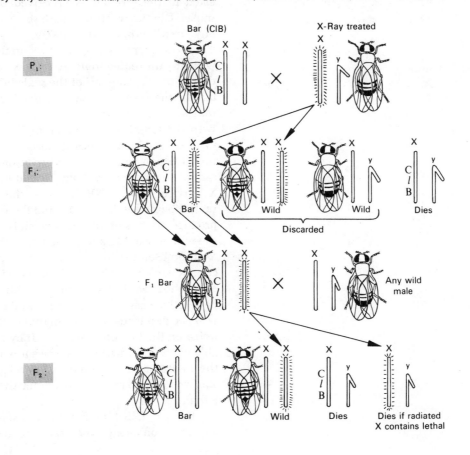

an inverted segment (C) which acts as a suppressor of crossing over in that part of the X between B and l. Such females are called "ClB" females and are phenotypically Bar eyed.

In the mutation study, two series of crosses were made with these females. In one series, the ClB flies were crossed with males which had received known doses of radiation. The other group, the controls, were mated with ordinary, untreated males. It can be seen from Figure 15-4 that the X chromosome of the treated P_1 males will go to all their daughters, both the wild and the ClB daughters. The latter can be recognized by their Bar phenotype. One-half of the F_1 male offspring will die because they receive an X chromosome which carries a lethal (the ClB) from their mothers. The remaining males, as well as the wild daughters, are rejected. Only the F_1 Bar-eyed daughters are saved for further mating. The reason for this is that we know that these Bar-eyed flies must contain the original lethal, because the Bar locus is linked to it, and crossing over is suppressed. We are now interested in the "treated X" received from the P_1 males. Did the treatment with the X-rays induce any "new" lethal mutations?

To answer this, each F_1 Bar female is crossed with any untreated male. It is known in advance that in the F_2, one-half of the male offspring will die for the same reason that one-half of the F_1 males died; they receive the ClB chromosome with the lethal. Now, the other X chromosome in these F_1 Bar females is derived from the P_1 males which were subjected to radiation. If the radiation produced a "new" lethal on the other X carried by the ClB females, the remaining male offspring will also die, and the females will produce no sons at all. The reason is simply that the female would have a different lethal on both X chromosomes; the males, being hemizygous, cannot survive. The F_1 females will survive, even with two lethals, one on each X, for any new lethal arising by chance would almost always be at a locus other than that of the original lethal on the ClB chromosome. If by the improbable chance it should be at the same locus, then the F_1 Bar female would not develop, because she would be homozygous (ll) for the recessive lethal.

Each F_1 ClB female was bred separately. If any male offspring were seen in the breeding bottle, it was immediately known that no new mutation had been produced. If, however, a C*lB* female failed to produce sons, then a new lethal mutation must have arisen on the other X. Since spontaneous mutations would also be occurring, the results on the X-ray treated series had to be compared with the nonexposed controls. The method also permitted the detection of any visibles produced on the X by the radiation. The procedure gives very clear-cut results; when sex-linked lethals are followed, one class, the male sex, is either present or absent. There is no source of error through difficulty in recognizing a particular phenotype.

Several thousand F_1 flies were followed in both the treated and untreated groups. The data were conclusive from the start. Similar results were later obtained by other investigators (Fig. 15-5). It was clear that the number of sex-linked lethals which arose was directly proportional to X-ray dosage as measured in roentgen units (1 roentgen (R) is equal to an amount of radiation dosage sufficient to produce 2×10^9 pairs of ions in 1 cc of air). A dose of approximately 60 R was found to be enough to double the spontaneous production of sex-linked lethals. A given amount of radiation would result in a certain percentage of lethals. Doubling or tripling the dosage would cause a corresponding increase in the number of lethals obtained. Of course, there is a limit to the dosage which can be applied. At higher levels, the viability of the flies is affected, so that a strict linear relationship begins to fall off. Moreover, at higher doses there is an increased probability that more than one lethal will be induced in the same X chromosome. Because two or more lethals would be counted as one, the direct relationship would tend to fall off for this reason as well. The direct relationship which does exist between radiation dosage and genetic effect has been found to apply generally to the production of other kinds of mutations besides sex-linked lethals. It also holds for wavelengths other than those of X-rays.

The corpuscular radiations, neutrons and protons, cause ionizations, and these were shown to be effective in the production of mutations. However, the distribution of their ionizations is much more condensed than that produced by electromagnetic radiations, so that certain differences are apparent in the relation-

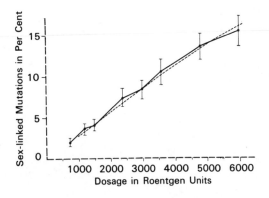

FIG. 15-5. Sex-linked mutations and X-ray dosage in *Drosophila*. The number of point mutations is directly proportional to the dosage of X-rays as measured in roentgens. At higher doses, the relationship falls off as the radiation decreases the viability of the flies. (Reprinted with permission from E. W. Sinnott, L. C. Dunn, and T. Dobzhansky, *Principles of Genetics*, 5th ed., McGraw-Hill, New York, 1958.)

ship of the dosage applied to the number of mutations produced. Nevertheless, all the ionizing radiations can induce genetic change. It must be kept in mind, however, that treatment with any of the mutagenic radiations is not specific for a particular gene and does not ensure that one certain kind of gene mutation will be produced in preference to another. There are no more visibles of a given kind produced by this or that agent. What *does* happen is that the entire spectrum of mutation is increased in frequency. Nothing new is produced. The mutation rate is elevated for all the loci. But with the increase, the loci still mutate with the same *relative* frequency. A locus with a spontaneous rate double that of another will still show this same relationship to the second one, even when the frequency of both of them has been raised.

SUGGESTED MECHANISMS OF RADIATION ACTION ON THE GENETIC MATERIAL. The observation that the production of point mutations is generally proportional to the dose applied led to the idea that induced mutations result from ionizations which are produced within the gene itself. According to this concept, the *target theory*, the gene itself must be struck if a mutation is to occur, just as if it were a target being hit by bullets. With twice the number of bullets, there would be twice the chance of hitting the target (the gene) and of causing ionizations within it.

Also, it shouldn't matter how the dose is distributed. If small doses were given over an extended period of time, there should be the same outcome as if one large dose had been applied. This would be similar to using a few bullets over a longer time or firing them all in one burst. The chance of hitting some things would be related to dose, not to distribution of the dose. The target theory seemed to apply to the production of point mutations in *Drosophila*.

Besides their ability to induce gene mutations, ionizing radiations were also found to be very effective in the production of structural chromosome changes. The frequency of all types, deletions, inversions, and translocations, etc., was increased by application of radiations. The role of direct proportion between mutation and dose of radiation, however, did not apply for the gross structural changes. This makes sense, when we consider the fact that two breaks are necessary for a structural change to take place. A given dosage (Fig. 15-6) would correspond to a certain probability for a break to occur in a chromosome. The chance of two breaks occurring together would be equal to the product of the chances of the single breaks. Therefore, the production of gross structural chromosome aberrations would increase *not* in a linear relationship to dose, but in proportion to the square of the dose. The higher the dose, the greater the chance of producing chromosome aberrations. Very small deletions and inversions might be caused by just one hit; these would be proportional to the dose, as seems to be the case. And the intensity of the dose seems to play a role in the production of structural changes. More of

FIG. 15-6. Radiation dosage and production of chromosome breaks. Single breaks in a chromosome show a direct proportion to dosage applied as is true for the production of point mutations. The chance of two breaks occurring in the same nucleus would be equal to the product of two single breaks. Consequently, large structural chromosome alterations increase in proportion to the square of the dose. This means, as seen on the right, that at lower doses more single breaks are produced, but with increased dosage, there is a greater chance for two breaks and hence structural aberrations to take place.

Radiation dose (arbitrary units)	Chance of one break	Chance of two breaks together
1	1/10	1/100
2	2/10	4/100
3	3/10	9/100
4	4/10	16/100

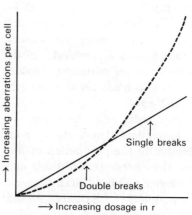

them are obtained if a large dose is given at one time rather than if it is distributed in smaller packages. This can be explained by assuming that when a dose is weak, there is less probability for two breaks to be produced in the same nucleus. The break would tend to be repaired before another one was formed. So if an intense dose is applied at one time, there will be more single breaks overall at one time which can engage in new arrangements.

Later work with mice has shown that these animals react differently from *Drosophila* in response to the intensity of the radiation. The intensity exerts an effect in the production of point mutations, which is similar to that in the production of structural aberrations. Such differences between species in response to radiation are one of several observations which suggest that the mechanism of inducing gene mutation is not as direct as the target theory states.

INTERACTION OF RADIATION AND OTHER FACTORS IN THE PRODUCTION OF MUTATIONS. For years, several factors have been known which can cause the number of mutations associated with a given dose of radiation to vary, even in the same organism. In *Drosophila*, a given dose will produce more point mutations if mature sperms are radiated rather than earlier stages. The chromosomes of mature sperms are also much more sensitive to chromosome breakage than are those in immature germ cells. Generally, cells with chromosomes in a condensed state, such as at nuclear division, are also much more susceptible to chromosome damage than those in the nondividing nucleus. This accounts in part for the greater susceptibility of cancer cells to radiation damage than the normal cells around them which would not be dividing as actively. Moreover, the number of mutations which arise after a given radiation dose may vary depending on the stock or strain. In fruit flies, just as some stocks have a higher spontaneous mutation frequency, so do some stocks vary in their sensitivity to the radiations. That the surrounding environmental conditions may also influence the yield of induced mutations has been definitely demonstrated by the effect of oxygen. Fewer mutations for a given dose are found at lower oxygen tensions. Among microorganisms, those forms which can exist either aerobically or anaerobically undergo a higher rate of both spontaneous and induced mutation under aerobic conditions.

The temperature of the surroundings at the time of applying the radiation has an influence which is probably related to the oxygen effect. Under colder conditions, the induction of chromosome alterations, as well as gene mutations, is strengthened; this is most likely associated with the presence of more oxygen at lower temperatures. Also connected with the amount of oxygen is the action of certain reducing agents when present in the environment at the time of irradiation. The number of chromosome breaks is reduced in the presence of —SH compounds, alcohols, and various strong reducers. These compounds may afford a certain amount of protection by removing oxygen dissolved in tissues.

ULTRAVIOLET AND THE INDUCTION OF MUTATIONS. We noted earlier in this chapter that the amount of energy associated with the wavelengths of ultraviolet is too small to produce ionizations. The relatively long waves are not highly penetrating and so do not usually reach those cells which will give rise to gametes. Nonetheless, it has been demonstrated in several higher organisms that ultraviolet has a definite mutagenic effect once it does reach the DNA. Microorganisms can be readily exposed to an ultraviolet source and have provided a wealth of information on its mutagenic action. Many years ago it was observed that the mutagenic activity seemed to be due to the fact that the hereditary material itself actually absorbs the ultraviolet. This was shown by comparing the effectiveness of the different wavelengths in their ability to induce mutations. Some wavelengths were found to be more effective than others, those of 2600 Å being the most mutagenic. When a curve is plotted showing these differences, it is found to agree with a curve which plots the ability of DNA to absorb wavelengths of ultraviolet. In other words, those wavelengths which cause more mutations are those which are more strongly absorbed by the genetic material. Little activity is found by those wavelengths only weakly picked up by the DNA. Later work with isolated DNA demonstrated changes which ultraviolet can produce in nucleic acid. It is the bases in the DNA which are primarily responsible for the absorption of the ultraviolet. The pyrimidine bases thymine and cytosine are particularly sensitive. A major consequence of the absorption is the formation of stable bonds between two thymine bases or between two units of cytosine. Such associations between two identical molecules or units of a molecule are called *dimers*. Production of thymine dimers seems to be a very important way in which ultraviolet alters the DNA (Fig. 15-7). Ultraviolet absorption may also cause stable bond formation between two different pyrimidine bases giving a "mixed dimer" (such as cytosine-thymine). When dimer formation occurs, it is two pyrimidine bases adjacent to each other on a DNA strand which become stably bonded. At very high doses, ultraviolet has been shown to be capable of breaking the sugar-phosphate backbone of isolated DNA. Whether or not it operates this way in the intact cell, ultraviolet can cause chromosome breakage in the fruit fly and some other species. Its ability to induce breakage, however, is much less than that of the high-energy ionizing radiations.

From our knowledge of the Watson-Crick model of DNA, we can surmise how ultraviolet may cause some of its mutagenic effects. Duplication of the DNA depends on clean separation of the two chains, each of which acts as a template in the construction of new, complementary chains. Obviously, any process which interferes with this, such as the formation of stable bonds by dimerization, could very possibly result in a DNA with altered activity or a DNA unable to transmit the information stored within it. One unusual aspect of ultraviolet is that its mutagenic effect may be reduced by white light. A simple way to demonstrate this (Fig. 15-8) is to spread each of two agar plates with the same number of bacteria and subject each to a dose of ultraviolet known to have killing activity. The plates are then incubated, one of them completely in the dark but the other after a 1/2 hr

FIG. 15-7. Dimerization of thymine. Ultraviolet light may cause stable bonds to form between two identical molecules of thymine or cytosine. The result is a dimer. If dimers form between two bases located on opposite strands or even along the same strand of a stretch of DNA, replication of the DNA molecule could become impaired.

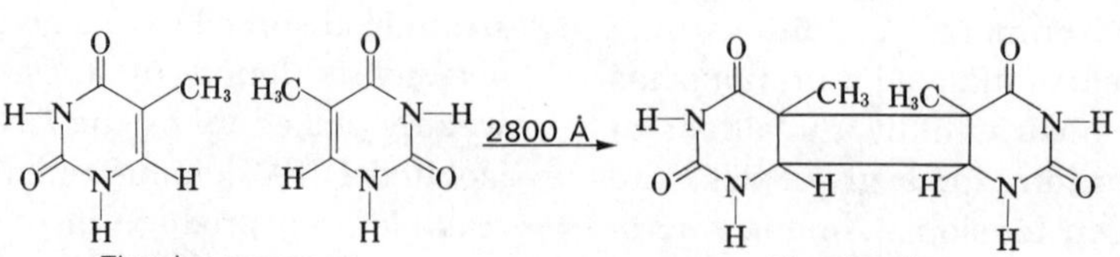

Thymine monomers $\xrightarrow{2800\ \text{Å}}$ Thymine dimer

exposure to a strong source of white light. Observation of the plates the next day will show a much better growth on the plate which was treated with the white light. The killing effect of the ultraviolet is reduced so that only about 40% of the original killing activity remains. Therefore, if a certain dosage can kill one-half the bacteria on a plate, exposure to white light can reduce this lethal action from 50% to 20% (0.50 × 0.40), so that 80% of the bacteria will survive. This reversal of the lethal effect of ultraviolet light by white light is called *photoreactivation* and has also been demonstrated in *Drosophila* and higher plants. The mechanism of photoreactivation remained a mystery until the discovery of "repair enzymes" which can undo damage to the DNA. One of these repair enzymes has been shown to be activated by light (actually the blue wavelengths) and to be able to break the bonds formed in dimerization.

Besides its direct effect on the DNA, there is evidence that some of the mutagenic activity of the ultraviolet may be through its action on other parts of the cell, which in turn exert a mutagenic effect. This is clearly indicated by experiments in which culture medium alone is exposed to the ultraviolet and then later inoculated with microorganisms. A definite increase in mutation is noted, even though the microorganisms were plated without direct contact with the ultraviolet light. The mutagenic effect has been linked to the formation of hydrogn peroxide (H_2O_2) in the medium by the radiation. Treatment of amino acids with peroxides was shown to make them capable of inducing mutation. It would thus seem that the ability of radiations to induce mutations (both point and chromosomal) is not dependent solely on ionizations directly within the gene. Indeed, a great many of the observations we have just discussed

FIG. 15-8. Photoreactivation. Two plates spread with the same number of bacteria are subjected to a dose of ultraviolet radiation. One is placed immediately in the dark. The other is kept in the light for at least 1/2 hr. After incubation of the plates, the light-treated one is found to have a much larger number of bacterial colonies than the plate placed immediately in the dark. The

number may approach that found on a plate which was not exposed to the ultraviolet at all. This untreated plate can be placed in the dark or given a light treatment, but the number of colonies on it will not be altered. The photoreactivation operates on the damage induced by the ultraviolet.

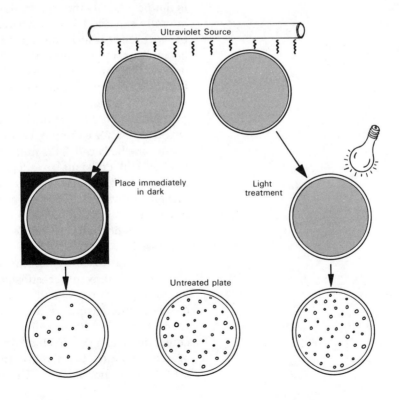

suggest that mutation frequency might be altered by chemical conditions surrounding the hereditary material.

CHEMICAL MUTAGENS. It was not suprising when the first chemical mutagen was discovered in 1941 by Auerbach and Robson. These investigators turned their attention to mustard gas and related compounds as possible mutagenic agents because their burning effects on somatic tissue were similar to those of high-energy radiations. And it was found that like high-energy radiations, the mustard compounds could produce a high incidence of both gene mutations and structural aberrations. The same types of gene mutations result as with the radiations (detrimentals, lethals, etc.). The mutagenic action was demonstrated on a variety of plant and animal species. The mustard compounds can also inflict high damage on rapidly growing tissue by their production of chromosome changes and have been used as well as radiations in the treatment of malignancies.

The mutagenic action of mustard gas, however, differs from that of radiations in a very important respect. The effect of treatment with mustard gas is frequently delayed; a mutation is not necessarily induced in the cell which was actually exposed but may arise in a descendant of that cell. It is as if the treated DNA somehow became unstable. The actual mutation may be delayed for two generations. And when the mutation does occur, it may arise in a body cell as well as in a germ cell. This means that if the mutation should occur during the development of a fruit fly, for example, that it can take place in just one body cell. This mutant cell then gives rise by cell divisions to a patch of cells surrounded by nonmutant ones. The result is a mosaic individual. This delayed effect makes mustard gas in many ways a more serious mutagenic agent than one whose action is directly expressed.

After the discovery of the first mutagenic chemicals, many other substances were also found to posses mutagenic activity, and the list has been increasing to this very day. Few of them have the strong effect of the mustard compounds, and the mechanism by which many of them operate to increase the frequency of mutations is not known. There is nothing in common about the chemical structures of the many mutagens. Furthermore, a substance found to be mutagenic in one organism may be ineffective in another. Caffeine, for example, is highly mutagenic in some organisms, but is apparently broken down in others before it can exert any action. Formaldehyde markedly increases the mutation rate in the fruit fly, but only if applied to the food of the larval male. It is ineffective at other stages of the life cycle and is entirely nonmutagenic in the female! Most of the chemical mutagens have also been found to be capable of inducing structural chromosome changes as well as point mutations.

The interesting discovery was made that some substances have an antimutagenic action, that is, they tend to suppress the mutagenic activity of other agents. Antimutagens were first detected by Novick, working with bacteria. The strong mutagenic action of purines, such as caffeine and adenine, was suppressed by the presence in the medium of purine ribonucleosides, such as adenosine and guanosine. The antimutagens were ineffective toward the mutagenic action of ultraviolet and gamma radiations. Even the spontaneous mutation rate was shown to be reduced when the antimutagens were present. It is significant that some purine nucleosides were found to have a high antimutagenic activity and that many of the nontoxic chemical mutagens were purines or their derivatives. Indeed, the mutation rate seemed to depend in part on the physiological state of the organism. All of these observations suggested that the study of mutation might be approached on a biochemical basis, even if the various mutagenic agents cause genetic change in different ways. Once the chemical structure of the hereditary material was clarified, such an attack on the problem became feasible.

REPAIR OF DAMAGED DNA. Although it is unclear how certain nucleosides effect their antimutagenic activity, several observations lead to the conclusion that the spontaneous mutation rate, which is characteristic for a species and which is geared to the generation time, results from an interaction between mutagenic substances and antimutagens which arise during the course of the metabolic activities of the cell. Moreover, we are now aware of the existence

within cells of several enzymes which can engage in the repair of damage caused to the DNA. The enzyme which is involved in photoreactivation is one of these. But in addition, there are others which repair ultraviolet-damaged DNA in a different manner and which do not depend on light. For example, the damage to DNA resulting from the formation of dimers can be repaired in the dark following a series of steps which involves the interaction of several enzymes.

These enzymes are representative of several groups of enzymes which play important roles in a variety of cellular activities. Two of the enzymes involved in DNA repair are called *nucleases*, since they can cut the phosphodiester bonds which link together nucleotides in a polynucleotide chain. A nuclease can attack either the 3' or the 5' end of a phosphodiester linkage. One group of nucleases is known as the *endonucleases*. These enzymes can attack only those bonds which occur in the interior of a polynucleotide chain. In essence, an endonuclease can produce a nick internally in a chain of a double helix. Another class of nucleases has been designated *exonucleases*. These too attack 3',5' phosphodiester bonds, but they attack only from the free end of a phosphodiester chain, either the 3' end or the 5' end.

Also involved in DNA repair is a type of enzyme called a *DNA ligase*. Such an enzyme is able to catalyze phosphodiester linkages. Thus, a ligase can restore a double helix to its intact condition following the production of an internal nick caused by an endonuclease. DNA polymerase is also essential to DNA repair. Its ability to catalyze the linkage of nucleotides on a DNA template was discussed at the end of Chapter 11.

The initial step in the repair entails an endonuclease which is able to recognize the damaged region, probably as a result of the change in the configuration of the DNA brought about by the presence of the dimers (Fig. 15-9). The endonuclease produces a nick in the region of the damage, causing the strand to break. The free end which arises is then recognized by an exonuclease which digests away a part of the strand including the dimers. Once the damaged portion is removed, the enzyme DNA polymerase synthesizes the missing segment, using as a template the complementary region which lies exposed on the other strand. Finally, the DNA ligase catalyzes a phosphodiester linkage between the end of the newly made segment and the portion of the original strand from which the damaged segment has been removed.

Although much of the information on repair enzymes has come from experiments with bacteria, it is known that eukaryotes also possess DNA repair mechanisms and that dimers may also be repaired in them by photoreactivation. Moreover, nuclei of human cells have been shown to contain an enzyme which can operate without light and which participates in the repair of ultraviolet-damaged DNA in much the way that the bacterial enzyme recognizes a damaged region of the DNA and initiates its removal.

The significance of repair enzymes to the organism cannot be overestimated. Failure of repair mechanisms in a cell may account for the increased mutation rate found in older seeds, sperms, etc. Conceivably, repair enzymes are needed in all kinds of cells for protection against the mutagenic action of factors which are found in the average environment. Indeed, this is strongly suggested by the observation that per-

FIG. 15-9. Steps in repair of damaged DNA without a requirement for light. A stretch of DNA (1) is exposed to ultraviolet light, and a thymine dimer forms, distorting the DNA (2). An endonuclease recognizes this region and produces a nick in the vicinity of the damage (3). A second enzyme, an exonuclease, recognizes the free end and digests away a segment of the DNA strand which includes the damage (4). DNA polymerase in a 5′ to 3′ direction then synthesizes the portion which has been eliminated, making a linkage to the 3′ end of the original strand (5). Finally, DNA ligase links the newly formed segment to the original strand, and the DNA is restored to normal (6).

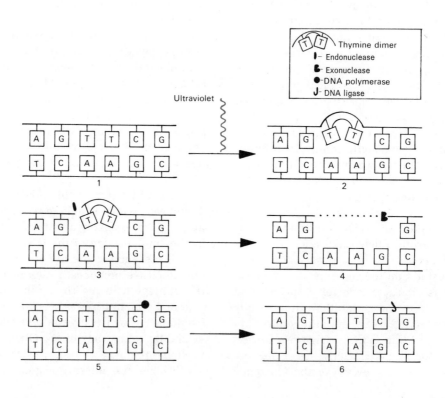

sons expressing a serious, inherited skin disorder (*xeroderma pigmentosum*) which entails malignancies of the skin have a lesser amount of the repair enzyme mentioned above which has been isolated from normal human cells and which can remove ultraviolet-induced thymine dimers. Repair enzymes such as this could be essential to the maintenance of normal cell activities by preventing the accumulation of DNA damage which may stem from various causes, both intracellular and extracellular. Failure of efficient repair mechanisms may contribute significantly to the "normal" process of aging, during which damage to the DNA may continue to accumulate, leading eventually to the death of the cell. We will return to a consideration of this point in Chapters 19 and 21.

THE EFFECT OF CHEMICAL MUTAGENS ON THE MOLECULAR LEVEL. Recall that the Wat-son-Crick model implies that the information stored in the gene is coded in the sequence of the nucleotide pairs which compose the DNA. The model predicts that a mistake in base pairing at the time of gene replication can lead to a point mutation. Such a mistake results in turn from the occurrence of a rare tautomeric form of one of the purine or pyrimidine bases (Chap. 11). Such a form arises from a shift or redistribution of electrons and protons in the molecule. Consequently, the usual pairing relationships among the bases may be altered. Instead of adenine pairing with its complementary base, thymine, the rare form of adenine may pair with cytosine to give an A:C pair instead of the normal A:T. Conversely, the occurrence of a rare tautomeric form of cytosine may result in a C:A pair instead of C:G. Likewise (Fig. 15-10), instead of T:A or G:C, the unusual pairs C:A and A:C may form. The ultimate effect of these pairing errors is

FIG. 15-10. Transitions. A stretch of four nucleotides with normal pairing is shown above. Abnormal pairs may form at the time of replication due to the occurrence of a rare tautomeric state of one of the bases. If adenine and cytosine form a pair, instead of the normal A:T, the result is a transition, a substitution of one purine-pyrimidine pair by another or the substitution of one pyrimidine-purine pair by another. In this example, we see: (1) G:C substitutes for A:T, (2) T:A for C:G; (3) C:G for T:A; and (4) A:T for C:G.

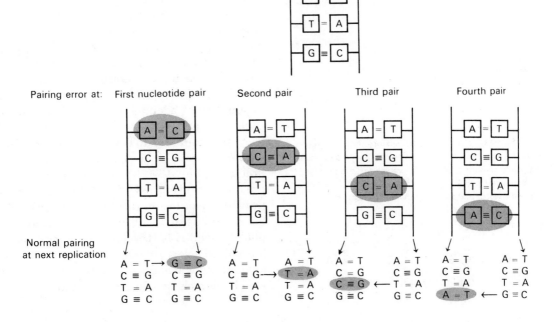

shown in Figure 15-10. It can be seen that the A:C error results in the formation of a G:C nucleotide pair. The original position of the purine, adenine, on one of the two strands of the double helix is now occupied by the purine, guanine. In other words, one purine:pyrimidine pair is now represented by a different purine:pyrimidine pair. It can be seen from the figure that in each case, the purine:pyrimidine relationship is maintained, one purine:pyrimidine pair substitutes for another *or* one pyrimidine:purine pair substitutes for another. Base pair substitutions of this type have been called *transitions*. In contrast to this type of change is the *transversion*, the substitution of a purine:pyrimidine pair by a pyrimidine:purine pair (or vice versa; Fig. 15-11). In the transversion, there is a new orientation of purine and pyrimidines. A purine on one strand is replaced by a pyrimidine; a pyrimidine on the complementary strand is replaced by a purine. The significant fact is that in the case of either a transition or a transversion, a different sequence arises at a site on the DNA. This alteration would thus modify the genetic information at that point and could cause a mutant effect as a consequence.

It was reasoned that if spontaneous mutation did actually result from such transitions and tranversions, then it should be possible to induce mutations with chemicals whose activities are known. Such a chemical, for example, is nitrous acid, which has the ability to remove free $—NH_2$ groups from the bases found in nucleic acids (Fig. 15-12*A*). After deamination by nitrous acid, adenine is converted to another base, hypoxanthine. In hypoxanthine, the arrangement of atoms involved in base pairing is different from the adenine (A), but similar to guanine (G). This means that if adenine is changed to hypoxanthine (H) at a point on the double helix, a transition from A:T to G:C can occur. By deamination, nitrous acid converts cytosine (C) to uracil (U, Fig. 15-12*B*), whose pairing characteristics are like those of thymine (T). Again a transition can result, this time from C:G to T:A. Since nitrous acid can cause such changes in nucleic acid, it is therefore not surprising that it is indeed a very powerful mutagenic chemical. Several other chemical agents are also known which can cause alterations in the bases of

FIG. 15-11. Summary of types of transversions. In an alteration of this kind, the end result is a substitution of a purine-pyrimidine pair by a pyrimidine-purine pair (or vice versa). For example, in *1*, the purine-pyrimidine pair A:T is replaced by the pyrimidine-purine pair T:A (or vice versa) and so on for the other substitutions.

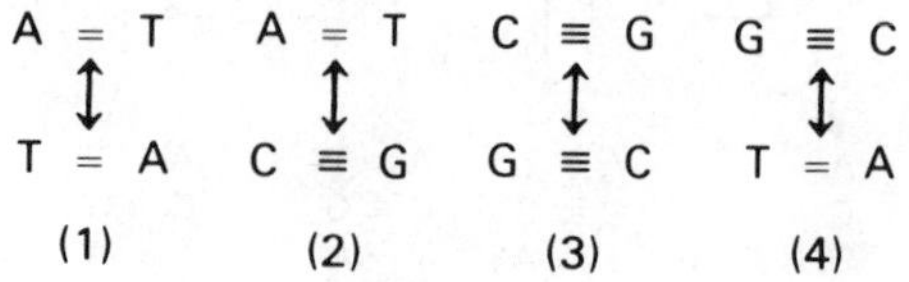

$$
\begin{array}{cccc}
A = T & A = T & C \equiv G & G \equiv C \\
\updownarrow & \updownarrow & \updownarrow & \updownarrow \\
T = A & C \equiv G & G \equiv C & T = A \\
(1) & (2) & (3) & (4)
\end{array}
$$

nucleic acids, and these too have been found to have mutagenic activity. Most of these investigations have been performed on microorganisms. Data from phages strongly support the concept of induced mutation arising from changes in base pairing at the time of gene replication. Very detailed biochemical analyses of certain bacterial viruses have shown that treatment with particular mutagenic substances whose chemical activity is known leads to alteration in protein products. These protein alterations are readily explained on the basis of transitions in the DNA caused by the mutagenic chemical.

Another class of mutagenic substances is known as the "base analogues." They are so called because they resemble the normally occurring nucleic acid bases both in structure and activity. For example, thymine has a base analogue, 5-bromouracil (5-BU). It is so similar to thymine that it can substitute in place of thymine. This means it can be attracted to adenine and form hydrogen bonds with adenine in much the same manner as thymine. One important difference between thymine and 5-BU, however, is that the latter undergoes tautomeric shifts more often than the thymine. In its rarer form, 5-BU behaves like cytosine and tends to pair with guanine. Therefore, the 5-BU could be in its "cytosine-like" state when it is being taken into the DNA as shown in Fig. 15-13. Or it may be in this form later at the time of replication *after* it has been incorporated into a strand of the nucleic acid. This is called a mistake in replication; the former is a mistake in incorporation. Again the end result is a transition, G:C to A:T (from the error of incorporation) or A:T to G:C (from the error of replication). Highly mutagenic base analogues such as 5-BU undoubtedly exert their mutagenic activity in this manner.

Other mutagenic chemicals are known which affect nucleic acid in other ways but the consequences still are base pair substitutions. The sulfur and nitrogen mustards can affect the base guanine in such a way that the linkage of the guanine base to the sugar backbone becomes unstable. The guanine may eventually be eliminated from the DNA (Fig. 15-14). This elimination doesn't cause the DNA to fall apart, but

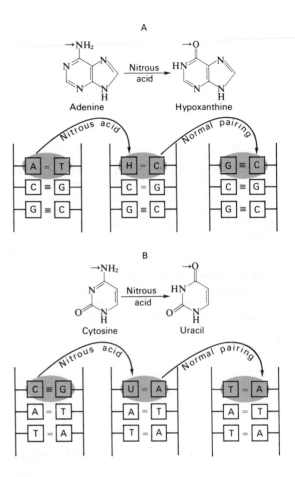

FIG. 15-12. Mutation induction by nitrous acid. Nitrous acid is able to remove free amino groups from bases in nucleic acid as seen above in the conversion of adenine to hypoxanthine. Hypoxanthine has pairing properties similar to guanine, and so it tends to pair with cytosine (A). The end result is a transition, the substitution here of G:C for the original A:T. B. Cytosine is converted to uracil which behaves like thymine and so pairs with adenine. The result again is a transition, in this case a change from C:G to T:A. Nitrous acid can also cause deamination of guanine and its conversion to xanthine. This product, however, behaves like guanine in its pairing preference for cytosine.

the missing guanine may be replaced by *any* one of the four bases present in the cellular environment. A transition or transversion could ensue. Knowledge of this type of action on the molecular level by the mustards gives us an insight into an understanding of the delayed effect which is characteristic of the mustard group.

Acridine dyes are mutagenic substances frequently used in mutation analysis. The exact manner in which they achieve their mutagenic action has not been established. However, it is known that acridines can insert themselves into the double helix between two adjacent nucleotides. Following exposure to acridines, there is an increase in a type of mutation in which there is either the addition of an extra base—called a (+) mutation—or the deletion of a base, a (−) mutation. Such (+) and (−) mutations are also known to occur spontaneously and following exposure to ultraviolet light. Mutations of this type, as we have seen, have been very important in deciphering the genetic code (Chap. 12). No matter what the exact mode of action, all the known chemical mutagens apparently bring about the same end result, a change in the coded information through some type of alteration in nucleic acid sequence.

Although we know that certain chemicals

FIG. 15-13. 5-bromouracil as a mutagen. This chemical resembles thymine closely in structure (*above*), and it can substitute for it at the time of DNA replication. The upper part of the figure shows this. At a following replication, it may be in its rarer form and behave like cytosine. At the next replication, the incorporated guanine pairs with cytosine. The result is a substitution of a G:C pair for the original A:T. The lower part of the figure indicates that 5-bromouracil may be in its rarer "C-like" state before it enters a DNA chain. It can thus pair with guanine. At the next replication, it would behave typically (like thymine) and attract adenine. At the next replication the adenine attracts thymine. The result is a substitution of an A:T pair for the original G:C.

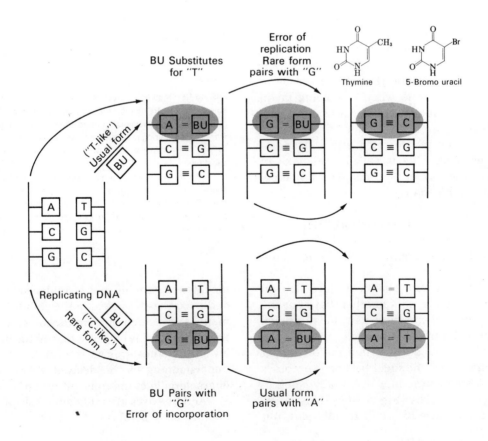

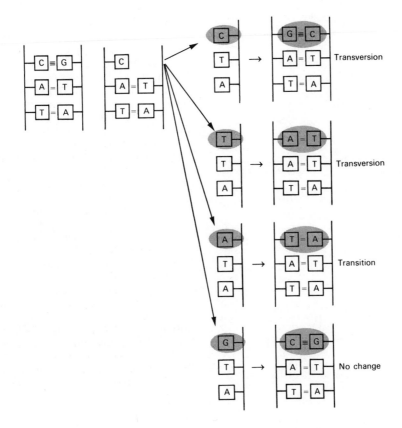

FIG. 15-14. Effect of mustard compounds. Sulfur and nitrogen mustards can cause the base guanine to be eliminated from a strand of DNA. The missing base may later be replaced by any one of the four bases. When this strand replicates, a transition or a transversion may occur, depending on which base substitutes for the original guanine.

can cause characteristic molecular changes, this information must not be misconstrued to mean that the investigator can bring about a specific kind of gene mutation by treating with a specific chemical. For these agents raise the entire mutation rate, and the four bases, adenine, thymine, guanine, and cytosine are found throughout the DNA. It is known from work with microorganisms that one mutagen may be more likely than another to induce a certain *class* of mutation. For example, Novick found that ultraviolet treatment of *E. coli* caused a higher mutational

change to resistance to a kind of virus, phage T/6. In contrast to this, certain purine mutagens caused a greater probability of mutation to resistance against a different virus, T/5. Comparison of some chemical mutagens has shown that some genes may be more susceptible to mutation by one chemical than to a second one. The chemical 2-amino purine favors a certain kind of back mutation (from mutant to wild) at a site in a gene controlling the enzyme tryptophan synthetase in *E. coli*. However, no mutagen is known which is specific for just one kind of gene.

REFERENCES

Auerbach, C., and B. J. Kilbey. Mutation in eukaryotes. *Ann. Rev. Genet., 5:* 163, 1971.

Bloom, A. D. Induced chromosomal aberrations in man. *Adv. Human Genet., 3:* 99, 1972.

Casarette, A. P. *Radiation Biology.* Prentice-Hall, Englewood Cliffs, N.J., 1968.

Drake, J. W. Mutagenic mechanisms. *Ann. Rev. Genet., 3:* 247, 1969.

Harris, M. Mutagenicity of chemicals and drugs. *Science, 171:* 51, 1971.

Muller, H. J. Artificial transmutation of the gene. *Science, 66:* 84, 1927.

Novick, A., and L. Szilard. Experiments on spontaneous and chemically induced mutations of bacteria growing in the chemostat. *Cold Spring Harbor Sympos. Quant. Biol., 16:* 337, 1951.

Setlow, R. B., J. D. Reagan, and J. German *et al.* Evidence that xeroderma pigmentosum cells do not perform the first step in the repair of ultraviolet damage to their DNA. *Proc. Natl. Acad. Sci., 64:* 1035, 1969.

Sutton, H., and M. I. Harris, (eds.), *Mutagenic Effects of Environmental Contaminants.* Academic Press, New York, 1972.

Witkin, E. M., Ultraviolet-induced mutation and DNA repair. *Ann. Rev. Genet., 3:* 525, 1969.

REVIEW QUESTIONS

1. The allele W for chlorophyll development is dominant to the recessive w for albino in a certain plant species. As a result of spontaneous mutation, which of the following plants would be the most likely to exhibit mosaicism if mutation arises in cells of a developing leaf? Which would be the least likely?

    (a) WW,

    (b) Ww,

    (c) ww,

    (d) WWWw, a tetraploid.

2. Suppose that a certain human disorder which is inherited as a dominant trait is associated with the allele B. The normal allele is the recessive b. It is found that in a given period of time five abnormal children are born among a total of 100,000 births to normal parents with no family histories of the disorder. From these fig-

ures, what appears to be the mutation rate per germ cell per generation from b to B?

3. Suppose a wild-type male fruit fly is treated with X-rays. It is then crossed to an ordinary, wild female carrying no known mutant alleles. If a recessive, X-linked lethal is induced in any of the radiated X chromosomes, what would be the effect on the sex ratio:

   A. In the first generation?
   B. In the second generation?

4. Answer the following questions which pertain to Muller's classic ClB method:

   A. If no inversion were present, what effect would this have in the P_1 female which would influence the results?
   B. Answer the same question in regard to the ClB F_1 females.

5. In the ClB experiment:

   A. What would be the phenotype of the males in the F_1 and the F_2 generations in regard to eye shape?
   B. Answer the same question, assuming no inversion in the P_1 female and in an F_1 ClB Bar female.

6. A certain dose of radiation applied to a culture of actively growing cells is found to induce single chromosome breaks at random with a frequency of 1/50.

   A. How many single breaks would be expected in the cell culture by increasing the dosage three times and then four times?
   B. How many structural chromosome changes (inversions, reciprocal translocations, etc.) would be expected by increasing the dosage three times and then four times?

7. Assume that a bacterial suspension is diluted so that a given volume will contain approximately 200 cells. This same amount of suspen-

sion is placed on several plates divided into three groups: A, B, and C. Nothing is done to group A. Groups B and C are exposed to a dosage of ultraviolet light sufficient to kill 30% of the cells. Group C plates are given a half-hour exposure to strong white light. The three groups are then placed in an incubator. Since about 60% of the killing effect of ultraviolet can be neutralized by photoreactivation, on the average about how many colonies would you expect to see the next day on plates of groups A, B, and C?

8. H. J. Muller estimated that the total background radiation from natural sources is equivalent to about 0.045r for a four-week period. Would you expect this background radiation to be more significant in the production of spontaneous mutations for a species like the human or one like the fruit fly with a life span of about four weeks? Why?

9. Give the specific role of each of the four kinds of enzymes involved in the repair of DNA damaged due to ultraviolet light.

10. Consider the following sequence of nucleotide pairs along a molecule of DNA: A G T . T C A
Exposure to nitrogen mustard causes G to become less firmly linked to the sugar-phosphate backbone, and it is eventually eliminated. Give the possible base sequences after the missing base is replaced and the strand replicates. Classify each change as a transition or a transversion.

11. Assume that the polynucleotide strand with the sequence A G T is replicating but that it has been stretched between G and T due to an acridine dye. On the growing complementary strand a nucleotide with thymine is inserted in the stretched region. What nucleotide pairs will result when this complementary strand has undergone a replication producing its own complementary strand?

12. A polynucleotide strand with the segment

A G T is exposed to nitrous acid, and the base
T C A
cytosine is changed to uracil.

A. What will be the strand formed which is complementary to this one in which the change has occurred?

B. What will be the base pairs at the next replication, and will this outcome be a transition or a transversion?

Select the correct answer or answers, if there are any, in the following three multiple choices:

13. The following can be said about mutations in general:

A. The commonest kinds of mutations are those giving conspicuous, visible effects.

B. Since they are usually beneficial, they further evolutionary progress.

C. Slightly harmful mutations which impart just a slight impairment to those possessing them are of no consequence to a population.

D. The different genes in an organism all have the same spontaneous mutation rate.

E. Gene mutations can produce only one allelic form for each genetic locus.

14. Ultraviolet:

A. Is a corpuscular radiation.

B. Is highly ionizing.

C. May cause the production of mutagenic chemicals.

D. Causes the gene to become unstable, resulting in mosaics.

E. Has a longer wavelength than X-rays.

15. The following is (are) true about radiations:

A. They are a valuable tool permitting the investigator to produce the specific kinds of gene mutations for a given genetic analysis.

B. High-energy radiations can impart excess kinetic energy to single atomic particles.

C. Electromagnetic radiations of longer wavelength have more quantum energy associated with the photons than do those of shorter wavelength.
D. The effect of radiations on the genetic material is not influenced by other environmental factors.
E. High-energy radiations do not cause structural chromosome changes.

16. Besides radiations and chemicals, give three or more other factors which have been found to affect the mutation rate in plants and animals.

17. Offer a suggestion for the fact that exposure to excess radiations can cause severe anemia due to blood cell destruction as well as a lowering of male fertility.

GENES, POPULATIONS, AND EVOLUTIONARY CHANGE

POPULATIONS AS UNITS OF EVOLUTIONARY CHANGE. Genetic discussions often concentrate on inheritance patterns found in families. However, it must be kept in mind that the various allelic forms of genes within a species are ultimately of consequence, not just to a few individuals, but to the entire population. In turn, mutations and changes in frequencies of alleles within populations influence the very course of evolution of the species. This is so, because the forces of evolution do not operate solely on separate genes or individuals. They act primarily on collections of individuals which compose a population. It is the population, not the individual, which is the main unit of evolution.

Although any population is made up of individuals, the size will vary from one population to another. The individuals within a population have become adapted to a set of environmental conditions under the direction of natural selection, the prime force of evolution. Thus, population members tend to exist in a defined location, somewhat different from that of any related group. One characteristic of a population is that its members tend to mate with one another more frequently than with individuals from other populations. Therefore, population members share more gene forms in common with individuals belonging to their group than they do with those in other related populations.

Most plant and animal species are composed of more than one population. The different populations which constitute a species may all be very similar and exhibit only minor variations from one to the other. On the other hand, they may be so different that within the species one can recognize varieties, races, or subspecies (Fig. 16-1A). Nevertheless, an important feature of all the different populations comprising a species is that they have the potential to interbreed with one another when brought together. Although it is true that population members usually mate within their own group, they are still sufficiently similar in their genetic composition to cross with members from other populations within the species. We are certainly aware of this in the human species, which is composed of many populations. Groups of human populations may be so different that we call them varieties or races, but still, any human from any population can exchange genetic material freely with any other when the two are brought together, no matter how different they may appear phenotypically. On the other hand, the inability of members from a certain group or population to cross with those in another would suggest that the two populations belong to two different species (Fig. 16-1B). Once barriers to crossing arise between two groups, perhaps after a long period of isolation between them, they are then considered members of two distinct species. They can no longer be recognized as subunits of a larger group, the species. For within a species, *all* members can exchange genetic material, even if they are separated into phenotypically distinct populations.

Within any one of the populations which form part of a species, the individuals tend to share a collection of information coded in deoxyribonucleic acid (DNA). This genetic information occurs as genes distributed among all the members. For any one pair of alleles ("A" and its allele "a"), the information may be present in homozygous (AA and aa) or heterozygous (Aa) condition. All the pairs of alleles within a population form a pool which population members share in common. Within any generation, the gene forms in the gene pool will undergo

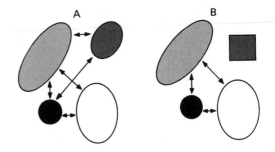

FIG. 16-1. Populations and species. *A.* One species. A species may be composed of populations of different sizes. These populations may be quite distinct in appearance so that we can recognize races or subspecies. Nevertheless (as shown by arrows), all members of a species can freely exchange genes, so that when members of the different subspecific population groups contact they can mate and produce viable offspring. *B.* Two species. If members of a certain population cannot exchange genes with individuals in certain other populations, more than one species can be recognized. In this example, we see two species, one of which is composed of three subspecies.

independent assortment and will enter into many different combinations through crossing over. But nevertheless, members of one population will generally pass these various arrangements of the genetic information back into the pool of their own population.

Although independent assortment and crossing over bring about a great deal of variation within a population, these processes by themselves do not contribute new genetic information to the population gene pool. New information, different allelic forms of the genes, may be added through matings of individuals with members of a different population. In addition, genetic changes can also be added to a population without contact with other populations through the origin of spontaneous point mutations or structural aberrations. As a matter of fact, it is an accumulation of these changes which tends to make one population distinct from another. Any population at a given time is the result of natural selection operating to achieve the highest level of adaptability to the specific environment. Frequencies of alleles may change in response to selection. Those individuals better adapted at one time will pass their favorable combinations of alleles to their offspring. Conversely, those with combinations which are less suitable or even deleterious under a given set of environmental conditions will leave fewer offspring, and the frequency of these alleles will fall. A population, then, is a dynamic group. Even without contact with another population, levels of alleles change from one generation to the next as certain allelic combinations impart more adaptive value than others at a given time.

EQUILIBRIUM FREQUENCY OF GENES IN POPULATIONS. To appreciate more fully the interaction of these factors, it is essential to grasp a concept which is fundamental to an understanding of population genetics. We have mentioned throughout the text that a dominant gene form is not necessarily superior to a recessive one, nor does the fact of dominance by itself make an allele more frequent than the recessive. An allele may increase in frequency if it imparts a reproductive advantage to those who carry it. But this is not directly a consequence of its being dominant or recessive. If the allele, dominant or recessive, is very beneficial, those who carry it will leave more progeny, and it will spread through the population. A deleterious genetic factor will be kept at a low frequency because it restricts in some way the number of offspring left by its carriers. Eventually, an allele will approach an equilibrium frequency in a population when the force of natural selection, for or against it, is balanced by the force of mutation which adds it (Fig. 16-2). A wild-type allele and its mutant form will thus reach an equilibrium which will tend to be maintained from one generation to the next.

To illustrate this idea in another way, consider a pair of alleles (the dominant "A" and its recessive form "a"). Suppose a group of heterozygotes becomes isolated and forms the basis for the origin of a new population. Individuals can mate only with others of the same genotype, Aa. The mating in the population (Fig. 16-3) is similar to that between any two heterozygous individuals. However, in this case, the alleles "A" and "a" represent genetic information in the gene pool of the population, not solely that in just one pair of parents. We see from the illustration that the original allele frequencies are 50% "A" and 50% "a." The genotypic frequency is 100% Aa, and the phenotypic is 100% "A," because "A" is dominant to "a." After one generation of mating, notice that the allele frequency is still the same: 50% "A" and 50% "a."

FIG. 16-2. Equilibrium frequency. Any allele in a population tends to reach a certain frequency which is determined largely by the reproductive advantage it imparts. More of it is being continually added to the population over a period of time due to mutation. However, at equilibrium, the same number tends to be eliminated in the same period of time, largely as a result of natural selection.

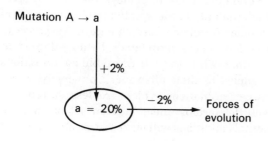

However, the genotypic frequency has now changed to 1AA:2Aa:1aa and the phenotypic to 3A:1a.

Let us assume that these alleles, "A" and "a," are neutral in their effect on the reproductive capacity of the individual. Also assume that persons of phenotype "A" or "a" are just as apt to mate with the opposite phenotype as they are with each other—that is, mating is at random; "A" doesn't prefer either type "A" or "a" over each other and likewise for "a." From Figure 16-3 and Table 16-1, it can be seen that the random mating will again result in a genotypic ratio of 1:2:1 and a phenotypic ratio of 3:1, exactly the same as in the preceding generation. The frequencies of "A" and "a" are still 50% and 50% Nothing about the gene pool has changed. It is as if the mating were still: Aa × Aa. We see that genetic equilibrium has been reached for this pair of alleles. And so it will continue generation after generation, as long as mating stays at random and no selective advantage or disadvantage is suddenly imparted to any one of the three genotypes. It doesn't matter whether "A" is dominant to "a" or not. Mendelian principles do not dictate that the frequency of one allele will increase over another. This point was expressed independently in 1908 by two different men, Hardy and Weinberg, in the simple algebraic expression, $p^2:2pq:q^2$. This Hardy-Weinberg law

TABLE 16-1 Random matings*

MATING	GENOTYPES			PHENO-TYPES	
AA × AA	AA			A	
AA × Aa	AA	Aa		2A	
Aa × AA	AA	Aa		2A	
AA × aa		Aa		A	
aa × AA		Aa		A	
aa × Aa		Aa	aa	A	a
Aa × aa		Aa	aa	A	a
Aa × Aa	AA	2Aa	aa	3A	a
aa × aa			aa		a
Total	4AA	8Aa	4aa	12A	4a
	1AA	2Aa	1aa	3A	1a

* If the three genotypes AA, Aa, and aa show no mating preference, various combinations can occur. All the possibilities are shown here. The outcome is that the genotypic and phenotypic frequencies will not change from generation to generation because the proportion of "A" vs. "a" alleles in the gene pool remains the same (also see Fig. 16-3).

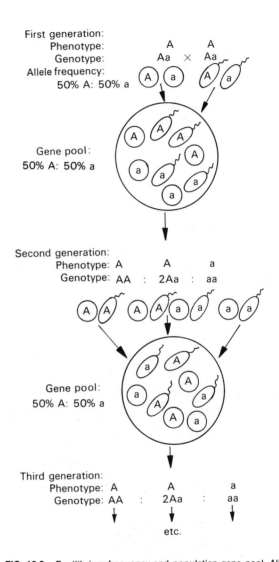

FIG. 16-3. Equilibrium frequency and population gene pool. All members of a population share genes from the same pool. Starting out with all individuals of genotype Aa, the frequency of the two alleles is 50:50 in the population, and they are contributed in that amount to the gene pool which will give rise to the next generation. As a result of random combinations of the alleles, the next generation will fall into a genotypic frequency of 1:2:1. However, the frequency of the alleles "A" vs. "a" in the population has not changed. It is still 50:50, and this same number will be contributed to the gene pool for the next generation. The population has reached equilibrium. The frequency of genotypes and phenotypes will not change for this pair of alleles as long as random mating persists.

or formula is nothing more than an expansion of the binomial $(p + q)^2$, but its implications are great for the genetic analysis of populations. We see here (Fig. 16-4) that AA = p^2 (25%), Aa = 2pq (50%), and aa = q^2 (25%). By using this simple expression, we can thus estimate the frequency of an allele in a population, as well as the frequency of genotypes. For at equilibrium, the Hardy-Weinberg formula tells us that for any pair of alleles, the genotypes will tend to occur in a ratio of $p^2 : 2pq : q^2$.

APPLICATION OF THE HARDY-WEINBERG LAW TO GENE FREQUENCIES IN POPULATIONS. In the example given, we assumed that the frequencies of "A" and "a" were both 50% or 1/2. Obviously if there are just two alternatives at a certain locus, a pair of alleles such as "A" and "a," then A + a must equal 100% or 1 (50% A + 50%a or 1/2 + 1/2). We can summarize this by the expression: p + q = 1. It is also evident that the frequencies of the three genotypes, AA (p^2), Aa (2pq), and aa (q^2), must total 1 (25% + 50% + 25%). Therefore, $p^2 + 2pq + q^2 = 1$. By using just these simple facts, a great deal of valuable information can be deduced on populations.

Obviously, not every pair of alleles occurs in a frequency of 50:50. More often, one gene form has a higher frequency in a population than its allele. But this does not upset anything which has been said. For example, suppose that in a certain population of Indians, whose members within the tribe share a common gene pool, the frequencies of the M and N blood group alleles (L^M and L^N) are 70% and 30%, respectively. With the Hardy-Weinberg expression, we can calculate the approximate frequency of the three genotypes in the population. Representing L^M and L^N by p and q, we know that at equilibrium the three genotypes will occur with frequencies of p^2 ($L^M L^M$):2 pq ($L^M L^N$):q^2 ($L^N L^N$). Simple arithmetic gives us 49% $L^M L^M$ (0.70 × 0.70):42% $L^M L^N$ (2 × 0.70 × 0.30):9% $L^N L^N$ (0.30 × 0.30), and we see that $L^M L^M + L^M L^N + L^N L^N = 1$ (0.49 + 0.42 + 0.09 = 1). Exactly the same principles hold for any pair of alleles, no matter what their frequencies.

In this example, the frequencies of L^M and L^N were given at the outset. Such information is, of course, not automatically known. But in the

FIG. 16-4. The Hardy-Weinberg formula. A pair of alleles in a population gene pool will tend to come together with a frequency of p^2: 2pq: q^2. This is nothing more than the familiar monohybrid ratio 1AA:2Aa:1aa in which A = p and a = q.

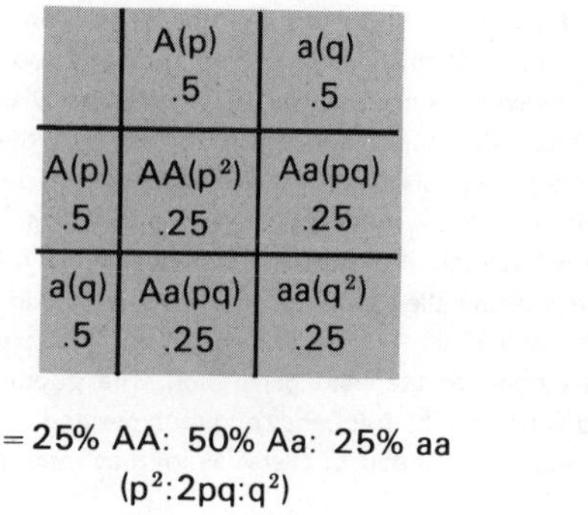

Aa × Aa

$$A = p$$
$$a = q$$

(A) (a)   (A) (a)

	A(p) .5	a(q) .5
A(p) .5	AA(p^2) .25	Aa(pq) .25
a(q) .5	Aa(pq) .25	aa(q^2) .25

= 25% AA: 50% Aa: 25% aa
(p^2:2pq:q^2)

case of alleles which are incompletely dominant or codominant such as these, the allelic frequencies can be directly calculated. The two homozygous classes can be distinguished from the heterozygotes. Finding that the frequencies of the M, MN, and N blood groups are respectively 49:42:9, we can reliably estimate the frequencies of the L^M and L^N alleles. For we know (Table 16-2) that 49% represents $L^M \times L^M$ (p^2); 42% indicates $2 \times L^M \times L^N$ (2pq); and 9% is the product of $L^N \times L^N$. The frequency of allele L^M must be 49% + 1/2 42% or 70%. This is so because *only* allele L^M is present in blood type M, whereas only one-half of the alleles in MN types are allele L^M. By using the same reasoning the frequency of allele L^N is 9% + 1/2 42% or 30%.

When dominancé is operating in the case of a pair of alleles we cannot recognize the heterozygotes (Aa) from the homozygotes (AA). So we cannot estimate the frequencies of "A" and "a" by counting the phenotypes directly. However, the Hardy-Weinberg formula enables us to achieve the same result. Suppose a population geneticist wishes to estimate the frequency of the recessive "a" for albinism in a certain island population. He is paticularly concerned about the number of heterozygous carriers (Aa) who appear normally pigmented and cannot be dis-

tinguished from the homozygotes of the genotype (AA). To solve his problem, the genticist must focus his attention on the double recessives (aa) who *can* be recognized by their albino phenotype. Counts of the number of such individuals must be obtained. Suppose they are found to occur with a frequency of 1/10,000. Simple arithmetic again tells us the approximate frequencies of the alleles "A" and "a," as well as the frequencies of the three genotypes AA, Aa, and aa. For the number 1/10,000 represents q^2 if we allow p to represent "A" and q to represent "a". The figure is nothing more than the product of two separate probabilities (Table 16-3A). Two gametes came together with a frequency of 1/10,000. This number is the product of the chance of a male gamete and that of a female gamete occurring separately in the gene pool of the population. Therefore, we take the square root of 1/10,000 which is 1/100. We now know the frequency of allele "a." Since p + q = 1, we also know the frequency of allele "A" (99/100). The frequencies of the three genotypes AA, Aa, and aa are, respectively, 9801/10,000; 198/10,000; and 1/10,000 (refer to Table 16-3A). And their sum is again 100%, which in this example is 10,000. We can of course work just as well using percentages instead of raw figures as was done in this example. Expressed in percentage, the wild allele is found with a frequency of 99%. It is very important to note that although the homozygous recessive makes up only 0.01% of the population, the heterozygous carriers are found with a frequency of almost 2% (Table 16-3B). Applying the Hardy-Weinberg formula in this way tells us that an appreciable number of mutant alleles accumulates in the population, *mainly* in heterozygotes. In effect, there is a pool of *mutant* alleles in any population, and this mutant gene pool constitutes an important portion of the total gene pool for the population. We see that the number of heterozygous individuals in a population at equilibrium will always be greater than the homozygous recessives in all cases of rare recessive alleles (2pq vs. q^2).

We can employ the same reasoning to estimate the frequencies of sex-linked alleles. Since males, in the case of mammals, are the heterogametic sex, they can carry only one of any two possible X-linked alleles. The number of males expressing the wild phenotype vs. those express-

TABLE 16-2 Estimation of gene frequency when alleles are codominant*

BLOOD GROUP	GENOTYPE	FREQUENCY IN POPULATION
M	$L^M L^M$ ($L^M \times L^M$)	49% (p^2)
MN	$L^M L^N$ ($L^M \times L^N$)	42% (2pq)
N	$L^N L^N$ ($L^N \times L^N$)	9% (q^2)
	49%	42%	9%
	(p^2)	(2pq)	(q^2)
	$L^M L^M$	$L^M L^N$	$L^N L^N$
Frequency allele L^M = 49% + 1/2(42%)		= 70%
Frequency allele L^N =	1/2(42%) + 9%	= 30%

* If there is lack of dominance, the frequencies of the two alleles in the population can be easily estimated because the three phenotypes and hence the three genotypes can be counted directly. In a population at equilibrium the genotypes tend to occur in a frequency of $p^2:2pq:q^2$. Therefore, the frequency of allele L^M here must equal the number of "M" individuals plus one-half the number of "MN." Similarly, the frequency of allele L^N must equal the number of "N" persons plus one-half the number of "MN."

ing a sex-linked recessive directly gives the frequencies of the alleles in the population for *both* males and females. For example, suppose in a certain human population the incidence of deutan color blindness is found to occur among 5% of the males (Table 16-4). This consequently means that 5% of the males carry the defective allele, d, and 95% the normal allele, D. If the entire population is at equilibrium, then the frequency of these alleles among the females must be the same as for the males, 95% and 5% for D and d, respectively. Treating the females as a separate population, the frequency of the three genotypes (DD, Dd, and dd) among them must be as follows: 0.95×0.95 (p^2) = 90.25%; $2 \times 0.95 \times 0.05$ (2 pq) = 9.50%; and 0.05×0.05 (q^2) = 0.25%. Males, of course, are either DY (95%) or dY (5%) which are the values of "p" and "q." So although five different genotypic frequencies will exist in a population for X-linked alleles, there is no contradiction to the Hardy-Weinberg formula. The same frequency of alleles is found among both sexes at equilibrium. Only because the male is hemizygous will the genotypic and phenotypic classes show a difference in relationship to sex.

In our use of the Hardy-Weinberg formula, we are assuming that each population has reached equilibrium for the alleles under consideration. Certain other assumptions are also being made: that mating is at random; that any gene form or genotype is not conferring a great advantage; that members of the population are crossing primarily with one another and are not leaving the population; and that there is no influx of individuals from other populations to add new genetic information. Obviously, any one of these or any combination of them can alter the genetic equilibrium. If, for example, a mutant allele suddenly gives an advantage in a changed environment, it will increase in frequency and spread through the population. The formerly "wild type" allele will fall correspondingly. Similarly, if mating habits suddenly cause members of the population to prefer certain phenotypes while rejecting others, fluctuations will take place in the genes affecting those phenotypes. Obviously, any population, being dynamic, will not conform to all the assumptions made when applying the Hardy-Weinberg formula. Nevertheless, even though one or more

TABLE 16-3   Estimation of gene frequency when one allele is dominant*

A. Calculation of frequency of allele from actual numbers†

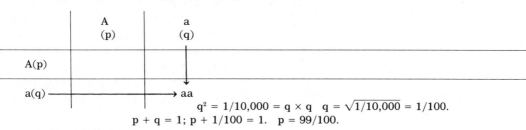

$$q^2 = 1/10,000 = q \times q \quad q = \sqrt{1/10,000} = 1/100.$$
$$p + q = 1; p + 1/100 = 1. \quad p = 99/100.$$

Genotype	Frequency	
AA	$99/100 \times 99/100$ =	9801/10,000
Aa	$2 \times 99/100 \times 1/100$ =	198/10,000
aa	$1/100 \times 1/100$ =	1/10,000
		10,000/10,000 = 1

B. Calculation of frequency in percentage†

Frequency of allele A (p) = 99/100 = 0.99
Frequency of allele a (q) =   1/100 = 0.01

Frequency of genotype:	AA (p^2)	Aa $(2pq)$	aa (q^2)
	0.99×0.99 =	$2 \times 0.99 \times 0.01$ =	0.01×0.01 =
	0.9801	+ 0.0198	+ 0.0001 = 1

* Normal = A = p; albino = a = q.
† Note that p + q always equals 1, as does the sum of AA + 2Aa + aa.

---

exceptions may pertain, the Hardy-Weinberg expression can be used to give good estimates of allele frequencies at any one time. It is the most important of the mathematical concepts in population genetics.

LETHAL AND DETRIMENTAL ALLELES IN POPULATIONS.   The formula can be put to very good use when approximating the number of rare, fully lethal alleles in a population. A frequently asked question is, "How can a gene form which kills before reproductive age still persist in a population? Since it kills all those who carry it while they are immature, shouldn't it be eliminated?" Yet we all know that lethal or semilethal alleles (such as those for Tay-Sachs disease, phenylketonuria, and hemophilia) persist in the populaton. The reason for this is simply that any genetic locus has a characteristic rate at which it undergoes spontaneous mutation. Therefore, when a lethal mutant is thrown out

TABLE 16-4   Estimation of frequency of a sex linked recessive allele*

GENOTYPE AND FREQUENCY		ALLELE AND FREQUENCY	
Males			
Normal	DY = 95%	D† = p =	95%
Color blind	dY = 5%	d† = q =	5%
	100%		100% = p + q

Genotype	Frequency		
Females			
DD	p^2 = 0.95×0.95	=	90.25%
Dd	$2pq$ = $2 \times 0.95 \times 0.05$	=	9.50%
dd	q^2 = 0.05×0.05	=	0.25%
			100.00%

* In a population at equilibrium, the frequency of the trait in the heterogametic sex indicates directly the frequency of the allele in the population as a whole. The frequency of the genotypes in the homogametic sex is calculated by applying the Hardy-Weinberg formula and treating it as a separate population.
† D, normal; d, color blind.

of the population through the death of an individual, it is also added as a result of mutation which arises in a germ cell and will be carried by a new individual. The elimination through death and the gain through recurrent mutation eventually reach an equilibrium value. The number of deaths from the genetic disorder will be balanced by the number of births (Fig. 16-5). Let us say that a certain lethal allele causes the death of individuals with a frequency of 1/100,000. The defective allele is thus being eliminated with this frequency; yet it does not disappear. This means that 1/100,000 victims are being born if an equilibrium exists. So the number of newly arisen mutant alleles balances the number lost. Therefore for a rare, fully lethal allele such as this in a population which is at equilibrium with respect to the trait, we can approximate the mutation rate, if we know the number of deaths it causes (or the number of individuals born with the defect). In this case the mutation rate would be in the vicinity of 1/100,000. This reasoning also assumes that there is no advantage conferred by the lethal allele on the heterozygote, as occurs in the case of certain alleles which are lethal to the homozygote, as in the case of sickle-cell anemia (later, this chapter).

The Hardy-Weinberg formula tells us that the number of heterozygotes in the population will be still more frequent than the number of victims: the square root of $1/100,000 = 1/316$ = the frequency of the lethal, "l"; thus 315/316 = the frequency of the wild allele, "L." Consequently, the number of heterozygotes must be: $2 \times 1/316 \times 315/316 = 1/158$. Again we see that the frequency of heterozygotes, even for a rare recessive lethal, will far outnumber the homozygous recessives in the population. The accumulation of lethal alleles in the mutant gene pool is appreciable and, as we shall see, of great significance to a population.

In the last chapter, it was stressed that most gene mutations impart just a slight impairment to the individual. Now let us examine the consequences of these detrimentals, alleles which decrease the chances of survival by just a small amount. Instead of killing outright, as does a full lethal, a detrimental will cause the elimination of only a certain number of those carrying it. Any one individual is being hampered in some

FIG. 16-5. Equilibrium of a fully lethal allele. A genetic factor which kills before reproductive age still persists in the population at a certain frequency. This equilibrium is reached when the number of lethal alleles eliminated is balanced by the number added to the gene pool through gene mutation (*above*). If the lethal were not replenished by mutation (*below*), it would eventually disappear, because the homozygote cannot reproduce, and the alleles are continually being lost from the population. This, however, does not happen due to the force of mutation pressure (*above*).

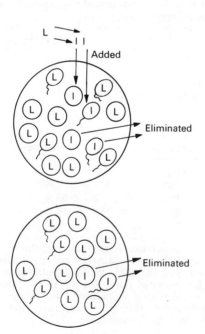

way by the detrimental but still has a chance to live to reproductive age and pass it down. Let us assume that an allele imparts a 10% disadvantage to the individual. We would say that the impairment is 10%. If the individual carrying it passes it down, again another individual arises who has this same 10% disadvantage. This second person may in turn pass the detrimental down. The process will continue until one of those receiving the allele is killed by its effects. *On the average,* an allele with an impairment of 10% will tend to pass through 10 individuals before it is eliminated from the population. At first, this might seem inconsequential and not as important to the population as a full lethal. However, H. J. Muller called attention to the importance of such detrimentals to the population, no matter how slight their impairment to the individual. For in the case of a 100% lethal, the one individual is removed at once, because the entire burden is placed on that person. Muller referred to any burden imparted by mutant genes as the *genetic load.* In the case of the detrimentals, he pointed out that the load, instead of falling on the one individual, was distributed in smaller amounts. However, this did not make them any less important to the population as a whole. For an impairment (i) of 10% would mean an average persistance of 10 generations (p = 1/i = 1/0.10 = 10). Not one, but 10 persons on the average would suffer some ill effects of the allele. The load is being shared by all of them, perhaps in just a small way, but each individual is having his chances of survival decreased by a real amount. And in the long run, a genetic death will result. The genetic death occurred at once for the full lethal. On the other hand, the detrimental passed through more individuals, but the outcome was the same: a genetic death. Genetic death does not necessarily imply the actual death of the individual. Any allele causes a genetic death if it prevents the individual from reproducing. In terms of genetic consequences, loss of reproductive capacity due to deleterious genetic factors is equivalent to an outright killing. In both cases, the individual fails to contribute alleles to the population gene pool and is therefore eliminated from passing genetic information to the next generation. Inability to reproduce is thus a death in the genetic sense.

GENETIC LOAD AND THE MUTANT GENE POOL.   The average individual in the population possesses not just one detrimental allele. Muller and others have calculated that a human carries in the heterozygous condition the equivalent of three to five of them. All populations thus possess a genetic load in the form of a mutant gene pool which accumulates mainly in the heterozygote. In our discussion of impairment, we did not consider whether the detrimentals were dominant or recessive. Recall that very few alleles are ever completely recessive. Usually some effect is expressed in the heterozygote, and this is true in the case of detrimentals as well. Muller pointed out that the amount of impairment in the heterozygote is usually what is significant. The damage that detrimentals do in the heterozygous condition, no matter how slight, is usually sufficient to eliminate them *before* they have a chance to become homozygous. Muller used the expression "effectively dominant" to describe this deleterious effect in the heterozygous state of alleles whose impairment in the homozygous condition would be even greater. So if an allele is highly disadvantageous when homozygous, say the aa state gives an impairment of 10%, its effective dominance may give it an impairment of 4% in the heterozygous condition, Aa. This means that *on the average,* 10 "aa" individuals would survive before a genetic death would result. On the other hand, 25 Aa persons (1/0.04 = 25) would arise before an elimination of the allele would occur. The important point is that because the allele "a" is relatively rare in the population compared to the wild form "A," the chances of its becoming homozygous are not too great. Therefore, most of the elimination will take place in the heterozygotes. So again, we see the importance of the accumulation of mutant alleles in the heterozygous condition.

Moreover, we see from every point of view that detrimentals, although they are less apt to kill an individual outright, nevertheless exert a real effect and will eventually be responsible for genetic deaths. If such detrimentals continue to be added to the human population, more and more people will suffer greater *overall* impairment due to the accumulation of many different detrimentals, each one with its own slight amount of impairment to the heterozygote. Muller was extremely concerned about the dangers inherent in adding detrimentals to the human population. For the human species is the only one which can tamper with the rate of mutation and allow mutant alleles to accumulate. Today we expose ourselves to an ever increasing list of chemicals, some known to be mutagenic, others with uncertain effects. Exposure to radiations produced by fallout may be adding more and more recessive lethals and detrimentals to the population. The consequences may remain unknown until accumulation reaches a point where almost every individual is burdened by a vast number of little genetic loads. These little deleterious effects may all add up to a significant burden which ensures a genetic death. Muller was alarmed for the human population if such a trend were not reversed. The exact rate of spontaneous mutation in humans has not yet been determined, but it certainly does not seem to be less than that for species which have been well studied in this respect. And we do not know how much extra genetic load we are actually adding through contact with mutagenic agents. Perhaps most significant is the fact that we have no idea how great a genetic burden the human population can tolerate without a survival crisis. Therefore, it would seem sensible to avoid adding extra mutant gene forms to the population through unnecessary contact with known or suspected mutagens.

INTERFERENCE WITH NATURAL SELECTION. Another way in which mutant alleles may accumulate in human populations is through man's interference with natural selection. This takes place whenever an individual who would normally suffer a genetic death is enabled to reproduce. Such a person passes on genetic information which natural selection would have eliminated. Let us consider a fully lethal allele which kills before reproductive age. Assume that victims are born with a frequency of one in a million and that the population is at equilibrium in respect to the trait under consideration. This frequency also reflects the death rate and the mutation rate. The frequency of the lethal allele is the square root of $1/1,000,000$ which equals $1/1,000$. Application of the Hardy-Weinberg formula tells us that the mutant gene pool in the population is: $2 \times 1/1{,}000 \times 999/1{,}000 = 1{,}998/1{,}000{,}000$ or approximately 1/500.

Now assume that science finds a way to prevent the death of many homozygous recessives and reduces the allele from a full lethal to a detrimental so that only 1 out of 10 homozygotes will die. However, the spontaneous mutation rate has not changed and still remains 1/1,000,000. Now more of the defective genes are allowed to accumulate in the population, a process which will continue until a new equilibrium is reached. This will occur when *10* per million are born instead of the original 1 per million. Just what benefits for the population are derived from this (Table 16-5)? The number of persons with the disorder has risen from 1 in a million to 10 in a million. More afflicteds are now in the population, and most are reproducing. Consequently, at the new gene equilibrium, the frequency of the mutant allele has risen from 1/1,000 to roughly 1/300 (the square root of 10/1,000,000). A corresponding increase has taken place in the mutant gene pool, from approximately 1/500 to approximately 1/150. But what about the decrease in the number of deaths directly attributed to the affliction? Ironically, the *total* number stemming from the disorder is the same! It is true that originally 1 in a million was dying from it and that at the new equilibrium

only 1 out of 10 million suffers a genetic death. In effect, we have allowed 10 more afflicteds to accumulate. But when this happens, we still get a genetic death. So the *total* number is the same. We have an increased birth rate: 10 where there was originally 1, but now the *accumulation* of 10 gives 1 genetic death. Originally, 10 defective births would result in 10 deaths. *Now* for 10 deaths, we accumulate 100 afflicted individuals, but still a total of 10 deaths from the disorder has been realized. We may well ask what benefits are truly realized by such manipulations of natural selection. The population is being burdened by extra afflicteds. These in turn may suffer from their disorder. But most important for the population from the long-term point of view is the additional load of defective alleles contributed to the gene pool. This must be considered along with loads which are being introduced through other channels. Such a continued process, Muller feared, could result in a real crisis for human survival.

THE SICKLE-CELL ALLELE IN POPULATIONS. Another important aspect of a genetic burden on a population is well illustrated by the familiar hereditary defect, sickle-cell anemia. In this disorder, the presence of a recessive allele results in the substitution of just one amino acid at a

---

TABLE 16-5   Calculation of effects of increase in incidence of a once fully lethal allele

Original situation
Frequency of victims	1/1,000,000
Frequency of deaths	1/1,000,000
Mutation rate	1/1,000,000
Frequency of allele	$\sqrt{1/1{,}000{,}000} = 1/1{,}000$
Size of gene pool	$= 2pq = 2 \times 1/1{,}000 \times 999/1{,}000 = 1{,}998/1{,}000{,}000 =$ approximately 1/500

New equilibrium:
Frequency of victims	10/1,000,000
Frequency of deaths	1/10,000,000
Mutation rate	1/1,000,000
Frequency of allele	$\sqrt{10/1{,}000{,}000} = \sqrt{1/100{,}000} = 1/316$
Size of gene pool	$= 2pq = 2 \times 1/316 \times 315/316 = 1/158$

Summary

	Mutation rate	Frequency of victims	Allele frequency	Gene pool	Total no. deaths
Old	1/1,000,000	1/1,000,000	1/1,000	1/500	10 for 10 births
New	1/1,000,000	10/1,000,000	1/316	1/158	10 for 100 births

GENES, POPULATIONS, AND EVOLUTIONARY CHANGE

critical position in the hemoglobin molecule. This change is sufficient to cause an assortment of physical disorders lethal to the homozygote. Because the effects of this mutant allele are so harmful, we would expect natural selection to operate against it, keeping it at a very low level in the population. Only by mutation pressure would we expect it to be maintained at all. Surprisingly, however, the frequency of the mutant allele is alarmingly high on the west coast of Africa, where it may reach a prevalence of approximately 30% in certain populations. Such a high value of a very deleterious genetic factor was quite puzzling until it was discovered that the frequency of the allele is correlated with the distribution of malaria. Where the incidence of this disease is low or absent, a corresponding drop is seen in the occurrence of the mutant allele and consequently of sickle-cell anemia. The high level of the mutant allele in some populations indicated that somehow natural selection was actually operating to maintain it in certain environments. Obviously, those homozygous for the mutant allele were not living long enough to reproduce; so the defective allele was not being added back to the gene pool through the victims of the disorder. Any selection in favor of the defective allele must therefore be on the heterozygote, who carries both a normal and a mutant gene form. These heterozygotes were shown to possess a certain number of abnormal, sickled, red blood cells. The mutant allele is actually a codominant, because its effect in the heterozygote can be detected. It was finally demonstrated that the heterozygote possesses an advantage in childhood in his or her natural environment over persons homozygous for the normal allele due to an increased resistance to the malarial parasite. Protected in this way, the heterozygote can survive a malarial attack as a young child and build up antibodies which provide resistance against any later attack. The heterozygote has a sufficient amount of normal hemoglobin in the red blood cells to prevent severe anemia. At most, the person will experience only a mild form, and this may not even become evident. The unfortunate result of the story is the maintenance of the deleterious allele at a high level in the population, because the heterozygotes will have a higher survival rate than the homozygous normals in areas of high

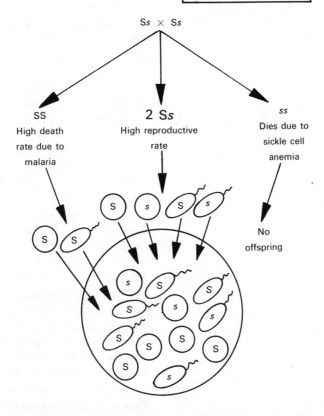

FIG. 16-6. Effect of selection on the heterozygote for sickle-cell anemia. Although the recessive allele causes death of the homozygote, it imparts an advantage to the carrier in regions of high malaria incidence. The homozygous normal individuals leave fewer offspring than the superior heterozygotes who keep adding recessives as well as dominant alleles to the gene pool. The number of harmful recessives therefore stays at a relatively high frequency in the population. (The "s" allele, strictly speaking, is a codominant.)

malaria incidence and as a group will leave more offspring (Fig. 16-6). The heterozygotes are thus superior in regard to this characteristic. But the population is being burdened by the production of homozygous recessives who suffer from the sickle-cell anemia. A certain balance is eventually reached between the frequency of the wild allele and its mutant form in different environments. In those where malaria is no longer a problem, the heterozygotes have no advantage over the homozygous normals, and so the frequency of the mutant allele can be expected to fall. And this seems to be the case in regions of Africa where the malarial parasite is being eliminated. Those heterozygotes who are descended from African populations and who are now living in malaria-free regions obviously possess no re-

productive advantage. These heterozygotes can be easily detected today by simple blood tests. Genetic counseling of known carriers can help to reduce the level of this very burdensome allele. Eventually it may be eliminated or fall to a low frequency which is maintained only through the recurrence of rare spontaneous mutation at the specific gene locus.

HYBRID VIGOR AND ITS BASIS.   The sickle-cell story dramatically demonstrates a well-known genetic effect, hybrid vigor or *heterosis*. In the sickle-cell case, the maintenance of the mutant allele entails very serious consequences for the homozygote. However, this is by no means the rule. Geneticists have known for a long time that inbred strains, when crossed to other inbred strains, produce offspring which are frequently more robust than members of either parental strain. This can be explained in part by the fact that inbreeding will increase the degree of homozygosity (Chap. 7), and this in turn increases the chances of bringing together in the homozygous condition alleles which are detrimental in some way. Outbreeding would thus decrease the frequency of harmful homozygous combinations, and it would generate heterozygotes which would be more robust. More recently, significant observations have been made on the extent of heterozygosity in populations of a few species, including fruit flies and even the human. When we consider any organism from a genetic point of view, we tend to envision a genotype composed of a preponderance of normal (wild type) alleles. Except for a few loci which may be homozygous or heterozygous for certain mutant alleles, we think of most of the loci in an individual as homozygous wild. This concept is now undergoing change, because this picture does not appear to be the correct one in natural populations. Indeed, it seems that a very significant proportion of the loci may be heterozygous. For it is now evident that the terms "wild-type allele" or "standard allele" are not very precise. This is so, because there may actually be two or more "wild-type" alleles possible at a locus. This may sound impossible, because we usually think of *a* wild-type allele and its mutant forms or alleles. However, chemical analyses of many so-called "normal" or "wild" protein gene products in the fruit fly, mice, and man have illustrated that the "normal" protein actually may be a mixture of two very similar, but still detectably different, proteins. For example, individuals thought to be homozygous wild in regard to a certain locus can be shown to be actually heterozygous. By refined chemical analysis, the protein product governed by the locus in question often can be shown to consist of slightly different molecules. These may differ somewhat in electric charge as the result of a difference in just one or a few amino acids. The individual is apparently "wild" at the locus, but strictly speaking is heterozygous, because each allele is directing a slightly different polypeptide. Therefore, two allelic forms of one gene are present, not just the one "wild allele" in the homozygous condition. Which form of the gene is the "true wild" or normal and which is the mutant becomes an almost meaningless debate. The important point is that a great deal more heterozygosity apparently exists in populations than has been suspected. Different forms of "wild alleles," often called *isoalleles*, may exist at many loci in the individuals composing the population. Since heterozygotes apparently are more vigorous in general than homozygotes, this would mean that being heterozygous for different forms of the same "wild" allele may be more advantageous than being homozygous for any *one* of them. Studies of different fruit fly populations strongly suggest that isoalleles are maintained in the population due to advantages they may confer, probably to the heterozygotes.

How can this hybrid vigor or heterozygote superiority be explained? Definite answers cannot be given, but our knowledge of the relationship between genes, and enzymes at the molecular level permits the formulation of some plausible explanations. We know that what often were considered to be single enzymes are actually mixtures of different, but very closely related, molecules. They are concerned with the same catalytic reaction but differ in response to temperature, pH, and other environmental changes. Such similar but slightly different enzymes are called *isoenzymes* or *isozymes*. Heterozygosity at a certain locus which controls a metabolic step would cause the production of not just one single type of enzyme but at least two forms of one enzyme, two isozymes. The presence of the two forms could very possibly confer an advantage on an individual. This advantage could be the ability to carry out the particular metabolic step more efficiently over a wider range of environmental fluctations (Fig. 16-7). The homozygote, having only one form of the enzyme, would be more limited. Therefore, heterozygosity for isoalleles at a number of loci might result in the presence of more isozymes for a number of metabolic steps. This could endow the individual with a versatility to cope with a broader spectrum of environmental situations. Unlike the sickle-cell victims, these individuals homozygous at a particular locus for one of the isoalleles would not suffer dire consequences. They would leave fewer offspring *on the average* and would thus be less vigorous than the heterozygotes due to limitations placed on the ability to adapt to a variety of conditions. The superiority of the heterozygote, however, might not always be evident. It is quite possible to imagine the homozygous and heterozygous conditions being equally effective under one

FIG. 16-7. An advantage of isoalleles. Two forms of a gene may govern the formation of two very similar but slightly different polypeptide chains, each of which acts efficiently as an enzyme concerned with the same metabolic step (conversion of substance X to Y). One form of the enzyme may be most efficient at a pH range of 6.9–7.1, whereas the other may operate best within a pH range of 7.1–7.4. The heterozygote possesses both forms of the enzyme and hence has a wider range of pH's over which the metabolic step can be fully carried out. This extension in range gives the heterozygote an advantage in case of environmental fluctuation which can alter the pH of the system.

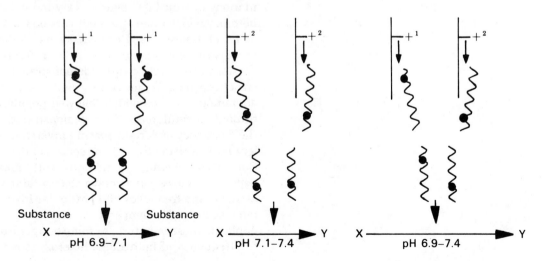

given set of environmental situations. We might think of the heterozygote having an advantage at particular temperatures or just certain parts of the life cycle. Homozygotes would be present constantly in the population and would keep contributing to the gene pool. So a balance would eventually be reached in the frequency of the isoallelic forms of a gene at a given locus. In the population, there would be a balance between homozygosity and heterozygosity in relationship to any locus, but this equilibrium value could change with a varying environment. Natural populations, therefore, are not composed of individuals that are mostly all homozygous for wild alleles. The population is dynamic and diversified, as natural selection continues to give the advantage to those best adapted to a set of environmental conditions.

These points have been well illustrated by Dobzhansky in his studies of natural populations of *Drosophila*. In Chapter 10, we discussed the high frequency of inversion heterozygotes in natural populations of fruit fly species. The presence of an inverted segment in one chromosome acts to decrease the amount of crossing over and hence the recombination of alleles in that chromosome region. Favorable gene combinations of alleles can thus be held together. The loci in the affected segment can therefore be maintained in a heterozygous state; consequently, this can preserve any advantages which hybrid vigor may confer. The presence of inversions is easily detected in *Drosophila*, because they can be recognized by alterations in the giant polytene chromosomes (Chap. 10). In some natural populations of fruit flies, nearly every individual is heterozygous for one or more inversions on each of its chromosomes. Dobzhansky has followed the distribution and frequency of different inversions and has clearly demonstrated that structural heterozygosity can confer a decided advantage. Studies which followed populations through several years have demonstrated that the frequency of certain specific inversions varies characteristically with the season of the year. Controlled laboratory conditions with flies in population cages have shown that similar variations in the frequency of a particular inversion type occur with temperature changes which duplicate those in nature. At some temperatures, the frequency of individuals heterozygous for a specific inversion and of those homozygous for the corresponding normal arrangement remained the same, indicating that the inversion heterozygote did not always have an advantage over the homozygote. These studies give a clear-cut demonstration that the fluctuation of genetic types within a population often is correlated with selective advantage. Although the inversion represents a structural change affecting a group of alleles in a certain combination, the same sort of frequency changes occur in the case of many individual loci with their slightly different isoallelic forms, as well as with their distinctly different mutant forms.

GENE EQUILIBRIUM AND EVOLUTION. Now that it is clear that populations are far from static, we must reexamine the Hardy-Weinberg formula. It was mentioned that an equilibrium value for a pair of alleles is reached only if several assumptions hold true: random mating and absence of migration, etc. No natural population conforms to these specifications, but with the Hardy-Weinberg expression we can still approximate the frequency of an allele in a population at a given time. Further consideration of the Hardy-Weinberg law will tell us that no real population *could* satisfy all the assumptions, for that would mean that evolutionary progress is no longer possible in the group. For the Hardy-Weinberg law demonstrates that Mendelian inheritance by itself does *not* lead to changes in allele frequencies in populations. Evolution requires alterations in the equilibrium frequency of alleles. For it is populations, not individuals, which evolve and give rise to different varieties and species. The population is the unit of evolutionary change upon which the forces of evolution act. These evolutionary forces constantly operate to upset the theoretical equilibrium value indicated by the Hardy-Weinberg expression. If equilibrium persists for all the pairs of alleles in the gene pool of a population, evolution would reach a standstill, because evolution requires change. Natural selection is the major force which upsets this equilibrium, and it operates on a population to achieve the highest level of adaptiveness in a given environment. We have seen how the adaptive value of a gene form may favor it over its allele and cause it to increase at a given time. At its simplest level,

evolution actually may be nothing more than just such minor fluctuations in the gene pool of a population from one generation to the next. But even these minor changes will lead to changes in the frequency of certain genotypes and phenotypes in the population. The population of one generation is not genetically identical to the one preceding it, due to modifications in the gene pool from one generation to the next. Evolution has taken place. Over a given period of time, the slight fluctuations in allele frequency may cause no significant differences between the original population and the later, descendant one. On the other hand, even minor changes in a given direction over generations can lead to a population quite distinct from the ancestral one.

INTERACTION OF MUTATION, NATURAL SELECTION, AND DRIFT. If a changed environment imparts a great selective advantage to an allele, natural selection will favor it, and the beneficial gene form will spread. It can be shown experimentally and mathematically that just a little selection is sufficient to cause the frequency of the favored allele to increase. Gene mutation by itself is obviously an evolutionary force, because the ultimate change in genetic information depends on it. Just by constant gene mutation $(A \rightarrow a)$, the equilibrium will be upset. So gene mutation (mutation pressure) is a driving force in evolution. However, by itself, gene mutation in one direction would be relatively insignificant. For if "A" continues to mutate to "a," eventually only "a" would be in the population. This does not occur, however, because of reverse mutation, "a" to "A." Nevertheless, gene mutation alone, although it causes changes in allele frequency, would eventually just lead to a new equilibrium. This in turn would have to be upset by subsequent mutation at the locus if any further change were to take place. Evolutionary progress would advance at a very slow pace if mutation pressure alone were the only driving factor. However, in natural populations, gene mutations provide different allelic forms of a gene, and these are spread throughout the population by sexual reproduction which entails independent assortment and recombination through crossing over. This makes possible many different combinations of newly arisen mutant gene forms with each other and with those already established in the population gene pool. The effect of gene mutation is amplified through the construction of new genetic combinations by independent assortment and crossing over. These processes operate in the population to produce variation. And variation in the form of new combinations of the genetic information is the raw material for the operation of natural selection. Natural selection changes the gene pool of the population by giving a reproductive advantage to those individuals with favored combinations of alleles, those combinations giving an advantage in a certain environment. Some genetic factors increase in frequency in the gene pool over their allelic forms. The effect of natural selection, selection pressure, is to steer the population toward the greatest level of adaptability and efficiency with the environment.

In addition to mutation and natural selection, there is another evolutionary force which operates on populations, and its effect may actually counter that of natural selection. This is the force of *drift*, fluctuations of allele frequencies within a population through random processes. This force can be appreciated in light of what we have learned about probability. We have noted that when we are dealing with small numbers, chance fluctuations may arise, such as a run of six heads in the tosses of a coin or the births of four boy babies in a row in a family. Ideal genetic ratios may not be realized when dealing with just a small sample. In a similar way, it is possible in a small population for certain alleles to become established or to become lost by chance, without any relationship to their adaptive value. Suppose that in a population "A" mutates to "A$_1$," an isoallele which can confer an advantage. However, the beneficial allele may be lost if the population is small and if only a few members come to acquire it. These individuals with the advantageous genetic factor may not be in the proper environment for the value of the allele to be expressed, or perhaps they may not even reproduce and would thus fail to add the beneficial gene form to the gene pool (Fig. 16-8A).

Conversely, a deleterious form could actually get established. Suppose by chance a few individuals carrying the same defective allele become isolated and form the basis of a new population (Fig. 16-8B). Without mutation or the introduction of a better allele from the outside, the defective allele can become established in the population. Its presence is due just to the chance that it happened to be in those who formed the basis of the population. Such a chance effect can actually be seen in some human populations. The founders of certain orders or sects, such as the Amish in some localities, just happened to carry particular detrimental alleles, for example, a recessive for a type of polydactyly. The frequency of the recessive is quite high in the descendant populations. Establishment of certain mutant alleles in this way has been termed the "founder effect." It illustrates the role of chance in fixing or establishing genetic factors regardless of any benefits or disadvantages they may entail. Although drift plays its greatest role in small populations, it may have been important in the evolution of many species. This is so, because the *effective* breeding population of a species may become greatly reduced at times. Drift could exert its effect at such times, even on species which eventually come to exist as very large populations. Several plant and animal groups display characteristics which have no apparent value and which may even entail certain disadvantages. These have very possibly become established through the random force of drift.

ROLE OF GENETIC ISOLATION IN SPECIATION. As the evolutionary forces of mutation, natural selection, and drift continue to operate on a single population generation after generation, the descendant population may become so distinct from the ancestral one that it can be recognized as a different species. In Chapter 10, we discussed another aspect of evolution, the origin of two distinct species from one original group. Evolution of this type demands reproductive isolation between the two segments of the original population. Operating under two different sets of environmental circumstances, natural selection may favor certain genetic combinations in one of the segments; these combinations may be selected against in the other segment where a different assemblage of genetic factors is favored. Moreover, different spontaneous gene mutations may occur by chance in the two separated groups. In effect, there are two separate populations which are

undergoing changes in their respective gene pools. If the gene pools become sufficiently different, individuals from the two populations may be so different genetically that they cannot cross. Two different species will have arisen from what was originally one (Fig. 16-9A). The original isolation preventing exchange of genetic information between the two segments is essential. If this reproductive isolation is broken down

FIG. 16-8. Effect of drift. *A.* Loss of a valuable allele. Suppose that the mutant allele "A_1" can be beneficial to an organism living in a region of drought. The valuable allele, however, may be carried only in a few individuals who never become exposed to the conditions where the benefits may be expressed. Therefore, these individuals would not leave any more offspring than the majority, and the allele would not increase in frequency. It may actually be lost if by chance the few individuals carrying it fail to reproduce for some reason completely unrelated to the gene in question. *B.* Establishment of a deleterious recessive. A deleterious recessive may be present in a few individuals in a population. Expression of the allele would be low, because most members of the population are homozygous dominant. If, however, a few carriers become isolated and form the basis of an isolated population, the frequency of the allele in the new population will be greater than in the original, and the recessive will thus come to expression more often.

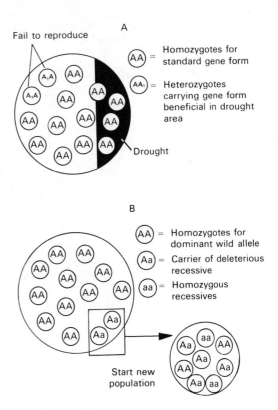

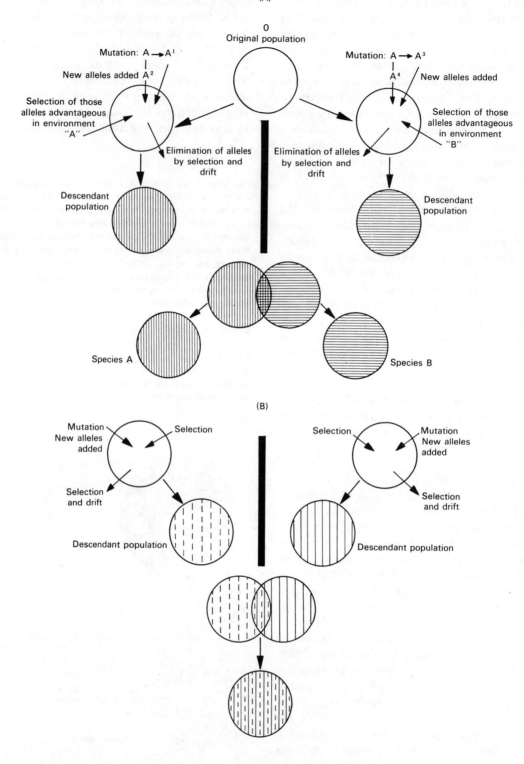

(A)

0
Original population

Mutation: A → A¹
New alleles added A²

Selection of those
alleles advantageous
in environment
"A"

Elimination of alleles
by selection and
drift

Descendant
population

Mutation: A → A³
A⁴
New alleles added

Selection of those
alleles advantageous
in environment
"B"

Elimination of alleles
by selection and
drift

Descendant
population

Species A

Species B

(B)

Mutation
New alleles
added

Selection

Selection
and drift

Descendant population

Selection

Mutation
New alleles
added

Selection
and drift

Descendant population

FIG. 16-9. Isolation and the evolution of populations. *A.* Population O (original) may become separated into two segments existing under different sets of environmental conditions. The two segments do not exchange genes due to an isolating barrier. Some of the spontaneous mutations arising in the two populations may represent different allelic forms of the original gene. Very different allele combinations may become established in the two separated groups as a result of the forces of natural selection and drift. The eventual descendant population may become so different as a result of these processes that members of the two populations cannot exchange genetic material when the isolating barrier is removed. Two separate species can be recognized. *B.* The barrier may be removed before the process has resulted in the establishment of very different allele combinations in the two groups. When brought together, genetic isolation may not be complete. If interbreeding continues, eventually all may become part of one new population.

---

before genetic isolation is completed, the members of the two populations may be nothing more than races or subspecies of the same species. Once free gene exchange takes place between the two groups, the effect will be to make them members of one interbreeding population again (Fig. 16-9B). At the beginning of this chapter, we noted that geographic isolation in the human species has resulted in various races. But no human population has been completely isolated genetically from all others. Sufficient genetic exchange has occurred among groups so that enough common genetic information has flowed from one population to the next to keep all groups of man members of one species. There have been established no genetic barriers to crossing. So we may think of the human species in one sense as a human population composed of smaller ones. As more and more free genetic exchange goes on, more variation will spread through the human population as a whole. Natural selection will work to favor the highest degree of adaptability as the environment of the planet changes. However, we must not forget the genetic loads which are present, largely from mutation pressure. And we must concern ourselves ever more today with the prevention of an increase in this load through the production of unnecessary mutation by mutagenic agents. The human alone among the many species of living things has the potential to direct his populations toward ever increasing levels of efficiency. Yet at the same time, he may through his own devices cause it to deteriorate.

REFERENCES

Avers, C. J. *Evolution.* Harper and Row, New York, 1974.

Cavalli-Sforza, L. L. Some current problems of human population genetics. *Am. J. Human Genet.,* 25: 82, 1973.

Cavalli-Sforza, L. L. The genetics of human populations. *Sci. Am. Sept:* 80, 1974.

Cavalli-Sforza, L. L., and W. F. Bodmer. *The Genetics of Human Populations.* W. H. Freeman, San Francisco, 1971.

Cold Spring Harbor Symposium on Quantitative Biology. *Population genetics: the nature and causes of genetic variability in populations.* Cold Spring Harbor Laboratory, New York, 1955.

Crow, J. F., and M. Kimura, *An Introduction to Population Genetics Theory.* Harper and Row, New York, 1970.

Eckhardt, R. B. Population genetics and human origins. *Sci. Am.,* (June): 94, 1972.

Green, E. Genetic effects of radiations on mammalian populations. *Ann. Rev. Genet.,* 2: 87, 1968.

King, J. C. *The Biology of Race.* Harcourt, Brace, Jovanovich, New York, 1971.

Mettler, L. E., and T. G. Gregg, *Population Genetics and Evolution.* Foundations of Modern Genetics Series. Prentice-Hall, Englewood Cliffs, N.J., 1969.

Muller, H. J. Radiation damage to the genetic material. *Am. Sci.,* 38: 33, 1950.

Stansfield, W. D. *The Science of Evolution.* Macmillan, New York, 1977.

Volpe, E. P. *Understanding Evolution,* 2nd ed. William C. Brown Co., Dubuque, Iowa, 1970.

Wills, C. Genetic load. *Sci. Am.* (March): 98, 1970.

REVIEW QUESTIONS

1. In a certain Indian tribe, the distribution of the MN blood types (which are based on the codominant alleles, L^M and L^N) is found to be: Type M, 64%; Type MN, 32%; Type N, 4%. What are the frequencies of the L^M and of the L^N alleles in this population?

(In Questions 2–7, assume that equilibrium exists for the alleles involved in each case.)

2. A farmer finds that among his flock of about 8100 birds, one is affected with an undesirable trait which affects the feathers. This trait is known to be inherited as an autosomal recessive. How many birds in the flock of 8100 could be expected to be carriers of the undesirable allele?

3. Approximately 70% of Americans can taste the bitter chemical PTC. This ability depends on the dominant, T, whereas inability to taste it results from homozygosity for the recessive allele, t. The pair of alleles has been found to exert no other detectable effect.

    A. Is the frequency of the dominant much higher in the population than the recessive?
    B. Can the dominant allele be expected to increase further?

4. Suppose that on a certain island the frequency of persons homozygous for the autosomal recessive (a) for albinism is 1 in 2500.

    A. How many persons in the population can be expected to be heterozygotes, and how many homozygotes for the wild-type allele?
    B. What is the probability that a normally pigmented person with no history of albinism in the family is a heterozygote?

5. Assume that in a certain population five males in 100 are found to be color blind due to the X-linked recessive, d.

    A. What is the frequency of the alleles D and d in the population?
    B. What is the frequency of carrier women in the population?
    C. What is the frequency of color-blind women in the population?

6. Assume that a certain autosomal recessive, a, causes severe anemia and death before the age of five years. In the population, one infant in 200,000 is born with the disorder.

    A. What appears to be the mutation rate at the "a" locus?

B. What is the frequency of the recessive in the population? (Round off to the nearest 10.)
C. Approximately how many persons in the population would be heterozygous for this recessive?

7. Assume that medical science is able to save many of the persons who are homozygous for the recessive "a" mentioned in Question 6. The survivors can reproduce, and eventually a new equilibrium is reached in which homozygotes are ten times as numerous:

A. What is the frequency of the gene in the population at the new equilibrium?
B. Approximately how many persons in the population would be heterozygous for the recessive at the new equilibrium?
C. What is the mutation rate at the new equilibrium?

8. In a population, a certain allele exerts a fully lethal effect. A second one is a detrimental and doesn't necessarily cause a genetic death in the person who carries it. However, it has an impairment of 5%. A third allele has an impairment of only 2%.

A. Through how many persons on the average will each of the three kinds of alleles tend to pass?
B. Which will cause the fewest number of genetic deaths?

9. A rare allele, "m," gives an impairment of 20% in the homozygous condition, but only a 1% impairment when its dominant allele "M" is present along with it.

A. Through how many homozygous persons will "m" tend to pass?
B. Through how many heterozygotes will it tend to pass?
C. Will most of the elimination occur in the heterozygotes or in the homozygotes?

10. Assume that two human populations, one in the United States and the other in Australia, both originated from migrants who left the same European population several generations ago. In the American population there is a high incidence of a recessive disorder which causes deformed skeletons. The disorder is unknown in the Australian population but is found in a very low frequency in the European population. Offer an explanation.

MICROORGANISMS: BACTERIA AS GENETIC TOOLS

VARIATIONS IN MICROORGANISMS. Early work with microorganisms showed that these lower forms are very diverse. For example, even within a given kind of bacterium, specific strains may be recognized by colony type, host specificity, pigment formation, drug susceptibility, nutritional requirements, and other distinctions. Although investigators before 1940 had demonstrated many such variations in microorganisms, the results of their studies were equivocal and did not permit a definite answer to the question, "Are the variations in microorganisms hereditary?" It was not even certain that true genetic mutations occur in bacteria as they do in higher organisms.

As more and more drugs were studied for their bacterial killing properties, a variety of responses by microorganisms was noted. A typical picture to emerge was as follows: a particular drug dosage might destroy most of a bacterial culture but would leave a few surviving members. Upon isolation, these resistant cells would be able to form the basis for a strain completely resistant to that first drug dosage. Treating cultures from this derived strain with a still higher dosage could result in the isolation of survivors able to withstand the higher level.

Repeating the procedure could lead to the establishment of strains restistant to a drug concentration which would have killed cells outright in the original culture.

The correct interpretation of such observations remained in doubt as long as work failed to distinguish between genetic and environmental factors. Two explanations were equally feasible. According to the one, the reaction of the cells to the drug stems directly from contact between the cells and the drug. In other words, an external agent can *cause* cells to adapt or change in direct response to its presence in the environment. A contrasting idea, incompatible with the first, is that spontaneous mutations take place in the bacterial population. Since these arise at random, some of the genetic changes result in a small number of cells which possess an increased drug tolerance. These changes in the cellular ability to withstand higher drug levels occur *independently* of the presence of the drug. The drug would be acting as a screening agent, selecting out those cells *already* immune to a certain drug level even *before* it was present in the environment. The two conflicting viewpoints are reminiscent of the controversy between the Lamarckian idea of acquired characters and the neo-Darwinian concept. According to the former, the environment directly alters individuals and brings about hereditary change. Adherents of the second school of thought would say that development of a new hereditary strain depends on random gene mutation followed by selection of those forms best suited to the immediate environment

SPONTANEOUS MUTATION IN MICROORGANISMS AND THE FLUCTUATION TEST. Since it was not even clear that true spontaneous mutations arise in microorganisms, no well-founded arguments were available to resolve the role of the environment in the origin of new strains. It remained until 1943 for the matter to be settled through the elegant experimental design of the "fluctuation test" developed by Luria and Delbruck. These pioneers in microbial genetics presented a methodology to answer the question once and for all. With the publication of the work, the foundation was laid for all later research in the discipline of the genetics of microorganisms.

The procedure was designed to distinguish between changes in population structure through: (1) direct modification by an environmental agent or (2) spontaneous mutation and subsequent selection by an environmental agent. The specific bacterium used was *Escherichia coli* strain B, which is sensitive to the bacteriophage, T/1. If the bacteria are exposed to the T/1 virus on solid media, most of the cells are destroyed, but a few colonies manage to develop. In broth, a clearing of the turbid culture takes place which is later followed by a clouding. These observations indicate that lysis is rampant at first, killing a majority of the cells. But a few survivors, resistant to the virus, remain and grow. When these are in turn exposed, they show resistance to the virus. This is the same situation as that resulting from exposure of bacterial cells to a drug, and it raises the same questions: "Did the agent (the virus here) cause some of the cells to change and become resistant by direct contact? Or were some cells already resistant to the virus before it was applied so that only these cells were able to survive and form the basis of a resistant culture?"

It was known that in a culture of 10^8 *E. coli* cells, approximately 100, on the average, could be expected to show phage resistance. To answer the question of the origin of these resistants, the experiment was set up in two different ways. In the first (Fig. 17-1A), a culture containing a large number of cells was divided into 10 parts. For the sake of simplicity, let us say that a culture of 10^8 was subdivided into 10 parts of 10^7 cells each. Therefore, in this part of the investigation, a culture of cells was simply subdivided into equal parts and then tested for any virus-resistant cells which might be present. No further growth of the samples from the original large culture was allowed up to this point. The subdivisions were plated on phage-impregnated media and then incubated to permit growth of any resistant cells. The results showed that a small portion, approximately 100 cells per 10^8 (or 10 cells on the average in each subdivision), were phage resistant. How could this be explained?

When the results were subjected to statistical analysis, it was found that the *mean* number of resistants in the subdivisions was of the same order of magnitude as the variance. The vari-

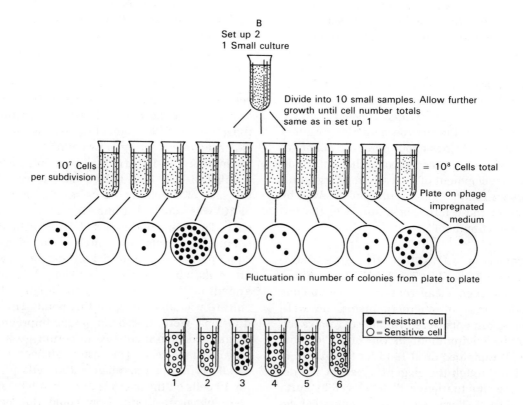

A

Set up 1
Culture of 10^8 cells

Subdivide into 10 equal sized samples

Allow no further growth

10^7 Cells per subdivision

= 10^8 Cells total

Plate on phage impregnated medium

Comparable number of colonies on each plate

B

Set up 2
1 Small culture

Divide into 10 small samples. Allow further growth until cell number totals same as in set up 1

10^7 Cells per subdivision

= 10^8 Cells total

Plate on phage impregnated medium

Fluctuation in number of colonies from plate to plate

C

● = Resistant cell
○ = Sensitive cell

1   2   3   4   5   6

FIG. 17-1. The fluctuation test. *A.* In setup 1, a large culture of cells, 10^8 in this example, is not permitted further growth. It is subdivided into 10 smaller samples of 10^7 each. Each of these samples is then plated on medium containing phage. Colonies which are phage resistant develop, and the number from one plate to the next is comparable. In one experiment, the mean number of resistant colonies was 10.7 and the variance 15, a figure close to the mean. Repetitions of this set up give comparable results. The observations can be explained on either the adaptation hypothesis or the spontaneous mutation theory. *B.* In setup 2, a small culture of cells is subdivided, and each is allowed to grow further in the separate tubes until the number of cells is the same as in the first setup. Thus, in both parts of the experiment, there are 10 tubes of 10^7 cells giving a total of 10^8. However, in setup 2, the results on the phage-impregnated plates are quite different. Certain plates may have no resistant colonies; some few; others many. There is a great deal of variation from one to the next, and the variance is far from the mean number. *C.* This fluctuation from one plate to the next is to be expected only on the basis of the spontaneous mutation theory. According to this concept, mutation can occur at any time in any one of the separately grown cultures. Certain tubes (1 and 6) may have no mutant cells at all. Others may contain a mutant cell which arose later in the growth of the culture (tube 2) and did not have the chance to divide further. Other mutant cells may arise earlier and at different times, giving different numbers of mutant cells at the end of the growth period. It is possible for two or more mutant cells to arise in one culture. When plated on phage-impregnated medium, the number of resistant colonies will vary from one plate to the other, reflecting the spontaneous nature of the mutations to phage resistance.

ance, it will be recalled in Chapter 7, is equal to the square of the standard deviation. Statistical analysis has demonstrated that in any random distribution, the value of the variance is close to that of the mean. Therefore, the results here are to be expected and can be explained by either of the two contrasting ideas. One could argue that in the culture of 10^8 cells, spontaneous mutations occur from time to time, and some of these genetic changes result in phage resistance. Some resistant cells have recently arisen in the culture, whereas others have descended from mutant cells which arose earlier. When the culture is subdivided and tested for mutants, one would obtain a random sampling of the culture and therefore a distribution of mutants with a variance equal to the mean. Alternatively, one could argue that no mutations existed in the culture as it grew to the original total of 10^8 cells. When the culture of 10^8 was subdivided and plated out in contact with the phage, most of the cells lyse. A few, however, adapt or are induced to mutate

by the phage. This occurs at random in the samples, and so the variance should be close to the mean.

Therefore, this first set up does not permit a distinction between the two hypotheses. However, the second part of the experimental design was crucial and allowed a clear choice between the two arguments. A small culture of bacteria was taken and divided (Fig. 17-1*B*). Each of these contained 50–500 cells and was then allowed to grow until the total number of cells was equal to the total number in the first set up, in our example 10^8 cells. Note the difference between the two set ups. In the first, a culture of 10^8 cells was subdivided and allowed no further growth before being brought into contact with the phage. In the second approach, each subdivision grew up independently from a small sample until a total of 10^8 cells was reached. Now, according to the adaptation hypothesis, there is little difference between the two set ups. In both cases, we have 10^8 cells. It shouldn't matter that in the one case a big culture of 10^8 was subdivided into 10 smaller ones and in the other case 10 smaller cultures were grown up separately until the same total of 10^8 was reached. There would be no relevant difference between them, because the adaptation (or induced mutation) hypothesis assumes that the resistants appear *only* when exposed to the phage. So when the samples are tested, the variance would reflect the randomness and should be about the same as the mean.

According to the spontaneous mutation concept, however, we would expect a wide distribution in the number of mutants in the separately grown cultures (Fig. 17-1*C*). This is so, because the separately grown cultures will differ in the number of mutations which arose within each of them during growth, and they will also differ in the stages of culture growth at which the mutations occurred. If a single mutation arises at the end of the growth of the culture, there will be fewer mutant cells than if the mutation occurs early. In the latter case, many descendants of the mutant cell would be present in the culture, because the bacteria multiply exponentially. So even in cultures in which the same number of mutations took place, the number of resistants recovered would be very different. This is *not* the same as taking random samples from one big culture of cells, and one would not expect the variance to be equal to the mean.

The results of many separate experiments of this type were comparable and gave an order of variance very far from the mean. Since the number of resistants fluctuates in the second set up from one separate culture to the next, the procedure has been named the "fluctuation test." Its application to investigations of variations in microorganisms leaves no doubt that spontaneous mutations do occur in these lower forms and that inherited variations can arise independently of an external agent.

THE REPLICA PLATING TECHNIQUE. After the fluctuation test, other procedures were developed which also established that spontaneous mutations arise in microorganisms. One which is now routinely used in laboratory investigations is the technique of replica plating, devised by Lederberg and Lederberg. Figure 17-2 demonstrates the procedure in reference to streptomycin resistance in bacteria. Many strains of *E. coli* cells are sensitive to streptomycin, and colonies developing from them will not grow on a solid medium which contains the drug. However, cells may arise, at times, which produce colonies which are resistant. It can be shown by the replica plating that this change to drug resistance is completely independent of the drug itself. A piece of velvet is secured over a block. The velvet is gently applied to the surface of the colonies growing on the plate. The nap of the velvet picks up sufficient cells which can then be transferred to other plates which contain various media. Therefore, imprints or replicas of the original colonies can be easily made onto a plate containing streptomycin. This second plate must be oriented in the same way as the original one so that any colonies which develop on it can be related back to colonies on the first plate. A colony may arise on the plate with the drug. Since the two plates are oriented, the colony on the first plate which produced this drug-resistant colony can be identified. Cells taken from the colony on the plate without the drug can then be exposed to a tube of broth containing streptomycin. These cells will grow in the presence of the drug, whereas other cells from neighboring colonies will not. Because no streptomycin was present on the first plate, the cells had never

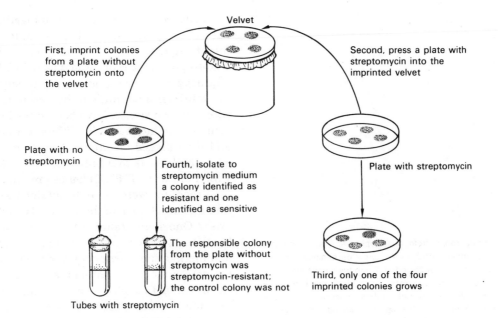

Velvet

First, imprint colonies from a plate without streptomycin onto the velvet

Second, press a plate with streptomycin into the imprinted velvet

Plate with no streptomycin

Fourth, isolate to streptomycin medium a colony identified as resistant and one identified as sensitive

Plate with streptomycin

The responsible colony from the plate without streptomycin was streptomycin-resistant; the control colony was not

Third, only one of the four imprinted colonies grows

Tubes with streptomycin

FIG. 17-2. The replica plating technique. This procedure, in addition to its other applications, also demonstrates that spontaneous mutations arise in microorganisms. See text for further details. (Reprinted with permission from R. Sager, and F. Ryan, *Cell Heredity*, Wiley, 1961.)

come into contact with the drug until they were placed in the broth. We would not have known that streptomycin-resistant cells had arisen and were present unless we had identified the colonies by replicating onto the second plate which has the drug. The procedure leaves no doubt that the drug-resistant cells were there on the first plate before any contact had been made with the streptomycin.

DETECTION OF SEXUAL RECOMBINATION IN BACTERIA. Once mutation in microorganisms was firmly established by such demonstrations, it became possible to utilize bacteria and viruses as tools in genetic analysis. Before the mid-1940's, bacteria were thought to lack any mechanism akin to a sexual process which could give rise to genetic recombination. The only way for new combinations of genes to arise in such organisms was thought to be by mutation alone. Therefore, these lower forms would be totally asexual and would be at a disadvantage in that beneficial gene combinations would arise more slowly than in sexual organisms.

Several early investigations were attempted to explore the possibility of a sexual mechanism among bacteria. However, none of these were

precise enough to preclude explanations of the observations on other grounds, such as spontaneous mutation. Until the mid-1940's there was no clear-cut proof of the contact of two genetically diverse bacterial cells followed by the segregation of genetic recombinants. The first unambiguous studies were those of Lederberg and Tatum who worked with multiply mutant strains of *E. coli.* These strains had been obtained through the mutagenic action of ultraviolet light and were deficient biochemically in one or more ways. Any wild type or normal bacterial cell can grow on a basic or minimal medium which contains just a few simple substances from which the cell can manufacture all the complex organic molecules needed for its metabolism. Such cells which thrive on a minimal medium are called *prototrophs*. In contrast, other cells may be nutritional deficients in that they lack the ability to make one or more essential growth factors from the minimal source. Such deficient cells are designated *auxotrophs* and must be provided with the growth factors they cannot synthesize. These nutritional mutants can grow only on a supplemental medium, a minimal medium to which one or more defined factors have been added.

Lederberg and Tatum mixed together different auxotrophic strains. For example, strain 1 might require biotin and methionine. Such an auxotroph can be represented as B⁻M⁻. Strain 2, requiring threonine and proline to supplement the minimal medium, can be designated T⁻P⁻. After the two strains have been mixed together, samples are incubated for different time periods and then placed on minimal medium (Fig. 17-3A). The logic is that only those cells will grow which are B⁺M⁺T⁺P⁺. Colonies growing on minimal medium were actually obtained after mixture and incubation of the strains. How did these arise? One suggestion is through sexual recombination. Strain 1, although deficient for biotin and methionine, carries the genetic information to manufacture threonine and proline (B⁻M⁻T⁺P⁺). Likewise, strain 2, although unable to manufacture threonine and proline, has the wild genes for biotin and methionine (B⁺M⁺T⁻P⁻). The mixture can be represented as a cross and the appearance of the wild colonies explained on the basis of cell fusion followed by recombinations (Fig. 17-3B).

If we assume that the loci are linked, we can imagine the operation of some process similar to

FIG. 17-3. Prototrophs from auxotrophs. A. An auxotrophic strain, such as strain 1 or 2, cannot grow on minimal medium. Essential substances must be added. When the strains are mixed and samples of the mixture plated on minimal medium, some plates develop colonies. The cells in these must be prototrophs, because no essential metabolites have been added to the medium. B. The mixture can be depicted as a cross of bacterial strains in which the genes are all linked. The prototrophs could arise through a crossover event. Cells with the reciprocal crossover product cannot develop, because they need to be supplemented with four essential metabolites. Other crossovers are possible, but these cannot develop because the cells will require one or more essential substances. Only the prototrophs (B⁺M⁺T⁺P⁺) can grow on the minimal medium.

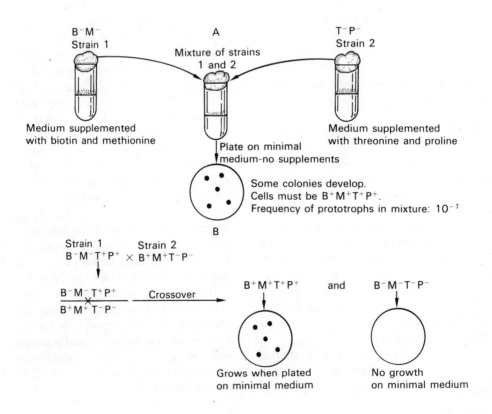

UNDERSTANDING GENETICS

crossing over which is able to produce the prototrophic recombinants, $B^+M^+T^+P^+$. The parental cells from strains 1 and 2 cannot grow on the minimal medium nor can any recombinants which lack the ability to carry out all the biochemical steps required for nutrition. The results, however, can be interpreted on the basis of mechanisms other than sexuality. A strong argument could be made for the occurrence of the prototrophs through spontaneous mutation. Of the cells in the mixture, only a very small number, approximately 1 in 10^7, is prototrophic. Perhaps these are not recombinants at all but are revertants, wild cells which have arisen through back mutations: $B^-M^-T^+P^+ \rightarrow B^+M^+$-$T^+P^+$ (strain 1) and $B^+M^+T^-P^- \rightarrow B^+M^+T^+P^+$ (strain 2). To answer this objection to the sexuality hypothesis, a second mixture was studied, this time employing two strains which were mutant for three factors (Fig. 17-4A). The prototrophs occurred in this mixture with a frequency of approximately 10^{-7}, just as they did in the first one. From the second mixture, several other types were also recognized. In addition to the prototrophs which were isolated on minimal medium (because the parental types can't grow on it), other colonies were recovered on media supplemented with one or more of the essential growth requirements. A variety of types was found: some with single requirements, some with double (including one requirement from each parent). Again these results can be explained on the basis of recombination through genetic exchange (Fig. 17-4B). A strong argument favoring this sexuality hypothesis is based on the frequency (10^{-7}) with which the prototrophs arise in the two separate experiments. It had been noted that reversion to wild (that is, back mutation) for any *one* of the requirements occurred with a frequency of 10^{-7}) in 1 day. So if the prototrophs in both of the experiments are arising through back mutation and not through sexual recombination, we would expect the following. (1) In the first mixture between the strains which are *doubly* deficient, prototrophs should arise with a frequency of approximately 10^{-14}. This would be so, because a doubly deficient cell such as $B^-M^-T^+P^+$ would require two back mutations, each with a probability of 10^{-7} ($10^{-7} \times 10^{-7} = 10^{-14}$). (2) In the second mixture between the *triply* deficient cells, prototrophs

should arise with a frequency of 10^{-21}, for a triply mutant cell such as $B^+Pa^+C^+T^-L^-B^-$ would require three independent mutations, each with a frequency of 10^{-7}. However, prototrophs arise with the same frequency when doubly and triply deficient strains are mixed, and this is the same frequency with which a single reversion takes place. These observations provide a strong argument against the origin of the prototrophs through spontaneous mutation and favor the sexuality concept. Other objections were still to be answered, but it is sufficient here to say that additional evidence showed that cell contact was essential for the origin of prototrophs after mixing multiply mutant strains. Experimental work eliminated the possibility of transformation (more later, this chapter), symbiotic association of two different cell types, and other phenomena as explanations for the origin of recombinants. Only the sexuality hypothesis was upheld.

THE FERTILITY FACTOR. Once the presence of a sexual process was established in *E. coli*, investigations concentrated on the nature of the mechanisms involved. More refined analyses indicated that the various markers (mutant alleles for auxotrophy as well as those which did not affect the nutritional requirements) were linked into just one group, in contrast to the two or more linkage groups which are typical of higher forms. Evidence for linkage and gene order was based on reasoning similar to that used in classical genetic analysis. Those genes closer together would undergo less recombination than those farther apart on the chromosome. Double crossovers would be less common than singles, and triples even less frequent than the doubles. So by crossing various multiply mutant strains, counting the resultant recombinant colonies (in the procedure each colony is derived from one cell), and scoring their phenotypic features, a single linkage map emerged.

However, during the course of the linkage analysis, it became apparent that the *E. coli* sexual system possessed several features which departed from the picture which is typical of sexual recombination in higher forms. One unexpected observation indicated that the production of recombinants depended on the continued survival of just *one* of the parents. For example, suppose strain 1 is $B^-M^-T^+P^+$ and is also strep-

A

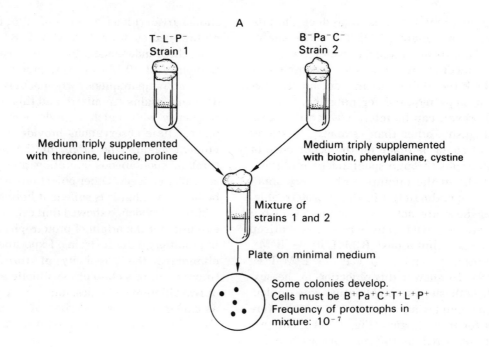

T⁻L⁻P⁻
Strain 1

B⁻Pa⁻C⁻
Strain 2

Medium triply supplemented
with threonine, leucine, proline

Medium triply supplemented
with biotin, phenylalanine, cystine

Mixture of
strains 1 and 2

Plate on minimal medium

Some colonies develop.
Cells must be B⁺Pa⁺C⁺T⁺L⁺P⁺
Frequency of prototrophs in
mixture: 10⁻⁷

B

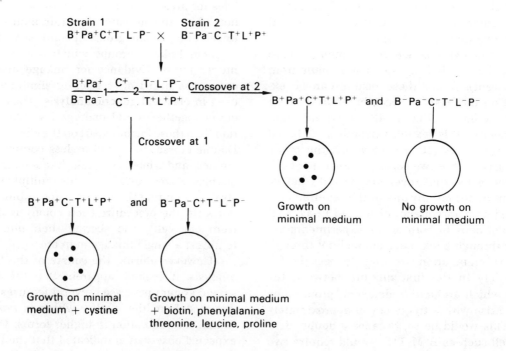

Strain 1
$B^+Pa^+C^+T^-L^-P^-$ ×

Strain 2
$B^-Pa^-C^-T^+L^+P^+$

$\dfrac{B^+Pa^+}{B^-Pa^-}$ 1 $\dfrac{C^+}{C^-}$ 2 $\dfrac{T^-L^-P^-}{T^+L^+P^+}$ Crossover at 2

$B^+Pa^+C^+T^+L^+P^+$ and $B^-Pa^-C^-T^-L^-P^-$

Crossover at 1

Growth on
minimal medium

No growth on
minimal medium

$B^+Pa^+C^-T^+L^+P^+$ and $B^-Pa^-C^+T^-L^-P^-$

Growth on minimal
medium + cystine

Growth on minimal medium
+ biotin, phenylalanine
threonine, leucine, proline

FIG. 17-4. Isolation of recombinants. *A.* When two strains are mixed, each of which is triply mutant for growth requirements, prototrophs may be isolated on the minimal medium. These prototrophs are found at a frequency of 10^{-7}, which is the same as that in a cross of two strains which are doubly mutant (Fig. 17-3A). This argues for sexual recombination vs. the origin of the prototrophs through back mutations. *B.* The mixture can be diagrammed as a cross. If the genes are all linked, a crossover at one point (point 2 in the figure) will produce recombinants which are prototrophs, as well as those which require all six growth requirements. Various other recombinant types would also be expected, such as those arising from a crossover at point 1 in the diagram. These will not grow on minimal medium but need to be supplemented with one or more essential metabolites. Such cells do arise and can be isolated. Their occurrence is to be expected on the basis of sexual recombination.

tomycin sensitive. Strain 2 is $B^+M^+T^-P^-$ but is resistant to the drug. The two strains are mixed and finally prototrophs are isolated on minimal medium containing streptomycin. Now suppose that exactly the same procedure and the same markers are followed in a second cross but that now strain 1 is streptomycin resistant, whereas strain 2 is streptomycin sensitive (Fig. 17-5). In such a cross, absolutely no recombinants form. Crosses of this type, in which the survival of only one of the parental strains was guaranteed, repeatedly demonstrated that only one of the parents was giving rise to the recombinants. In our example, they are coming from strain 2, for if it is killed, no recombinants are formed. Strain 1 is transferring genetic material to strain 2, but once the transfer has been achieved, it is no longer needed for the formation of recombinants. Further studies revealed that the appearance of recombinants definitely depends on two types of cells, a donor and a recipient. Genetic material is transferred from the donor cell to the recipient,

FIG. 17-5. Survival of one parent. In order to obtain any recombinants, the survival of one of the parental strains must be ensured. One way to demonstrate this is by performing two types of crosses between two auxotrophic strains. The only difference between the crosses is that a parental strain is sensitive to streptomycin in one

cross and resistant to it in the other. As the diagram shows, there is one of the two strains which must not be killed if recombinants are to be obtained (strain 2 must survive here). This would appear to be the recipient strain. Strain 1 would be a donor which transfers genetic material to a recipient.

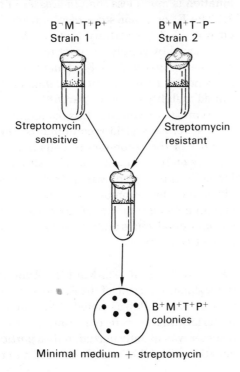

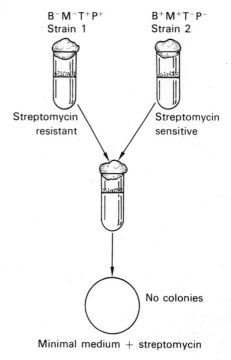

which then acts as a "zygote" whose survival is necessary for the segregation of recombinant cells. It was also discovered that at times cells could arise in a culture of donor types which no longer had the ability to act as donors. They lacked the capacity for transferring genetic material to the recipient. It appeared that some factor was required for a cell to act as a donor type. If it were lost, a cell originally from a donor culture could now be used as a recipient. This fertility factor needed to confer donor capacity was designated "F." Donor cells were thus called "F⁺" (with fertility factor) and recipient cells "F⁻" (lacking fertility factor).

Observations then showed that the fertility factor (F) could be transferred with relative ease to F⁻ cells, thus converting them to the donor state. Simply bringing donor and recipient cells into contact for 1 hr or so could change the F⁻ cells to the F⁺ state. The following will summarize what was now evident about the sexual system in *E. coli*. (1) Sexual recombination depends on the presence of two cell types, a donor or male (F⁺) and a recipient or female (F⁻). F⁻ × F⁻ crosses are ineffective. Mixing together F⁺ cultures with other F⁺ cultures yields some recombinants due to the presence of F⁻ cells which spontaneously arise, but the amount of recombination is much less than in crosses of F⁺ × F⁻. (2) The F⁺ or donor state results from the presence of "F," the fertility factor, which can be transferred by cellular contact to F⁻ cells which are then converted to F⁺ donors. (3) The appearance of recombinants depends on the continued viability of the F⁻ parent.

It must be realized at this point that although F⁺ × F⁻ crosses yield recombinants, the actual number of the recombinant cells is low, in the vicinity of 10^{-6}. In other words, although the F factor is transferred with a high frequency or efficiency, the chromosome itself is transferred only in a small minority of the cells. Hence, the chromosomal genes undergo recombination with a low frequency.

PARTIAL GENETIC TRANSFER. Another unorthodox characteristic of the *E. coli* sexual system was revealed. It was noted that when genetic material was transferred from F⁺ to F⁻, the transfer was usually partial; only a portion of the genetic material of the F⁺ seemed to enter the

F⁻. Certain genes were not transferred at all from donor to recipient. Evidence for this was obtained from reciprocal crosses between strains. For example, in *both* strain 1 and strain 2, F⁺ and F⁻ types were isolated. This made possible reciprocal crosses, and the results of these proved to be quite different (Fig. 17-6). Only some of the genetic material of the F⁺ parent entered the F⁻ cell to contribute to the offspring. The F⁻ parent, on the other hand, contributes all of its genetic material. Recombinants, therefore, will contain a greater percentage of their genes from the F⁻ cells than from the F⁺. This is quite a different situation from that in higher forms where both parents make equal donations of the genetic material to the zygote.

HIGH-FREQUENCY TYPES AND THE TRANSFER OF GENETIC MARKERS. Some clarification of the unusual features of recombination in *E. coli* was provided through the discovery of still another donor or male type. These donors behave as "supermales." When F⁻ cells are crossed with them instead of with ordinary F⁺ males, the percentage of recombination is increased approximately one thousand times. Therefore, the "supermales" were termed "high frequency" ("Hfr"), denoting their ability to bring about genetic transfer. Various Hfr males arose independently in cultures of ordinary F⁺ donors. When different strains of Hfr were examined, it was noted that they could become F⁺ again. This suggested that the sex factor (F) was present in the Hfr cell. But some puzzling facts emerged when different Hfr × F⁻ crosses were compared with one another and also with ordinary F⁺ × F⁻ crosses. In the latter type of cross, it will be recalled that the F factor itself is transferred quite readily to the F⁻ cells, con-

FIG. 17-6. Partial transfer. F⁺ and F⁻ types were found within strains. This makes possible reciprocal crosses. In this example, strains 1 and 2 are both auxotrophs, because each lacks the ability to manufacture a specific amino acid (histidine in 1 and arginine in 2). The other loci indicated are for traits which do not influence the ability to grow on minimal medium. When the strains are crossed and the cells placed on minimal medium, only the prototrophs (A⁺H⁺) will grow. These prototrophs can be tested for the other markers. It is found that the reciprocal crosses differ. Although various combinations are possible, most of the alleles in the A⁺H⁺ prototrophs will have been derived from the F⁻ parent. The transfer is only partial from the F⁺ to the F⁻.

Cross #1

Strain 1 F⁺ (auxotrophic-requires histidine) Strain 2 F⁻ (auxotrophic-requires arginine)

A⁺B⁺C⁺D⁺E⁺G⁺H⁻ × A⁻B⁻C⁻D⁻E⁻G⁻H⁺

↓ Place on minimal medium

A⁺B⁻C⁻D⁻E⁻G⁻H⁺- Prototrophs
Genes mainly from F⁻ strain 2

Cross #2

Strain 1 F⁻ (auxotrophic-requires histidine) Strain 2 F⁺ (auxotrophic-requires arginine)

A⁺B⁺C⁺D⁺E⁺G⁺H⁻ × A⁻B⁻C⁻D⁻E⁻G⁻H⁺

↓ Place on minimal medium

A⁺B⁺C⁺D⁺E⁺G⁺H⁺- Prototrophs
Genes mainly from F⁻ strain 1

| A⁺ makes arginine |
| A⁻ requires arginine |
| H⁺ makes histidine |
| H⁻ requires histidine |
| C,D,E,G,- other genes not influencing basic nutrition |

verting them into donor types. And although a low frequency of recombinants is obtained in F$^+$ × F$^-$ crosses, *any* portion of the genetic material can be transferred from F$^+$ to F$^-$. In contrast to this, when any specific Hfr strain is used as the donor parent, only a restricted portion of its linkage group is readily transferred. Some genes rarely go from the Hfr to the F$^-$ parent. Moreover, the F factor which must be present in the Hfr cell usually is *not* transferred. Consequently, in a cross of Hfr × F$^-$, the offspring are almost always F$^-$. For example, consider the cross in Figure 17-7. After the strains are mixed, samples are plated out on a medium lacking threonine and leucine but containing streptomycin. By doing this, we are selecting prototrophs which are resistant to streptomycin. Therefore, only T$^+$L$^+$ colonies will grow. These cells must be recombinant, because the drug will have killed the Hfr parent; the F$^-$ parent, being an auxotroph (T$^-$L$^-$), cannot grow on minimal medium. The recombinant colonies on the plates will thus all be T$^+$L$^+$. They can subsequently be tested for any other markers transferred by the Hfr parent.

In one strain of Hfr, the streptomycin marker is rarely transmitted, whereas the segment of genes T through lambda is transmitted with a high frequency. Figure 17-7 shows that very curious ratios are found among the recombinants. Of the T$^+$L$^+$ offspring, 90% are azide sensitive (meaning that 90% received the Az

FIG. 17-7. Cross of Hfr and F$^-$. Prototrophs are selected by plating on minimal medium. These cells will all possess the T$^+$L$^+$ markers which come from the prototrophic Hfr parent. The Hfr parent, however, is destroyed by the presence of the streptomycin in the minimal medium; so all the prototrophic colonies are truly recombinant. The F$^-$ parent, being an auxotroph, cannot grow on the minimal medium. These recombinant offspring are then tested for the presence of the other markers from the F$^+$ parent. It is seen that the markers have very different chances of being transferred to the F$^-$ cells to enter into the formation of recombinants.

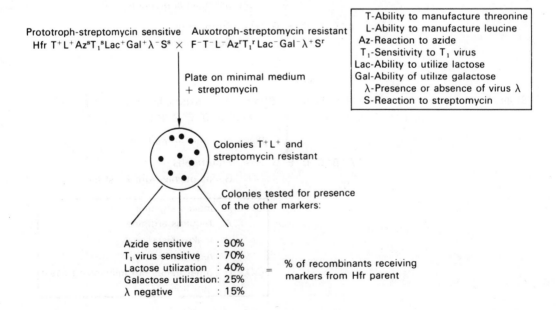

Prototroph-streptomycin sensitive   Auxotroph-streptomycin resistant
Hfr T$^+$L$^+$Az^{s}T$_1^s$Lac$^+$Gal$^+$λ$^-$S^s × F$^-$T$^-$L$^-$Az^{r}T$_1^r$Lac$^-$Gal$^-$λ$^+$S^r

| T-Ability to manufacture threonine |
| L-Ability to manufacture leucine |
| Az-Reaction to azide |
| T$_1$-Sensitivity to T$_1$ virus |
| Lac-Ability to utilize lactose |
| Gal-Ability of utilize galactose |
| λ-Presence or absence of virus λ |
| S-Reaction to streptomycin |

Plate on minimal medium
+ streptomycin

Colonies T$^+$L$^+$ and
streptomycin resistant

Colonies tested for presence
of the other markers:

Azide sensitive   : 90%
T$_1$ virus sensitive   : 70%
Lactose utilization   : 40%   =   % of recombinants receiving
Galactose utilization: 25%   markers from Hfr parent
λ negative   : 15%

marker from the Hfr parent), 70% received the T_1 marker, and so on for the others. It would seem that the marker alleles did not all have the same chance of entering the F⁻ cell from the Hfr. Moreover, the recombinants are usually F⁻, not F⁺ or Hfr as would be expected if the F factor had a good chance of entering the recipient cell.

An insight into the basis of this odd mating system was provided by an ingenious procedure employed by Wollman and Jacob. Using strains similar to the ones just discussed, they mixed together Hfr and F⁻ and then interrupted the mating at various time intervals. They did this by subjecting the mixture to a Waring blender, which can pull apart pairs of cells without damaging them. They then plated out samples and tested for the presence of Hfr markers in the recombinant cells. They found (Fig. 17-8) that no prototrophs were formed before an 8-min period. Then the threonine marker was transferred and shortly thereafter the leucine marker. Although only 9 min of contact between cells was needed for entry of the Az marker, 25 and 26 min were necessary for Gal and lambda. If

mating was interrupted at 18 min, for example, some cells would have received Lac, but none would be Gal⁺. This suggested to Wollman and Jacob that the genetic material or chromosome of the Hfr parent entered the F⁻ cell at a definite rate and in an established order (Fig. 17-9A), in this case, the order T through lambda. If the mating was interrupted and the chromosome broken at a certain time, some genes would not have yet been able to penetrate into the F⁻ cell. The breaking of the chromosome doesn't prevent the integration of the donor genes into the recipient by recombination. The peculiar recombination ratios which occur under normal conditions of mating can be explained by the spontaneous breakage of the donor chromosome as it is being transferred. The closer a gene is to the origin, that chromosome end which penetrates the recipient first, the greater its chance of being transferred. The spontaneous breakage point would vary from one mating pair to the next. The result would be different percentages of recombination for the different genetic markers carried by the male parent. The bacterial

FIG. 17-8. Interrupted mating. The same two strains shown in Fig. 17-7 may be allowed to mate for different periods of time. Samples of a mating mixture are removed at specific times and subjected to a Waring blender which pulls the cells apart. The sample is then tested for recombinants. It is found in this cross that no markers enter if the cells are separated before 8 minutes. The markers being followed in the cross are then found to enter at different characteristic times.

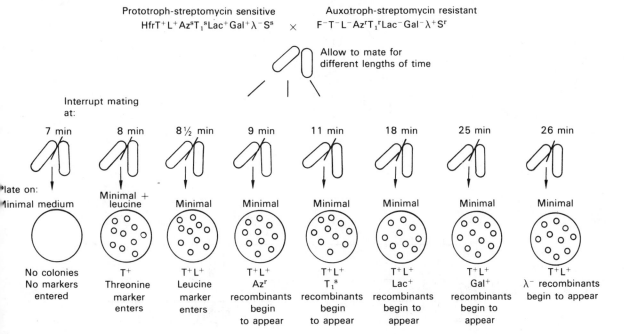

mating system, therefore, has unusual features which permit us to map genes in two ways: by recombination percentages and by time of entry into the female parent (Fig. 17-9B).

RELATIONSHIP AMONG THE *E. COLI* MATING TYPES. Still unsettled was the fundamental difference between Hfr and ordinary F⁺. Why did the latter show a high frequency of transfer of the F factor but a low frequency of transfer of the marker alleles? Since Hfr originated from F⁺ cultures, it was considered possible that the Hfr were mutants and that *only* these mutants of F⁺ were actually able to transmit the chromosome. The F⁺ cells would really be ineffective in bringing about transfer of genes from male to female cells. Wollman and Jacob applied the fluctuation test and showed that the origin of Hfr through mutation of F⁺ was a correct concept. They also showed this through application of the replica plating technique, which in addition enabled them to isolate different strains of Hfr in pure culture. Recall that replica plating enables us to obtain copies of an original plate and thus permits study of the same bacterial cells or colonies under a variety of conditions. Using the replica

FIG. 17-9. Entry of Hfr chromosome. *A.* The results shown in Figs. 17-7 and 17-8 can be explained as shown here. The chromosome of *E. coli* is known to be circular. Apparently, the circle in the Hfr cell gives rise to a linear chromosome which penetrates into the F⁻ cell at a specific rate and in a definite order. The Hfr chromosome may become linear at any point, so that only portions of the chromosome enter the F⁻ cell and engage in recombination. The closer a locus to the "beginning" of the Hfr chromosome, the greater the chance that it will be transferred. The T and L loci as shown here have a greater chance of entry than those following them. Some loci, such as the one for streptomycin tolerance, are so far toward the end that breakage usually occurs in front of them, and they rarely enter. *B.* On this concept, the chromosome may be mapped in either of two ways, by recombination percentages or by times of entry. In this example, all of the cells would be T⁺L⁺, because only prototrophs were selected in the cross. Different percentages of these prototrophs carry the other markers. We see that 90% of them will receive the azide marker, only 25% the Gal marker. The difference is explained due to the different times of entry, 9 minutes vs. 25. The sooner a marker enters, the closer it is to the beginning and the greater the number of cells which will receive it. The whole chromosome may be mapped in time units. At 37°C., approximately 90 minutes are required if any cells are to receive the entire Hfr chromosome.

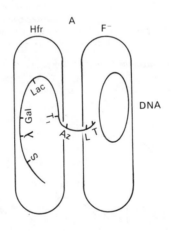

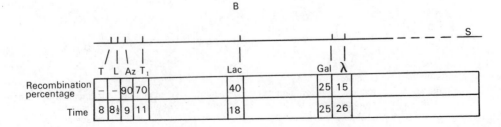

	T	L	Az	T₁		Lac		Gal	λ	
Recombination percentage	–	–	90	70		40		25	15	
Time	8	8½	9	11		18		25	26	

plating method, Wollman and Jacob started with a population of F⁺ cells which were T⁻L⁻ (Fig. 17-10). They grew discrete colonies on plates supplemented with threonine and leucine. A second set of plates contained minimal medium, and these were spread with F⁻ cells which were M⁻B⁻. Imprints of the F⁺ colonies were then made onto the second set of plates. The original plate was oriented in exactly the same way as the replica plates. After transferring cells in this way from the F⁺ plate to the plates containing the F⁻, it was found that some colonies formed on the latter. These must be recombinant prototrophs, because only they can grow on minimal medium. The positions of the colonies on the second set of plates were then related back to the positions of certain F⁺ colonies on the first plate. Cells from these F⁺ colonies which had been picked up by the velvet must have contacted the F⁻ cells on the second plate and transferred genetic material to them. After genetic recombination, prototrophs arose from the F⁻ cells and were selected on the minimal medium, where they formed colonies. The replica plating tech-

nique thus permitted the identification of just those few F⁺ colonies which contained cells capable of acting as donor or male cells and which could transfer chromosome markers to F⁻. Each of the F⁺ colonies had arisen on the plate from a single cell and, therefore, each donor colony represented a cell which was different from the other cells on the plate which did not behave as donors. Each chromosome donor in the original F⁺ population is in the minority and must have arisen independently by spontaneous mutation.

After the donor colonies were identified in this way, they were isolated to establish different, pure Hfr strains. Crosses were made between these several Hfr types and F⁻ cells. Further analysis employing the interrupted mating method contributed information of fundamental importance to bacterial genetics. The results clearly showed that different Hfr types had a different gene at the anterior end of the chromosome (Fig. 17-11A). Moreover, some blocks of genes which were *not* transferred with a high frequency in one strain were transferred readily

FIG. 17-10. Origin and isolation of Hfr strains. The replica plating procedure demonstrated that only a few cells in an F⁺ population are capable of transferring genetic material to the F⁻. These are

Hfr cells which arise spontaneously from F⁺. Identification of Hfr colonies growing among F⁺ colonies on the same plate permitted the isolation of pure Hfr cultures.

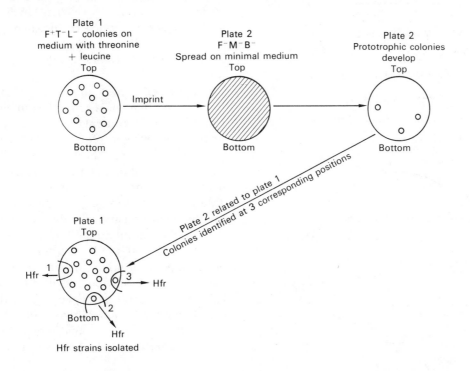

in another (the segment M through S is transferred in strain 1 but not in strain H; the segment Az through lambda is transferred with high frequency in strain H but not in strain 1).

When the different strains are compared in this way, an unexpected picture emerges. All of the observations indicate only one chromosome or linkage group, and this linkage group appears to have no beginning or end. It can only be interpreted in the form of a circle (Fig. 17-11B). Comparison of the Hfr strains shows that the arrangement of their genes is the same; however, the order in which the genes are transferred by the several strains to the F⁻ parent differs. In addition, for each Hfr type, there is a characteristic segment of the circular chromosome which is usually not transmitted at all.

Most of the peculiarities of the system have been clarified and now make perfect sense in light of the findings. In the F⁻ cell, the circular chromosome remains as such, and no linear chromosome arises to be transferred. This is generally the case in F⁺ cells. However, these contain F, the fertility factor. This element may exist free in a cell, as it does in the F⁺ cell, or it may insert itself into the circular chromosome, and thus convert the cell to a Hfr type. The F factor has been shown to be an entity, also circular in nature in the F⁺ cell and composed of double-stranded DNA. When an F⁺ cell and an F⁻ make contact, the freely existing F⁺ DNA is replicated and transferred to the F⁻ cell which then becomes F⁺ (more details later this chapter).

A Hfr cell carries an inserted F factor, and the inserted element is generally replicated right along with the bacterial chromosome, just as if it were a native part of it. However, when a Hfr contacts an F⁻ cell, the F factor somehow becomes activated and can now trigger DNA replication and chromosome transfer. Note that the different Hfr strains simply represent cells with the sex factor integrated at different positions in the circular chromosome (Fig. 17-12).

Before transfer of DNA from Hfr to F⁻ takes place, a point of origin, or lead point, is established, and this obviously will differ from one Hfr strain to another, depending on the site of insertion of F. A nick or break occurs within the integrated F factor (in just one strand of the double helix, as we shall see shortly). One end

FIG. 17-11. Hfr types and the linkage map. *A.* Several Hfr strains were isolated (only six are shown here). When mated to F⁻, the strains were found to differ in relationship to the genes which were transferred with a high frequency and to the order in which the markers enter the F⁻ cell. *B.* The observations on the six Hfr types can be explained on the basis of a circular linkage group. When genetic material is transferred to the F⁻ cell, a linear chromosome is derived from the circle. The anterior end which is established for the linear chromosome would differ among the Hfr strains so that the first loci to enter the F⁻ would vary from one Hfr to the other. The arrangement of the loci, however, remains the same; only the order of entry is different. Those genes near the leading end will be transferred with a high frequency. Those farther back will go over to the F⁻ with a low frequency or not at all.

A

Hfr Type	Leading End
H	T L Az T₁ Pro Lac T₆ Gal λ
1	L T B₁ M Mtol Xyl Mal S
2	T₁ Az L T B₁ M Mtol Xyl Mal S
3	T₆ Lac Pro T₁ Az L T B₁ M Mtol Xyl Mal S
4	B₁ M Mtol Xyl Mal S λ Gal
5	M B₁ T L Az T₁ Pro Lac T₆ Gal λ

B

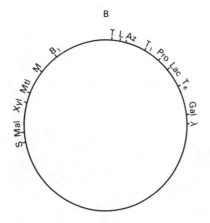

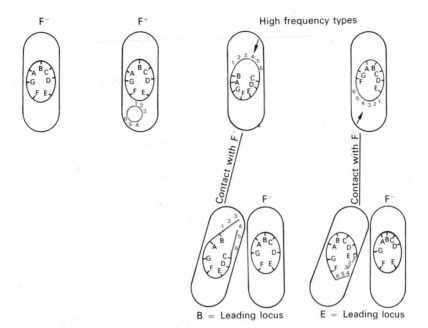

Contact with F

F⁻

Contact with F

F⁻

B = Leading locus   E = Leading locus

FIG. 17-12. Differences in mating type. The F⁻ cell contains a circular chromosome, as do the F⁺ and Hfr types. In the F⁺, however, the fertility factor (F) is present as a free entity. The Hfr type contains F in an integrated state on the circular chromosome. The various Hfr types differ in the position at which F is located (*arrows*). Contact with an F⁻ cell triggers the formation of a linear chromosome in the Hfr, and a break is produced within F. This establishes a leading point and an end point. Because F can be located at different positions on the chromosome, the order of entry of the genes will vary from one strain to the next.

of the broken F factor is then transferred to the F⁻ through a conjugation tube which forms between the cells. Since the break takes place within the F factor, this element is consequently split. As conjugation starts, some of its genes are at the "front end" and so enter the F⁻ cell first. The remaining portion of F can be transferred to the recipient cell only if all the rest of the DNA including the entire Hfr chromosome is transferred. Since breakage usually occurs somewhere before the entire amount of DNA is transferred, the F⁻ cell usually receives only a part of the F factor and so remains F⁻. Those Hfr genes in the region near the end of the linear chromosomal DNA will also have a low chance of entering the recipient. This end region will differ from one Hfr type to another and explains why a given Hfr usually fails to transfer some given chromosome segment.

When the circular chromosome map was first proposed, the actual physical shape of the *E. coli* chromosome was unknown. As discussed in Chapter 11, the refined procedures of Cairns clearly demonstrated the actual circular nature of the *E. coli* chromosome. This represents another superb correlation between genetic analysis and cytology.

Once a portion of the Hfr chromosome has penetrated the F⁻ parent, the chromosomal alleles which it contains may undergo recombination with the genetic material in the F⁻ cell. The chromosome material contributed by the Hfr parent does not always enter into the formation of recombinants when it *is* present in the F⁻ cell. It is believed that the introduced Hfr segment must pair with the homologous region in the F⁻ cell and that recombination takes place by a crossover event (Fig. 17-13). Genes are thus reciprocally exchanged between the recipient chromosome and the introduced fragment. The free fragment which remains is apparently discarded from the recombinant cell.

When a recombinant cell is found which has received an allele of a gene that is typically *not* transmitted by a particular Hfr strain, the recombinant often proves to be Hfr. This is understandable, because the entry of a gene at the extreme end implies that the entire chromosome

MICROORGANISMS: BACTERIA AS GENETIC TOOLS   **491**

may have been transferred, and this would include all portions of the sex factor. The Hfr recombinants derived in this way may also revert to the F⁺ state, indicating that the F factor may exist in a free state or attached to the chromosome.

An ordinary population of F⁺ cells contains occasional Hfr cells which have arisen independently by insertion of the F factor at different positions. So when F⁺ and F⁻ populations are mixed, it is only the occasional Hfr types in the population which are able to transfer the chromosome. But because the point of insertion of the F factor varies from one Hfr type to another, the linear chromosomes which arise in the several Hfr types at the time of conjugation will have different anterior ends and different terminal ones. Consequently, different segments will be transferred from one mating pair to another. The F⁺ population is thus a mixed one containing an assortment of Hfr types, in contrast to an isolated population pure for one Hfr strain. So when F⁺ and F⁻ are mixed, it seems as if *all* genes are being transferrd at a low frequency. The F particle itself goes over at a high frequency because it exists free in most of the cells in the F⁺ population.

THE F FACTOR AS AN EPISOME. Jacob and Wollman coined the term *episome* to designate any factor with a behavior such as the F factor. An episome is a hereditary entity which may exist free in a cell or attached to the chromosome. Such a particle may have several remarkable attributes, as seen by the fertility factor of *E. coli*. In this bacterium, the F factor makes a cell a potential donor. Its presence in a cell has also

FIG. 17-13. Formation of recombinants. Once a segment of the Hfr (shown here as *BAG*) enters the F⁻ parent, it may or may not enter into the formation of recombinants. If recombinants are to arise, the introduced fragment must pair with the homologous region on the F⁻ chromosome. A reciprocal crossover event is believed to take place. The free fragment (*B'A'G'*) which arises is eventually lost or discarded.

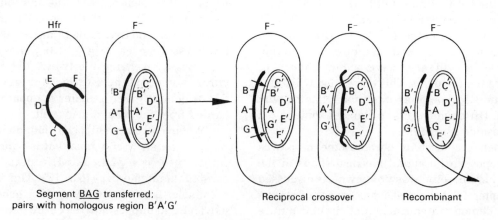

Segment <u>BAG</u> transferred;
pairs with homologous region B'A'G'

Reciprocal crossover

Recombinant

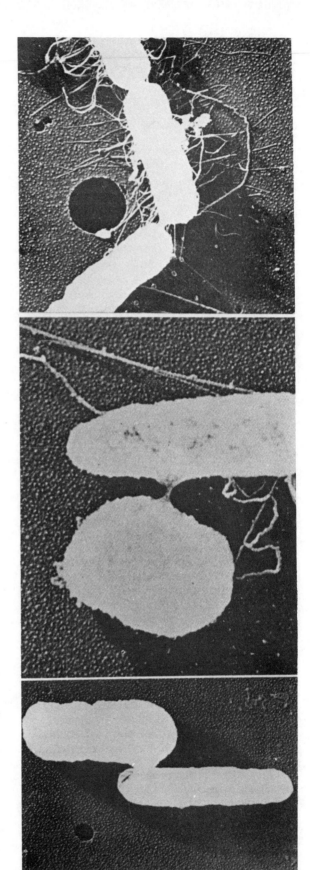

been shown to cause the production of a particular type of appendage. These have been termed *sex pili* or *sex fimbriae* and have been shown to be necessary for the cells to conjugate. Electron micrographs of mating cells reveal cytoplasmic connections or bridges between them. Although it still has not been definitely established, each of these bridges may represent one of the sex fimbriae which may be required to establish a connection between mating cells which is essential for the transfer of the chromosome (Fig. 17-14).

Different techniques have clearly demonstrated that the F factor is an entity composed of deoxyribonucleic acid (DNA) (more detail in next section). Isolation of sex factors transferred to F^- cells indicates that they are approximately 2% of the size of the bacterial chromosome. Various procedures indicate that only one sex factor exists within any F^+ cell for each nuclear region, so that the relationship is one F factor for each circular chromosome. Almost one-half of the DNA composing the F factor has been shown to be able to hybridize with chromosomal DNA. Therefore, a significant portion of the episome apparently contains regions homologous to portions of the chromosome. As noted, there is evidence that the fertility factor, like the chromosome, is a circular DNA element. The F factor can apparently pair with certain chromosome segments. This pairing between the F factor and the chromosome evidently results from regions of homology which exist between short stretches of DNA in the episome and corresponding regions in the chromosome. After pairing, the episome may then insert itself into the chromosome through the process of crossing over. By a reversal of this, it can again become free, converting the cell back to an F^+ (Fig. 17-15*A*). However at rare times, a portion of the chromosome may become associated with the F factor and carried away by it (Fig. 17-15*B*). As a result, an F factor can originate which contains a marker allele, such as Gal$^+$. Harboring pieces of genetic information in this way, F factors can enter F$^-$ cells. This means that genes may be transferred from one cell to another by an episome, which can thus confer a new genetic property on a recipient cell. Suppose (Fig. 17-16) that an F$^-$ cell is Gal$^-$ and thus cannot ferment galactose. This cell may contact an F$^+$ cell which contains an F factor carrying a Gal$^+$ allele. The F$^-$, Gal$^-$ cell in turn can be converted to an F$^+$, Gal$^+$ type by receiving the fertility factor, which not only makes it a potential donor cell but which also gives it the capacity to ferment galactose. The transfer of genetic material in this way from one cell to another by way of a sex factor has been called *sexduction*. The term *F-genote* has been coined to designate a sex factor which has incorporated a portion of the chromosomal DNA. Those bacterial strains which carry F-genotes are known as F-prime (F') strains. A bacterial cell therefore can acquire the ability to perform a function as a result of an allele contained in an episome which it harbors and not as a result of an allele on its chromosome. A cell such as this is haploid except for a small chromosome portion. In this short region, it can possess a pair of alleles and would therefore be partially diploid

FIG. 17-15. Insertion and removal of F factor. *A.* The F factor possesses certain regions which are homologous to segments of the chromosome. This enables the fertility factor to pair with the chromosome at different points. A process akin to crossing over is believed responsible for the insertion of F into the chromosome of an F$^+$ cell, thereby converting it to Hfr. *B.* At times, the F factor may remove itself from its integrated state and become free, converting the cell to an F$^+$. Normally, this would occur by a reversal of the process shown in (*A*). Homologous regions would again pair and a crossover would take place. However, at rare times, the F factor may carry away a portion of the chromosome. This is believed to come about when small homologous portions of the chromosome itself on either side of the inserted F factor pair and then a crossover event takes place.

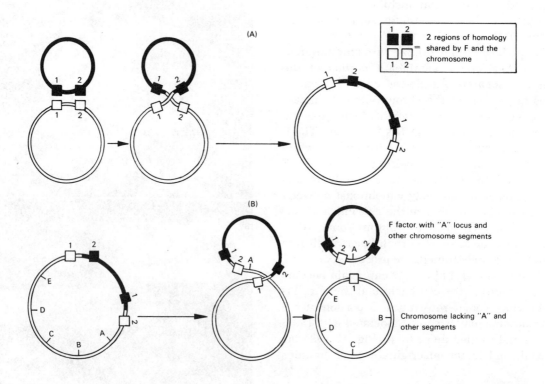

as well as a partial heterozygote. The expression *heterogenote* describes a cell heterozygous for just a segment of the genetic material, as opposed to "heterozygote," which is reserved for diploid cells.

EVENTS IN CHROMOSOME TRANSFER FROM Hfr TO F⁻. When a Hfr cell transfers chromosomal material to an F⁻, the Hfr parent remains viable because the Hfr parent does not actually lose genetic material. The chromosome transfer has been shown to be related to the replication of the Hfr chromosome as has been demonstrated in various ways. If any genetic material is to be donated by the Hfr cell, a cycle of DNA replication must be started. Inhibitors of DNA synthesis may be applied to cells at the time of conjugation. If this is done and if the initiation of DNA synthesis is prevented, no gentic markers will be transferred from Hfr to F⁻. Moreover, as we noted, only one strand of the double-stranded DNA of the donor cells moves into the F⁻ as the DNA synthesis takes place (Fig. 17-17A). This has been illustrated by the use of labeled DNA precursors. For example, Hfr cells may be grown on a medium containing only heavy carbon and nitrogen (C^{14}, N^{15}) so that their DNA is made heavy. At the time of mating, these cells may be transferred to "light" medium along with the F⁻ recipients. The DNA which enters the F⁻ can be isolated. After such procedures, it has been found that the donated DNA is composed of one heavy and one light strand. This indicates that only one strand of the double helix entered the F⁻ cell from the Hfr parent. Once inside the F⁻ cell, this single donor strand is used as a template for the synthesis of a complementary strand. A double helix representing a portion of the Hfr genetic material is thus formed in the F⁻ cell. It is this double-helical DNA which may then engage in the crossing over and recombination with the F⁻ chromosome.

Recall that a nick occurs within the inserted F factor itself at the start of chromosome transfer. The nick affects just one of the two strands of the DNA of the F element as indicated by the discussion above. Once the strand is split, there is a free 5′ and a free 3′ end. The lead point of the F factor DNA which enters the F⁻ cells is the 5′ end. As the single strand enters the recipient, it is copied in the F⁻ cell in a 5′ to 3′

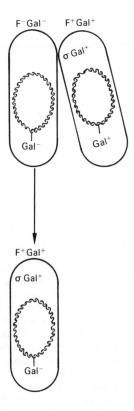

FIG. 17-16. Sexduction. This diagram shows an F⁻ cell which cannot ferment galactose and an F⁺ cell which carries the allele for galactose utilization on its F factor and also on its chromosome. After the two cells make contact, the fertility factor may be transferred to the F⁻, which is therefore converted to an F⁺ with the ability to utilize galactose. It carries the allele for this ability on its F factor, whereas its chromosome carries the allele which does not enable the cell to use the sugar.

direction. At the same time in the Hfr donor, a complementary copy is made of the DNA strand which remains and which acts as a template. Therefore, the double-helical nature of the donor DNA is maintained, because a copy of the transferred strand is made during conjugation. We see that new DNA synthesis takes place in both the Hfr and the F⁻ parents and that the Hfr cell does not lose any of its genetic material during conjugation.

The transfer of the sex factor (F) from F⁺ to F⁻ cells entails the same steps described here for chromosome transfer (Fig. 17-17B). Conjugation triggers production of a nick in one of the strands of F as well as a new cycle of DNA synthesis in the sex factor as it exists in the F⁺ cell, separate from the chromosome. The nicked strand with the 5′ end leading moves into the F⁻

cell where synthesis of the complementary strand produces a double helix. The F⁻ cell has now been converted to F⁺. The original F⁺ parent remains F⁺ because the strand which remained is used as a template to preserve the double helix of the F particle.

TRANSFORMATION IN RELATIONSHIP TO RECOMBINATION. The discovery of the many remarkable properties of the F factor focused attention on the possible occurrence of episomes or similar particles in other kinds of cells. The episome concept has inspired investigations which have led to the recognition of still other kinds of genetic elements which may exist in the cell independently of the chromosome. The genetics of microorganisms has also raised provocative questions concerning the nature of viruses and their role in the transfer of genetic information from one cell to another. The following chapter will deal with aspects of these points. However, before leaving this discussion of bacteria, let us reexamine one method of genetic change in bacteria which may have implications for higher forms of life. In Chapter 11, the phenomenon of transformation in *Pneumococ-*

FIG. 17-17. Transfer of a single DNA strand. *A.* The transfer of genetic material from Hfr to F⁻ is associated with DNA replication. Hfr cells can be grown on a heavy medium, so that all the donor DNA is heavy (*1*). These heavy donors may be mated to F⁻ cells with light DNA. The mating is done on a light medium (*2*). The segment which is donated can be isolated and shown to consist of one heavy and one light strand (*3*). This indicates that only one strand of DNA entered the F⁻ from the Hfr parent and that a complement to it was made in the F⁻ parent. This strand would be light because the cells were mated on light medium. The Hfr cell remains viable, because the strand which passes into the F⁻ is also replicated at the same time in the Hfr cell itself (*2* and *3*). *B.* An F⁻ cell is converted to an F⁺ by contact with an F⁺. The F factor is believed to be circular. In the process of conversion to the F⁺ state, a single strand of F factor DNA moves into the F⁻. A complement of this is built in the F⁻ cell which now becomes F⁺ with an F factor composed of two strands. The strand which remained in the original F⁺ donor also builds a complementary one, and the donor remains a donor.

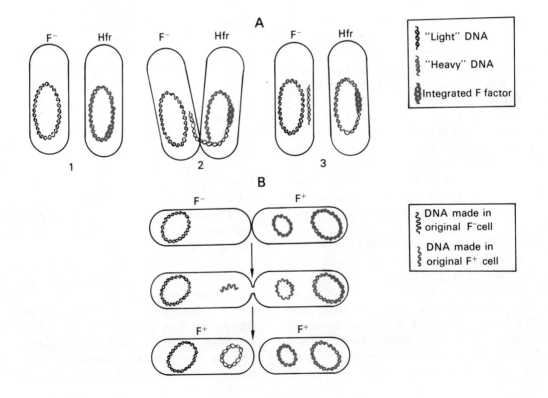

cus was discussed in relationship to the identification of the chemical nature of the gene. It will be recalled that in transformation, DNA preparations from a donor strain may be applied to a recipient. After a period of incubation, the latter may incorporate some of the donor DNA and acquire a new trait. This transformation when completed, is stable and represents a true genetic alteration. At first, transformation was considered more or less a curiosity in the pneumonia organism, but it now seems to be rather widespread. Transformation has been recognized in at least 17 bacterial species, among them such well known groups as *Streptococcus, Bacillus,* and *Hemophilus.* Many characteristics of these microorganisms are known to be susceptible to transformation: drug resistance and ability to synthesize specific enzymes, etc. It seems that any genetic locus may be transferred from donor to recipient in transformation. It must be emphasized that the DNA donated to the recipient does not require cell contact; pure DNA preparations extracted from donor cells are effective. Nor does the transfer depend on any vector, such as an episome or virus. Bacterial cells appear to have an affinity for uptake of DNA from the environment during a certain phase of the growth cycle of the bacterial population, near the end of the log phase. This receptive stage is of brief duration in the recipient cells and entails physiological changes which are not completely understood. A cell in the receptive stage which can take up extraneous DNA and be transformed by it is said to be *competent. Competence* may also refer to the portion of the growth cycle of the bacterial population when most of the cells are physiologically capable of transformation.

Throughout the past decade, several very ingenious experiments have revealed many facts about transformation which are of general biological significance. In the early 1960's, studies were conducted in which labeled donor DNA was followed from the moment of contact with recipient cells until the latter were transformed. It was found that immediately after entering the recipient cells, some of the donor DNA is converted to a single stranded state, while the rest is degraded. This labeled, single-stranded donor DNA is then inserted into the recipient DNA (Fig. 17-18). Once inserted, it has replaced the

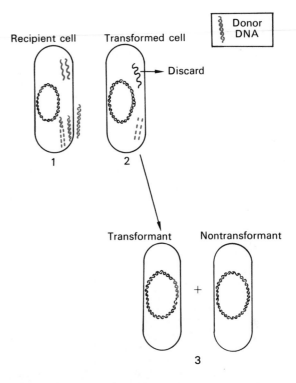

FIG. 17-18. Steps in transformation. In order to bring about transformation, the donor DNA must be double stranded; otherwise it will not penetrate the recipient cell. Once inside the recipient, some of the donor DNA is degraded and some transformed to the single-stranded condition (*1*). This single-stranded form may associate with the region on the chromosome with which it is homologous. It may then replace a single-stranded portion of the recipient DNA (*2*). This recipient segment is eventually lost. The cell, now carrying the hybrid DNA, is transformed. After it divides (*3*), it will give rise to *nontransformed* and to transformed cells. In the latter, the region involved in the transformation has both strands identical to the original donor DNA. This results from the fact that the inserted donor strand will form a complementary copy of itself at the next replication.

homologous segment of the recipient DNA, and this is then discarded. It is this newly inserted DNA which is responsible for the transformation. It is able to effect a genetic change and become a part of the genetic material of the recipient cell. This cell in turn may be used as a donor in a subsequent transformation. Thus, the labeled DNA introduced in the first transformation can be followed further to see how it behaves in later transformations or in later cell divisions. Extremely valuable information has come from following DNA throughout all phases of transformation in a single cell and throughout two or more divisions of transformed cells. One of the most important discoveries, with great implications for genetic thought, was the demonstration that the introduced donor DNA is physically integrated into the recipient DNA. This observation supports the concept of crossing over as a result of synapsis followed by breakage and reunion, much as Janssens pictured it in 1909, a matter which had been controversial for decades. The studies with the labeled donor DNA clearly showed that the introduced transforming DNA penetrates the cells and becomes associated with the homologous DNA segment of the recipient. It is then incorporated into the recipient chromosome by an actual physical insertion which is accompanied by the genetic transformation of the cell.

As noted, only one strand of the donor DNA participates, and the evidence indicates that *either* strand of the donor double helix can bring about the transformation. It is interesting that the DNA which is picked up by the recipient cell in the first place *must* be double stranded. It must also be above a certain minimal molecular weight (5×10^5). Single-stranded DNA is completely ineffective in transformation. But once inside the recipient cell, the double stranded donor DNA is altered, and it is only a single strand which engages in the transformation process. This means that when this single donor strand replaces one of the two strands of the recipient, a hybrid DNA duplex results. One strand of the double helix in the involved region would be a strand of the recipient, the other the strand of the donor. And this has been found to be true. Therefore, just one strand, either one of the two of the original double helix, can cause a transformation. At the cell divisions which follow the transformation, the DNA of the transformed cells will come to have both strands of the *original* donor DNA in the region of the transformation (Fig. 17-18).

TRANSFORMATION IN HIGHER LIFE FORMS. Although most studies of transformation employ DNA preparations which are extracted from donor cells by an investigator, it is now known that transformation may take place spontaneously. Aging cultures of some bacterial strains contain cells which are undergoing self-digestion and which release DNA into the medium. This

DNA can cause transformations of other cells in the population. *Pneumococcus* transformations may occur spontaneously when two different bacterial strains are injected into mice. Such observations raise the possibility of transformation in higher species. Actually, this has been achieved, as exemplified by the fruit fly. Eggs of *Drosphila* containing very young embryos have been treated with DNA from other stocks carrying certain marker alleles. The treated embryos have given rise to stocks of flies which are true genetic transformants, whose cells have incorporated some of the genetic material applied to them. However, one significant difference is known to exist between transformation in flies and in bacteria. In bacteria, the introduced DNA replaces the homologous DNA of the recipient (Fig. 17-18). In the flies, the donor DNA is incorporated into the cell and associated with the homologous locus. However, it does not replace the resident DNA of the recipient chromosome but manages somehow to persist along with it in the cell. As a result, the transformed flies are always mosaics. Their bodies are not wholly transformed for any introduced allele; only segments express the introduced genetic information. Within a transformed stock of flies, different degrees of mosaicism exist. Apparently, the donor DNA does not always undergo transcription. Whether the information contained in the DNA of the donor or in that of the host will be transcribed seems to vary from cell to cell so that mosaicism results. The mosaic flies pass down to their offspring the information which was introduced from the donor stock, but these offspring are in turn mosaics. The appearance of the mosaicism actually may be delayed one or two generations. Nevertheless, the fact that most of the offspring continue to pass down the donor information, as seen by the mosaicism, even though the degree of mosaicism varies, tells us that most or all of the cells of a transformed fly contain the extra information. The expression of this additional genetic material, however, is sporadic.

Certain reports hold implications for still higher forms of life. For example, fibroblast cells of the Chinese hamster and certain cells of the mouse were raised separately *in vitro*. The mouse cells were from a line which, due to a gene mutation, lacks the ability to manufacture a specific enzyme, a transferase. The hamster cells carried the information for this protein. In the procedure, metaphase chromosomes of the hamster were actually isolated from the cells; these isolated chromosomes were then incubated with mouse cells (Fig. 17-19). Previous studies with labeled DNA had shown that cells in tissue culture will take up and ingest isolated metaphase chromosomes and that some of their DNA becomes distributed in the chromosomes of the cell. In the hamster-mouse experiment, a definite gene marker was being followed, the allele for the production of the transferase protein. The mouse cells incubated with the hamster chromosomes were exposed to a medium which would allow only those cells to survive which possessed the ability to make the transferase. Colonies did develop on the selective medium.

FIG. 17-19. *In vitro* transformation of mammalian cells. Mouse cells lacking the ability to manufacture a specific transferase can be incubated with the isolated chromosomes from a hamster strain which does have the ability to manufacture it. Cells can then be screened for their ability to produce the transferase by plating on a medium which allows only those cells to grow which can make it. Colonies of mouse cells were isolated on the medium, and these had acquired the ability to make the enzyme. They were not spontaneous mutants, because the enzyme was found to be hamster type.

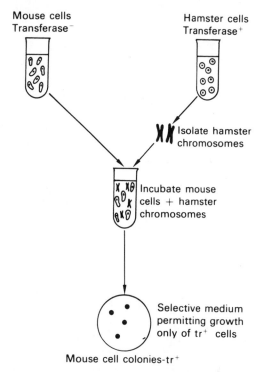

The cells of the colonies were mouse cells, of course, but they had acquired the lost ability to make the enzyme transferase. The cells which gave rise to the colonies were very few in number, and one might argue that they represented revertants, mutant cells with back mutations to wild. However, the enzyme was isolated from the cells and was shown to be hamster transferase and *not* the mouse enzyme which can be distinguished from it. Clearly, a genetic transformation had occurred. For the first time in mammals it was thus demonstrated that a known genetic marker could be acquired by transformation and that it could also be reproduced and handed down to progeny cells. Although this work deals with *in vitro* techniques, the questions and implications it raises about the occurrence and use of transformation as a mechanism of genetic exchange are staggering. The introduction of microorganisms as genetic tools continues to offer insights into biological phenomena which hold great significance for all living things.

SOME DIFFERENCES BETWEEN PROKARYOTES AND EUKARYOTES. Knowledge of molecular dynamics in microorganisms has built a foundation for our understanding of transcription, translation, and replication and has given us an insight into the nature of the gene, its expression, and its regulation. The information from microorganisms has provided a basis for investigations into related phenomena in higher cell types. Since most of the work has involved prokaryotes, bacteria especially, we must not reach premature conclusions about eukaryotes and assume that the identical situation pertains. It is appropriate here to point out some differences on the molecular level which have emerged between prokaryotes and eukaryotes as research progresses on the molecular genetics of higher cells. Certain differences have already been encountered, and others will be noted in remaining discussions.

Since prokaryotic cells lack a membrane-bounded nucleus and hence a nucleolus, processing of the ribosomes is different, and the S values of the ribosomes are distinct in the two groups (Chap. 13). The enzyme RNA polymerase has been described for prokaryotes (Chap. 12).

In eukaryotes, however, not just one kind but three distinct types of RNA polymerase have been recognized. One of these occurs in the nucleolus, the rRNA polymerase which is concerned with the synthesis of ribosomal RNA. The other two kinds are found in the nuclear sap. One of these is reserved for the synthesis of transfer RNA, the role of the other being the synthesis of messenger RNA. Each of the three RNA polymerases is composed of several polypeptide chains. The precise structure of the three types and the manner in which they operate on a DNA template still await clarification.

The mRNA of eukaryotes possesses several features not found in that of prokaryotes, and the reason for the differences is not known. For example, the mRNAs from eukaryotes, after being formed by transcription and then released into the rest of the nucleus, are acted upon by an enzyme which can recognize mRNA. The enzyme adds to the mRNA transcript very long stretches of nucleotides, approximately 200, each containing the base adenine. The mRNA then becomes linked to protein units, at least two types being involved. Complexed with protein, it moves out of the nucleus into the cytoplasm.

Not too much is known about the lifetime of mRNA in a eukaryotic cell, but the *relative* lifetimes in prokaryotes and eukaryotes seem to be comparable. The average is about 3 hr for eukaryotic cells which are rapidly growing, but it is much longer in highly differentiated cells such as the red blood cell which produces no new RNA during hemoglobin formation.

In eukaryotes, ribosomes are apparently broken down and have a lifetime of only about 100 hr, so that a turnover of rRNA is occurring. In prokaryotes, no ribosomal breakdown appears to take place in normal cells; it only occurs under unfavorable conditions.

A surprising fact relating to translation is that while the eukaryote possesses two types of tRNA for methionine, and while one of these acts as initiating tRNA as does tRNA$^{f met}$ in prokaryotes, ordinary methionine is always the first amino acid placed in the polypeptide chain. No formylation occurs!

Among the most interesting distinctions between prokaryotes and eukaryotes is in relation to their DNA. There is, of course, much more DNA in a eukaryotic cell than in a prokaryotic one. In Chapter 11, we noted that the percentage of G:C in lower forms, especially in prokaryotes, shows great variability from one species to the next, whereas in eukaryotes, the value is much more stabilized, being about 40% in mammals.

Another distinction between prokaryotic and eukaryotic DNA is seen when DNA is isolated from cells, fragmented into segments of comparable size, and then subjected to density gradient ultracentrifugation. Handled in this way, the prokaryotic DNA yields one band following centrifugation, indicating that the separate fragments of comparable size are approximately the same in density and hence in their base composition. When eukaryotic DNA is processed in this way, however, centrifugation typically reveals a main band (including the bulk of the DNA) and one or more satellite bands (Chap. 13). This means that the G:C and A:T pairs are not uniformly distributed on the fragments but may be clustered in certain regions yielding DNA fragments of quite different densities. Some of the satellite DNA represents chromosomal DNA, as seen in the case of nucleolar DNA in *Xenopus*. However, some satellite bands represent the DNA which is found outside the nucleus in organelles such as the mitochondria and chloroplasts (Chap. 20).

A most unusual feature of eukaryotic DNA is its inclusion of base sequences of varying lengths which are repeated throughout the bulk of the DNA. The difference between prokaryotes and eukaryotes in this respect is shown as follows. DNA extracted from a prokaryotic cell may be fragmented into pieces all approximately the same size, perhaps 500 nucleotide base pairs long. The DNA is then subjected to a high temperature so that the DNA strands separate (denature), and single-stranded DNA fragments arise. The material is then cooled, permitting the single strands to rejoin and form doublehelical DNA segments (renaturation). If only one copy of a stretch of a nucleotide sequence is present, then renaturation can occur only when a single-stranded DNA fragment encounters its original partner or the same complementary stretch from another cell. This means that renaturation will tend to occur slowly. In *E. coli*, for example, renaturation progresses at a constant rate for the first 10 minutes and then slows down even more. This is so because as strands slowly find their complements, fewer and fewer single strands are present, and the chance that a complementary single strand encounter will occur is rarer (Fig. 17-20*A*).

When this same procedure is followed with DNA from eukaryotic cells, a different pictures emerges. At first a very fast renaturation takes place. This tends to level off and then renaturation starts to take place more rapidly again, giving a curve such as that in Figure 17-20*B*. It is clear that eukaryotic DNA is heterogeneous in the sense that it contains some DNA stretches which are repeated over and over again in a cell. Three major classes of eukaryotic chromosomal DNA are recognized. Most of the DNA is non-repetitive, meaning that when the chromosomes are fragmented, a given stretch of nucleotides in the DNA will tend not to be repeated elsewhere in the cell's DNA. When the DNA is isolated from many cells, such a fragment can unite only with its original complementary partner or with that same complementary stretch from another cell. This DNA would renature slowly, since most chance encounters would not bring together complementary fragments.

A highly repetitive class of eukaryotic DNA is recognized as making up as much as 10% of the DNA in some species. This DNA has base sequences that may be repeated as many as a million times! The repeated sequences may be short, such as a 6 nucleotide sequence repeating over and over or as long as 300 nucleotides whose sequence occurs again and again. This DNA will renature quickly, since chance collisions are more apt to bring together complementary sequences.

A moderately repetitive DNA fraction is also recognized in which a given fragment (say 500 base pairs long) has a sequence quite similar to that present in other DNA fragments from the same nucleus. In the moderately repetitive fragments, the nucleotide sequences are not identical but have stretches that are sufficiently alike to permit a single strand to pair not just with its exact complement but also with other strands which bear similar, though not completely iden-

FIG. 17-20. DNA renaturation curves. *A.* Generalized renaturation curve expected for DNA in which stretches of nucleotide sequences occur only in single copies. Curves for prokaryotes, such as *E. coli*, approach this idealized one. *B.* DNA renaturation curves of *E. coli* and a typical eukaryote compared.

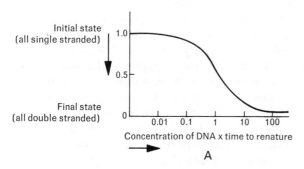

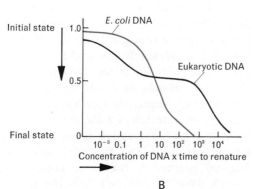

tical, sequences. This DNA renatures at a rate in between that of the other two classes.

The DNA of eukaryotes is thus shown to contain significant stretches of repetitive DNA in contrast to prokaryotes which do not tend to contain any which are even moderately repetitive. What some of this repeated DNA may represent will be referred to in later discussions (Chap. 21).

REFERENCES

Adelberg, E. A., and S. N. Burns. Genetic variation in the sex factor of *Escherichia coli*. *J. Bact., 79:* 321, 1960.

Britten, R. J., and D. E. Kohne. Repeated segments of DNA. *Sci. Am.* (April): 24, 1970.

Campbell, A. *Episomes*. Harper and Row, New York, 1969.

Fox, A. S., S. B. Yoon, and W. M. Gelbart. DNA-induced transformation in *Drosophila*: genetic analysis of transformed stocks. *Proc. Natl. Acad. Sci., 68:* 342, 1971.

Gurney, T., Jr., and M. S. Fox. Physical and genetic hybrids in bacterial transformation. *J. Mol. Biol., 32:* 83, 1968.

Hayes, W. *The Genetics of Bacteria and Their Viruses*, 2nd ed. Wiley, New York, 1968.

Lacks, S. Molecular fate of DNA in genetic transformation of *Pneumococcus*. *J. Mol. Biol., 5:* 119, 1962.

Lederberg, J., and E. L. Tatum. Gene recombination in *Escherichia coli*. *Nature, 158:* 558, 1946.

Luria, S. E., and M. Delbruck. Mutations of bacteria from virus sensitivity to virus resistance. *Genetics, 28:* 491, 1943.

McBride, O. W., and H. L. Ozer. Transfer of genetic information by purified metaphase chromosomes. *Proc. Natl. Acad. Sci., 70:* 1973b.

Taylor, A. L., and C. D. Trotter. Linkage map of *E. coli* K12. *Bact. Rev., 36:* 504, 1972.

Vapnek, D., and W. D. Rupp. Identification of individual sex factor DNA strands and their replication during conjugation in thermosensitive DNA mutants of *Escherichia coli*. *J. Mol. Biol., 60:* 413, 1971.

Watson, J. D. *Molecular Biology of the Gene*. Benjamin, Menlo Park, Calif., 1976.

Wollman, E. L., F. Jacob, and W. Hayes. Conjugation and genetic recombination in *Escherichia coli* K-12. *Cold Spring Harbor Symp. Quant. Biol., 21:* 141, 1962.

REVIEW QUESTIONS

For each of the following four multiple-choice questions, select the correct answer or answers, if there are any:

1. The replica plating technique:

   A. Can be used routinely to transfer bacterial colonies from a plate with minimal medium to others containing different kinds of media.

   B. Has demonstrated that spontaneous mutations occur at random without the presence of a selective agent.

   C. Has shown that the production of a new mutation to drug resistance depends upon prior exposure to that drug.

   D. Can be used to identify colonies derived from single cells which differ genetically from the majority of cells in the culture.

   E. Has provided information on mutation in microorganisms which supports that obtained from the fluctuation test.

2. In the *E. coli* sexual system:

   A. The unattached F factors are not usually transmitted to the F⁻ cells.

   B. Cells are considered to be "female" or "recipient" when they lack a sex factor.

   C. Recombinant cells come from divisions of the F⁺ cells after mating.

   D. The F factor may carry within it a gene normally found on the chromosome.

   E. The F factor is composed of DNA.

3. In the *E. coli* sexual system:

   A. Only the Hfr cells can actively transmit the chromosome to the F⁻ cells.

   B. When Hfr and F⁻ cells are mated, most of the F⁻ becomes Hfr or F⁺

C. The point of origin in the various Hfr types is the same.
D. One parent does not usually transmit to the other parent an entire complement of genetic material.
E. The various markers of the donor all enter the recipient cell at the same time.

4. *E. coli* cells:

A. Are typically diploid.
B. May be diploid for only a few genetic regions.
C. Contain several linkage groups.
D. Contain DNA complexed with protein.
E. Do not contain ribosomes.

5. In one part of an experiment with a strain of penicillin-sensitive cells, a sample of bacteria is taken and allowed no further growth. The sample is divided into several equal parts and then each is plated immediately on a medium containing penicillin. In another part of the experiment, a small sample of the sensitive culture is taken and divided further into several subdivisions. Each subdivision is then allowed to grow until each contains the same number of cells as does each of the divisions in the other part of the experiment. From the data below, tell which part of the experiment is represented by the data in column A and in column B. Explain.

	A	B
Total number of cells	10^9	10^9
Mean number resistant cells	22	15
Variance	4480	17

6. Let A^+, B^+, C^+, D^+, E^+ represent markers enabling an *E. coli* cell to manufacture amino acids required for growth on a minimal medium. *E. coli* strain No. 1 is $A^+B^+C^-D^+E^-$. Strain No. 2 is $A^-B^-C^+D^-E^+$. The two strains are mixed and incubated together. After plating on minimal medium, a small percentage of prototrophs is selected.

A. What would be the genotype of the prototrophs in respect to these markers?
B. Two other *E. coli* strains, strains No. 3

and No. 4, can manufacture all the amino acids required for growth on minimal medium. Strain No. 3 is Lac$^+$ and is consequently able to utilize lactose, a sugar which can be used as an accessory energy source. Strain No. 4 is Lac$^-$ and cannot utilize lactose when this sugar is provided. Which of these strains is auxotrophic? Explain.

7. Two reciprocal crosses are made:

       Strain No. 1   Strain No. 2

(1) A$^+$B$^+$C$^-$D$^+$E$^-$ × A$^-$B$^-$C$^+$D$^-$E$^+$
   streptomycin-   streptomycin-
   resistant   sensitive

(2) A$^+$B$^+$C$^-$D$^+$E$^-$ × A$^-$B$^-$C$^+$D$^-$E$^+$
   streptomycin-   streptomycin-
   sensitive   resistant

After incubation, plating is made on minimal medium containing streptomycin. From cross (1) colonies develop, but none at all arise from cross (2). Explain.

8. Let A$^+$ and B$^+$ represent markers for essential growth substances needed for survival on minimal medium. The other markers, L$^+$, M$^+$, etc., are for traits which do not affect the minimal nutritional requirements. From the cross and the information given below, arrange the marker alleles in the proper order of entry.

Hfr A$^+$B$^+$L$^+$M$^+$N$^+$O$^+$Q$^+$ × F$^-$ A$^-$B$^-$L$^-$M$^-$N$^-$O$^-$Q$^-$
strep.-sensitive   strep.-resistant

Colonies on minimal medium with strep.:

    A$^+$—100%; B$^+$—100%; L$^+$—25%; M$^+$—88%;
    N$^+$—none; O$^+$—73%; Q$^+$—35%

9. Markers are transferred from five Hfr strains to F$^-$ cells in the following order:

Hfr strain	Order of entry
1	B K A R M
2	D L Q E O C
3	O E Q L D N
4	M C O E Q L D N
5	R A K B N

A. Draw a map showing the sequence of these markers on the chromosome.

B. For each strain, indicate on the map the site of insertion of the fertility factor by placing an arrowhead so that the first gene to be transferred by a strain is behind the arrowhead.

10. Assume that recombinants are desired which are Hfr and that crosses are made with the Hfr strains in Question 9. For each strain, indicate which marker should be selected in order to derive the highest number of recombinants which will also be Hfr. Explain.

11. Assume a DNA of an Hfr E. coli has the following arrangement of nucleotides:

    5′   A G C T A T   3′
    3′   T C G A T A   5′

Assume the Hfr cells have been growing on medium supplied with C^{14} and N^{15} so that all its DNA is heavy. Mating of the Hfr with F$^-$ cells is performed on ordinary medium. In the mating, the lower DNA strand is transferred from its 5′ end.

A. Show this segment of the double helix in the donor cell after DNA replication, indicating light (L) and heavy (H) strands.

B. Do the same for the recipient, assuming this whole strand is incorporated.

12. A certain Hfr strain normally transmits the lac$^+$ marker as the last one during conjugation. In a cross of this strain with an F$^-$lac$^-$, some recombinant lac$^+$ cells received the lac marker too early in the mating process. When these lac$^+$ cells are mixed with F$^-$lac$^-$ cells, the majority of the latter are converted to F$^+$lac$^+$, but other markers are not transferred. Explain. What can you say about the genotypes of the cells from the F$^-$ strain which became lac$^+$?

13. Suppose Strain A of *Pneumococcus* is streptomycin resistant and Strain B is streptomycin sensitive. In a transformation experiment, competent B cells are incubated with DNA from

Strain A. DNA of Strain A carrying the resistant marker enters some B cells and associates with the B DNA. The latter cells become transformed. Answer the following:

A. Will the DNA of the transformed recipients have A or B DNA for the marker region?

B. When any transformed cell divides, what will be the nature of its two cell products in regard to streptomycin resistance, and what will the DNA of these cells be like regarding the A and the B DNA?

14. DNA analyses of three different mammalian species shows the following base percentages:

	A	T	G	C
Species A—	29.6	29.6	20.4	20.4
Species B—	29.5	29.5	20.5	20.5
Species C—	29.8	29.8	20.2	20.2

Denaturation-renaturation experiments are performed with two species at a time. In such an experiment, the separate DNA chains of one species are left intact; in the other, they are fragmented. It is found that almost 100% of the DNA from species A and C hybridize. Only about 5% of the DNA of species B hybridizes with A or C. Offer an explanation.

15. When DNA-RNA hybridization experiments are performed with eukaryotes, would there be any difference in the rate of hybridization between those RNA transcripts formed from single-copy DNA sequences and those RNA transcripts formed from repetitive DNA sequences? Explain.

16. Next to each of the following statements, place a "P" if it is more applicable to prokaryotes, an "E" if more applicable to eurkaryotes, and "P + E" if about equally applicable to both.

(1) ____ Little variation G:C value
(2) ____ 5' to 3' growth of DNA chains
(3) ____ 3 types of RNA polymerase
(4) ____ n-formyl methionine

(5) ___ mRNAs terminating in stretches of A-containing nucleotides

(6) ___ two initiating types of tRNA

(7) ___ satellite DNA

(8) ___ very fast renaturing DNA fraction

(9) ___ partial genetic transfer in sexual reproduction

(10) ___ spontaneous mutation independent of a selective agent

(11) ___ colinearity

(12) ___ ribosomal processing in nucleolus

(13) ___ degenerate code

(14) ___ initiation of DNA synthesis required for transfer of genetic markers

VIRUSES AS GENETIC TOOLS

Bacteria and the viruses which infect them (bacteriophages) have proved to be extremely valuable tools for the molecular biologist. Because of their rapid reproductive rate, microorganisms enable the geneticist to examine in a very short period of time many generations consisting of extremely large numbers of individuals. This has made possible a refinement of genetic analysis which would be impossible with higher forms. In addition, studies of viruses, phages in particular, have disclosed mechanisms of genetic recombination and control which were unsuspected by the classical geneticist. Phages and their interactions with the bacterial cell have also focused attention on cellular particles which may be of paramount importance to an understanding of differentiation and of pathology in higher organisms.

EVENTS FOLLOWING INFECTION BY A VIRULENT PHAGE. It was noted in several chapters that phages may destroy the bacterial hosts which they infect. These are *virulent* phages which must encounter a supply of host cells in order to continue their reproductive cycle. We have seen that it is the nucleic acid of the phage which enters the host cell; the protein envelope is left behind (refer to Fig. 11-4). Once inside the cell, the phage particle takes control of the cellular machinery. One of the first "new" substances to appear after infection is a mRNA which is complementary to the deoxyribonucleic acid (DNA) of the virus, *not* to that of the host. This is followed by the appearance of proteins which represent certain enzymes needed for the activities of the virus. Meanwhile, the normal activities of the host cell cease. The bacterial DNA becomes degraded and enters a pool in the cell. Substances from the cell medium are added to this pool, from which virus-specific substances will be made. In short, after entering the host cell, the genetic material of the virus subverts that of the bacterium (refer to Fig. 12-6). By the ability of its nucleic acid to undergo transcription in the host cell, virus-specific mRNA is formed. When this is translated by the ribosomes of the host, enzymes are constructed which are needed to build new viruses. Therefore, products now form in the host cell which are unique and foreign to the cell. For example, the nucleic acid of some viruses contains a variant form of cytosine (5-hydroxy methyl cytosine) instead of the typical cytosine. Not only is this unusual substance made in the host cell, but the enzymes which are required for its synthesis must be made first. All of this is accomplished by the ability of the virus DNA to be transcribed and its mRNA translated in the host cell.

In the first few minutes after entry, no mature virus particles can be detected in the bacterial cell because the preliminary activities needed for virus construction are going on. We say that the virus is in *eclipse* at this time. After the degradation of the nucleic acid of the host, construction of the phage DNA begins. This occurs approximately 6 minutes after infection. After the appearance of the protein which has an enzymatic function, another class of protein begins to appear. This is the protein which will form the envelope of the completed viruses. Finished phage particles may appear in the cell approximately 12 minutes after the infection, and their number will increase up to the lysis of the cell. The exact time of this cell burst depends on the particular bacterial host and the strain of the virus. Approximately 1/2 hr after infection by a phage, the cell will burst and release 100 or so mature phage particles.

The lysis results not just from pressure of the virus particles in the cell, but apparently from the activity of an enzyme (lysozyme) in the tail of the virus which can break down the cell wall. This same enzyme would also enable the phage to inject its DNA into the cell. After lysis, the released phage particles are free to enter other susceptible cells and repeat the process. It is evident that the virus is much more than just a simple association of nucleic acid and protein. Indeed, its ability to use cell machinery for the complex reactions necessary for its own reproduction implies a high degree of specialization. This must have involved a long history in which the forces of evolution acted on both the virus parasite and the host. Figure 11-3 shows a very well known virus which infects *Escherichia coli*, phage T/4. Although not cellular in structure, it is much more complex than was first suspected. The DNA is stuffed into the protein of the head. Exactly how the long double helix is packaged into this headpiece during maturation of the phage in the host cell is still not understood. The tailpiece which is needed to attach the infecting virus to the cell is associated with a plate equipped with prongs. The contraction of the tailpiece causes the viral DNA to pass through the hollow core into the bacterial cell.

TEMPERATE VIRUSES AND PROPHAGE. Not all phages, however, are virulent, as just described. Many are able to infect cells and take up residence in them without causing lysis or cell destruction. Such a nonvirulent virus residing in a bacterium is called a *symbiotic* or *temperate phage*. The bacterial cell remains unharmed by the presence of the phage and may actually receive some benefits from the particle it contains. If we artificially burst open a bacterial cell which harbors a temperate phage, we cannot detect any mature phage particles. The virus seems to have disappeared. Yet its presence in the cell can still be demonstrated. Assume that we have a strain of bacterial cells which contains a temperate virus. Although we find no mature virus in the cells, the fluid in which the cells are growing will always contain some mature phage particles. The reason for this is that in a few rare cells of the bacterial strain, the virus becomes virulent and causes lysis of its host. We can detect these released particles if we bring the culture fluid in contact with an indicator strain of bacteria (Fig. 18-1). The latter would be a strain which is susceptible to the phage. If a drop of the culture fluid is added to a cloudy tube containing the susceptible cells in broth, a clearing of the broth takes place due to the destruction of the sensitive cells by the virus. Of course, we might not have discovered that the first bacterial strain is harboring a phage unless we had the indicator strain. Therefore, many bacteria normally contain unsuspected virus of some kind whose presence awaits detection by contact with a sensitive strain.

We see, therefore, that bacterial cells which harbor a virus give no visible evidence of the presence of the virus. Yet the virus *is* present, and the cell contains genetic information to make virus, as evidenced by the presence of virus paricles in the culture fluid. The reduced state of the temperate virus in the host cell is called *prophage*. The cell contains only one intact prophage of a particular kind, because the presence of a virus in a cell makes the cell immune to further infection by the same kind of virus. In its prophage state, the virus is integrated into the bacterial chromosome (Fig. 18-2A). It apparently inserts itself by a process similar to crossing over, just as the "F" factor of *E. coli* becomes inserted in high frequency (Hfr) strains (see Fig. 17-15). In its integrated state, the virus in every way behaves like a gene of the bacterium, replicating along with the DNA of the cell. Occasionally the virus, by a reversal of the process of insertion, may come out of the chromosome (Fig. 18-2B). It can do this spontaneously, or it can be stimulated to do so by certain agents, such as ultraviolet light. Normally, the temperate virus is prevented from becoming infectious and detrimental to the host because of a repressor substance in the cell (next chapter). If this is somehow destroyed, as by ultraviolet radiation, the genes of the virus may be expressed. The virus can then subvert the cell and cause new mature virus to be made (Fig. 18-2C). Lysis results, and mature particles enter the culture fluid. They may then take up residence in similar host cells or they may destroy cells which are sensitive to them.

Since the bacterial strain with the temperate phage contains information to make virus which can cause lysis of sensitive cells, it is called

FIG. 18-1. Lysis of a cell culture. If drops of culture fluid are exchanged between certain strains (1 and 2 here), a clearing of the culture fluid may occur, indicating that cells of the strain have been destroyed. In this case, strain 1 causes lysis of strain 2. The latter acts as an indicator strain for the presence of phage in strain 1.

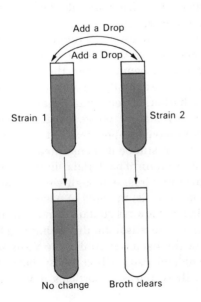

Add a Drop

Add a Drop

Strain 1   Strain 2

No change   Broth clears

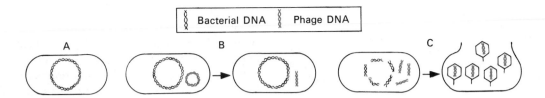

FIG. 18-2. Prophage and lysogenic bacteria. *A.* When a temperate phage enters a bacterium, it may become integrated into the chromosome of the cell. It behaves very much like a bacterial gene and is passed on from one cell division to the next. In this prophage state, it remains undetected. *B.* In an occasional cell, the phage DNA comes out of the chromosome. It is believed to be in a circular form which then becomes linear. *C.* The virus then subverts the cell, bringing about production of new viral DNA and protein with the eventual lysis of the bacterium and the release of mature viruses. These, in turn, may inject their DNA into other bacterial cells and take up residence as prophage.

lysogenic. This means that it can bring about lysis of an indicator strain. On the level of bacteria and their viruses, we find many different strains of cells and phages which vary, especially, in their susceptibility and their virulence. A spontaneous mutation may arise which can change a sensitive bacterial cell to one which is now resistant to a specific virus. In turn, mutation in the virus can make it more virulent to a bacterium, or else it may change it to a symbiotic phage. The establishment of a symbiotic relationship between a bacterium and its temperate phage is the product of mutation operating under the force of natural selection. The closeness of the phage-bacteria relationship is seen when we realize that viruses, in their prophage state, reside *not* at just any location on the chromosome but rather at very specific sites. The chromosome map in Figure 17-11*B* indicates the location of phage lambda. This virus resides near the bacterial marker Gal (for the ability to produce an enzyme needed for utilization of galactose). Phage lambda may or may not be present in a cell, depending on the bacterial strain. A Hfr cell that is lambda$^-$ can transmit the absence of lambda to a lambda$^+$ cell, just as if the lambda state (presence or absence of lambda phage) were being inherited as a specific bacterial gene (Fig. 18-3*A*).

If a Hfr lambda$^+$ cell is mated to an F$^-$ lambda$^-$, the chromosome of the former may penetrate into the F$^-$ cell and transfer prophage. However, the prophage is immediately activated when transmitted in this way to an F$^-$ cell (Fig. 18-3*B*). The virus removes itself from the chromosome and causes destruction of the F$^-$. Thus, some phages can be induced to become virulent,

not just by exposing the cells to certain environmental conditions but also as a result of transfer during conjugation to cells lacking them. Not all prophages, however, may be induced. A host cell, then, may contain one or more different kinds of viruses in their prophage state, each inserted at a characteristic location on the chromosome and each behaving as if it were a bacterial locus.

PHAGES AND TRANSDUCTION. When the virus in its prophage state removes itself from the chromosome, it may in so doing carry with it an adjacent piece of the host chromosome. Phage lambda may on rare occasions pick up a Gal$^+$ marker from a cell (follow Fig. 18-4 in this discussion). As it does this, it leaves behind a portion of itself which remains inserted in the host chromosome. The virus, now containing a bacterial gene, may be released into the medium where it is free to enter an appropriate cell which contains no lambda prophage. The virus may insert itself as prophage in the second cell. As a result, it may confer a new property on the host. If the latter, for example, is Gal$^-$ on the chromosome, it may acquire the ability to utilize galactose due to the presence of the Gal$^+$ allele associated with the prophage that it is now harboring. This transfer of genetic material from one cell to another by way of a virus is called *transduction*. The phenomenon of transduction was discovered in 1952 by Zinder and Lederberg working with another species of bacterium, *Salmonella typhimurium*.

It is now well known that variations occur in transduction, depending on the host and the phage under consideration. Phage lambda,

VIRUSES AS GENETIC TOOLS

FIG. 18-3. Transfer of lambda (λ) prophage. *A.* A prophage occupies a definite location on the bacterial chromosome and may behave like an ordinary bacterial gene. The phage λ occupies a position adjacent to the marker, Gal. In conjugation, the absence of λ (λ⁻) state may be transferred to the F⁻ cell, just as any other marker. Recombinants can arise which are λ⁻. *B.* The prophage cannot be transferred from the Hfr which is in the λ⁺ condition without causing lysis of the F⁻ cell. This is due to zygotic induction in which the phage, upon entering the F⁻ cell, becomes activated to manufacture more λ virus. Lysis of the cell results with release of mature λ phage.

which has been extensively studied in *E. coli*, illustrates "specialized transduction." The behavior of this virus brings to light several interactions which can take place between the host cell and the virus. When transduction takes place by a virus such as lambda which has existed as a prophage in a previous host cell, the second host becomes diploid for a very small segment of the genetic material, in our example, the Gal region. The cell is a partial diploid, heterozygous for just a small portion of the genome. We again

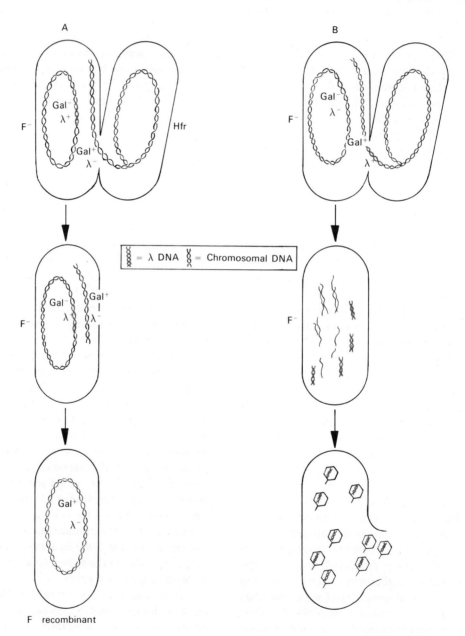

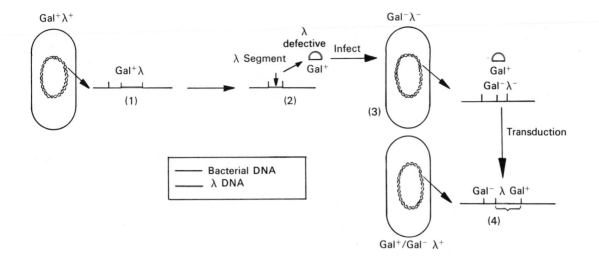

FIG. 18-4. Prophage and transduction. Phage lambda (λ) resides adjacent to the Gal locus (*1*) in a λ+ cell. When it comes out of the prophage state and deintegrates from the chromosome, it may carry the Gal marker with it (*2*), but in so doing, it leaves a portion of itself behind. The phage is now defective, having lost some of its abilities. If the defective virus, carrying a Gal+ marker infects a Gal− cell which does not contain λ (*3*), it may insert itself in the Gal region. The cell becomes transduced (*4*) and contains the defective λ plus the Gal locus in the diploid condition. The cell is a heterogenote. The defective nature of λ is seen in the fact that by itself, it cannot remove itself from the chromosome and reproduce. This function was lost when the phage left a segment of itself behind (*2*) and incorporated the Gal locus.

use the term "heterogenote" for such a condition. Since the virus left a part of itself inserted in the chromosome of the first host as it picked up the Gal marker, we might expect the virus to show some alteration in its behavior. And indeed, the virus is now changed. This is seen in its inability to become infective again; it cannot cause cell lysis or become released into the cellular environment. The defective prophage *can* be liberated again, however, if the cell also contains an *intact* lambda. A transduced bacterial cell (one which has received a marker by way of a virus) may contain both the defective transducing phage with the marker allele as well as an intact phage which contains all of its own genetic information (Fig. 18-5). The latter is able to supply the information missing in the former and can bring about the lysis of an occasional cell. Bacterial cells from a strain which has been transduced may thus undergo lysis and contribute to the culture medium both kinds of phages, the defective ones carrying a bacterial gene and the intact ones with all the needed virus information to effect lysis. Both types of particles would be able in turn to enter other cells of the same host strain and again take up residence.

It should be apparent that transduction by a virus clearly resembles sexduction, transfer of a bacterial gene by a sex factor. Indeed, the two processes are so similar that they may be considered aspects of the same phenomenon. The "F" factor, as well as the phage, may also leave a piece of itself behind when it picks up a bacterial gene. It is the presence of these small segments of "F" factor or prophage which may account for the specific point of origin in a particular Hfr strain and also for the characteristic point of residence of a prophage in a bacterial cell (Fig. 18-6). We may consider the phage to be an episome, because it satisfies the definition in the same way as the "F" factor.

It has been demonstrated that the very presence of a phage in a bacterium may impart a new characteristic to the cell, one which is *not* due to a gene introduced from another bacterium. For example, certain nontoxic strains of *Corynebacterium diphtheriae* gain the ability to produce the virulent diphtheria toxin when they are infected with particular phages. The production of the toxin by the cells does not result from the presence of any bacterial genes which have been carried by the virus; rather, it

FIG. 18-5. Double infection. A bacterial cell may become infected simultaneously with an intact virus and a defective one carrying a marker (1). Both viruses take up residence on the chromosome, and the cell is transduced and becomes a heterogenote (2). In a rare cell, the two viruses may deintegrate (3). This is possible, only because the normal, intact virus is present to supply the function missing in the defective phage. Both types of phages are able to reproduce, and the cell lyses (4) with release of the two types of particles.

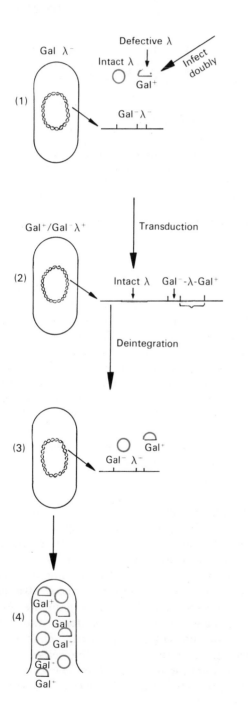

depends on both the genetic machinery of the cell and that of the phage. If the phage is lost from the cell, so is the ability to form the toxin. This is an example of *conversion*, the acquisition of an inherited cell property as the result of the interaction of host and virus genes.

RECOMBINATION AND THE GENETIC SYSTEM OF PHAGES. It must be kept in mind that the bacteriophage is not a simple entity which just wanders around from one host cell to another. For the virus itself has a set of genetic instructions which determines its ability to infect this or that host strain. The genes of the virus also enable it to direct its own DNA replication and the assembly of its parts. Although the number of genes in a virus may be small when compared to that in a higher form, they nevertheless enable the virus to engage in various activities which require coordinated interactions. Mutations occur in viral genes just as they do in other forms. As in higher organisms, the study of the inheritance of a mutant allele and its standard form, a pair of alleles, permits a genetic analysis of the virus and the mapping of its genes.

What kinds of characters can be followed when the genetic system of the virus itself is being investigated? We have noted that viruses grow only in certain host strains. Phages thus have characteristic host ranges. A mutation, however, may widen or narrow that range. For example, the standard "T-even" phages can infect and lyse *E. coli*. A variant form of these viruses can infect and lyse strain B of *E. coli* but cannot cause lysis of strain K. Host range is thus a genetic characteristic which can be recognized in viruses. Another common one is the type of plaque produced by a virus when it infects a lawn of bacteria spread on an agar dish (Fig. 18-7). The wild type plaque is quite distinct for a type of virus. It may be small with smooth edges, whereas a mutant form may be large. Some plaques are cloudy; others are clear. Another trait which can be easily followed is the rapidity with which the virus bursts the host cells.

Studies conducted in the late 1940's by Hershey and Doermann and several others brought out the features which are typical of phage crosses. Suppose host range and speed of lysis are being followed. The wild allele (h+) enables the virus to grow on *E. coli* strain B but not on

E. coli B/2. The mutant allele, (h) increases the host range to include the B/2 strain. The wild allele (r^+) determines small plaques, whereas its allele (r) results in large plaques as a consequence of very rapid lysis. To determine whether or not recombination can take place between the loci "h" and "r," we may proceed as follows (Fig. 18-8A). The two bacterial strains, B and B/2, are mixed and then spread on agar plates. We know that if a virus is "h^+r," it can grow only on strain B and will produce large plaques. We can easily identify plaques produced on a plate containing *both* B and B/2 bacteria. The virus will lyse only the B cells, but the large plaques will be turbid or cloudy because they will contain unlysed cells of strain B/2. On the other hand, a virus that is "hr^+" will produce small clear plaques due to the fact that r^+ determines slower (wild) lysis and that the "h" allele enables the phage to lyse both B and B/2 cells. The resulting plaques will contain no intact cells and therefore will be clear.

A mixture of the two parental virus types (h^+r and hr^+) is allowed to infect a liquid culture of strain B host cells (Fig. 18-8B). Both parental viruses can infect and lyse the B cells. This B culture is then diluted by liquid broth and allowed to undergo lysis by the phages. The viruses which are released after lysis of the B cells are then plated on a mixture of fresh host cells, this time both B and B/2. By plating on this mixture, we will be able to detect virus offspring with normal host range (turbid plaques) and mutant (clear plaques). The plaques which form are then examined for their size and clearness. The two types caused by the two parental phages are seen (large, turbid ones and small, clear ones).

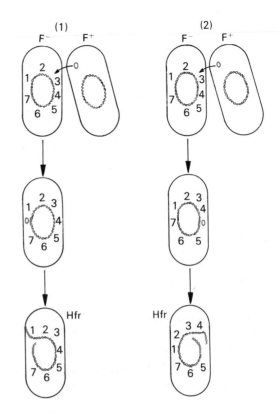

FIG. 18-6. Origin of Hfr types. When the F factor picks up a marker allele from a bacterial chromosome, it may leave a segment of itself behind, An F⁻ cell may later arise carrying this small portion of "F." Different F⁻ cells may carry segments of F at different locations on the chromosome (1 and 2). When such F⁻ cells receive a fertility factor in conjugation, the F factor will be attracted to the site of the segment, because the latter contains regions homologous to it. Hfr types can thus arise which will have different genes at the anterior end of the chromosome when the cell undergoes conjugation. A similar picture is believed to hold for the insertion of viruses at particular prophage sites, because transducing phages may leave portions of their own DNA behind in the chromosome.

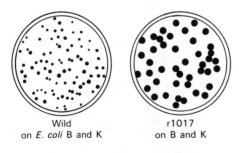

Wild
on *E. coli* B and K

r1017
on B and K

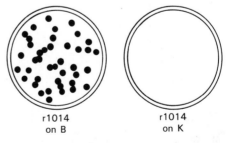

r1014
on B

r1014
on K

FIG. 18-7. Plaque formation. Viruses may differ in the type of plaque they form and also in the range of hosts in which they can reproduce. Phage T/4 produces the small, wild type plaques on

E. coli strains B and K. Certain mutants (r mutants) produce large plaques on B and K. Still others, the rII mutants produce the large mutant plaques on B but no plaques at all on K.

But in addition, we also find some which are large and clear and others which are small and cloudy. These latter two types must result from the presence of viruses with the respective genotypes, hr and h⁺r⁺. This tells us that recombination has occurred. We can express the percentage of recombination by using the same reasoning as was used to determine crossover frequency in higher organisms. The proportion of the number of recombinant offspring, hr and h⁺r⁺ (as measured by features of the plaques), to the number of total progeny (the total of all the plaques) is obtained. The percentage value tells the recombination frequency, just as it would if we were studying eye color and body color in fruit flies.

The classical genetic principles used in making three point testcrosses (Chap. 8) may also be applied to phages. Suppose that three loci are being mapped: m, r, and tu (Table 18-1). Again plaque characteristics are examined. From the data, it can be seen that the most numerous type of plaque and the least numerous represent the parental types and the double crossovers, re-

FIG. 18-8. Recombination in phages. *A.* Mixture of strains B and B/2 can be used to identify the kinds of phages present. The h⁺r type produces large, turbid plaques, because it produced rapid lysis on strain B but cannot grow in B/2. Therefore, the plaques are cloudy due to the unlysed B/2 cells. The hr⁺ produce wild type (small) plaques, but they are clear, because both B and B/2 cells are lysed. *B.* One can determine the amount of recombination in this procedure by looking at the type of plaques. Again a mixture of B and B/2 is used as an indicator, and the four types of plaques are easily identified. The amount of crossing over is determined by counting the total number of plaques and expressing the number of recombinants as a percentage.

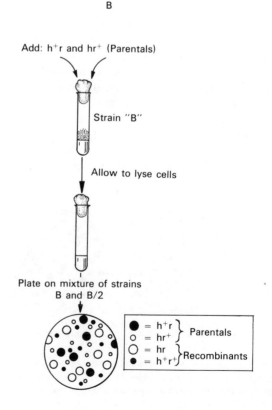

TABLE 18-1* Progeny from three-factor linked cross in phage T/4†

CATEGORY	CLASS	GENO-TYPE	TOTAL PLAQUES	PER-CENT-AGE
Parental	A	m r tu	3467	33.5
Parental	A	m⁺ r⁺ tu⁺	3729	36.1
Single crossover	B	m r⁺ tu⁺	520	5.0
Single crossover	B	m⁺ r tu	474	4.6
Single crossover	C	m r tu⁺	853	8.2
Single crossover	C	m⁺ r⁺ tu	965	9.3
Double crossover	D	m r⁺ tu	162	1.6
Double crossover	D	m⁺ r tu⁺	172	1.7

* Reprinted with permission from A. H. Doermann, *Cold Spring Harbor Symp. Quant. Biol.,* XVIII: 4–11, 1953.

† A, noncrossovers; B, crossover between m and r; C, crossover between r and tu; D, double crossover; m, minute; r, rapid lysis; and tu, turbid.

spectively. The locus "r" is established as the one in the middle, because it appears to have switched position. Establishment of regions 1 and 2 allows us to determine the corresponding crossovers in the two regions. Again we must add the frequency of doubles to each class (3.3 here). And so the map becomes:

m   12.9   r   20.8   tu.

By applying detailed linkage analysis to phage T/2 of *E. coli*, it was found that all of the genes fall into one linkage group and that the map is circular. The circularity of the map, however, does not mean the DNA of the virus "chromosome" is necessarily in the form of a circle as in *E. coli*. Rather, the isolated DNA of phage T/2 seems to be linear. The DNA of some viruses (φX 174 which will be discussed shortly) has been shown to be definitely circular. On the other hand, the "chromosome" of phage lambda seems to exist sometimes as a circle and other times in a linear form. Perhaps one form is necessary at the time of replication (maybe the circular one) and the other form at the time of infection when the viral DNA is injected.

REPEATED MATING IN PHAGES. Although genetic maps may be built up employing the fa-

miliar logic applied to higher organisms, close inspection reveals several unusual features which are peculiar to phage systems. When the crosses are performed in the manner as described above, a tremendous number of bacterial cells and phage particles is obviously involved. Techniques are available, however, which enable us to make critical dilutions. From these, we may study the offspring phages coming from the burst of a *single* bacterial cell which had been infected by two types of phage particles, let us say h⁺ r and hr⁺. When such dilutions are made, it is found that the recombinant classes, hr and h⁺r⁺, are *not* present in equal amounts. On the basis of our understanding of crossing over, we would expect the recombinant classes to be reciprocal and hence equal in number. In the case of phages, these are usually found to be very unequal. Morever, if a cell is infected with three different kinds of virus particles, such as m r⁺ t⁺, m⁺ r t⁺, and m⁺r⁺ t, we might find offspring that are m r t. For such a combination to arise, three parents must be involved! How can such an observation be explained?

Some clarification is provided by bursting the infected cells artifically several minutes before they would lyse naturally. When this is done, it can be demonstrated that recombinant viruses can exist in the cell many minutes before the natural cell burst. As time increases toward natural burst, the number of recombinants also increases. In other words, the acts involved in recombination start well in advance of the bursting of the host cell. This means that before lysis, more than one round of mating and recombination can take place. According to the theory of "repeated mating," the phages enter the cell and then multiply. The phage units in the cell can now mate at random repeatedly in pairs so that several rounds of mating and multiplication can take place. Exchange of genetic material by crossing over takes place and produces recombinants. If reciprocal recombinants are formed early in the mating process, they may be changed by participating in later mating. Therefore, the phage offspring coming from a lysed cell represent several generations of multiplication. We can now understand how three parents may be involved in the production of a phage genotype (Fig. 18-9). A mating of m⁺r t⁺ with m r⁺t⁺ in one round of multiplication can give m⁺r⁺t⁺ and m r t⁺. If the latter mates with m⁺r⁺t in the next round of mating, the genotype m r t can arise.

When we study the products of many cell bursts after cell infection by two different phage types (h⁺r and h r⁺), we are unaware that multiple rounds of mating have taken place. Equal numbers of reciprocal recombinants are found among the collection of offspring from the bursting of the *entire* population of cells as a result of randomness. Since an extremely large number of phage units is involved, the recombinant types will be produced in comparable amounts.

PHAGE SYSTEMS AND THE RECOGNITION OF THE CISTRON. One of the greatest contributions of phage analysis is the revised concept it has given us on the nature of the gene. Previous to the 1950's, classical ideas held the gene to be an indivisible unit. It had a certain function which it performed. As a unit, the gene could mutate and engage in crossing over. When the gene mutated, all of it was somehow changed. Mutation at different sites within the gene was not considered possible. Nor was crossing over within a gene recognized. Because it was an intact unit, breakage and crossing over could occur on either side of it but not within it. Contradictory cases arose in the fruit fly as well

FIG. 18-9. Repeated mating. When phage DNA enters a cell, it multiplies. The resulting phage units may undergo several rounds of mating which are at random and during which recombination can take place. If a cell is infected with phages of three different genotypes as shown here, offspring may arise which carry contributions from three separate phage units (in this case, m, r, and t were originally in three separate particles). This is easily understood when it is realized that more than one mating takes place in the infected cell before mature viruses are released. In this example, one of the products of the first round of mating (m r t⁺) participates in a second mating. The genotype mrt can thus be found in a mature phage released upon lysis of the cell.

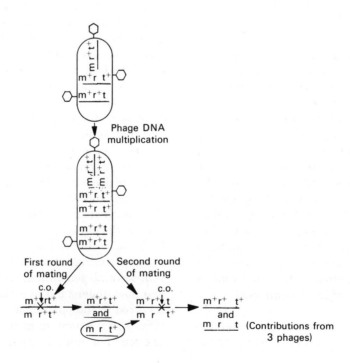

as in other well-studied organisms, but these apparent exceptions, as in the case of lozenge (Chap. 14), were explained by coining new terms and adapting them to fit the classic, accepted idea of the gene. The work of Benzer, undertaken in the 1950's, revolutionized the picture of the gene as the indivisible unit of function, mutation, and recombination. Benzer worked with the T-even viruses which infect *E. coli*. He was one of the first investigators to try to relate genetic units, such as units of crossing over and units of mutation, to the actual physical basis of DNA. This procedure requires the construction of very detailed chromosome maps. In turn, constructing such maps relies on the examination of extremely large numbers of offspring. We have seen that microorganisms are ideal for such purposes, especially phages, where millions of progeny can be examined in a short time.

Benzer studied in detail the genetics of plaque formation in phage T/4 of *E. coli*. The wild-type plaque, as seen on an agar plate, is a small clear area with rough edges. These wild-type plaques are produced when phage T/4 is plated on either strain B or strain K of *E. coli*. Certain mutants arise in phage T/4 which are called "rII" mutants. When plated on strain B, they form a so-called "r" plaque, one which is larger than the wild type (r^+) and which has smooth edges. These rII mutants, however, produce no plaques on strain K. (Reference back to Fig. 18-7 will summarize these points.) The rII mutants infect K cells and start the synthesis of phage DNA and other viral products, but they cannot produce infective offspring. As a result of this defect, they cannot cause cells of strain K to lyse. This growth defect of the rII viruses on strain K provided Benzer with a very critical tool. One of his goals was to take different rII mutants which had arisen independently and cross them in various combinations. The point was to determine whether the numerous independent mutants represented separately occurring mutations at the same or at different sites on the genetic material. For example (Fig. 18-10A) assume that two "r" mutations, r_1 and r_2, are actually alterations at two different positions within the DNA. Being rII mutants, the phages produce a mutant plaque on strain B and *none* on strain K. However, a strain-B cell, infected with the two types, could give rise by recombi-

nation to wild, $r_1{}^+r_2{}^+$. This is possible, because the r_1 mutant is carrying a normal r_2 region and would have the genotype $r_1 r_2{}^+$. Similar reasoning tells us that the r_2 mutant would be $r_1{}^+r_2$. Recombinant wild-type phages, $r_1{}^+r_2{}^+$, could arise by crossing over. The resulting wild phage would be able to produce wild plaques on both strains B and K. But if the two mutant sites, r_1 and r_2, are sites *within* the same gene and are therefore extremely closely linked, the wild recombinants would be exceedingly rare. This would mean that the investigator would be forced to search through perhaps millions of mutant plaques on strain B in order to find the one wild plaque. This is a feat which would be truly impossible. Fortunately, the growth defect of the rII virus on strain K can avoid the need to do this. For it can act as a screen to uncover the rare, recombinant among the millions of others.

In the procedure, rII mutants are therefore recognized by the "r" plaques which they produce on strain B and their inability to produce any plaques on K. Strain B is then infected with two rII mutants which arose independently (Fig. 18-10B). The infected *E. coli* B are allowed to lyse. Samples of the resulting phages are then plated on agar dishes containing strain K. Only the wild recombinants will form plaques. The number of them will be few, because the recombinants will be rare. But this number of wild plaques on K can now be compared to the number of "r" plaques produced by samples plated on strain B. From these counts, the frequency of recombination is easily determined; the logic is again the same as for the estimation of crossover frequency in higher organisms.

Benzer's method was so sensitive that he could have detected one recombinant among 10^8. The wild plaque which this recombinant can form would be lost among the "r" plaques if only strain B were available. The rare recombinant would never be found utilizing only strain B, and so, no maps could be constructed. With his method, Benzer found that the rII mutations cluster in a small portion of the T/4 map. Within that small region, the positions of different rII mutations could be established. Actually, 300 separate sites have been recognized within the small map segment (Fig. 18-11). Well over 2000 separate mutations were studied, and it was found that some of these were mutations at the

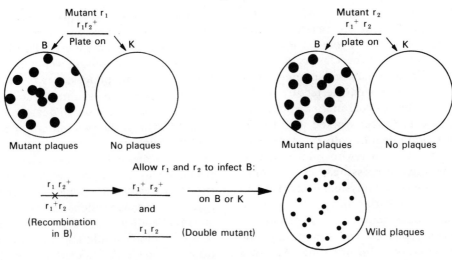

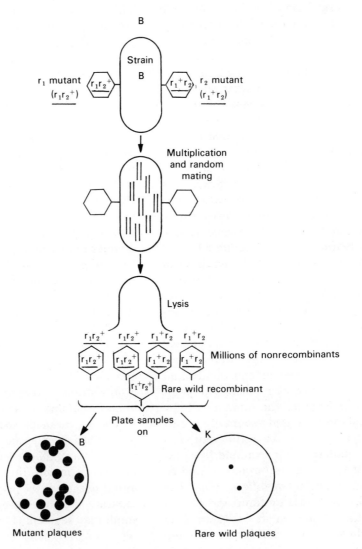

FIG. 18-10. Crossing of rII mutants. *A.* Recognition of phage types. rII mutants are recognized by the fact that they produce a mutant plaque (large) on *E. coli* strain B and no plaques on strain K. If strain B is infected with two rII mutants which have alterations at different sites within the gene, a very rare recombinant may be formed $r_1^+ r_2^+$. This can produce wild (small) plaques on either B or K. On strain B, both the rII mutants and the wild type can grow. The rare, wild plaque would be lost among the mutant ones. Strain K, however, acts as a screen, because only the wild type can grow on it. *B.* Recombination frequency in the rII region. The different rII mutants are recognized as shown in *A* above. Two types of rII mutants are allowed to infect sensitive strain B. When B cells lyse, they will release phages which are primarily mutant, because recombinants within the short rII region under study would be rare. However, by plating on strain K and observing for wild plaques, the number of these can be established. These can then be compared with the number of mutant plaques produced on strain B. A comparison of the number of mutant plaques on B to the number of wild ones on K allows a calculation of the frequency of crossing over between the two mutant sites. (The figure does not attempt to reflect actual numbers of plaques.)

very same site. As a matter of fact, Benzer recognized "hot spots," sites where large numbers of the mutations map in contrast to other locations where only one or a few mutations have arisen. The smallest recombination frequency was approximately 0.02%, meaning that two different rII mutant sites may be only 0.02 map units apart. Benzer translated this into terms of nucleotides along a stretch of DNA. Making certain assumptions, the 0.02% value indicated that a distance of not more than six nucleotide pairs would separate two mutant sites along a strand of DNA. From the wealth of genetic analyses of microorganisms which followed Benzer's studies, we now know that two mutant sites may be adjacent to each other, representing two nucleotide pairs which are side by side.

Does the fact that two "r" mutant sites may be so close together mean that there are different mutant sites within the gene and that they can undergo recombination by crossing over within the gene? And consequently, is the unit of function different from the unit of mutation and

FIG. 18-11. Map of rII region. A large number of different mutations which originated independently have been assigned to the rII region. The exact order of the mutations in a small segment is not known in most cases. Each square represents an independent mutation whose site has been mapped. It can be seen that more mutations occur at certain sites than at others. (Reprinted with permission from S. Benzer, *Proc. Natl. Acad. Sci., U.S.A., 47:* 410, 1961.)

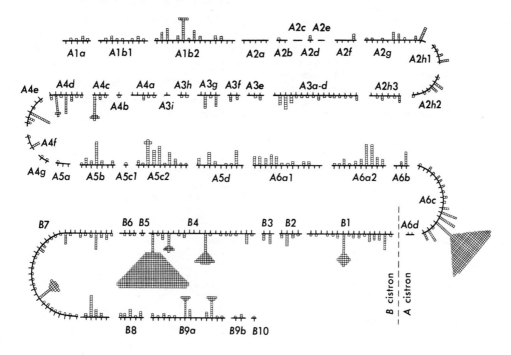

recombination? We have already discussed these points in relationship to the gene and molecular interactions (Chap. 14), but it was Benzer's work which reformed thought on the nature of the units of heredity. Utilizing strain K of *E. coli*, Benzer demonstrated that *two* genes of function are present in the rII region of phage T/4. He discovered that a mixed infection of strain K with a wild-type phage plus any rII mutant results in the lysis of the cell and the appearance of *both* wild type and mutant viruses (Fig. 18-12). This means that the presence of the wild phage makes possible the reproduction of the mutant type. In a sense, it is as if the wild (+) form is dominant over a recessive "r" allele. Benzer then infected strain K with different pairs of rII mutants (follow Fig. 18-12). From the mapping which he had done, he knew how far apart any two rII mutants would be. He now wanted to see if two rII mutants could interact in some way to cause lysis of strain K with the production of offspring phages; this is an effect which no rII mutant can bring about on its own. Benzer found that certain pairs of rII mutants could interact to effect lysis on strain K, whereas others could not. The small rII region was shown to be divisible into two subregions, A and B. A combination of any two mutants in the same region, both in A or both in B, cannot cause lysis. However, a combination of two mutant types, one with a mutation anywhere in A and the other with one anywhere in B can cause lysis. This observation told Benzer that all of those sites within subregion A had more in common with one another than with any site in B. The same relationship would hold for the B sites. The common feature which distinguishes all of the

FIG. 18-12. Demonstration of the functional unit. Strain K can be infected with wild-type phages plus one of the rII mutant types. Any combination of a wild and a mutant results in lysis of strain K (*1* and *2*), indicating that the wild-type phage is able to carry out the function which is deficient in the mutant (inability to grow on strain K by itself). Certain combinations of two different rII mutants on strain K result in lysis (*3* and *6*), whereas other combinations do not (*4* and *5*). It is found that the rII region can be divided into two segments (indicated by the dotted line). Any combination of mutants whose mutant sites are within the same region fails to cause lysis. This indicates that two such mutants carry defects in the same functional region, even though the defects are not at the exact same site. When the defects are in separate regions, the one mutant contains a normal, functional unit which the other lacks (*3* and *6*). Lysis occurs because the two functional units needed for growth in K are present in the cell.

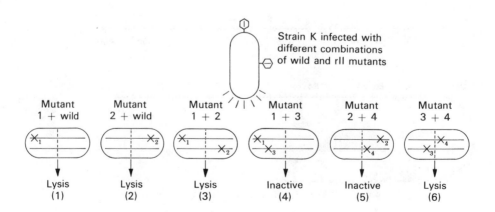

sites within one of the subregions is evidently a specific function to which all of the sites are related. In this case, the function is one that is necessary for cell lysis and the production of phage. We have then in the rII region two subregions, A and B, which are units of function. In order for lysis to occur, *each* unit must be carrying out its specific function. When a wild phage is crossed with any rII mutant, the wild particle contributes two normal functional units, A and B, and so lysis can occur. Similarly, a cross of any A mutant with any B produces lysis, because a mutant which is defective in A contributes an intact B and vice versa. However, if a cell of strain K is infected with two B mutants, there is no normal, functional B unit in the cell (and likewise for any two A mutants). Because one of the two functions is lacking, lysis cannot occur. The work with rII phage clearly demonstrates that the unit of function is larger than the unit of mutation and crossing over. Benzer coined the term "cistron" to designate the unit of function. And the concept of the cistron, as we have seen, has been shown to be applicable to higher species as well as to microorganisms.

In Chapter 14, the cistron, or gene of function, was discussed in relationship to its control of a polypeptide chain. In the light of the knowledge assembled by molecular biologists since Benzer's original work, the concept of the gene as a unit of function (a stretch of DNA) composed of smaller units (nucleotide pairs which may mutate independently and engage in crossing over) agrees perfectly with the known physical basis of the gene itself. As stressed in Chapter 14, there need be no confusion about use of the terms "cistron" and "gene." They are equivalent if we picture the "familiar gene" as a stretch of DNA controlling a specific function (formation of a specific polypeptide chain) and composed of smaller sites which may mutate independently and engage in crossing over (the nucleotide pairs). The expressions "muton" and "recon" were also coined by Benzer to refer to these sites of mutation and crossing over within a cistron. However, they are now less frequently used, because they are known to represent the same thing in physical terms, one nucleotide pair.

A cistron, or gene of function, can be recognized in any group of organisms by the performance of a *cis-trans* test (Chap. 14). When any two mutants are crossed, the mutant sites in the F_1 offspring are found in the *trans* arrangement. In our discussion of allelism in higher forms, we saw that a cross of two recessive mutants will produce mutant offspring if the two parents are carrying mutations which are within the same gene. Benzer was actually performing *cis-trans* tests in his work with strain K and the rII mutants. For a cross of any rII mutants gives a *trans* arrangement (Fig. 18-12). If the resulting phenotype is wild (here lysis and plaque formation), the mutations complement each other and hence belong to separate genes or functional units. If a mutant phenotype (no lysis) follows from a mixed infection by two rII mutants, then the mutations do not complement each other but are in the same gene or cistron.

PHAGE WITH SINGLE-STRANDED DNA. Far from being just simple little packages of double-stranded DNA, viruses have proved to be quite diversified. Their variation has led to the clarification of many critical points related to the nature of gene action. One virus which has been used as a valuable tool is ϕX 174. After its discovery in 1958, chemical analysis of ϕX 174 revealed an unexpected feature. Although its DNA contains the familiar building blocks of phosphate, sugar, and bases, the proportion of adenine to thymine is not 1:1 nor is that of guanine to cytosine! At first, this seemed to contradict the rule of preferential base pairing. The ϕX 174 DNA was subjected to a variety of tests, including ultracentrifugation and exposure to enzymes. The data clearly showed that the DNA of this virus is single stranded rather than in the form of a double helix. Moreover, it exists in the form of a closed circle.

One question which arose concerned the manner in which such a species of DNA could replicate. It was found that the single-stranded form is infective, capable of invading its *E. coli* host. Once inside the cell, however, the single-stranded DNA can direct the formation of a complementary strand, producing a double-stranded form of the virus, the so-called "replicative form." The latter is thus composed of the infective strand (+) and its complementary strand (−) made inside the cell. The replicative form is now able to direct the formation of new, single-stranded virus (+ strands). In the process

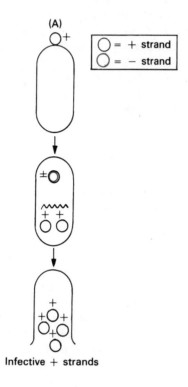

(A)

◯ = + strand
◯ = − strand

Infective + strands

(B)

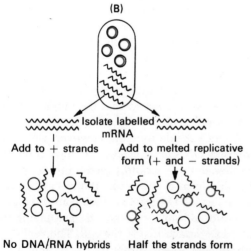

Isolate labelled mRNA

Add to + strands | Add to melted replicative form (+ and − strands)

No DNA/RNA hybrids | Half the strands form DNA/RNA hybrids

of replication, mRNA must be formed to direct the synthesis of essential proteins (Fig. 18-13A). Do both strands undergo transcription when the required mRNA is formed? To answer this, *E. coli* cells were supplied with P^{32} and then infected with ϕX 174. Any newly formed mRNA would thus be labeled. Such P^{32}-labeled mRNA was isolated and mixed with the strands of the infective form (+) of the virus (Fig. 18-13B). The labeled mRNA did not form DNA-RNA hybrids with the "+" strands, indicating that it was not transcribed from it. The double-stranded, replicative form was then subjected to a melting temperature sufficient to separate the double-stranded DNA into "+" and "−" strands. When the labeled mRNA was mixed with these, exactly one-half of the DNA strands formed DNA-RNA hybrids. These observations clearly demonstrated that the mRNA was a transcription of only one of the two strands of the replicative form. It had to be the "−" strand, because no hybrids were formed when the labeled mRNA was mixed with the "+" strands alone. Subsequent chemical analysis of the mRNA later confirmed that it was complementary only to the "−" strand. We now have other lines of evidence from other viruses and bacteria which also indicate that only one of the two strands of the DNA composing a gene normally undergoes transcription into a mRNA, which is then later translated into protein. This strand has been termed the "sense strand" to distinguish it from the strand which is not transcribed into mRNA, the "nonsense strand."

OVERLAPPING GENES. Later investigations with the phage ϕX 174 have yielded some un-

FIG. 18-13. Replication of ϕX 174. *A.* In its infective state outside the *E. coli* cell, the virus is single stranded. Once inside, a complementary strand is formed, producing the replicative double-stranded form of the virus. Messenger RNA is sent out, which directs the formation of infective, single-stranded virus of the original type. *B.* Under experimental conditions, labeled mRNA formed by the replicative form may be isolated from the cell. When this is added to a fresh supply of infective (+) strands, no DNA-RNA hybrids form. When a supply of the replicative form is separated by melting into the two strands, "+" and "−," it is found that one-half of the viral strands form hybrids with the DNA. Obviously, transcription of mRNA came from only one strand, the "−" strand.

expected results which demand a new look at the nature of the gene and its reading. The single DNA strand of the phage has been subjected to intense investigations, many involving the cleavage of the DNA with enzymes and then separation of the resulting DNA fragments. The length of the ϕX 174 DNA has been found to be 5400 nucleotides, and their exact sequence has been completely determined. This is quite a feat unto itself, the sequencing of the entire hereditary material of a genetic element. A dilemma, however, presented itself regarding the length of the DNA of the phage and the number of proteins for which the DNA is apparently coded. The virus can bring about the production of nine different proteins whose molecular weights are known. The amount of DNA required to code for these is far in excess of the amount of DNA which seems to be present in the phage. The answer to the problem was quite unanticipated. Research teams in the laboratory of F. Sanger in Cambridge, England studied in detail several genetic regions of ϕX 174, one of which contains two genes, genes D and E, which code for proteins D and E, respectively. The D protein, whose exact amino acid sequence was known, is produced in great amounts in the host cell and is required for the production of single-stranded viral DNA, although its exact role in the process is unclear. The E protein is required for the lysis of the host cell when hundreds of new viruses have been produced and are to be released. The genes D and E on all counts seem like any other two separate and distinct genes. Nonsense mutations are known to occur in each one, without apparently affecting the function of the other.

The exciting finding was made that a certain stretch of the phage DNA contains the coded information for both the D and the E proteins and that the D and E genes overlap! The start of the E gene is in the middle of the D gene. Both genes end close to the same spot. However, the E gene is not just a segment of the D gene or just the end part of the D gene. The two proteins, D and E, are very different in amino acid sequence. The DNA, however, in the D and E genetic region is read in different frames. Recall frame shift mutations (Chap. 12) which were recognized in Crick's pioneer studies on the triplet code. Reading of a gene begins at a fixed starting point, three nucleotides at a time;

A "−" or a "+" mutation may cause the entire reading of a gene to shift. In the case of the D and E genes, the starting sequence for the E gene was found to be within the D gene. The D gene and the E gene within it, however, are read in different frames (Fig. 18-14).

Another interesting finding is that the terminating codon of the D gene overlaps by one nucleotide the start codon of the next gene, gene J. Moreover, overlapping of reading frames does not seem to be confined to just this one region of ϕX 174 DNA. Gene A, which contains coded information for another protein needed for viral construction, overlaps with gene B, coded for a protein which produces a nick in the DNA when new copies of the DNA are needed. Note in Figure 18-14 that the triplets indicated have the same sense as the mRNA. This is so, since this strand which is shown and the one which was analyzed is the one which is replicated in the host. It is the complementary strand which is transcribed. Hence, shown here as "stop" is the triplet TAA which will have the sequence UAA (chain terminating) in the mRNA. Similarly, ATG will be AUG in the messenger.

The discovery of the overlapping of genes D and E is quite an exception to the general picture of the gene. As discussed in Chap. 12, the genetic code has been found to be nonoverlapping, each single nucleotide being part of only one code word. We see here in ϕX 174 that a single nucleotide may be part of two code words, depending on the reading of the frame. The finding of the overlapping genes helps solve the problem of fitting the 9 genes into a supposedly less-than-adequate length of DNA. The overlapping may have evolved in ϕX 174 as a device to accommodate a higher number of genes than would otherwise be possible in so small an amount of DNA. The findings in ϕX 174 will surely focus attention on overlapping genes and the possibility of the phenomenon in other organisms. The discovery also raises intriguing questions concerning gene control and expression.

USE OF VIRUSES IN THE ISOLATION OF A GENE. Although atypical in nature, ϕX 174 is now known to be only one of several DNA viruses which exist primarily in a single-stranded form. The recognition of sense and nonsense strands has aided one group of investigators to perform

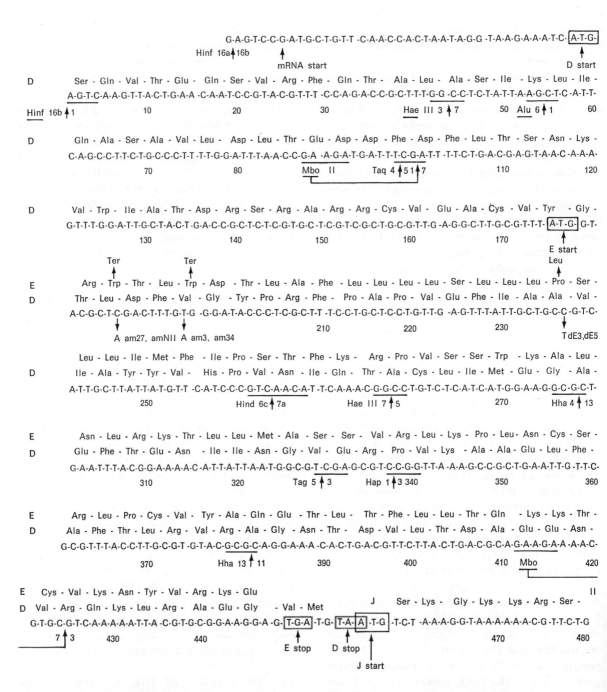

FIG. 18-14. Overlapping genes in φX 174. The starting sequence and the end sequence of the E gene is within the D gene. In the overlapping genes, codons are read in different frames. The amino acid sequences of the D and E proteins, as well as that of the J protein (product of the gene following gene D) are aligned with the DNA sequences. Note the overlapping between the terminating codon of gene D and the start codon of gene J. The DNA triplets shown here have the same sense as the RNA triplets of the messenger RNA, since the strand shown here is not the one which undergoes transcription. The strand complementary to this is the one which acts as template for the messenger RNA.

a very exacting feat, the actual isolation of a specific gene found in E. coli! The particular gene which was isolated was "lac," which determines the ability of the cell to utilize the sugar lactose. At the beginning of this chapter, we discussed the ability of certain viruses to pick up pieces of genetic material from bacterial chromosomes and to integrate these pieces with their own DNA. The phage may then carry the integrated gene to another cell (transduction). In their isolation of the lac gene, Beckwith and his group employed two different kinds of viruses (lambda plac 5 and φ80 plac 1), each of which had incorporated the lac marker of E. coli. (Actually, two small regions closely associated with the lac marker were also involved, the "promoter" and the "operator," both of which are discussed fully in the next chapter. However, they need not complicate our discussion here.)

The team took advantage of certain special features of these transducing phages when they selected them for their work. In each phage, the base composition of the DNA is such that one of the two strands is heavier than the other. Thus, each phage normally possesses DNA composed of a heavy strand and a light one. Moreover, the bacterial DNA carried by one phage was inserted in a direction opposite to that in the other phage. Now recall that of the two DNA strands, one of them is a sense strand and one is a nonsense strand. Because of the difference in orientation (Fig. 18-15A) of the bacterial DNA which is integrated into the two phages, the sense strand is attached to a heavy strand in one phage, whereas the nonsense (or complementary) strand is attached to the heavy strand in the other phage. This means that one strand of the lac gene is associated with a heavy strand of one of the phages, and its complementary strand is associated with the heavy strand of the other phage. If the two strands can be brought together and freed from the associated viral DNA, then the complete lac gene, composed of a sense and a nonsense strand, will be isolated by itself. In the actual procedure (Fig. 18-15B), the DNA of each virus was separated into two separate strands. The heavy strands could be easily isolated from the light ones in each case. The lac "sense" is on the heavy strand of one phage and the lac "nonsense" is on the heavy strand of the other. When the two strands are brought together, they form double helical DNA, because they are complementary. However, they now have "tails" associated with them. These, of course, represent the DNA of the phages which had incorporated the bacterial DNA. The DNA strands of the phages are not complementary; they are completely different DNA segments from two different viruses. Therefore, the DNA strands of the two viruses do not associate to form a double helix. These single-stranded tails were removed by treatment with an enzyme which is specific for single-stranded DNA but which will not harm double stranded. Removal of the tails by the enzyme leaves double helices, and these represent pure lac genes. The isolated gene has been visualized by the electron microscope (Fig. 18-16) and has been shown to be 1.4 mμ in length. This first isolation of a gene has opened up new pathways for the study of exactly how single genes undergo transcription. Using other transducing phages and similar procedures, it should be possible to isolate other known genes. The implications of such a feat are staggering and present promise for genetic engineering, as well as grave ethical problems for the molecular geneticist.

TOBACCO MOSAIC VIRUS—AN RNA VIRUS. The study of viruses has also shown that the hereditary material is not universally DNA. Long before the foundations of molecular genetics were established, it was known that tobacco mosaic virus (TMV) is composed of a protein coat enclosing a core of ribonucleic acid (RNA). The RNA was later shown conclusively to be the hereditary material of this plant virus, which can also attack members of the sunflower and nightshade families. The protein coat of TMV is necessary to protect the RNA; without it, there is a decrease in the ability of the virus to enter the plant cells. The protein coat and the RNA can be separated, and it has been clearly demonstrated that the coat alone cannot infect a cell and bring about either the production of new virus protein or virus RNA. On the other hand, TMV RNA by itself can do this. Experiments have been performed using different strains of the virus. For example, in addition to the standard strain (S), there is a more virulent strain (H). After separating the protein coats from the RNA of viruses of each strain, hybrid virus particles can be reconstituted, those with H-RNA and S coat and those with S-RNA and H-coat protein. When these are allowed to infect plant cells, it is found that the progeny viruses which form are always typical of the strain which provided the RNA. As seen in Figure 18-17, H-RNA with S coat governs the formation of typical H viruses which have both H-RNA and H protein.

Not only does this show that the RNA is the genetic material of these viruses but that it governs the formation of more RNA and also of virus-specific protein, much as DNA governs the formation of more DNA and protein. The coat of the TMV is composed of many protein subunits. Each of these, however, is identical and is in turn made up of 158 amino acids. Since the genetic code is a triplet one, this means that 474 nucleotides are needed to translate the protein of the TMV. Surprisingly, analysis of the TMV RNA shows it to be much longer, 6400 nucleotides long and apparently in a linear rather than a circular form. There is thus a great deal more

FIG. 18-15. Isolation of the lac gene. *A.* The two phages selected for the isolation possess certain special features. In each, the DNA is composed of one strand which is heavy and one which is light. This enables them to be easily separated under laboratory conditions. Both phages had incorporated the bacterial gene, lac, but the gene is inserted in the opposite direction in each case. Consequently, the sense strand of the lac gene is associated with the heavy strand in one phage and with the light strand in the other. The same is obviously true for the nonsense strand. *B.* The two strands of each virus can be separated by centrifugation after melting of the double-stranded DNA. The heavy strands of each are then brought together. The DNA strands of the lac gene, incorporated into the phage DNA, form a double helix, because they are complementary. The tails of the DNA of the two different phages remain attached. Enzyme digestion removes the phage DNA which has remained single stranded, because the strands are from different phages and are not complementary. This leaves isolated the complete lac gene, composed of a sense (+) and a nonsense (−) strand.

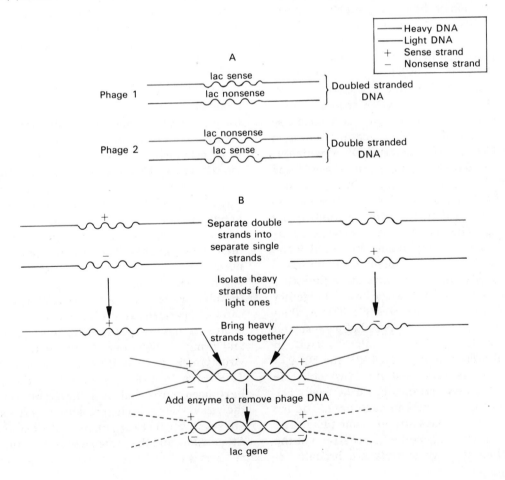

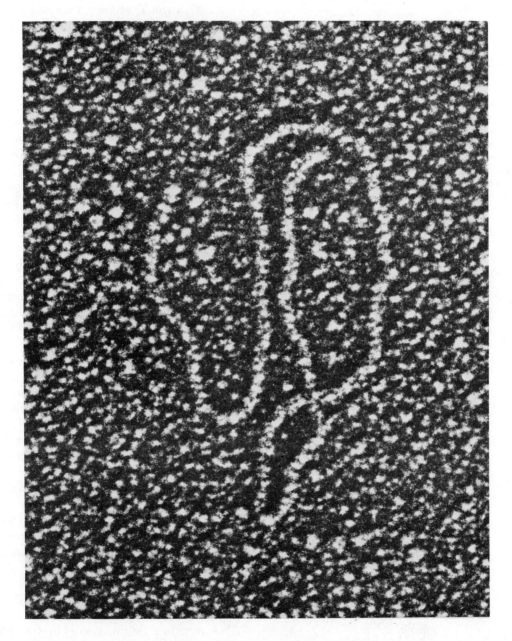

FIG. 18-16. Electron micrograph of lac gene isolated by procedure summarized in Fig. 18-15. (Reprinted with permission from J. Shapiro, L. MacHattie, L. Eron, *et. al. Nature, 224:* 768–774, 1969.)

genetic information present than that needed for the production of the single type of protein found in the TMV coat. When the RNA of the virus enters a host cell, it acts as a template for the synthesis of more viral RNA, and it also governs the formation of the coat protein which then combines with the RNA. These processes undoubtedly involve several enzymes, proteins which would not be present in the completed coat. This would account for some of the other messages which are evidently coded in the viral RNA.

RNA VIRUSES AND THE DIRECTION OF FLOW OF GENETIC INFORMATION. RNA viruses are not peculiar to plant hosts. A number of them is known to attack bacterial and animal cells, among them those responsible for polio, influenza, and the common cold. The RNA of many of the animal viruses, as well as the RNA of many bacteriophages, infects the cell as a single RNA strand surrounded by a protein coat. After entry, the single RNA strand of a phage, the "+" strand, can act as a template for the synthesis of a complementary strand, the "−" strand. In turn, the "−" strand may act as a template for the formation of new "+" strands. Any "+" strand may also act as messenger RNA and attach to ribosomes (Fig. 18-18A). Indeed, after entry of the "+" strands into a cell following infection by virus particles, "+" strands at first act as mRNA. This is essential for the synthesis of the enzyme RNA replicase, required for the formation of RNA on an RNA template. This enzyme is *not* produced in an uninfected cell. An uninfected cell never contains RNA molecules capable of serving as templates for RNA formation. The "+" strands are coded for the formation of this enzyme needed for the replication of the virus. They also contain the information for the coat proteins of the infective virus. These coat proteins eventually surround "+" strands which were formed in the cell on the "−" strand templates. Finally, an infected cell bursts, releasing infective viruses, each with a single "+" strand.

As mentioned in the discussion of TMV, this process is similar to the manner in which the DNA transfers information (Fig. 18-18A). The flow of genetic information pictured in Figure 18-18B was first proposed by Crick. It seemed to

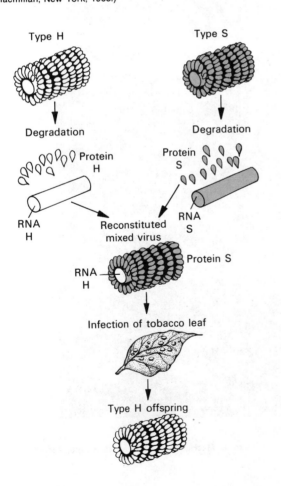

FIG. 18-17. Reconstitution of tobacco mosaic virus. The RNA of one strain of virus can be combined with the protein coat of another strain. When such hybrids infect the tobacco plant, the viruses which are later isolated are typical of the one from the strain which supplied the nucleic acid. The RNA of the virus must therefore direct the production of more specific RNA and also of the specific protein which surrounds it. (Modified slightly and reprinted with permission from M. W. Strickberger, *Genetics*, Macmillan, New York, 1968.)

UNDERSTANDING GENETICS

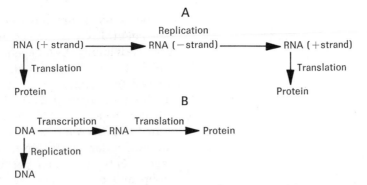

FIG. 18-18. Flow of genetic information. *A.* In many RNA viruses, the nucleic acid can bring about the formation of more RNA. Some of the RNA can act as mRNA which is translated into protein. *B.* DNA contains information which can guide the formation of more DNA or of RNA. The latter contains information imparted to it by the DNA, and this is then translated into protein. The concept of the flow of genetic information only in one direction, from DNA to RNA to protein, is known as the "central dogma."

be the universal process for transferring coded information from DNA, and so the concept became established as the "central dogma" of molecular biology. The central dogma was taken by many biologists to mean that the flow of genetic information from DNA to protein could occur only in the one direction, as indicated in the figure. Protein could not impart information to other protein or to RNA. RNA could not impart information to DNA. Any such "reverse flow" of information, as from RNA to DNA, would be a violation of the central dogma.

Therefore, it was with a great deal of excitement that certain findings were reported from studies with a group of RNA viruses; these appeared to violate the central dogma. The particular assortment of viruses includes some which are capable of inducing tumors in several animal species: rats, mice, and chickens, *et al.* The Rous sarcoma virus has been one of the most widely studied members. After infecting a chicken cell, the virus can transform it into a cancer cell which can in turn divide and produce new virus particles. The first clue to the unexpected way in which this virus replicates came through the investigations of Temin, who worked with the antibiotic, actinomycin D, as part of his analytical procedure. This substance had been shown to prevent transcription of DNA into RNA by blocking the DNA as a template. Therefore, it has become a commonly used tool for the determination of DNA-dependent RNA synthesis. If applied to a cell infected with an RNA virus, RNA would still be produced (Fig. 18-19*A*). This is so, because the RNA template

FIG. 18-19. Effect of actinomycin D. *A.* The drug blocks DNA as a template, so that no RNA transcript can be formed when it is applied to a cell (*above*). When applied to a cell harboring an RNA virus (*below*), more viral RNA is made, because the drug cannot block an RNA template. *B.* Reproduction of the RNA of the Rous sarcoma virus entails DNA which evidently depends on the RNA template. This DNA in turn gives rise to more viral RNA. If actinomycin D is applied, production of the viral RNA ceases, because the DNA template is blocked and is needed to guide the formation of the RNA of the virus.

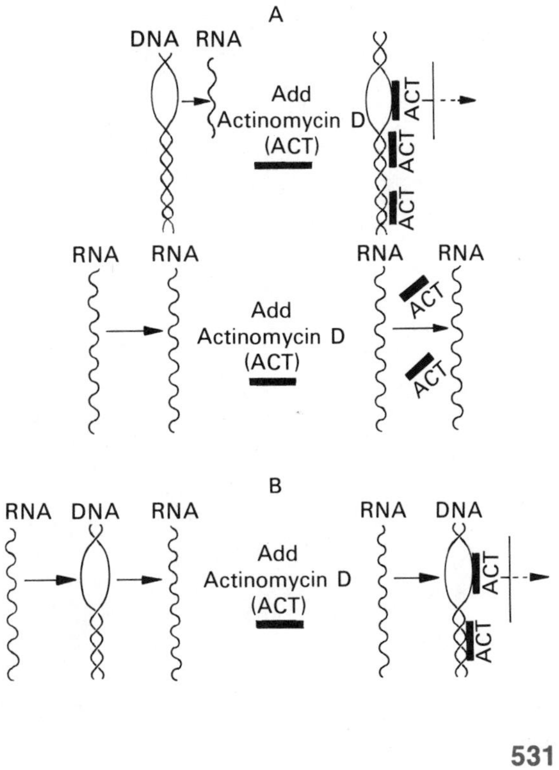

of the virus is not blocked. Therefore, RNA-dependent RNA synthesis would still go on, because it is only DNA-dependent RNA synthesis which is shut off by the antibiotic. When Temin applied the drug to cell cultures infected with the Rous sarcoma virus, he found that all RNA production ceased. This was perplexing and suggested that DNA might somehow be involved in the production of the RNA virus. A variety of experiments was then performed by Temin's group and others, and the data strongly supported the idea that *new* DNA is produced in a cell infected with the virus. This DNA is actually viral DNA and is formed on an RNA template, the RNA of the infective virus (Fig. 18-19B). If actinomycin D is applied to a cell containing the Rous virus, the viral DNA formed from the viral RNA template cannot undergo transcription into RNA, and so no new viral RNA is made by the cell.

Temin's group succeeded in demonstrating the presence of a DNA polymerase in the Rous sarcoma virus particles as they occur *outside* the cell. This DNA polymerase can form DNA from an RNA template as well as from a DNA template. It has been designated "RNA directed DNA polymerase." In popular usage, this polymerase is often called "reverse transcriptase." This term is somewhat unfortunate, because it has now been demonstrated that DNA polymerases which usually use DNA as a template can also, under certain conditions, use RNA as a template to make DNA. In the Rous sarcoma virus, however, the polymerase preferentially uses RNA as the template for DNA, whereas most familiar DNA polymerases use a DNA template. This RNA-directed DNA polymerase of the Rous sarcoma virus has been shown to produce a very small, double-stranded DNA molecule which contains base sequences complementary to the RNA. The virus DNA polymerase thus uses the RNA of the virus to build up a corresponding DNA strand. A DNA strand complementary to this one is then formed, giving double-helical DNA.

IMPLICATIONS OF THE FLOW OF INFORMATION FROM RNA TO DNA. The unexpected findings with the Rous virus hold implications for various cellular phenomena. It is apparent that a strict interpretation of the central dogma of molecular biology must be modified to include reverse flow of information from RNA to DNA. The findings also demand changes in many ideas on the interactions between the genetic material of a virus and the host cell. In addition to the DNA polymerase, the Rous virus also contains other enzymes. Among them is a ligase, which can repair breaks in the DNA. Its presence suggests a way in which an RNA virus can seem to disappear in an infected cell, only to reappear at some later time. After entering the cell, the RNA virus, under the influence of the DNA polymerase, could transfer its information to a bit of DNA (Fig. 18-20A). This in turn could become inserted into the chromosome of the host cell through the activity of the repair enzyme. Although the RNA form of the virus disappears, all of the information to make RNA virus later is present in the cell and is carried along with the cell's own genetic material. Earlier in this chapter, we noted that a symbiotic virus may be activated by an environmental agent. Some such trigger could cause the integrated viral DNA to become activated suddenly and to guide the formation of RNA viruses in the cell (Fig. 18-20B). Conceivably, such activation could result in a transformation of the cell from normal to malignant. Although these ideas are somewhat speculative, they afford a new outlook on the nature of viruses and their relationship to the genetics of the cell. We have seen that the virus can be considered an episome. In a higher cell, perhaps episomelike entities are transmitted from one generation to the next after their integration with the genetic material of the host cell. They may become capable of expressing themselves, given the appropriate set of conditions. The borderline between a gene of the host and the virus itself is seen to be a very vague one.

The result of the virus activities may be detrimental to the cell, causing it to become abnormal in some way or even malignant. However, there are still other possibilities to consider. Conceivably, such a virus particle could impart a benefit to the host. The cells of a higher organism might harbor entities comparable to the prophages of lysogenic bacteria, which act very much like genes of the organism itself. The

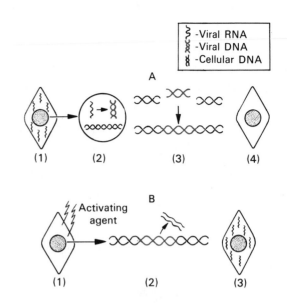

FIG. 18-20. RNA virus in a cell. *A.* Disappearance of viral RNA. Evidence suggests that an RNA virus may enter a cell (*1*) and bring about the production of viral DNA (*2*). This DNA may insert into the chromosome with the aid of an enzyme. The viral DNA would then replicate along with the DNA of the cell (*3*). The cell appears to be virus free (*4*), although it contains the information in its nucleus to manufacture viral RNA. *B.* Reappearance of viral RNA. It is conceivable that the viral DNA may become activated if the cell is subjected to certain environmental influences (*1*). The viral DNA may then undergo transcription into viral RNA (*2*). This viral RNA would then be detectable in the cell and could cause some cellular abnormality (*3*).

genetic information for some special ability could reside in such elements and not in the genes of the host.

RNA tumor viruses all contain an enzyme system similar to the Rous virus; however, all viruses with these enzymes do not produce tumors. Since DNA polymerases in general may be induced to use RNA as a template, there is the possibility that the flow of genetic information from RNA to DNA may occur in normal cells as a part of normal cell metabolism. Temin suggests that a chromosomal gene may undergo transcription into RNA. This RNA in turn may act as a template for the synthesis of the gene again (Fig. 18-21). If this newly constructed gene becomes inserted into the chromosome, the particular gene is now present in an extra dose. In this way, genes could become amplified in different cell types; such a process could be important in normal cell change or differentiation.

AN RNA VIRUS GENE SEQUENCE. Research with RNA viruses is contributing information which holds great significance for molecular biology, as seen by the accomplishment of Fiers and his group who have reported the exact makeup of an RNA gene. This team employed another phage of *E. coli,* the MS_2 virus, one of the simplest viruses known, and determined the complete nucleotide sequence of one of its genes. The specific gene is one which governs the formation of the protein in the coat of the virus. The protein contains 129 amino acids. The exact order of these amino acids had been established employing procedures similar to those used for the determination of amino acid sequence in hemoglobin (Chap. 13). The RNA of the MS_2 infective virus serves both as messenger RNA and as the genetic material of the virus itself (the "+" strand). This RNA was fragmented by enzyme treatment, and the sequence of nucleotides in the fragments was then determined. Since the amino acid sequence of the coat protein was known, the workers could relate this sequence to the sequence of bases in the RNA. For example, a fragment of RNA containing a sequence of nucleotides will code for a corresponding stretch of amino acids in a polypeptide. Such a fragment of RNA could be identified as part of the coat protein, because it could be related to a portion of the coat protein (Fig. 18-22).

In the procedure, Fiers and his team used RNase to digest the viral RNA into fragments. These fragments were isolated on gels and separated by fingerprinting. The sequence of the nucleotides in a fragment was then compared to amino acid stretches in the coat protein. By doing this, it could be determined whether or not the particular RNA fragment was in the coat protein gene. The work demanded analysis of even larger, overlapping fragments of the coat protein gene. Finally, the group ended up with four long fragments. These were not overlapping. And the nucleotide sequence of the fragments corresponded exactly to the polypeptide stretches of the coat protein. It was possible to place the fragments in the proper order by referring to the amino acid sequence of the coat protein. In this way, the order of the nucleotides of the entire coat protein gene was determined along with nucleotide sequences which are not usually translated. Included among these are

FIG. 18-21. Gene amplification. It is possible that the flow of information from RNA to DNA may play a role in the normal cell nucleus. Conceivably, a gene (gene "C" in the figure) could undergo transcription to RNA. This RNA in turn could act as a template to guide the formation of the specific gene. With the aid of an enzyme, this duplicated gene could become inserted into the chromosome, which would now possess a double dose of the particular gene.

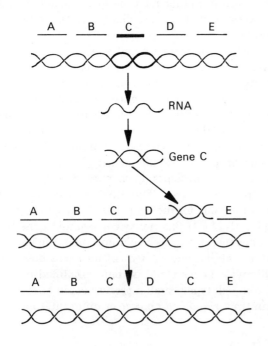

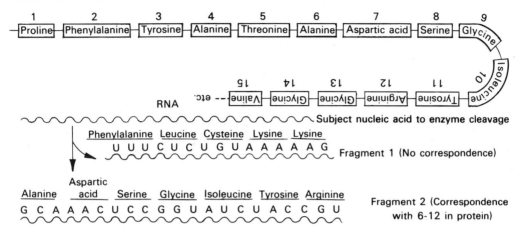

FIG. 18-22. Determination of nucleotide sequence in a viral RNA gene. The sequence of the amino acids in the protein governed by the particular gene is known and is used as a reference. The nucleic acid of the virus is fragmented by enzyme treatment. The nucleotide sequence within a fragment is determined. The sequence corresponds to codons which designate amino acids. The sequence of the codons can be related to the protein by comparing the amino acids designated by the codons to the known amino acids in the protein. If a sequence (as in fragment 1) does not correspond to any portion of the protein, it is discarded, because it is not part of the gene which determines the amino acid sequence of that protein. If a stretch of nucleotides (and consequently the amino acids they designate) does correspond to a stretch within the protein (fragment 2), it may be a part of the gene which governs the protein. As more and more fragments of different sizes are examined and related to the protein, the entire sequence of nucleotides within the specific gene can be determined.

sequences which serve as punctuation marks by providing "stop" and "start" signals.

Since the coat protein contains 129 amino acids, this means that a stretch of 387 nucleotides must determine the amino acid sequence. Once this RNA stretch which represents the coat protein gene was recognized, an opportunity was available for the first time to compare directly each amino acid with each code word or codon. The actual comparison *directly* confirmed the genetic dictionary, as it had been established by other approaches.

Fiers and his group showed that the RNA of the coat protein gene is not digested at random by enzymes. This observation, in addition to data from electrophoretic studies, provided strong evidence that the coat protein gene has a certain secondary structure and is not simply a linear stretch of nucleotides. According to their "flower model" (Fig. 18-23), the most stable configuration, with the greatest base pairing, is one in which there is a long double-stranded stem associated with a set of double-stranded loops or "petals." The base pairing maintains these loops and hence the secondary structure of the gene. The suggestion has been made that when the gene is not being transcribed, it may be folded and opens up as it is being read.

The determination of the exact nucleotide sequence of a gene of an RNA virus has profound implications for the subject of malignant transformation. Most of the viruses known to cause cancer in animals are RNA viruses. If the genes of such viruses can be isolated along with the "stop" and "start" signals, there is the hope of learning how to turn off permanently any cancer-causing virus which might be present in a cell. The problem of normal cell change or differentiation also involves "stop" and "start" signals and is the main topic of the next chapter. We will see how information from microorganisms has laid a foundation for some understanding of this critical process. Current studies with virus-host systems may force reinterpretations of various other well-established ideas in addition to the central dogma. They are almost certain to provide clues to the solution of the riddle of normal and pathological cell changes.

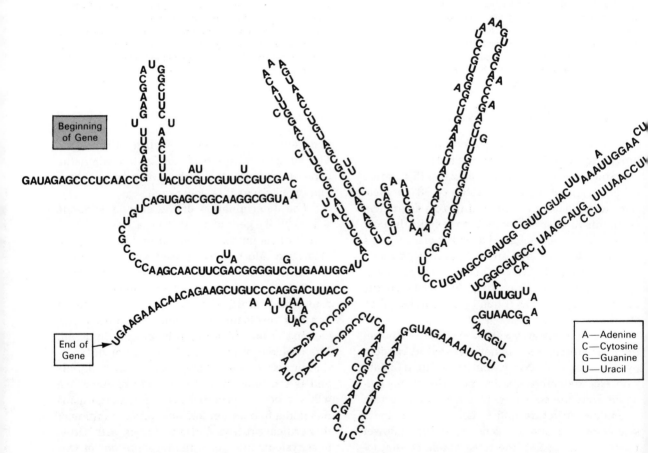

FIG. 18-23. Flower model of the gene. Data suggest that the coat protein gene of the MS$_2$ virus possesses a secondary folded structure which roughly resembles that of a flower on a stalk. Base pairing is responsible for holding the gene in this configuration. (Reprinted with permission from W. Min Jou, G. Haegeman, M. Ysebaert et al. Nature (Lond.), 237: 82–88, 1972.)

REFERENCES

Benzer, S. On the topology of the genetic fine structure. *Proc. Natl. Acad. Sci., 45:* 1607, 1959.

Campbell, A. M. How viruses insert their DNA into the DNA of the host cell. *Sci. Am.* (*Dec.*): 1976.

Cold Spring Harbor Symposium on Quantitative Biology. *Replication of DNA in microorganisms*, vol. 33, 1968.

Cold Spring Harbor Symposium on Quantitative Biology. *Transcription of Genetic Material*, vol. 35, 1970.

Fiddes, J. C. The nucleotide sequence of a viral DNA. *Sci. Am.* (*Dec*): 54, 1977.

Fraenkel-Conrat, H., and B. Singer. Virus reconstitution II. Combination of protein and nucleic acid from different strains. *Biochim. Biophys. Acta, 24:* 540, 1957.

Gajdusek, D. C. Unconventional viruses and the origin and disappearance of kuru. *Science, 197:* 943, 1977.

Hayes, W. *The Genetics of Bacteria and Their Viruses: Studies in Basic Genetics and Molecular Biology,* 2nd ed. Wiley, New York, 1968.

Holland, J. J. Viruses, slow, inapparent, and recurrent. *Sci. Am.* (*Feb*): 32, 1974.

Kamine, J., and J. M. Buchanan. Cell-free synthesis of two proteins unique to RNA of transforming virions of Rous sarcoma virus. *Proc. Natl. Acad. Sci., 74:* 2011, 1977.

Luria, S. E., and J. E. Darnell. *General virology,* 2nd ed. Wiley, New York, 1968.

Marx, J. L. RNA tumour viruses: getting a handle on transformation. *Science, 199:* 161, 1978.

Min Jou, W. G. Haegeman, and M. Ysebaert et al. Nucleotide sequence of the gene coding for the bacteriophage MS_2 coat protein. *Nature, 237:* 82, 1972.

Okada, Y., Y. Nozu, and T. Ohno. Demonstration of the universality of the genetic code *in vivo* by comparison of the coat proteins synthesized in different plants by tobacco mosaic virus RNA. *Proc. Natl. Acad. Sci., 63:* 1189, 1969.

Pratt, D. Single stranded DNA bacteriophages. *Ann. Rev. Genet., 3:* 343, 1969.

Sanger, F., G. M. Air, B. G. Barrell, and N. L. Brown et al. Nucleotide sequence of bacteriophage ϕX174 DNA. *Nature, 265:* 687, 1977.

Smith, M., N. L. Brown, G. M. Air, and B. G. Barrell et al. DNA sequence at the C termini of the overlapping genes A and B in bacteriophage ϕX174. *Nature, 265:* 702, 1977.

Stent, G. S. *Molecular Biology of Bacterial Viruses.* Freeman, San Francisco, 1963.

Temin, H. M. RNA-directed DNA synthesis. *Sci. Am.* (*Jan*): 24, 1972.

REVIEW QUESTIONS

1. Let A, B, and C represent three strains of *E. coli*. Drops of culture fluid are interchanged as shown below. The recipient tubes are examined in each case for clearing (+) or no change (−). Explain the results.

Drop derived from:	Recipient	Clearing (+)
A	B	+
A	C	+
B	A	+
B	C	+
C	A	−
C	B	−

2. The order of entry of markers in a certain Hfr strain is:

$$\overset{\longleftarrow}{\underline{\quad T \quad L \quad Lac \quad \lambda \quad S \quad}}$$

The three crosses below are made. Answer the questions which follow, and offer an explanation.

(1) Hfr T^+ L^+ Lac^+ Gal^+ λ^+ $S^s \times$
$\qquad\qquad \times F^-$ T^- L^- Lac^- Gal^- λ^- S^r

(2) Hfr T^+ L^+ Lac^+ Gal^+ λ^- S^s
$\qquad\qquad \times F^-$ T^- L^- Lac^- Gal^- λ^+ S^r

(3) Hfr T^+ L^+ Lac^+ Gal^+ λ^+ S^s
$\qquad\qquad \times F^-$ T^- L^- Lac^- Gal^- λ^+ S^r

A. From which of the crosses would it be possible to obtain recombinants which are Lac^+ Gal^+ λ^-?

B. From which of the crosses would it be possible to obtain recombinants which are Gal^+ λ^+?

C. From which of the crosses would it be possible to obtain rare recombinants which are S^s?

3. A. Give some ways in which transformation differs from specialized transduction.

   B. How does conversion resemble and differ from transduction?

4. A culture of *E. coli* cells which are λ⁻ Gal⁻ is infected with lambda phages which have been residing in cells which were λ⁺ Gal⁺. The concentration of virus used to infect the *E. coli* is very low. A small percentage of Gal⁺ cells is isolated. However, none of these transduced cells is found to be lysogenic, and none can be made to liberate virus. Explain.

5. *E. coli* cells which are λ⁻ Gal⁻ may become transduced by phage lambda to Gal⁺ cells. These cells can then give rise to colonies. It is then found that 1% to 10% of the cells in most of the colonies cannot utilize galactose. Offer an explanation.

6. One type of T/2 virus is h⁺r⁺ and another is h r. The two types are allowed to infect Strain B of *E. coli*. Plating of viruses from the lysate is made on a mixture of Strains B and B/2 of *E. coli*. Plaques are then scored for size and cloudiness. From the information below give the percentage of recombination and the viral genotypes represented by the plaques.

Small, clear	42
Small, cloudy	2195
Large, clear	2230
Large, cloudy	33

7. When an experiment is performed such as that described in Question 6, the virus partricles emerging from a single cell may be followed further. It is often found that one recombinant class, h⁺r or h r⁺, may not be found at all. Offer an explanation.

8. Assume that a DNA segment of the infective form of the virus φX 174 has the nucleotide sequence: *A A G T T A C C A*. The single-stranded DNA infects an *E. coli* and mRNA is formed by the replicative form. Give the sequence of bases in the mRNA which would relate to the segment given above.

9. The sequence in the gene below has the same sense as the mRNA. Consulting a genetic dictionary, if needed, determine where it is possible in the sequence for two genes to overlap, assuming a difference in reading frames.

AAATTAGCTCCCGGAGCGTGATGTCTAAAGGT

10. If a DNA strand has the same sense as the mRNA, then the triplet GTT corresponds to the amino acid valine.

    A. What is the corresponding DNA sequence and its polarity in the strand which underwent transcription?

    B. Write the corresponding anticodon showing the polarity.

    C. Show the polarity of the triplet GTT mentioned above.

11. Assume that the "+" strand of an RNA virus enters a bacterial cell. A stretch of the RNA in the "+" strand is as follows: AACAG-GACGCAG.

    A. What will be the nucleotide sequence in the nucleic acid which attaches to the ribosomes of the host?

    B. What RNA sequence is required for rellication of the sequence shown above?

    C. Name the enzyme required for replication of the viral RNA.

12. In various kinds of experiments in molecular biology, it is possible to isolate mRNA of a given kind. It is often desirable for accuracy of the procedure to have many copies of the sense DNA strand from which the RNA was transcribed. Suggest how this might be solved and give the logic.

    Circle the correct answer or answers, if any, for each of the following:

13. Strain K of *E. coli* is subjected to infection with two different rII mutants of phage which are separable by crossing over. Following infection, lysis occurs. The following would appear to be true.

    A. Either of the rII mutants by itself can obviously lyse strain K.

    B. These two rII mutants must be in different functional segments.

    C. These two rII mutants are part of the same cistron.

    D. These two rII mutants are recons within the same cistron.

    E. These two rII mutants are allelic to each other.

14. Assume that a, b, and c are recessive mutations and that they are very closely linked. Suppose the following double heterozygotes have been constructed and that the resulting phenotypes are as indicated here:

$$\frac{a\ b^+}{a^+\ b} = \text{mutant} \qquad \frac{a\ b}{a^+\ b^+} = \text{wild} \qquad \frac{b\ c^+}{b^+\ c} = \text{wild}$$

$$\frac{b\ c}{b^+\ c^+} = \text{wild} \qquad \frac{a\ c^+}{a^+\ c} = \text{wild} \qquad \frac{a\ c}{a^+\ c^+} = \text{wild}$$

The following would appear to be true:

    A. "b" and "c" are engaged in different functions.

    B. All three are engaged in the same function.

    C. "a" and "b" are complementary to each other.

    D. "a" and "b" are alleles.

    E. The three sites are all part of the same cistron.

    F. "b" and "c" are alleles.

15. In the isolation of the lactose gene

    A. The lactose sense strand was attached to the heavy DNA strand of both of the transducing phages.

    B. The transducing phages carried single-stranded DNA, and one of them carried a sense strand; the other one a nonsense strand.

    C. The DNA components of the two viruses were fully complementary and formed double-helical DNA when brought together.

    D. The lactose sense and nonsense strands formed double-helical DNA resistant to the enzyme digestion used to isolate the gene by itself.

E. Both sense strands were separated from the phages and then brought together as a double helix, thus isolating the complete lactose gene.

16. In tobacco mosaic virus:

A. The nucleic acid component is single-stranded DNA.

B. The protein coat is composed of many subunits, each differing from the next.

C. There is little genetic information present beyond that required to code for the coat protein.

D. When a hybrid virus, composed of nucleic acid from one strain and protein coat from another, is allowed to infect a plant, the offspring viruses have a protein coat typical of the strain which provided the nucleic acid.

E. RNA of the virus governs the formation of more RNA and protein.

17. The work of Temin and others with the Rous sarcoma virus has shown.

A. All RNA production ceases if actinomycin D is applied after infection with the virus.

B. That the virus carries into the cell an enzyme which carries on exclusively RNA-dependent RNA synthesis.

C. RNA may serve as a template for the formation of a complementary DNA strand.

D. A double-stranded DNA arises in an infected cell which has base sequences complementary to the viral DNA.

E. The virus carries enzymes which can repair breaks in the DNA.

F. All animal RNA viruses are tumor-inducing.

18. In the work which gave the exact nucleotide sequence of a gene of an RNA virus.

A. The virus employed has double-stranded RNA.

B. The amino acid sequence of the coat

protein was deduced from studying fragments of the viral nucleic acid.

C. The nucleotide sequence of the "−" strand was fragmented and related to the amino acids in the coat protein.

D. The concept that the genetic code is degenerate was violated.

E. Evidence was obtained that a gene is folded when it is not being transcribed.

19

CONTROL MECHANISMS AND DIFFERENTIATION

CELL SPECIALIZATION AND THE GENETIC IN-
FORMATION. In higher organisms, genes in the
many kinds of cells are continually undergoing
transcription, which is then followed by translation essential for normal cellular function. The
diverse cell types which compose a multicellular
organism are engaged in a host of activities, but
each type of cell has a specific role which it
performs for the benefit of the individual. Higher
forms of life are differentiated into different kinds
of cells which compose distinct tissues. In humans, a cell of the bone marrow engages primarily in synthesis of hemoglobin, whereas a
pancreatic cell produces insulin. We do not expect skin tissue to secrete gastric juice any more
than we expect stomach cells to form pigments
which screen out light rays. Each kind of differentiated cell carries out a special function, and
this is reflected mainly in the specific proteins
which it produces.

A moment's thought tells us, however, that
the genetic information in all the cells of the
body is the same. From the zygote onward, a
human arises through repeated nuclear divisions; except for cells in the germ line, all of
these are mitotic. The first few mitotic divisions
of the embryo produce cells which are all essen-

tially the same, but then something happens to cause them to diverge and give rise to different cell types. If specialization or differentiation had never occurred in living things, only masses of identical cells could form, and life as we know it would not exist. Not only is differentiation required for the orderly development of a normal organism, it must not be upset once it is established. A pathological disorder, such as a malignancy, reflects some type of change in the cell which causes it to depart from its normally defined role. Differentiation and its maintenance depend mainly on cell mechanisms which control the formation of specific protein products. Obviously, only a part of the total genetic information is expressed in any one cell. A liver cell will not translate information pertinent only to a cell of the eye, even though all of the information is there in the liver cell nucleus. How do diverse cell types arise from a population of identical cells? What devices permit only certain genes to function in a given kind of cell? Although these questions are still largely unsettled, genetic studies with microorganisms have contributed important information which has great bearing on differentiation and cell change in higher species.

GENETIC CONTROL OF ENZYME INDUCTION IN BACTERIA. The critical investigations have depended on the isolation from *Escherichia coli* of certain mutants in which the control mechanisms have gone awry. For example, a normal *E. coli* cell can utilize the sugar, lactose, when it is supplied in place of glucose. The ability to split the lactose, a disaccharide, into the monosaccharides, glucose and galactose, depends on a specific enzyme, β-galactosidase, which the normal cell can manufacture. However, in the absence of lactose, this enzyme is not needed, and it is not produced. Indeed, it would be a waste of energy for the cell to produce any protein which it does not require. Normally, therefore, the enzyme is made only when lactose, the substrate of the enzyme, is supplied. Such a cellular reaction is called *induction,* the formation of a specific enzyme in response to the presence of the substrate. However, in populations of bacteria, certain mutant cells arise in which enzyme is formed regardless of the presence of the substrate. These are called *consti-*

tutive mutants, and they produce large amounts of the enzyme both in the absence and the presence of substrate (Fig. 19-1).

In the metabolism of lactose, two enzymes besides the β-galactosidase are also needed (β-galactoside permease and galactoside transacetylase). These are involved in the localization and concentration of the lactose within the cell before it is cleaved by the galactosidase into glucose and galactose. *All three* enzymes are formed by the constitutive mutants in the absence of the substrate, lactose. Chemical analysis of these three protein products in the constitutive mutants shows them to be normal. Therefore, nothing has gone wrong to alter the composition of the protein. Instead, a mutation has interfered with the cell's ability to produce the three enzymes only when they are needed. The constitutive mutant cell cannot control the level of enzyme production; it produces normal

FIG. 19-1. Wild-type cultures and constitutive mutants. Normal *E. coli* cells do not waste energy by producing the enzyme beta-galactosidase in the absence of the enzyme's substrate (1). A culture of the wild *E. coli* cells can be induced to produce the enzyme when lactose is supplied as the sugar source. In contrast, constitutive mutants produce the enzyme both in the absence and presence of the lactose (2). These mutants have some defect in a control mechanism.

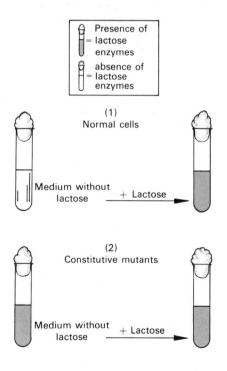

enzyme indiscriminately, a process which wastes cell energy.

The lactose system is just one of several in which mutations are known which can upset normal controls, but it has been studied in the greatest detail. The information gained from it seems to apply to the other systems as well and has permitted Jacob and Monod to formulate a model of genetic control. This model is based on the results of crossing various kinds of mutants in different combinations. As in the case of any other proteins, there are also mutations which *can* alter the structure of the three enzymes of the lactose system. From crossing cells containing mutations of this sort, it has been possible to map the sites of the genes which control the formation of the three enzymes. Mapping has shown (Fig. 19-2A) that the three loci are clustered together in a sequence on the chromosome (z, y, and x for the galactosidase, the permease, and the transacetylase, respectively). A mutation in any one of them can alter the structure of its specific protein product. These genes and all others which direct amino acid sequences in enzymes or any protein which is not a regulatory substance is called a *structural* gene.

It is important to understand that mutations which affect the regulation of enzymes and thus produce constitutive mutants do *not* occur in the structural genes. Instead, they map in other locations. One kind of constitutive mutation maps in a portion of the chromosome which is a distance away from the three clustered structural genes. However, these mutations result in the unregulated production of the three enzymes of the lactose system. Such mutations have enabled Jacob and Monod to identify a *regulator* gene, a gene concerned not with the formation of an enzyme but rather with a regulatory substance (Fig. 19-2B). But the regulator gene is not the only one involved in control of the three structural genes of the lactose system. A gene called the *operator* has also been recognized as a result of mutations which have arisen within it. It has been mapped and shown to be located adjacent to gene z. Like the regulator gene, it is concerned with control and not with a finished structural protein product.

Jacob and Monod were able to show that the regulator gene is responsible for the production of a repressor substance, R (Fig. 19-3A). This

FIG. 19-2. The lactose operon. *A.* Three enzymes are involved in the metabolism of lactose. Each is controlled by a gene, and the three genes map together in a portion of the *E. coli* chromosome. These genes undergo transcription to form mRNA when the cell is induced by the presence of lactose. A mutation within any one of the genes can alter the structure of the protein it governs. Genes such as these which dictate the structure of enzymes are called "structural genes" and are the ones most familiar to us. *B.* Other genes, the operator and the regulator, can affect the lactose system. However, a mutation within either of these does not alter the structure of any one of the three enzymes. Instead, the mutation can result in a failure of the enzymes to be produced at the appropriate time. The operator gene is located adjacent to the structural genes. The promoter region to which RNA polymerase attaches to begin transcription is next to the operator. The structural genes plus the operator and promoter compose an operon. All the genes and sites within an operon are transcribed on the same mRNA. The regulator gene (*R*) maps on the other side of the promoter.

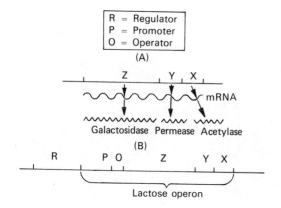

substance is produced by the regulator in the *absence* of the substrate, lactose. When lactose is not present in the cell, the repressor can combine with the operator. As a consequence, RNA polymerase which attaches to the promoter site cannot transcribe the information of the three genes into messenger RNA, even in the presence of cyclic AMP and CAP. Transcription is thus effectively blocked by the interaction of the repressor with the operator site.

Now let us suppose that lactose is added to the medium and enters the cells (Fig. 19-3B). The substrate, lactose, has the ability to act as an *effector* which binds the repressor molecules. The latter would normally be present only in a small amount. As a result of the interaction between repressor and effector (lactose), the operator becomes freed. The RNA polymerase bound at the promoter encounters no barrier and can proceed with transcription. However, if a constitutive mutation arises in the regulator gene, the regulator may be unable to produce a normal repressor. This altered repressor may be unable to combine with the operator. Consequently, the RNA polymerase is free to transcribe the structural genes at any time. Therefore, transcription can even take place in the absence of the substrate when the three enzymes are not needed (Fig. 19-3C).

If a constitutive mutation occurs in the operator gene, again the enzymes are produced without regard to the presence of substrate. This is so because the mutation in the operator renders it incapable of binding with the normal repressor (Fig. 19-3D). Consequently, the RNA polymerase is again free to transcribe the structural genes. It is to be noted that the three structural genes are undergoing transcription together on the same stretch of messenger RNA. Thus, the operator and the genes adjacent to it define a unit, a unit of transcription called the *operon*. Any operon is composed of an operator, a promoter, and one or more structural genes (or cistrons) which are all transcribed on the same piece of mRNA. A piece of mRNA which contains the coded information for two or more genes is polycistronic; it carries the information which can direct the formation of more then one protein. In transcription of the operon, the promoter region is apparently not included in the transcript, but the operator or a portion of it may

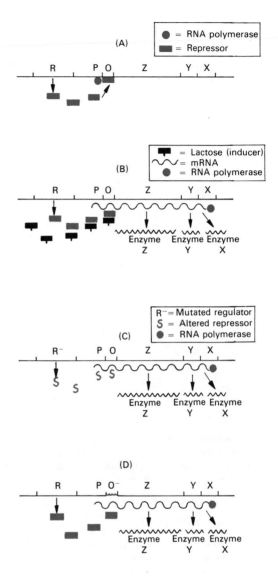

FIG. 19-3. The repressor. *A.* The regulator genes produce a substance which can combine with the operator. In the absence of the substrate, lactose in this example, combination of the repressor and the operator prevents the RNA polymerase from moving along the DNA to bring about transcription. *B.* When an inducer is added (the lactose molecules), it binds to the repressor. The operator site becomes freed. The polymerase may now move along the DNA to transcribe it into mRNA. *C.* If a mutation arises in the regulator gene, the repressor may become altered in structure so that it cannot combine with the operator. Therefore, even in the absence of the lactose, the RNA polymerase is free to transcribe genes in the operon. The mRNA is then translated, and the three enzymes are formed in the cell, even when they are not needed. *D.* A mutation may affect the operator in such a way that it can no longer combine with the normal repressor. The result again is freedom of the RNA polymerase to transcribe genes in the operon. The three enzymes are again formed constitutively, without regard to presence of the substrate.

be. We can now appreciate more fully the need for punctuation codons in the mRNA to distinguish between two or more polypeptide chains controlled by two or more different genes in an operon. The RNA codons, UAA, UAG, and UGA, along the messenger are needed to prevent a run-on protein. Without them, the ribosome would be unable to complete one protein and start the next. The three enzymes in the lactose operon would all be hooked together and would not exist as separate functional entities.

The model of the operon explains beautifully the control of inducible enzymes. Jacob and Monod did not identify the repressor, but they had sufficient evidence to propose its existence. For example, it was possible through sexduction (Chap. 17) to obtain *E. coli* cells which were diploid in the lactose region of the chromosome (Fig. 19-4). Partial heterozygotes were derived which possessed a normal regulator and a normal operator on the sexduced fragment; the chromosome contained a defective regulator. Such a cell behaves normally; it is not constitutive but produces enzyme only when needed. Figure 19-4 shows clearly why this is so. The defective repressor cannot bind with the operator. If the cell were haploid, lacking the F factor with the chromosome segment, enzyme would be indiscriminately produced (Fig. 19-3C). However, the presence of a normal regulator anywhere in the cell, even on the episome, gives rise to the production of normal repressor. This is a substance which can diffuse throughout the cell and bind with *both* normal operators, the one on the chromosome and the one carried by the sex factor. The structural genes on both DNAs are therefore not transcribed unless substrate can tie up the normal repressor. In effect, the constitutive regulator mutations are behaving as recessives to the normal gene form.

In contrast to this, the constitutive operator mutations were shown to be dominant ones. A study of Figure 19-5 shows why this would be expected. In the diploid state, one of the DNA segments has a normal operator (the F factor in the figure). The repressor substance produced by either normal regulator can bind to it. But this is not so with the operator which is mutant (the one on the chromosome). Since it cannot bind with the regulator, the structural genes adjacent

FIG. 19-4. A heterogenote for the regulator. A bacterial cell may possess a fertility factor which carries a lac operon and a regulator gene. The regulator on the chromosome may be defective and produce ineffective repressor. However, if a normal regulator is present in the sex factor, it will cause normal repressor to be formed. This will combine with the operator on the chromosome and also with the operator associated with the lac genes on the F factor itself. The cell consequently behaves normally and does not produce enzyme in absence of the substrate. The regulator mutation is thus behaving as a recessive to the normal.

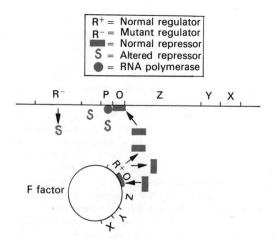

R⁺ = Normal regulator
R⁻ = Mutant regulator
▬ = Normal repressor
S = Altered repressor
● = RNA polymerase

to it (in the *cis* arrangement) are free to be transcribed in *both* the absence and the presence of the substrate.

REPRESSORS AND THEIR INTERACTIONS. Results such as these from various arrangements of mutant and normal alleles in operator, regulator, and structural genes allowed Jacob and Monod to propose the model of the operon in 1961. However, it was not until later in the decade that genetic repressors were actually isolated by the team of Gilbert and Muller-Hill working with the lactose operon and by Ptashne studying a repressor of phage in *E. coli*. It was expected that the repressors would contain an RNA component to allow them to recognize complementary DNA sequences in specific operators. Surprisingly, the repressor turned out to lack an RNA component and to be a rather ordinary acidic protein. The lactose repressor is an aggregate and consists of four identical polypeptide subunits. Each is composed of 147 amino acids whose exact sequence is known.

In one of the experiments, isolated repressor was made radioactive and mixed with the appropriate DNA. For example, lac repressor was placed in a centrifuge tube along with DNA known to contain the lactose genes. When spun in a density gradient, the repressor bound tightly to the DNA and sedimented with it. In contrast, when the repressor was spun in the centrifuge tube with DNA from cells containing lactose constitutive mutations, the repressor did not bind. These results provided strong support for the model of the operon and the concept of a repressor which physically blocks genetic transcription.

Extensive investigations have been made on the lactose operon throughout the years. The exact nucleotide sequence of the operator region is now known. This has been accomplished by subjecting *E. coli* DNA to nuclease treatment. Lactose repressor protein, when added to the DNA before exposure to the nuclease, protects the operator region to which it binds from digestion. Analysis of the protected portion has shown it to consist of 21 base pairs and to contain two regions where there is a degree of symmetry. This means that starting at a given point on a DNA strand and reading in the 5′ to 3′ direction,

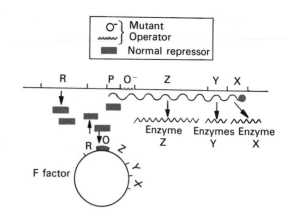

FIG. 19-5. A heterogenote for the operator. If a mutation occurs in an operator gene, it may not bind with the normal repressor. This means that those structural genes adjacent to it (in the *cis* arrangement) will be freely transcribed. The figure shows a normal operator in the sex factor combining with the repressor, which is being produced by both regulators, one on the chromosome and one on the episome. The operator on the chromosome does not combine with the repressor, so that the three enzymes will be produced in the presence or absence of substrate. The operator mutation therefore behaves as a dominant.

a corresponding region can be recognized on the complementary strand which reads the same in the 5′ to 3′ direction. For example:

```
5′  A A T T G T̲‾‾‾‾‾‾‾‾ A C A A T T  3′
3′  T T A A C A̲_____T̲G̲T̲T̲A̲A̲  5′
```

These regions may be segments which are recognized by the repressor and to which two subunits of the repressor bind. Not more than 20 copies of the lac repressor are usually found in a cell.

The 21 base pairs of the operator region are now know to contain the starting point for synthesis of lac messenger RNA. It will be recalled (Chap. 13) that the length of the promoter region for the lactose operon is 80 base pairs long and contains a CAP site as well as an RNA polymerase binding site. It has been shown that a bit of overlapping of nucleotides exists between the end of the lac promoter region and the beginning of the lac operator. We can now picture how the bound repressor prevents transcription in this operon. It would do so by preventing RNA polymerase from binding to the appropriate signals in the promoter region where the overlap occurs (Fig. 19-6).

REPRESSION OF ENZYME SYNTHESIS. The operon concept applies to other types of genetic regulation in bacteria, such as the *repression* of enzyme synthesis. This phenomenon is illustrated in *E. coli,* where the enzyme tryptophan synthetase is produced by cells growing in an environment which lacks the amino acid, tryptophan. The bacteria need tryptophan for growth and usually must synthesize it. Therefore, they manufacture tryptophan synthetase, an essential catalyst in synthesis of the amino acid. However, if tryptophan is supplied to the medium, the enzyme production is repressed. Again this is economical, because it would be wasteful for the cell to make both the enzyme and its end product, tryptophan, when there is a ready supply of the required substance. The operon model explains this repression in a manner similar to that used for the lactose operon (Fig. 19-7). The only difference between the tryptophan and lactose control systems is that the operator in the former does *not* react with the repressor in the absence of the effector, in this case the tryptophan. But when tryptophan is added to the cell, it acts as a *corepressor* and alters the repressor. It is this changed repressor which can then bind to the operator and block transcription. Therefore, in a repressible system, such as seen in the example of tryptophan, the effector (the end product in a reaction) alters the repressor so that it *can* combine with the operator and block transcription. On the other hand, in inducible systems, as in the example of the lactose operon, the effector (substrate of an enzyme) alters the repressor so that it *cannot* combine with the operator, thus allowing transcription to take place (see Fig. 19-3*B*).

The type of regulation in the examples of tryptophan synthetase and the lactose operon is

FIG. 19-6. Blockage of transcription by repressor in the lac operon. The lac promoter has been found to contain a CAP site to which the CAP-CAMP complex binds before transcription can proceed. The promoter also contains a region containing signals for the binding of RNA polymerase. At the end of this region is the site at which RNA polymerase starts transcription. This site overlaps with the beginning of the operator. If repressor is bound to the operator, this site is blocked to RNA polymerase and transcription cannot proceed.

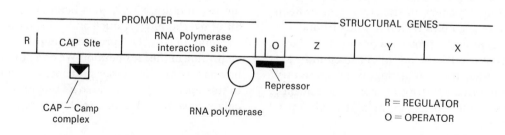

one of *negative control.* This is so since in cases such as these, the product of the regulator acts as a repressor, either alone or with a corepressor. Negative controls such as this, however, are by no means the only kind of control involved in gene regulation. Indeed, both cyclic AMP and CAP act as *positive control* elements, substances required for the activation of transcription.

Moreover, operons are known in bacteria (the L-arabinose and L-maltose operons, for example) in which a specific regulator gene associated with a specific operon produces a substance required for activation of the structural genes in the operon. Both kinds of control elements, negative and positive, are undoubtedly involved in the regulation of most structural genes.

HISTONES AND GENETIC CONTROL IN EUKARYOTES. Genetic control in the prokaryotic cell has undoubtedly evolved under the force of natural selection, for it guarantees a high level of efficiency with a minimum of wasted energy. Comparable mechanisms must operate in higher species to maintain efficient cell activities and to bring about cell differentiation by repressing and activating specific genes during development. It must be remembered that the genetic control mechanism of the prokaryotes is an "open" one; the DNA is not complexed with the other macromolecules which are typical of eukaryotic chromosomes. The genes of eukaryotes, on the other hand, are not exposed directly to the cellular environment. Any sort of repressor would not be free to combine with the DNA without encountering other molecules. Besides the DNA, it is the histone protein which predominates in chromatin. Histones occur only in eukaryotes, all of which have well-defined nuclei with classical chromosomes (see Chap. 2). The histones are distributed over the entire length of the chromosomal DNA and are bound to it, forming a nucleohistone complex.

It was suspected that the histones themselves might in some way perform a regulatory role. Histones are actually relatively small proteins or polypeptides (M.W. 10,000–20,000). They are very basic in nature due to the inclusion of large amounts of the basic amino acids. This basic property enables them to react with the

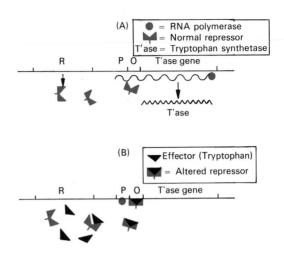

FIG. 19-7. Repression of enzyme synthesis. *A.* In the absence of tryptophan, the *E. coli* cell manufactures this necessary amino acid. The regulator is producing a repressor, but it cannot combine with the operator, so that transcription is allowed, and the required enzyme, tryptophan synthetase, is formed. *B.* If tryptophan is added to the cell medium, the enzyme is no longer necessary. The tryptophan itself acts as an effector which alters the repressor. This altered repressor can now combine with the operator and block transcription.

negative charges on the phosphates of the DNA and to form salt complexes. Histones may be removed from their combination with DNA by subjecting the chromatin to a 1M salt solution. The DNA, free of histone, can then be extracted. Following such a procedure, Huang and Bonner measured the ability of naked DNA to undergo transcription *in vitro.* They compared this with the ability of this same DNA to support RNA synthesis when it is associated with histones (Fig. 19-8). Working with chromatin from cells of the pea embryo, they found that *whole* chromatin was much less effective in guiding the formation of RNA than was the DNA which had been stripped of its histone. The deproteinized DNA supported RNA synthesis at five times the rate of the intact chromatin. The work was so designed that there was no doubt that both the naked DNA and the whole chromatin were supplied with sufficient RNA polymerase for transcription. The inhibitory effect on RNA synthesis was definitely due to the presence of the histone protein.

Bonner and his colleagues, as well as other teams, extended studies of histone inhibition to an assortment of cells from various sources. It was found that most of the chromatin in any kind of cell is actually repressed. Not more than 20% of the DNA of a cell is usually active in the synthesis of mRNA. Indeed, in some highly differentiated cells which are specialized for one main activity (such as hemoglobin synthesis by erythrocytes), almost all of the genetic material is repressed. Studies of a variety of plant and animal cells leave no doubt that the histones of the chromatin repress most of the genes in the cell. They do this somehow by preventing genetic loci from acting as templates in the formation of mRNA. And the repression is specific; the DNA which is repressed in one kind of cell differs from that which is turned off in another. Therefore, the genes which are active vary from one cell or tissue type to another.

Just how are the histones able to do this? How can they recognize the precise genes which are to be repressed? One might suspect that many different kinds of histones exist and that they would vary greatly from one tissue to another and from one species to another. But this is definitely not the case. The histone composition is amazingly similar in all kinds of eukaryotic cells, and there is no evidence of a specific histone type for a specific gene. It is not known how histones identify the correct genes nor how they bind to cause repression. However, many ideas have been presented on the subject. It is known that histones, when added to DNA, can cause it to assume a packed or more condensed configuration. Conversely, extracting the protein from condensed chromatin causes the latter to become more diffuse. These observations are significant, becaue many studies have shown that chromatin which is in an extended state (as in the nondividing nucleus) is much more active in synthesizing RNA than this same chromatin when it is in a packed state (as in the mitotic chromosomes or in the sperm). Recall that the Barr body is condensed chromatin of one X chromosome which is inactive in the cell.

Histones appear to exert their regulatory effect by inducing packed and unpacked states in the chromatin. They accomplish this somehow without actually coming off the chromatin and then later rejoining it. This is seen by the fact that the percentage of histone which is bound to both packed and unpacked chromatin does not vary greatly. There is good evidence that chem-

FIG. 19-8. Transcription by whole chromatin and deproteinized chromatin. Chromatin can be extracted from cell nuclei. This can then be added to a test-tube system containing the other required ingredients for RNA synthesis (see Fig. 12-4). The chromatin is able to generate the formation of RNA. If this same chromatin is stripped of its protein, to yield naked DNA, it is found that approximately five times as much mRNA is then generated *in vitro*.

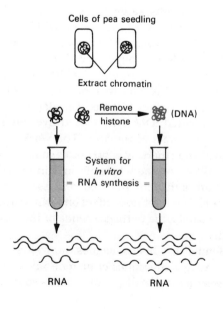

Cells of pea seedling

Extract chromatin

Remove histone → (DNA)

System for *in vitro* = RNA synthesis =

RNA   RNA

ical alterations take place in the histone as gene activity increases. As inhibition by histone becomes less, the number of acetyl groups and the degree of phosphorylation of the histone proteins are increased. This indicates that the histones are being modified somehow as their effectiveness as genetic inhibitors decreases.

Hormones are chemical messengers which can stimulate the synthesis of certain proteins in target cells, those cells which are receptive to the hormones. When given to animals *in vivo,* steroid hormones increase the rate of RNA synthesis. Chromatin taken from hormone-treated tissue is more active as a template in RNA synthesis than chromatin which is untreated. These and other observations have led some biologists to propose that the hormones act as inducers and that they do so by antagonizing gene repressors, perhaps by blocking histone binding in some way. Viewed in this light, certain hormones would act as inducers in much the same way as the inducers in the Jacob and Monod model and would resemble the effect of lactose, as it triggers the induction of the enzymes in the lactose operon (Fig. 19-9). It is known that the action of most steroids depends on their binding to very specific protein receptors in the cell cytoplasm. After the arrival of a steroid hormone, a receptor-hormone complex is formed, and this complex moves to the nucleus and eventually binds to the chromatin. Transcription is then somehow stimulated. Certain of the nonhistone proteins described below may act as acceptors in the chromatin for the binding of the steroid-receptor complex.

THE NONHISTONE PROTEINS AND REGULATION. It will be recalled that in addition to the histones, other proteins are also associated with the chromatin (Chap. 11). These are known collectively as the "nonhistone proteins," an assortment which includes different kinds of functional proteins (enzymes such as the RNA and DNA polymerases) and various acidic proteins. In contrast to the histones, which are quite constant in amount and composition from one cell type to another (and even from one species to another), the nonhistones occur in increased amounts in the chromatin of cells which are active. It has also been demonstrated that they are very variable in kind and are specific for a

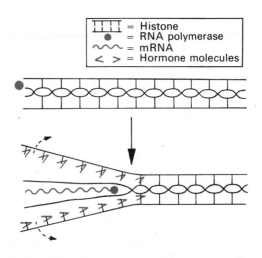

FIG. 19-9. Hormones as inducers. In this speculative diagram, the histones are pictured as binding the DNA and thus blocking transcription by RNA polymerase. Certain hormones could conceivably remove the blocking action in some way and free portions of the DNA so that it can unwind. RNA polymerase could then select one strand for transcription.

particular type of tissue within a species. Moreover, they show variation from one species to the next.

Several lines of experimental evidence implicate the nonhistone proteins (mainly the acidic ones) with the regulation of genetic activity. They may even be involved with the activation of that part of the genetic material which carries the information needed for mitosis and the very replication of the DNA itself. This is suggested by the fact that when cells are stimulated to divide (such as chemical stimulation of white blood cells), there is an increased uptake of amino acids into the nonhistone proteins *before* DNA synthesis starts.

The nonhistone proteins may in some way regulate transcription during different stages of the cell cycle, particularly in those cells which are continuously dividing. Certain techniques enable the investigator to selectively remove proteins from the chromatin, to add specific protein fractions back to denuded DNA (to reconstitute the chromatin), and to compare the amount of transcription under the various conditions (Fig. 19-10A). When nonhistone proteins from *inactive* cells are specifically used to reconstitue chromatin, the rate of transcription is *less* than when the chromatin is reconstituted with nonhistone proteins taken from active cells. Nonhistone protein can be obtained from nuclei which are in interphase and also from chromosomes which are in the mitotic cycle. The nonhistone proteins taken from cells in different parts of the cell cycle can then be used to reconstitute chromatin. A comparison is therefore possible between chromatin which has been reconstituted with nonhistone protein taken from interphase and chromatin which has been reconstituted from mitotic nonhistone protein. The results show that the latter type of reconstituted chromatin possesses a template activity which is less than that of chromatin reconstituted with the nonhistone protein of interphase (Fig. 19-10B). These observations agree with the fact that transcription is reduced at the time of nuclear division when the chromosomes are condensed for their orderly distribution; they also implicate the nonhistone proteins as agents which alter the amount of transcription at different times during the cell cycle. It is highly significant that chromatin which has been reconstituted using *histones* taken from the different stages of the cell cycle or from cells which

FIG. 19-10. Chromatin reconstitution. *A.* The procedure can be manipulated to associate DNA and chromosomal proteins obtained from cells in different stages of the cell cycle. One kind of reconstituted chromatin may contain DNA and nonhistone protein from interphase complexed with histone from mitotic cells (*left*). Another kind (*right*) may contain DNA and nonhistone protein from mitotic cells complexed with histone obtained from interphase. Chromatins reconstituted in various ways can then be compared for their ability to generate mRNA. *B.* Chromatin reconstituted with nonhistone protein from interphase has greater template activity than that chromatin reconstituted with nonhistone proteins from mitotic cells. (Fig. 19-10 continues on opposite page.)

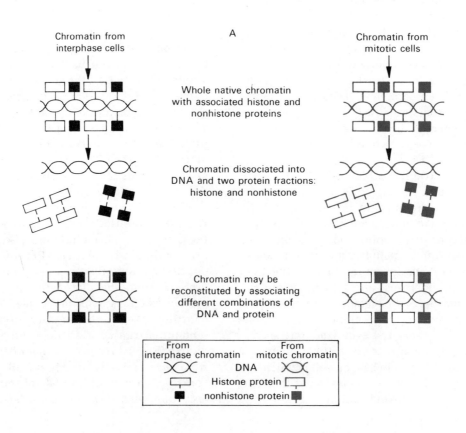

vary in their metabolic activities *does not* show these differences in template activity.

Although it appears that the nonhistone proteins do exert some regulatory role in the control of gene expression, the mechanism involved is unclear. Some recent observations indicate they may interact with the histones. It has been found that chromatin from mitotic cells (which shows a decrease in amount of transcription as compared to that of interphase chromatin) is bound more tightly to the histones than is the chromatin of interphase. Since chromatin reconstituted with nonhistone chromosomal proteins of mitotic cells has a decreased template activity in comparison with chromatin reconstituted with interphase nonhistone protein, it is conceivable that the nonhistone proteins can influence the degree or manner of binding of the histones to the DNA and in this way affect transcription. There are other lines of evidence which also support this concept.

There is also experimental work which indicates that the nonhistone chromosomal proteins play an acceptor role in binding steroid hormones, such as progesterone. As a result of the binding, the chromatin would become altered and the capacity of the chromatin for transcription could in turn be changed.

Phosphorylation of the nonhistone proteins seems to play an important role in gene activity. Many observations show that when chromatin becomes activated for RNA synthesis (such as the stimulation of gene activity in the prostate by testosterone), there is an associated increase in the rate of phosphorylation of nonhistone proteins. And significantly, each tissue shows the presence of a unique phosphoprotein component. We see again that differences in nonhistone protein occur from one tissue to another, unlike the histones which do not vary. There have also been reports that when phosphorylated nonhistone chromosomal proteins are added to cell-free systems, there is a stimulation of RNA synthesis.

Much evidence has led to the suggestion that phosphorylated nonhistone chromosomal proteins may be able to recognize specific DNA sequences and in this way exert an effect on gene control. Perhaps they may then interact with histones, causing them to bind less tightly to the DNA. However, it has been shown that phosphorylated nonhistone proteins can stimulate RNA synthesis when histones are *not* complexed with the DNA.

Although the mechanism of their action is still not clear, the nonhistone chromosomal proteins must be recognized as possible key factors in the regulation of gene activity. It is very likely that these proteins may cause changes in gene activity in a number of different ways. Perhaps they may interact at some times with the histones; at others they may interact directly with enzymes, such as the polymerases, which are also components of the chromatin.

GENE REGULATION AND EVOLUTION. Besides the importance of genetic regulatory mechanisms in differentiation and normal cell activity, there is good reason to believe that mutation in regulatory mechanisms may be the basis for anatomical evolution in higher life forms. For example, the human and the chimpanzee differ considerably in anatomical aspects. However, when their body proteins (such as blood proteins) are examined, it is found that very little difference exists. The protein differences between the two do not seem to account for the anatomical distinctions. Mutation in regulatory systems

FIG. 19-10. (Continued).

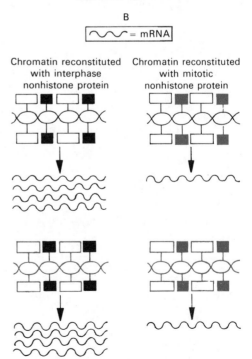

B

⌐\~\~\~ = mRNA

Chromatin reconstituted with interphase nonhistone protein

Chromatin reconstituted with mitotic nonhistone protein

would appear to be a more likely explanation. Different regulatory activities in two groups which possess similar proteins could affect the timing of gene expression and the pattern of development of the organisms. Creatures with very similar proteins would be unable to hybridize if they were members of separate groups in which the regulatory mechanisms operating during development are very different.

Mutation in regulatory genes may also account for the very rapid evolution which placental mammals have undergone in their anatomy in only 75 million years. The rate of protein evolution has not proceeded at a comparable pace. The proteins in this diverse group show fewer distinctions than would be expected on the basis of the anatomical differences which are found. Lower vertebrates, such as the thousands of frog species, show significant protein differences when compared. Yet the various frog species are very similar anatomically. Some biologists recognize two types of evolution which proceed independently and which can have different rates, protein evolution and anatomical evolution, the latter due largely to changes in regulatory mechanisms and responsible for the great diversification seen in the anatomy of higher vertebrates.

GIANT CHROMOSOMES AND GENE ACTIVITY. Among the best studies of differentiation are those which have made use of giant polytene chromosomes described in Chapter 2. It will be recalled that such giant chromosomes can be seen in the salivary gland nuclei of the larval *Drosophila* (Fig. 2-13). They are also found in other fly genera and can be seen in cells of other organs as well: the midgut, rectum, and excretory organs. These unusual structures provide a unique opportunity to study gene action. Each giant chromosome is a linear body composed of about 2^{10} single-unit chromosome threads. As a result of the juxtaposition of so many individual chromosomes, the polytene structure reflects features of the single chromosome which would otherwise go unnoticed. For example, the bands or chromomeres along the salivary chromosomes are regions where DNA appears to be concentrated due to tight packing of the chromosome fiber; the interbands on the other hand are more extended. We can see the band and interband arrangement, because more than 1000 separate fibers are closely associated next to each other. The nuclei in which these giant chromosomes occur are in a permanent interphase state. They are not preparing for cell division, but are concerned only with the synthesis of products needed in the nondividing nucleus. They are thus in a more active, stretched out condition than they would be if they were preparing to divide. Structural chromosome changes (Chap. 10) have enabled us to associate some of the bands with specific genes. An inversion, duplication, or deletion often can be related to a comparable change in a portion of one of the giant chromosomes. Each band appears to represent a site occupied by one gene. Besides DNA, the chromosomes contain all of the other molecules typically associated with chromatin, such as RNA and protein.

In the early 1950's, Beerman, working with *Chironomus,* noted that some of the bands of the polytene chromosomes may appear in a swollen or puffed condition. (These puffs are also found associated with giant chromosomes of several fly species including *Drosophila.*) Comparison of the giant chromosomes in different tissues of the same individual also showed puffing, but significantly, the pattern of puffing differed from one kind of cell to another. Moreover, the pattern seen in a certain tissue is characteristic of a certain stage of development. It changes and assumes another pattern at a later stage (Fig. 19-11).

Cells with polytene chromosomes were supplied with labeled amino acids and labeled RNA precursors. The results clearly showed that any swollen band is actively engaged in RNA synthesis but is not incorporating amino acids at an increased rate. All of the evidence taken together strongly suggested that the bands are sites of genes and that the puffed state of a band indicates a gene or genes which have become active. The gene activity is reflected both by the puffing and the increased rate of RNA formation. Observations with the electron microscope indicate that the puffing itself results from an unwinding of chromosome fibers in the more condensed regions of the bands. This unpacking exposes a much greater length of the individual chromosome fibers and this makes them available as templates for RNA transcription (Fig. 19-12).

FIG. 19-11. Chromosome puffs in *Rhynchosciara angelae.* The chromosome shows puffs which were not present at earlier stages of larval development but which appear at a characteristic time. (Reprinted with permission from W. V. Browh, *Textbook of Cytogenetics,* p. 27, C. V. Mosby Co., St. Louis, 1972.)

Beerman and his colleagues have actually been able to isolate RNA associated with different puffs. If the RNA formed at separate puffs represents different kinds of mRNA coded by different genes, we would expect the base composition of the isolated RNA samples to vary. And this is what has been found. Not only have different types of RNA been demonstrated, but the various puffs show different rates of RNA synthesis. There is little doubt that the polytene chromosomes with their puffing patterns and associated species of RNA afford an exceptional opportunity to study differential gene activity at various stages of development.

If the puffs represent genes which have been swtiched on at specific times by an unpacking of the DNA, an important question arises concerning the nature of the mechanism which accomplishes this. A clue was provided by the discovery that a specific hormone, the insect molting hormone ecdysone, can induce the formation of very specific puffs. When ecdysone is released by the larva at a certain stage in development, the puffs which were already present decrease in size, and new ones quickly appear, eventually over 100 of them until the pupal stage is reached. Each puff has a characteristic time of appearance and duration.

The hormone ecdysone is a steroid, and it is believed that the first puffs appearing after its release are caused directly by the hormone's ability to promote transcription. Various steroid hormones, as noted earlier in this discussion, are believed to exert their effects by promoting directly the transcription of specific genes. Ecdysone will bring about formation of the earlier puffs even if protein inhibitors are applied. This indicates that synthesis of new proteins is not essential for the formation of these early puffs but that the hormone is acting directly on the chromatin. Later puffs, however, do seem to depend on proteins made as a result of the earlier puffing, since later puffs will not arise if protein synthesis is blocked. The hormone level also seems to be important, because several puffs may appear at ecdysone concentrations which bring about no changes in other bands along the length of the chromosome. Certain puffs may actually decrease in size, whereas others increase in response to the hormone when it is given experimentally. The observations indicate that the various genetic regions along the length of the chromosome may react quite differently to a given hormone level. Although the changes in puffing patterns are taking place, the histone content remains constant, again implying that these proteins are not removed from the DNA when it becomes active in transcription. The exact changes which take place in histones or other components of the chromatin of the polytenes and which permit transcription at a specific time await clarification. It is possible that nonhistone proteins in the chromatin may accept an ecdysone-receptor complex and that this complex may interact with histones, removing the blocking action.

THE CYTOPLASM AND EXTERNAL ENVIRONMENT IN CELLULAR DIFFERENTIATION. The importance of the cytoplasm must not be overlooked in considerations of cellular differentiation. Very striking experiments which show the involvement of the cytoplasm are those which entail the transplantation of nuclei in amphibians. In such an experiment, the nucleus is removed from an egg cell. Another nucleus is then excised from a different kind of cell and placed into the egg cytoplasm (Fig. 19-13). Shortly after this is done, synthesis of RNA and DNA takes place. This is so even when the nucleus introduced into the egg is from a highly differentiated cell type. A brain cell nucleus, for example, does not normally divide again; yet

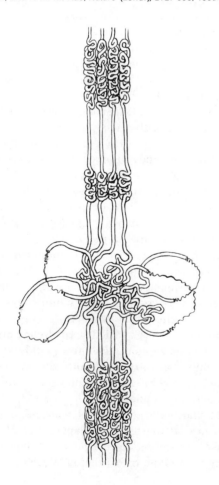

FIG. 19-12. Nature of a polytene chromosome. The diagram shows that the giant chromosome consists of many chromosome threads side by side. Tight folding in all the threads at a specific site produces a band. A puff arises when the threads at a site become unpacked. The threads in their extended state can then engage in RNA synthesis. RNA isolated from puffs at different sites has been shown to be different in its composition, indicating that the RNA is mRNA. Different bands along a chromosome would thus include different genes which form different mRNA's during the puffed state. (Reprinted with permission from E. J. DuPraw, and P. M. M. Rae, *Nature* (*Lond.*), *212:* 598, 1966.)

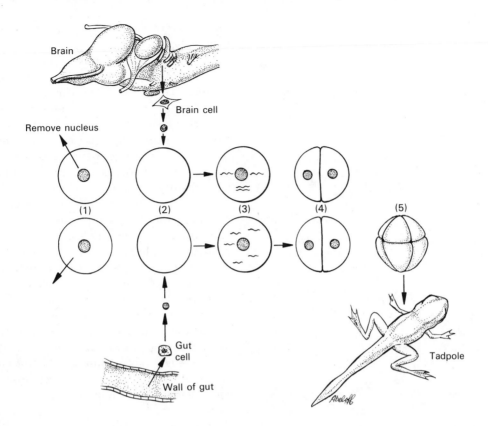

FIG. 19-13. Transplantation of nuclei in amphibians. The nucleus may be removed from an unfertilized egg (1). Another nucleus may be transplanted to such an egg from a cell which is differentiated, such as one from the brain or from the gut (2). After the transplant, DNA and RNA synthesis can be detected in the cell with the transplanted nucleus (3). The cell may divide several times (4). The nuclei taken from some cells trigger more divisions than others (5). Some result in the formation of normal tadpoles.

when inserted into the egg, it is able to reacquire this ability. Nuclei taken from the gut may not only divide but may give rise to normal tadpoles! Such results demonstrate that the genetic material in differentiated as well as nondifferentiated cells possesses all of the information needed to direct the development of a complete individual. Cytoplasmic substances undoubtedly play a regulatory role by signaling the expression of specific genes at the proper time.

Superb experiments with plants have shown both the ability of a single differentiated cell to develop into a new individual and also the importance of the external environment in cellular control. Most cells in the root of a carrot or of the underground stem (tuber) of the potato no longer divide when the plant reaches a mature state. However, it is common knowledge that such quiescent cells express their capacity for cell division when the organ is wounded by cutting

and the cells become exposed. Once the wound is healed, division again comes to a halt. Evidently, the information for cell division and growth is present in these cells. There must be some kind of signal which indicates when these processes are to start and stop. From the observations with healing, it seemed likely that some signal could come from outside the cell itself. To test this idea, Steward and his associates obtained minute pieces of carrot root tissue. These fragments, composed of adult, nongrowing cells, were exposed to a variety of substances in order to determine whether growth could be stimulated. The most potent stimulator was found to be coconut milk, the liquid endosperm of the seed that normally supplies nutrient to the coconut embryo. Fragments of carrot root exposed to the milk increased in weight about eighty times in approximately 3 weeks. The coconut milk must contain something which triggers the mechanism for division in these typically quiescent cells.

The potential for growth apparently can be switched on or off by various regulatory substances. This was shown through experiments using cells from other sources. For example, buds of certain trees become dormant late in the summer. They must be exposed to the cold if they are to resume growth. Cells from dormant maple buds did not respond to coconut milk. And significantly, extracts from these dormant cells counteracted the stimulating effect of the coconut milk on the carrot cells. However, after exposure to the cold, the same cell-free extracts from the maple buds no longer possessed their inhibitory effect. The cold must have destroyed the responsible inhibitor or rendered it inactive (Fig. 19-14). In contrast to inhibition, Steward found other substances which, when added to the coconut milk, actually increased its power to stimulate growth many fold.

An even more graphic illustration of the potential of a mature cell to express all of its coded genetic information was shown when individual root cells were freed completely from their neighbors. Apparently, surrounding cells in the intact root can restrict the growth of other cells in the vicinity. Separate root cells cultured in the coconut milk may develop into complete, normal plants! This is clear-cut evidence that all of the genetic information for directing the formation of another individual resides in the nucleus of the mature root cell. In the intact root tissue, however, the ability to express all of this information is suppressed by surrounding cells. When freed from this inhibitory effect and subjected to substances which can stimulate growth, the full potential of the mature nucleus is realized. Another striking demonstration of this was seen when separate cells were obtained from an intact carrot embryo and then were cultured in coconut milk. Thousands of such individual embryo cells developed into thousands of other embryos, just as if each separate cell had regained the power of the fertilized egg (Fig. 19-15). Moreover, the stages of development through which these secondary embryos passed were similar to those of embryos which develop in the usual way in the intact plant.

We see that while an adult, differentiated cell contains all the coded information of the

FIG. 19-14. Growth inhibitor. An extract from the cells of a dormant bud inhibits the growth of carrot fragments in coconut milk. If a bud is cold treated, the extract obtained from it no longer inhibits the growth of the carrot fragments which continue to increase in weight in the coconut milk.

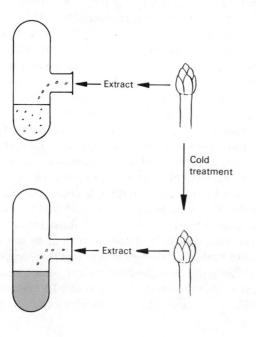

Extract

Cold treatment

Extract

fertilized egg, it is normally restricted from expressing the entire set of instructions. Apparently, this suppression takes place during normal development by the inhibitory effect of certain environmental substances, as well as by the influence of surrounding cells. Other substances may trigger the information for division and growth at the proper time or "turn on" the transcription of the appropriate genes when their products are needed. By a series of coordinated inhibitory and stimulatory signals, the genetic information is so controlled that essential gene products are present at the needed time. When not required, superfluous gene products could be detrimental and would waste valuable cell energy, as in the case of the constitutive bacterial mutants.

It is, of course, imperative that cells do *not* express all of their genetic information, because the result would be just masses of identical cells. Normal differentiation from the fertilized egg onward requires that only a fraction of the total coded information be translated. Expression of extra information in a growing cell or a mature cell may cause chaotic or abnormal growth. Other works of Steward showed that abnormalities could be brought about by the exposure of mature plant cells to certain environmental agents. A type of tumor occurs on stems of plants in the genus *Kalanchoe*. Tumor induction requires the presence of a virus, but once formed, the tumor may be perpetuated independently of the virus. Virus-free tumor extracts were applied to fragments of carrot tissue. Substances in the extract stimulated them to grow. However, the result was an abnormal growth of unspecialized cells, unlike the organized tissue which formed under the stimulation of the coconut milk. Such observations pose questions on the origin of abnormal growths in general. The undirected growth of tumor tissue implies that some regulatory function of the cell has been upset. The malignant cell, unlike a normal cell, is capable of almost limitless growth and is not restricted from dividing by the influence of its neighbors. The disorganized activity suggests that certain genes, normally repressed in a specific cell type, have now been released from their repression and are "switched on." The ensuing transcription and translation would bring about the production of products foreign to that cell type.

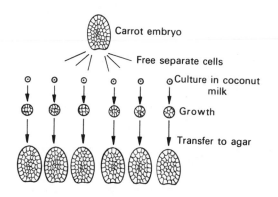

FIG. 19-15. Embryos from embryos. Separate cells can be freed from a growing carrot embryo. When these separate cells are cultured in coconut milk, they may multiply to form a small group of cells. Each of these in turn can form a carrot embryo when raised on agar medium.

Released from its restrictions, the cell may divide repeatedly to form disorganized tissue which burdens and kills the normal tissues by depriving them of their nutrient requirements.

IMPLICATIONS OF GENETIC CONTROL IN PATHOLOGICAL CELL CHANGES. Although we do not know exactly how any malignant cell arises, genetic knowledge can provide us with models which may offer points of departure for experimental attacks on the problem. We can conceive of some external agent which can act as a cancer inducer by causing the destruction of a repressor in the cell, thus allowing indiscriminate gene transcription. It is known that ultraviolet light can activate prophage in the host cell, evidently by destroying a viral repressor which is required to maintain the virus in its prophage state. Once the repressor is removed, the virus can detach itself from its characteristic location on the chromosome (Fig. 19-16). Its genes can now undergo transcription; the result is rapid virus production and the death of the cell.

The ability of a virus to take up residence at all on the chromosome of a bacterial host and become prophage (to lysogenize) depends on the production of a repressor under the direction of the viral genes (Fig. 19-17A). Normally, phage lambda can lysogenize *E. coli*. However, certain mutations are known which prevent it from doing so. The phage is no longer temperate.

Instead, its structural genes are all expressed, and the host cell is consequently destroyed. These mutations are in every way similar to constitutive mutations in *E. coli*, and their effect on the cell can be explained by the Jacob-Monod model (Fig. 19-17*B*). A mutant regulator produces a defective repressor due to the mutation. The phage genes are not "turned off" and so the phage cannot exist as prophage. These mutations are recessive because the presence of a normal viral regulator allele on another particle such as the "F" factor causes production of normal repressor which inhibits transcription of the genes of the phage.

Although these examples deal with microorganisms, we can imagine similar upsets in cells of higher species. Repressor substances may become ineffective after gene mutation or exposure to certain external agents. We do not know of any virus which can definitely generate a malignancy in man, but it is conceivable that viruses or episome-like particles could play a role. We can imagine prophage-like entities inserted into the chromosomes of many normal cells and inherited from one generation to the next. Possibly an external factor could activate them by destruction of a certain repressor and thus set off abnormal growth. In other cases of abnormal growth, viruses or episomes may not even be involved. A mutation in the cell's own genetic material could alter a repressor, which normally prevents the indiscriminate transcription of certain genes in a differentiated cell. Or outside agents could somehow change one or more repressors and thus cause uncontrolled cell

FIG. 19-16. Maintenance of prophage. In a cell infected with a temperate phage, the virus is kept in the prophage state by a repressor which is formed by the expression of certain genes of the virus (*above*). RNA polymerase cannot transcribe the DNA of the virus. If the cell is subjected to certain environmental influences, such as ultraviolet (*below*), the repressor is destroyed. RNA polymerase can now transcribe the viral DNA. This brings about the formation of viral proteins and then the viral DNA. The cell is eventually lysed, and mature viruses emerge. Conceivably, certain influences could destroy repressors in higher cells and cause transcription of certain genetic elements which might evoke pathological changes.

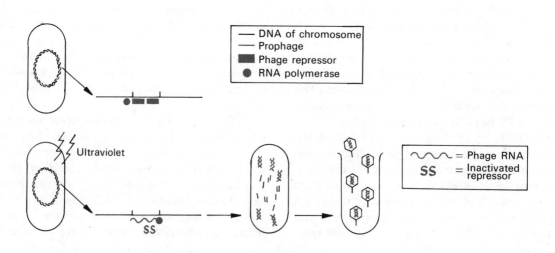

activites leading to the deterioration of normal development.

DIFFERENTIATION IN RELATIONSHIP TO TIME AND AGING. We must remember that differentiation in a many-celled organism begins with the development of the fertilized egg. And it does not cease upon completion of the individual but continues throughout adulthood until death terminates the process. Perhaps the very earliest sign of differentiation occurs *even before* zygote formation, with the appearance in the oocyte of DNA polymerase. No active form of this enzyme can be found in the oocyte until the latter is ready for fertilization. Once the oocyte becomes receptive, the polymerase appears in the cytoplasm and moves to the nucleus where it can now activate the synthesis of DNA.

In the zygote, most of the structural genes are repressed. During ontogeny, only certain genes will be activated at certain times by signals from cytoplasmic substances and from the outside environment. There is a correct time for specific genes to be called to expression. Such genes as those for development of limbs and digits would be turned on at the appropriate embryonic stage. After birth, the DNA segments coding for mature eye pigments will undergo transcription. At puberty, the information for beard development in the male or breast development in the female is switched on. Throughout the normal life span, the whole array of genes may have come to expression in the appropriate cells at the correct time. Even when mutant gene forms are present in place of the normal alleles, we can see that they too have characteristic times of expression. The dominant for polydactyly will affect digit development in the embryo. The recessive allele for Tay-Sachs disorder will exert its effect in early infancy. A particular type of muscular dystrophy appears in early adolescence when the responsible defective allele is transcribed. The genetic factor for Huntington's disease (progressive mental and nervous deterioration) usually will not express itself until after the age of 30. There are thus characteristic times for the action of both normal and mutant alleles from the stage of the fertilized egg to death. Indeed, aging (which also begins with the zygote) and life span itself may be a genetic characteristic of all living things.

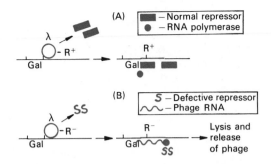

FIG. 19-17. Ability of a phage to lysogenize a cell. *A.* A temperate phage such as lambda (λ) is able to take up residence as prophage, because it normally carries a regulator gene which causes production of a repressor. This product prevents the phage genes from being transcribed. *B.* If a phage carries a mutation in the regulator, it may not be able to maintain itself as prophage. The defective repressor does not prevent transcription of the phage DNA. After transcription, more phage DNA is produced and eventually cell lysis as shown in Fig. 19-16.

We all realize that as one gets older, particularly into middle age, that the chances of death increase. A person of 50 has a greater probability of death than one of age 30. The 20-year-old has a lesser probability than the other two. Certainly more people reach an older age today than was true years ago. But has the true limit to the span of life really been increased? Or have our modern medical practices and technology allowed more people to reach an age which is an upper limit that cannot be exceeded? An intriguing series of experiments by Hayflick suggests that the latter idea may be so and that life span and aging are genetic characteristics, just as much as eye color or blood groups.

Hayflick has employed cell cultures of human fibroblasts, cells which produce collagen and fibrin and which continue to divide in the body of the adult. If there is no restriction on the capacity of these cells to divide, then one might expect them to continue on forever in tissue culture from one cell generation to the next. However, immortality does not seem to be a property of normal human cells. Fibroblasts taken from 4-month-old embryos divided over and over again for approximately 50 cell generations. Then the population died. An indication that the number of doublings is somehow genetically controlled was suggested by the amazing fact that the *total number* of doublings was not changed, even if cell division was interrupted for different periods of time. For example, cells were allowed to undergo 30 population doublings and were then suspended in cold storage for years. Upon thawing, they went on to divide an average of 20 times more!

Fibroblasts were taken from persons in different age groups. These divided fewer times than the average of 50 which is typical of embryos. Moreover, cells from older donors underwent fewer doublings than those from younger persons. A limit to length of life span through genetic restriction on cell division was strongly suggested by observations made on cells from species with short life spans. Fibroblasts taken from rat embryos undergo no more than 15 doublings in cell culture, as opposed to the 50 divisions for the human. And those from the adult rat undergo even fewer.

Single human embryonic cells were isolated and allowed to divide. Any cells arising from a common cell would all be genetically identical. From these, groups of genetically identical cells, clones, can be established (Fig. 19-18). The ability of these genetically identical populations to divide was then followed. It was shown that a gradual decline in the capacity for cell division took place. More and more of the clones lost the ability to divide as the fiftieth doubling generation was approached. Finally, division ceased in all the clones. The clonal studies strongly supported the idea that aging was determined largely by factors internal to the cell, rather than by the occurrence of substances in the outside environment.

It is known that in vertebrates, certain kinds of body cells cease active division as a normal part of their development or maturation. Such a restriction on division is seen in the pronephros or metanephros during embryology. It is also known that at old age, many organs may weigh less and contain fewer cells. Perhaps this is a reflection of a limit to the number of cell doublings somehow coded in the genetic material. If so, the DNA would even determine the limit to the life span of normal cells. Therefore, modern science would be unable to avoid the eventual fate set down in the genetic program. We can speculate that the scientist might still push back the limit once he knows exactly what is taking place on the molecular level as aging

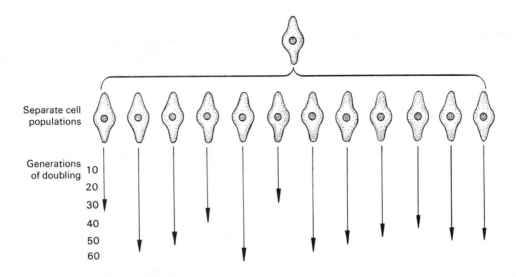

Separate cell populations

Generations of doubling

10
20
30
40
50
60

FIG. 19-18. Clonal aging. Groups of cells which trace back to the same ancestral cells are genetically identical and compose a clone. When populations of identical cells are followed through cell divisions, it is found that the separate populations cease division at different times but that more and more of them lose the capacity for division as they approach the fiftieth doubling generation. Eventually all the cell populations cease to divide further and die.

progresses. Hayflick's observations on chromosomes show an increase with age in the number of structural aberrations as cells grow in culture, but it is still not certain that this takes place in the body during aging. It is also possible that aging involves mistakes during replication of the DNA or even in transcription or translation of the genetic message. An accumulation of such errors with time could increase the number of critical metabolic pathways which become blocked. Added to an accumulation of chromosome aberrations, these could cause a deterioration of chemical activities necessary to maintain normal cell function.

If such factors are clearly shown someday to be involved in the aging process, perhaps techniques may be devised to correct them, for enzymes are known which can repair damage to the DNA (Chap. 15). An assortment of animals and plants apparently contains such enzymes. Perhaps those animals with longer life spans have evolved more efficient repair mechanisms than those with shorter ones, as Hayflick suggests. If these repair systems become precisely understood, perhaps science may supplement the normal mechanism by injecting the active enzymes into cells. We may even learn how to correct faulty messages which arise as copy errors at transcription or at the time of gene replication. The accumulation of genetic blocks could be prevented or at least decreased in number. In the final chapter, we shall speculate further on approaches to immortality for the human species.

REFERENCES

Bahl, C. P., R. Wu, J. Stawinski, and S. A. Narana. Minimal length of the lactose operator sequence for the specific recognition by the lactose repressor. *Proc. Natl. Acad. Sci.*, 74: 966, 1977.

Beckwith, J. R. Regulation of the lac operon. *Science*, 156: 597, 1967.

Beerman, W., and U. Clever. Chromosome puffs. *Sci. Am.* (Apr): 210, 1964.

Bonner, J. *The Molecular Biology of Development.* Oxford University Press, London, 1965.

Clever, U. Regulation of chromosome function. *Ann. Rev. Genet.*, 2: 11, 1968.

De Robertis, E. M., and J. B. Gurdon. Gene activation in somatic nuclei after injection into amphibian oocytes. *Proc. Natl. Acad. Sci.*, 74: 2470, 1977.

Dickson, R. C., J. Abelson, W. M. Barnes, and W. S.

Reznikoff. The lac control region. *Science, 187:* 27, 1975.

Gilbert, W., N. Maizels, and A. Maxam. Sequences of controlling regions of the lactose operon. *Cold Spring Harb. Symp. Quant. Biol., 38:* 845, 1974.

Gilbert, W., and B. Muller-Hill. Isolation of the lac repressor. *Proc. Natl. Acad. Sci., 56:* 1891, 1966.

Gurdon, J. B. Transplanted nuclei and cell differentiation. *Sci. Am. (Dec):* 24, 1968.

Hayflick, L. The limited *in vitro* lifetime of human diploid cell strains. *Exp. Cell. Res., 37:* 614, 1965.

Hayflick, L. Human cells and aging. *Sci. Am. (Mar):* 32, 1968.

Jacob, F., and J. Monod. Genetic regulator mechanisms in the synthesis of a protein. *J. Mol. Biol., 3:* 318, 1961.

Jamrick, M., A. L. Greenleaf, and E. K. F. Bautz. Localization of RNA polymerase in polytene chromosomes of *Drosophila melanogaster*. Proc. Natl. Acad. Sci., 74:2079, 1977.

Jansing, R. L., J. L. Stein, and G. S. Stein. Activation of histone gene transcription by nonhistone chromosomal proteins in WI-38 human diploid fibroblasts. *Proc. Natl. Acad. Sci., 74:* 173, 1977.

Maniatis, T., and M. Ptashne. A DNA operator-repressor system. *Sci. Am. (Jan):* 64, 1976.

Marx, J. L. Viral messenger structure: some surprising new developments. *Science, 197:* 853, 1977.

O'Malley, B. W., and W. T. Schrader. The receptors of steroid hormones. *Sci. Am. (Feb):* 32, 1976.

Ptashne, M., and W. Gilbert. Genetic repressors. *Sci. Am. (June):* 36, 1970.

Stein, G. S., T. C. Spelsberg, and L. J. Kleinsmith. Nonhistone chromosomal proteins and gene regulation. *Science, 183:* 817, 1974.

Stein, G. S., J. S. Stein, and L. J. Kleinsmith. Chromosomal proteins and gene regulation. *Sci. Am. (Feb):* 46, 1975.

Steward, F. C. The control of growth in plant cells. *Sci. Am. (Apr):* 104, 1963.

Tomkins, G., and D. W. Martin. Hormones and gene expression. *Ann. Rev. Genet., 4:* 91, 1970.

REVIEW QUESTIONS

In the following, allow R^+ and O^+ to stand for normal regulator and operator and R^- and O^- to represent their deficient alleles. Z^+, Y^+, and X^+ represent normal alleles of structural genes for enzyme production and Z^-, Y^-, X^- their alleles for absence of enzyme. In normal cell types, the enzymes are inducible.

1. In cells of the following genotypes, indicate whether enzyme production will be inductive or constitutive:

    A. $\underline{R^+\ O^+\ Z^+\ Y^+\ X^+}$ B. $\underline{R^+\ O^-\ Z^+\ Y^+\ X^+}$

C.
$$\frac{R^+\ O^+\ Z^+\ Y^+\ X^+}{R^+\ O^-\ Z^+\ Y^+\ X^+}$$
D.
$$\frac{R^+\ O^-\ Z^+\ Y^+\ X^+}{R^+\ O^-\ Z^+\ Y^+\ X^+}$$

E.
$$\frac{R^+\ O^+\ Z^+\ Y^+\ X^+}{R^-\ O^+\ Z^+\ Y^+\ X^+}$$
F.
$$\frac{R^-\ O^+\ Z^+\ Y^+\ X^+}{R^-\ O^+\ Z^+\ Y^+\ X^+}$$

2. For each of the following, tell which enzymes will be produced constitutively and which by induction:

A.
$$\frac{R^+\ O^-\ Z^+\ Y^-\ X^-}{R^-\ O^+\ Z^-\ Y^+\ X^+}$$
B.
$$\frac{R^+\ O^+\ Z^-\ Y^+\ X^+}{R^-\ O^+\ Z^+\ Y^-\ X^-}$$

3. Let O^o represent an operator mutation causing the operator to bind irreversibly with the normal repressor. Let R^s represent a mutation in the regulator which causes the formation of an altered repressor which cannot react with the inducer even though it can bind with the normal operator. For each of the following, tell which enzymes will be produced constitutively and which by induction.

A. $R^s\ O^+\ Z^+\ Y^+\ X^-$ B. $R^+\ O^o\ Z^+\ Y^+\ X^+$

C.
$$\frac{R^+\ O^o\ Z^+\ Y^+\ X^-}{R^-\ O^+\ Z^-\ Y^-\ X^+}$$
D.
$$\frac{R^-\ O^+\ Z^+\ Y^+\ X^-}{R^s\ O^-\ Z^-\ Y^-\ X^+}$$

4. Assume that A^+ and B^+ govern enzyme formation in an operon which is a repressible system such as in the case of tryptophan synthetase. In such a system, the effector (end product) must combine with the product of the regulator in order for the operator to block transcription. In each of the following, tell what enzymes would be expected from the cells in the absence of effector and in the presence of effector.

A. $\underline{R^+\ O^+\ A^+\ B^+}$ B. $\underline{R^+\ O^-\ A^+\ B^+}$

C. $\underline{R^-\ O^+\ A^+\ B^+}$ D.
$$\frac{R^+\ O^+\ A^+\ B^-}{R^+\ O^-\ A^-\ B^+}$$

5. Two regions are found in a segment of DNA where there is a symmetrical arrangement of base pairs. One of these regions has the following sequence on one strand, reading in the 5' to 3' direction: GGCCAG. Show the nucleotide pairs and the polarity in the region which corresponds to this segment.

For each of the following select the correct answer or answers, if any.

6. The operon repressor in inducible systems:

A. Has been shown to be a histone protein.
B. Has been shown to contain a significant amount of RNA which enables it to recognize a specific operator.
C. Binds directly to specific regions of DNA.
D. Is produced by the operator which lies adjacent to the structural genes.
E. Is able to repress the synthesis of all the enzymes controlled by all the structural genes within the operon.

7. In an inducible system such as the lactose operon:

A. A constitutive mutation in the regulator acts as a recessive to wild.
B. A constitutive mutation in the regulator acts as a dominant to wild and causes production of the enzymes controlled by both of the DNAs in a heterogenote.
C. A constitutive operator mutation behaves as a recessive to wild.
D. A constitutive operator mutation permits those genes adjacent to it on the same DNA strand to be transcribed in a heterogenote.
E. A constitutive operator mutation permits transcription in the partial diploid of the structural genes on both DNA strands.

8. In a repressible system such as the tryptophan synthetase operon:

A. The enzymes governed by the structural genes within the operon are produced in the absence of the end product whose synthesis they control.
B. The repressor does not react with the operator in the absence of the end product.
C. The end product is required to bind the repressor so that transcription of the structural genes can take place.
D. Addition of the end product results in increased production of the enzymes involved in the synthesis of that end product.
E. Addition of the end product has no effect on the level of the enzymes involved in the synthesis of that end product.

9. Histones:

   A. Are very basic in nature.
   B. Are molecules of very high molecular weight.
   C. Do not occur in association with the nucleic acid of prokaryotes.
   D. Vary greatly in kind from one species of organism to the next.
   E. Become completely dissociated with chromatin and decrease in amount when transcription is taking place.

10. The following can be said about the giant chromosomes found in the salivary glands, midgut, rectum, etc., of certain insects.

   A. They are actually in an interphase condition and are not preparing for cell division.
   B. The puffs are very active in the incorporation of amino acids.
   C. The RNAs of the different puffs have been shown to be different.
   D. The puff pattern is identical in cells of different tissues.
   E. Each giant chromosome is composed of a thousand or more separate chromatid fibers which are synapsed.
   F. The giant chromosomes contain interbands which are richer in DNA concentration than are the bands.

11. Which is (are) true regarding work with chromosomal proteins and chromatin reconstitution?

   A. Nonhistone proteins vary in amount and kind from one tissue to the next.
   B. Chromatin reconstituted with mitotic-phase nonhistone protein has a greater template activity than that reconstituted with interphase nonhistone protein.
   C. Chromatin reconstituted with interphase histone protein has a greater template activity than that reconstituted with mitotic phase histone protein.
   D. Histones are probably bound more tightly in chromatin reconstituted with interphase nonhistone protein.
   E. Phosphorylated nonhistone proteins can stimulate RNA synthesis even in the absence of histones.

12. Which of the following has (have) been shown experimentally from various studies of differentiation?

 A. A nucleus from a highly differentiated animal cell can no longer be stimulated to engage in DNA and RNA synthesis following transplantation.
 B. The genetic material in a differentiated animal cell nucleus lacks the information needed for the development of a complete individual.
 C. The presence of neighboring cells can limit the mitotic activity of a given cell.
 D. Normal development from a zygote requires that only a fraction of the coded genetic information be transcribed in a given kind of cell.

13. Which of the following has (have) been suggested from studies of differentiation?

 A. The cytoplasm is ineffective in regulating transcription and replication in a nucleus from a differentiated cell.
 B. In malignant cells, certain genes which are normally inactive have been called to expression.
 C. Hormones may act as inducers.
 D. Chromatin in a packed state is more active in transcription than chromatin in an extended condition.
 E. Cells taken from young individuals and those taken from older persons undergo the same number of doublings in tissue culture.

20

NONCHROMOSOMAL GENETIC INFORMATION

Genetic investigations with microorganisms have clearly pointed out that inheritance may entail more than the information which resides in the chromosome complement of the cell. Episomes such as the fertility factor ("F") and the F-genote derived from it may exist apart from the bacterial chromosome. Nevertheless, they provide the cell with extra genetic information and can confer properties upon it which do not depend on chromosomal genes for their expression. Viruses, which resemble episomes, may infect a cell and change it by introducing new genetic material. Bacterial and viral systems clearly demonstrate that nonchromosomal genetic information may be present within a cell. We now know that this is true as well for eukaryotes, from unicellular forms to the highest plant and animal species.

LIFE CYCLE OF *PARAMECIUM*. Investigations performed by Sonneborn and his colleagues with the protozoan, *Paramecium aurelia*, were among the first to emphasize the importance of the cytoplasm in inheritance and to focus attention on the existence of nonchromosomal genetic determinants. An acquaintance with a few features of the life cycle of *Paramecium* is essential for an understanding of the genetic implications of certain inheritance patterns in this organism. The nondividing *Paramecium* contains a large macronucleus and two tiny micronuclei (Fig. 20-1A). Each of these divides when the cell divides; however, the nuclear membranes remain intact. Each micronucleus contains the diploid number of chromosomes, and these can be seen at mitosis. The macronucleus, on the other hand, never shows chromosomes or a spindle; only granules are evident within it. However, a macronucleus is essential for the life of the cell. Therefore, the essential information which it contains is somehow equally distributed at cell division. The micronuclei are dispensable to a cell but are necessary for sexual reproduction; it is these which demand our attention in this discussion.

As Figure 20-1B shows, when two cells of proper mating type conjugate, the animals touch on the mouth side. Each micronucleus undergoes meiosis so that eight haploid micronuclei arise in a cell. Seven of these eight micronuclei degenerate, leaving only one haploid nucleus in each conjugating cell. The remaining nucleus then undergoes a *mitosis*, producing two identical haploid nuclei per cell. Of these two, one is capable of migrating to the other cell by way of a small connection which forms between the mating cells. The result is that each member receives a haploid nucleus from the other. The migratory nucleus then unites with the stationary nucleus of the cell, restoring the diploid condition. This fertilization nucleus in each conjugating cell contains equal amounts of genetic material from each of the two parents. The two conjugants thus become identical in respect to their chromosomal information.

While these events are taking place, the macronucleus in each cell breaks down. The fertilization nucleus then divides to produce two diploid nuclei, each of which divides in turn to give four per cell. Two of these four begin to enlarge. When the exconjugant cell divides mitotically after having mated, the two daughter cells receive two micronuclei and one of the two enlarging ones. This latter will become the new macronucleus. The important point is that after conjugation and fertilization, each parent cell contains new genetic information which it has received from its mate, and both of the two parent cells are identical for their chromosomal complements. The diploid condition is present again in the micronuclei. If a *Paramecium* is prevented from engaging in conjugation, it may undergo a self-fertilization process known as autogamy (Fig. 20-1C). The sequence of steps in this process is identical to those just described for conjugation, except for the fact that no nuclear exchange takes place between cells. The two haploid nuclei present after meiosis simply unite with each other. This means that after autogamy, a cell will be homozygous for *all* of its chromosomal genes, because the two haploid nuclei which unite in the self-fertilization process are identical.

THE GENETICS OF THE KILLER TRAIT. In the 1930's, Sonneborn observed that at times, when two stocks of *Paramecium* are mixed together, some of the animals become abnormal and die. It was easy to mark either one of the stocks by feeding the animals colored food and to demonstrate that one of the stocks in each case was responsible for the killing effect. Such stocks were designated "killers," and those affected by them were called "sensitives." Studies were undertaken to clarify the nature of the killing action. Examination of the fluid in which killers had lived demonstrated that the cell-free filtrate contained killing action. Evidently, something was being liberated from the bodies of the killers which could destroy sensitive cells swimming in the fluid. The killing agent was shown to be quite different from other known antibiotic agents. For example, sensitive cells may be exposed to a given amount of medium in which a known number of killers has lived. For this particular amount of medium, only a certain number of sensitives will die. Say that 1000 sensitives are added to 1 cc of filtrate and that 10 animals are killed. If 2000 or 3000 or just 100 are added to other 1-cc portions of the same filtrate, 10 animals on the average will be killed in each case. If sensitives are exposed to the same volume of fluid in which a larger number of killers has lived, then a proportionately larger number of sensitives will be killed. In other words, the number of sensitives killed is directly proportional to the number of killers. Such observations suggested that the killing agent exists in the form of discrete, lethal particles. A certain

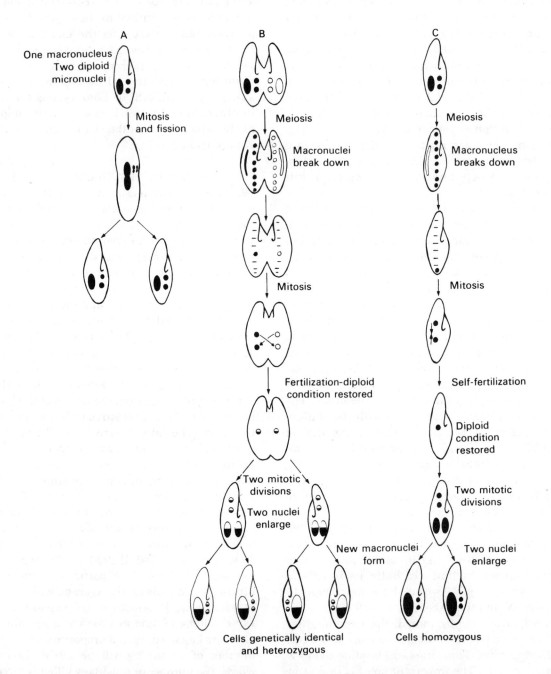

FIG. 20-1. Life cycle of *Paramecium aurelia*. A. Reproduction by cell division and mitosis. B. Conjugation and exchange of micronuclei. Note that the outcome is the production of cells which are genetically identical due to the fact that the mating cells exchanged a haploid nucleus. C. Autogamy or self-fertilization. Note that this process results in homozygosity. (Follow text for details.)

number of killers will liberate a certain number of particles in a given period of time, each of which can kill a sensitive. This idea was clearly demonstrated in various ways, such as the following. Immediately after fission, a single killer may be placed in a given amount of fluid (Fig. 20-2A). At the end of 1 hr, it may be transferred to a fresh quantity in another container. This process may be repeated so that the cell remains 1 hr in each of several separate equal volumes. The outcome is that at the end of 5 hr, which is about the length of a fission cycle, one of the volumes will contain enough killing agent to kill one sensitive cell. If two killers had been used (Fig. 20-2B), then there would be two killing particles liberated into one or the other of the five volumes. This clearly illustrates that a discrete particle is suddenly liberated from the body of a killer into the medium.

Investigations were undertaken to clarify the genetic basis for the formation of this unusual killing agent. Fortunately, sensitive stocks are immune to the killing action during conjugation, so that matings may be performed between killers and sensitives. Let us examine such a cross, one between killer stock 51 and sensitive stock 47 (Fig. 20-3). After conjugation, each of the parent cells can be isolated and followed through subsequent divisions. It is found that the exconjugant cell which had been the killer cell gives rise only to killers by fission, and the killer trait persists throughout further self-fertilizations (autogamies). The sensitive exconjugant produces only sensitive cells when it is followed in the same way. This immediately tells us that something other than Mendelian inheritance is involved in the transmission of the killer trait. Both exconjugant cells must be identical for chromosomal genes, because they exchanged micronuclei. They and the cells they produce by fission are F_1's resulting from a cross of two pure-breeding stocks. Any difference between them must be due to something other than the expression of genetic information in their nuclei. Apparently, the killer trait is following the cytoplasm in this cross. If a parent cell was sensitive, all of the cells arising from it after conjugation with a killer are also sensitive, even

FIG. 20-2. Particulate nature of killing action. *A.* One killer cell may be transferred from one to the other of several containers after a stay of 1 hr in each. The amount of fluid is fixed. When a given number of sensitives is added to each, it is found that one cell will be killed in one of the first five containers. *B.* If the same procedure is followed with two killers, then two sensitives will be killed, either in the same or different containers.

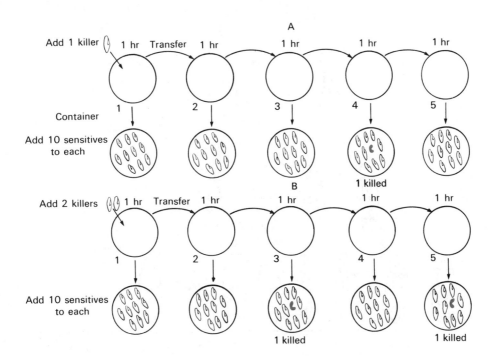

though it must contain the same chromosomal information as the killer.

The importance of the cytoplasm becomes even more apparent when conjugating cells exchange an appreciable amount of cytoplasm during mating. This exchange takes place at times when the mating period is prolonged. A sizable bridge which can be seen with the aid of a dissecting microscope may form between the mating cells. When conjugation is completed and the exconjugants followed, the results of a cross between killer stock 51 and sensitive stock 47 are now different (Fig. 20-4) from those of an ordinary mating. We see that *both* the original killer and the original sensitive give rise to killers through subsequent fissions and autogamies. Obviously, something has passed from the cytoplasm of the killer through the cytoplasmic bridge to the body of the sensitive. This factor, which can convert a sensitive to a killer and which resides in the cytoplasm was named *kappa*. Additional studies showed that if a potential killer cell divides very rapidly, it may lose kappa. A cell can regain it and become a killer only by obtaining kappa from another cell.

FIG. 20-3. A cross of killer stock 51 with sensitive stock 47. After conjugation, the killer cell remains a killer and the sensitive cell remains sensitive. The killer and sensitive traits persist throughout further cell divisions and autogamies. Since all the cells are identical for nuclear genes, the genetic basis for the killing must entail a cytoplasmic element.

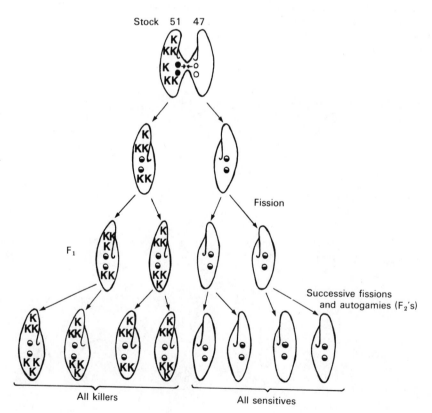

Stock 51 47

F₁

Fission

Successive fissions
and autogamies (F₂'s)

All killers

All sensitives

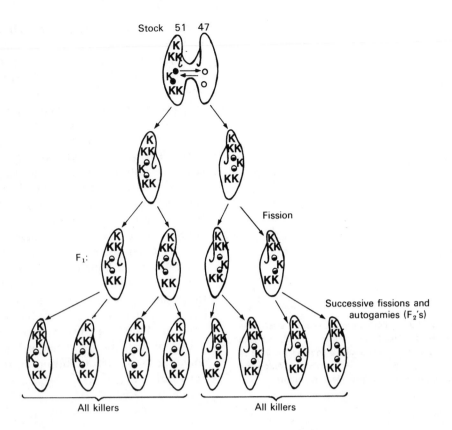

Stock 51 47

F₁:

Fission

Successive fissions and
autogamies (F₂'s)

All killers   All killers

FIG. 20-4. A cross of killer stock 51 with sensitive stock 47—large bridge. When a large cytoplasmic connection forms between the two conjugating animals, the sensitive cell is then converted to a killer. Both exconjugant cells now give rise to killers, and the killing trait persists through further divisions and autogamies. This cross is the same as that in Fig. 20-3 except for the amount of cytoplasm exchanged.

Further crosses between killer and sensitive stocks showed that more than the cytoplasm is involved in the inheritance of kappa. Let us next examine a cross between our killer stock 51 and another sensitive, stock 29. In an ordinary mating where no large cytoplasmic bridge forms between the cells, it is found that the two exconjugants retain their original phenotypes, the killer giving rise to killers by fission and the sensitive to sensitives (Fig. 20-5). This is similar to the F₁ results shown in Figure 20-3. However, a significant difference is seen when the F₂ results are compared. In this cross of 51 × 29, the sensitive F₁'s continue to produce sensitives after self-fertilization, but the killer F₁'s can give rise to either killers or to sensitives. A population of F₁ killers such as these will produce killer and sensitives in a ratio of 1:1. This 1:1 segregation of killer and sensitives is also seen when a large amount of cytoplasm is exchanged between two

parents (Fig. 20-6). Only now, the original sensitive parent is converted to a killer and produces killer F₁'s by fission; it also gives rise to an F₂ in which there is a 1:1 segregation of killer and sensitive.

What is the explanation of the difference between the cross of this same killer parent (stock 51) with the two different sensitive stocks, 47 (the first example) and 29 (the second cross)? The 1:1 ratio produced in the second cross tells us that a pair of alleles is segregating. Apparently, the cytoplasmic factor, kappa, depends for its maintenance on a dominant ("K"); its recessive allele ("k") cannot support kappa. Killer stock 51 must be homozygous ("KK") and sensitive stock 29 homozygous recessive ("kk") (Fig. 20-7A; compare with Figs. 20-5 and 20-6). When the two are crossed, the F₁'s are heterozygotes which can support kappa, if it is present in the cytoplasm. At autogamy, however, meiosis occurs, and the alleles segregate. A haploid nucleus may receive either "K" or "k." This means that one-half of the F_1 cells in a population will become "KK" after self-fertilization and will be able to support kappa; the other one-half become "kk" and cannot maintain kappa. Sensitive stock 29 differs in its genic constitution from sensitive stock 47 in the first cross. Stock 47, like killer stock 51, is homozygous "KK" (Fig. 20-7B). It is sensitive only because it lacks kappa. Once it acquires kappa through a cytoplasmic bridge, it can maintain it and is converted into a killer. We see from these crosses that an inherited characteristic may have a basis which is both chromosomal and cytoplasmic. To be a killer, a cell must have kappa in its cytoplasm. It also requires an allele K to support kappa, but this same allele does not enable the cell to make kappa. Kappa must be introduced into the cytoplasm. When it is present, it somehow converts a cell of the proper genotype to a killer by enabling it to liberate killing particles into the environment.

FIG. 20-5. A cross of killer stock 51 with sensitive stock 29— small bridge. The results of the F_1 are the same as in the cross of 51×47; the killer cell gives rise to killers and the sensitive to sensitives. When the F_1 cells are allowed to undergo autogamy, however, one-half of the killer F_1's give rise to sensitives.

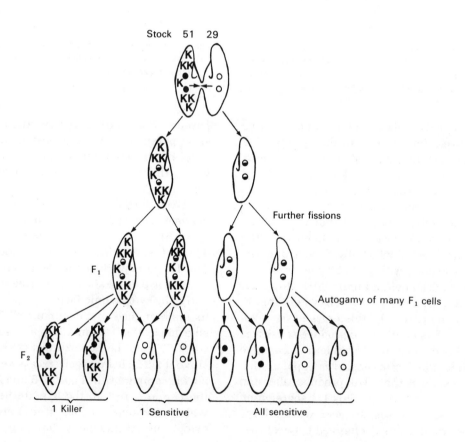

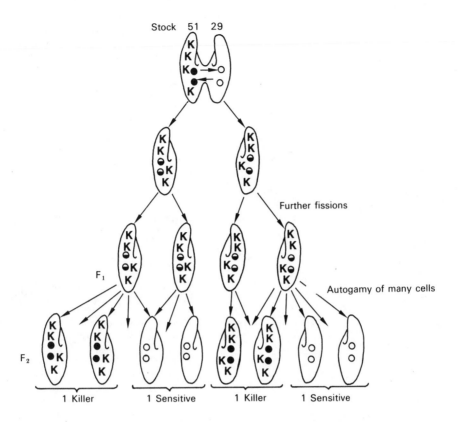

FIG. 20-6. A cross of killer stock 51 with sensitive stock 29—large bridge. When a large amount of cytoplasm is exchanged, the sensitive is converted to a killer, as is true when stocks 51 and 47 exchange a lot of cytoplasm. In this case, however, at autogamy, there is now a 1:1 segregation of killer to sensitives when the F_1's from both the original killer and the original sensitive are allowed to undergo the self-fertilization process.

THE NATURE OF KAPPA AND RELATED PARTI-CLES. Various studies were undertaken to elucidate the nature of kappa. Preer found that X-rays could inactivate kappa when they were applied to killer cells. The dosage required to do this indicated that kappa should be of a size which can be resolved by the light microscope. Killer and sensitive cells were stained by various procedures which revealed the presence of large numbers of particles in the cytoplasm of killer stocks (Fig. 20-8A). These particles were never seen in sensitive stocks, and all the observations indicated that the cytoplasmic bodies were indeed kappa. Detailed examination later showed that the kappa particles in a killer are not all the same but that two kinds exist, "brights" and "nonbrights" (Fig. 20-8B). The two are very similar except that the former type contains a refractile body which is evident with bright-phase contrast. Both the brights and nonbrights are bounded by a double membrane and contain appreciable amounts of deoxyribonucleic acid (DNA) distributed throughout. Killer cells have been followed cytologically throughout all phases of their life cycle. The observations have shown that the maintenance of the killing activity depends on both kinds of kappa particles. The brights seem to develop from the nonbrights. The nonbrights alone have the ability to divide; if they are lost, neither nonbrights nor brights can reappear. It is the bright, however, which is evidently responsibile for the killing activity. Cells have arisen in which mutations have taken place in the kappa particles. One kind of mutant kappa is called "pi." Cells carrying pi possess no killing activity and are also sensitive. Cytological examination shows that the pi particles all resemble the nonbrights; no brights are present. It seems that in these cells, brights cannot develop from the nonbrights. In ordinary killer

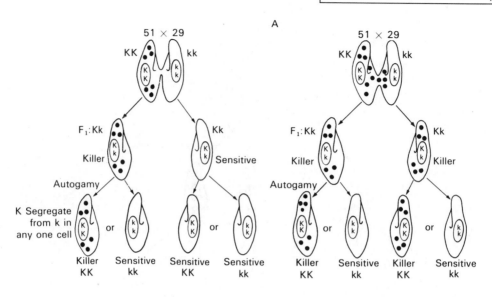

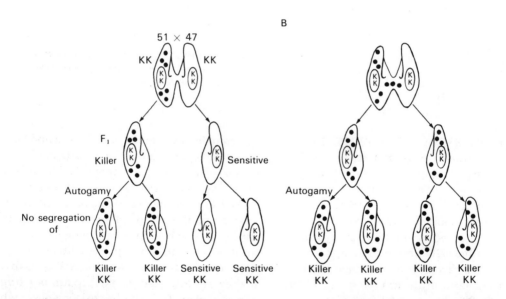

FIG. 20-7. Genetic basis for inheritance of the killer trait. *A.* Stock 51 × 29—small bridge (*left*). Stock 51 is a killer because it possesses kappa in the cytoplasm and carries in its nuclear material allele K needed to support it. Stock 29 is sensitive, carrying the allele k which cannot maintain kappa. A cross between the two makes all of the F₁ cells heterozygotes (Kk). However, if no kappa goes over to the sensitive mate, it will remain sensitive even though it has the "K" to support kappa. The F₁ cells from the killer cells produce a ratio of 1 killer to 1 sensitive at autogamy, because in any one cell which undergoes the self-fertilization, there is a 50:50 chance whether the two haploid nuclei which unite carry K or k. Therefore,

one-half of the F₂ becomes killer (KK) and one-half sensitive (kk). When a large amount of cytoplasm is exchanged (*right*), kappa converts the sensitive to a killer. Kappa is supported, because the original sensitive cell is now genetically heterozygous (Kk). At autogamy, there will be a 1:1 segregation of killer to sensitive from this cell line, as well as for the other. The reason is the same as just explained above. *B.* Stock 51 × 47. Stock 47 is genetically identical to stock 51. It is sensitive only because it lacks kappa. Once kappa is received through a large bridge, all the cells remain killers. There is no recessive k present to give rise to sensitives if kappa is also present in the cytoplasm.

cells, the disappearance of brights is followed by their reappearance. Many observations of this kind indicate that the bright particle is actually liberated from the body of the killer and that this particle is responsible for the killing action on a sensitive cell.

We can now understand some of the unusual features of the killing action which seemed so puzzling when it was first analyzed. The killing event is due to the liberation of a discrete entity, a bright. This develops from a nonbright in the cytoplasm of a killer. But the killer must have a nuclear genetic factor ("K") to support the cytoplasmic kappa. These unusual features have posed many questions concerning the origin of kappa. It is certainly a genetic entity containing DNA. However, kappa is not essential to the life of the cells and is absent from most paramecia. This suggests that it is not native to the species but has arisen from another source, an idea supported by the fact that it cannot be made by a cell but must be introduced. As shown by the electron microscope, the structure of kappa bears many resemblances to bacterial cells. Moreover, cytochemical tests have demonstrated that kappa is a Gram-negative body and that it contains cytochromes which resemble those of bacteria and not those of any eukaryote, including *Paramecium* itself. Most authorities agree that kappa is an infective agent with definite relationships to bacteria.

Further aspects of the story raise additional questions about the significance of such an extra genetic element in the cytoplasm of a eukaryotic cell. For kappa has been found to be but one of numerous kinds of different entities which may infect *Paramecium*. We have already mentioned pi, which appears to be a mutant form of kappa. Another type of killer, the mate killer was later discovered. In contrast to the original killer type, the mate killer exerts its effect only if it has prolonged contact with a sensitive cell. The latter is not affected by the fluid in which mate killers have lived; no killing particle is liberated. Cytological examination shows that mate killers possess kappa-like particles, but that they are all of the nonbright type. This supports the concept that brights must be formed from nonbrights and liberated from the cell if the original type of killing action is to occur. These particles in the mate killers, believed to be another mutant form

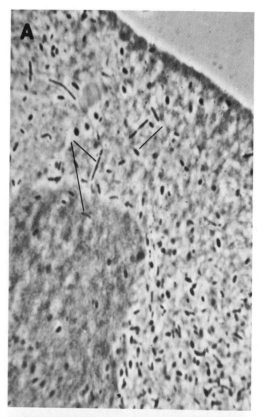

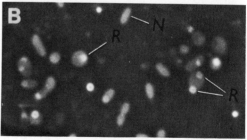

FIG. 20-8. Kappa particles. *A.* Kappa in a stained *Paramecium* killer cell. The nucleus is in lower left of picture and appears slightly darker than the cytoplasm. The kappa particles are very numerous in the cytoplasm. Those in the nuclear area are actually on the surface of the nucleus. *B.* Brights and nonbrights. These particles are from a freshly broken animal which has not been stained. The phase contrast microscopy reveals the refractile body (*R*) in the bright particle which is absent from the nonbright (*N*). (Reprinted with permission from *J. Cell. Sci., 5:*65–91, 1969.) (Courtesy of Dr. J. R. Preer, Jr.)

of kappa, have been named "mu" particles and are responsible for the production of a toxin. After conjugation between a mate killer and a sensitive, the mate killer remains a mate killer, but the sensitive dies. It may survive long enough to produce a few abnormal progeny which also become distorted and die. If cytoplasm is exchanged between a mate killer and a sensitive, the latter may become a mate killer or it may die. Mate killers are not protected against the killer types which carry kappa. Similarly, killers with kappa are sensitive to mate killers. Although mu particles may make a cell a mate killer, their maintenance depends on the presence of *either one* of two unlinked dominants of the host cell, M_1 or M_2 (Table 20-1). Homozygous double recessives cannot support mu, and they become sensitive. M_1 or M_2 cannot bring about the manufacture of mu; it must be introduced. These facts are reminiscent of kappa. Additional examples could be given of other types of similar particles in *Paramecium*. Indeed, any *Paramecium* may be infected by some kind of extraneous entity whose maintenance depends on the genotype of the host cell. This fact reflects the close association which has evolved in the relationship between the cell and the infective agent. It suggests that some sort of advantage may be conferred on the host cell by the presence of kappa and related particles, all of which may be thought of as symbionts. If the presence of the symbiont does confer an advantage, any genetic changes in the host to support the infective particles would be selected by the forces of evolution. However, no advantage from these

TABLE 20-1 Summary of relationships in expression of the mate killer phenotype*

GENOTYPE OF NUCLEUS				PARTICLES IN CYTOPLASM	PHENO-TYPE
M_1 ____		M_2 ____		Mu	Mate killer
M_1 ____		M_2 ____		None	Sensitive
M_1 ____		m_2	m_2	Mu	Mate killer
M_1 ____		m_2	m_2	None	Sensitive
m_1	m_1	M_2 ____		Mu	Mate killer
m_1	m_1	M_2 ____		None	Sensitive
m_1	m_1	m_2	m_2	None	Sensitive

* Either one of the unlinked dominants, M_1 or M_2, must be present to support the mu particle and make the cell a mate killer. These alleles, however, cannot bring about production of the mu particle, which must be introduced into the cell.

particles to *Paramecium* has yet been demonstrated other than the fact that cells with a symbiont are immune to any toxic effects which it may produce. Most cells, however, would not seem vulnerable to the lethal effects, because a killing particle or a liberated toxin would find little chance in the natural environment of contacting a sensitive cell swimming by. It is possible that some still unknown advantage is provided by the cytoplasmic particles. Whatever the reason is for such a variety of kappa-like entities in the cells of *Paramecium*, it poses interesting questions relating to the establishment of host-symbiont relationships and the benefits which can be derived from the presence of chromosomal and infective DNA in the same cell. We will see that such questions have significance even for the human species.

EXTRANUCLEAR GENETIC INFORMATION IN HIGHER SPECIES. The presence of nonchromosomal genetic information in a cell is not just a peculiarity of lower species. The fact is that all eukaryotic cells, including those of mammals, contain DNA outside the cell nucleus. All animal and plant sources which have been examined show that DNA exists in the mitochondria. Moreover, the chloroplasts of plants, from algae through the flowering plants, contain amounts of DNA even larger than that found in the mitochondria. The DNA of both the chloroplasts and the mitochondria is double stranded and occurs unassociated with histone protein and unbounded by any membrane. The organelle DNA usually differs in its adenine: thymine-/guanine:cytosine (A:T/G:C) base ratio from that of nuclear DNA. The difference permits the organelle DNA to be recognized as a distinct satellite band when DNA is extracted from a cell and subjected to density-gradient ultracentrifugation. It also makes possible the isolation of this DNA from the nuclear DNA so that it can be used in further studies.

In addition to their DNA, mitochondria and chloroplasts also possess components required for the synthesis of protein. Both kinds of organelles contain ribosomes, and these have been isolated and characterized. Their physical properties bear striking similarities to those of prokaryotes. As in bacteria, the sedimentation value of the chloroplast and mitochondrial ribosomes is 70S, as opposed to 80S for the cytoplasmic ribosomes of eukaryotes (see Fig. 13-1). The 70S ribosomes of the organelles have been dissociated and the composition of their subunits analyzed in some detail. This procedure has shown that the ribonucleic acid (RNA) of each subunit also resembles that of bacteria, rather than the RNA of the cytoplasmic ribosomes. Isolated mitochondria and chloroplasts can be supplied in the test tube with labeled amino acids. When this is done, the organelles incorporate amino acid units which can be found later in certain protein fractions. We see, therefore, that the mitochondria, essential to all eukaryotes and the chloroplasts, essential to all life on earth, contain genetic information and the machinery to manufacture protein. These observations suggest that a certain degree of autonomy may reside in these cell parts. The facts also raise interesting questions concerning the very origin of mitochondria and chloroplasts during the evolution of eukaryotes. Before discussing these points as well as details about the choloroplast, let us first concentrate on the mitochondrion and certain genetic phenomena which seem to be related to it.

FEATURES OF MITOCHONDRIA. The mitochondrial DNA of all the animal species examined has proved to be circular and to measure approximately 5 μ in circumference. The DNA in the mitochondria of some protozoans and plants appears to be linear. In general, the mitochondrial DNA of plants and eukaryotic microorganisms is appreciably longer than that in animal cells. In addition to their DNA and prokaryotic type of ribosome, the mitochondria also contain the following components associated with replication and information transfer: a DNA polymerase, different tRNAs which correspond to the 20 kinds of amino acids, amino acid activating enzymes (the aminoacyl synthetases), and DNA-dependent RNA polymerase. We see, therefore, that most of the elements which permit independent replication and autonomous protein synthesis are present in the mitochondria. Moreover, each of these components has been shown by chemical and physical procedures to be quite distinct from its counterpart in the cytoplasm. A mitochondrion, therefore, contains its own unique enzymes and tRNAs. This is also demonstrated by the fact that the activating synthetases of the cytoplasm cannot hook amino acids to mitochondrial tRNAs. A mitochondrial tRNA requires a specific mitochondrial synthetase. Since all the observations indicate a special mechanism of information transfer in the mitochondria, we may well evaluate the significance of such a system in the cell. Hybridization studies have been performed between the DNA and the RNA of the mitochondria, and these have yielded some information concerning the informational role of the mitochondrial DNA. There are various lines of evidence for mRNA which is a transcript of the mitochondrial DNA, although no such specific mRNA has been directly identified. It is clear that the mitochondrion's tRNA, as well as the RNA component of its ribosomes, are transcripts of the DNA of the organelle itself. When mitochondria are isolated and supplied with labeled amino acids, they incorporate them at a low rate; the label then appears in the membrane fraction of the organelle, probably the inner membrane. This suggests that some of the genetic information in a mitochondrion may code for certain structural parts of the mitochondrion itself. However, the greatest part of the mitochondrial protein (approximately 95%) has been shown to be formed outside the organelle in the surrounding cytoplasm and to be coded for by genetic information in the nucleus of the cell. The evidence indicates that the ribosome machinery of the mitochondrion, as well as that of the cytoplasm, contributes to the formation of mitochondrial protein but that the larger contribution is made by the cytoplasm of the cell.

When the total DNA content of the mitochondria of animal cells is considered, it is seen to be quite low. We have noted that the tRNA and the rRNA of the mitochondria are transcribed from this organelle DNA. This means that the amount of information left in it to code for protein is very small, only about 20-30%. In some eukaryotic microorganisms, such as the mold *Neurospora,* the amount of mitochondrial DNA is much greater. In such cases, more protein could be encoded by the organelle DNA than by the mitochondria of the average animal species. At the moment, we are faced with facts which pose many unanswered questions about the mitochondria. We see that a unique DNA as well as unique components of a protein-synthesizing mechanism is present in the organelle, but yet we find that most of the protein of the mitochondrion depends on information which is coded in the cell nucleus and then translated by the ribosomes of the cytoplasm. There is evidence, however, which shows that some of the mitochondrial protein depends on translation of mRNA by the ribosomes inside the organelle itself. It is nevertheless conceivable that certain information stored in the DNA of the nucleus undergoes transcription to mRNA and that *this* then travels to the mitochondrion where it is translated. In like fashion, transcripts of the mitochondrial DNA could move from the organelle and be translated by the ribosomes of the cytoplasm (Fig. 20-9). Such interactions between organelle ribosomes and those of the cytoplasm have not, however, been conclusively demonstrated.

THE PETITE MUTATIONS IN YEAST. Though we still don't know the full significance of the extrachromosomal information in the mitochondrion, we do have excellent genetic evidence that mutations within the mitochondrial DNA are responsible for certain traits which are inherited in a non-Mendelian fashion. The most thoroughly studied of these are the petite mutations in yeast, *Saccharomyces.* Cells carrying a petite mutation form tiny colonies on agar, in contrast to the large ones of the wild phenotype. These petites have been shown to be deficient in certain cytochromes. They require a substrate containing a fermentable product such as glucose which they use in the presence of oxygen as if they were growing anaerobically. In order to understand the inheritance of the petite traits, we must be familiar with the very simple life cycle of yeast (Fig. 20-10). Like *Neurospora,* yeast is an ascomycete, or sac fungus. It is a unicellular organism, and the haploid cells can be classified into either of two mating types, "+" or "−." The diploid zygote formed from fusion of a "+" and a "−" cell may grow by budding to produce a diploid colony. The diploid cells can also be stimulated to undergo meiosis. The cell then enlarges and forms four haploid nuclei, each of which becomes the nucleus of a spore. The meiotic cells behave like an ascus or sac and so the four spores are considered to be

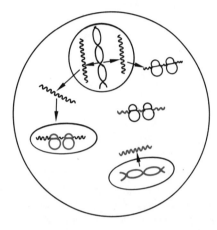

⬭	Mitochondrion
✖	Mitochondrial DNA
∿∿	Mitochondrial transcript
⊖	Mitochondrial ribosome
∿∿	Nuclear transcript
⊖	Cytoplasmic ribosome

FIG. 20-9. Interactions between cytoplasm and organelle. In this hypothetical scheme, some of the mRNA of the nucleus travels to the mitochondrion and is translated on the ribosomes of the organelle. The presence of a protein inside the organelle, therefore, would not mean that the protein was coded by information in the organelle itself. Similarly, transcription of mitochondrial DNA could produce messenger which travels to the cytoplasm and which is translated on the ribosomes there.

ascospores. It can be seen from Figure 20-10 that two of the ascospores will be mating type "+" and two will be "−." This 2:2 segregation indicates that the genetic determinants of mating type are nuclear and exist as one pair of alleles which segregate at the meiosis which precedes ascospore formation. Similarly, many other characteristics are known in yeast which behave in the same way, such as the ability to produce adenine or some other essential metabolite. In all of these cases, 2:2 segregations are found when a single pair of alleles is followed. Linkage as well as independent assortment is easily detected in yeast and has permitted the construction of chromosome maps using the same reasoning followed for other organisms.

Mutations to the petite phenotype have arisen in both mating types, "+" and "−." Some of these petites behave in the expected Mendelian fashion: a cross of a petite with a wild produces wild diploid cells; the ascospores yield a 2:2 segregation of wild type to petite. Petites such as these are called "nuclear petites." In contrast to these are others which are decidedly non-Mendelian in their pattern of inheritance. One kind of nonchromosomal petite is known as the "neutral" or "recessive" petite. A cross of a neutral petite with a wild (Fig. 20-11) produces diploid cells which are normal in phenotype. When sporulation is induced, the ascus yields spores which produce only wild-type cells. The segregation is thus 4:0. The petite phenotype has disappeared and does not reappear when these wild cells are followed further in similar crosses. Clearly, the neutral petite is not behaving as a Mendelian trait, as indicated in the departure from the 2:2 segregation in the ascus.

Experiments were performed in which yeast was treated with acriflavine. (It has been found that acridine dyes can eliminate "F" factors from *Escherichia coli* cells.) The results showed that almost a whole population of normal cells could be transformed to petite after exposure. No known mutagen can affect nuclear genes to such an extent that every exposed cell contains an induced mutation. Other observations of this type strongly indicated that the determinants for the petite phenotype reside in the cytoplasm. Another class of petites was discovered in which the trait also behaved in a non-Mendelian fashion, but it differed in its pattern from the neutral or recessive petite. This is the "suppressive petite." When it is crossed with a wild type, the results depend on when the zygotes are induced to sporulate (Fig. 20-12). If ascospore formation takes place very soon after the zygote forms, it is found that most of the asci will give a segregation of 0:4; that is, all the spores will give rise to petites. The zygotes, if immediately plated out on agar after the mating, form diploid colonies which are also petite. In contrast to the neutral petite, it is as if the wild type were tending to disappear. However, different results can be obtained from the same cross. Instead of being induced to sporulate immediately, the zygotes may be subcultured in liquid medium for a period of time. If the diploid colonies are then plated out, they are almost all wild type! Similarly, if

FIG. 20-10. Life cycle of *Saccharomyces.* Haploid cells of opposite mating types ("+" and "−") fuse to form a diploid zygote. This can give rise to a colony of diploid cells by budding. Diploid cells may also be stimulated to undergo meiosis. The cell enlarges, and four haploid nuclei result, each the nucleus of an ascospore. The mating type as well as other nuclear alleles segregate 2:2 from the ascus.

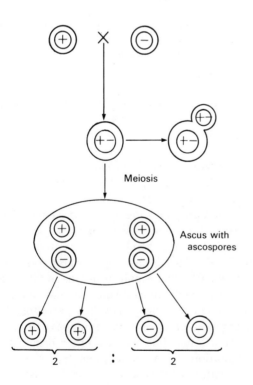

Meiosis

Ascus with ascospores

2 : 2

these zygotes are induced to sporulate *after* being subcultured, the ascus gives a 4:0 segregation, four wild to no petite. So depending on the treatment after mating, two very different pictures may emerge. The appearance of the wild type after subculture in the liquid can be explained by the fact that the liquid culture would favor the survival of wild-type cells. Any cells of normal respiratory phenotype which are present immediately after mating would be selected in the liquid medium, because it is not conducive to the growth of the petites. Therefore, wild-type cells would increase and take over the culture, giving normal diploids upon plating and 4:0 segregation after sporulation.

There are many other features of the petite story, but it should be evident from this short discussion that the inheritance of the neutral petite and of the suppressive petite is decidely non-Mendelian. It had been considered likely for many years that the petite mutation is due to some change in the mitochondrion, because the respiratory capacity of the petite cell is deficient. It has now been demonstrated by density gradient ultracentrifugation that the mitochondrial DNA of a cytoplasmic petite (neutral or suppressive) differs in its buoyant density from the mitochondrial DNA of the wild yeast. The mitochondrial DNAs from various cytoplasmic petite strains also differ from one another when they are compared. Detailed analyses of certain strains leave no doubt that the mitochondrial DNA has been physically altered with the change from wild type to petite.

INHERITANCE OF DRUG RESISTANCE IN YEAST. In the past 5 years, another class of non-Mendelian mutations has been recognized in yeast. These bring about resistance to various antibiotics, such as erythromycin and chloramphenicol. From Figure 20-13, it can be seen that resistance to the former drug is decidedly non-Mendelian. After the cross of sensitive to resistant, diploids can be isolated which are of the parental types. Upon sporulation, all resistant spores or all sensitive ones are obtained. Resistance to chloramphenicol follows the same pattern. If a cross is made between two strains, each resistant to one of the two drugs, it is seen (Fig. 20-14) that recombination can occur to give

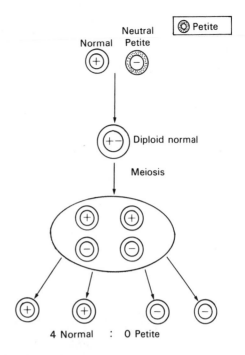

FIG. 20-11. Wild type yeast cells and neutral petite. In a cross between the two types, diploids arise, and these are normal in phenotype. After meiosis, the mating types segregate 2:2 as expected, but all the resulting ascospores produce wild type cells. The petite trait seems to have disappeared and is not behaving in a Mendelian fashion.

asci containing all doubly resistant spores and to asci with all doubly sensitive spores. This tells us that recombination of nonchromosomal genetic determinants can occur as well as recombination of nuclear alleles. Such a process adds even more diversity to the combinations of inherited material which are possible after sexual reproduction, and its full significance in all species is yet to be assessed. There is good evidence that the determinants of drug resistance in yeast also reside in the mitochondrial DNA. For example, when mutations occur to petite and produce a physical alteration in the mitochondrial DNA, markers for drug resistance may be lost. This implies that the physical change in the organelle DNA, possibly the elimination of a segment, has brought about the petite phenotype and the simultaneous loss of the resistance determinants. The inheritance in yeast of the petite trait and of the resistance to certain drugs offers the best evidence for the localization of specific nonchromosomal genes in the mitochondrion.

FEATURES OF THE CHLOROPLAST. We have already noted in this chapter that the chloroplast as well as the mitochondrion contains the essentials for protein synthesis and that isolated chloroplasts can incorporate labeled amino acids. The DNA of the chloroplasts exists in a circular form, but in contrast to mitochondria, which contain very small amounts, the quantity of DNA in the chloroplast is appreciable. In certain algae (*Chlamydomonas*, for example), the amount is about equivalent to that in a bacterial cell. The information to manufacture perhaps several hundred proteins, much more than in the mitochondrion, may reside in the chloroplast. Some of the plastid DNA has been shown to be used for the transcription of the RNA of chloroplast ribosomes and tRNAs. One important enzyme involved in photosynthesis (ribulose 1,5-diphosphate carboxylase) apparently forms as a result of mRNA translation in the plastid, but as indicated before in the discussion of mitochondria, the occurrence of a protein in an organelle does not mean that it or its mRNA was formed there. Actually, there is evidence that the mRNA which is translated to carboxylase is a transcript of nuclear DNA and that this nuclear mRNA moves to the plastid where it is translated (review Fig. 20-9). Moreover, the DNA polymerase

FIG. 20-12. Wild type yeast × suppressive petite. The suppressive petite, unlike the neutral one, tends to express itself over the wild. The diploids are petite, and if diploid cells are induced to sporulate immediately, all the spores give rise to petite colonies. However, if diploids are subcultured and later sporulated, then all the spores give rise to wild colonies.

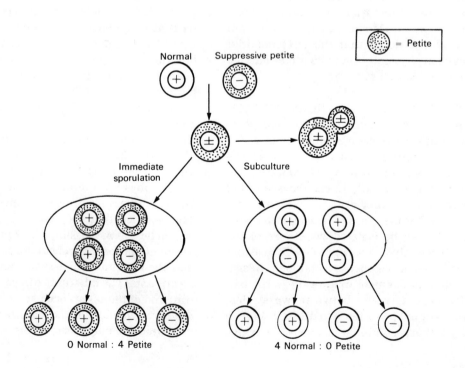

which replicates the DNA of the plastid does not arise by translation of mRNA in the plastid. Actually, the polymerase forms in the cytoplasm as a result of translation of mRNA by the cytoplasmic ribosomes. This can be shown in the following way. Cells may be treated with the drug, rifampicin, which prevents the activity of the RNA polymerase of the plastid. This interference actually causes a loss of the ribosomes of the plastid, because no rRNA can be transcribed from the DNA of the plastid. The drug does not inhibit the RNA polymerase of the nucleus, and so cytoplasmic ribosomes are not affected by it. When ribosomes are lost from the plastids after treatment with rifampicin, the plastids nevertheless still manage to replicate their DNA. Therefore, the DNA polymerase needed for this replication must enter the plastid, because it is a protein and cannot be made there in the plastid if no ribosomes are present (Fig. 20-15).

As in the case of the mitochondrion, the full significance of the genetic material and the protein-synthesizing machinery of the chloroplast remains unknown. The presence of appreciable amounts of hereditary material in the plastid certainly suggests a degree of independence from the nucleus. In certain algae, it is very evident that the plastids divide and are continous from one cell generation to the next. In unicellular algae, the plastids divide in synchrony with the nucleus and are carried along through the gametes where they even remain green. In higher plants, chloroplasts increase in number by simple division, which is not correlated with the division of the cell. No clear-cut plastids are seen in embryonic cells; however, little bodies known as proplastids may be recognized, and it is from these that the chloroplasts arise in the cell. There are actually no plastids in the gametes of a higher plant species, and it is believed that they are passed from one generation to the next in the form of these undeveloped bodies, the proplastids. Thus, the chloroplast is thought to be continuous in higher as well as in lower plant species, arising only from other plastids or from proplastids.

Many nuclear genes are known which control the development of the plastid from the proplastid. In corn, several hundred are known. Mutant forms of many of these nuclear genes

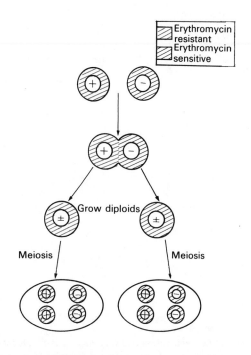

FIG. 20-13. Erythromycin resistance in yeast. After the cross of resistant × sensitive, diploid cells may be established from growth of the zygotes. Upon sporulation, the diploids give rise either to all resistant cells or to all sensitive ones. (The two kinds of diploid cells do not necessarily occur in a 1:1 ratio.)

have been shown to block specific steps in plastid differentiation. Depending on the stage which the nuclear gene controls, a plant with a mutant allele may be unable to develop any chlorophyll at all, or it may form reduced levels of the green pigment. Such a plant may die when the food reserves from the seed are exhausted. There is thus no doubt that the nucleus exerts control over the plastid; however, this is not necessarily absolute, as the following two examples illustrate.

PLASTID AUTONOMY IN HIGHER PLANTS. Rhoades studied the inheritance of a gene in corn known as "iojap." When a plant is a homozygous recessive ("ij ij"), there is a tendency for plastids to become so altered that chlorophyll is not produced. Not all the plastids necessarily undergo this change, but the tendency is sufficient to produce streaks of white tissue, giving the plant a striped or variegated appearance. Some seedlings appear colorless and do not survive because the majority of their plastids have been altered and lack chlorophyll. The results of a cross between a striped "iojap" plant and a normal green one differ, depending on the way the cross is made. Figure 20-16A shows that the inheritance is strictly maternal and seems to depend on what the female gamete contributed. If the female parent is normal green (Ij Ij) and the male parent is striped (ij ij), all of the F₁ plants will be normal in appearance. On the other hand, if the female parent in a cross is striped (ij ij) and the male parent is green (Ij Ij), three phenotypes occur among the F₁: normal green, striped, and colorless. It thus seems that the genotype of the male parent is ineffective in the expression of the iojap trait. The inheritance is maternal when one of the parents is of the genotype "ij ij."

Particularly significant is the result of backcrossing (Fig. 20-16B). In a backcross, if a striped F₁ is used as the female parent and is crossed with a homozygous wild (Ij Ij), three classes of progeny again arise: green, striped, and colorless. This means that plants heterozygous or homozygous for the wild type allele (Ij ij or Ij Ij) may still show the mutant effect. Some plastids have been altered, but they are obviously reproduced and carried along in this changed condition. The original mutation (ij) occurred in a nuclear gene, but once the alteration in the plastid has arisen, it becomes established and

FIG. 20-14. Two-factor cross involving drug resistance. If chloramphenicol resistance is followed alone, it shows the type of non-Mendelian inheritance illustrated in Fig. 20-13 for erythromycin. If chloramphenicol and erythromycin resistances are followed together in a two-factor cross, it can be demonstrated that recombination of the non-Mendelian factors is able to take place. After the mating, diploid cells are cultured. From these, four types can be isolated (not necessarily in a 1:1:1:1 ratio). Each gives rise to ascospores, which are either all parental or all recombinant. The recombinants here are the $E^R C^R$ and $E^S C^S$. (Modified slightly and reprinted with permission from R. Sager, *Cytoplasmic Genes and Organelles*, p. 132, Academic Press, New York, 1972.)

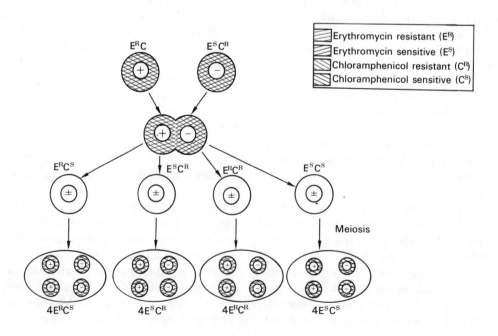

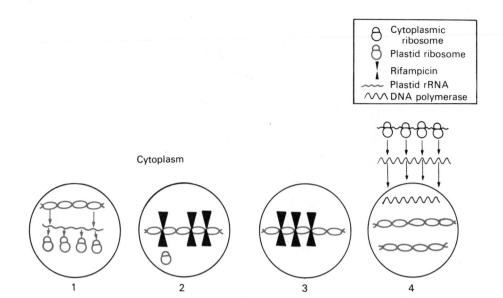

Cytoplasmic ribosome
Plastid ribosome
Rifampicin
Plastid rRNA
DNA polymerase

Cytoplasm

1   2   3   4

FIG. 20-15. Interaction of chloroplast and cytoplasm. *1.* The normal chloroplast contains among other things its own DNA, RNA, and ribosomes. Some of the chloroplast DNA may undergo transcription to form the rRNA of the ribosomes of the plastid. *2.* The drug rifampicin blocks transcription of the plastid DNA. It does not block transcription of the nuclear DNA. *3.* Plastids arise which lack ribosomes, because no ribosomal RNA is formed in presence of the drug. *4.* In the plastids lacking the ribosomes, the DNA manages to replicate. This requires the enzyme DNA polymerase. The polymerase could not be formed in these plastids because they lack ribosomes, and these are needed for protein formation. The polymerase must have entered the plastid from the cytoplasm. The cytoplasm contains ribosomes because the drug does not prevent RNA formation from the nuclear DNA.

persists; it no longer depends for its perpetuation on the recessive allele. There is obviously some degree of plastid autonomy. This is clearly indicated by the striped and colorless phenotypes which persist in homozygous wild plants (Ij Ij). The genus *Oenothera* also has given evidence for a certain amount of plastid autonomy, although nuclear genes also exert an effect on plastid development.

NONCHROMOSOMAL GENES IN CHLAMYDO-MONAS. All of the evidence indicates that there is genetic information in the chloroplast which influences the reproduction and development of the organelle. Since a significant amount of DNA is present, there is the possibility that the chloroplast also contains genetic information which can affect a characteristic of the plant *other than* the plastid itself. The best evidence for this comes from studies of Sager with the unicellular green alga, *Chlamydomonas*. The life cycle of this species is very simple and is comparable to that of yeast. Again two mating types, "+" and "−," are recognized. Mating between cells of opposite mating type produces a zygote (Fig. 20-17). When the zygote matures, meiosis occurs, and four zoospores (the motile products of the meiotic division) arise. One-half of these are mating type + and one-half are "−." The zoospores may undergo mitotic division to form clones of cells. Again the mating type is seen to depend on a pair of nuclear alleles inherited in typical Mendelian fashion. Many characteristics in *Chlamydomonas* are known to depend on nuclear genes: spore color and requirements for certain metabolites, etc. Both linkage and independent assortment have been demonstrated for these genes associated with Mendelian traits. There are, however, some genetic determinants in *Chlamydomonas* which are definitely nonchromosomal. Sager has worked extensively with certain streptomycin-resistant strains which exhibit non-Mendelian inheritance. As Figure 20-18 shows, the results of a cross between streptomycin resistant and streptomycin sensitive depends on the phenotype of the mating type "+" parent. Although mating type itself segregates among the zoospores in the Mendelian 2:

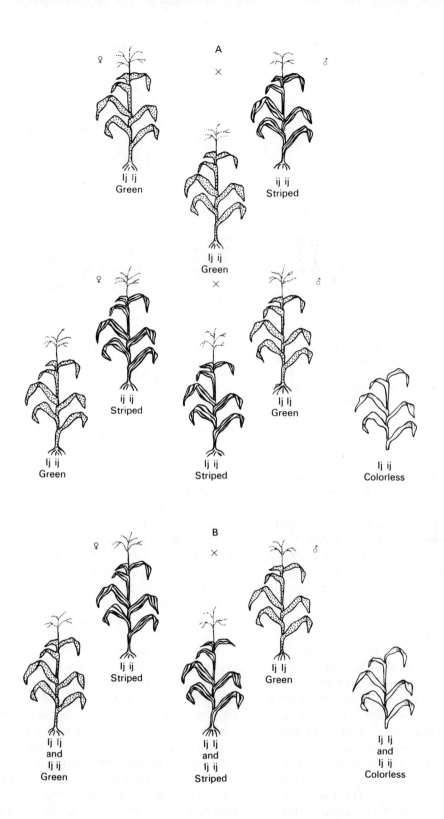

FIG. 20-16. Iojap mutation and plastid inheritance. *A.* Plants that are homozygous for the recessive "iojap" will carry altered plastids and will appear variegated or striped with white streaks. If an iojap plant is used as a male parent (*above*), the offspring are green, but the reciprocal cross produces three classes of offspring (*below*). The altered plastid condition can evidently be transmitted by the female gamete but not by the male. *B.* When a striped heterozygous plant is used as the female parent and crossed with a homozygous normal, offspring arise which have altered plastids. This means that individuals homozygous for the wild allele (Ij Ij) are expressing the mutant effect and may be either striped or colorless. The original plastid alteration resulted from the presence of the mutant nuclear allele in homozygous condition (ij ij), but the plastid, once changed, can still reproduce itself even though the wild genotype is present. The plastid thus has a certain amount of independence.

2 fashion, the zoospores always have the streptomycin trait which was shown by the mating type "+" parent. This suggests maternal inheritance, but here, the "+" and the "−" cells are identical. Neither class of gamete contributes more cytoplasm and hence more cytoplasmic genes than the other. Something is operating to prevent the transmission of the nonchromosomal genes of the mating type "−" parent from the zygote to the meiotic products. Since the discovery of the nonchromosomal streptomycin determinants, many others have been found. Some of these were induced by mutagens, particularly by streptomycin itself, which in *Chlamydomonas* is able to induce mutations in nonchromosomal genes but not in those of the nucleus.

It was later found that rare exceptions occur to this "mating type +" pattern of inheritance. In less than 1% of the zygotes, exceptions arise in which both sets of nonchromosomal genes, the set from the mating type "−" as well as the one from the mating type "+" parent, are transmitted. Sager also found that if the mating type "+" parent is treated with a dose of ultraviolet

FIG. 20-17. Life cycle of *Chlamydomonas.* Two mating types occur. After fusion of two cells of opposite type, meiosis takes place, and the mating type segregates in a Mendelian fashion. The motile cells can divide mitotically to form clones of other identical cells.

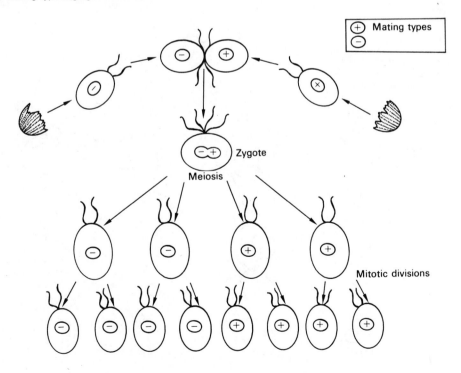

light before mating, 50% of the zygotes exhibit biparental inheritance. All of the offspring arising from such a zygote contain a complete set of nonchromosomal genes from each parent. Such cells are called *cytohets*, indicating that they are heterozygous for cytoplasmic genetic determinants (Fig. 20-19A and B). It is possible to follow more than one pair of cytoplasmic markers in biparental inheritance. The cytoplasmic markers do *not* segregate when meiosis takes place. The haploid zoospores are therefore heterozygous for the cytoplasmic genes. When a heterozygous zoospore divides mitotically and forms a clone, the alternative forms of the cytoplasmic genes segregate and undergo recombination. This segregation and recombination continues with each additional mitotic division of the zoospores until there are no more heterozygous markers to detect. The recombination frequency of these non-Mendelian genes has been measured. This has permitted the construction of a map and the assignment of several nonchromosomal genes to definite positions. It appears that the cytoplasmic genes are all part of a single linkage group which is circular in nature. Sager has presented several lines of evidence which support the hypothesis that the physical basis of the nonchromosomal genes

FIG. 20-18. Inheritance of streptomycin resistance in *Chlamydomonas.* It can be seen that inheritance of sensitivity or resistance to streptomycin depends on the phenotype of the parent which was mating type "+." While mating type is segregating as expected, the cells always are like those of the "+" parent in relationship to tolerance of streptomycin.

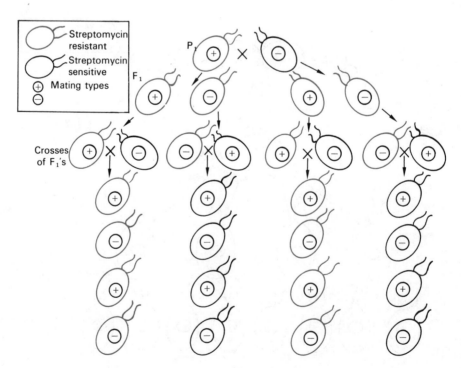

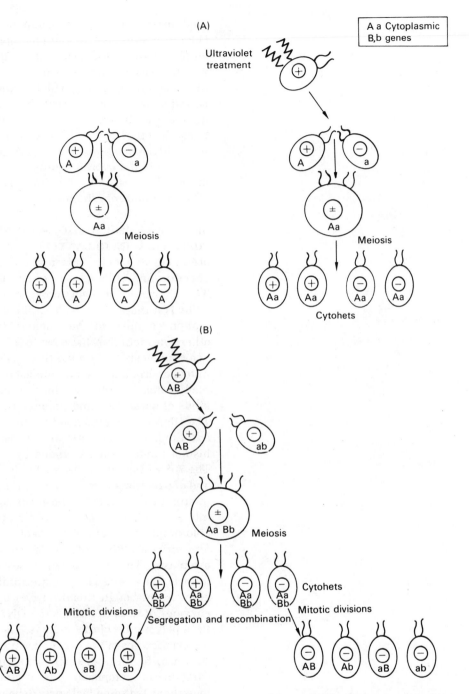

FIG. 20-19. Biparental inheritance of cytoplasmic genes. *A.* Typically, the cytoplasmic genes of the mating type "−" parent are not transmitted from the zygote, so that the haploid products of meiosis, the zoospores, show the phenotype of the "+" parent (*left*). If, however, the mating type "+" parent is treated with ultraviolet and then mated, the cytoplasmic genes from both parents may be transmitted to the zoospores (*right*). Although the nuclear factors (such as "+" and "−") segregate at meiosis, the cytoplasmic ones do not. The F₁ cells thus have two complete sets of cytoplasmic genes and are called "cytohets." *B.* Two or more pairs of cytoplasmic factors may be followed in cases of biparental inheritance. The haploid zoospores are cytohets. If each zoospore is then allowed to divide further to form a clone, segregation of the alleles takes place, as does recombination among the members of the different allelic pairs. Thus, *nuclear alleles in Chlamydomonas segregate and undergo recombination at meiosis, whereas these processes do not take place for the cytoplasmic genes until the mitotic divisions of the haploid zoospores, after meiosis has occurred.*

(and hence the linkage group to which they belong) is the DNA of the chloroplast. For example, most of the cytoplasmic determinants which have been mapped were caused to mutate by streptomycin, a drug which is known to exert its influence almost exclusively on the development of the chloroplast and which has no direct effect on other parts of the cell. Further work with *Chlamydomonas* promises to give a greater insight into the overall significance of nonchromosomal genes, as well as into the mechanism of uniparental inheritance.

BIOLOGICAL IMPLICATIONS OF GENETIC INFORMATION IN ORGANELLES. Although the presence of genetic information in the mitochondria and chloroplasts is now firmly established, it raises some very provocative questions. "Why is it there? Why isn't all the genetic information confined to the nucleus?" These and other considerations have led to the formulation of a hypothesis concerning the very origin of the mitochondria and the chloroplasts. We have already noted several resemblances between both kinds of organelles and prokaryotic organisms. The DNA of the mitochondria, the chloroplasts, and the DNA of bacteria are unassociated with histones and are not bounded by a membrane. The 70S ribosomes of the two kinds of organelles and of bacteria appear to be very much the same in structure. Moreover, these ribosomes are inhibited by chemicals such as streptomycin and chloramphenicol, antibiotics which do not interfere with the activity of the 80S ribosomes in the cytoplasm. The RNA polymerases of bacteria, mitochondria, and plastids are inhibited by rifampicin. Although this drug can bind to these enzymes, it does not affect the DNA-dependent RNA polymerase of the nucleus. The tRNAs and the synthetases of the organelles have also been shown to be quite distinct from their counterparts in the cytoplasm. Observations such as these have led some biologists to believe that the mitochondria and the chloroplasts in eukaryotic cells represent the descendants of ancient symbiotic organisms which took up residence in ancient prokaryotic host cells. The mitochondrial ancestor would have been a primitive type of aerobic bacterium and the ancestral chloroplast a primitive type of blue-green alga. According to the idea, their residence gave a decided advantage to the primitive prokaryotic host. Those with the mitochondrial ancestor would have been provided with an efficient energy source. The host cytoplasm would contain the machinery for anaerobic respiration, whereas the symbiont would contain the aerobic mechanism. As a result of the benefits conferred to the host, other features of the eukaryotic cell became established: the nuclear membrane, mitotic apparatus, and endoplasmic reticulum, etc. Those cells which also carried the primitive blue-green algae evolved into eukaryotic plant cells which had acquired the ability to trap the energy of light.

Thus, according to this idea, all eukaryotic cells have arisen from a host-symbiont relationship, and we see today in the cells of every eukaryote the descendants of the original symbionts in the form of mitochondria or plastids. During the courses of evolution, natural selection would favor any mutations in the genetic material of the host cell which would help to preserve the relationship or make it still more efficient. And so, the host cell would eventually take over some of the functions of the symbiont as the relationship became more and more advantageous. Such a picture is reminiscent of *Paramecium* with kappa and the related particles which most probably *do* represent symbionts. We saw that the host cells carried chromosomal genes which affect the maintenance and development of the particles. So it is conceivable that the eukaryotic cell organelles arose in a similar fashion and that today, most of the properties of the symbionts are now controlled by nuclear genes of the host. Most of the proteins and the developmental stages of the mitochondria and plastids are certainly under the control of nuclear genes. And most of the essential structural components of the organelles are made outside of them in the cytoplasm of the cell. The amount of information in the mitochondrion is so small that not very many proteins could be coded for. And so, one can argue that during the course of evolution, most of the original mitochondrial functions were assumed by the cell nucleus because mutations in this direction were favored by natural selection. But the annoying question still remains as to why any genetic information at all continues to reside in the mitochondrion or plastid. If the symbiont hypothesis is correct, the relationship was es-

tablished countless ages ago, long before the origin of the simplest eukaryotic cell. If it were beneficial for the nucleus to assume control over the organelles, why has it not been complete? The force of evolution usually weeds out any superfluous features of an organism. Therefore, we would have to conclude that the presence of the extra genetic material in the organelles is not superfluous. Perhaps the information is controlling a few very important functions which are more efficiently handled by the mitochondrion or the chloroplast than by the nucleus. Or possibly, there may be some interrelationship between the nuclear and the nonchromosomal genetic information which has yet to be revealed. Definitive answers may someday be provided through investigations of nonchromosomal genetic determinants in a variety of species.

THE REPLICON. The behavior of extrachromosomal genetic particles in bacteria has brought to light several basic points which relate to nonchromosomal genetic information in general. Foremost among these are the requirements essential to the replication and persistance of any extra chromosomal genetic material. In our discussion of chromosome aberrations (Chap. 10), we noted that any chromosome fragments in a cell, such as those resulting from crossing over within an inversion, do not persist; they disintegrate in the cytoplasm or are otherwise lost from the cell. An entire extra chromosome, on the other hand, may survive and replicate, as the many examples of aneuploidy illustrate. When we look at bacteria, we see a comparable picture. A piece of the bacterial chromosome by itself will not survive in the bacterial cytoplasm. An example of this was seen in the case of transformation. If a piece is not integrated into the bacterial chromosome, it will not function in transforming the cell and will disappear. However, if a piece of the chromosome, which cannot survive by itself, is incorporated into an "F" factor of a virus, it *can* persist. The F-prime factors carry pieces of the bacterial chromosome which replicate along with the extrachromosomal "F" particle.

A transducing phage, such as lambda, may carry a chromosomal gene or fragment which may be replicated independently of the chromosome when it becomes part of the virus.

Viruses themselves are able to replicate autonomously when in a host cell. Compared to the size of the host genetic complement, they are extremely small, tinier than most chromosome fragments which cannot replicate by themselves. However, if the virus inserts into the bacterial chromosome and takes up residence as prophage, it then replicates right along with the host chromosome, as if it were a native part of it. Such observations tell us that an extra chromosomal entity must possess certain features if it is to survive autonomously and that this is not simply related to its size. To account for all the facts, a model has been proposed by Jacob, Monod, and Brenner. Any genetic element which can carry on independent replication is called a *replicon*. A complete chromosome would be a replicon, as would an "F" factor or a virus. A replicon is thus a unit of replication, and it must possess at least two essential determinants. One of these is the replicator, the specific genetic point at which the double helix of the DNA unwinds so that each strand can act as a template. The entire replicon is then duplicated once the DNA is opened up at the replicator and exposed to the enzyme DNA polymerase. However, before this can happen, a stimulus is needed. This is provided by a cytoplasmic factor, the initiator, which acts on the replicator (Fig. 20-20). Initiator molecules are produced by a specific genetic region of the replicon, a region which acts as a regulator. Close similarities can be seen between the model of the replicon and that of the operon. Many phenomena related to replication can be explained on this idea. A chromosome fragment, no matter what its size, cannot replicate autonomously unless it has a

FIG. 20-20. The replicon. The figure shows a unit of replication. This may be a whole chromosome, an F factor, or a virus. Before the DNA can be replicated, the double helix must open up so that the DNA polymerase can copy each strand. In order for the DNA to unwind, a stimulus is needed. This is provided by the initiator molecules whose formation is governed by a regulator. The initiator causes the DNA to unwind at a specific region, the replicator. DNA polymerase then copies both strands, and the entire replicon is duplicated. If a DNA element, no matter what its size, lacks a replicator or regulator, it is not able to replicate independently because the DNA will not unwind.

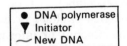

● DNA polymerase
▼ Initiator
⁓ New DNA

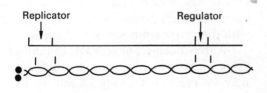

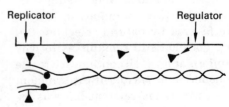

UNDERSTANDING GENETICS

replicator and the required region to provide the stimulating (initiator) molecules. A fragment would usually lack one or both of these. A tiny virus or "F" factor, on the other hand, contains both. If the episome is carrying a piece of the chromosome, the latter is replicated right along with the genetic material of the episome, for the fragment is now a part of a replicon and is being controlled by the regulator and initiator of the episome.

When a temperate phage inserts into a bacterial chromosome, its own replicator does not function (Fig. 20-21A). This is so, because its own initiator molecules are inhibited by a repressor whose formation in the cell is governed by a gene of the phage itself. This repressor prevents the initiator molecules of the virus from acting on the virus replicator. As a result, the replication of the host chromosome, controlled by the host initiator, determines the replication of both the chromosome and the virus which is integrated into it. If the repressor of the virus initiator is rendered ineffective (and this can happen by mutation of the specific viral gene or as a result of an environmental factor such as ultraviolet), the virus can again replicate independently (Fig. 20-21B). It may come out of the chromosome, replicate repeatedly, and lyse the cell. A virulent virus is not repressed when it enters a cell. Its initiator molecules allow it to replicate over and over, independently of the chromosome, and the result is lysis. The concept of the replicon pertains to all kinds of cells, prokaryotic and eukaryotic. We must now recognize the existence in eukaryotes of extrachromosomal replicons in the mitochondria and plastids, major cell components essential to all higher life forms.

RESISTANCE TRANSFER FACTORS. Of ever increasing importance to humans are certain replicons which have been recognized in bac-

FIG. 20-21. Phage as a replicon. *A.* A temperate phage can insert itself into the chromosome of the cell and become prophage. The phage possesses its own replicator and regulator. However, genes of the temperate phage produce a repressor which reacts with the phage's initiator molecules. This prevents the phage initiator from acting on the phage replicator. The initiator molecules of the chromosome are not influenced by the phage repressor. The DNA can be unwound at the site of the chromosome replicator. The DNA polymerase will then start replicating DNA from that point, including all the DNA of the virus in its prophage state. *B.* A mutation in the phage may cause defective repressor or no repressor to form. The phage initiator can then act on the phage replicator. DNA polymerase will then bring about reproduction of the phage DNA independently of the chromosome. The phage is now virulent. Any further replication of the chromosomal DNA will cease, and the cell will be lysed.

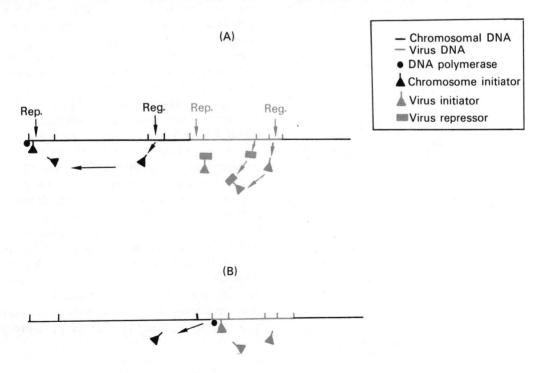

(A)

— Chromosomal DNA
- Virus DNA
● DNA polymerase
▲ Chromosome initiator
△ Virus initiator
▪ Virus repressor

(B)

teria and which can confer drug resistance on the prokaryotic cell. These nonchromosomal genetic entities are like episomes because they may be absent from a cell but when present can replicate autonomously. However, they have not all been shown to integrate with the chromosome of the cell as do "F" factors and temperate phage. The more general term *plasmid* is used to define any nonessential genetic particle which may exist independently of the chromosome. Episomes would be plasmids which have the ability to integrate with the chromosome.

Within the past 10 years or so, the number of bacterial strains which have become resistant to one or more drugs has increased with amazing rapidity. There are several loci on the bacterial chromosome which are known to control drug sensitivity or resistance. Antibiotics such as chloramphenicol commonly exert their lethal effect on the bacterial cell by association with the ribosomes and consequently interfering with translation. Gene mutations to drug resistance generally bring about changes in the ribosome itself so that the antibiotic is no longer able to combine with it, thus permitting normal translation and protein synthesis in the presence of the drug. Another important class of drug-resistance determinants is now known, and these do not represent genetic information on the chromosome. Instead, they are genes which exist outside the chromosome and which depend for their transfer on the presence of a plasmid, known as a *resistance transfer factor* (*RTF*). This factor has certain features in common with "F," the sex factor. Like "F," it can bring about conjugation between two cells and so promote its own transfer to a cell which lacks it. This transfer is independent of the presence of "F" which can coexist in a cell with the resistance-transfer factor (Fig. 20-22). Since an RTF may carry several genetic determinants for drug resistance, conjugation can confer multiple drug resistance on a previously sensitive cell. The RTF, like "F," is known to be a DNA element. RTFs in certain bacteria such as *Proteus* have an A:T/G:C ratio quite distinct from that of the DNA of the chromosome. This has made possible the isolation of the DNA of the RTF by density gradient ultracentrifugation. Special procedures have permitted the separation of individual

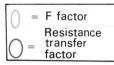

$\bigcirc$	= F factor
$\bigcirc$	= Resistance transfer factor

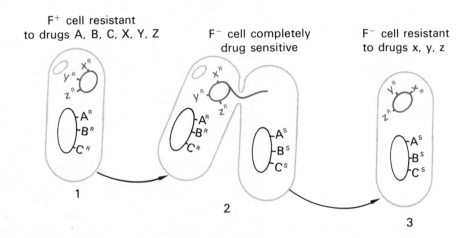

F$^+$ cell resistant to drugs A, B, C, X, Y, Z

F$^-$ cell completely drug sensitive

F$^-$ cell resistant to drugs x, y, z

FIG. 20-22 Resistance transfer factors (RTFs). A bacterial cell may carry genetic determinants which make it resistant to certain drugs (*1*). Some of these determinants may be on the chromosome, but others may be on an RTF. The latter can exist in a cell independently of the "F" factor. It can also promote conjugation with a cell which lacks an RTF (*2*). In this way, a previously drug sensitive cell can be converted to one which is multiply drug resistant if it receives an RTF which carries resistance to two or more drugs (*3*). It is possible for a cell to carry a resistance determinant to a specific drug, say streptomycin, on either its chromosome or on its RTF or other plasmid. However, the gene on the plasmid is not the same as the one on the chromosome. They bring about resistance to the specific drug in different ways. See text for more details.

RTFs. These have been visualized with the electron microscope and have been shown to be circular molecules.

We noted that the "F" factor causes the production by the cell of a specific type of pilus or sex fimbria, which is believed to be essential for conjugation. These sex fimbriae also make an F$^+$ or high-frequency (Hfr) cell sensitive to certain RNA phages which can adsorb to them and cause destruction of the cell. Some of the RTFs have also been shown to cause a cell to produce sex fimbriae which resemble those under the control of the "F" factor. The cells with these RTF-induced fimbriae also become sensitive to the RNA phages which can lyse and destroy male cells. Both the RTF and the "F" factor may be removed from cells after treatment with acridines. This loss is accompanied by the inability of the cells to produce sex fimbriae and to adsorb the male-specific phages. We see from

this that these two types of replicons, the "F" factor and the RTF, have many features in common. The fertility factor, however, serves mainly to bring about transfer of the chromosome once it becomes integrated, whereas the RTF is responsible for its own transfer and for that of the resistance genes which it may be carrying. It has been shown that while certain nonchromosomal genes for drug resistance may exist as part of an RTF, others may exist independently of the RTF in a cell. For example, in *E. coli*, a determinant which confers tetracycline resistance is integrated with an RTF which regularly transmits itself and the gene for tetracycline resistance. However, this RTF is not integrated with the determinants for resistance to streptomycin and to ampicillin. These latter two genetic elements exist and replicate independently of the RTF. By themselves, however, they cannot be transferred from cell to cell. If the RTF is present, the streptomycin determinant and the ampicillin factor may be transferred either together or separately (Fig. 20-23). This means that these two genetic elements occur on plasmids and can confer properties on a cell, the property of specific drug resistance. They are not infectious, however, and cannot promote their own transfer without the presence of an RTF. An RTF in turn *is* infectious, and it may be joined with certain determinants. There appears to be a variety of different RTFs which carry various combinations of resistance determinants and which may react in various ways with other plasmids in the cell. Indeed, there is evidence that besides interacting with other plasmids which confer drug resistance (such as the ones for streptomycin and ampicillin resistance), the RTF may also exert some inhibiting effect on the transfer of "F" from an F^+ to an F^- cell.

The existence of such an assortment of resistance-determining plasmids raises questions concerning their origin. Unlike the chromosomal genes which can confer resistance by altering the structure of the ribosome, the nonchromosomal genes exert their effect by directing the synthesis of products which act directly on the drug itself and inactivate it. The ribosome remains unaltered. This very different method of conferring drug resistance suggests that the nonchromosomal determinants are quite distinct from their chromosomal counterparts. Note that all the plasmids are replicons; they possess all the genetic information required for their independent replication. Such evidence strongly suggests that these plasmids have actually originated independently of the cell. They would thus represent symbionts whose presence confers certain properties on a cell. This reminds us of a symbiotic virus which can carry into a cell genetic information picked up from another. We have seen that the distinction between a virus and an F factor is a vague one and that F also has many properties of a symbiont. The first topic in this chapter dealt with the more obvious symbionts of *Paramecium,* and questions were raised on the implications of the relationship. Now we realize that all higher species contain extrachromosomal information in the DNA of their mitochondria and that green plants carry even more in their plastids. Studies of extrachromosomal elements have very practical ap-

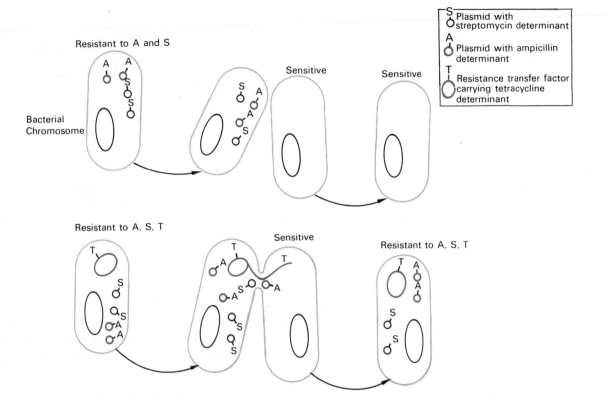

FIG. 20-23. RTFs and other plasmids. Certain determinants to drug resistance do not reside on an RTF but instead are found on plasmids which can replicate independently. They cannot bring about their own transfer to another cell (*above*). An RTF, carrying one or more resistance determinants, can bring about its own transfer. When this takes place (*below*), the other plasmids, if present in the same cell, may also be transferred.

plications in the control of drug-resistant pathogens. They may also provide clues regarding the factors involved in pathological cell changes. At the same time, we may gain a fuller appreciation of the origins of the eukaryotic cell and the interrelationships of higher and lower species. Not least of all is the use of RTFs in genetic manipulation (Chap. 21).

REFERENCES

Clowes, R. C. Molecular structure of bacterial plasmids. *Bacter. Rev.*, 36: 361, 1972.

Clowes, R. C. The molecule of infectious drug resistance. *Sci. Am. (Apr)*: 18, 1973.

Goodenough, U. W., and R. P. Levine. The genetic activity of mitochondria and chloroplasts. *Sci. Am. (Nov)*: 22, 1970.

Margulis, L. Symbiosis and evolution. *Sci. Am. (Aug)*: 48, 1971.

Preer, J. Extrachromosomal inheritance: hereditary symbionts, mitochondria, chloroplasts. *Ann. Rev. Genet.*, 5: 361, 1971.

Sager, R. *Cytoplasmic Genes and Organelles.* Academic Press, New York, 1972.

Scragg, A. H. A mitochondrial DNA-directed RNA polymerase from yeast mitochondria. In *The Biogenesis of Mitochondria*, A. M. Kroon and C. Saccone (eds.), p. 47. Academic Press, New York, 1974.

Schwartz, R. D., and M. O. Dayhoff. Origins of prokaryotes, eukaryotes, mitochondria, and chloroplasts. *Science:* 199: 395, 1978.

Shumway, L. K., and T. E. Weier. The chloroplast structure of *iojap* maize. *Am. J. Bot.*, 54: 773, 1967.

Sonneborn, T. M. Kappa and related particles in *Paramecium. Adv. Virus Res.*, 6: 229, 1959.

Sonneborn, T. M. The gene and cell differentiation. *Proc. Natl. Acad. Sci.*, *46:* 149, 1960.

Wurtz, E. A., J. E. Boynton, and N. W. Gillham. Perturbation of chloroplast DNA amounts and chloroplast gene transmission in *Chlamydomonas reinhardii* by 5-fluorodeoxyuridine. *Proc. Natl. Acad. Sci.*, *74:* 4552, 1977.

REVIEW QUESTIONS

1. Assume that each of the following kinds of *Paramecium* cells undergoes autogamy. Give the genotypes of the two cells which will arise. Indicate whether they are killer or sensitive.

    A. KK with kappa.
    B. KK without kappa.
    C. Kk with kappa.
    D. Kk without kappa.

2. In each of the following, give the genotypes of the two stocks or strains of *Paramecium* which, when crossed, produce the results described:

    A. A killer stock and a sensitive stock whose F_2 are all killers following cytoplasmic exchange.

    B. A killer stock and a sensitive stock whose F_2 are killers and sensitives in a 1:1 ratio following cytoplasmic exchange.

    C. Two killer strains whose cells, when mated, give an F_1 producing killers and sensitives in a 3:1 ratio with or without cytoplasmic exchange.

3. A. In Question 2A, assume there is no cytoplasmic exchange. What would be the genotype and the phenotype of the F_1 cells arising after conjugation from the P_1 sensitive? What would be the result from the P_1 killer?

    B. In Question 2B, assume there is no cytoplasmic exchange. What would be the genotype and the phenotype of the F_1 cells arising after conjugation from the P_1 sensitive? What would be the result from the P_1 killer?

4. Give at least five facts which can be used to support the argument that the mitochondria and the chloroplasts are descendants of ancestral

prokaryotes which took up residence in a host cell.

5. A. In yeast, would you expect the wild or the petite condition in the zygotes and from sporulation of the zygotes after crossing two neutral petites?

B. Answer the same question for the cross of a neutral petite and a nuclear petite.

C. How would you expect mating type to segregate in each of the above crosses?

6. In *Chlamydomonas*, assume the nuclear alleles C, c and D, d are being followed along with the cytoplasmic determinants P, p and Q, q. Give the genotype expected in the zygote and in the products of the zygote in each of the following crosses:

A. Mating type "+"; CD; Pq × mating type "–"; cd; pQ.

B. The same cross as in the above, assuming that ultraviolet light was applied to the "+" parent.

7. Allow "A" and "B" to represent determinants for drug resistance which are found on an RTF in certain *E. coli* strains. "C" and "D" represent drug-resistant determinants which occur on two different noninfectious plasmids. "E" and "F" are genetic drug-resistance determinants found on the chromosome. For each of the situations given below, tell what is likely to be the maximum number of resistance determinants transferred to any cell which completely lacks them.

A. F$^+$ ABCDEF mixed with F$^-$, completely resistant

B. F$^-$ ABCDEF mixed with F$^-$, completely resistant

C. F$^-$ CDEF mixed with F$^-$, completely resistant

D. Hfr ABCDEF mixed with F$^-$, completely resistant

For each of the following, select the correct answer or answers, if any:

8. In *Paramecium*, a cell can act as a killer:

A. If it carries kappa particles and the dominant "K" to support them.

B. If it receives through an ordinary conjugation bridge the dominant "K" which can bring about the formation of kappa particles.

C. If it carries kappa in the cytoplasm and is either genotype "KK" or "kk" following autogamy of an F$_1$.

D. If it arises following the conjugation of any two cells of genotype "KK."

E. As long as both of its parents were killer cells.

9. Which of the following is (are) true about kappa and the killing action?

A. Nonbrights are required for the maintenance of brights and the killing action.

B. Nonbrights are eliminated from the cell and cause the killing action.

C. DNA is found in both the brights and the nonbrights.

D. The numbers of sensitive cells killed is directly proportional to the number of killer cells.

E. Kappa is essential for the survival of the cell.

F. Kappa cannot mutate independently of the nucleus.

10. In the case of the mate killer trait:

A. A mu particle is liberated into the surrounding fluid.

B. A mate killer also has protection against the killing action of kappa.

C. A sensitive cell is usually transformed to a mate killer after mating with a mate killer.

D. A cell must carry both dominants, M$_1$ and M$_2$, if it is to be a mate killer.

E. A cell lacking mu particles will manufacture them if the genotype includes either M$_1$ or M$_2$.

11. Which of the following is (are) true about the petite story in yeast?

A. Following the cross of a nuclear petite with a normal, the segregation of ascospores from the zygote is 2 normal:2 petite.

B. Following the cross of a neutral petite

with a normal, the diploid cells are petite and give rise to all petites following ascospore formation.

C. Following the cross of a suppressive petite with a normal, the diploids will be petite if they are plated on agar soon after mating.

D. Following the cross of normal with suppressive petite, wild-type cells are favored if subculturing precedes the plating on agar.

E. Petites cannot be induced by any known mutagen.

12. In yeast:

A. Following the cross of an erythromycin-sensitive cell with a resistant one, the sensitive trait disappears, and all cells become resistant.

B. Following the cross of an erythromycin-resistant, chloramphenicol-sensitive cell ($E^R C^S$) with an erythromycin-sensitive, chloramphenicol-resistant cell ($E^S C^R$) asci are produced which contain spores only of the parental types.

C. The cross $E^R C^S \times E^S C^R$ produces four kinds of asci: $E^R C^S$, $E^S C^R$, $E^R C^R$, and $E^S C^S$.

D. The cross $E^R C^S \times E^S C^R$ produces asci, each of which produces four different kinds of spores, the two parentals and the two recombinant types.

E. The determinants for drug resistance may be associated with a physical alteration in mitochondrial DNA.

13. In higher plants:

A. Nuclear genes do not seem to influence the development of the plastid from the proplastid.

B. There is more DNA in a chloroplast than in a mitochondrion.

C. All of the enzymes involved in photosynthesis are apparently formed from transcripts of the plastid DNA.

D. The RNA of the ribosomes of the plastid arises from transcription of the plastid DNA.

E. Replication of the plastid DNA depends on the formation of plastid DNA polymerase on the ribosomes of the plastid.

14. In corn:

  A. The iojap mutation is due to an alteration in the DNA of the chloroplast.
  B. Altered plastids due to iojap become normal if they are transmitted to a plant carrying the dominant allele "Ij."
  C. An individual homozygous for the wild allele (Ij Ij) may have mutant plastids.
  D. If a normal, fully green plant (Ij Ij) is used as a female, the offspring will be all green, even though the male parent is striped and is genotype "ij ij."
  E. If a striped plant used as female parent (ij ij) is crossed to a fully green one (Ij Ij), the F_1 may consist of three types of plants: green, striped, and nongreen.

15. In bacteria:

  A. Mutations to drug resistance do not arise on the chromosomes but are restricted to plasmids.
  B. The resistance transfer factor (RTF) can bring about conjugation between cells.
  C. The RTF cannot be transmitted unless a cell is F^+.
  D. Some of the nonchromosomal determinants for drug resistance are not found on the RTF.
  E. The RTF must integrate with the chromosome if it is to be transferred.
  F. Both the chromosomal and the nonchromosomal determinants for drug resistance alter the structure of the ribosomes and in this way cause the resistance.

GENETIC RESEARCH: IMPACT AND CHALLENGE

SYNTHETIC DNA. An important goal of the geneticist is to help prevent the suffering of those afflicted with a genetic disease and to effect a cure. Certain steps in this direction have already been taken, as we will see in the following sections.

In the mid 1950's, Kornberg isolated the enzyme DNA polymerase from *E. coli*. Shortly thereafter, Kornberg and his associates were able to synthesize DNA *in vitro* by using this polymerase along with other essential components. Intact cells were not even necessary, but the cell-free systems had to include the four different kinds of deoxynucleotides in an activated form, the triphosphate form. Also required was some DNA to act as a primer or template for the formation of any new DNA. It was found that just about any preformed DNA could be employed as a primer, that of *E. coli*, that of a virus, or even of a higher form. The preformed DNA was able to direct *in vitro* synthesis of DNA which was identical to it. For example, when *E. coli* DNA was used as primer, a double-stranded DNA was formed under its direction which possessed the same base ratio (A:T/G: C = 1) as the primer DNA. In each case, the base ratio of the synthesized DNA was the same as that of the DNA used as the template. There was little doubt that a DNA primer in the test tube could direct the manufacture of additional DNA similar to it in chemical composition.

But a very important question remained. Was this synthesized DNA capable of the same biological activity as the original DNA which acted as its template? Perhaps the test-tube DNA did not possess the exact physical structure of the original and was biologically inert. To ascertain whether any biologically active DNA could be synthesized, Kornberg realized the need to work with very simple DNA. One obvious reason is that DNA from a higher form has a complex organization in the chromosome and can be easily damaged in the procedures used. Such damage could prevent any test-tube DNA from showing the activity of related intact DNA of the cell. For this reason, Kornberg selected the bacteriophage ϕX 174 which can attack *E. coli*. This virus contains only about a half dozen genes arranged on a circular, single-stranded DNA molecule which acts as its "chromosome." The virus, in its single-stranded state, can infect a bacterial cell. Inside the cell, the single strand directs the formation of a complementary DNA strand, thus giving a form of the virus which is double stranded. The single-stranded DNA is composed of 5400 nucleotides and can be extracted from the virus without damage.

In an experiment of superb design reported in 1967, the intact circular DNA was introduced into a cell-free system containing the four kinds of deoxynucleotides and the other essentials (Fig. 21-1). Among the latter were the DNA polymerase required to link the activated nucleotides and also another enzyme, the "joining enzyme, a DNA ligase." The latter is needed because DNA polymerase can join the nucleotides in a linear fashion, but it cannot form a closed loop. Being circular, the DNA of ϕX 174 requires a joining enzyme. Without it, any new DNA formed *in vitro* would remain linear. In the experimental set up, the single stranded, circular DNA used as a primer directed the formation of a complementary strand. The latter strand was synthesized by the DNA polymerase using the circular template and resulted in a double-stranded stage of the virus comparable to that which is formed in *E. coli* after infection. This double-stranded form is known as the "replica-

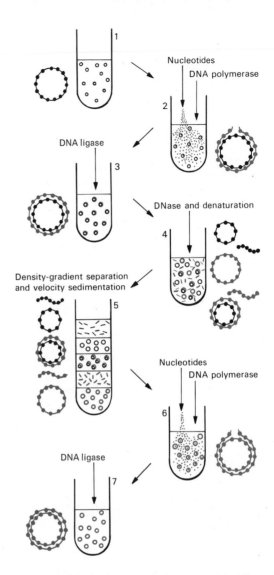

FIG. 21-1. Synthesis of ϕX 174 DNA. *1.* The template DNA was circular and single stranded. It was tagged with tritium so that it could be followed and identified throughout the experiment. *2.* Activated deoxyribonucleotides were added. However, instead of nucleotides with thymine, those containing 5'- bromouracil were employed. This is an analogue of thymine, but it is heavier. A molecule carrying the 5'-bromouracil can be separated from one carrying thymine. This was important, as will be seen in step 5. The required enzyme DNA polymerase was also added. *3.* The ligase was needed to close the loop. *4.* An amount of DNase was added which was just enough to break about one strand in approximately one-half of the double stranded loops. *5.* The mixture of material was subjected to density-gradient ultracentrifugation. The synthetic loops made on the template would be heavy because they contain 5'-bromouracil and so they separate from the other strands. *6.* These synthetic loops were isolated and were used in turn as templates. *7.* Completely synthetic two-stranded loops were formed. (Reprinted with permission from A. Kornberg, The synthesis of DNA. Copyright © 1968 by Scientific American, Inc. All rights reserved.)

tive form." The joining enzyme was responsible for making the new strand circular. Now the critical question arose. Was this newly synthesized strand capable of biological activity? Could it infect and lyse *E. coli* as the native DNA can do? The design of the experiment allowed Kornberg to isolate some of these synthetic circular DNA molecules (Fig. 21-1). They were then incubated with *E. coli* and were shown to be able to infect the bacterial cells in the same way as the naturally occurring DNA. The synthetic DNA was thus shown to be active biologically, capable of attacking *E. coli*. It was even demonstrated that these synthetic loops could direct the formation of other circular loops which were complementary to them. A synthetic loop could thus produce a replicative form of the virus which was identical to the natural replicative form. And the *second* synthetic strand was also able to infect *E. coli* cells. Complete biological activity of the DNA made in the test tube had been conclusively demonstrated.

A COMPLETELY ARTIFICIAL GENE.   In 1977, the announcement was made that Khorana and his associates at MIT had succeeded in synthesizing the first truly artificial gene, one which can function in a living cell. Unlike Kornberg and other investigators who used a naturally occurring DNA as the template for the production of a complementary copy of DNA, Khorana's group synthesized a gene by chemical procedures alone without relying on a natural DNA template. The gene which was synthesized is one found in a bacteriophage, phage lambda (Chap. 18), which can infect *E. coli* cells. The naturally occurring viral gene codes for a new kind of tRNA which appears in the bacterial cells after viral entry, a tRNA specific for the amino acid tyrosine. (It is now known that certain viruses carry genes which code for new types of tRNA. These new tRNA species may be necessary for viral replication in the cell, since some viruses, after entry, are known to alter the ribosomes of the host. Perhaps the new types of tRNA are needed to work efficiently with the altered ribosomes.)

The sequence of nucleotides in the gene proper (the part which undergoes transcription into RNA) includes 126 nucleotide pairs. This sequence had been deduced by English investigators from analyses of the RNA transcript. For reasons still unknown, the transcript is cleaved by enzyme action after its formation. Forty nucleotides are removed in the host cell, leaving a tRNA composed of 86 nucleotide pairs. In addition to the gene proper which is transcribed into RNA, the gene has a promoter region of 56 nucleotide pairs and at its other end, a stop signal which is 25 nucleotide pairs long. Khorana and his group determined the nucleotide sequences of these other gene parts. They also artificially synthesized short gene fragments, each containing a dozen or so nucleotide pairs. In the procedure, each segment which was used bore a short piece of single-stranded DNA extending from each end (Fig. 21-2). These extensions act as joints or splints which enable them to complex more readily as a result of the complementarity of bases in the extensions. The segments are then joined together by DNA ligase. Without the extensions to act as splints, the short segments would only fall into proper position by chance, rendering the procedure too inefficient for practicality.

By joining the short segments, the entire gene (gene proper, promoter, and terminator regions) composed of 207 nucleotide pairs was constructed without the use of any natural DNA product. Khorana then tested the ability of the synthetic gene to function. A certain mutant strain of phage lambda cannot produce the new tyrosine tRNA in the host cell due to a defect in the gene. The defect appears to be in the terminator region. As a result of it and hence the defective tRNA, the mutant strain produces short, nonfunctional proteins in the cell and cannot multiply in the host. The artificial gene was incorporated into mutant viruses, and these were then allowed to infect *E. coli* cells. The originally defective viruses could now multiply in the host cells just as well as normal viruses. This conclusively showed that the artificial gene was functioning. The new and necessary tyrosine tRNA was proven to be present in the host, since normal translation took place with the formation of normal viral proteins.

The construction of totally artificial genes opens up many new pathways in genetic research. One extremely important one can lead to a better understanding of a gene, how it is regulated, and how it is expressed. The investigator can produce at will alterations in the gene and observe any ensuing effect. Information of this type is of profound significance in studies of cell differentiation and pathological cell changes (Chap. 19).

Accomplishments such as those of Kornberg and Khorana with microorganisms are quite a distance from a comparable feat with higher species. However, the incredible strides which are being made with lower forms suggest that similar manipulation of human DNA may be more than speculative. Artificial synthesis of certain human genes which control the formation of valuable substances such as insulin could pave the way for the insertion of these genes into cells of microorganisms which in turn might produce abundant quantities of the desired substance in cell cultures. The introduction of for-

FIG. 21-2. Joining of DNA fragments. Each fragment is synthesized bearing single stranded projections to act as splints so that stretches of complementary bases on the extensions are exposed and can complex, bringing the segments together. After the segments are properly positioned, DNA ligase activity (arrows) joins them together by forming phosphodiester linkages. (The sequence shown here is purely illustrative.)

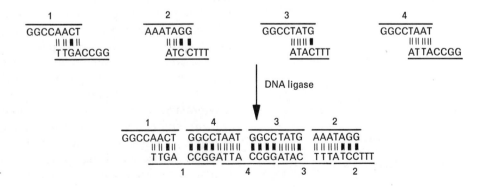

eign DNA into a cell forms the basis of the evolving science of genetic engineering, a discipline which will be faced with seemingly insurmountable problems of both a scientific and moral nature.

ACHIEVEMENTS IN GENETIC MANIPULATION. At the forefront of the most dramatic scientific research in modern times have been the accomplishments made in genetic manipulation in which genetic materials from different sources are combined. We now realize that genetic engineering is no longer in the realm of speculation but is a decided fact which is receiving the fullest attention from large segments of the scientific and lay communities alike.

The procedures now used in genetic manipulation depend to a large extent on microorganisms, plasmids, and the activities of various kinds of enzymes, especially the ligases and endonucleases. Recall that DNA ligase is able to synthesize phosphodiester bonds between nucleotides. On the other hand, an endonuclease is capable of breaking an internal phosphodiester bond and can produce a nick in a DNA chain. A ligase can repair a break made by an endonuclease. The endonucleases utilized in genetic manipulation constitute a special class known as *restriction endonucleases*. These enzymes occur naturally in microorganisms and afford a protection against the entry of foreign DNA. For example, certain viral DNA, after entering a bacterial cell, may be recognized and cleaved by the cell's restriction enzymes. The restriction enzymes have been selected in genetic research for certain of their unusual properties. Unlike most endonucleases, which are nonspecific as to where they cleave a DNA strand, restriction endonucleases produce internal nicks only within certain specific nucleotide sequences which they can recognize. These DNA sequences all contain a point or axis around which a symmetrical arrangement of base pairs occurs (Fig. 21-3A). Note that in such a stretch of DNA the reading in the 5' to 3' direction on either paired strand produces the same base sequence. Such a symmetrical arrangement is also known as a *palindrome*, comparable to words or sentences which read the same starting at either end. In Chapter 19, we noted that certain symmetrical regions have been found in the lactose operator

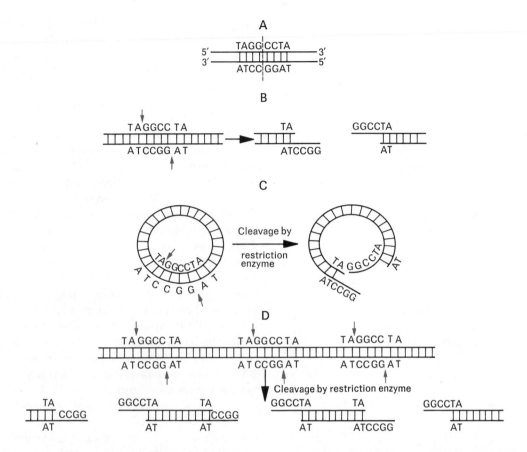

FIG. 21-3. Restriction enzymes and genetic manipulation. *A.* A palindrome, a region containing a point around which there is a symmetrical arrangement of nucleotide pairs, may occur in a stretch of DNA. A specific restriction enzyme recognizes a particular region of symmetry. *B.* Certain restriction enzymes, when they act on a region of axial symmetry, produce breaks (arrows) which are several nucleotides apart on each complementary strand, in this case a distance of 4 nucleotides. The cleavage produces fragments which have single stranded projections. *C.* If a circular plasmid contains only one palindrome which can be recognized by a particular restriction enzyme, the plasmid is broken at a specific site in each strand (arrows) amd becomes linear, with single stranded ends projecting. *D.* If the same palindrome repeats more than once in DNA, the DNA will be cleaved at a specific site in each region (arrows) with the production of many fragments, each with projecting single stranded ends. Note that such extensions in the fragments are complementary to projections in the plasmid (*C*). Such a fragment can base pair with exposed bases in the plasmid. DNA ligase can then join the ends together by forming phophodiester linkages.

and may represent binding sites which can be recognized by the repressor.

Some of the restriction enzymes produce breaks in complementary DNA strands at points which are actually several nucleotides apart in each strand (Fig. 21-3*B*). As a result of this property, DNA regions are generated which have single-stranded projections. Khorana recognized the need for such projections to act as splints in joining separate DNA fragments together, and he utilized this in the synthesis of the first completely artificial gene. As we will see, these projections are also essential to genetic engineering.

Remember that fragments or segments of DNA do not usually possess the ability to replicate unless they are part of an intact replicon (see Fig. 20-20). Therefore, if a bit of foreign DNA is to be introduced into a cell it must be part of a suitable replicon. The vehicle selected to carry foreign DNA into a bacterial cell was the resistance transfer factor (RTF), the replicon which may contain factors providing antibiotic resistance. RTF plasmids can be isolated from cells utilizing density gradient ultracentrifugation. It was discovered that if bacteria are treated with calcium chloride, their cell membranes are rendered permeable to extraneous DNA. When salt-treated cells are exposed to RTFs, about one cell in a million becomes transformed by taking up an RTF.

With this information and the available techniques, the stage was set for combining DNAs from different sources. In one kind of genetic engineering, plasmid DNA from two distinct species of bacteria are combined. For example, E. coli and Staphylococcus aureus are different species and cannot exchange genetic material. Either one may carry its own kind of RTF. Assume the S. aureus RTF carries a factor for penicillin resistance and that the E. coli plasmid carries one for resistance to tetracycline. From cultures of cells carrying known RTFs the DNA may be extracted and subjected to density-gradient ultracentrifugation. Certain chemical treatments used impart a greater density to the RTFs which are then easily separated in the gradient from the chromosomal DNA.

Once isolated, the two kinds of plasmids are mixed and subjected to a restriction endonuclease which can recognize a specific palindrome and cleave the DNA in the region of axial symmetry (Fig. 21-3B). The E. coli plasmid used as a carrier in the experiments has only one such site so that the circular plasmid becomes linear and is not broken into short fragments (Fig. 21-3C). Fragmentation occurs in a plasmid or a stretch of DNA if the palindrome is repeated in various regions, since the enzyme will attack each one of them. The same palindrome may occur in DNAs from different sources. The particular E. coli plasmid used as carrier has the same palindrome as the S. aureus RTF; but in the latter the palindrome may occur more than once so that fragments are produced from the S. aureus RTF as a result of the enzyme activity (Fig. 21-3D). Note that the enzyme action will also cause any broken DNA to have projecting single-stranded ends and that these ends will be complementary. Complementary projections on any DNA pieces, no matter what the source, will permit base pairing. If this pairing occurs between DNA pieces from different sources, exposure to DNA ligase can join them by forming phosphodiester links.

Plasmids treated in this way are then mixed with *E. coli* cells which have been treated with calcium chloride to render them permeable to DNA. Since the majority of salt-treated cells will not become transformed and since some of them which have been transformed will contain two plasmids of the same kind joined together, it is necessary to screen out cells which carry a composite plasmid, composed of DNA from the two bacterial species. It is possible to do this by growing cells in the presence of the antibiotics for which the two types of plasmids carry resistance factors, in our example penicillin and tetracycline. Only doubly resistant cells can grow and therefore must carry antibiotic resistance factors from both species.

Following these steps, such cells carrying composite DNA have been isolated. In these cells, the integrated foreign DNA combined with the native DNA continues to replicate as if it were native to the cell. The process bridges the natural barrier to crossing or genetic exchange between two species. In the example given, most of the genetic information which is expressed is typical of *E. coli*, but the cells also carry DNA derived from a genetically separate species, and this DNA is also expressed in *E. coli*!

An even more amazing and wider bridge between reproductively isolated species was spanned when *E. coli* DNA was combined with DNA from a vertebrate, the toad *Xenopus*. In the following section, techniques will be discussed which permit the isolation of specific genes from *Xenopus*, genes which code for ribosomal RNA. We will see that these genes have certain properties which are quite characteristic and which permit their ready identification. The toad DNA was subjected to restriction enzymes which break it into pieces having characteristic size and base composition. The steps followed in the toad-bacterium work were basically the same as those described for the two bacterial species. The outcome was the production of a composite DNA composed of the *E. coli* RTF and fragments of toad DNA. However, the toad DNA carries no resistance factors to enable their quick identification. Still, any cells carrying composite toad-*E. coli* DNA can be screened out since any transformed cells would be resistant to the antibiotic for which the RTF carries factors. The DNA from transformed cells is then extracted and analyzed further for the presence of those features unique to the toad DNA. It has actually been demonstrated that the animal DNA not only had combined with the *E. coli* plasmid DNA but that the animal genes were also replicating in the bacterial cells and undergoing transcription into mRNA!

The full potential of the process of genetic manipulation would seem to be almost limitless in its scope and is at the moment a challenge to the imagination. One eventual goal of such research is the introduction into bacterial DNA of genes known to control the formation of some rare but vital substance, such as insulin. If the introduced gene can function and produce the valuable product in the cells, vast amounts of costly and rare substances might become easily available to help alleviate human suffering. Conceivably ways might be found to introduce into human cells which carry a defective allele some kind of carrier DNA bearing the functional gene form. If successful, certain genetic diseases might be alleviated by correcting the genetic defect in the body cells and thus enabling the individual to pursue a normal existence with no need for continued treatment. Similarly, plants or animals so treated could be given new powers of disease resistance or some other desirable attribute as a result of the introduction of DNA from a different strain or species.

While such speculations on genetic manipulation are exciting to ponder, they nevertheless raise controversial questions more profound than those arising from most scientific research. Several scientific congresses have met to recommend guidelines to be followed in conducting genetic engineering. Many fears have arisen concerning the posssible construction and accidental escape of some cell or virus whose powers might prove lethal for the human or some valued plant or animal group. Fears concerning the misuse of the procedure are also being voiced. Many of the points raised deserve serious appraisal and consideration. Those involved in genetic manipulation utilize strains of nutritionally deficient cells which cannot survive outside of test tube conditions. Other extreme precautions are also being followed. Still the need for continued surveillance is recommended by many. Perhaps genetic engineering may bring about the unique coalition of teams composed of members of the scientific and lay communities which will oversee this type of research as it probes even deeper into the unknown.

ISOLATION OF VERTEBRATE GENES. Insertion of ribosomal RNA genes of the toad *Xenopus* into bacterial plasmids testifies to the progress being made in the isolation of specific genes in higher organisms. The selection of ribosomal genes, rDNA, as targets of isolation was based on several features of these genes which make them good candidates for such a process. One of the attributes of rDNA is its unusual base composition which distinguishes it from the rest of the nuclear DNA. The ribosomal DNA is especially rich in its G + C content. In *Xenopus,* the genes for 18S, 28S, and 5S RNA contain about 53%, 63%, and 57% G + C, respectively, in contrast to the rest of the DNA which has a G + C content of about 40%. The G + C content of the ribosomal DNA was deduced from analysis of the nucleotide content of the three species of ribosomal RNA. Since cellular RNA is a transcript of DNA, the base content of the former must be complementary to the latter. This illustrates another important feature of ribosomal genes, the ready availability in pure form of their RNA transcripts. The ease with which rRNA can be obtained is put to still another use in gene isolation, as we will see shortly.

The higher the G + C content of a specific kind of DNA, the greater its density, making it possible to separate it from a mixture of DNAs by density-gradient ultracentrifugation. When nuclear DNA is fractionated into pieces of a given size and centrifuged in a salt solution, the denser pieces with the higher G + C content separate as a band distinct from the rest of the less dense DNA (see Fig. 11-16). The DNA in a given band may be isolated and then tested with known RNA to identify it, utilizing the technique of DNA-RNA hybridization (see Fig. 13-9). Since ribosomal RNA is abundant in the cell, it is relatively easy to determine whether or not a given sample of DNA represents ribosomal genes, rDNA. This is so since only the rDNA strands will hybridize with rRNA, the transcripts of the rDNA. In this way, genes for 18S and 28S rRNA, were isolated by density-gradient ultracentrifugation and then identified by DNA-RNA hybridization.

The genes for the 5S RNA, (the 5S DNA) separated at a different position in the density gradient but could also be isolated and identified in the same way. Since the rDNA for 18S RNA and the rDNA for 28S RNA band together in the density gradient, they must be on the same DNA molecule, separate from the 5S genes which have a different density and so form a distinct band of their own.

The availability of the ribosomal RNA, the product of the ribosomal genes, permitted an estimate to be made of the number of ribosomal genes present in a set of *Xenopus* chromosomes. Again this was made possible by DNA-RNA hybridization. A sample of cellular DNA of a given amount is allowed to hybridize with a known quantity of pure ribosomal RNA of one of the three RNA species. The greater the amount of cellular DNA which hybridizes with the rRNA, the greater the number of ribosomal genes which must be present in the bulk DNA sample. In the process (similar to that shown in Fig. 13-9), The radioactive rRNA is added to single-stranded DNA. The DNA-RNA hybrid molecules are trapped on filter paper, the remaining unbound RNA being washed away. The amount of bound RNA in the form of hybrid molecules can be measured since the RNA is radioactive. This hybridization procedure thus acts as an assay and reveals that about 450 copies of each 18S and each 28S gene are present in a haploid set of *Xenopus* chromosomes! We see here a third feature of ribosomal genes which made them

very suitable for isolation, their presence in many copies in a body cell.

Once isolated, the ribosomal genes were subjected to very detailed analyses using an assortment of chemical procedures, as well as electron microscopy. As a result, we now have a very precise picture of the genes for ribosomal RNA. A ribosomal DNA region which governs the synthesis of the 18S and 28S RNAs was found to consist of repeating units of a given length, each unit containing certain recognizable sequences (Fig. 21-4A). The unit includes a gene for 18S RNA and a gene for 28S RNA. On either side of the 18S gene are two small regions that are transcribed along with them on the same RNA and are called the "transcribed spacers." This portion of the rDNA (the 18S and 28S genes along with their spacers) is transcribed into a 40S RNA. It is this 40S RNA which is processed by enzymatic cleavage in the nucleolus (Fig. 13-10). About 20% of the RNA is then discarded, the RNA of the spacer region. The remainder goes into one 18S RNA molecule and one 28S RNA molecule.

Also present in a ribosomal DNA region is a large "untranscribed spacer" which has never been found to hybridize with any cellular RNA. This portion of the ribosomal DNA can actually be observed with the electron microscope as a region between stretches of ribosomal DNA to which RNA transcripts are attached (Fig. 21-5). The function of this untranscribed spacer between the repeating units is unknown.

FIG. 21-4. Ribosomal RNA genes (rDNA) in *Xenopus*. *A.* The genes for the 18S and 28S RNA occur in repeating units along with spacer regions. The transcript of each unit is a 40S RNA which is processed in the nucleolus. The transcribed regions are separated by untranscribed spacers. *B.* The genes for 5S RNA also occur in units which repeat, the 5S genes composing only a small portion of the unit and separated by large untranscribed spacers. The latter differ in size from one species to the next.

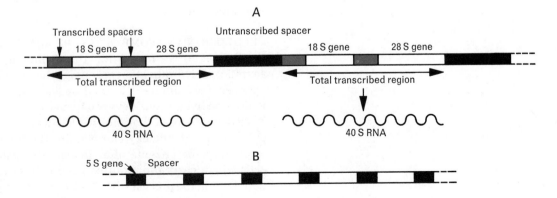

A

Transcribed spacers   Untranscribed spacer
18 S gene   28 S gene   18 S gene   28 S gene
Total transcribed region   Total transcribed region
40 S RNA   40 S RNA

B
5 S gene   Spacer

Analyses of the ribosomal DNA molecules for the 5S RNA have shown that the 5S DNA also includes units which repeat (Fig. 21-4*B*). The portion which is transcribed into 5S RNA makes up only about 1/7 of a unit and is separated from the next unit by a large untranscribed spacer. The latter differs greatly in length from one species of toad to the next. However, the size of the 5S gene itself is the same.

Ribosomal RNA of *Xenopus* has been tested with ribosomal genes of various plants and animals using DNA-RNA hybridization to determine whether similarity in base sequence exists. The results show that the ribosomal genes of the very dissimilar species are amazingly alike, indicating that the evolution of ribosomal genes has progressed at a very slow pace. This conservative evolution of ribosomal genes may reflect the fact that they determine in large part the structure of the ribosomes, which are a fundamental, essential component of all cellular forms. However, when spacer regions between units are compared among different species, a great deal of variation is found in nucleotide sequence. No good explanation for this rapid evolution of the spacers has been offered.

The isolation of ribosomal DNA was facilitated by the special features it possesses, and its analysis has revealed many unexpected facts about genes which are essential to any normal cell activity. New insights into molecular interactions have also resulted from the observations. As techniques become still more refined, certain characteristics of other eukaryotic genes may be revealed and these special features may permit the development of procedures to isolate them. Isolation of a spectrum of genes may offer ways to solve certain genetic problems, but they are certain to raise certain controversial points at the same time.

CELL FUSION AND SOME OF ITS APPLICATIONS. In Chapter 9, we noted that the technique of somatic cell hybridization permits fusion of cells between very different animal species. Observations on hybrid cells between mouse and chick have demonstrated that mouse cells with a specific enzyme deficiency can acquire the ability to produce the enzyme after fusion with chick cells which carry the required genetic information. During the process, the chick chromosomes

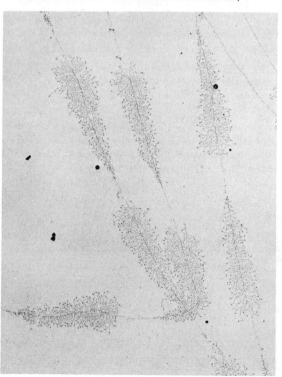

FIG. 21-5. Visualization of ribosomal DNA from oocyte nucleoli of the spotted newt, *Triturus viridescens*. The linearly repeated ribosomal genes (rDNA) are separated by untranscribed spacer segments. Each ribosomal gene actively transcribes rRNA molecules. The shorter transcripts are the more recently synthesized and are closer to the beginning of each gene. The difference in their length accounts for the feathery, arrowhead appearance. (Courtesy O. L. Miller, Jr., from Miller, O. L., Jr., and B. Beatty, *Science:164*:955–957, Fig. 2, 1969. Copyright © 1969 by the American Association for the Advancement of Science.)

become fragmented, but some of the pieces become permanently established in the surviving nuclei (Fig. 21-6). A very small percentage of these contain pieces which include the gene controlling the enzyme in question. These few cells were then used to derive a line of cells which are fundamentally mouse cells and which have gained the ability to make the enzyme. These fusion procedures offer hope that human mutant cells which carry genetic defects may become supplied with the correct information. Suppose that a person suffers from an enzyme deficiency due to a genetic defect. Cells from the affected organ might be stimulated to grow in tissue culture along with normal ones from a donor. After infection of the culture with a harmless virus to alter the cell membranes, hybrid cells could be isolated. Successive loss of chromosomes in the hybrid cells could eventually produce a line which has the normal chromosome number and contains the necessary allele. Should this intraspecific fusion meet with failure, perhaps fusion between human cells and those of an animal carrying a gene which controls the enzyme may accomplish the same end. The cells of the animal might fragment in the fusion nucleus, and as in the case of the mouse, a line might be derived which is essentially a normal human cell line containing just a piece of needed genetic information from the other species. Such "corrected" cells could then simply be injected into the bloodstream. It is known that certain cell types, when so injected, will travel and become incorporated into the appropriate organ or tissue. Bone marrow cells, for example, will become established and grow at the proper site after their introduction into the bloodstream. Thus, cell fusion procedures could give rise to normal cell colonies at the appropriate organ of the body.

Results from cell fusion studies in mice have also provided another insight into the problem of malignancy. After fusion between normal and malignant cells, hybrid cells have been produced which are not malignant. This nonmalignant hybrid can, however, give rise to some cancer cells after later cell divisions in which there is a loss of chromosomes (Fig. 21-7). In addition, fusion between malignant cells of different types has also resulted in normal hybrid cells. This observation suggests that the return to a non-

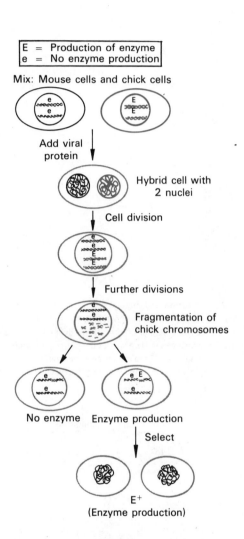

FIG. 21-6. Interspecific cell fusion and transfer of genes. Mouse cells, lacking the ability to produce a specific enzyme, may be cultured with chick cells which carry a gene permitting enzyme production. The cells may fuse after addition of Sendai virus protein. The contents of the two nuclei become incorporated into one nucleus after cell division. In further divisions of this hybrid cell, the chick chromosomes tend to fragment and become lost. Occasional pieces of these may become incorporated into some of the mouse chromosomes. In a few cells, the gene for enzyme production may become inserted. These cells can be selected and give rise to cells which are essentially mouse cells and which can produce the enzyme.

malignant condition results from the fact that the hybrid cells contain necessary genetic information which the malignant cells lack. These experiments indicate therefore that loss of certain genetic information may trigger a malignancy. In the body, cell fusions may conceivably occur spontaneously. This in turn could produce cell lines which lack particular chromosomes. The absence of a critical genetic message or messages might result in confusion of information transfer. If a specific repressor protein is no longer made by the cell, perhaps a latent virus or episome could then be released from repression and effect a malignant transformation. Although still inconclusive, the cell fusion results point out the need to examine loss of genetic information as a condition predisposing a cell to malignancy.

Besides its application to genetic engineering in animals, techniques applying cell fusion to plants offer exciting possibilities for the derivation of superior hybrids. Mature, fertile, interspecific hybrids have actually been obtained through cell fusion methods which completely bypass the normal, sexual process. The feat involved two wild tobacco species, *Nicotiana glauca* (2n = 24) and *Nicotiana langsdorfii* (2n = 18). Leaf cells were subjected to enzyme digestion to free them from their cellulose walls. More than 10^7 protoplasts of each species were plated in a solution which stimulated them to

FIG. 21-7. Cell fusion and malignancy in the mouse. If normal cells and malignant ones undergo fusion (*left*), hybrid cells can arise which are nonmalignant. When these hybrids undergo cell divisions, chromosomes may be lost. Some of the cells which arise become malignant. Highly malignant cells of different ancestries may be fused (*right*). Some of the resulting hybrid cells are normal.

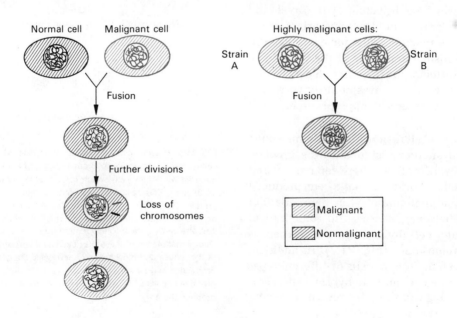

fuse (Fig. 21-8). Approximately 25% of the cells underwent fusion in culture. Some fusions were between two cells of the same species, but others were hybrid fusions. These latter cells were then selected by plating all the cells on a medium that permits growth *only* of those containing the genetic information of both parent species. After further culture, the isolates developed rudimentary shoots and leaves but failed to form roots. However, when grafted to other tobacco roots, they developed into mature plants, which were identical in chromosome number (2n = 42), and

in all other characteristics, to hybrid plants produced by conventional cross-pollination. Moreover, their seeds gave rise to offspring identical to them in all ways, illustrating that these "fusion hybrids" can breed true to type. In the typical production of an interspecific polyploid such as this (Chap. 10), chromosome doubling is required to produce fertile hybrid offspring from highly infertile F_1 plants. The cell fusion method, called *parasexual hybridization* because the normal sexual process is circumvented, avoids this difficulty and others involved

FIG. 21-8. Parasexual hybridization in tobacco. Cells of tobacco leaves are enzyme treated to release free cells. Cells from two different species which differ in chromosome number are cultured together. Cellular fusions occur; some of these are fusions between the same kinds of cells; others are hybrid cells between the species. These are selected by the growth medium. The hybrid cells develop into plantlets lacking roots. The small plants are grafted onto other roots and eventually develop into mature plants bearing flowers and producing seeds. The chromosome numbers as well as all the other characteristics of the plants derived in this way are identical to the interspecific hybrid produced by crossing the two species and doubling the chromosome number to obtain a fertile polyploid.

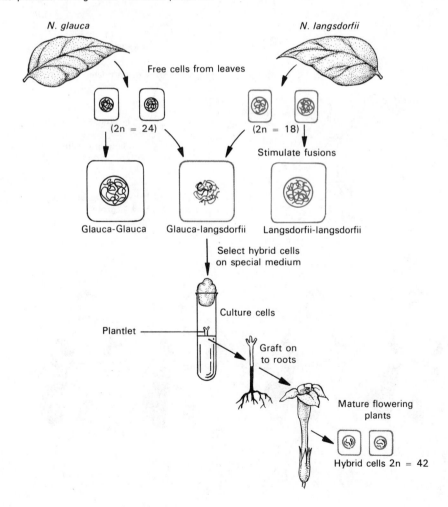

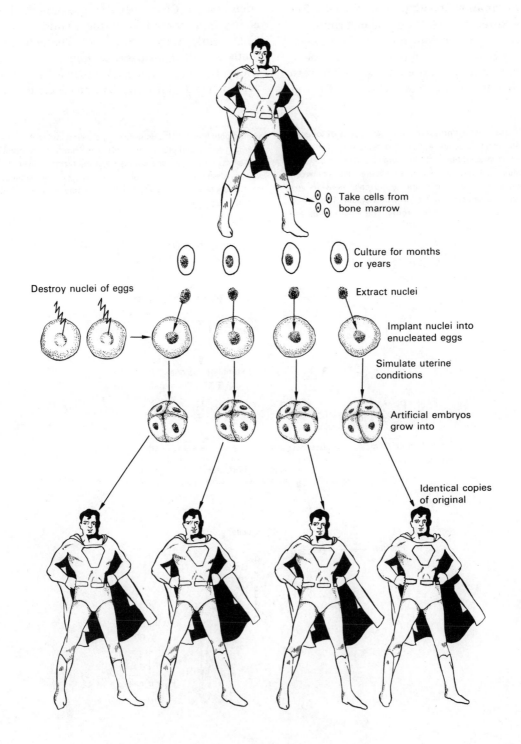

FIG. 21-9. Hypothetical scheme for cloning a human.

Take cells from bone marrow

Culture for months or years

Destroy nuclei of eggs

Extract nuclei

Implant nuclei into enucleated eggs

Simulate uterine conditions

Artificial embryos grow into

Identical copies of original

UNDERSTANDING GENETICS

in crossing species. The parasexual process holds great promise as a way to combine the genetic information from plant species which would otherwise be unable to cross due to reproductive barriers. Hybrids derived parasexually may very well enable the plant breeder to produce vigorous new species which combine the most desirable features of each parent, such as increased disease resistance or increased yield of an edible product. Attempts to produce plant hybrids through cell fusion depend largely on selective techniques which make possible the isolation of only the hybrid cells which would be scattered among the total percentage of cells which have fused. It seems quite likely that this problem can be solved once differences in growth requirements between the parental species are fully known. The parasexual process holds great promise in the near future for increased food production in needed areas by the derivation of hardier and more nutritious plant species.

DELAY OF AGING AND CLONING. The correction of hereditary defects in humans through cell fusion or some other device of genetic engineering may allow the individual to survive to his maximum possible age. There is evidence, however (Chap. 19), that there may be an upper limit to this which is species specific. For aging and death may involve the accumulation of a critical number of genetic blocks which have arisen in body cells as a result of chromosome aberrations and errors in transcription and translation. If the specific biochemical blocks are detected, a tissue may be cured by supplying it with the lost genetic information utilizing the methods of transduction or cell fusion. Future genetic analysis may yet reveal the loci of genes controlling the formation of repair enzymes. Perhaps aging of cells may be delayed by amplifying those genes which control the production of repair enzymes. Extra sets of these genes introduced into cells would provide an increased supply of enzyme to correct damage to the DNA itself arising spontaneously or from an assortment of mutagenic agents.

A very imaginative way to prevent the death of a particular human genotype might involve cloning. Since all members of a clone trace their ancestry back to a single cell, they are thus identical genotypically. The body of any multi-cellular organism represents a clone, because the genetic information in each cell of the body is the same. Different genes are suppressed in the various cell types, but suppression may be overcome through removal of histones and release of cells from the influence of neighbors, etc. (Chap. 19). In much the same way that amphibian nuclei from differentiated cells are transplanted into enucleated eggs, a cloning technique could be perfected in humans. Fibroblasts, erythroblasts, or some other readily obtainable cell type could be taken from an individual and raised in tissue culture. These could provide the source of the nuclei to be transplanted into human eggs whose nuclei had been purposely destroyed. A large number of eggs, each implanted with a nucleus from the same individual, could then be stimulated to divide and raised under conditions simulating those normally found *in utero*. If these survive to term and pass through normal development, the result would be an assemblage of individuals, genetically as identical as any identical twins. They would all be members of a clone, because their nuclei trace back to the transplant nuclei which came from one individual. The donor may age and die in one sense, but his exact genotypic copy could go on indefinitely (Fig. 21-9).

Speculations such as these seem staggering; yet equally overwhelming are the accomplishments of the molecular biologist in just a little over 20 years. Such achievements, however, entail challenges for the human who must strive to utilize such information constructively for the benefit of the human and other species.

DRUGS AND THE GENETIC MATERIAL. Today, drugs are playing a wider role in medicine and, unfortunately, among larger segments of society. Indiscriminate drug use is a matter of concern for many reasons. The potential damage to the genetic material attributable to drugs has not been assessed, and there is no reason to exempt them from suspicion. They should be treated with the same reserve held for any untested chemical. Champions of this or that substance are usually ignorant of such matters as mutagenesis, information transfer, or induction of latent viruses. Few will question the carcinogenic effect of tobacco smoking; yet many assume marijuana is harmless in this regard. Just how tobacco smoke increases the risk of malignancy is unknown. We could suggest that smoke itself somehow alters a protein repressor substance. Its destruction could release a virus-like particle from repression. Such an activation of a latent agent could trigger a malignant transformation. The smoke of marijuana would not be exempt from such a triggering role. Although these are admittedly speculations, it is just as valid to propose them as to assert the completely innocuous nature of substances which have become fads.

On the other hand, we do know with certainty that several drugs of therapeutic value can induce violent reactions in certain individuals. Moreover, such sensitivity to specific chemicals has a definite genetic basis. This is clearly illustrated by the differences in response to the harmless chemical, phenylthiocarbamide (PTC). Most persons carry a dominant allele enabling them to taste PTC when they chew a bit of paper treated with the substance. Those homozygous for the recessive allele taste nothing at all. The tasters, however, experience a variety of sensations. Some detect it as extremely bitter, some mildly so. Others describe it as sweet or salty. Although perhaps nothing more than a curiosity, the different reactions to PTC emphasize the fact that inherited physiological differences exist which may elicit one of several possible responses to a given drug or chemical. This is demonstrated in the reaction of some persons to primaquine, a drug used for years to treat malaria. After administration of primaquine, certain patients develop an anemia which follows from a breakdown of red blood cells.

FIG. 21-10. Waltzing behavior in mice. This aberrant waltzing trait is inherited as a simple Mendelian recessive.

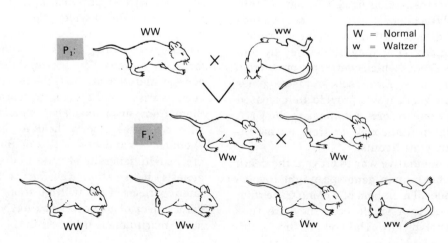

These cells in sensitive persons have been shown to be normal until they contact primaquine. Sensitive persons also lack the enzyme glucose-6-phosphate dehydrogenase, although the relationship between this enzyme and the fragility of the red cells in unknown. The condition is inherited as a sex-linked recessive and also predisposes the individual to an increased sensitivity to still other drugs, including sulfanilamide.

It is significant that those showing this sensitivity are otherwise completely normal. Only an encounter with the specific drugs reveals any difference from the majority of persons. The reaction of the sensitives points out the caution which should be exercised when experimenting with drugs for any reason. Wide differences in response resulting from genetic factors could have disastrous effects for some individuals. This is vividly seen in the violent reaction to barbiturates which can result in the death of susceptible persons. These barbiturate-sensitive individuals are also usually very sensitive to sunlight, a symptom of a condition known as "porphyria." The basis for these two reactions is definitely genetic, but the responsible genetic factors vary from one pedigree to another. A most significant point is that the genetic disorder usually is not severe unless barbiturates are encountered. Again we see that exposure to a drug may trigger a very unexpected reaction by an apparently normal individual. The effects of long-term drug usage on transmission of genetic material to the next generation is still unknown and demands the same attention as that given to chemicals which surround us in our daily environment.

GENETIC FACTORS AND BEHAVIOR. Perhaps the greatest challenge to man will be information yet to be uncovered concerning the genetics of behavior. That behavior which is typical of a species may well prove to be the most complex of all phenotypic characteristics. Like any other aspect of the phenotype, behavior has both genetic and environmental components which interact in complex and very subtle ways. To ask which is the more important becomes meaningless, because the development of the behavior which is typical or normal for any animal group depends on both. Characteristic behavior patterns of most species depend on many genes whose expression may be altered at various developmental stages by environmental influences. There are a few specific behavioral traits in some animals for which the genetic basis has been recognized. In mice, an aberrant form of behavior occurs which has been well understood for years. This is the waltzing trait characterized by pointless, irregular twirling motions which leave the animal exhausted. When a cross is made between waltzers and wild stock mice, the F_1 exhibit normal behavior, but the F_2 generation produces normal and waltzers in a ratio of 3:1. Clearly, the trait is inherited as an autosomal recessive (Fig. 21-10). The resulting behavior indicates a nervous disorder, and this has been shown to be associated with deafness and various abnormalities of the inner ear. In homozygous condition, there is undoubtedly a lack of correct genetic information needed for normal development of the inner ear and associated nervous tissue. The result of the block manifests itself in the waltzing behavior.

Another example of simple Mendelian inheritance as the basis of a behavioral trait is seen in honeybees. Bees are particularly interesting in behavior studies, because workers and queens, which are alike genetically, come to display very different types of behavior as a result of environmental influences operating during larval development. In addition, there are great differences in behavior among strains of bees. Some colonies exhibit "hygienic" behavior, which means that the workers remove dead larvae and pupae from the cells of the hive, an act requiring uncapping the cell followed by physical removal of the dead remains. These hygienic colonies are also resistant to a bacterial infection, American foulbrood, and their resistance stems from the removal of the dead bees which may harbor the responsible agent. Other bee colonies may be "nonhygienic"; no attempt is made by the workers to remove the dead from their cells. Such colonies are thus sensitive to the bacterial disease.

A cross between hygienic and nonhygienic strains produces an F_1 in which the workers display nonhygienic behavior. It will be recalled that worker bees are sterile females; only the queen engages in sexual reproduction. The drones, or males, are all haploid and develop parthenogenetically from unfertilized eggs. Male offspring of the F_1 queen (whose workers are nonhygienic) can be mated to queens from other colonies which are inbred or pure breeding for either the hygienic or nonhygienic behavior. When these drones are mated to inbred hygienic queens (Fig. 21-11), the results give a 1:1:1:1 ratio of hygienic bees: nonhygienic ones: bees which just uncap the cell but leave the remains: bees which do not uncap cells but which remove remains when the uncapping is performed for them. This typical testcross ratio clearly indicates a two-gene difference between hygienic and nonhygienic behavior.

The example of behavior in bees also illustrates a very significant point. The behavior of the hygienic strain is of definite value to the colony. It is important to bear in mind that the general behavior which is typical of a species is a phenotypic characteristic which has undergone a long history of natural selection. Behavioral patterns have been selected for their survival value, just as any other part of the total phenotype.

BEHAVIOR DISTURBANCES IN THE HUMAN. Even in man, the genetic basis is known for certain types of behavior. We have discussed the autosomal recessive responsible for the production of phenylketonuria, a condition which can cause severe mental retardation. However, dietary restrictions can prevent the accumulation of phenylalanine in the blood of afflicted babies and in turn prevent toxic damage to the central nervous system. We see here the interaction of environmental and genetic components in the expression of a phenotype. An unfortunate behavioral outcome may be avoided if the proper environment (low amounts of phenylalanine) is present at a crucial time. This latter point is to be stressed, for the mental retardation cannot be prevented if the dietary regimen is undertaken too late. Nor will higher levels of phenylalanine in the diet impair the mental functions of individuals homozygous for the allele who were given the proper diet in infancy. As in many cases of phenotypic expression, behavioral or otherwise, the environmental influence at a specific time or period of development may be critical.

Other examples of genetic alterations which

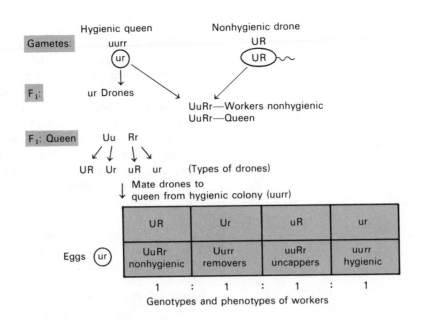

FIG. 21-11. Inheritance of hygienic behavior in worker bees. Workers are sterile females, but certain of their behavioral characteristics can be studied genetically by crossing queens and drones of various stocks. The male, or drone, develops from an unfertilized egg of the queen. Thus, males are haploid, and in effect represent the kinds of gametes produced by the queen. The difference between hygienic and nonhygienic behavior in the workers has been shown as follows to be based on two pairs of Mendelian alleles. If a queen from a hygienic colony is mated to a drone from a colony that is nonhygienic, all the F_1 workers exhibit nonhygienic behavior. The F_1 drones are all haploid and reflect the genotype of the hygienic queen. The F_1 queen and workers are dihybrids, but only the workers exhibit the behavior characteristic under study. Because the workers are sterile, only the F_1 queen can be crossed. The dihybrid queen will produce four kinds of gametes. There will thus be four different kinds of drones representing these gametes. If these drones are mated to a queen from a hygienic colony, four different kinds of workers result in a ratio 1:1:1:1, showing that a two gene difference is involved. (If the drones from the F_1 queen are mated with a queen from a nonhygienic colony, all the workers will be nonhygienic, because the two dominant alleles, "U" and "R," would be contributed by the nonhygienic queen.)

can affect human behavior have also been encountered in various chapters (cri-du-chat babies, Tay-Sachs infants). There is also the need to investigate more fully the possibility of extra Y chromosomes as genotypic factors predisposing individuals to aggressive and antisocial behavior.

One of the most bizarre behavior disturbances in the human is associated with a single genetic defect, a sex-linked recessive. This is a type of cerebral palsy known as the Lesch-Nyhan syndrome in which afflicted children, all males, develop a compulsion for self mutilation. While seeming to wish protection against his own self-destructive behavior, a Lesch-Nyhan child will nevertheless bite his tongue, lips, and fingers. Such an unfortunate boy must be fitted with restraints to protect him from inflicting serious damage on his own body. He also demonstrates very aggressive behavior toward others, and when upset will kick and bite those caring for him and in addition, may vomit over them.

The responsible sex-linked recessive has been shown to produce a molecular defect, the deficiency of a certain enzyme which is required for normal purine metabolism. Lacking the enzyme, the person accumulates products which produce a high metabolic rate. Lesch-Nyhan victims die before puberty, usually as a result of kidney damage and frequently after developing gout. Persons who are gout sufferers also have a biochemical upset similar to that found in the Lesch-Nyhan syndrome. However, those with gout do not experience nervous derangements.

It is obvious that the Lesch-Nyhan syndrome is extremely relevant to studies of aggressive behavior and compulsions of various kinds. A complete understanding of this syndrome could provide valuable information on behavioral disturbances in general, the biochemical upsets which accompany them, and the genetic and environmental factors which may interact to cause them. Directly related to such investigations is the problem of mental illness. There is good evidence that certain psychoses, the most serious of the mental illnesses, have a genetic as well as an environmental component. Schizophrenia, the most frequently occurring psychosis, is characterized by very unpredictable and unusual forms of behavior associated with the individual's complete withdrawal into a world of his or her own creation. Controversies have been waged up to the present time on the causes of schizophrenia. Many psychiatrists subscribe to the thesis that environmental stresses alone trigger the disturbance. However, present data assembled from studies of identical twins and families in which the illness occurs indicate that genetic as well as environmental factors are involved. As to the genetic component, again there is no consensus. Some favor a single dominant allele hypothesis, others an interpretation based on a recessive. Still others favor an explanation based on polygenes or multiple factors.

All those who recognize a genetic component also implicate the environment in the expression of schizophrenia. There are those who feel that genetic factors may predispose a person to expression of the illness so that he or she has a greater risk than those lacking such factors of developing the disturbance. However, also involved are unknown environmental stresses which may be required before a predisposed person can develop the illness. Consequently, such a person could remain healthy. On the other hand, a person lacking the predisposing genetic elements could develop the disorder as the result of subjection to certain stresses.

While still unsettled, data indicate that genetic factors play a role in even other behavioral disturbances, such as the manic-depressive psychosis. It is hoped that a more complete understanding of the various mental illnesses could yield information useful for the treatment and prevention of the disorders, as well as the clarification of many aspects of other behavior patterns.

BEHAVIOR AS A COMPLEX PHENOTYPIC CHARACTERISTIC. The basis of most types of behavior is so complex that the genetic and environmental components are only remotely appreciated. We expect animals to behave in certain ways which are typical of their species, and we sense that such characteristic behavior must somehow be influenced by genetic information which all members of a species inherit in common. Immediately after hatching, baby ducks will walk, follow their mother, and even swim. Human babies, in contrast, are quite helpless, but at a certain age, they will start to speak a language, something no other animal can do, no matter how precocious at birth. Surely the child has the potential to speak, because the DNA transmitted to him has the encoded information required to equip him with a complex brain, voice box, and other necessary anatomical structures. Animals without the set of genetic instructions needed to guide orderly construction of the features essential to speech will never utter an intelligible word. Does his mean that speaking a language is an inherited trait? Is the precocity of the duckling in recognizing its mother an inherited trait for that species? An unqualified "yes" to either of these would indicate a failure to remember that no phenotypic characteristic is inherited as such. There is no gene for "speaking" any more than there is a special gene for eye color. In addition to his structural equipment, the child will begin to speak at a certain period in his development only if his environment surrounds him with the proper sound incentives. Whether he will speak English, Chinese, or some other tongue will depend on the sounds he hears. The genetic endowment of a child empowers him with the remarkable ability to assemble separate sound symbols and arrange them in meaningful order. Without ever being taught the rules of grammar, he becomes able to understand the meaning of sentences and to express himself by manipulating the symbols of speech. The magnitude of this feat is appreciated when we note that those children surrounded by the unrelated sound stimuli of more than one language can, without difficulty, sort them out in their proper relationships and so achieve expertise in more than one tongue. However, at about the age of sexual maturity, this talent begins to fade, so that mastery of a language from that time on becomes an effort. It is the rare adult who achieves the perfection and facility with a newly studied tongue which a child acquires so effortlessly. So there is a critical period during development when environmental factors interact with those genetically determined. The former are the proper sounds, the latter the structures needed to receive, interpret, and associate symbols. Should a child be deprived of the proper stimuli at a critical developmental stage, he may never acquire the same language facility as one who was surrounded by the sounds which are meaningful in his cultural environment. Therefore, group members who are isolated from infancy from appropriate language stimuli would be unable to communicate verbally in any known language. After a series of generations, we might expect descendants of this still isolated group to have developed some means of verbal communication which follows certain rules that give it a structure and grammar. For speaking a language is an aspect of behavior, a potential which is peculiar to the human zygote. Moreover, it is a phenotypic character with survival value. Speech enables the members of the group to transfer an ever growing amount of cultural information from one generation to the next. And so the members, building on information from the past, may continue to devise new ways of coping with different environments, enabling members of their species to survive under a range of diverse conditions. Natural selection has undoubtedly favored this aspect of man's behavior which helps set him apart from other creatures.

Of course, the ability to speak a language is by no means the only type of communication. An individual born with defective vocal apparatus can still communicate by symbols, as can a person deaf from birth. The latter, although possessing perfect vocal cords, will never speak with the clarity of the more fortunate person who has been able to receive the appropriate sound stimuli during his development. The complex interaction of hereditary and environmental elements is very evident here. Hereditary deafness can deprive the child from reacting with the appropriate environmental factors. However, if he is genetically normal otherwise, such a child will have the kind of nervous system which enables him to associate and manipulate nonverbal abstract symbols that he can substitute for the spoken word. On the other hand, deafness or faulty speech may be environmentally caused. The proper genetic information coded in DNA may be present in all the cells, but suppose a virus crosses the placenta at a critical point in embryonic development. Its interference with the formation of a normal anatomy can prevent the individual from ever speaking clearly, even though he is later given the most conducive environment.

We must remember that speaking is just one form of behavior that permits communication and is certainly not the only kind of language. This is graphically illustrated by the gifted chimpanzee, Sarah, who has achieved a certain ability to read and write by manipulating plastic pieces of various shapes and colors. These plastic bits can be used as symbols to represent parts of speech: nouns, verbs, and adjectives. The chimpanzee was taught to arrange approximately 130 "words" to form meaningful sentences, a feat which has indicated the animal's talent to associate an object with a symbol and also to describe and classify it as well. Sarah also learned to recognize sentences written in plastic symbols and to respond correctly to them, such as following a "written" order to place a banana in a dish. There is every reason to believe that the chimpanzee eventually became able to associate a symbol with an object, even in the absence of the object. If this proves to be correct, it would mean that the animal can *think* with language and has the ability to represent mentally in terms of symbols objects which are not present. This is perhaps the most important feature of language, as it enables individuals to communicate about things which are not physically there.

Investigations of this type are very important in understanding language and human behavior. A chimpanzee cannot verbally speak a language because its genetic makeup does not endow it with a potential for forming the anatomy required for speech. However, it evidently does possess the genetic information enabling it to grasp certain concepts and to communicate in symbols. The investigators correctly point out that studies such as theirs may clarify the very nature of language. For we can see that language is not unique to humans; human language is just one form of a larger and more general system of communication. Moreover, certain aspects of language which are often considered "human" may yet prove to be part of a larger system which man shares with other animals. Further studies may help us to recognize some of the specific genetic factors which influence language ability and to appreciate their selective value in the evolution of man and his relatives.

SPECIES SPECIFIC BEHAVIOR. Arguments have raged for years over the inheritance of "innate" behavior, that behavior so typical of a species that it may seem inborn and unalterable. After hatching, the duckling will typically follow its mother faithfully; a female rat will properly attend to her offspring after their birth. These animals seem to "know" enough to do such things without having in some way been taught. Are there segments of DNA which program this type of behavior? Again, this sounds like the concept of the inheritance of a specific character. From what we have been discussing, we should expect such innate behavior to reflect an interplay of environmental and genetic factors.

The mere fact that we can recognize a behavior which is typical of a species implies that there is some genetic basis. That much more is also involved can be well illustrated from studies performed with ducks and geese. Years ago, the observation was made that a duckling, which normally forms a rigid attachment for its mother, will form a bond toward a human if it sees that person before it sees its mother. Even artificial objects were shown to serve as "parents" if isolated birds were exposed to the inanimate objects before having any type of social contact. This phenomenon was designated "imprinting" and has been taken by many students of behavior to represent some form of learning. Lorenz, who called attention to imprinting, considered it a different process, because it happened so quickly and seemed to have a permanent effect. Recent observations by Hess clearly show that imprinting can be reversed. Laboratory hatched duck-

lings which were exposed to humans formed attachments to them, but these "human-imprinted" birds, when given to a female duck, would then accept her and behave just as other ducklings which had been normally reared by their mother. Hess's observations caution against definitive statements on behavior which are based solely on laboratory observations. Essential to any interpretation are studies of the animal's behavior in its natural habitat. Rearing an animal in some artificial environment may evoke certain behavioral responses which actually tell little about the normal behavior pattern. Hess's studies of natural imprinting indicate that normal bonding between mother and duckling involves an interplay of sounds between the mother and her offspring, an interaction which starts well in advance of hatching! The unhatched birds emit noises to which the mother reacts by clucking. The stimuli provided by sound, along with tactile stimulation of the progeny (squeezing and scratching), are normal factors in the development of the species typical bond between ducklings and mother. The process of natural imprinting ensures a very strong filial-maternal attachment. These observations again illustrate that many environmental stimuli may be needed for the development of a normal behavior pattern. And the best place to study this is in nature. The innate behavior is not just encoded in the zygote's DNA so that it will emerge intact at a given moment.

Still other influences must be considered before reaching conclusions about innate behavior. The female rat apparently does not need to learn to act as a mother. Normally, she cares for her offspring and will do so with her first litter, even without prior experience or opportunity to observe another mother rat. On the surface, it might appear that this maternal behavior is inherited as such. However, it has been demonstrated that a female rat will show hostility toward her offspring, avoid them, or even kill them if she is deprived of certain stimuli. Normally, a rat licks and smells its own body as it develops into adulthood. If a newborn female is fitted with a collar of sufficient size, she will be unable to examine herself in this way. When she bears her young, even after removal of the collar, she will fail to display the species-specific maternal behavior. This example shows that a normal environmental stimulus may be something quite unsuspected. In this case "seeing" maternal behavior is not a necessary prerequisite. Instead, quite a different experience is required, the familiarity of the female rat with her own body.

The lessons learned from animal studies caution us about reaching premature conclusions concerning human behavior. Gaining deeper insights into why he acts the way he does may enable man to elevate himself to undreamed of heights. But erroneous conclusions based on failure or reluctance to recognize the subtleties in the interactions of genetics and environment may result in his self-destruction.

REFERENCES

Abelson, J. Recombinant DNA: examples of present-day research. *Science, 196:* 159, 1977.

Athwal, R. S. and O. W. McBride. Serial transfer of a human gene to rodent cells by sequential chromosome-mediated gene transfer. *Proc. Natl. Acad. Sci., 74:* 2943, 1977.

Baldessarini, R. J. Schizophrenia. *New Eng. J. Med., 297:* 988, 1977.

Brown, D. The isolation of genes, *Sci. Am. (Aug):* 20, 1973.

Carlson, P. S., H. H. Smith, and R. G. Dearing. Parasexual interspecific plant hybridization. *Proc. Natl. Acad. Sci., 69:* 2292, 1972.

Cohen, S. N. The manipulation of genes. *Sci. Am. (July):* 24, 1975.

Cohen, S. N. Recombinant DNA: fact and fiction. *Science, 195:* 654, 1977.

Grobstein, C. The recombinant DNA debate. *Sci. Am. (July):* 22, 1977.

Hess, E. H. Imprinting in a natural laboratory. *Sci. Am. (Aug):* 24, 1972.

Kidd, K. K., and L. L. Cavalli-Sforza. Analysis of the genetics of schizophrenia. *Social Biology, 20:* 254, 1973.

Kornberg, A. The synthesis of DNA. *Sci. Am. (Oct):* 64, 1968.

Maugh, T. H. II. The artificial gene: it's synthesized and it works in cells. *Science, 194:* 44, 1977.

McClearn, G. E. Behavioral genetics. *Ann. Rev. Genet., 4:* 437, 1970.

Mertz, J. E., and J. B. Gurgon. Purified DNAs are transcribed after microinjection into *Xenopus* oocyte. *Proc. Natl. Acad. Sci. : 74:* 1501, 1977.

Miller, O. L. The visualization of genes in action. *Sci. Am., (Mar):* 34, 1973.

Miller, O. L., and B. R. Beatty. Visualization of nucleolar genes. *Science, 164:* 955, 1969.

Premack, A. J., and D. Premack. Teaching language to an ape. *Sci. Am. (Oct):* 92, 1972.

Rambach, A., and D. S. Hogness. Translation of *Drosophila melanogaster* sequence in *Escherichia coli. Proc. Natl. Acad. Sci.,* 74: 5041, 1977.

Rothenbuhler, W. C., J. Kulincevic, and W. E. Kerr. Bee genetics. *Ann. Rev. Genet.,* 2: 413, 1968.

Seegmiller, J. E. Lesch-Nyhan syndrome and the X-linked uric acidurias. In *Medical Genetics.* V. A.

McKusick and R. Claiborne (eds.), pp. 101–119. HP Publishing Co., New York, 1973.

Weiner, F., E. M. Fenyö, and G. Klein. Fusion of tumour cells with host cells. *Nature (New Biol.),* 238, 1972.

Winston, M. L., and C. D. Michener. Dual origin of highly social behavior among bees. *Proc. Natl. Acad. Sci.,* 74: 1135, 1977.

REVIEW QUESTIONS

1. Below is a segment of DNA containing a palindrome. A restriction enzyme which recognizes this produces breaks four nucleotides apart from the axis of symmetry in each strand. The break is to the left in the top strand and to the right in the lower one. Show the fragments which can be generated with complementary single-stranded projections.

$$\underline{\overline{\text{C T A C G A A T T C G T C T}}}$$
$$\underline{\text{G A T G C T T A A G C A G A}}$$

2. Below is a segment of a circular plasmid. A restriction enzyme specific for the palindrome present produces breaks four nucleotides apart from the axis of symmetry, to the left in the inner or upper strand, and to the right in the outer, or lower one.

   A. Show the ends which can be generated.
   B. Could the fragments generated in Question 1 be inserted? Explain.

$$\underline{\overline{\text{G A C T A T T A C G T A A T A G C A}}}$$
$$\underline{\text{C T G A T A A T G C A T T A T C G T}}$$

3. When pure DNA of the nucleolar organizer is hybridized with 18S and 28S RNA from the same cell type, only about 25% of the DNA present hybridizes with the RNA. Offer an explanation.

4. List at least three features of the rRNA genes which made their isolation feasible. Briefly cite the advantages of each feature.

5. The evolution of rDNA has been very conservative as shown when comparing very different species. Nevertheless, the untranscribed spacer regions show great variation from one species to the next. Still, within one species, these repeated untranscribed spacer regions are the same. Can you recognize the dilemma here from what you know about duplicated genetic material? (Refer to Chap. 10).

6. Khorana's artificial gene is a DNA unit 207 nucleotide pairs long. Suppose this DNA were hybridized with the tRNA for which it is coded. About what percentage of the DNA would you expect to form hybrids with the RNA? Why?

7. Assume that plant species A has a diploid chromosome number of 22 and that species B has a chromosome number of 16. The two species can be crossed with difficulty and produce sterile hybrids.

   A. What is the probable chromosome number of an F_1 hybrid?
   B. How could one derive a fertile plant from an F_1, and what would the chromosome number of the derivative most likely be?
   C. What would be the chromosome number of a plant derived from the two species utilizing somatic cells and the technique of parasexual hybridization?

8. Suppose one individual is homozygous for a nuclear genetic factor "A" and that a second individual is homozygous for the recessive allele "a." A clone is derived by transplanting cell nuclei from individual one into enucleated cells of individual two. However, individual one carries the determinant C in the mitochondria, whereas individual two carries D in the mitochondria. What would be the constitution of cells composing the clone?

9. A. Diagram a cross between a queen bee from a nonhygienic hive and a drone from a hygienic colony. Give the genotypes of the F_1 queen and the workers. What is the workers' behavior?
  B. What are the genotypes of the drones produced by the P_1 queen?
  C. What are the genotypes of the drones produced by the F_1 queen?
  D. The drones from the F_1 queen are mated to a queen from a nonhygienic colony. Give the genotypes and the behavior of the resulting workers.

Select the correct answer or answers, if any, in each of the following:

10. Cell hybridization studies have shown that:

  A. The nucleic acid of the Sendai virus is required to cause cell fusion.
  B. Fusions between human cells and those of the mouse usually give rise to a stable line of cells with complete chromosome complements of both species.
  C. Fusions can result in the incorporation of a chromosome fragment of one species into the chromosome of another species.
  D. Fusions between malignant cells from different lines can give rise only to malignant cells.
  E. Fusions between malignant cells and normal cells give rise only to malignant cells.

11. Species-specific behavior in higher animals:

  A. Is a phenotypic characteristic.
  B. Has evolved under the force of natural selection.
  C. Depends on environmental stimuli for its normal development.
  D. Requires that the individual be taught by observing the typical behavior.
  E. Is inborn in those cases where the animal displays the typical behavior pattern without having seen it at all.

12. Below is a list of human disorders which have a decided genetic component. In the blank

next to each, place a "C" if the condition is associated with a chromosome anomaly. Place "A-R" if it is inherited as an autosomal recessive, and "A-D" if it is an autosomal dominant. Place an "X" if the condition is X-linked.

A. *cri-du-chat* syndrome ———
B. myotonic dystrophy ———
C. Christmas disease ———
D. Down's syndrome ———
E. phenylketonuria ———
F. primaquine sensitivity ———
G. Tay-Sachs disorder ———
H. Klinefelter syndrome ———
I. sickle-cell anemia ———
J. Turner syndrome ———
K. albinism ———
L. Lesch-Nyhan syndrome ———

13. Many scientists and scientific teams have made outstanding contributions to genetic progress. Associate each term on the left with the number of the investigator or team on the right most closely associated with it.

A. Deciphering the genetic code
B. Upset of the central dogma
C. Biologically active, synthetic DNA using DNA primer
D. Intragenic mapping and the cistron
E. Very first studies of DNA
F. Mutagenic effect of radiation
G. Continuity of the germ plasm
H. First demonstration of DNA as genetic material
I. First to clearly demonstrate sex in bacteria
J. Amino acid sequence of hemoglobins
K. Discovery of Rh antigen
L. Discovery of linkage, crossing over, sex linkage
M. Gene control and regulation
N. First completely artificial gene using no DNA primer
O. Segregation and recombination of cytoplasmic genes

1. Avery, McCarty, and MacLeod
2. Beckwith
3. Benzer
4. Ingram
5. Jacob and Monod
6. Khorana
7. Kornberg
8. Landsteiner and Wiener
9. Lederberg and Tatum
10. Miescher
11. Morgan
12. Muller
13. Nirenberg
14. Sager
15. Sonneborn
16. Temin
17. Watson and Crick
18. Weismann
19. Wilkins

GLOSSARY

acentric. An abnormal chromosome or chromosome fragment lacking a centromere.

acquired characteristics. Those features which an organism takes on during its lifetime through the effect of the environment on somatic tissue and which are not transferred to the next generation.

acrocentric. A chromosome with its centromere very close to one end, giving it one very short arm. The centromere located near one end of a chromosome.

adenine. A purine base found in DNA and RNA.

adenosine triphosphate (ATP). A nucleoside triphosphate which provides a source of high energy for metabolic processes requiring energy transfers. The main energy-storing molecule in the cell.

allele. A given form of a gene which occupies a specific position (locus) on a specific chromosome. Alternative forms of a gene occur and can thus occupy that specific locus. These variant forms are said to be *alleles* or to be allelic to one another.

allopolyploidy. A polyploid condition in which the extra chromosome sets are derived from different groups or species following hybridization.

amino group. The —NH$_2$ chemical group, basic in nature, and characteristic of amino acids.

aminoacyl-tRNA synthetase. One of a group of enamniotic fluid is removed from a pregnant woman amino acid and links it to its specific kind of tRNA molecule.

amniocentesis. A procedure in which a sample of amniotic fluid is removed from a pregnant woman so that the fluid and fetal cells present in it can be subjected to various analyses.

anaphase. The stage of nuclear division at which the chromosomes move to opposite poles.

aneuploidy. A chromosome anomaly in which the number of chromosomes in a cell or individual departs from the normal diploid number by less than a whole haploid set.

Angstrom (Å). A unit of measurement equal to 10^{-8} cm.

antibody. A protein molecule, also known as an immunoglobulin, which is formed in response to a specific antigen and which can recognize and react with it.

anticodon. A sequence of three nucleotides in a tRNA molecule which pairs with a specific codon in the mRNA and thus enables the tRNA to position its amino acid properly in a growing polypeptide chain.

antigen. Any substance or large molecule which can stimulate the production of antibodies following entrance into the tissues of a vertebrate.

aster. The region which marks the poles of dividing cells and which includes rays of microtubules surrounding a clear area within which two centrioles are located.

autopolyploidy. A polyploid condition in which the additional chromosome set or sets have been derived from within the same group or species.

autoradiography. A technique which permits the localizing of sites occupied by a radioactive substance in biological material by covering slide preparations of the material with a photographic emulsion which is later developed to reveal the darkened points at which radioactive emanations have caused decay in the emulsion.

autosomes. All the chromosomes in the complement other than the sex chromosomes.

auxotroph. A nutritional deficient which must be supplied with one or more nutrient factors which it cannot manufacture from a minimal medium.

axial symmetry. (See *palindrome*).

backcross. A cross of a member of the F_1 generation to a member of one of the parental lines or to an individual which is a parental type.

back mutation. A mutation in a mutant allele which results in a change back to the wild or standard form of the gene.

bacteriophage. (See *phage*).

Barr body. (See *sex chromatin*).

bivalent. An intimate association of two homologous chromosomes seen at first meiotic division.

CAP (catabolite gene activator protein). A positive control element which complexes with cyclic AMP and which must bind to a region of the promoter before transcription of certain genes can begin.

carboxyl group. The —COOH chemical grouping, acidic in nature, and found in all amino acids.

centriole. An organelle marking the poles of the spindle in certain types of dividing cells.

centromere. That part of the chromosome responsible for chromosome movement and to which the spindle fibers attach.

characteristic. A general attribute of an organism. In some characteristics, variation is recognizable as distinct, contrasting traits. For others, the variation is continuous, showing a gradual transition from one extreme to the other.

chi-square. A value which indicates the probability that the figures in a set of numerical data are compatible with those ideally expected for a certain ratio.

chiasma. A cross-like figure seen at first meiotic division and associated with the exchange between two homologous chromatids.

chiasmatype theory. Interpretation of a crossover event as a result of breakage of two homologous chromatids in a bivalent followed by reciprocal reunion of chromatid ends and the appearance of a chiasma.

chromatid. A strand of a chromosome which has replicated. A chromosome is composed of two chromatids joined at the centromere region during mitotic prophase and metaphase.

chromatin. The chromosome material composed of DNA and associated proteins.

chromomere. A region of a chromosome thread which appears as a bead-like structure.

chromonema. The chromosome when it is seen as an extremely thin thread.

chromosome. A structure in the cell nucleus composed of chromatin and which stores and transmits genetic information.

chromosome aberration (chromosome anomaly). Any modification which alters the morphology or number of chromosomes in the complement.

cis arrangement. That arrangement in a doubly heterozygous individual in which two linked wild genes occur together on one homologue and the two recessive mutant genes occur on the other.

cis–trans effect. A position effect in which two individuals, dihybrid in respect to the same genetic sites, show a difference in phenotype as a consequence of the arrangement of the mutant sites on the chromosome, the *cis* arrangement producing a more normal phenotype than the *trans*.

cistron. The gene considered as a unit of function; a segment of DNA which is coded for one polypeptide.

clone. A population of genetically identical cells.

codominance. The expression in the heterozygote of both alleles. Codominance is generally used to refer to those cases in which two products can be detected in the heterozygote, each associated with one of the allelic forms. Often used interchangeably with incomplete dominance, as the distinction between the two is often not clear.

codon. A sequence of three adjacent nucleotides (an RNA or a DNA triplet) which designates a specific amino acid or which indicates that translation is to be terminated.

coefficient of coincidence. A figure which takes into consideration interference and is used to indicate the chance that a double crossover will occur within a certain map distance.

colinearity. The correlation between the sequence of codons in the DNA of a cistron (gene) and the sequence of amino acids in the polypeptide which it specifies.

continuous variation. That phenotypic variation shown by some characteristics in which there is a gradation from one extreme to the other without any delineated categories or distinct traits.

corepressor. A small molecule needed to interact with a specific repressor before the latter can combine with an operator to block transcription.

covalent bond. A strong chemical interaction in which atoms share electrons.

crossing over. The reciprocal exchange of segments between two homologous chromatids which results in the recombination of linked alleles.

cyclic AMP (3',5'-cyclic adenylic acid). A small molecule derived from ATP which interacts with specific hormones in cells and which must complex with CAP before transcription of certain genes can take place.

cytokinesis. Division of the cytoplasm producing two daughter cells from a parent cell.

cytoplasm. All the living parts of the cell outside the nucleus.

cytosine. A pyrimidine base found in DNA and RNA.

degenerate code. The genetic code which is characterized by the fact that two or more codons may designate the same amino acid.

deletion. A chromosome alteration in which a portion of a chromosome has been lost; loss of a part of a DNA molecule from the genetic complement.

denaturation (of DNA). The separation of the two strands of the double helix as a result of hydrogen bond disruption following exposure to high temperature or chemical treatment.

density gradient centrifugation. A technique in which a mixture of substances is spun at high speed in an appropriate solution until each component of the mixture reaches an equilibrium position where its density matches the density of the solute molecules at that position in the centrifuge tube.

detrimental allele. Any gene form which imparts a disadvantage to the individual carrying it and decreases the carrier's chances of reproduction or survival with respect to an individual not possessing the allele.

diakinesis. Last stage of first meiotic prophase during which the bivalents are extremely condensed and well separated from one another.

dicentric. An abnormal chromosome having two centromeres.

dihybrid (cross). A cross in which two pairs of alleles are being followed. Also any individual carrying two pairs of alleles and thus heterozygous at two loci which are under consideration.

dimer. A compound formed between two identical molecules and thus having the same percentage composition but twice the molecular weight of one of the original molecules.

diploid. Having two complete haploid sets of chromosomes typical for the species or group.

diplonema. Stage of first meiotic prophase during which homologous chromosome threads repel and chiasmata appear.

discontinuous variation. That phenotypic variation shown by some characteristics which can be recognized by clear-cut differences or distinct traits.

disjunction. The separation of chromosomes at anaphase of a mitotic or meiotic division.

DNA polymerase. An enzyme which catalyzes the synthesis of DNA on a DNA template from deoxyribonucleotide triphosphate precursors.

dominant. Any gene form which expresses itself in some way in the presence of its allele in the heterozygote. Also any trait which is phenotypically expressed when the responsible gene form is present singly with its allele in the heterozygote.

drift. Random or chance fluctuation in the frequencies of alleles in a population independent of natural selection and mutation.

duplication. An extra copy of a gene or genetic region in the complement. A duplication refers to any situation in which one or more additional copies of a genetic region are present.

endonuclease. An enzyme which can cut phosphodiester bonds which occur internally in a DNA chain.

endoplasmic reticulum (ER). System of membranes distributed within the cytoplasm of eukaryotes which provides sites of protein synthesis and intracellular channels of transport.

episome. A nonessential hereditary factor which may exist either free within the cell or in a state in which it is integrated with a chromosome.

epistasis. A type of genic interaction in which a gene form at a given locus masks the expression of another genetic factor which is located at a different locus and is therefore not its allele.

euchromatin. Those chromosome regions which are noncondensed and active in the interphase nucleus.

eukaryote. A cell or organism having a distinct membrane-bound nucleus as well as one or more membranous subcellular components (such as mitochondria, Golgi) which are specialized for the performance of certain specific cellular functions (see prokaryote).

exonuclease. An enzyme which attacks either the 3' or the 5' end of a polynucleotide chain and cuts phosphodiester bonds.

expressivity. The variation in the phenotypic expression of a specific allele or genotype when that genetic constitution is penetrant.

F_1 (see *first filial generation*).

F-genote. A fertility factor which carries a portion of DNA which is typically found in the chromosome of a bacterium.

fingerprinting. A technique in which a protein is subjected to enzyme digestion and the resulting peptide fragments separated by a combination of paper chromatography and electrophoresis.

first filial generation (F_1). The generation of individuals produced by the first parental generation (P_1) or the first parents being considered.

first parental generation (P_1). The first set of parents considered in any pedigree or mating.

5' end. That end of a nucleic acid chain terminating in a free phosphate group.

5' to 3' growth. The synthesis of a nucleic acid chain by the joining of the 5' (PO_3) end of a nucleotide being added to the 3' (OH) end of the last nucleotide already in the uncompleted chain.

forward mutation. A mutation which results in the formation of a mutant gene form from the wild or standard allele.

frame shift mutation. A deletion or duplication of a base in the DNA which causes a portion of the coded information in the gene to go out of phase.

gene. A stretch of nucleotides or nucleotide pairs along a nucleic acid molecule which contains coded information for some functional product such as RNA or a polypeptide.

gene pool. The total of all the genes carried by all the individuals in a population.

genetic engineering. Manipulation in which the genetic material is purposefully altered in some way through combining hereditary materials from different sources or by the removal of a native segment and the insertion of one from another source.

genetic load. The total in a population of the mutant alleles having a detrimental effect and accumulating mainly in heterozygotes.

genotype. The genetic constitution of a cell or of an individual.

gynandromorph. A sex mosaic; an individual with certain body segments that can be recognized as phenotypically male and others that are female.

guanine. A purine base found in DNA and RNA.

haploid. Having one complete set of chromosomes typical for the species or group.

Hardy-Weinberg law. A mathematical expression which shows that the frequency of alleles and the resulting genotypes will remain constant in a population from one generation to the next and will not change as a result of dominance or the way in which the alleles are transmitted.

hemizygous. Having certain loci present in a single rather than a double dose, as the X-linked genes in a mammalian male.

heterochromatin. The chromatin or chromosomes which tend to remain condensed in the interphase nucleus.

heterogenote. A partially diploid haploid cell heterozygous for just a segment of the genetic material.

heteropycnosis. The differential reactivity to staining displayed by a chromosome or a part of a chromosome in relation to the rest of the complement.

heterosis. Hybrid vigor; the more robust nature of a heterozygote as opposed to that of the more inbred parental lines.

heterozygous. Having alternative forms of a gene (alleles) at a given locus, one allelic form on each of the two homologues.

histocompatibility loci. Those genetic locations occupied by genes which produce tissue antigens that can be recognized by the immune system of a host, resulting in tissue rejection following a transplant.

histones. Basic proteins of low molecular weight which are complexed with the DNA of eukaryotes.

holandric. Any allele whose locus is on the Y chromosome; any trait associated only with the Y chromosome.

homogametic sex. That sex in a given species which produces a single kind of gamete with respect to the type of sex chromosome. The mammalian female is the homogametic sex, since every normal egg carries an X chromosome.

homologous. Having corresponding genetic loci.

homozygous. Having the same allele (gene form) present at a given locus on both homologous chromosomes.

hydrogen bond. A weak chemical interaction between an electronegative atom and a hydrogen atom that is covalently linked to another atom.

incomplete dominance. The expression in the heterozygote of both allelic forms. Generally used to refer to those cases where the phenotypic effect of one gene form in the heterozygote appears more pronounced than the other. Incomplete dominance and codominance are frequently used interchangeably, since the distinction between them is often not clearly defined.

incomplete sex linkage. The situation in which a specific locus is found on both the X and the Y chromosomes in that portion of the sex chromosomes which have homologous regions.

independent assortment. The segregation at meiosis of a pair of alleles independently of other allelic pairs whose loci are found on different chromosomes.

induction (of enzyme synthesis). The stimulation of specific enzyme synthesis in a cell when the specific substrate of the enzyme is supplied.

initiating tRNA. A special form of tRNA which can recognize the triplet AUG at the site in the mRNA which marks the start of translation.

interference. The effect exerted by a crossover which decreases the probability that a second crossover will occur within a certain map distance in the chromosome.

interphase. The condition of the nucleus which is not undergoing mitotic or meiotic division.

inversion. A chromosome aberration which entails two breaks in a chromosome followed by a reversal of the segment and consequently of the gene sequence in the segment. Pericentric inversions include the centromere in the inverted segment, whereas paracentric inversions do not. In an overlapping inversion a point of breakage in a particular inversion occurs in a segment which was reversed in a previous inversion.

ionizing radiation. Any radiation possessing sufficient energy to cause an outright separation of electric charges in the material it strikes.

isoenzymes. Two or more enzymes capable of catalyzing the same reaction but differing slightly in their structures and consequently in their efficiencies under various environmental conditions.

kappa. A DNA-containing cytoplasmic particle found in certain strains of *Paramecium* which is capable of self-reproduction and which can exert a killing effect on sensitive cells when it is liberated in the culture medium.

karyotype. The chromosome constitution of a cell or individual.

leptonema. The first stage of first meiotic prophase before synapsis and during which each chromosome is in an extremely extended state.

lethal allele. Any allele which can cause the death of the individual who possesses it. A complete lethal removes the individual sometime before the age of reproduction.

ligase. An enzyme which can catalyze phosphodiester linkages and can thus restore intact a polynucleotide chain which contained one or more nicks.

linkage. The association of genes or genetic loci on the same chromosome. Any genes at linked loci will tend to be transmitted together.

locus. The specific position occupied by a given gene on a chromosome. At a particular locus, any one of the variant forms of a gene may be present.

lysogenic. A strain of bacteria harboring prophage and which can thus cause the lysis of a strain sensitive to that virus.

map unit. One percent of crossing over on a linkage map.

mean. The numerical average; the sum of all the values of a group of figures divided by the number of individual measurements in the group.

megaspore. A haploid plant cell derived from meiotic division of a megaspore mother cell and which may give rise to a female gamete following mitotic divisions.

meiosis. The nuclear process which includes two divisions and which results in the reduction of chromosome number from diploid to haploid.

messenger RNA (mRNA). The complementary RNA copy of DNA formed on the DNA template during transcription and carrying coded information for the amino acid sequence of a polypeptide.

metacentric. A chromosome with its centromere in a median position, thus having two arms of equal length; the median location of the centromere in the chromosome.

metaphase. That stage of nuclear divisions when the chromosomes are arranged at the equatorial plane of the spindle.

microspore. A haploid plant cell derived from the meiotic division of a microspore mother cell and which may give rise to male gametes following mitotic divisions.

missense mutation. A point mutation in which a codon becomes changed to a different codon designating a different amino acid.

mitochondrion. A double-membrane cytoplasmic organelle found in eukaryotes and the site of aerobic respiration.

mitosis. The nuclear division which results in the accurate distribution of the genetic material and produces two nuclei exactly like the parent nucleus in chromosome content.

modifier (modifying gene). A gene whose expression affects the expression of a gene or genes at other loci which are thus not allelic to it. A modifier often has no other known effect.

monohybrid. An individual heterozygous at a locus under consideration, thus having a pair of alleles at that locus. Also, a cross in which a pair of alternative gene forms (alleles) is being followed.

monosomy. A chromosome aberration in which only one of a given kind of chromosome is present instead of the normal two.

mosaicism. A situation in which the body of an individual is composed of two or more genetically distinct cell types.

multiple alleles. Three or more forms of a gene, any one of which can occur at a given locus on a specific chromosome.

multiple-factor inheritance. Inheritance which entails many nonallelic genes and the complex interaction of environmental factors, many of which are unknown (see quantitative inheritance).

mutagen. Any agent which can cause an increase in the rate of mutation in an organism.

mutation. A sudden inheritable change which includes gene (point) mutation and chromosome aberrations in its broadest sense.

mutation pressure. Spontaneous gene mutation occurring continuously in a population.

natural selection. The differential reproduction of alleles which occurs in a population from one generation to the next and which results in an increase in the frequency of certain alleles and a decrease in others.

negative control. The type of genetic regulation in which the product of the regulator gene acts (alone or with a corepressor) to repress transcription of specific structural genes.

nondisjunction. The failure of homologous chromosomes or sister chromatids to separate at mitotic or meiotic division, resulting in aneuploid cells or individuals.

nonhistone protein. An assortment of proteins, mostly acidic in nature, which vary in kind with the species and cell type and which appear to function in the control of transcription in eukaryotes.

nonsense codon. One of three codons whose sequence of three nucleotides indicates that translation is to be halted and thus the assembly of amino acids into a growing polypeptide chain is to be terminated.

nonsense mutation. A point mutation in which a codon specific for an amino acid is changed to a codon which indicates that translation is to be halted.

nonsense strand. That one of the two strands of DNA composing a gene which does not undergo transcription.

nuclease. Any enzyme which can break the phosphodiester bonds which join together nucleotides in a polynucleotide chain.

nucleolar organizer. That region of a specific chromosome in the complement (the nucleolar organizing chromosome) which contains genes for the RNA of the ribosome and which is associated with the formation of the nucleolus.

nucleolus. A dense body within the nucleus produced by the nucleolar organizer and associated with the processing of the ribosomes.

nucleoside. A molecule composed of a 5-carbon sugar linked to a nitrogen base.

nucleotide. A nucleic acid unit composed of a 5-carbon sugar joined to a phosphate grouping and a nitrogen base.

oogenesis. The entire series of events in a female in which a gamete is produced from the maturation of an immature germ cell.

oogonium. An immature germ cell in the female which can give rise to a primary oocyte.

operator. That segment of DNA associated with a structural gene and which interacts with the product of a regulator in the control of transcription.

operon. A unit of transcription composed of an operator and one or more structural genes associated with it.

organelle. A subcellular component containing particular enzymes and specialized for a certain cell role.

overlapping inversion. (See *inversion*).

P_1 (See *first parental generation*).

pachynema. Stage of meiotic prophase I after completion of synapsis and characterized by a detectable thickening of the chromosome threads.

palindrome. In DNA, a region in which a symmetrical arrangement of bases occurs around a point in the molecule with the result that the base sequence is the same on either side of it, reading either of the two paired strands in the 5' to 3' direction.

paracentric inversion. (See *inversion*).

penetrance. The ability of a specific allele or genotype to express itself in any way at all when present in an organism.

pericentric inversion. (See *inversion*).

phage. A virus having a bacterial cell as its host.

phenocopy. A nongenetic, environmentally induced imitation of the effects of a specific genotype.

phenotype. Any detectable feature of a living organism. The phenotype is a product of the interaction between the genetic material and the environment.

photoreactivation. The reversal of the killing effect of ultraviolet radiation by white light.

plasmid. Any nonessential genetic entity which may exist in the cell independently of the chromosomes.

640

pleiotropy. The multiple phenotypic effects of an allele.

polar body. A tiny cell which receives little cytoplasm, produced by the division of a primary or a secondary oocyte.

polygenic inheritance. Inheritance involving alleles at many genetic loci which interact with environmental factors. "Polygenic" refers to the many genetic factors as opposed to the environmental component. (Also see *quantitative inheritance*).

polypeptide. A single long chain composed of amino acid units joined by peptide linkages.

polyploidy. A condition in a cell or individual in which one or more whole haploid sets of chromosomes are present in addition to the normal diploid number typical for the species or group.

polysome. An assembly of ribosomes active in translation and connected by the same mRNA strand.

polytene chromosome. A chromosome composed of many intimately associated strands or unit chromatids formed as a result of endoreduplication.

population. An interbreeding group of organisms of the same species.

position effect. A phenotypic change resulting solely from the placement of a gene or genetic region in a new location in the chromosome complement without any change within the newly located genetic material itself.

positive control. The type of genetic regulation in which a substance produced in the cell is required to activate transcription of specific structural genes.

primary oocyte. A cell in the female which is derived from an oogonium and which undergoes first meiotic division.

primary spermatocyte. A cell in the testis derived from a spermatogonium and which undergoes first meiotic division.

prokaryote. Organisms or cells which lack a nucleus bounded by a membrane as well as specialized membrane bound organelles. (Also see *eukaryotes*).

promoter. The segment of DNA containing start signals which can be recognized by RNA polymerase for the start of transcription of a gene.

prophage. A temperate virus devoid of its protein components which is integrated with the DNA of the host and behaves like a genetic factor of the infected cell.

prophase. That phase of nuclear divisions after DNA replication and during which the chromosomes become evident as they gradually shorten and thicken.

prototroph. A microorganism which is capable of supplying its own nutrient requirements from a minimal medium.

purine. A nitrogen-containing compound with a double ring structure and the parent compound of adenine and guanine.

pyrimidine. A nitrogen-containing compound with a single ring structure and the parent compound of cytosine, thymine, and uracil.

quantitative inheritance (variation). That type of hereditary transmission or variation which involves many alleles at different genetic loci, each allele producing some measurable effect (often used synonomously with polygenic and multiple factor inheritance).

recessive. Any gene form which is not expressed in the presence of its allele in the heterozygote. Also any trait which is expressed phenotypically only when the responsible gene form is present in double dose in the homozygote.

recombination. A new combination of alleles resulting from rearrangement following crossing over or independent assortment. (Some restrict usage of the term only to new combinations produced by crossing over).

regulator gene. A DNA sequence or cistron which codes for the production of a nonenzyme protein which regulates the transcription of another gene.

renaturation (of DNA). The reconstitution of double helical DNA by the reassociation of single strands derived from it following melting.

repetitive DNA. Repeated nucleotide sequences which may occur in hundreds, thousands, or more copies in the chromosome complement of a eukaryote.

replication. The synthesis of a macromolecule identical to and under the guidance of a parent or template macromolecule.

replication fork. The Y-shaped region within a double-stranded DNA molecule which is undergoing replication and which marks the site at which complementary strands are being synthesized at that time.

replicon. Any genetic entity which can replicate itself independently.

repression (of enzyme synthesis). The cessation of production of a specific enzyme by a cell when the end product of the enzyme reaction is supplied.

repressor. The protein product of a regulator gene which can combine with a specific operator and block transcription of the structural genes in an operon.

resistance transfer factor (RTF). A nonchromosomal entity or replicon which carries one or more factors capable of conferring drug resistance on a bacterial cell.

restriction enzyme. A type of endonuclease which produces internal cuts within a specific region of axial symmetry in the DNA.

reverse transcriptase. (See *RNA-dependent DNA polymerase*).

rho factor. A protein necessary to halt the transcription of certain genes.

ribosomal DNA (rDNA). Those chromosomal genes whose transcripts are processed into the RNA components of the ribosomes.

ribosome. A complex cellular component composed of RNA and protein which interacts with mRNA and tRNA to join together amino acids into a polypeptide chain. Found in the cytoplasm, mitochondria, and chloroplasts.

RNA. (See "types of," for example, *messenger RNA*).

RNA-dependent DNA polymerase (reverse transcriptase). A polymerase which preferentially assembles a DNA strand using a strand of RNA as the template.

RNA polymerase. An enzyme which catalyzes the synthesis of RNA from ribonucleoside triphosphate precursors.

RNA replicase. An enzyme required for the assembly of RNA strands on an RNA template.

S value. A sedimentation coefficient or constant expressed in Svdeberg units.

satellite. A terminal end of a chromosome arm produced by a secondary constriction, connected to the rest of the chromosome by a very narrow region, and typically associated with nucleolus formation.

satellite DNA. That DNA of a eukaryotic cell which

has a different density than the bulk of the cellular DNA and which equilibrates at a different position from the main band following density gradient centrifugation.

secondary oocyte. The larger of the two cells produced by the division of a primary oocyte.

secondary spermatocyte. One of two cells derived from the division of a primary spermatocyte which has completed first meiotic division.

selection pressure. The operation of natural selection on the allele frequency in a population resulting in an increase in the frequency of certain alleles and a decrease in the frequency of others.

semiconservative replication. The manner in which double-stranded DNA is synthesized and which produces from the original DNA two double-stranded molecules each containing one parental strand and one strand which is newly formed.

sense strand. That one of the two DNA strands composing a gene which undergoes transcription.

sex bivalent. An association composed of the paired X and Y chromosomes in a primary spermatocyte.

sex chromatin. A condensed or inactivated X chromosome (or part of an X) in the interphase nucleus seen as a densely staining body in cells containing more than one X chromosome.

sex chromosomes. Those chromosomes involved in sex determination (such as the mammalian X and Y) and which show a difference in morphology or number between the sexes.

sex-influenced allele. Any allele which is expressed as a dominant in one sex and a recessive in the other.

sex-limited allele. Any allele which expresses itself in only one of the sexes.

sex-linked (X-linked) allele. Any gene form having its locus on the X chromosome. Also any trait associated with such a genetic factor.

sexduction. The transfer of a chromosomal gene from one cell to another by way of a sex factor, such as F, the fertility factor.

shift. A chromosome aberration in which a chromosome segment is transposed to a new location in the same or in a different chromosome. The shift is a simple type of translocation.

sigma factor. One of the polypeptide chains composing RNA polymerase and which is essential for the recognition of signals in the DNA for the start of transcription.

silent mutation. A point mutation in which a codon specific for a given amino acid is changed to a different codon which designates the same amino acid.

somatic cell. A cell which is part of the body as opposed to a reproductive cell in the germ line.

spermatid. One of two cells derived from the division of a secondary spermatocyte which has completed second meiotic division.

spermatogenesis. The series of events starting with the origin of a primary spermatocyte and culminating with the production of sperms.

spermatogonium. A type of cell found in the wall of the testis which can give rise to a primary spermatocyte.

spermiogenesis. The portion of spermatogenesis during which a spermatid is transformed into a mature sperm.

spindle. An aggregation of microtubules essential for the positioning and distribution of the chromosomes at nuclear divisions.

spindle fiber. A microtubule composed of an assemblage of protein units.

standard deviation. A statistic which describes the amount of variation on either side of the mean in a sample and which can be used to calculate the standard error of the mean.

standard error of the mean. A statistic which indicates the probability that the mean calculated from a sample is close to the true mean for the entire population.

statistics. Measurements or values obtained from samples rather than those measurements made on an entire population, the parameters.

steroid. One of an assortment of complex lipids, composed of four interlocking rings of carbon atoms, which are often biologically important compounds such as vertebrate male and female sex hormones.

structural gene. A segment of DNA which is coded for a polypeptide which serves as an enzyme or as some nonregulatory substance.

structural heterozygote. An individual having a normal chromosome and some kind of change in the homologue which affects its morphology or structure.

suppressor. A gene which prevents the expression of a mutant gene found at a different genetic locus, which is therefore not allelic to it.

symbiosis. An intimate association of two organisms of different species living together, usually for mutual benefits.

synapsis. The intimate pairing, locus for locus, of homologous chromosomes at first meiotic prophase.

syndrome. The collection or group of symptoms or features associated with a specific disease or abnormality.

synteny. The association of genes or genetic loci on the same chromosome. Often used synonomously with linkage. Some restrict the term "linkage" only to those genes which clearly do not undergo independent assortment as shown by genetic analysis. "Syntenic" is used to indicate that procedures such as somatic-cell hybridization have shown two genes are on the same chromosome, even if genetic analysis has not yet demonstrated this.

T cell (T lymphocyte). A white blood cell which becomes differentiated in the thymus gland and which can recognize foreign antigens on cell surfaces and bring about the removal of the foreign cells.

t test. A statistic which indicates the chance that the values between two separately calculated means are significantly different.

tautomerism. The reversible shifting of the location of a proton in a molecule which changes certain chemical properties of the molecule.

telocentric. A chromosome with its centromere appearing to be in a terminal location, thus having only one evident arm. The centromere located at one end of the chromosome.

telomere. A region of a chromosome marking the extreme end of a chromosome arm. Every normal chromosome possesses two telomeres.

telophase. The stage of nuclear division during which the nuclear membrane reforms and the chromosomes gradually become less and less evident.

temperate (symbiotic) phage. A bacterial virus which tends to take up residence in the host cell without destroying it.

template. A macromolecule which serves as a blueprint or mold for the synthesis of another macromolecule.

testcross. The cross of an individual to one which expresses a specific recessive trait or traits under consideration.

tetrad. The four nuclei or four cells which are the immediate results of meiosis in a parent cell. (Sometimes also used to designate a bivalent).

tetraploid. A cell or organism having four complete haploid sets of chromosomes instead of the typical two.

3′ end. That end of a nucleic acid chain terminating in a free—OH group.

thymine. A pyrimidine base found in DNA but not in RNA.

trait. A distinct alternative form of a characteristic.

trans arrangement. That linkage arrangement in a double heterozygote in which a wild allele and a recessive mutant gene form occur together on one chromosome and the corresponding recessive and wild alleles occur together on the homologue.

transcription. The assembly of a complementary single-stranded molecule of RNA on a DNA template.

transduction. The transfer of genetic material from one cell to another by a viral vector.

transfer RNA (tRNA). A small RNA molecule which recognizes a specific amino acid, transports it to a specific codon in the mRNA and positions it properly in a growing polypeptide chain.

transformation. A genetic change effected in a cell as the result of the incorporation of DNA from a virus or some genetically different cell type.

transgressive variation. The quantitative variation in a characteristic of the offspring which exceeds that shown by the same characteristic in either parent.

transition. A point mutation in which a purine is replaced by a different purine or a pyrimidine by a different pyrimidine.

translation. The assembly of a polypeptide chain from the coded information in the mRNA which directs the amino acid sequence of the chain.

translocation (reciprocal). A chromosome alteration in which a chromosome segment or arm is transposed to a new location. A reciprocal translocation involves a mutual exchange of chromosome segments or arms between two nonhomologous chromosomes.

transversion. A point mutation in which a purine is replaced by a pyrimidine or a pyrimidine is replaced by a purine.

triploid. A cell or organism having three complete haploid sets of chromosomes instead of the normal two.

trisomy. A condition in a cell or individual in which a particular chromosome is present in three doses instead of the normal two.

uracil. A pyrimidine base found in RNA but not in DNA.

variance. A statistic which measures the amount of variability in a sample and which can be broken up into separate components to ascertain the role of responsible genetic and environmental factors in the variation; the square of the standard deviation.

virulent phage. A bacterial virus which typically destroys the host cell which it infects.

wild type. The form of a gene or of an individual which is considered the standard one typically found in nature.

wobble hypothesis. The concept that the base at the 5′ end of the anticodon is somewhat free to move about spatially and can thus form hydrogen bonds with more than one kind of base at the 3′ end of a codon in the mRNA.

X linked. See sex linked.

zygonema. The stage of first meiotic prophase during which homologous chromosomes pair.

Answers to Review Questions

CHAPTER 1

1. A. OO
   B. oo
   C. Oo
   D. oo

2. A. 1 OO:1 Oo. All round
   B. 1 OO: 2 Oo: 1 oo. 3 round: 1 oblong
   C. oo. All oblong
   D. 1Oo: 1 oo. 1 round: 1 oblong

3. Short hair is the dominant trait. Genotypes: female No. 1 is Ss; female No. 2 is SS. The male is ss.

4. A. Horned and hornless in a ratio of 1:1
   B. One-fourth
   C. Three-fourths

5. A. Red (RR), roan (Rr), white (rr) in a ratio of 1:2:1
   B. One roan (Rr); one white (rr)

6. A. hhrr
   B. H__Rr
   C. HhRR

7. A. All with black fur and short hair.
    B. One out of four. BBLL—black short.
       BBll—black long. bbLL—albino short.
       bbll-albino long.

8. A. All blue and mildly frizzled.
    B. 1 blue, mildly frizzled: 1 white, mildly
       frizzled.
    C. 1 blue, mildly frizzled: 1 blue frizzled:
       1 white, mildly frizzled: 1 white friz-
       zled.
    D. 1 black frizzled: 2 blue frizzled: 1 white
       frizzled: 2 black, mildly frizzled: 4 blue,
       mildly frizzled: 2 white, mildly frizzled:
       1 black straight: 2 blue straight: 1
       white straight.

9. A. Aamm
    B. A_Mm
    C. aamm

10. A. Parental genotype: AaMm. Gametes:
       AM, Am, aM, am
       Parental genotype: Aamm. Gametes:
       Am, am.
    B. Offspring: nonalbino with migraine,
       nonalbino without migraine, albino
       with migraine, albino without mi-
       graine in a ratio of 3:3:1:1

11. A. No chance
    B. One-fourth
    C. Two-thirds

12. All the children will be genotype "ss,"
    homozygous for the defective allele, just as
    the parents are. The children would require
    treatment, since the environment has not
    altered the genetic material which is trans-
    mitted through the gametes. A Lamarckian
    would claim that the treatment has pro-
    duced the normal condition and this trait
    can now be transmitted so that the children
    would be normal eventually regardless of
    treatment.

13. A. (1) ffSs (2) FfSs
    B. Free, no sickling; free, some sickling;
       free, with anemia; attached, no sick-
       ling; attached, some sickling; attached
       with anemia. Ratio 1:2:1:2:1.

14. A. An allele does not increase in frequency
       simply because it is dominant. The
       frequency is the result of the interac-
       tion of many factors. Dominance by
       itself does not make an allele increase.
       Dominance means that a gene form
       will express itself in the presence of its
       allele.
    B. The person has a 50:50 chance of de-
       veloping the disorder. The afflicted par-
       ent is most likely genotype Hh, since
       the allele is so rare.

15. A. 4
    B. 16
    C. 16

16. A. 16
    B. 8
    C. 32

17. A. 81
    B. 27
    C. 243

18. (1) TG, tG
    (2) TGW, TGw, tGW, tGw
    (3) TGW, TgW, tGW, tgW, TGw, Tgw, tGw,
        tgw

19. (1) 1 tall yellow: 1 dwarf yellow
    (2) 1 tall yellow round: 1 tall yellow wrin-
        kled: 1 dwarf yellow round: 1 dwarf
        yellow wrinkled
    (3) 1 tall yellow round: 1 tall green round:
        1 dwarf yellow round: 1 dwarf green
        round: 1 tall yellow wrinkled: 1 tall
        green wrinkled: 1 dwarf yellow wrin-
        kled: 1 dwarf green wrinkled

20. 9 tall yellow round: 3 dwarf yellow round:
    3 tall yellow wrinkled: 1 dwarf yellow wrin-
    kled

CHAPTER 2

1. A. Prophase
    B. Telophase
    C. Metaphase
    D. Prophase

E. Interphase

F. Anaphase

2. A. (1) 92
      (2) 46
      (3) 92

   B. (1) 46
      (2) 46
      (3) 46

3. A. 22 pairs
   B. 22 pairs
   C. 2
   D. 2
   E. 7

4. A. AAXX
   B. AAXY

5. 1. B
   2. E
   3. F
   4. A
   5. E
   6. C
   7. E
   8. H
   9. G

6. Prokaryotes lack mitochondria, a nuclear envelope, chloroplasts, endoplasmic reticulum, centrioles, mitotic spindles (additional distinctions are made in later chapters).

7. (1) locus
   (2) homologous
   (3) homozygote or homozygous
   (4) alleles
   (5) DNA
   (6) metacentric
   (7) polytene
   (8) karyotype
   (9) satellite
   (10) acrocentric
   (11) colchicine

CHAPTER 3

1. A. Diplonema
   B. Zygonema
   C. Diakinesis

D. Leptonema
E. Anaphase II
F. Metaphase II

2.

	Chromatid Number	Chromo- some Number	Number of Bivalents
(1)	92	46	23
(2)	92	46	23
(3)	92	46	23
(4)	46	23	0
(5)	46	23	0
(6)	23	23	0

3. A. At mitotic metaphase single chromosomes are oriented at the equator, whereas bivalents are at the equator of first meiotic metaphase.

B. In mitosis, each chromosome moving poleward is composed of a single chromatid; at meiotic first anaphase, each is composed of two chromatids.

C. No bivalents would be present at mitotic prophase. At the earliest meiotic prophase, the chromosome threads are much more extended than at mitosis and show obvious chromomeres.

4. A. Only AaBb
B. AB, Ab, aB, ab

5. A. 46
B. 46
C. 23
D. 23
E. 23
F. 23

6. A. AX
B. AX and AY
C. AAXY
D. AAXX

7. Only in the primary oocyte.

8. A. (1) 4000 (2) 2000 (3) 1000 (4) 8000
B. (1) 1000 (2) 1000 (3) 1000 (5) 250

9. A. 10
B. 10

C. 11
D. 6
E. 5

10. A. 10
B. 20
C. 20
D. 20
E. 10
F. 10

CHAPTER 4

1. The dominant "E" shows variable expressivity and reduced penetrance. The man with the extra toe on each foot definitely carries allele "E" and is most likely genotype "Ee," since the allele is rare. The wife is most probably genotype "ee," since the allele is rare, and she does not come from her husband's family line. Both their children are genotype "Ee," even the one of normal phenotype. The normal offspring's wife is almost certainly "ee," again because the allele is rare, but the offspring here is again "Ee," having received the dominant from the phenotypically normal parent who carried it but did not express it.

2. 60%.

3. Pleiotropy.

4. A. CcPp × Ccpp
B. White and colored in a ratio of 5:3

5. A. Red
B. White
C. Yellow
D. White

6. A. Brown × white. Offspring: 2 white: 1 black: 1 brown
B. White × black. Offspring: 2 white: 1 black: 1 brown
C. White × white. Offspring: 3 white: 1 black

7. A. Deaf × deaf. No chance of a deaf child.
B. Normal × normal. No chance of a deaf child.

    C. Normal × normal. One chance out of four.

    D. Normal × normal. One chance out of four.

8. A. Pea × rose. Offspring: all walnut

    B. Walnut × walnut. Offspring: 9 walnut: 3 rose: 3 pea: 1 single

    C. Walnut × single. Offspring: 1 walnut: 1 rose: 1 pea: 1 single.

9. A. RrPp × rrpp

    B. RRpp × rrPp

    C. Rrpp × RrPp

10. A. RRppBB × rrPPbb. F_1 birds are walnut, blue: RrPpBb

    B. RrPpBb × rrppBB. F_2: Black birds of the four different types of combs and blue birds of the four different types of combs.

11. A. BBCCoo × BBccOO. F_1: BBCcOo.

    B. BBCcOo × BBCcOo. F_1: 9 BBC__O__; 3 BBC__oo; 3 BBccO__; 1 BBccoo. The ratio is 9:7.

    C. BbCCOO × BbCCOO. F_1: BBCCOO; BbCCOO; bbCCOO. The ratio is approximately 1:2:1.

12. The F_2 plants fall into a ratio of 9:7, a modified dihybrid ratio. Two pairs of alleles are involved, say Aa and Bb. High cyanide content requires both dominants, A and B. All those plants lacking either A or B will have a low cyanide content.

13. The ratio among the F_2's is approximately 15:1, indicating a modified dihybrid ratio. Assuming the allelic pairs Aa and Bb, the F_1 parents must have been dihybrids AaBb. The P_1's would have been AAbb and aaBB. Only double recessives, aabb, have unfeathered legs. Duplicate dominant epistasis is involved so that the presence of either dominant, A or B, will give feathers on the legs.

14. Mexican hairless dogs are all heterozygotes, carrying an allele which can be represented as "H." This acts as a dominant

in respect to the hairless condition, but it also has a recessive lethal effect. Other breeds of dogs are homozygous for the recessive allele "h" which permits hair growth. They thus lack the lethal. Genotypes are: Mexican hairless, Hh; dogs with hair, hh; dead pups, HH.

15. A. Hhww
     B. hhww
     C. hhWW

16. A. 1 hairless: 1 wire-haired
     B. 4 hairless: 3 wire-haired: 1 straight-haired

17. When two plants are crossed and produce albino offspring, it is known they are heterozygotes. These plants of known genotype Ww can be crossed to those whose genotype is still unknown. If the cross of a known heterozygote and an unknown produces a large number of offspring (say, over 20) without the appearance of any albinos, it would be most likely that the unknown is genotype WW. Recognizing WW plants in this way followed by their exclusive use in further crosses would eliminate the lethal.

18. A. 1 black: 1 albino
     B. Same results as above since ovaries are aa.
     C. All black offspring.

19. A, C, D

20. A, B

CHAPTER 5

1. A. AaPp
    B. AapY
    C. AaPY

2. A. aP, ap
    B. Ap, AY, ap, aY
    C. AP, Ap, aP, ap

3. A. Daughters all carriers with normal iris (Ii). Sons all with normal iris (IY).

    B. Daughters all carriers with normal iris (Ii). Sons all with cleft iris (iY).
    C. Daughters: half of them carriers with normal iris (Ii) and half of them with cleft iris (ii). Sons: half of them with normal iris (IY) and half of them with cleft iris (iy).

4. Girl: Mmii; mother: mmIi; father: M__iY

5. Migraine, normal iris; migraine, cleft iris; no migraine, normal iris; no migraine, cleft iris in a ratio of 1:1:1:1 among both sons and daughters.

6. A. ppTT × PYtt. F₁ PpTt (daughters with normal vision); pYTt (sons with red-green color blindness).
    B. Normal vision, red-green color blind, and totally color blind in a ratio of 3:3:2 among both sons and daughters.

7. A. All daughters with rickets (Rr); all sons normal (rY).
    B. Half of the sons and half of the daughters with rickets (RY;Rr) and half of the sons and half of the daughters normal (rY; rr).

8. A. Females: bY (non-Bar); Males: Bb (Bar)
    B. Females: BY (Bar) and bY (non-Bar); Males: Bb (Bar) and bb (non-Bar)

9. A. The females carrying the lethal would produce only one-half the expected number of male offspring, since no male could be born with the allele. The 1:1 sex ratio would become distorted to 2:1 in favor of females.
    B. The males carrying the lethal would produce only half the expected number of female offspring. No hen could carry the allele. The sex ratio of 1:1 would become distorted in favor of males.

10. To have the affliction, a female must receive an X chromosome bearing the recessive from both parents. Since males with this affliction do not live to reproduce, any offspring receives the recessive only from

a carrier mother. Since a male has only one X chromosome, a male receiving the allele will express it.

11. The butterfly must be female, since the allele for whiteness can only be expressed in the female of this species. The genotype could be WW or Ww. The tortoise kitten must be female, since the genotype for tortoise must be Bb, and this requires two X chromosomes. The sex of the yellow kitten is in doubt. It could be a male (bY) or a female (bb).

12. Noncolor-blind female with three X chromosomes (PPp); noncolor-blind Turner female (P0); noncolor-blind Klinefelter male (PpY). The Y0 zygote would not survive.

13. Noncolor-blind Turner female (P0); noncolor-blind Klinefelter male (PPY); color-blind Turner female (p0); noncolor-blind Klinefelter male (PpY).

14. (1) 100 (2) 100 (3) 50 (4) None
    (5) None (6) 50 (7) 100 (8) 100

15. This male must be carrying certain alleles which cause reduction in milk yield. These would be sex-limited alleles which do not come to expression in the male. However, in a female these alleles can come to expression whether received from a male or a female parent. In this example the bull is carrying them and transmitting them to female offspring where they come to expression causing a reduction in milk yield.

16. A. Half of the sons nonbald (bb) and half of them bald (Bb). Daughters all nonbald (Bb and bb).
    B. Sons all bald (Bb); daughters all nonbald (Bb).
    C. Same answer as in part A.

17. Daughters: all nonbald, but half of them will be color blind. Sons: equal chances for bald and normal vision; bald and color blind; nonbald and normal vision; nonbald and color blind.

18. D, E

19. A

20. C, E

CHAPTER 6

Answers—Pedigree I:

2. There is no chance, because the allele is dominant. If neither parent shows the trait, they are not carrying the allele and cannot pass it down.

3. The answer is the same as given above. As far as this trait is concerned, the fact that they are cousins doesn't matter because neither shows it and so neither carries the responsible allele.

Pedigree II:

5. The chance is 1/3 that V-1 is a heterozygote. The chance is 2/3 that her mother is a heterozygote. If she is, there is a chance of 1/2 that she will pass the recessive to an offspring. Thus: $2/3 \times 1/2 = 1/3$.

6. The answer is 1/6. The chance is 1/3 that she is a carrier. If a carrier marries an affected (Aa × aa), the chance is 1/2 for an affected child. The chance for the events to happen together is $1/3 \times 1/2$ or 1/6.

Pedigree III:

7. The answer is 1/2 because it will be passed to all the males.

8. No chance, because person V-5 is not carrying it on his Y chromosome.

9. 100% or 1, because all of the male children will be affected.

Pedigree IV:

11. The answer is 1/8. The chance that V-3 is a carrier is 1/2. Her mother must be "Aa," because person V-1, her brother, shows the sex-linked trait. There is a chance of 1/2 that the mother passed it to V-3. *If* she is a heterozygote, the cross is Aa × aY and 1/2 of the children will be affected. This means the chance of an affected child is $1/2 \times 1/2$ or 1/4. Because there is a chance of 1/2 for a boy, the probability for an affected son is 1/8.

12. There is no chance. Since V-6's father doesn't have the rare sex-linked trait, she hasn't received it from him. Her mother is not a blood relative of the father and wouldn't be carrying it. Although person IV-10 has the trait, the cross would be AA × aY, and no children will show it.

13. $(1/2)^6$

14. 1/2

15. (1) 1/64 (2) 1/64 (3) 1/32 (4) 5/16

16. A. 1/8
    B. 3/8
    C. $3/8 + 1/8 = 1/2$

17. A. $3/8 \times 1/8 \times 1/8 \times 3/8 \times 3/8 = 27/32,768$
    B. $10 (3/4)^3 \times (1/4)^2 = 270/1024 = 135/512$
    C. The chances are 3 out of 4 that the child will be normally pigmented and 1 out of 4 that the child will be albino.

18. The answer is 1/3. The reasoning is as follows: In a family of two children, the probability is 1/4 for two girls, 1/2 for a boy and a girl, and 1/4 for two boys. You can eliminate the two boys in this case, since you know one of the children is a girl. This leaves the following possibilities: a girl, then a girl; a girl, then a boy; or a boy, then a girl. Since two of the remaining possibilities are boy-girl combinations, one of the three is girl-girl.

19. A. $(1/3)^3 = 1/27$
    B. $(2/3)^3 = 8/27$

20. $2/3 \times 1/2 = 1/3$

21. A. 3/64
    B. $4 \times (1/4)^3 \times 3/4 = 12/256 = 3/64$

22. A. 1/2
    B. 1/4
    C. 1/4
    D. 1/2

23. A. $3/16 \times 9/16 \times 3/16 = 81/4096$
    B. $1/4 \times 1/2 \times 1/4 = 1/32$

24. $\dfrac{4!}{2!\,1!\,1!} \times (3/16)^2(3/16)(1/16) =$
    324/65,536

25. Calculation of χ^2:

o	e	d	d^2	d^2/e
33	37	4	16	0.43
74	74	0	0	0
41	37	4	16	0.43
				$0.86 = \chi^2$

P is between 0.50 and 0.80 (Table 6-5). χ^2 does not tell you that it *is* a certain ratio. In this case, the figures are compatible with the hypothesis of a ratio of 1:2:1 which has become distorted by chance factors.

26. If genes are assorting independently, a ratio of 1:1:1:1 is expected in a dihybrid testcross. The χ^2 value on this hypothesis turns out to be 53.06 in this case as shown by the calculations given below. Such a high χ^2 gives a P value much less than 0.01 for 3 degrees of freedom. Therefore, from these data one cannot say that the genes are assorting independently, since the data are incompatible with a ratio of 1:1:1:1. The probability that chance factors have been operating to distort the ratio is extremely low. Calculation of χ^2:

o	e	d	d^2	d^2/e
539	600	61	3721	6.20
659	600	59	3481	5.80
712	600	112	12544	20.90
490	600	110	12100	20.16
				$53.06 = \chi^2$

CHAPTER 7

1. A. AABb or AaBB
   B. AaBb
   C. aabb

2. The white parent is aabb. The intermediate is probably either AAbb or aaBB. This is so, since intermediates with genotype AaBb would make it possible for children of various shades to arise.

3. Intermediate: AaBb
   Light: Aabb and aaBb
   White: aabb

4. The two parents in the first case are probably not heterozygous at both loci. Both could be AAbb; both could be aaBB; or one could be AAbb and the other aaBB. The second two parents are probably both AaBb.

5. A. $2^6 = 64$
   B. 1/4096 red and 1/4096 white.

6. A. Mean = 10 in.
   B. Variance = 1.68
   C. Standard deviation = 1.30
   D. Standard error = 0.29.

7. A. 69 in. to 72 in.
   B. 67.5 in. to 73.5 in.

8. A. 0.15
   B. The chance is 68% that the true mean for the whole population lies within the values 70.50 ± 0.15 (70.35 and 70.65 in.).

9. A. Mean = 107. Calculated as follows:

v	f	fv
90	1	90
93	2	186
94	2	188
97	1	97
101	4	404
108	7	756
111	1	111
115	1	115
119	3	357
125	3	375
	25	2679

Mean = 2679/25 = 107

B. Variance = 114.75. Calculated as follows:

d	d²	f	fd²
17	289	1	289
14	196	2	392
13	169	2	338
10	100	1	100
6	36	4	144
1	1	7	7
4	16	1	16
8	64	1	64
12	144	3	432
18	324	3	972
			2754 = Σ fd²

Variance = 2754/24 = 114.75

C. Standard deviation = $\sqrt{114.75}$ = 10.72

10. A. Mean = 110. Calculated as follows:

v	f	fv
99	6	594
103	2	206
110	3	330
111	4	444
112	3	336
118	1	118
119	2	238
120	2	240
126	2	252
	25	2758

mean = 2758/25 = 110

B. Variance = 74.08. Calculated as follows:

d	d²	f	fd²
11	121	6	726
7	49	2	98
0	0	3	0
1	1	4	4
2	4	3	12
8	64	1	64
9	81	2	162
10	100	2	200
16	256	2	512
			1778 = Σ fd²

Variance = 1778/24 = 74.08

C. Standard deviation = $\sqrt{74.08}$ = 8.60

11. Standard error of first group = 2.14. Standard error of second group = 1.72. The value of "t" is approximately 1.09. The "t" value is far too low to consider the means significantly different.

12. Four pairs of alleles. Each effective allele contributes 1/4 in. to ear length.

13. Three pairs of alleles. Each effective allele contributes 1 g.

14. Four pairs of alleles are involved. Each effective allele contributes 1 mm.

15. The two parents are obviously not extreme types. Several genotypes are possible. For example, each could be AaBbCcDd. One other possibility is that one parent is genotype AABbCcdd and the other parent AABbccDd.

CHAPTER 8

1. A. $\underline{r^+ \ s}$ and $\underline{r \ s^+}$
   B. $\underline{r^+ \ s^+}$ and $\underline{r \ s}$

2. A. $\underline{t^+ \ u}$ (35%); $\underline{t \ u^+}$ (35%); $\underline{t^+ \ u^+}$ (15%); $\underline{t \ u}$ (15%)
   B. $\underline{t^+ \ u^+}$ (35%); $\underline{t \ u}$ (35%); $\underline{t^+ \ u}$ (15%); $\underline{t \ u^+}$ (15%)

3. $\underline{t^+ \ u^+}$ and $\underline{t \ u}$, the original combinations from the dihybrid $\dfrac{t^+ \ u^+}{t \ \ u}$.

4. A. $\dfrac{pr \ \ vg^+}{pr^+ \ \ vg} \times \dfrac{pr \ \ vg}{pr \ \ vg}$ B. 15%

5. Purple flies and vestigial flies in a ratio of 1:1.

6. A. $\dfrac{o^+ \ s}{o \ \ s^+} \times \dfrac{o \ \ s}{o \ \ s}$ B. 20%

7. A. $\underline{o^+ \ s^+}$ (40%); $\underline{o^+ \ \ s}$ (10%); $\underline{o \ \ s^+}$ (10%) $\underline{o \ \ s}$ (40%)

656

B. Round, simple (66%); round, branched (9%); long, simple (9%); long, branched (16%).

C. $1000 \times 9\% = 90$

8. P_1: $\dfrac{I^+ \ F}{I^+ \ F} \times \dfrac{I \ F^+}{I \ F^+}$; F_1: $\dfrac{I^+ \ F}{I \ F^+}$

9. A. $\dfrac{I^+ \ F}{I \ F^+} \times \dfrac{I^+ \ F^+}{I^+ \ F^+}$

   B. $\dfrac{I \quad F^+}{I^+ \quad F^+}$ (white, normal);

   $\dfrac{I^+ \ F}{I^+ \quad F^+}$ (colored, mildly brittle)

   $\dfrac{I^+ \quad F^+}{I^+ \quad F^+}$ (colored, normal);

   $\dfrac{I \quad F}{I^+ \quad F^+}$ (white, mildly brittle)

   C. 20%

10. P_1: $\dfrac{w \ y^+}{w \ y^+} \times \dfrac{w^+ \ y}{Y}$:

   F_1: $\dfrac{w \ y^+}{w^+ \ y}$ (red-eyed, gray-bodied females) and

   $\dfrac{w \ y^+}{Y}$ (white-eyed, gray-bodied males)

11. 15/685 + approx. 2%. Consider only the male offspring.

12. A. $\dfrac{d \ p^+}{d \ p^+} \times \dfrac{d^+ \ p}{Y}$

   offspring: $\dfrac{d \ p^+}{d^+ \ p}$ (noncolor-blind daughters) and

   $\dfrac{d \ p^+}{Y}$ (color-blind sons)

   B. $\dfrac{d^+ \ p}{d \ p^+} \times \dfrac{d^+ \ p^+}{Y}$;

   offspring:

   $\dfrac{d^+ \ p}{d^+ \ p^+}$ and $\dfrac{d \quad p^+}{d^+ \ p^+}$ (noncolor-blind daughters)

   $\dfrac{d^+ \ p}{Y}$ and $\dfrac{d \quad p^+}{Y}$ (sons color blind)

   C. Daughters all with normal vision:

   $\dfrac{d^+ \ p}{d^+ \ p^+}$; $\dfrac{d \quad p^+}{d^+ \ p^+}$;

   $\dfrac{d^+ \ p^+}{d^+ \ p^+}$; $\dfrac{d \ p}{d^+ \ p^+}$

   Sons: $\dfrac{d^+ \ p}{Y}$; $\dfrac{d \ p^+}{Y}$; $\dfrac{d \ p}{Y}$ (color blind)

   $\dfrac{d^+ \ p^+}{Y}$ (normal vision)

13. A. $\dfrac{o \ d^+}{o \ d^+} \times \dfrac{o^+ \ d}{Y}$

   offspring: $\dfrac{o \ d^+}{o^+ \ d}$ (daughters normal)

   $\dfrac{o \ d^+}{Y}$ (sons with ocular albinism)

   B. $\dfrac{o^+ \ d^+}{o \ d} \times \dfrac{o^+ \ d^+}{Y}$

   offspring: $\dfrac{o^+ \ d^+}{o^+ \ d^+}$ and

   $\dfrac{o^+ \ d^+}{o \ d}$ (daughters all normal)

   $\dfrac{o^+ \ d^+}{Y}$ and $\dfrac{o \ d}{Y}$ (sons normal and sons with both ocular albinism and color blindness)

14. $\underline{e \quad 4 \quad r \quad 7 \quad s}$ or $\underline{s \quad 7 \quad r \quad 4 \quad e}$ (both maps are the same)

15. A. $0.18 \times 0.14 = 0.025$ (expected);

   coincidence $= \dfrac{0.015}{0.025} = 0.6$

   B. The locus cannot be placed with certainty on the basis of the information given. It could be located 13 units to the right of "p" or 13 units to the left. Information is needed on crosses involving genes at the "s" locus and those at "m" or "d."

16. A. $\dfrac{v^+ \ 1 \ b^+}{v \ 1^+ \ b}$ B. $\underline{v \quad 17.4 \quad 1 \quad 26.7 \quad b}$

   C. $.174 \times .267 = 5\%$ (approx.)

   coincidence $= \dfrac{40 \, (\text{obtained})}{50 \, (\text{expected})} = 0.8$

17. $\underline{1 \quad 11.8 \quad r \quad 33.7 \quad g}$

18. A. P_1 females: $\dfrac{x^+ \ z^+ \ y}{x \ z \ y^+}$ males: $\dfrac{x^+ \ z^+ \ y^+}{Y}$

   B. $\underline{x \quad 6.4 \quad z \quad 7.3 \quad y}$

   C. Little, if any, interference, since five doubles were obtained, and this is about what is to be expected, assuming no interference.

19. A. Would expect one-half of the flies to be females. Therefore, approximately 2000 would be wild females, $r^+s^+ \ t^+$.

B. Would expect 2000 males. Approximately six should be r^+ s t. This is a double crossover type. The total number of doubles expected on the basis of the map is $2000 \times 1.2\%$ or 24. However, the coincidence is 0.5, and so a total of 12 doubles is expected. Of these, 6 would be r^+st and 6 would be r s^+ t^+.

C. These are parental types. They are determined by subtracting the sum of the doubles (12) and the singles in region I (228) and the singles in region II (188). The singles in region I, for example, are determined in this way: $12\% \times 2000 = 240$ expected from the map information. However, the doubles (12) must be subtracted, since they were included in the construction of the map. This gives 228. The same reasoning gives 188 for region II. The parental types among the male flies, r^+s^+t and r s t^+, would total about 1572 (200 − 428, the sum of the doubles and the singles in I and II).

20. A. 2 wild (red eyes, straight wings):1 sepia-eyed, straight wings:1 red-eyed, curled wings.

B. $\dfrac{se\ \ e^+}{se\ \ e^+} \times \dfrac{se^+\ \ e}{se^+\ \ e}$; F_1: $\dfrac{se\ \ e^+}{se^+\ \ e} \times \dfrac{se\ \ e^+}{se^+\ \ e}$

	se e⁺	se⁺ e	se⁺ e⁺	se e
	se e⁺	se⁺ e	se⁺ e⁺	se e
se e⁺	se e⁺	se e⁺	se e⁺	se e⁺
	(sepia)	(wild)	(wild)	(sepia)
	se e⁺	se⁺ e	se⁺ e⁺	se e
se⁺ e	se⁺ e	se⁺ e	se⁺ e	se⁺ e
	(wild)	(ebony)	(wild)	(ebony)

The amount of crossing over cannot be estimated from a cross in *Drosophila* in which the linked alleles are in the *trans* arrangement in the F_1's. This is so since there is no crossing over in the male fruit fly. Consequently, both the crossover and noncrossover offspring will arise in the F_2 in a ratio of 2 wild:1 mutant:1 mutant.

CHAPTER 9

1. 2%

2. 90%

3. $\dfrac{gd^+ \quad P}{gd \quad p} \times \dfrac{gd^+ \quad P}{Y}$.

Offspring:

Females—All enzyme producers with normal vision

Males—Enzyme producer, normal vision, 47%

No enzyme, color blind, 47%

Enzyme producer, color blind, 3%

No enzyme, normal vision, 3%

4. A. $\underline{pro^+ \quad leu^+}$ (2); $\underline{pro^+ \quad leu}$ (2);

$\underline{pro \quad leu^+}$ (2); $\underline{pro \quad leu}$ (2)

B. $\underline{pro^+ \quad leu^+}$ (4); $\underline{pro \quad leu}$ (4)

C. $\underline{pro^+ \quad leu}$ (4); $\underline{pro \quad leu^+}$ (4)

D. $\underline{pro^+ \quad leu^+}$ (2); $\underline{pro \quad leu}$ (2);

$\underline{pro^+ \quad leu^+}$ (2); $\underline{pro \quad leu}$ (2)

5. A. $\underline{pro^+ \quad ser^+ \quad leu^+}$ (4); $\underline{pro \quad ser \quad leu}$ (4)

B. $\underline{pro^+ \quad ser^+ \quad leu^+}$ (2); $\underline{pro^+ \quad ser \quad leu}$ (2);

$\underline{pro \quad ser^+ \quad leu^+}$ (2); $\underline{pro \quad ser \quad leu}$ (2)

C. $\underline{pro^+ \quad ser^+ \quad leu^+}$ (2); $\underline{pro^+ \quad ser^+ \quad leu}$ (2);

$\underline{pro \quad ser \quad leu^+}$ (2); $\underline{pro \quad ser \quad leu}$ (2)

D. $\underline{pro^+ \quad ser^+ \quad leu^+}$ (2); $\underline{pro^+ \quad ser \quad leu^+}$ (2);

$\underline{pro \quad ser^+ \quad leu}$ (2); $\underline{pro \quad ser \quad leu}$ (2)

6. The crossover will produce two chromosomes in each of which the two chromatids are different genetically:

$\dfrac{sn \qquad y^+}{sn^+ \qquad y}$ and $\dfrac{sn \qquad y^+}{sn^+ \qquad y}$

An arrangement on the spindle as shown here in some of the cells at mitosis will give rise to cells which are homozygous for "y" and cells which are homozygous for "sn." These cells in turn will give rise to "yellow" and "singed" patches among the cells which show the gray and the normal bristle traits.

7. The "h" locus must be on chromosome III, linked to sepia. The reason is that no double recessives can be found among the F_2 offspring of a cross in *Drosophila* when two mutant stocks are crossed and the recessives being followed are linked. The F_1 dihybrid offspring would have alleles in the *trans* arrangement: $\dfrac{se \quad h^+}{se^+ \quad h}$. Since there is no crossing over in the male fruit fly, only parental combinations can be passed to the gametes: se h^+ and se^+ h. When these fertilize the four kinds of gametes formed by the female, a 2:1:1 ratio results. No double recessive can be produced.

8. Chromosome 8

9. A chromosome alteration such as a shift or translocation should be suspected. A karyotype analysis should be performed using some techniques such as fluorescent staining to bring out the chromosome banding pattern. This person's karyotype should then be compared to that of typical persons. If a band normally present in Chromosome 1 is absent from that chromosome of the atypical person and if a similar band is now present in Chromosome 5 where it is not typically seen, then good evidence can be offered for a shift. The locus governing the enzyme's production can now be more precisely assigned to a specific region of Chromosome 1.

10. All are correct.

11. C

12. B, C

Chapter 10

1. The dicentric $\underline{\quad}_0 \underline{ABCdefba}_0\underline{\quad}$ and the acentric $\underline{g\ c\ D\ E\ F\ G}$

2. $\underline{\quad}_0\underline{ABCdeFG}$ and $\underline{\quad}_0\underline{abfEDcg}$

3. $\underline{HIJ}_0\underline{Klh}$ and $\underline{nmij}_0\underline{kLMN}$

4. $\underline{HIj}_0\underline{kLMN}$ $\underline{hlK}_0\underline{Jimn}$

5. At meiosis, each chromosome will form a loop which includes the genes T U and V.

A crossover anywhere between the loops, before W or before X, will lead to:

—o‾R‾S‾T‾U‾V‾W‾X‾T‾U‾V‾Y‾Z‾

(a duplication) and —o‾R‾S‾W‾X‾Y‾Z‾

(a deficiency)

6. A. ‾g‾—o—‾h‾   ‾m‾—o—‾n‾
      ‾g‾—o—‾n‾   ‾m‾—o—‾h‾

   B. g-h, m-h and m-n, g-n (one possibility)
      m-h, m-n and g-n, g-h (the other possibility)

   C. g-h, m-n and g-n, m-h

7. alpha-alpha (dies); alpha-beta (survives); beta-beta (dies).

8. A. One-half of the male embryos would die so that the sex ratio would be 2 females: 1 male. One-half of the females carry the deletion.

   B. One-fourth of the offspring from two carrier parents will fail to survive, just as if a recessive lethal gene were present.

9. From eggs with exact pairing: $\dfrac{B\ B}{B\ B}$

   (Bar females) and $\dfrac{B\ B}{Y}$ (Bar males)

   From exceptional eggs: $\dfrac{B\ B\ B}{B\ B}$ (very

   narrow Bar females) $\dfrac{B\ B}{B}$ (wide-Bar

   females)

   $\dfrac{B\ B\ B}{Y}$ (ultra-Bar males) $\dfrac{B}{Y}$ (normal-

   eyed males)

10. A. I I II II II III III IV IV V V
    B. I I I II II II III III III IV IV IV V V V
    C. I I II II III III IV V V
    D. I I I I II II II II III III III III IV IV IV
       IV V V V V

11. A. 24
    B. 48

12. A. 24
    B. 17
    C. 32
    D. 15

13. Two species can be recognized. Populations B and C, while phenotypically distinct, have not accumulated genetic differences which limit free gene exchange. The two populations may be considered varieties of the same species (let us say, Species B). Population A is genetically isolated from members of the other two populations (members of Species B), as seen by the lack of fertile hybrids when A members are crossed to B and C members. Population A is thus composed of individuals of a different species (let us say, Species A).

14. A. Nondisjunction of X chromosomes at first anaphase of oogenesis giving an XX egg which becomes fertilized by a Y-bearing sperm. Also nondisjunction of the X and Y chromosomes at first anaphase of spermatogenesis, giving an XY sperm which fertilizes an egg.
    B. Nondisjunction of sex chromosomes at spermatogenesis or oogenesis, producing gametes with no sex chromosomes. Also loss of one X chromosome at first mitotic anaphase of an XX zygote.
    C. Nondisjunction of Y chromosome at second anaphase of spermatogenesis.
    D. A pericentric inversion.
    E. Complete failure of anaphase movement at meiosis producing an unreduced sperm or egg which unites with a haploid gamete.
    F. A translocation.
    G. Nondisjunction of X chromosomes at oogenesis producing an XX egg which becomes fertilized by an X-bearing sperm.
    H. Loss of an X chromosome during mitotic divisions of an XX embryo.
    I. Loss of a Y chromosome during mitotic divisions of an XXY embryo.
    J. Nondisjunction of Y chromosome during mitotic divisions of an XY embryo.

15. Turner syndrome: 2, 3, 8
    Klinefelter syndrome: 2, 8
    Down's syndrome: 2, 6, 8
    cri-du-chat syndrome: 7
    Philadelphia chromosome: 6
      (previously considered a deletion)
    XYY male: 2, 8

16. A metacentric chromosome may be formed following two breaks, one in each of two acrocentrics. In one of the acrocentrics, the break is just before the centromere, in the long arm of the chromosome. In the other, the break is just behind the centromere, in the short arm. The latter long piece with the centromere joins the other long arm piece which lacks a centromere. The tiny pieces left will generally consist largely of heterochromatin. These may become lost or may unite to form a tiny chromosome which is later lost.

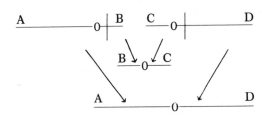

17. Testis determination seems to depend on very few (possibly just one) Y-linked loci. This small region could have become inserted into an X chromosome at the time of chromosome pairing at meiotic prophase. The rest of the Y chromosome could have been lost following this at meiotic divisions. The outcome could be a gamete lacking a visible Y but containing the tiny, testis-determining region.

18. C

19. B, C, D

20. C

21. B, D

22. E

23. E

24. A, C, D, E

CHAPTER 11

1. A. Capsulated, type III
   B. Streptomycin sensitive
   C. Penicillin resistant

2. Most will be in the cell-free fluid, since the radioactive sulfur was in the protein coats which didn't enter the host cells.

3. A. Phosphoric acid, purine or pyrimidine base, sugar component.
   B. Purine or pyrimidine base, sugar component.

4. A. DNA contains thymine, never uracil. RNA contains uracil, never thymine.
   B. The deoxyribose of DNA has one less oxygen at the No. 2 position in the sugar ring than does the ribose of RNA.
   C. DNA is confined mainly to the chromosomes, whereas RNA is much more widely distributed in the nucleus and the cytoplasm.

5. Adenine, 33%; guanine, 17%; cytosine, 17%.

6. 3′ A-T-A-A-T-C-T-A-G-G-T-A

7. T-A-G paired with A-T-C (the original sequence of base pairs) and T-G-G paired with A-C-C (a G-C pair now replaces the original A-T).

8. (A) The one with guanine.
   (B) The one with thymine.
   (C) The 3′ end or —OH end.
   (D) Adenine.

9. No hybrid DNA band would form. At 0 generations, one heavy band. At one generation, two bands, one heavy (N^{15}/N^{15}) and one light (N^{14}/N^{14}). At two generations, two bands, one heavy and one light, but the latter about three times as wide as the former. At three generations, almost all light.

10. B, C, E

11. B, C, D

12. A and E

CHAPTER 12

1. A. 5′ *A T G T T T A G A G T A A C A T A T C C T* 3′
   B. 5′ *A U G U U U A G A G U A A C A U A U C C U* 3′
   C. 7

2. A. Translation would halt prematurely.
   B. Nonsense.

3. A. No effect on the protein, since both code for phenylalanine.
   B. Silent.
   C. Cysteine would be substituted for phenylalanine.
   D. Missense.

4. A. The codons from the point of the first deletion will go out of phase and will remain out. Likely outcome is a completely nonfunctional protein because of amino acid substitutions and also possibly one which is incomplete because of the formation of a nonsense codon.
   B. Reading will come back into phase with codon 5. Polypeptide may function if first four amino acid substitutions are not too detrimental.
   C. Reading will come back into phase with codon 4. Polypeptide may function.
   D. Frame shift mutations.

5. Tryptophan, since it is designated by only one codon, whereas arginine is represented by six.

6. All possess a carboxyl group, an amino group, and an "R" group.

7. ATP, aminoacyl synthetases, tRNA, mRNA, ribosomes.

CHAPTER 13

1. A. 1   D. 2
   B. 4   E. 5
   C. 7

2. A. 1   E. 2
   B. 6   F. 11
   C. 4   G. 9
   D. 5   H. 3, 13

3. A. tRNA and rRNA
   B. rDNA (DNA of nucleolar organizer)
   C. rRNA (40S and 60S rRNA)
   D. mRNA
   E. 5S rRNA

4. 50% at most, since only one strand of each gene undergoes transcription.

5. A. 7
   B. *U A C   A A A   U C U   C A U   U G U   A U A   G G A*

6. A. RNA codons: 5′ AAU 3′
      DNA codons: 3′ TTA 5′
   B. RNA codons: 5′ AGA 3′ and 5′ AGG 3′
      DNA codons: 3′ TCT 5′ and 3′ TCC 5′
   C. RNA codons: 5′ UUA 3′ and 5′ UUG 3′
      DNA codons: 3′ AAT 5′ and 3′ AAC 5′
   D. RNA codons: 5′ GGA 3′; 5′ GGU 3′; 5′ GGC 3′
      DNA codons: 3′ CCT 5′; 3′ CCA 5′; 3′ CCT 5′

7. (A) asparagine (B) arginine (C) leucine (D) glycine

8. A minimum of three would be needed. Four of the degenerate codons have the same bases in the first two positions; however no single anticodon can recognize four different codons. The base "G" at the wobble position can recognize "U" or "C" at the 3′ position in the codon. Similarly, "U" at the wobble position can recognize "A" or "G." So, two tRNAs can suffice for these four. The other two codons for serine can be recognized by one transfer RNA for the same reasons.

9. Three in each case.

10. A. The fifth position
    B. 7

11. A. Asparagine
    B. Tryptophan
    C. Methionine
    D. Asparagine
    E. Methionine

12. A. 80S
    B. 30S
    C. 50S
    D. 60S
    E. Mg^{++}

CHAPTER 14

1. Three loci are being followed: adenine 1, adenine 3, and the locus at which the adenine 8 and adenine 16 mutant sites have occurred. The ad^8 and ad^{16} mutations do not complement each other, and so the ad^8 and ad^{16} sites are allelic.

2. Complementation is seen between "b" and "c." However, "a" and "b" are alleles. Therefore, "a" and "c" will most likely complement each other in a cross.

3. A. Apricot males and intermediate apricot females.
   B. Apricot males and females intermediate between cherry and apricot.
   C. Cherry males and red-eyed females.
   D. Males: 1 wild: 1 cherry. Females: 1 red: 1 intermediate between cherry and apricot.

4. A. S_1S_4; S_1S_5; S_2S_4; S_2S_5
   B. S_3S_5; S_4S_5
   C. No offspring can result, since all the pollen will abort.
   D. S_3S_4; S_3S_5

5. A. $c^{ch}c^h$ (light gray) and $c^{ch}c$ (light gray)
   B. c^+c (wild) and $c^{ch}c$ (light gray)
   C. $c^{ch}c^{ch}$ (chinchilla); $c^{ch}c^h$ (light gray); $c^{ch}c$ (light gray); c^hc (Himalayan)

6. A. I^Ai and ii
   B. I^AI^B and I^Ai
   C. I^AI^A and I^BI^B

7. A. $I^{A1}I^{A2}$ (Type A_1); $I^{A1}i$ (Type A_1);
        $I^{A2}i$ (Type A_2); ii (Type O)
    B. $I^{A1}I^{A2}$ (Type A_1); $I^{A2}I^B$ (Type A_2B);
        $I^{A1}i$ (Type A_1); I^Bi (Type B)
    C. $I^{A1}I^{A2}$ (Type A_1); $I^{A1}I^B$ (Type A_1B);
        $I^{A2}I^B$ (Type A_2B); I^BI^B (Type B)
    D. $I^{A1}i$ (Type A_1); I^Bi (Type B)

8. Neither. A type O person such as man No.
    1 cannot have AB offspring. The second
    man, being blood type A_2, cannot contribute
    an I^{A1} allele required for the child's $I^{A1}I^B$
    type.

9. A. I^Ai Se se (will secrete A antigen) and I^Bi
        Se se (will secrete B antigen).
    B. I^Ai Se Se; I^Ai Se se (both types secrete A
        antigen). I^Bi Se Se; I^Bi Se se (both types
        secrete B antigen). I^Ai se se; I^Bi se se
        (both types are nonsecretors).
    C. I^AI^B Se Se (secretes antigens A and B);
        I^AI^B Se se (secretes antigens A and B);
        I^Bi Se Se (secretes B); I^Bi Se se (secretes
        B); I^AI^B se se (nonsecretor); I^Bi se se
        (nonsecretor); I^Ai Se Se (secretes A); I^Ai
        Se se (secretes A); I^Ai se se (nonsecre-
        tor); ii Se Se; ii Se se; ii se se (all nonse-
        cretors, since no A or B antigen can be
        produced with O type).

10. A. 3 type A; 3 type B; 6 type AB; 4 type O
    B. 3 type A; 5 type O
    C. 5 type O; 1 A; 1 B; 1 AB

11. Man No. 1 cannot be the father, since the
    father must contribute the L^N allele. Man
    No. 2 could be the father, but one cannot
    say that he is simply because he has the
    appropriate genotype.

12. A. O Rh$^+$ and O Rh$^-$
    B. AB Rh$^+$; AB Rh$^-$; A Rh$^+$; A Rh$^-$; B Rh$^+$;
        B Rh$^-$; O Rh$^+$; O Rh$^-$
    C. A Rh$^+$; A Rh$^-$; B Rh$^+$; B Rh$^-$

13. Both men could be.

14. There would be a risk in matings B and C,
    since the female is Rh$^-$ (dd) and the male
    is Rh$^+$, possessing allele D. The risk is

greater in mating B, since the male is
homozygous for allele D.

15. A. $\dfrac{A^1\ S^2\ B^1\ D^2}{A^3\ S^4\ B^3\ D^4}$; $\dfrac{A^1\ S^2\ B^1\ D^2}{A^4\ S^3\ B^4\ D^3}$;

        $\dfrac{A^2\ S^1\ B^2\ D^1}{A^3\ S^4\ B^3\ D^4}$; $\dfrac{A^2\ S^1\ B^2\ D^1}{A^4\ S^3\ B^4\ D^3}$

    B. They can neither donate to nor accept
        from the parents.
    C. The boy will be able to accept from
        either parent, since he is deficient in
        the immune response, but he will not
        be able to donate tissue, since his tissue
        carries antigens which will trigger an
        antibody response in the parents.
    D. 25%

16. A. All normally pigmented, genotype
        AaBb.
    B. Chance for an albino is 7 out of 16.

17. The enzyme involved is composed of two
    different polypeptide chains, each con-
    trolled by a separate cistron. Mutant sites
    a^1 and a^4 are in the same cistron which is
    a unit of function separate from the cistron
    containing both a^2 and a^3. The crosses are:

(1)	$\dfrac{a^1}{a^1}$	(2) $\dfrac{a^1}{a^4}$
(3)	$\dfrac{a^1}{\quad a^2}$	(4) $\dfrac{a^1}{\quad a^3}$
(5)	$\dfrac{\quad a^2}{\quad a^3}$	(6) $\dfrac{a^1\ a^4}{\quad}$
(7)	$\dfrac{a^1\ \ a^2}{\quad}$	(8) $\dfrac{\quad}{a^2\ a^3}$

18. The two different chains of hemoglobin A
    are coded by 2 different genes. The female
    parent may be represented as genotype
    aaBB (defect in the gene for the alpha
    chain), and the male parent as AAbb (defect
    in the gene for the beta chain). The offspring
    would be genotype AaBb, having received
    a functional alpha gene from one parent
    and a functional beta gene from the other.

Sufficient hemoglobin A which is normal may form. Complementation has taken place.

CHAPTER 15

1. Plants of genotype Ww will have the greatest probability, since mutation is rare and is usually to an allele which is recessive to normal or standard. The tetraploid is the least likely to show the effect, since two dominant alleles would still be present should one of them mutate in this case.

2. 1/40,000, since 5 children out of 100,000 = 1/20,000 afflicted individuals. Each individual represents two sex cells, and the new mutation could have arisen in either the sperm or the egg.

3. A. Since the P_1 females are carrying no X-linked recessives, the sex ratio should be 1:1, even if any treated male X chromosomes carry a recessive, because no F_1 male would receive an X from a P_1 male.

   B. If a sex-linked lethal has been induced, any F_1 female carrying the recessive and crossed to an F_1 male will produce female:male in a ratio of 2:1.

4. A. The dominant B serves as a marker to indicate that any Bar-eyed female carries one recessive lethal on the X with Bar. Without the inversion, crossing over could separate the Bar marker from the lethal and place the B marker on the X chromosome without the lethal. These F_1 flies would be assumed, however, to carry the lethal on the X along with Bar. Breeding them would produce a sex ratio of 1:1 if no lethal were produced in the P_1 male and a sex ratio of 2 females: 1 male, even if a lethal were produced. Seeing any males at all would be counted as "no induced lethals," even though there was one.

   B. If crossing over occurs in an F_1 ClB female, it can produce some males, even if a lethal is on the treated X. If a lethal has been induced, crossing over could thus produce an X which carries two lethals, the original one and the induced. The other X would then lack the lethals and could give rise to males.

5. A. Must be non-Bar, since the Bar trait is carried on the X with the lethal.
   B. Crossing over in the P_1 female could separate B and l and would give rise to F_1 males which are Bar-eyed. Crossing over in the F_1 ClB would give rise to some Bar-eyed males, whether or not the treated X carried a lethal.

6. A. 3/50 and 4/50
   B. 9/2500 and 16/2500

7. About 200 on plates of Group A; about 140 on B, and about 176 on C. (About 40% of the killing action remains; therefore, 12% killing action is left following photoreactivation.)

8. More significant to the human with his longer life span, since the effects of radiation are cumulative. The average fruit fly would be getting very little natural radiation. A human with a longer life span would accumulate about 15r in thirty years.

9. 1. Endonuclease needed to recognize the damaged region and produce a free end to the segment with the dimer.
   2. Exonuclease needed to recognize the free end and remove the damaged DNA.
   3. DNA polymerase needed to synthesize the digested segment using the complementary strand as a template.
   4. DNA ligase needed to link the ends of the newly synthesized segment to the free ends of the original strand.

10. A G T (base replaced with guanine,
    T C A   no change)

    A C T (base replaced with cytosine,
    T G A   a transversion)

A T T (base replaced with thymine,
T A A   a transversion)
A A T (base replaced with adenine,
T T A   a transition)

11. A G A T
    T C T A

12. A. A A T   B. A A T (a transition)
                     T T A

13. No answer.

14. C and E

15. B

16. Temperature extremes, aging, nutrition, mutator genes.

17. Cells in the bone marrow and in the testes are actively dividing. In such cells, the chromosomes are in a more condensed state than in nondividing cells. With their condensed chromosomes, these cells are much more susceptible to destruction by mutagens, since the chromosomes can be more readily damaged.

CHAPTER 16

1. M = 0.8; N = 0.2

2. 178 out of 8100

3. A. The frequency of the recessive allele must be equal to $\sqrt{.30}$ or about 54.8%; that of the dominant must therefore be less, about 45.2%.
   B. The allele will not increase simply because it is dominant. Unless some selective value is imparted to T or to t, the frequencies of the alleles will remain the same except for chance fluctuations.

4. A. 98/2500 heterozygotes and 2401/2500 homozygotes for the wild allele.

B. 98/2500 divided by
98/2500 + 2401/2500 = 2/51

5. A. $d = 1/20$; $D = 19/20$
   B. 38/400
   C. 1/400

6. A. 1/200,000
   B. 1/450
   C. $2 \times 1/450 \times 449/450$ = about 1/220

7. A. 1/141
   B. $2 \times 140/141 \times 1/141 = 1/71$ (approx.)
   C. 1/200,000

8. A. The lethal will not be passed down but will exert its full effect on the one individual. The other genes will pass through 1/.05 or 20 and 1/.02 or 50 persons.
   B. Each one of the three will result in a genetic death, although it may take longer for the detrimentals to do so.

9. A. Through 5 homozygotes.
   B. Through 100 heterozygotes.
   C. Most of the elimination will be in the heterozygotes, since the gene is rare, and homozygotes will be much less frequent than the heterozygotes.

10. This can be explained on the basis of the Founder effect. The undesirable allele, while low in the ancestral European population, was undoubtedly present in some of the few members who founded the American population. This would have been a chance event. Those who founded the Australian population happened to be free of the allele.

CHAPTER 17

1. A, B, D, and E

2. B, D, and E

3. A and D

4. B

5. The fluctuation from the mean is very great in *A*, as shown by the large difference between the mean and the variance. The *A* data must refer to the second part of the experiment, since spontaneous mutations will result in fluctuation from one sample to the next due to the fact that their occurrence is at random.

6. A. $A^+B^+C^+D^+E^+$
   B. Neither. Both are prototrophs, since both can grow on minimal medium.

7. Strain No. 2 is the donor strain. The recipient strain is strain No. 1, and it must survive since the recombinants are derived from the F⁻ cells.

8. M-O-Q-L-N. Since the procedure selects only A^+ and B^+ cells, the order of these cannot be established here.

9. A and B:

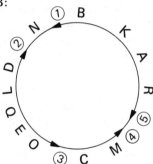

10. The last known marker to enter should be selected in each case. Select for M in the case of strain 1, C in the case of strain 2, etc. Since the site of the fertility factor will be the last to enter, chances are greater that selection for a marker near the terminus of the transferred donor chromosome will screen out cells which have received the entire donor strand along with the fertility factor.

11. A. (H) 5′ A G C T A T 3′
       (L) 3′ T C G A T A 5′
    B. (L) 5′ A G C T A T 3′
       (H) 3′ T C G A T A 5′

12. The cells which first received lac⁺ too early received the sex factor which had removed itself from its site of insertion on the chromosome along with the adjacent lac⁺ marker. These cells then acted as donor cells, transferring F⁺lac⁺ to F⁻lac⁻ with a high frequency. This is an example of sexduction. The derived F⁺lac⁺ cells are heterogenotes. They carry lac⁻ on the chromosome and lac⁺ on the sex factor which is existing free in the cell.

13. A. The DNA will be hybrid: one strand A, one strand B.
    B. One cell will be streptomycin resistant and the DNA will be composed of two "A" type strands in the marker region. The other cell is sensitive and will have two "B" type strands.

14. Simply because the percentage values of the bases in two DNAs are the same, it doesn't necessarily follow that the distribution of the bases in the two DNAs is the same. Nucleotide sequences would be very different in species which are not closely related. Genes from more closely related species would be expected to be more similar and to be coded for related protein products. In this case, "B" apparently has many nucleotide sequence differences (and hence gene differences) from "A" and "C". The latter two species would appear to be closely related, since most DNA stretches are complementary.

15. Hybridization will occur much more rapidly between the repetitive DNA and its RNA transcripts, since any RNA derived from a gene which is present in many copies has a much greater chance of finding a DNA complement than does an RNA with only one complement in a cell.

16. (1) E (2) P + E (3) E (4) P (5) E (6) P + E (7) E (8) E (9) P (10) P + E (11) P + E (12) E (13) P + E (14) P

CHAPTER 18

1. A and B are both lysogenic. Each harbors a different prophage, so each can be lysed

by the other. C harbors no known prophage and is thus susceptible to both A and B but cannot lyse either of them.

2. A. From either (1) or (2), since phage lambda would not be transferred from the Hfr, and zygotic induction wouldn't occur.
    B. From either (2) or (3). In (2) the prophage is not transferred. In (3), both the Hfr and F$^-$ carry the prophage, so induction won't occur.
    C. From either (2) or (3). Impossible in (1), since site of lambda is well before the S locus and therefore lambda would be transferred before S, and lysis would follow.

3. A. Transformation requires no vector. Transformation can occur by uptake of pure donor DNA alone. Transduction requires a vector; the DNA is introduced into the cell by the vector and doesn't enter by itself. Transformation requires that the cell be in a state of "competence." This is not so in transduction. Transduction depends on the ability of a cell to harbor appropriate prophage. This is not required in transformation.
    B. Both depend on introduction of a virus into the bacterial cell, but in conversion the introduced virus does not necessarily carry any bacterial genes. The new features of the bacterial cell are not due to any introduced genes from another bacterial strain but are the result of the interaction of the genetic material of the virus and of the cell itself.

4. If the level of infection is very low, only one virus particle would be likely to infect a cell. Those which are transducing the cells are defective. Since no intact helper phage is present, the transducing phage cannot effect lysis.

5. The transduced cells are heterogenotes. Some of them may lose the Gal$^+$ marker and revert to the Gal$^-$ trait.

6. 1.7% recombination. Small clear h r⁺. Small cloudy, h⁺r⁺. Large clear, h r. Large cloudy, h⁺r.

7. Repeated mating occurs. One recombinant class formed early in the cycle would probably engage in later matings and thus not be represented at the time of burst.

8. *A A G U U A C C A*

9. Taken three nucleotides at a time from the left, the seventh triplet is TGA, which corresponds to UGA, a chain-terminating codon in the messenger. Starting a new frame with the A in this triplet produces ATG, which corresponds to the start signal AUG in mRNA. It is thus at this point that an overlapping could occur.

10. A. 3′ CAA 5′; B. 3′ CAA 5′; C. 5′ GTT 3′

11. (A) The same as the "+" strand, since the "+" strand acts as mRNA.
    (B) UUGUCCUGCGUC
    (C) RNA replicase

12. Supply the isolated RNA with the necessary deoxyribonucleotides and RNA-directed DNA polymerase (reverse transcriptase). DNA will be formed on the RNA template. The DNA would reflect the base sequence of the original DNA strand from which the mRNA was transcribed.

13. B

14. A and D

15. D

16. D and E

17. A, C, D, and E

18. E

CHAPTER 19

1. A. Inductive
   B. Constitutive
   C. Constitutive
   D. Constitutive
   E. Inductive
   F. Constitutive

2. A. Z constitutively; Y and X by induction.
   B. X, Y, and Z only by induction.

3. A. No enzymes produced constitutively or by induction.
   B. Same as A.
   C. No enzymes produced constitutively. Only X produced by induction.
   D. X produced constitutively. Nothing else produced.

4. A. A and B produced only in absence of effector.
   B. A and B produced in absence and in presence of effector.
   C. Same as B.
   D. A and B produced in absence of effector. B produced in presence of effector.

5. 5′ CTGGCC 3′
   3′ GACCGG 5′

6. C and E

7. A and D

8. A and B

9. A and C

10. A, C, and E

11. A and E

12. C and D

13. B and C

CHAPTER 20

1. A. KK, killers.
   B. KK, sensitives.
   C. Two KK killers or two kk sensitives.
   D. Two KK sensitives or two kk sensitives.

2. A. KK and KK
   B. KK and kk
   C. Kk and Kk

3. A. KK sensitives from the sensitive and KK killers from the killer.
   B. Kk sensitives from the sensitive and Kk killers from the killer.

4. Several facts are:
   (1) DNA of the organelles is unassociated with histones as is also true of prokaryotes.
   (2) DNA of the organelles is unbounded by a membrane as is also true of prokaryotes.
   (3) Organelle ribosomes have S values similar to prokaryotic ones, differing from the cytoplasmic ribosomes.
   (4) Organelle ribosomes and those of prokaryotes are inhibited by drugs which do not affect the cytoplasmic ribosomes.
   (5) RNA polymerases of the organelles and of bacteria are affected by rifampicin which doesn't affect the RNA polymerase of the nucleus.
   (6) The tRNAs and the synthetases of the organelles are distinct from their cytoplasmic counterparts.

5. A. Zygotes and ascospores petite.
   B. Zygotes wild. Ascospores 2 wild: 2 petite.
   C. $2^+:2^-$

6. A. Zygote: CcDdPpQq. Products of zygote: (1) CDPq (2) CdPq (3) cDPq (4) cdPq. Each cell type will produce cells of the same type.
   B. Zygote: CcDdPpQq. Products of zygote: (1) CDPpQq (2) CdPpQq (3) cDPpQq (4) cdPpQq. Type (1) will give rise to: CDPQ, CDPq; CDpQ; CDpq. Type (2) will give rise to CdPQ; CdPq; CdpQ; Cdpq. Type (3) will give rise to: cDPQ; cDPq; cDpQ; cDpq. Type (4) will give rise to: cdPQ; cdPq; cdpQ; cdpq.

7. A. ABCD, since RTF and F factor can't transfer the chromosome.

672

B. Same answer as above.

C. No determinants will be transferred.

D. EF, since integrated F factor brings about chromosome transfer mainly.

8. A

9. A, C, and D

10. None

11. A, C, and D

12. C and E

13. B and D

14. C, D, and E

15. B and D

CHAPTER 21

1. C T A

G A T G C T T A A G C   and

        C G A A T T C G T C T

               A G A

2. A. G A C T A

    C T G A T A A T G C A T T   and

    T T A C G T A A T A G C A

             A T C G T

B. No. Different palindromes are involved and the ends generated in the plasmid are not complementary to those in the DNA segments of Question 1.

3. Not all the DNA of the nucleolar organizer is complementary to the 18S and 28S RNA, since a considerable amount is spacer which is not transcribed at all. The transcribed spacer would not form a part of the final rRNA, and this would also account for some of the DNA which does not hybridize with the RNA of the ribosomes.

4. (1) Base composition of the rDNA which gives it a different buoyant density from the bulk of the DNA, facilitating the separation of the ribosomal RNA genes in a density gradient.

(2) Abundance of the products of the genes, the transcripts, making identification of the specific genes easier in any sample of isolated DNA. Also permits estimate of gene number to be made in DNA-RNA hybridization experiments.

(3) Presence of rDNA in many copies making the genes themselves more abundant and thus facilitating their isolation, their assay, and their detailed analysis.

5. Duplication is believed to be an important way of adding different genes to the genome. Once a duplication exists, independent mutation events will cause the once identical segments to diverge so that eventually they can be recognized as separate genes. In the case of the spacers, this has not happened within a species. The spacers within a species remain the same. Since they vary greatly from one species to the next, it would appear there is no selective advantage for any particular arrangement of nucleotides. No satisfactory explanation has been found for the fact that a given arrangement within a species stays the same.

6. Somewhat under 20%. This is so, since only half of the DNA would be involved in transcription. Of this amount, not all the DNA would be transcribed, such as the promoter and the terminator. Of that part that is transcribed, a portion is cleaved from the final RNA product. Only 86 nucleotides are present in the final product.

7. A. 19

B. Double the chromosome number using colchicine. 38 chromosomes.

C. 38

8. AA in the nucleus. D in the cytoplasm.

9. A. (nonhygienic queen) UURR × ur (hy-

gienic drone). UuRr—both workers and queen.

Workers' behavior nonhygienic.

B. UR drones.

C. UR, Ur, uR, ur

D. UURR, UURr, UuRR, UuRr—all non-hygienic.

10. C

11. A, B, and C

12. A. C; B. A-D; C. X; D. C; E. A-R; F. X; G. A-R; H. C; I. A-R; J. C; K. A-R; L. X

13. A. 13; B. 16; C. 7; D. 3; E. 10; F. 12; G. 18; H. 1; I. 9; J. 4; K. 8; L. 11; M. 5; N. 6; O. 14

INDEX

Mutator genes, 427

Nail-patella syndrome, 243, 245
Natural selection, 467–69, 471
   interference with, 460–61
Neurospora, crossover analysis and, 251–53
n-formyl methionine, 372, 373, 374
Nicotiana, cell fusion and, 616–17
Nilsson-Ehle, H., 176
Nirenberg, Marshall W., 354, 356
Nitrous acid, 442
Nondisjunction, 117
   in humans, 123, 127
Nonhistone protein, 556
   regulation and, 551–53
Nonsense strand, 524, 527
Nonsense triplet, 352–53, 358. *See also* Chain-terminating
   codon
Novick, A., 439, 445
Nucleolar organizer, 37–38, 377
Nucleolus, 37–38
   ribosomes and, 375–76, 378
Nucleosides, kinds, 320
Nucleotides, kinds, 319
Nutrition, gene expression and, 92, 97

Ochoa, S., 354
Oenothera. *See* Evening primrose
One gene-one enzyme theory, 339, 378–79
One gene-one polypeptide theory, 378–79
Oogenesis, 68–70
Operator, 544, 547
Operon, 545

Pachynema, 54
Palindrome, 608
Paramecium
   killer trait in, 569–77
   life cycle of, 568–69
   mate killer, 577–78
Parasexual hybridization, 617–18
Particulate theory of inheritance, 6, 13
Pedigree analysis, 155–63, 241–43
Penetrance, 89–91, 92
Peptide linkage, 339
Petite trait, in yeast, 581–83
Phage, 316–17. *See also* Bacteriophage; *specific phage types*
   recombination in, 514–17
   repeated mating in, 517–18
   single stranded, 523
   temperate, 509–11
   transduction and, 511–13
   virulent, 508–9
Phage ϕX, 174, 517, 523–24
   overlapping genes in, 524–25
   synthetic DNA and, 605–6
Phage lambda, 511–13
Phage T$_2$, 517
Phage T$_4$, 509
Phenocopy, 94
Phenotype, 11, 74–75

Phenylalanine, metabolism of, 87–88
Phenylketonuria (PKU), 86–89, 92, 97
Philadelphia chromosome, 283, 293
Phosphodiester linkage, 322
Photoreactivation, 436–37
Pi particles, 575
Plasmids, 596–98
   in genetic manipulation, 609–11
Pleiotropy, 86, 91, 92
Pneumococcus transformation, 313–14
Polarity, of codons, 369
Polydactyly, 89
Polygenic inheritance. *See* Quantitative inheritance
Polynucleotide phosphorylase, 354
Polyploidy, 296. *See also* Autopolyploidy;
   Allopolyploidy
   in animals, 302–4
   in humans, 304
Polysome (polyribosome), 368
Polytene chromosomes, gene activity and, 554–56
Population(s)
   allelic frequencies in, 452–56
   deleterious alleles in, 457–60
   evolution and, 450–52, 467–71
Position effect
   Bar locus and, 287–88
   heterochromatin and, 288–89
Poultry, 79–81, 82–83, 143
Prenatal diagnosis, 245–47
Probability (ies), 148–55
   combining, 149–55
   independent events, 150–51
   more than two alternatives, 154–55
   mutually exclusive events, 149–50
Prokaryotes
   cells of, 33–34
   contrast with eukaryotes, 500–3
Promoter, 547–48
Prophage, 510
Prophase
   first meiotic, 53–57
   mitotic, 43
Proteins, general characteristics of, 339
Ptashne, M., 547
Puffing, of polytene chromosomes, 555–56

Quantitative inheritance, 176–85
   calculation of gene effect, 181–83

Rabbit
   coat color in, 93–94
   fat color in, 92–93
   multiple alleles in, 402–3
Race(s), 451
Race, R. R., 412
Radiation
   chromosome anomalies and, 434
   general effects, 428–31
   interaction with other factors, 435
   mutation and, 431–33
   types, 428–31

Ratio(s)
interpretation of, 17–18
modified, 79–84, 99
rDNA. *See* Ribosomal DNA
Recessive factors, 11
Regulation, evolution and, 553–54
Regulator gene, 544
Release factors, 374
Renaturation kinetics, 501–3
Repair enzymes, 437, 439–41
Repeated mating, in phage, 517–18
Replica plating, 478–79, 488
Replication fork, 332–33
Replicon model, 593–95
Repression, of enzyme synthesis, 548–49
Repressor(s), 544, 545, 547
Resistance transfer factors (RTF), 595–98
in genetic manipulation, 609–11
Restriction enzyme(s), 608, 609, 610, 611
Reverse transcriptase, 532
Rh blood grouping
general features, 409
genetic interpretation, 411–13
pregnancy and, 409–11
Rh locus, 243, 245
Rho factor, 364
Rhoades, M. M., 585
Ribosomal DNA, 377
isolation of, 612–14
Ribosome(s), 34, 38, 341
characteristics, 366–68
during translation, 369–71
in eukaryotes, 501
processing, 375–78
Rous sarcoma virus, 531, 532, 533
RNA, 341. *See also specific types*
chemistry, 319–20
as genetic material, 527–28, 530
polarity, 346–47
soluble. *See* tRNA
RNA polymerase, 345, 362, 363–64
in eukaryotes, 501
RNA virus, 527–28, 530–32, 533. *See also specific types*
gene sequencing, 534–35
replication, 530

Saccharomyces. See Yeast
Sager, R., 587, 589, 590
Sanger, R., 412, 525
Satellite, of chromosome, 37, 378
Satellite DNA, 376, 377, 501
Schizophrenia, 624
Schleiden, M. S., 3
Schwann, T., 3
Secretor trait, 243, 246–47, 406
Semiconservative replication, 326–27, 329–30
Sense strand, 524, 527
Sex bivalent, 123
Sex change, in humans, 142–43
Sex chromatin, 132–34, 289
Sex chromosome(s)

anomalies in humans, 121–23
in *Drosophila*, 108–9
in humans, 128–29
Sex determination
in *Drosophila*, 118–21
in humans, 128–29
Sex-influenced alleles, 139–41
Sex-limited genes, 141
Sex linkage
chromosome theory and, 115–18
discovery, 109–15
in humans, 129–32
Sex-linked alleles, equilibrium frequency, 455–56
Sex mosaic, 123, 126
Sex pili, 493
Sex ratio, 129
Sexduction, 494, 513, 546
Sexual reproduction, gene combinations and, 79
Shift, 268
Sickle-cell anemia, 15–16, 381, 382–84
in populations, 461–63
Sickle-cell hemoglobin, 381–83
Sigma chain (factor), 362
Skin pigmentation, 178–79
Somatic cell hybridization, 243–45
Sonneborn, T. M., 569
Spacer DNA, 613
Speciation
isolation and, 469–71
polyploidy and, 301–2
Species, population and, 450–51
Spermatogenesis, 64, 66–68
Spermiogenesis, 64
Spindle, 43
Stahl, F. W., 330–31
Standard deviation, 186–90
Standard error of the mean, 191–92
Statistics, 185–95
Sterility alleles, 403, 404
Stern, C., 249–51
Steroid hormones, 551, 555–56
Steward, F. C., 557, 559
Strasburger, E., 3, 52
Sturtevant, A. H., 286
Sutton, W., 28, 53
Superfemale, 119
Supermale, 119
Suppressor genes, 91
Synapsis, 54
Synteny, 231

t test, 192–93
T_2 bacteriophage, 316
T_4 bacteriophage, 519
Target theory, 433–35
Tatum, E. L., 479
Tautomerism, 327, 328
Tay-Sachs disease, 99, 385
Telomere, 263
Telophase
first meiotic, 58